Graduate Texts in Mathematics 241

*Editorial Board*
S. Axler K.A. Ribet

# Graduate Texts in Mathematics

1 TAKEUTI/ZARING. Introduction to Axiomatic Set Theory. 2nd ed.
2 OXTOBY. Measure and Category. 2nd ed.
3 SCHAEFER. Topological Vector Spaces. 2nd ed.
4 HILTON/STAMMBACH. A Course in Homological Algebra. 2nd ed.
5 MAC LANE. Categories for the Working Mathematician. 2nd ed.
6 HUGHES/PIPER. Projective Planes.
7 J.-P. SERRE. A Course in Arithmetic.
8 TAKEUTI/ZARING. Axiomatic Set Theory.
9 HUMPHREYS. Introduction to Lie Algebras and Representation Theory.
10 COHEN. A Course in Simple Homotopy Theory.
11 CONWAY. Functions of One Complex Variable I. 2nd ed.
12 BEALS. Advanced Mathematical Analysis.
13 ANDERSON/FULLER. Rings and Categories of Modules. 2nd ed.
14 GOLUBITSKY/GUILLEMIN. Stable Mappings and Their Singularities.
15 BERBERIAN. Lectures in Functional Analysis and Operator Theory.
16 WINTER. The Structure of Fields.
17 ROSENBLATT. Random Processes. 2nd ed.
18 HALMOS. Measure Theory.
19 HALMOS. A Hilbert Space Problem Book. 2nd ed.
20 HUSEMOLLER. Fibre Bundles. 3rd ed.
21 HUMPHREYS. Linear Algebraic Groups.
22 BARNES/MACK. An Algebraic Introduction to Mathematical Logic.
23 GREUB. Linear Algebra. 4th ed.
24 HOLMES. Geometric Functional Analysis and Its Applications.
25 HEWITT/STROMBERG. Real and Abstract Analysis.
26 MANES. Algebraic Theories.
27 KELLEY. General Topology.
28 ZARISKI/SAMUEL. Commutative Algebra. Vol.I.
29 ZARISKI/SAMUEL. Commutative Algebra. Vol.II.
30 JACOBSON. Lectures in Abstract Algebra I. Basic Concepts.
31 JACOBSON. Lectures in Abstract Algebra II. Linear Algebra.
32 JACOBSON. Lectures in Abstract Algebra III. Theory of Fields and Galois Theory.
33 HIRSCH. Differential Topology.
34 SPITZER. Principles of Random Walk. 2nd ed.
35 ALEXANDER/WERMER. Several Complex Variables and Banach Algebras. 3rd ed.
36 KELLEY/NAMIOKA et al. Linear Topological Spaces.
37 MONK. Mathematical Logic.
38 GRAUERT/FRITZSCHE. Several Complex Variables.
39 ARVESON. An Invitation to $C^*$-Algebras.
40 KEMENY/SNELL/KNAPP. Denumerable Markov Chains. 2nd ed.
41 APOSTOL. Modular Functions and Dirichlet Series in Number Theory. 2nd ed.
42 J.-P. SERRE. Linear Representations of Finite Groups.
43 GILLMAN/JERISON. Rings of Continuous Functions.
44 KENDIG. Elementary Algebraic Geometry.
45 LOÈVE. Probability Theory I. 4th ed.
46 LOÈVE. Probability Theory II. 4th ed.
47 MOISE. Geometric Topology in Dimensions 2 and 3.
48 SACHS/WU. General Relativity for Mathematicians.
49 GRUENBERG/WEIR. Linear Geometry. 2nd ed.
50 EDWARDS. Fermat's Last Theorem.
51 KLINGENBERG. A Course in Differential Geometry.
52 HARTSHORNE. Algebraic Geometry.
53 MANIN. A Course in Mathematical Logic.
54 GRAVER/WATKINS. Combinatorics with Emphasis on the Theory of Graphs.
55 BROWN/PEARCY. Introduction to Operator Theory I: Elements of Functional Analysis.
56 MASSEY. Algebraic Topology: An Introduction.
57 CROWELL/FOX. Introduction to Knot Theory.
58 KOBLITZ. $p$-adic Numbers, $p$-adic Analysis, and Zeta-Functions. 2nd ed.
59 LANG. Cyclotomic Fields.
60 ARNOLD. Mathematical Methods in Classical Mechanics. 2nd ed.
61 WHITEHEAD. Elements of Homotopy Theory.
62 KARGAPOLOV/MERLZIAKOV. Fundamentals of the Theory of Groups.
63 BOLLOBAS. Graph Theory.

*(continued after index)*

**Joseph H. Silverman**

# The Arithmetic of Dynamical Systems

With 11 Illustrations

Springer

Joseph H. Silverman
Department of Mathematics
Brown University
Providence, RI 02912
USA
jhs@math.brown.edu

*Editorial Board*

S. Axler
Mathematics Department
San Francisco State University
San Francisco, CA 94132
USA
axler@sfsu.edu

K.A. Ribet
Mathematics Department
University of California, Berkeley
Berkeley, CA 94720-384
USA
ribet@math.berkeley.edu

Mathematics Subject Classification (2000): 11-01, 11G99, 14G99, 37-01, 37F10

ISBN 978-1-4419-2417-9 e-ISBN-13: 978-0-387-69904-2

Printed on acid-free paper.

© 2010 Springer Science+Business Media, LLC
All rights reserved. This work may not be translated or copied in whole or in part without the written permission of the publisher (Springer Science+Business Media, LLC, 233 Springer Street, New York, NY 10013, USA), except for brief excerpts in connection with reviews or scholarly analysis. Use in connection with any form of information storage and retrieval, electronic adaptation, computer software, or by similar or dissimilar methodology now known or hereafter developed is forbidden.
The use in this publication of trade names, trademarks, service marks, and similar terms, even if they are not identified as such, is not to be taken as an expression of opinion as to whether they are subject to proprietary rights.

9 8 7 6 5 4 3 2 1

springer.com

# Preface

This book is designed to provide a path for the reader into an amalgamation of two venerable areas of mathematics, Dynamical Systems and Number Theory. Many of the motivating theorems and conjectures in the new subject of Arithmetic Dynamics may be viewed as the transposition of classical results in the theory of Diophantine equations to the setting of discrete dynamical systems, especially to the iteration theory of maps on the projective line and other algebraic varieties. Although there is no precise dictionary connecting the two areas, the reader will gain a flavor of the correspondence from the following associations:

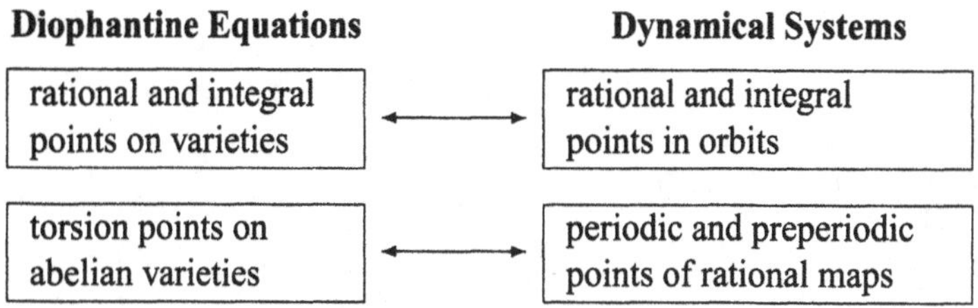


There are a variety of topics covered in this volume, but inevitably the choice reflects the author's tastes and interests. Many related areas that also fall under the heading of arithmetic or algebraic dynamics have been omitted in order to keep the book to a manageable length. A brief list of some of these omitted topics may be found in the introduction.

### Online Resources

The reader will find additonal material, references and errata at

`http://www.math.brown.edu/~jhs/ADSHome.html`

### Acknowledgments

The author has consulted a great many sources in writing this book. Every attempt has been made to give proper attribution for all but the most standard results. Much of the presentation is based on courses taught at Brown University in 2000 and 2004, and the exposition benefits greatly from the comments of the students in those courses. In addition, the author would like to thank the many students and mathematicians who read drafts and/or offered suggestions and corrections, including Matt Baker, Rob Benedetto, Paul Blanchard, Rex Cheung, Bob Devaney, Graham Everest, Liang-Chung Hsia, Rafe Jones, Daniel Katz, Shu Kawaguchi, Michelle Manes, Patrick Morton, Curt McMullen, Hee Oh, Giovanni Panti, Lucien Szpiro, Tom Tucker, Claude Viallet, Tom Ward, Xinyi Yuan, Shou-Wu Zhang. An especial thanks is due to Matt Baker, Rob Benedetto and Liang-Chung Hsia for their help in navigating the treachorous shoals of $p$-adic dynamics. The author would also like to express his appreciation to John Milnor for a spellbinding survey talk on dynamical systems at Union College in the mid-1980s that provided the initial spark leading eventually to the present volume. Finally, the author thanks his wife, Susan, for her support and patience during the many hours occupied in writing this book.

Joseph H. Silverman
*January 1, 2007*

# Contents

# Introduction

A *(discrete) dynamical system* consists of a set $S$ and a function $\phi : S \to S$ mapping the set $S$ to itself. This self-mapping permits iteration

$$\phi^n = \underbrace{\phi \circ \phi \circ \cdots \circ \phi}_{n \text{ times}} = n^{\text{th}} \text{ iterate of } \phi.$$

(By convention, $\phi^0$ denotes the identity map on $S$.)

For a given point $\alpha \in S$, the *(forward) orbit of* $\alpha$ is the set

$$\mathcal{O}_\phi(\alpha) = \mathcal{O}(\alpha) = \{\phi^n(\alpha) : n \geq 0\}.$$

The point $\alpha$ is *periodic* if $\phi^n(\alpha) = \alpha$ for some $n \geq 1$. The smallest such $n$ is called the *exact period of* $\alpha$. The point $\alpha$ is *preperiodic* if some iterate $\phi^m(\alpha)$ is periodic. The sets of periodic and preperiodic points of $\phi$ in $S$ are denoted respectively by

$$\begin{aligned} \operatorname{Per}(\phi, S) &= \{\alpha \in S : \phi^n(\alpha) = \alpha \text{ for some } n \geq 1\}, \\ \operatorname{PrePer}(\phi, S) &= \{\alpha \in S : \phi^{m+n}(\alpha) = \phi^m(\alpha) \text{ for some } n \geq 1, m \geq 0\} \\ &= \{\alpha \in S : \mathcal{O}_\phi(\alpha) \text{ is finite}\}. \end{aligned}$$

We write $\operatorname{Per}(\phi)$ and $\operatorname{PrePer}(\phi)$ when the set $S$ is fixed.

**Principal Goal of Dynamics**

Classify the points $\alpha$ in the set $S$ according to the behavior of their orbits $\mathcal{O}_\phi(\alpha)$.

If $S$ is simply a set with no additional structure, then typical problems are to describe the sets of periodic and preperiodic points and to describe the possible periods of periodic points. Usually, however, the set $S$ has some additional structure and one attempts to classify the points in $S$ according to the interaction of their orbits with that structure. There are many types of additional structures that may imposed, including algebraic, topological, metric, and analytic.

*Example* 0.1. (*Finite Sets*) Let $S$ be a finite set and $\phi : S \to S$ a function. Clearly every point of $S$ is preperiodic, so we might ask for the number of elements in the set of periodic points

$$\mathrm{Per}(\phi, S) = \{\alpha \in S : \phi^n(\alpha) = \alpha \text{ for some } n \geq 1\}.$$

An interesting class of sets and maps are finite fields $S = \mathbb{F}_p$ and maps $\phi : \mathbb{F}_p \to \mathbb{F}_p$ given by polynomials $\phi(z) \in \mathbb{F}_p[z]$. For example, Fermat's Little Theorem says that

$$\mathrm{Per}(z^p, \mathbb{F}_p) = \mathbb{F}_p \qquad \text{and} \qquad \mathrm{Per}(z^{p-1}, \mathbb{F}_p) = \{0, 1\},$$

which gives two extremes for the set of periodic points. A much harder question is to fix an integer $d \geq 2$ and ask for which primes $p$ there is a polynomial $\phi$ of degree $d$ satisfying $\mathrm{Per}(\phi, \mathbb{F}_p) = \mathbb{F}_p$. Similarly, one might fix a polynomial $\phi(z) \in \mathbb{Z}[z]$ and ask for which primes $p$ is it true that $\mathrm{Per}(\phi, \mathbb{F}_p) = \mathbb{F}_p$. In particular, are there infinitely many such primes?

In a similar, but more general, vein, one can look at a rational function $\phi \in \mathbb{F}_p(z)$ inducing a rational map $\phi : \mathbb{P}^1(\mathbb{F}_p) \to \mathbb{P}^1(\mathbb{F}_p)$. Even more generally, one can ask similar questions for a morphism $\phi : V(\mathbb{F}_p) \to V(\mathbb{F}_p)$ of any variety $V/\mathbb{F}_p$, for example $V = \mathbb{P}^N$.

*Example* 0.2. (*Groups*) Let $G$ be a group and let $\phi : G \to G$ be a homomorphism. Using the group structure, it is often possible to describe the periodic and preperiodic points of $\phi$ fairly explicitly. The following proposition describes a simple, but important, example. In order to state the proposition, we recall that the *torsion subgroup of an abelian group* $G$, denoted by $G_{\mathrm{tors}}$, is the set of elements of finite order in $G$,

$$G_{\mathrm{tors}} = \{\alpha \in G : \alpha^m = e \text{ for some } m \geq 1\},$$

where $e$ denotes the identity element of $G$.

**Proposition 0.3.** *Let $G$ be an abelian group, let $d \geq 2$ be an integer, and let $\phi : G \to G$ be the $d^{th}$ power map $\phi(\alpha) = \alpha^d$. Then*

$$\mathrm{PrePer}(\phi, G) = G_{\mathrm{tors}}.$$

*Proof.* The simple nature of the map $\phi$ allows us to give an explicit formula for its iterates,

$$\phi^n(\alpha) = \alpha^{d^n}.$$

Now suppose that $\alpha \in \mathrm{PrePer}(\phi_d, G)$. This means that $\phi^{m+n}(\alpha) = \phi^m(\alpha)$ for some $n \geq 1$ and $m \geq 0$, so $\alpha^{d^{m+n}} = \alpha^{d^m}$. But $G$ is a group, so we can multiply by $\alpha^{-d^m}$ to get $\alpha^{d^{m+n}-d^m} = e$. The assumptions on $d$, $m$, and $n$ imply that the exponent is positive, so $\alpha \in G_{\mathrm{tors}}$.

Next suppose that $\alpha \in G_{\mathrm{tors}}$, say $\alpha^m = e$, and consider the following sequence of integers modulo $m$:

$$d, d^2, d^3, d^4, \ldots \text{ modulo } m.$$

Since there are only finitely many residues modulo $m$, eventually the sequence has a repeated element, say $d^i \equiv d^j \pmod{m}$ with $i > j$. Then

$$\phi^i(\alpha) = \alpha^{d^i} = \alpha^{d^j} = \phi^j(\alpha), \quad \text{since } \alpha^m = e \text{ and } d^i \equiv d^j \pmod{m}.$$

Hence $\alpha \in \mathrm{PrePer}(\phi)$. □

*Example* 0.4. (*Topological Spaces*) Let $S$ be a topological space and let $\phi : S \to S$ be a continuous map. For a given $\alpha \in S$, one might ask for a description of the accumulation points of $\mathcal{O}_\phi(\alpha)$. For example, a point $\alpha$ is called *recurrent* if it is an accumulation point of $\mathcal{O}_\phi(\alpha)$. In other words, $\alpha$ is recurrent if there is a sequence of integers $n_1 < n_2 < n_3 < \cdots$ such that $\lim_{i\to\infty} \phi^{n_i}(\alpha) = \alpha$, so either $\alpha$ is periodic, or it eventually returns arbitrarily close to itself.

*Example* 0.5. (*Metric Spaces*) Let $(S, \rho)$ be a compact metric space. For example, $S$ could be the unit sphere sitting inside $\mathbb{R}^3$, and $\rho(\alpha, \beta)$ the usual Euclidean distance from $\alpha$ to $\beta$ in $\mathbb{R}^3$. The fundamental question in this setting is whether points that start off close to a given point $\alpha$ continue to remain close to one another under repeated iteration of $\phi$. If this is true, we say that $\phi$ is *equicontinuous* at $\alpha$; otherwise we say that $\phi$ is *chaotic* at $\alpha$. (See Section 1.4 for the formal definition of equicontinuity.) Thus if $\phi$ is equicontinuous at $\alpha$, we can approximate $\phi^n(\alpha)$ quite well by computing $\phi^n(\beta)$ for any point $\beta$ that is close to $\alpha$. But if $\phi$ is chaotic at $\alpha$, then no matter how close we choose $\alpha$ and $\beta$, eventually $\phi^n(\alpha)$ and $\phi^n(\beta)$ move away from each other.

*Example* 0.6. (*Arithmetic Sets*) An arithmetic set is a set such as $\mathbb{Z}$ or $\mathbb{Q}$ or a number field that is of number-theoretic interest, but doesn't have a natural underlying topology. More precisely, an arithmetic set tends to have a variety of interesting topologies; for example, $\mathbb{Q}$ has the archimedean topology induced by the inclusion $\mathbb{Q} \subset \mathbb{R}$ and the $p$-adic topologies induced by the inclusions $\mathbb{Q} \subset \mathbb{Q}_p$. In the arithmetic setting, the map $\phi$ is generally a polynomial or a rational map. Here are some typical arithmetical-dynamical questions, where we take $\phi(z) \in \mathbb{Q}(z)$ to be a rational function of degree $d \geq 2$ with rational coefficients:

- Let $\alpha \in \mathbb{Q}$ be a rational number. Under what conditions can the orbit $\mathcal{O}_\phi(\alpha)$ contain infinitely many integer values? In other words, when can $\mathcal{O}_\phi(\alpha) \cap \mathbb{Z}$ be an infinite set?
- Is the set $\operatorname{Per}(\phi, \mathbb{Q})$ of rational periodic points finite or infinite? If finite, how large can it be?
- Let $\alpha \in \operatorname{Per}(\phi)$ be a periodic point for $\phi$. It is clear that $\alpha$ is an algebraic number. What are the arithmetic properties of the field $\mathbb{Q}(\alpha)$, or more generally of the field generated by all of the periodic points of a given period?

**What is in this book**: We provide a brief summary of the material that is covered.

1. *An Introduction to Classical Dynamics*
   We begin in Chapter 1 with a short self-contained overview, without proofs, of classical complex dynamics on the projective line.
2. *Dynamics over Local Fields: Good Reduction*
   Chapter 2, which starts our study of arithmetic dynamics, considers rational maps $\phi(z)$ with coefficients in a local field $K$, for example, $K = \mathbb{Q}_p$. The emphasis in Chapter 2 is on maps that have "good reduction modulo $p$." The good

reduction property imples that many of the geometric properties of $\phi$ acting on the points of $K$ are preserved under reduction modulo $p$. In particular, the map $\phi$ is $p$-adically nonexpanding, and periodic points behave well when reduced modulo $p$. The remainder of the chapter gives applications exploiting these two key properties of good reduction.

3. *Dynamics over Global Fields*
   We move on in Chapter 3 to arithmetic dynamics over global fields such as $\mathbb{Q}$ and its finite extensions. Just as in the study of Diophantine equations over global fields, the theory of height functions plays a key role, and we develop this theory, including the construction of the canonical height associated to a rational map. We discuss rationality of preperiodic points and formulate a general uniform boundedness conjecture. Using classical results from the theory of Diophantine approximation, we describe exactly which rational maps $\phi$ can have orbits containing infinitely many integer points, and we give a more precise result saying that the numerator and denominator of $\phi^n(\alpha)$ grow at approximately the same rate. We consider the extension fields generated by periodic points and describe their Galois groups, ramification, and units.
4. *Families of Dynamical Systems*
   At this point we change our perspective, and rather than studying the dynamics of a single rational map, we consider families of rational maps and the variation of their dynamical properties. We construct various kinds of parameter and moduli spaces, including the space of quadratic polynomials with a point of exact period $N$ (which are analogues of the classical modular curves $X_1(N)$), the parameter space $\mathrm{Rat}_d$ of rational functions of degree $d$, and the moduli space $\mathcal{M}_d$ of rational functions of degree $d$ modulo the natural conjugation action by $\mathrm{PGL}_2$. In particular, we prove that $\mathcal{M}_2$ is an isomorphism to the affine plane $\mathbb{A}^2$. We also study twists of rational maps, analogous to the classical theory of twists of varieties, and the field-of-moduli versus field-of-definition problem.
5. *Dynamics over Local Fields: Bad Reduction*
   Chapter 5 returns to arithmetic dynamics over local fields, but now in the case of "bad reduction." It becomes necessary to work over an algebraically closed field, so we discuss the field $\mathbb{C}_p$ and give a brief introduction to nonarchimedean analysis and Newton polygons. Using these tools, we define the nonarchimedean Julia and Fatou sets and prove a version of Montel's theorem that is then used to study periodic points and wandering domains in the nonarchimedean setting. This is followed by the construction of $p$-adic Green functions and local canonical heights. The chapter concludes with a short introduction to dynamics on Berkovich space. The Berkovich projective line $\mathbb{P}^{\mathcal{B}}$ is path connected, compact, and Hausdorff, yet it naturally contains the totally disconnected, non-locally compact, non-Hausdorff space $\mathbb{P}^1(\mathbb{C}_p)$.
6. *Dynamics Associated to Algebraic Groups*
   There is a small collection of rational maps whose dynamics are much easier to understand than those of a general map. These special rational maps are associated to endomorphisms of algebraic groups. We devote Chapter 6 to the study of

these maps. The easiest ones are the power maps $M_d(z) = z^d$ and the Chebyshev polynomials $T_d(z)$ characterized by $T_d(2\cos\theta) = 2\cos(d\theta)$. They are associated to the multiplicative group. More interesting are the Lattès maps attached to elliptic curves. We give a short description, without proofs, of the theory of elliptic curves and then spend the remainder of the chapter discussing dynamical and arithmetic properties of Lattès maps.

7. *Dynamics in Dimension Greater Than One*
With a few exceptions, the results in Chapters 1–6 all deal with iteration of maps on the one-dimensional space $\mathbb{P}^1$, i.e., they are dynamics of one variable. In Chapter 7 we consider some of the issues that arise in studying dynamics in higher dimensions. We first study a class of rational maps $\phi : \mathbb{P}^N \to \mathbb{P}^N$ that are not everywhere defined. Even over $\mathbb{C}$, the geometry of dynamics of rational maps is imperfectly understood. We restrict attention to automorphisms $\phi : \mathbb{A}^N \to \mathbb{A}^N$ and study height functions and rationality of periodic points for such maps. We next consider morphisms $\phi : X \to X$ of varieties other than $\mathbb{P}^N$. In order to deal with higher-dimensional dynamics, we use tools from basic algebraic geometry and Weil's height machine, which we describe without proof. We then study arithmetic dynamics, heights, and periodic points on K3 surfaces admitting two noncommuting involutions $\iota_1$ and $\iota_2$. The composition $\phi = \iota_1 \circ \iota_2$ provides an automorphism $\phi : X \to X$ whose geometric and arithmetic dynamical properties are quite interesting.

**What's missing**: A book necessarily reflects the author's interests and tastes, while space considerations limit the amount of material that can be included. There are thus many omitted topics that naturally fit into the purview of arithmetic dynamics. Some of these are active areas of current mathematical research with their own literature, including introductory and advanced textbooks. Others are younger areas that deserve books of their own. Examples of both kinds include the following, some of which overlap with one another:

- *Dynamics over finite fields*
This includes general iteration of polynomial and rational maps acting on finite fields, see for example [41, 42, 102, 106, 109, 179, 220, 222, 275, 308, 336, 350, 385, 384, 400, 437, 444], and more specialized topics such as permutation polynomials [275, Chapter 7] that are fields of study in their own right.
- *Dynamics over function fields*
The study of function fields over finite fields has long provided a parallel theory to the study of number fields, but inseparability and wild ramification often lead to striking differences, while function fields of characteristic 0 present their own arithmetic challenges, e.g., they have infinitely many points of bounded height. The study of arithmetic dynamics over function fields is in its infancy. For a handful of results, see [20, 61, 85, 107, 207, 279, 354, 356, 415].
- *Iteration of power series*
There is an extensive literature, but no textbook, on the iteration properties of formal, $p$-adic, and Puiseux power series. Among the fundamental problems are

the classification of nontrivial commuting power series (to what extent do they come from formal groups?) and the description of preperiodic points. See for example [246, 266, 267, 268, 269, 270, 271, 272, 273, 274, 280, 281, 282, 283, 284, 389, 390].

- *Algebraic dynamics*
  There is no firm line between arithmetic dynamics and algebraic dynamics, and indeed much of the material in this book is quite algebraic. Some topics of an algebraic nature that we do not cover include irreducibility of iterates [6, 17, 113, 115, 345, 346, 425], formal transformations and algebraic identities [55, 80, 79], and various results of an algebro-geometric nature [140, 149, 163, 392].
- *Lie groups and homogeneous spaces, ergodic theory, and entropy*
  This is a beautiful and much studied area of mathematics in which geometry, analysis, and algebra interact. There are many results of a global arithmetic nature, including for example hard problems of Diophantine approximation, as well as an extensive $p$-adic theory. For an introduction to some of the main ideas and theorems in this area, see [47, 247, 304, 423], and for other arithmetic aspects of ergodic theory and entropy, including relations with height functions, ergodic theory in a nonarchimedean setting, and arithmetic properties of dynamics on solenoids, see for example [9, 31, 104, 130, 147, 160, 188, 229, 239, 241, 251, 277, 278, 351, 440, 441, 442, 447].
- *Equidistribution in arithmetic dynamics*
  There are many ways to measure (arithmetic) equidistribution, including via canonical heights, $p$-adic measures, and invariant measures on projective and Berkovich spaces. In Section 3.10 we summarize some basic equidistribution conjectures and theorems (without proof). For additional material, see [15, 24, 28, 98, 99, 168, 182, 209, 429, 432, 450].
- *Topology and arithmetic dynamics on foliated spaces*
  This surprising connection between these diverse areas of mathematics has been inverstigated by Deninger in a series of papers [124, 125, 126, 127, 128].
- *Dynamics on Drinfeld modules*
  It is natural to study local and global arithmetic dynamics in the setting of Drinfeld modules, although only a small amount of work has yet been done. See for example [180, 181, 182, 349, 395].
- *Number-theoretic iteration problems not arising as maps on varieties*
  A famous example of this type of problem is the notorious $3x+1$ problem, see [253] for an extensive bibliography. Another problem that people have studied is iteration of arithmetic functions such as Euler's $\varphi$ function; see for example [151, 342].
- *Realizability of integer sequences*
  A sequence $(a_n)$ of nonnegative integers is said to be realizable if there are a set $S$ and a function $\phi : S \to S$ with the property that for all $n$, the map $\phi$ has $a_n$ periodic points of order $n$. See [158] for an overview and [12, 144, 157, 362, 363, 419] for further material on realizable sequences.

**Prerequisites**: The principal prerequisite for reading this book is basic algebraic number theory (rings of integers, ideals and ideal class groups, units, valuations and absolute values, completions, ramification, etc.) as covered, for example, in the first section of Lang's *Algebraic Number Theory* [258]. We also assume some knowledge of elementary complex analysis as typically covered in an undergraduate course in the subject. No background in dynamics or algebraic geometry is required; we summarize and give references as necessary. In particular, to help make the book reasonably self-contained, we have included introduction/overview material on non-archimedean analysis in Section 5.2, elliptic curves in Section 6.3, and algebraic geometry in Section 7.2. However, previous familiarity with basic algebraic geometry will certainly be helpful in reading some parts of the book, especially Chapters 4 and 7.

**Cross-references and exercises**: Theorems, propositions, examples, etc. are numbered consecutively within each chapter and cross-references are given in full; for example, Proposition 3.2 refers to the second labeled item in Chapter 3. Exercises appear at the end of each chapter and are also numbered consecutively, so Exercise 5.7 is the seventh exercise in Chapter 5. There is an extensive bibliography, with reference numbers in the text given in square brackets.

This book contains a large number of exercises. Some of the exercises are marked with a single asterisk * , which indicates a hard problem. Others exercises are marked with a double asterisk ** , which means that the author does not know how to solve them. However, it should be noted that these "unsolved" problems are of varying degrees of difficulty, and in some cases their designation reflects only the author's lack of perspicacity. On the other hand, some of the unsolved problems are undoubtedly quite difficult. The author solicits solutions to the ** marked problems, as well as solutions to the exercises that are posed as questions, for inclusion in later editions. The reader will find additional notes and references for the exercises on page 441.

**Standard Notation**: Throughout this book we use the standard symbols

$$\mathbb{Z},\ \mathbb{Q},\ \mathbb{R},\ \mathbb{C},\ \mathbb{F}_q,\ \mathbb{Z}_p,\ \mathbb{A}^N,\ \text{and}\ \mathbb{P}^N$$

to represent the integers, rational numbers, real numbers, complex numbers, field with $q$ elements, ring of $p$-adic integers, $N$-dimensional affine space, and $N$-dimensional projective space, respectively. Additional notation is defined as it is introduced in the text. A detailed list of notation may be found on page 445.

# Exercises

**0.1.** Let $S$ be a set and $\phi : S \to S$ a function.

(a) If $S$ is a finite set, prove that $\phi$ is bijective if and only if $\mathrm{Per}(\phi, S) = S$.

(b) In general, prove that if $\mathrm{Per}(\phi, S) = S$, then $\phi$ is bijective.

(c) Give an example of an infinite set $S$ and map $\phi$ with the property that $\phi$ is bijective and $\mathrm{Per}(\phi, S) \neq S$.

(d) If $\phi$ is injective, prove that $\operatorname{PrePer}(\phi, S) = \operatorname{Per}(\phi, S)$.

**0.2.** Let $S$ be a set, let $\phi : S \to S$ and $\psi : S \to S$ be two maps of $S$ to itself, and suppose that $\phi$ and $\psi$ commute, i.e., assume that $\phi \circ \psi = \psi \circ \phi$.

(a) Prove that $\psi\bigl(\operatorname{PrePer}(\phi)\bigr) \subset \operatorname{PrePer}(\phi)$.

(b) Assume further that $\psi$ is a finite-to-one surjective map, i.e., $\psi(S) = S$, and for every $x \in S$, the inverse image $\psi^{-1}(x)$ is finite. Prove that $\psi\bigl(\operatorname{PrePer}(\phi)\bigr) = \operatorname{PrePer}(\phi)$.

(c) We say that a point $P \in S$ is an *isolated preperiodic point of* $\phi$ if there are integers $n > m$ such that $\phi^n(P) = \phi^m(P)$ and such that the set

$$\bigl\{Q \in S : \phi^n(Q) = \phi^m(Q)\bigr\}$$

is finite. Suppose that every preperiodic point of $\phi$ is isolated. Prove that

$$\operatorname{PrePer}(\phi) \subset \operatorname{PrePer}(\psi).$$

Conclude that if the commuting maps $\phi$ and $\psi$ both have isolated preperiodic points, then $\operatorname{PrePer}(\phi) = \operatorname{PrePer}(\psi)$.

**0.3.** Let $\phi(z) = z^d + a \in \mathbb{Z}[z]$ and let $p$ be a prime. Prove that $\operatorname{Per}(\phi, \mathbb{F}_p) = \mathbb{F}_p$ if and only if $\gcd(d, p-1) = 1$.

**0.4.** Let $G$ be a group and let $\phi : G \to G$ be a homomorphism.

(a) Prove that $\operatorname{Per}(\phi, G)$ is a subgroup of $G$.

(b) Is $\operatorname{PrePer}(\phi, G)$ a subgroup of $G$? Either prove that it is a subgroup or give a counterexample.

**0.5.** Let $G$ be a topological group, that is, $G$ is a topological space with a group structure such that the group composition and inversion laws are continuous maps. Let $\phi : G \to G$ be a continuous homomorphism. Exercise 0.4 says that $\operatorname{Per}(\phi, G)$ is a subgroup of $G$, so its topological closure $\overline{\operatorname{Per}(\phi, G)}$ is also a subgroup of $G$. Compute this topological closure for each of the following examples. (In each example, $d \ge 2$ is a fixed integer.)

(a) $G = \mathbb{C}^*$ and $\phi(\alpha) = \alpha^d$.

(b) $G = \mathbb{R}^*$ and $\phi(\alpha) = \alpha^d$.

(c) $G = \mathbb{R}^N/\mathbb{Z}^N$ and $\phi(\alpha) = d\alpha \bmod \mathbb{Z}^N$.

**0.6.** (a) Describe $\operatorname{Per}(\phi, \mathbb{Q})$ for the function $\phi(z) = z^2 + 1$.

(b) Describe $\operatorname{Per}(\phi, \mathbb{Q})$ for the function $\phi(z) = z^2 - 1$.

(c) Let $\phi(z) \in \mathbb{Z}[z]$ be a monic polynomial of degree at least two. Prove that $\operatorname{Per}(\phi, \mathbb{Q})$ is finite. (*Hint.* First prove that $\operatorname{Per}(\phi, \mathbb{Q}) \subset \mathbb{Z}$.)

(d) Same question as (c), but now $\phi(z) \in \mathbb{Q}[z]$ has rational coefficients and is not assumed to be monic.

**0.7.** Let $\phi(z) = z + 1/z$ and let $\alpha \in \mathbb{Q}^*$. Prove that $\mathcal{O}_\phi(\alpha) \cap \mathbb{Z}$ is finite. What is the largest number of points that it can contain?

# Chapter 1

# An Introduction to Classical Dynamics

Classically, the subject of discrete complex dynamical systems involves the study of iteration of polynomial and rational maps on $\mathbb{C}$ or $\mathbb{P}^1(\mathbb{C})$. The local theory of analytic iteration dates back to the 19th century, but the modern theory of global complex dynamics starts with the foundational works of Fatou [166, 167] and Julia [223] in 1918–1920. From that beginning has arisen a vast literature.

Our goal in this book is to study number-theoretic questions associated to iteration of polynomial and rational maps. A standard technique in number theory is to attempt to answer questions related to a number field $K$ by first studying analogous questions over each completion $K_v$ of $K$. In particular, the archimedean completions of $K$ lead back to classical dynamics over $\mathbb{R}$ or $\mathbb{C}$. In this chapter we describe some of the fundamental concepts and theorems in complex dynamics. These theorems provide both tools and templates for our subsequent study of dynamics over complete nonarchimedean fields in Chapters 2 and 5 and over number fields in Chapters 3 and 4. This chapter provides sufficient background and motivation for reading the remainder of this book, but the reader wishing to learn more about the beautiful subject of complex dynamics might look at one or more of the following texts:

- A. Beardon [43], *Iteration of Rational Functions*.
- L. Carleson and T. Gamelin [95], *Complex Dynamics*.
- R. Devaney [132], *An Introduction to Chaotic Dynamical Systems*.
- J. Milnor [302], *Dynamics in One Complex Variable*.

## 1.1 Rational Maps and the Projective Line

A *rational function* $\phi(z) \in \mathbb{C}(z)$ is a quotient of polynomials

$$\phi(z) = \frac{F(z)}{G(z)} = \frac{a_0 + a_1 z + \cdots + a_d z^d}{b_0 + b_1 z + \cdots + b_d z^d}$$

with no common factors.[1] The *degree of* $\phi$ is

$$\deg\phi = \max\{\deg F, \deg G\}.$$

So if we write $\phi$ as above with at least one of $a_d$ and $b_d$ nonzero, then $\phi$ has degree $d$. We generally consider maps with $\deg\phi \geq 2$.

A rational function of degree $d$ induces a *rational map* of the *complex projective line* $\mathbb{P}^1(\mathbb{C})$,

$$\phi : \mathbb{P}^1(\mathbb{C}) \longrightarrow \mathbb{P}^1(\mathbb{C}). \tag{1.1}$$

The map $\phi$ is a ramified $d$-to-1 covering. The complex projective line is also known as the *Riemann sphere*, since topologically, and indeed real analytically, $\mathbb{P}^1(\mathbb{C})$ is isomorphic to $S^2$. Two very important properties of rational maps (1.1) are that they are *continuous* and *open*.[2]

There are many equivalent (and useful) ways to construct the projective line, including

$$\mathbb{P}^1(\mathbb{C}) \cong \mathbb{C} \cup \{\infty\} \qquad \text{and} \qquad \mathbb{P}^1(\mathbb{C}) \cong \frac{\mathbb{C}^2 \smallsetminus \{(0,0)\}}{\sim},$$

where $[X,Y] \sim [X',Y']$ if there is a $u \in \mathbb{C}^*$ such that $X' = uX$ and $Y' = uY$. In terms of homogeneous coordinates, a rational map $\phi : \mathbb{P}^1 \to \mathbb{P}^1$ is given by a pair of homogeneous polynomials $\phi(X,Y) = [F^*(X,Y), G^*(X,Y)]$ of degree $d$. These are related to the description $\phi(z) = F(z)/G(z)$ by the relations

$$F^*(X,Y) = Y^d F(X/Y) \qquad \text{and} \qquad G^*(X,Y) = Y^d G(X/Y).$$

A *linear fractional transformation* (or *Möbius transformation*) is a map of the form

$$z \longmapsto \frac{az+b}{cz+d} \qquad \text{with } ad - bc \neq 0.$$

It defines an automorphism of $\mathbb{P}^1$, and composition of transformations corresponds to multiplication of the corresponding matrices $\left(\begin{smallmatrix} a & b \\ c & d \end{smallmatrix}\right)$. Two matrices give the same transformation if and only if they are scalar multiples of one another, and it is not hard to prove that these are the only automorphisms of $\mathbb{P}^1(\mathbb{C})$; see [198, Exercise I.6.6 and Remark II.7.1.1]. So we have

$$\operatorname{Aut}(\mathbb{P}^1(\mathbb{C})) = \mathrm{PGL}_2(\mathbb{C}) = \mathrm{GL}_2(\mathbb{C})/\mathbb{C}^*.$$

We observe that given any two triples $(\alpha, \alpha', \alpha'')$ and $(\beta, \beta', \beta'')$ of distinct points in $\mathbb{P}^1$, there exists a unique automorphism $f \in \mathrm{PGL}_2(\mathbb{C})$ satisfying

$$f(\alpha) = \beta, \quad f(\alpha') = \beta', \quad f(\alpha'') = \beta''.$$

---

[1]It is permissible to have $G(z) = 0$, in which case $F(z) = a_0$ is a nonzero constant and $\phi(z)$ maps every point to $\infty$. Similarly we may have $F(z) = 0$ and $G(z) = b_0 \neq 0$. Note that by convention we set $\deg 0 = -\infty$, so if $\phi(z)$ is constant, then $\deg\phi = 0$.

[2]Let $\phi : X \to Y$ be a map between topological spaces. We recall that $\phi$ is *continuous* if $U \subset Y$ open implies $\phi^{-1}(U) \subset X$ open, and $\phi$ is *open* if $V \subset X$ open implies $\phi(V) \subset Y$ open.

This is intuitively reasonable because $\mathrm{PGL}_2(\mathbb{C})$ is a three-dimensional space that acts more-or-less simply transitively on $\mathbb{P}^1 \times \mathbb{P}^1 \times \mathbb{P}^1$.

If $\phi : \mathbb{P}^1 \to \mathbb{P}^1$ is a rational map and $f \in \mathrm{PGL}_2(\mathbb{C})$, the *linear conjugate of $\phi$ by $f$* is the map

$$\phi^f = f^{-1} \circ \phi \circ f.$$

Linear conjugation corresponds to a change of variables on $\mathbb{P}^1$ as illustrated by the commutative diagram

$$\begin{array}{ccc} \mathbb{P}^1 & \xrightarrow{\phi^f} & \mathbb{P}^1 \\ {\scriptstyle f}\big\downarrow & & {\scriptstyle f}\big\downarrow \\ \mathbb{P}^1 & \xrightarrow{\phi} & \mathbb{P}^1 \end{array}$$

Two rational maps $\phi$ and $\psi$ are *linearly conjugate* if $\psi = \phi^f$ for some $f \in \mathrm{PGL}_2(\mathbb{C})$. This is clearly an equivalence relation. We also observe that

$$(\phi^f)^n = f^{-1} \circ \phi^n \circ f = (\phi^n)^f,$$

which shows that linear conjugation is a good operation to use when studying iteration.

A convenient metric on the projective line $\mathbb{P}^1(\mathbb{C})$ is the *chordal metric*, given in homogeneous coordinates by the formula

$$\rho([X_1, Y_1], [X_2, Y_2]) = \frac{|X_1Y_2 - X_2Y_1|}{\sqrt{|X_1|^2 + |Y_1|^2}\sqrt{|X_2|^2 + |Y_2|^2}}.$$

If neither point is the point at infinity (i.e., neither point is equal to $[1, 0]$), then using the substitution $z_i = X_i/Y_i$ to dehomogenize gives

$$\rho(z_1, z_2) = \frac{|z_1 - z_2|}{\sqrt{|z_1|^2 + 1}\sqrt{|z_2|^2 + 1}}.$$

Notice that $0 \le \rho \le 1$. Identifying $\mathbb{P}^1(\mathbb{C}) \cong \mathbb{C} \cup \{\infty\}$ with the sphere $S^2$ by drawing lines from the north pole of the sphere (see Figure 1.1), the chordal metric $\rho(z, w)$ is related to the Euclidean distance between the corresponding points on the sphere by the formula

$$\rho(z, w) = \frac{1}{2}|z^* - w^*|.$$

(See Exercise 1.1.) In particular, the triangle inequality in $\mathbb{R}^3$ implies the triangle inequality for the chordal metric

$$\rho(P_1, P_3) \le \rho(P_1, P_2) + \rho(P_2, P_3) \qquad \text{for all } P_1, P_2, P_3 \in \mathbb{P}^1(\mathbb{C}). \tag{1.2}$$

Rational maps $\phi : \mathbb{P}^1(\mathbb{C}) \to \mathbb{P}^1(\mathbb{C})$ also have the *Lipschitz property* relative to the chordal metric. This means that there is a constant $C(\phi)$ such that

$$\rho(\phi(\alpha), \phi(\beta)) \le C(\phi)\rho(\alpha, \beta) \qquad \text{for all } \alpha, \beta \in \mathbb{P}^1(\mathbb{C}). \tag{1.3}$$

(See Exercise 1.3.)

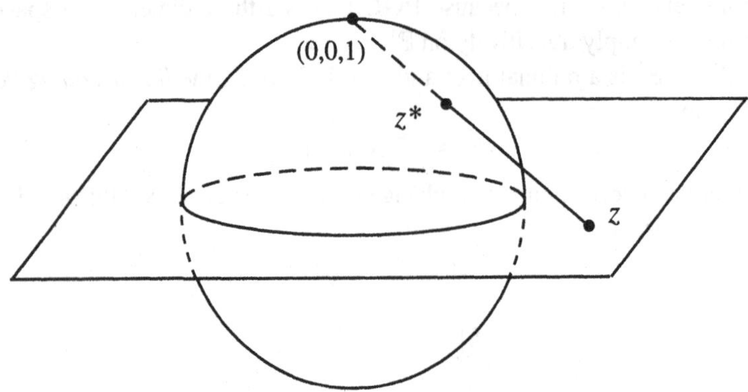


Figure 1.1: Identifying $\mathbb{C} \cup \infty$ with the Riemann sphere.

## 1.2 Critical Points and the Riemann–Hurwitz Formula

Given a rational function $\phi(z) = F(z)/G(z)$, the formal derivative

$$\phi'(z) = \frac{G(z)F'(z) - F(z)G'(z)}{G(z)^2}$$

is defined at any point $\alpha$ satisfying $\alpha \neq \infty$ and $G(\alpha) \neq 0$. (In order to compute the derivative at the excluded points, make a linear change of variables. See Exercise 1.5.)

Looking locally around $z = \alpha$, Taylor's theorem says that

$$\phi(z) = \phi(\alpha) + \phi'(\alpha)(z - \alpha) + O\big((z - \alpha)^2\big).$$

The map $\phi : \mathbb{P}^1(\mathbb{C}) \to \mathbb{P}^1(\mathbb{C})$ is *ramified* at $\alpha$ if $\phi'(\alpha) = 0$, in which case the point $\alpha$ is called a *critical point* or a *ramification point* of $\phi$. (Algebraic geometers talk about ramification points; dynamicists talk about critical points.) Note that $\phi$ is locally one-to-one around a noncritical point. As we will see, the orbits of the critical points play an important role in determining the behavior of the dynamical system.

Assuming as above that $\alpha \neq \infty$ and $\phi(\alpha) \neq \infty$, the *ramification index* of $\phi$ at $\alpha$ is

$$e_\alpha(\phi) = \operatorname{ord}_\alpha\big(\phi(z) - \phi(\alpha)\big).$$

In particular, $\phi$ is ramified at $\alpha$ if and only if $e_\alpha(\phi) \geq 2$. Locally, there is a constant $c \neq 0$ such that

$$\phi(z) = \phi(\alpha) + c(z - \alpha)^{e_\alpha(\phi)} + O\left((z - \alpha)^{e_\alpha(\phi)+1}\right),$$

and $\phi$ is locally $e_\alpha(\phi)$-to-1 in a neighborhood of $\alpha$. It is not hard to see that $e_\phi(\alpha) \leq \deg(\phi)$. If $e_\phi(\alpha) = \deg(\phi)$, then we say that $\phi$ is *totally ramified at $\alpha$*.

The next result gives a global relation satisfied by the locally defined ramification indices. We sketch two distinct proofs of this important theorem.

**Theorem 1.1.** (Riemann–Hurwitz Formula for $\mathbb{P}^1$)

$$2d-2=\sum_{\alpha\in\mathbb{P}^1}\bigl(e_\alpha(\phi)-1\bigr).$$

*Proof.* (Algebraic Proof) After a change of variables using a linear fractional transformation, we may assume that $\infty$ is neither a ramification point nor the image of a ramification point, and also that $\phi(\infty)=0$. This means that $\phi(z)$ has the form

$$\phi(z)=\frac{F(z)}{G(z)}=\frac{az^{d-1}+\cdots}{bz^d+\cdots}$$

with $ab\neq 0$. Then

$$\phi'(z)=\frac{G(z)F'(z)-F(z)G'(z)}{G(z)^2}=\frac{-abz^{2d-2}+\cdots}{G(z)^2}.$$

For any $\alpha\neq\infty$ such that $\phi(\alpha)\neq\infty$, i.e., such that $G(\alpha)\neq 0$, we have

$$\phi(z)=\phi(\alpha)+(z-\alpha)^{e_\alpha(\phi)}\psi(z)$$

for a function $\psi(z)\in\mathbb{C}(z)$ satisfying $\psi(\alpha)\neq 0,\infty$. Differentiating yields

$$\phi'(z)=e_\alpha(\phi)(z-\alpha)^{e_\alpha(\phi)-1}\psi(z)+(z-\alpha)^{e_\alpha(\phi)}\psi'(z),$$

from which we deduce

$$\operatorname{ord}_\alpha\phi'(z)=e_\alpha(\phi)-1.$$

(Notice we are using strongly the fact that $\mathbb{C}$ has characteristic 0.) Summing over $\alpha$ gives

$$\sum_{\alpha,\phi(\alpha)\neq\infty}\bigl(e_\alpha(\phi)-1\bigr)=\sum_{\alpha,\phi(\alpha)\neq\infty}\operatorname{ord}_\alpha\phi'(z)=\deg(-abz^{2d-2}+\cdots)=2d-2.$$

(Topological Proof) It is a well-known topological fact that if a sphere is triangulated with $V$ vertices, $E$ edges, and $F$ faces, then $V-E+F=2$. We take a (sufficiently small) triangulation of the sphere with the property that the image of every ramification point of $\phi$ is a vertex. Let

$$\{v_i:1\le i\le V\},\quad \{e_i:1\le i\le E\},\quad \{f_i:1\le i\le F\},$$

be the vertices, edges, and faces of the triangulation. The map $\phi$ is a local isomorphism except around the ramification points, so

$$\{\phi^{-1}(v_i):1\le i\le V\}\cup\{\phi^{-1}(e_i):1\le i\le E\}\cup\{\phi^{-1}(f_i):1\le i\le F\}$$

also gives a triangulation of the sphere. To ease notation, we let $V'$, $E'$, and $F'$ denote the numbers of vertices, edges, and faces in this inverse image triangulation.

The map $\phi$ is exactly $d$-to-1 except over the ramification points, so the set

$$\{\phi^{-1}(e_i) : 1 \le i \le E\}$$

consists of exactly $dE$ edges, and similarly the set

$$\{\phi^{-1}(f_i) : 1 \le i \le F\}$$

consists of exactly $dF$ faces. In other words, $E' = dE$ and $F' = dF$.

The vertices are a little more complicated. If $v$ is a vertex, then $\phi^{-1}(v)$ consists of $d$ points counted with appropriate multiplicities, where the multiplicity of $\alpha \in \phi^{-1}(v)$ is precisely $e_\alpha(\phi)$. In other words,

$$d = \sum_{\alpha \in \phi^{-1}(v)} e_\alpha(\phi),$$

or equivalently,

$$d - \#\phi^{-1}(v) = \sum_{\alpha \in \phi^{-1}(v)} (e_\alpha(\phi) - 1).$$

Summing over the vertices $v_i$ yields

$$dV - \sum_{i=1}^{V} \#\phi^{-1}(v_i) = \sum_{i=1}^{V} \sum_{\alpha \in \phi^{-1}(v_i)} (e_\alpha(\phi) - 1).$$

Note that the lefthand side is $dV - V'$. Further, since we chose the $v_i$'s to include the image of every critical point, we know that $e_\alpha(\phi) = 1$ for every point $\alpha$ not in any of the $\phi^{-1}(v_i)$'s, so we can extend the sum on the righthand side to include every $\alpha \in \mathbb{P}^1(\mathbb{C})$. This proves that

$$dV = V' + \sum_{\alpha \in \mathbb{P}^1(\mathbb{C})} (e_\alpha(\phi) - 1).$$

Next we take an alternating sum to obtain

$$d(V - E + F) = V' - E' + F' + \sum_{\alpha \in \mathbb{P}^1(\mathbb{C})} (e_\alpha(\phi) - 1).$$

Finally, we use the fact that

$$V - E + F = V' - E' + F' = 2$$

to obtain the Riemann–Hurwitz formula. □

**Corollary 1.2.** *A rational map of degree d has exactly* $2d-2$ *critical points, counted with appropriate multiplicities.*

The following weaker version of the Riemann–Hurwitz formula is often convenient for applications.

**Corollary 1.3.** *Let $\phi : \mathbb{P}^1(\mathbb{C}) \to \mathbb{P}^1(\mathbb{C})$ be a rational map of degree $d \ge 1$.*
(a) *Let $\alpha \in \mathbb{P}^1(\mathbb{C})$. Then*

$$\sum_{\beta \in \phi^{-1}(\alpha)} e_\beta(\phi) = d.$$

(b) [Weak Riemann–Hurwitz Formula]

$$2d - 2 = \sum_{\alpha \in \mathbb{P}^1(\mathbb{C})} \left(d - \#\phi^{-1}(\alpha)\right).$$

*Proof.* Making a change of coordinates reduces us to the case that $\alpha \ne \infty$ and $\infty \notin \phi^{-1}(\alpha)$. Writing $\phi(z) = F(z)/G(z)$ and $\phi^{-1}(\alpha) = \{\beta_1, \dots, \beta_r\}$, this means that there is a factorization

$$F(z) - \alpha G(z) = c(z - \beta_1)^{e_1}(z - \beta_2)^{e_2} \cdots (z - \beta_r)^{e_r}.$$

Notice that the exponents are exactly the ramification indices $e_i = e_{\beta_i}(\phi)$, so

$$\sum_{\beta \in \phi^{-1}(\alpha)} e_\beta(\phi) = \sum_{i=1}^{r} e_i = \deg(\phi) = d.$$

This proves (a). We then use the Riemann–Hurwitz formula (Theorem 1.1) to compute

$$\begin{aligned} 2d - 2 = \sum_{\beta \in \mathbb{P}^1} \left(e_\beta(\phi) - 1\right) &= \sum_{\alpha \in \mathbb{P}^1(\mathbb{C})} \sum_{\beta \in \phi^{-1}(\alpha)} \left(e_\beta(\phi) - 1\right) \\ &= \sum_{\alpha \in \mathbb{P}^1(\mathbb{C})} \left(d - \#\phi^{-1}(\alpha)\right). \end{aligned}$$

□

*Remark* 1.4. The Riemann–Hurwitz formula is an example of a local–global formula. The quantity $2d - 2$ is a global quantity, given in terms of how many times the map $\phi$ covers the Riemann sphere $\mathbb{P}^1(\mathbb{C})$ and reflecting the topology of the sphere. The general version of the Riemann–Hurwitz formula, which we now state, includes global information about the genera of the curves under consideration.

**Theorem 1.5.** (Riemann–Hurwitz Formula) *Let $C_1$ and $C_2$ be algebraic curves (Riemann surfaces) of genus $g_1$ and $g_2$, respectively, and let $\phi : C_1 \to C_2$ be a finite map of degree $d \ge 1$. Then*

$$2g_1 - 2 = d(2g_2 - 2) + \sum_{P \in C_1} \left(e_P(\phi) - 1\right).$$

*Proof.* See [198, IV §2]. □

In this general formulation, the quantity $2g_1 - 2 - d(2g_2 - 2)$ is the global part of the Riemann–Hurwitz formula. It reflects the topology of the two curves, since the genus of a curve is an intrinsic global quantity. Of course, for rational maps $\phi : \mathbb{P}^1 \to \mathbb{P}^1$, these genera are $g_1 = g_2 = 0$.

The ramification indices, on the other hand, are purely local quantities, since they can be computed in a neighborhood of a point. Thus the Riemann–Hurwitz formula computes a difference of global topological quantities in terms of quantities that can be computed purely locally. This helps to explain both its importance and its usefulness.

The next two results illustrate some of the power of the Riemann–Hurwitz formula.

**Theorem 1.6.** *Let $\phi : \mathbb{P}^1(\mathbb{C}) \to \mathbb{P}^1(\mathbb{C})$ be a rational map of degree $d \ge 2$, and let $E \subset \mathbb{P}^1(\mathbb{C})$ be a finite set satisfying $\phi^{-1}(E) = E$. Then $\#E \le 2$. Further:*

(a) *If $\#E = 1$, then there is an $f \in \mathrm{PGL}_2(\mathbb{C})$ such that*

$$E = \{f(\infty)\} \quad \text{and} \quad \phi^f(z) \in \mathbb{C}[z].$$

(b) *If $\#E = 2$, then there is an $f \in \mathrm{PGL}_2(\mathbb{C})$ such that*

$$E = \{f(0), f(\infty)\} \quad \text{and} \quad \phi^f(z) = z^d \text{ or } z^{-d}.$$

*Proof.* The assumption that $E$ is finite and satisfies $\phi^{-1}(E) = E$ combined with the fact that $\phi : \mathbb{P}^1(\mathbb{C}) \to \mathbb{P}^1(\mathbb{C})$ is surjective implies that $\phi$ acts as a permutation on $E$. Since $E$ is finite, some power of $\phi$ acts as the identity on $E$, say $\phi^n(\alpha) = \alpha$ for every $\alpha \in E$. Replacing $\phi$ by $\phi^n$, we find that every point $\alpha \in E$ satisfies $\phi^{-1}(\alpha) = \{\alpha\}$. Thus every $\alpha \in E$ is a totally ramified fixed point of $\phi$, i.e., $\phi(\alpha) = \alpha$ and $e_\alpha(\phi) = d$. Applying the Riemann–Hurwitz formula (Theorem 1.1) yields

$$\begin{aligned} 2d - 2 &= \sum_{\beta \in \mathbb{P}^1} (e_\beta(\phi) - 1) \\ &\ge \sum_{\alpha \in E} (e_\alpha(\phi) - 1) = \sum_{\alpha \in E} (d-1) = \#E \cdot (d-1). \end{aligned}$$

Since $d \ge 2$, we conclude that $2 \ge \#E$, which completes the proof of the first part of the theorem.

Suppose next that $\#E = 1$, say $E = \{P\}$. We conjugate $\phi$ by any $f \in \mathrm{PGL}_2(\mathbb{C})$ satisfying $f^{-1}(P) = \infty$. Then $(\phi^f)^{-1}\{\infty\} = \{\infty\}$, so the only pole of $\phi^f(z)$ is $z = \infty$. This proves that $\phi^f$ is a polynomial.

Finally, suppose that $\#E = 2$, say $E = \{P, Q\}$. We conjugate $\phi$ by some $f \in \mathrm{PGL}_2(\mathbb{C})$ satisfying $f^{-1}(P) = \infty$ and $f^{-1}(Q) = 0$. There are then two cases, depending on whether $\phi^f$ fixes or permutes the set $\{0, \infty\}$. Thus

$$\begin{aligned} (\phi^f)^{-1}\{\infty\} = \{\infty\} \quad &\text{and} \quad (\phi^f)^{-1}\{0\} = \{0\} \quad \Longrightarrow \quad \phi^f(z) = cz^d, \\ (\phi^f)^{-1}\{\infty\} = \{0\} \quad &\text{and} \quad (\phi^f)^{-1}\{0\} = \{\infty\} \quad \Longrightarrow \quad \phi^f(z) = cz^{-d}. \end{aligned}$$

A further conjugation by $g(z) = az$ for an appropriate value of $a$ removes the $c$, which completes the proof of Theorem 1.6. □

**Definition.** A rational map $\phi : \mathbb{P}^1 \to \mathbb{P}^1$ is a *polynomial map* if it has a totally ramified fixed point, that is, if there is a point $\alpha \in \mathbb{P}^1$ such that $\phi(\alpha) = \alpha$ and $e_\alpha(\phi) = \deg(\phi)$.

Notice that the polynomial $\phi(z) = a_0 + a_1 z + \cdots + a_d z^d$ has $\infty$ as a totally ramified fixed point. This is typical. If $\phi$ has $\alpha$ as a totally ramified fixed point, then after a change of variables using any $f \in \mathrm{PGL}_2(\mathbb{C})$ satisfying $f(\infty) = \alpha$, one finds that $\phi^f(z)$ is in $\mathbb{C}[z]$. See Exercise 1.9.

**Definition.** A set satisfying $\phi^{-1}(E) = E = \phi(E)$ is called a *completely invariant set* for $\phi$. We note that it suffices to require $\phi^{-1}(E) = E$, since then the surjectivity of $\phi : \mathbb{P}^1(\mathbb{C}) \to \mathbb{P}^1(\mathbb{C})$ automatically implies that $\phi(E) = E$. The classical term for a finite completely invariant set is an *exceptional set*. Thus Theorem 1.6 says that an exceptional set for $\phi$ contains at most two elements. Further, if the exceptional set contains two elements, then $\phi$ is conjugate to $z^{\pm d}$, and if the exceptional set contains one element, then $\phi$ is conjugate to a polynomial.

**Theorem 1.7.** *Let $\phi : \mathbb{P}^1(\mathbb{C}) \to \mathbb{P}^1(\mathbb{C})$ be a rational map of degree $d \ge 2$, and suppose that $\phi^n$ is a polynomial map for some $n \ge 1$. Then already $\phi^2$ is a polynomial map. Further, if $\phi$ itself is not a polynomial map, then $\phi$ is linearly conjugate to the function $1/z^d$.*

*Proof.* Let $\alpha$ be a totally ramified fixed point of $\phi^n$, so $(\phi^n)^{-1}(\alpha) = \{\alpha\}$. Consider the chain of maps

$$\{\alpha\} \xrightarrow{\ \phi\ } \{\phi\alpha\} \xrightarrow{\ \phi\ } \{\phi^2\alpha\} \xrightarrow{\ \phi\ } \cdots \xrightarrow{\ \phi\ } \{\phi^{n-1}\alpha\} \xrightarrow{\ \phi\ } \{\alpha\}.$$

The fact that $(\phi^n)^{-1}(\alpha)$ consists of the single point $\alpha$ implies that each set $\{\phi^i(\alpha)\}$ in the chain is the complete inverse image of the next set $\{\phi^{i+1}(\alpha)\}$ in the chain. It follows that the set of points

$$E = \{\alpha, \phi(\alpha), \phi^2(\alpha), \phi^3(\alpha), \ldots, \phi^{n-1}(\alpha)\}$$

satisfies $\phi^{-1}(E) = E$. Now Theorem 1.6 tells us that $\#E \le 2$, so there are two cases to consider.

First, if $\#E = 1$, then $\alpha = \phi(\alpha)$, and we see that $\phi$ is a polynomial map, since it has $\alpha$ as a totally ramified fixed point. Second, if $\#E = 2$, then $\phi^2(\alpha) = \alpha \ne \phi(\alpha)$, so $\phi^2$ is a polynomial map with $\alpha$ and $\phi(\alpha)$ as distinct totally ramified fixed points.

In the second case, we choose a fixed point $\beta$ of $\phi$ and perform a linear conjugation to move $\alpha, \phi(\alpha), \beta$ to $0, \infty, 1$ respectively. Then writing $\phi(z) = P(z)/Q(z)$ as a quotient of polynomials with no common roots, we have

$$\phi^{-1}(0) = \{\infty\} = \{z : P(z) = 0\} \quad \text{and} \quad \phi^{-1}(\infty) = \{0\} = \{z : Q(z) = 0\}.$$

Thus $P(z)$ has no zeros (in $\mathbb{C}$), so it is constant, while $Q(z)$ vanishes only at 0, so has the form $Q(z) = cz^d$. Thus $\phi(z) = 1/cz^d$, and the fact that $\phi(1) = 1$ gives $\phi(z) = 1/z^d$. □

*Remark* 1.8. There are easy counterexamples to Theorem 1.7 for fields of positive characteristic, but it remains true for separable maps; see Exercise 1.11. There is also a higher-dimensional version of Theorem 1.7; see Exercise 7.16.

## 1.3 Periodic Points and Multipliers

If $\alpha$ is a fixed point of $\phi$, then the *multiplier of $\phi$ at $\alpha$* is the derivative

$$\lambda_\alpha(\phi) = \phi'(\alpha).$$

(If $\alpha = \infty$, this formula needs to be modified; see Exercise 1.13.) In general, the value of a derivative depends on a choice of coordinates, since $\phi'(z)$ is specifically the derivative with respect to the variable $z$. However, it turns out that the derivative at a fixed point is coordinate-invariant.

**Proposition 1.9.** *Let $\phi \in \mathbb{C}(z)$ be a rational map and let $\alpha$ be a fixed point of $\phi$. Let $f \in \mathrm{PGL}_2(\mathbb{C})$ be a change of coordinates and set $\beta = f^{-1}(\alpha)$, so $\beta$ is a fixed point of the conjugate map $\phi^f = f^{-1} \circ \phi \circ f$. Then*

$$\phi'(\alpha) = (\phi^f)'(\beta).$$

*Proof.* Two applications of the chain rule yield

$$(\phi^f)'(w) = (f^{-1} \circ \phi \circ f)'(w) = (f^{-1})'(\phi(f(w))) \cdot \phi'(f(w)) \cdot f'(w). \tag{1.4}$$

Hence

$$\begin{aligned}
(\phi^f)'(\beta) &= (f^{-1})'(\phi(\alpha)) \cdot \phi'(\alpha) \cdot f'(\beta) && \text{evaluating (1.4) at } w = \beta = f^{-1}(\alpha),\\
&= (f^{-1})'(\alpha) \cdot \phi'(\alpha) \cdot f'(\beta) && \text{since } \phi(\alpha) = \alpha,\\
&= (f^{-1})'(f(\beta)) \cdot \phi'(\alpha) \cdot f'(\beta) && \text{since } \alpha = f(\beta),\\
&= \phi'(\alpha) \quad \text{since } (f^{-1})'(f(z)) \cdot f'(z) = (f^{-1} \circ f)'(z) = 1.
\end{aligned}$$

□

Thus the multiplier $\lambda_\alpha(\phi)$ of a fixed point $\alpha$ of $\phi$ is well-defined, independent of the choice of coordinates on $\mathbb{P}^1$.

We next consider points that are fixed by some iterate of $\phi$. Recall that a point $\alpha$ is called a *periodic point* for $\phi$ if $\phi^n(\alpha) = \alpha$ for some $n \geq 1$. The smallest such $n$ is called the *exact period of $\alpha$*.[3] (A point of exact period 1 is a *fixed point*.) We set

$$\begin{aligned}
\mathrm{Per}_n(\phi) &= \{\alpha \in \mathbb{P}^1(\mathbb{C}) : \phi^n(\alpha) = \alpha\},\\
\mathrm{Per}_n^{**}(\phi) &= \{\alpha \in \mathrm{Per}_n(\phi) : \alpha \text{ has exact period } n\}.
\end{aligned}$$

(The reason for this notation is that there is a weaker notion of *formal period* $n$, and we will later write $\mathrm{Per}_n^*(\phi)$ to denote the set of points having formal period $n$. See Exercise 1.19 and Section 4.1.) It is easy to see that $\mathrm{Per}_n(\phi)$ is the disjoint union of $\mathrm{Per}_m^{**}(\phi)$ over all $m \mid n$; see Exercise 1.14.

[3] In the literature one finds many terms for the smallest value of $n$ satisfying $\phi^n(\alpha) = \alpha$, including *exact period, least period, primitive period,* and *prime period.* We will use the first three interchangeably, but we eschew the fourth, since in arithmetic dynamics, the phrase "$\phi$ has prime period $n$ at $\alpha$" should mean that the integer $n$ is a prime number!

Let $\alpha \in \mathrm{Per}_n^{**}(\phi)$ be a point of exact period $n$ for $\phi$. In particular, $\alpha$ is a fixed point of $\phi^n$. We define the *multiplier of $\phi$ at $\alpha$* to be

$$\lambda_\alpha(\phi) = (\phi^n)'(\alpha).$$

Notice that this may be calculated using the chain rule as

$$(\phi^n)'(\alpha) = \phi'(\alpha) \cdot \phi'(\phi\alpha) \cdot \phi'(\phi^2\alpha) \cdots \phi'(\phi^{n-1}\alpha).$$

In other words, $\lambda_\alpha(\phi) = (\phi^n)'(\alpha)$ is the product of the values of $\phi'$ at each of the $n$ distinct points in the orbit of $\alpha$.

*Remark* 1.10. The set of multipliers

$$\{\lambda_\alpha(\phi) : \alpha \in \mathrm{Per}(\phi)\}$$

depends only on the conjugacy class $\{\phi^f : f \in \mathrm{PGL}_2(\mathbb{C})\}$ of $\phi$. It can thus be used to define useful dynamical invariants of the map $\phi$.

*Remark* 1.11. There is a more intrinsic way to define the multiplier of a fixed point $\alpha$ in terms of the space of differential one-forms $\Omega_\alpha^1$ on $\mathbb{P}^1$ at $\alpha$. This space has dimension 1, since $\mathbb{P}^1$ is nonsingular of dimension 1, so $\phi^* : \Omega_\alpha^1 \to \Omega_\alpha^1$ is an element of $\mathrm{GL}(\Omega_\alpha^1) = \mathbb{C}^*$. This element of $\mathbb{C}^*$ is $\lambda_\alpha(\phi)$. More concretely, taking any nonzero $\omega \in \Omega_\alpha^1$, we define $\lambda_\alpha(\phi)$ by the equation $\phi^*\omega = \lambda_\alpha(\phi)\omega$. For example, if $\omega = dz$ for some uniformizer $z$ at $\alpha$, then $\phi^*(dz) = d(\phi(z)) = \phi'(\alpha)dz$ in $\Omega_\alpha^1$, and we recover the earlier definition $\lambda_\alpha(\phi) = \phi'(\alpha)$.

If $|\lambda_\alpha(\phi)| < 1$, then a small neighborhood of $\alpha$ will shrink each time it returns to $\alpha$, while if $|\lambda_\alpha(\phi)| > 1$, then it will expand. This observation prompts the following definitions.

**Definition.** Let $\alpha$ be a periodic point for a rational function $\phi \in \mathbb{C}(z)$, and let $\lambda_\alpha(\phi)$ be the corresponding multiplier. Then $\alpha$ is called

$$\begin{aligned} \textit{superattracting} \quad & \text{if } \lambda_\alpha(\phi) = 0, \\ \textit{attracting} \quad & \text{if } |\lambda_\alpha(\phi)| < 1, \\ \textit{neutral} \quad & \text{if } |\lambda_\alpha(\phi)| = 1, \\ \textit{repelling} \quad & \text{if } |\lambda_\alpha(\phi)| > 1. \end{aligned}$$

(Neutral periodic points are also sometimes called *indifferent*.)

*Remark* 1.12. A periodic point is superattracting if its orbit contains a critical point, so Corollary 1.2 implies that a map of degree $d$ can have at most $2d - 2$ superattracting cycles. We will state a much stronger result below (Theorem 1.35(a)).

*Remark* 1.13. A neutral periodic point is called *rationally neutral* if its multiplier is a root of unity, that is, if $\lambda_\alpha(\phi)^k = 1$ for some $k \geq 1$. Otherwise, it is called *irrationally neutral*.

We next prove a useful formula giving a relation between the multipliers of the fixed points of a rational map. For an application of this formula to $p$-adic dynamics, see Corollary 5.19 and Proposition 5.20.

**Theorem 1.14.** *Let $K$ be an algebraically closed field and let $\phi(z) \in K(z)$ be a rational function of degree $d \geq 2$. Assume that*

$$\lambda_P(\phi) \neq 1 \qquad \textit{for all } P \in \operatorname{Fix}(\phi).$$

*Then*

$$\sum_{P \in \operatorname{Fix}(\phi)} \frac{1}{1 - \lambda_P(\phi)} = 1.$$

*Remark* 1.15. The condition that $\lambda_P(\phi) \neq 1$ is equivalent to the condition that the fixed point $P$ has multiplicity 1, so Theorem 1.14 holds provided $\phi$ has $d+1$ distinct fixed points. If some fixed point has $\lambda_P(\phi) = 1$, then a similar, but more complicated, formula involving a higher-order index is true; see Exercise 1.17 or [302, §12].

*Remark* 1.16. More generally, we can apply Theorem 1.14 to an iterate $\phi^n$ to obtain a relation among the multipliers of the $n$-periodic points of $\phi$ provided that those multipliers are not equal to 1.

*Example* 1.17. Let $\phi(z) = z^2 + c$. The fixed points of $\phi(z)$ are

$$\alpha = \frac{1 + \sqrt{1-4c}}{2}, \qquad \beta = \frac{1 - \sqrt{1-4c}}{2}, \qquad \text{and} \qquad \infty,$$

with corresponding multipliers $\lambda_\alpha(\phi) = 2\alpha$, $\lambda_\beta(\phi) = 2\beta$, and $\lambda_\infty(\phi) = 0$. (Note that the map $\phi$ is totally ramified at $\infty$, so in particular $\infty$ is a critical point.) The sum in Theorem 1.14 is

$$\frac{1}{1-2\alpha} + \frac{1}{1-2\beta} + \frac{1}{1-0} = \frac{1}{-\sqrt{1-4c}} + \frac{1}{\sqrt{1-4c}} + 1 = 1.$$

*Proof of Theorem* 1.14. We prove the theorem under the assumption that $K$ has characteristic 0, in which case the Lefschetz principle (see, e.g., [410, VI §6]) says that we can embed a sufficiently large subfield of $K$ into $\mathbb{C}$. It thus suffices to prove the theorem for $K = \mathbb{C}$. (The case of characteristic $p$ can be proven by reduction modulo $p$ from the characteristic 0 case or by developing the theory of residues in characteristic $p$ as described in Exercise 5.10.)

To simplify the proof, we make a change of variables (i.e., conjugate $\phi$ by an element of $\operatorname{PGL}_2(\mathbb{C})$) so that $\phi(\infty) \neq \infty$. For any rational function $\psi(z) \in \mathbb{C}(z)$, or more generally for any meromorphic function, the Cauchy residue theorem says that

$$\sum_{P \in \mathbb{P}^1(\mathbb{C})} \operatorname{Res}_P(\psi(z)\,dz) = 0.$$

For $P = \alpha \in \mathbb{C}$ in the affine plane, the residue is given by the usual formula

$$\operatorname{Res}_{z=\alpha}(\psi(z)\,dz) = \frac{1}{2\pi i} \int_{|z-\alpha|=\epsilon} \psi(z)\,dz \qquad \text{for sufficiently small } \epsilon > 0.$$

For $P = \infty$, we substitute $z = 1/w$ and use the same formula to compute the residue with respect to $w$ at $\alpha = 0$.

We apply the Cauchy residue formula to the rational function

$$\psi(z) = \frac{1}{\phi(z) - z},$$

whose poles are the fixed points of $\phi$. For any $\alpha \in \operatorname{Fix}(\phi)$ (note that $\alpha \neq \infty$) we observe that

$$\begin{aligned} \phi(z) - z &= \{\phi(\alpha) + \lambda_\alpha(\phi)(z-\alpha) + O(z-\alpha)^2\} - z && \text{Taylor expansion,} \\ &= \{\alpha + \lambda_\alpha(\phi)(z-\alpha) + O(z-\alpha)^2\} - z && \text{since } \phi(\alpha) = \alpha, \\ &= (z-\alpha)\{\lambda_\alpha(\phi) - 1 + O(z-\alpha)\}. \end{aligned}$$

Hence under the assumption that $\lambda_\alpha(\phi) \neq 1$, the function $\frac{1}{\phi(z)-z}$ has a simple pole at $z = \alpha$ and its residue is given by[4]

$$\begin{aligned} \operatorname*{Res}_{z=\alpha}\left(\frac{dz}{\phi(z)-z}\right) &= \frac{1}{2\pi i}\int_{|z-\alpha|=\epsilon} \frac{dz}{(z-\alpha)\{\lambda_\alpha(\phi) - 1 + O(z-\alpha)\}} \\ &= \frac{1}{\lambda_\alpha(\phi) - 1}. \end{aligned}$$

The differential $\frac{dz}{\phi(z)-z}$ has no other poles in $\mathbb{C}$. In order to compute its residue at $\infty$, we substitute $z = w^{-1}$ and use $d(w^{-1}) = -w^{-2}dw$ to compute

$$\operatorname*{Res}_{z=\infty}\left(\frac{dz}{\phi(z)-z}\right) = \operatorname*{Res}_{w=0}\left(\frac{d(w^{-1})}{\phi(w^{-1}) - w^{-1}}\right) = \operatorname*{Res}_{w=0}\left(\frac{-dw}{w\bigl(w\phi(w^{-1}) - 1\bigr)}\right).$$

The assumption that $\phi(\infty) \neq \infty$ implies that the function $w\phi(w^{-1})$ vanishes at $w = 0$, so this last differential has a simple pole at $w = 0$ and its residue is equal to

$$\operatorname*{Res}_{w=0}\left(\frac{-dw}{w\bigl(w\phi(w^{-1}) - 1\bigr)}\right) = \lim_{w\to 0} w\left(\frac{-1}{w\bigl(w\phi(w^{-1}) - 1\bigr)}\right) = 1.$$

Substituting these residue values into the Cauchy residue formula yields

$$\begin{aligned} 0 = \sum_{P\in\mathbb{P}^1(\mathbb{C})} \operatorname*{Res}_{P}\left(\frac{dz}{\phi(z)-z}\right) &= \sum_{\alpha\in\operatorname{Fix}(\phi)} \operatorname*{Res}_{z=\alpha}\left(\frac{dz}{\phi(z)-z}\right) + \operatorname*{Res}_{z=\infty}\left(\frac{dz}{\phi(z)-z}\right) \\ &= \sum_{\alpha\in\operatorname{Fix}(\phi)} \frac{1}{\lambda_\alpha(\phi) - 1} + 1. \end{aligned}$$

Finally, we move the sum to the other side, changing $\lambda - 1$ to $1 - \lambda$, to complete the proof of Theorem 1.14. □

[4]In general, if $\psi(z)$ has at most a simple pole at $\alpha$, then $\operatorname{Res}_\alpha(\psi(z)\,dz)$ equals $\lim_{z\to\alpha}(z-\alpha)\psi(z)$.

## 1.4 The Julia Set and the Fatou Set

The notion of equicontinuity is central to dynamics. We begin by recalling the classical definition of continuity. A function $\phi : (S_1, \rho_1) \to (S_2, \rho_2)$ between metric spaces is *continuous* at $\alpha \in S_1$ if for every $\epsilon > 0$ there exists a $\delta > 0$ such that

$$\rho_1(\alpha, \beta) < \delta \quad \Longrightarrow \quad \rho_2(\phi\alpha, \phi\beta) < \epsilon.$$

We extend this notion to a collection of functions $\Phi$ from $S_1$ to $S_2$ by requiring that a single $\delta$ work for every function in $\Phi$. This means that if $\beta$ is chosen sufficiently close to $\alpha$, then $\phi(\alpha)$ and $\phi(\beta)$ will be close to one another for every map $\phi$ in $\Phi$.

**Definition.** Let $(S_1, \rho_1)$ and $(S_2, \rho_2)$ be metric spaces, and let $\Phi$ be a collection of maps from $S_1$ to $S_2$. The collection $\Phi$ is said to be *equicontinuous* at a point $\alpha \in S_1$ if for every $\epsilon > 0$ there exists a $\delta > 0$ such that

$$\rho_1(\alpha, \beta) < \delta \quad \Longrightarrow \quad \rho_2(\phi\alpha, \phi\beta) < \epsilon \quad \text{for every } \phi \in \Phi.$$

The collection $\Phi$ is equicontinuous on a subset $U \subset S_1$ if it is equicontinuous at every point of $U$.

For an individual map $\phi : S \to S$ from a set $S$ to itself, we say that $\phi$ *is equicontinuous at* $\alpha$ if the collection of iterates $\{\phi^n : n \ge 1\}$ is equicontinuous at $\alpha$.

**Definition.** Let $\phi$ be a map from a metric space to itself. The *Fatou set of* $\phi$, denoted by $\mathcal{F}(\phi)$, is the maximal open set on which $\phi$ is equicontinuous. The *Julia set of* $\phi$, denoted by $\mathcal{J}(\phi)$, is the complement of the Fatou set. Note that equicontinuity is not, in general, an open condition (see Exercise 1.22), but that $\mathcal{F}(\phi)$ is an open set by definition.

Informally, one might say that points in the Julia set $\mathcal{J}(\phi)$ tend to wander away from one another as $\phi$ is iterated, so $\phi$ behaves *chaotically* on its Julia set. Nearby points in the Fatou set $\mathcal{F}(\phi)$, on the other hand, tend to stay together, so the behavior of $\phi$ on $\mathcal{F}(\phi)$ is considerably easier to analyze. Much of complex dynamics is devoted to studying the Fatou and Julia sets of rational maps. For further reading, see the textbooks cited at the beginning of this chapter and the two survey articles [74, 235].

*Example* 1.18. Let $\phi(z) \in \mathbb{C}(z)$ be a rational map. If $\alpha$ is an attracting periodic point of $\phi$, then points close to $\alpha$ are all attracted to the points in the orbit of $\alpha$, from which it is easy to conclude that they are in the Fatou set. Similarly, it is easy to check that repelling periodic points are in the Julia set. See Exercise 1.27. The behavior near the neutral points is more complicated.

*Example* 1.19. Consider the map $\phi : \mathbb{P}^1(\mathbb{C}) \to \mathbb{P}^1(\mathbb{C})$ given by $\phi(z) = z^d$. Then $\phi^n(z) = z^{d^n}$, so

$$\lim_{n\to\infty} \phi^n(\alpha) = \begin{cases} 0 & \text{if } |\alpha| < 1, \\ \infty & \text{if } |\alpha| > 1. \end{cases}$$

It is easy to see that if $|\alpha| \neq 1$, then $\alpha \in \mathcal{F}(\phi)$. In particular, the superattracting fixed points 0 and $\infty$ are both in $\mathcal{F}(\phi)$, and every point $\alpha$ with $|\alpha| \neq 1$ is attracted to one of these fixed points.

However, if $|\alpha| = 1$, then $|\phi^n \alpha| = 1$ for all $n \geq 1$, but there are points $\beta$ arbitrarily close to $\alpha$ satisfying $\phi^n \beta \to 0$ and other points arbitrarily close to $\alpha$ satisfying $\phi^n \beta \to \infty$. Therefore $\mathcal{J}(\phi)$ is equal to the unit circle $\{\alpha \in \mathbb{C} : |\alpha| = 1\}$. We also observe that, aside from 0 and $\infty$, the periodic points of $\phi$ are the $(d^n - 1)^{\text{th}}$-roots of unity

$$\operatorname{Per}(\phi) = \{e^{2\pi i k/(d^n-1)} : k \in \mathbb{Z}, n \geq 1\}.$$

If $\alpha$ has exact period $n$, then its multiplier is

$$\left|\lambda_\alpha(\phi)\right| = \left|(\phi^n)'(\alpha)\right| = \left|d^n \alpha^{d^n-1}\right| = d^n > 1,$$

so these periodic points are all repelling. Notice that $\operatorname{Per}(\phi)$ is dense in $\mathcal{J}(\phi)$, and indeed $\mathcal{J}(\phi)$ is the closure of the repelling periodic points of $\phi$.

*Example* 1.20. The Julia set of the polynomial $\phi(z) = z^2 - 2$ consists of the closed interval on the real axis between $-2$ and 2. See Exercise 1.28.

*Example* 1.21. The preceding two examples are actually quite misleading. We will see in Section 1.6 what makes them so special. A more typical example of dynamical behavior is exhibited by the polynomial $\phi(z) = z^2 - 1$. Its fractal-like Julia set is a connected set, while the Fatou set consists of an infinite number of connected components.

Another kind of behavior is seen for the polynomial $\phi(z) = z^2 + \frac{1}{2}$. The Julia set for this map is totally disconnected, although it also has the property that every point in $\mathcal{J}(\phi)$ is the limit of a sequence of distinct points in $\mathcal{J}(\phi)$, i.e., the Julia set is *perfect*.

*Remark* 1.22. It turns out that if the Julia set of a polynomial $\phi$ is neither a line segment nor a circle, then $\phi$ is more-or-less determined by its Julia set. More precisely, let $J \subset \mathbb{C}$ be the Julia set of some polynomial. Then there is a polynomial $\phi_J(z) \in \mathbb{C}[z]$ associated to $J$ with the following property: If $\phi(z) \in \mathbb{C}[z]$ satisfies $\mathcal{J}(\phi) = J$, then $\phi = f \circ \phi_J^n$ for some $n \geq 1$ and some rotation of the plane $f$ with $f(J) = J$. See [394].

*Remark* 1.23. The above examples show that Julia and Fatou sets can be extremely complicated, even for quadratic polynomials. The situation for polynomials of higher degree becomes increasingly complex, and rational functions lead to additional difficulties. Finally, although possibly of less interest from an arithmetic viewpoint, the dynamics of holomorphic or meromorphic functions $\mathbb{C} \to \mathbb{C}$ having an essential singularity at $\infty$ (e.g., $\phi(z) = e^z$) exhibit many further interesting dynamical properties.

We begin our description of the Fatou and Julia sets by showing that each is invariant under both $\phi$ and $\phi^{-1}$.

**Proposition 1.24.** *Let $\phi : \mathbb{P}^1(\mathbb{C}) \to \mathbb{P}^1(\mathbb{C})$ be a rational map of degree $d \geq 2$, and let $\mathcal{F}$ and $\mathcal{J}$ be the Fatou and Julia sets of $\phi$, respectively.*

(a) *The Fatou set $\mathcal{F}$ is completely invariant, i.e., $\phi^{-1}(\mathcal{F}) = \mathcal{F} = \phi(\mathcal{F})$.*
(b) *The Julia set $\mathcal{J}$ is completely invariant.*
(c) *The boundary $\partial\mathcal{J}$ of the Julia set is completely invariant.*

*Proof Sketch.* (a) Since $\phi$ is surjective, it suffices to prove that $\phi^{-1}(\mathcal{F}) = \mathcal{F}$. Suppose first that $\alpha \in \mathcal{F}$, and let $\phi(\beta) = \alpha$. For any point $\beta'$ that is close to $\beta$, the Lipschitz property (1.3) of $\phi$ says that

$$\rho\bigl(\alpha, \phi(\beta')\bigr) \le C(\phi)\rho(\beta, \beta').$$

In particular, $\phi(\beta')$ can be made arbitrarily close to $\alpha$ by taking $\beta'$ sufficiently close to $\beta$. Since $\alpha \in \mathcal{F}$, we know that $\phi^n(\alpha)$ stays close to $\phi^n(\phi\beta')$, and hence that $\phi^{n+1}(\beta)$ stays close to $\phi^{n+1}(\beta')$. Therefore $\beta \in \mathcal{F}$, which proves that $\phi^{-1}(\mathcal{F}) \subset \mathcal{F}$.

Next suppose that $\alpha \in \mathcal{F}$. We need to check that $\phi(\alpha) \in \mathcal{F}$. If $U$ is a small neighborhood of $\alpha$, then the open mapping property of $\phi$ implies that $\phi(U)$ is a (small) open neighborhood of $\phi(\alpha)$. In particular, if $\beta$ is sufficiently close to $\phi(\alpha)$, then $\beta$ will lie in $\phi(U)$, say $\beta = \phi(\beta')$ with $\beta' \in U$. Then the assumption that $\alpha \in \mathcal{F}$ implies that the iterates $\phi^n(\alpha)$ and $\phi^n(\beta')$ remain close to one another, and hence the same is true of the iterates $\phi^{n-1}(\phi\alpha)$ and $\phi^{n-1}(\beta)$. Therefore $\phi(\alpha) \in \mathcal{F}$, which completes the proof of the other inclusion $\mathcal{F} \subset \phi^{-1}(\mathcal{F})$. (You should fill in the details of this proof sketch and give a rigorous $(\delta, \epsilon)$ proof. See Exercise 1.23.)
(b) The Julia set is the complement of the Fatou set, so the complete invariance of $\mathcal{F}$ implies the complete invariance of $\mathcal{J}$.
(c) Let $\mathcal{J}^\circ$ be the interior of $\mathcal{J}$. The rational map $\phi$ is continuous, so $\phi^{-1}(\mathcal{J}^\circ)$ is open, and the complete invariance of $\mathcal{J}$ shows that it is contained in $\mathcal{J}$; hence $\phi^{-1}(\mathcal{J}^\circ) \subset \mathcal{J}^\circ$. For the other inclusion, we use the fact that $\phi$ is an open map, so $\phi(\mathcal{J}^\circ)$ is open. Again the complete invariance of $\mathcal{J}$ shows that it is contained in $\mathcal{J}$, and the fact that it is open shows that it is contained in $\mathcal{J}^\circ$. Hence

$$\mathcal{J}^\circ \subset \phi^{-1}\bigl(\phi(\mathcal{J}^\circ)\bigr) \subset \phi^{-1}(\mathcal{J}^\circ).$$

This proves that $\phi^{-1}(\mathcal{J}^\circ) = \mathcal{J}^\circ$, so the interior of $\mathcal{J}$ is completely invariant. Finally, the fact that $\mathcal{J}$ and $\mathcal{J}^\circ$ are completely invariant implies that the set difference $\partial\mathcal{J} = \mathcal{J} \smallsetminus \mathcal{J}^\circ$ is also completely invariant. □

**Proposition 1.25.** *For every integer $n \ge 1$,*

$$\mathcal{F}(\phi^n) = \mathcal{F}(\phi) \qquad \textit{and} \qquad \mathcal{J}(\phi^n) = \mathcal{J}(\phi).$$

*Proof.* Let $\psi = \phi^n$. It is clear from the definition that $\mathcal{F}(\phi) \subset \mathcal{F}(\psi)$, since if we know that iteration of $\phi$ maintains closeness of points, then the same is certainly true for $\phi^n$.

To prove the opposite inclusion, for each $0 \le i < n$ we consider the collection of maps

$$\Phi_i = \{\phi^i\psi^k : k \ge 0\}.$$

For any fixed $i$, the map $\phi^i$ satisfies a Lipschitz inequality (1.3), so the Fatou set $\mathcal{F}(\Phi_i)$ of $\Phi_i$ contains the Fatou set $\mathcal{F}(\psi)$ of $\psi$. Hence

$$\mathcal{F}(\Phi_0) \cap \mathcal{F}(\Phi_1) \cap \cdots \cap \mathcal{F}(\Phi_{n-1}) \supset \mathcal{F}(\psi).$$

But the intersection is clearly equal to $\mathcal{F}(\phi)$, since

$$\{\phi^j : j \geq 0\} = \bigcup_{0 \leq i < n} \{\phi^i \psi^k : k \geq 0\}.$$

This completes the proof that $\mathcal{F}(\phi^n) = \mathcal{F}(\phi)$. The corresponding fact for the Julia set is then clear. □

*Remark* 1.26. The idea of writing $\{\phi^j : j \geq 0\}$ as the union

$$\{\phi^j : j \geq 0\} = \bigcup_{0 \leq i < n} \{\phi^i (\phi^n)^k : k \geq 0\}$$

will reappear later in the proof of Theorem 3.48 when we need to spread out the ramification of a map.

The notions of equicontinuity and normality are closely related.

**Definition.** Let $(S_1, \rho_1)$ and $(S_2, \rho_2)$ be metric spaces, and let $\Phi$ be a collection of maps from $S_1$ to $S_2$. The collection $\Phi$ is said to be *normal*, or a *normal family*, on $S_1$ if every infinite sequence of functions from $\Phi$ contains a subsequence that converges locally uniformly on $S_1$.

The following two important theorems from complex analysis are used extensively in the study of dynamical systems on $\mathbb{P}^1(\mathbb{C})$.

**Theorem 1.27.** (Arzelà–Ascoli Theorem) *Let $U$ be an open subset of $\mathbb{P}^1(\mathbb{C})$, and let $\Phi$ be a collection of continuous maps $U \to \mathbb{P}^1(\mathbb{C})$. Then $\Phi$ is equicontinuous on $U$ if and only if it is a normal family on $U$.*

**Theorem 1.28.** (Montel's Theorem) *Let $U$ be an open subset of $\mathbb{P}^1(\mathbb{C})$, and let $\Phi$ be a collection of analytic maps $U \to \mathbb{P}^1(\mathbb{C})$. If $\bigcup_{\phi \in \Phi} \phi(U)$ omits three or more points of $\mathbb{P}^1(\mathbb{C})$, then $\Phi$ is a normal family on $U$.*

The next theorem summarizes a number of important properties of the Julia set.

**Theorem 1.29.** *Let $\phi(z) \in \mathbb{C}(z)$ be a rational function of degree $d \geq 2$.*
(a) *The Julia set $\mathcal{J}(\phi)$ is nonempty.*
(b) *Let $\alpha \in \mathcal{J}(\phi)$. Then the backward orbit $\mathcal{O}_\phi^-(\alpha) = \{\phi^{-n}(\alpha) : n \geq 1\}$ of $\alpha$ is dense in $\mathcal{J}(\phi)$.*
(c) *Let $U$ be a union of connected components of the Fatou set $\mathcal{F}(\phi)$ that is completely invariant, i.e., $\phi^{-1}(U) = U = \phi(U)$. Then $\mathcal{J}(\phi) = \partial U$.*
(d) *The Julia set $\mathcal{J}(\phi)$ contains no isolated points, i.e., $\mathcal{J}(\phi)$ is a perfect set.*

*Proof.* (a) See [43, Theorem 4.2.1] or [95, Theorem III.1.2].
(b) See [43, Theorem 4.2.7] or [95, Theorem III.1.6].
(c) See [95, Theorem III.1.7].
(d) See [43, Theorem 4.2.4] or [95, Theorem III.1.8]. □

Theorem 1.29(a) says that the Julia set is always nonempty. The Fatou set, on the other hand, may be empty. We state some conditions that characterize this phenomenon.

**Theorem 1.30.** *Let $\phi(z) \in \mathbb{C}(z)$ be a rational function of degree $d \geq 2$. Then the following are equivalent:*

(a) *The Julia set $\mathcal{J}(\phi)$ is equal to $\mathbb{P}^1(\mathbb{C})$.*

(b) *The Julia set $\mathcal{J}(\phi)$ has a nonempty interior.*

(c) *There exists a point $\alpha \in \mathbb{P}^1(\mathbb{C})$ whose (forward) orbit $\mathcal{O}_\phi(\alpha)$ is dense in $\mathbb{P}^1(\mathbb{C})$.*

*Further, if every critical point of $\phi$ is strictly preperiodic, i.e., preperiodic, but not periodic, then $\mathcal{J}(\phi) = \mathbb{P}^1(\mathbb{C})$.*

*Proof.* See [43, Theorems 4.2.3, 4.3.2, 9.4.4] or [95, Theorems III.1.9, V.1.2]. □

*Example* 1.31. Consider the map $\phi(z) = 1 - 2/z^2$. Its critical points are 0 and $\infty$. (See Exercise 1.25.) Applying $\phi$ repeatedly to 0 yields

$$0 \xrightarrow{\ \phi\ } \infty \xrightarrow{\ \phi\ } 1 \xrightarrow{\ \phi\ } -1 \xrightarrow{\ \phi\ } -1 \xrightarrow{\ \phi\ } \cdots,$$

so the critical points are strictly preperiodic. It follows from Theorem 1.30 that $\mathcal{J}(\phi) = \mathbb{P}^1(\mathbb{C})$.

*Remark* 1.32. Theorems 1.29 and 1.30 say that a rational map $\phi : \mathbb{P}^1(\mathbb{C}) \to \mathbb{P}^1(\mathbb{C})$ may have $\mathcal{J}(\phi) = \mathbb{P}^1(\mathbb{C})$, but that it is not possible to have $\mathcal{F}(\phi) = \mathbb{P}^1(\mathbb{C})$. The situation becomes exactly reversed if $\mathbb{C}$ is replaced by an algebraically closed field $K$ having a nonarchimedean absolute value. For example, if $K = \mathbb{C}_p$ (the completion of the algebraic closure of $\mathbb{Q}_p$), then the Fatou set of $\phi : \mathbb{P}^1(\mathbb{C}_p) \to \mathbb{P}^1(\mathbb{C}_p)$ is always nonempty, but it is possible, and indeed quite easy, to find maps with empty Julia set and $\mathcal{F}(\phi) = \mathbb{P}^1(\mathbb{C}_p)$. We will discuss this in more detail when we study dynamics over complete local fields in Chapters 2 and 5.

**Theorem 1.33.** *Let $\phi(z) \in \mathbb{C}[z]$ be a polynomial of degree $d \geq 2$.*

(a) *The Julia set $\mathcal{J}(\phi)$ is connected if and only if for every critical point $\alpha \neq \infty$, the orbit $\mathcal{O}_\phi(\alpha)$ is bounded in $\mathbb{C}$.*

(b) *If every critical point $\alpha$ of $\phi$ satisfies $\lim_{n\to\infty} \phi^n(\alpha) = \infty$, then the Julia set $\mathcal{J}(\phi)$ is totally disconnected.*

*Proof.* See [95, Theorems III.4.1, III.4.2]. □

*Remark* 1.34. A quadratic polynomial has only one finite critical point. Thus Theorem 1.33 covers all cases, so we can divide the set of quadratic polynomials into two classes according to the behavior of their finite critical point. Every quadratic polynomial is linearly conjugate to a unique polynomial of the form $\phi_c(z) = z^2 + c$, so we define a portion of the $c$-plane by

$$\begin{aligned} \mathcal{M} &= \{c \in \mathbb{C} : \phi^n(0) \text{ is bounded for } n \geq 1\} \\ &= \{c \in \mathbb{C} : \mathcal{J}(\phi_c) \text{ is connected}\}. \end{aligned}$$

This set $\mathcal{M}$ is the famous *Mandelbrot set*, illustrated in Figure 1.2. It is a dynamically determined subset of the moduli space of quadratic polynomial maps.

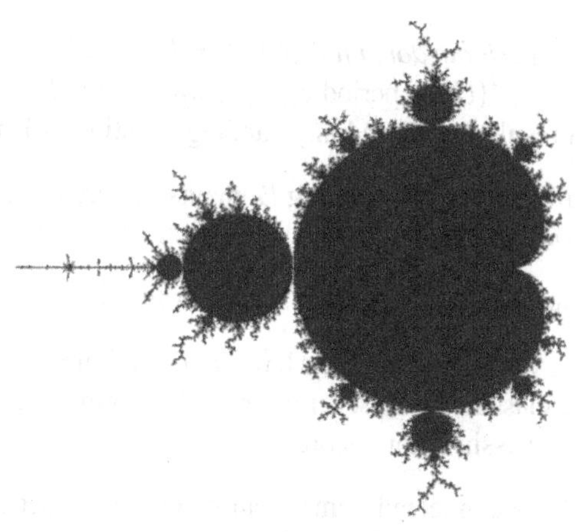

Figure 1.2: The Mandelbrot set $\mathcal{M}$.

## 1.5 Properties of Periodic Points

The dynamics of a rational map is influenced not only by the behavior of its critical points, but also by the behavior of its periodic points. The next result describes some of the properties of the periodic points and periodic cycles of a rational map.

**Theorem 1.35.** *Let $\phi(z) \in \mathbb{C}(z)$ be a rational function of degree $d \ge 2$.*

(a) *The map $\phi$ has at most $2d-2$ nonrepelling periodic cycles in $\mathbb{P}^1(\mathbb{C})$. If $\phi$ is a polynomial map, then it has at most $d-1$ nonrepelling periodic cycles in $\mathbb{C}$.*

(b) *The Julia set $\mathcal{J}(\phi)$ is equal to the closure of the repelling periodic points of $\phi$.*

(c) *Let $U \subset \mathbb{P}^1(\mathbb{C})$ be an open set such that $\mathcal{J}(\phi) \cap U \ne \emptyset$. Then there is an integer $n \ge 1$ such that $\phi^n(U \cap \mathcal{J}(\phi)) = \mathcal{J}(\phi)$.*

*Proof.* (a) The sharp bound of $2d-2$ for rational maps is due to Shishikura, see [43, Theorem 9.6]. Much earlier, Fatou gave a weaker bound that is sufficient for many applications, see [95, Theorem III.2.7]. The bound for polynomial maps is due to Douady, see [95, Theorem VI.1.2].
(b) This important result is due independently to Julia and Fatou. See [43, Theorem 6.9.2] or [95, Theorem III.3.1].
(c) See [95, Theorem III.3.2]. □

The Fatou set $\mathcal{F}(\phi)$ is an open subset of $\mathbb{P}^1(\mathbb{C})$, so it consists of one or more connected components. It is known that the number of components is equal to 0, 1, 2, or $\infty$; see [95, Theorem IV.1.2]. If $U$ is a connected component of $\mathcal{F}(\phi)$, then the open mapping property of $\phi$ implies that $\phi(U)$ is also a connected component of $\mathcal{F}(\phi)$. In this way we obtain a map

$$\text{Components}\big(\mathcal{F}(\phi)\big) \longrightarrow \text{Components}\big(\mathcal{F}(\phi)\big), \qquad U \longmapsto \phi(U),$$

and we say that $U$ is a *periodic domain* if $\phi^n(U) = U$ for some $n \ge 1$, a *preperiodic domain* if some iterate $\phi^n(U)$ is periodic, and a *wandering domain* otherwise. The following important result answers a long-standing question of Fatou and Julia.

**Theorem 1.36.** (Sullivan's No Wandering Domains Theorem) *A rational map $\phi \in \mathbb{C}(z)$ has no wandering domains.*

*Proof.* See [43, Chapter 8], [95, Theorem IV.1.3] or [426]. □

The possibility of wandering domains having been eliminated, the periodic components of $\mathcal{F}(\phi)$ are classified into several types. We begin with the necessary definitions, then state the classification theorem.

**Definition.** Let $U$ be a connected component of the Fatou set $\mathcal{F}(\phi)$, and assume that $U$ is forward invariant, i.e., $\phi(U) = U$.

- $U$ is *parabolic* if the boundary of $U$ contains a rationally neutral periodic point $\zeta$ such that $\lim_{n\to\infty} \phi^n(\alpha) = \zeta$ for every $\alpha \in U$.
- $U$ is a *Siegel disk* if there is an analytic isomorphism $f$ from the unit disk $\{w \in \mathbb{C} : |w| < 1\}$ to $U$ such that $\phi^f$ is a rotation of the unit disk.
- $U$ is a *Herman ring* if there is an analytic isomorphism $f$ from some annulus $\{w \in \mathbb{C} : a < |w| < b\}$ to $U$ such that $\phi^f$ is a rotation of the annulus.

Here a rotation is simply a map of the form $w \mapsto e^{i\theta} w$.

If $U$ is a periodic connected component of period $n$ of $\mathcal{F}(\phi)$, then $U$ is forward-invariant for $\phi^n$ and we say that $U$ is *parabolic*, a *Siegel disk*, or a *Herman ring* if it is of the appropriate type for $\phi^n$.

**Theorem 1.37.** *Let $U$ be a periodic connected component of the Fatou set of a rational map $\phi \in \mathbb{C}(z)$. Then $U$ fits into exactly one of the following categories:*
(a) *$U$ contains an attracting periodic point.*
(b) *$U$ is parabolic.*
(c) *$U$ is a Siegel disk.*
(d) *$U$ is a Herman ring.*

*Proof.* See [43, Theorem 7.1] or [95, Theorem IV.2.1]. We mention that it is nontrivial to prove that Siegel disks and Herman rings exist. In particular, the Fatou set of a polynomial map cannot contain a Herman ring. □

## 1.6 Dynamical Systems Associated to Endomorphisms of Algebraic Groups

In this section we study certain rational maps whose dynamical properties are comparatively easy to analyze due to the existence of an underlying group structure.

These maps thus provide a class of examples on which to make and test conjectures, albeit with the caution that they are rather special and in some ways atypical. Additional material on this interesting collection of maps may be found in Chapter 6.

### 1.6.1 The Multiplicative Group

The dynamics of the polynomial $\phi(z) = z^d$ is extremely easy to analyze, since we have the explicit formula $\phi^n(z) = z^{d^n}$. However, there is a more intrinsic reason that $z^d$ is such a nice map; namely, it is an endomorphism of the multiplicative group $\mathbb{C}^*$ of nonzero complex numbers. In other words, the map

$$\mathbb{C}^* \longrightarrow \mathbb{C}^*, \qquad z \longmapsto z^d,$$

is a homomorphism. It is the existence of the underlying group $\mathbb{C}^*$ that makes $z^d$ relatively easy to analyze.

*Remark* 1.38. In fancier language, the map $\phi(z) = z^d$ is an endomorphism of the algebraic group $\mathbb{G}_m$, where for any field $K$, the set of points of $\mathbb{G}_m$ is the multiplicative group $\mathbb{G}_m(K) = K^*$. The full endomorphism ring of $\mathbb{G}_m$ is $\mathbb{Z}$ via the identification

$$\mathbb{Z} \cong \operatorname{End}(\mathbb{G}_m), \qquad d \longleftrightarrow (z \mapsto z^d).$$

In even fancier language, $\mathbb{G}_m$ is a group scheme

$$\mathbb{G}_m = \operatorname{Spec} \mathbb{Z}[X,Y]/(XY-1)$$

called the *multiplicative group scheme*. It is characterized by the property that for every ring $R$, the group of $R$-valued points of $\mathbb{G}_m$ is the unit group $\mathbb{G}_m(R) \cong R^*$.

### 1.6.2 Chebyshev Polynomials

The multiplicative group $\mathbb{C}^*$ has a nontrivial automorphism given by the reciprocal map $z \mapsto z^{-1}$. We can take the quotient of $\mathbb{C}^*$ by this automorphism to obtain a new space that turns out to be isomorphic to $\mathbb{C}$ via the map

$$\mathbb{C}^*/\{z \sim z^{-1}\} \xrightarrow{\sim} \mathbb{C}, \qquad [z] \longmapsto z + z^{-1}.$$

The inverse isomorphism takes $w \in \mathbb{C}$ to the equivalence class of either of the roots of $z^2 - zw + 1 = 0$.

This inversion automorphism commutes with the $d^{\text{th}}$-power map, so when we take the quotient, the $d^{\text{th}}$-power map will descend to give a map on the quotient space. In other words, there is a unique map $T_d$ that makes the following diagram commute:

$$\begin{array}{ccc}
\mathbb{C}^* & \xrightarrow{z \to z^d} & \mathbb{C}^* \\
\downarrow & & \downarrow \\
\dfrac{\mathbb{C}^*}{\{z \sim z^{-1}\}} & \xrightarrow{z \to z^d} & \dfrac{\mathbb{C}^*}{\{z \sim z^{-1}\}} \\
{\scriptstyle z \downarrow z+z^{-1}} \downarrow & & \downarrow {\scriptstyle z \downarrow z+z^{-1}} \\
\mathbb{C} & \xrightarrow{w \to T_d(w)} & \mathbb{C}
\end{array}$$

The map $T_d$ is characterized by the equation

$$T_d(z + z^{-1}) = z^d + z^{-d} \qquad \text{for all } z \in \mathbb{C}^*. \tag{1.5}$$

It is not hard to see that $T_d$ is a polynomial; indeed, it is a monic polynomial of degree $d$ with integer coefficients called the $d^{\text{th}}$ *Chebyshev polynomial*. Writing $z = e^{it}$ with $t \in \mathbb{C}$, we see that the Chebyshev polynomial satisfies

$$T_d(2\cos t) = 2\cos(dt).$$

Using these formulas, it is not hard to prove that the Julia set of $T_d$ is given by $\mathcal{J}(T_d) = [-2, 2]$, the closed real interval from $-2$ to 2, and to derive many further properties of the Chebyshev polynomials. See Exercises 1.30 and 1.31, as well as Section 6.2.

*Remark* 1.39. Classically the Chebyshev polynomials are normalized to satisfy the identity $\widetilde{T}_d(\cos t) = \cos(dt)$, in which case the Julia set becomes $\mathcal{J}(\widetilde{T}_d) = [-1, 1]$. The two normalizations are related by the simple formula $\widetilde{T}_d(w) = \frac{1}{2}T_d(2w)$.

### 1.6.3 Rational Maps Arising from Elliptic Curves

There are three different types of (connected) algebraic groups of dimension one. First is the *additive group* $\mathbb{G}_a(\mathbb{C}) = \mathbb{C}^+$, whose endomorphisms are the maps $z \to dz$. The dynamical systems associated to endomorphisms of the additive group are relatively uninteresting. Second is the *multiplicative group* $\mathbb{G}_m(\mathbb{C}) = \mathbb{C}^*$, whose endomorphisms $z \mapsto z^d$ have interesting, but relatively easy to analyze, dynamical properties. The same is true of the Chebyshev polynomials, which come from the restriction of $z \mapsto z^d$ to a quotient of $\mathbb{G}_m$.

The third type of one-dimensional algebraic groups consists of a family of groups called *elliptic curves*. We do not have space to go into the theory very deeply, so we are content to make a few brief remarks here and later expand on this material in Chapter 6. For further material on the analytic, algebraic, and arithmetic theory of elliptic curves, see for example [1, 198, 254, 410, 420, 412]. We note that the reader may omit this section on first reading, since the material is not used other than as a source of examples until Chapter 6.

An elliptic curve $E(\mathbb{C})$ is the set of solutions $(x, y)$ to an equation of the form

$$y^2 = x^3 + ax + b$$

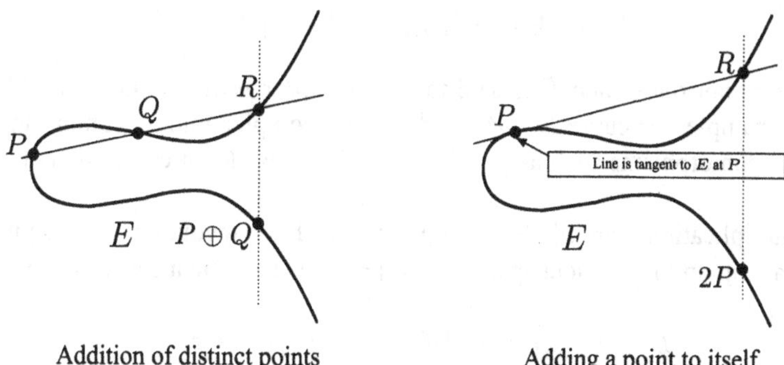


Addition of distinct points

Adding a point to itself

Figure 1.3: The addition law on an elliptic curve.

together with an extra point $\mathcal{O}$. We also require that $\Delta = 4a^3 + 27b^2 \neq 0$, which means that the cubic polynomial has distinct roots and ensures that the curve $E(\mathbb{C})$ is nonsingular.

Alternatively, $E(\mathbb{C})$ is the curve in the projective plane $\mathbb{P}^2(\mathbb{C})$ defined by the homogeneous equation

$$Y^2Z = X^3 + aXZ^2 + bZ^3,$$

with the extra point being $\mathcal{O} = [0,1,0]$. There is a natural automorphism on $E(\mathbb{C})$ given by $[-1](X,Y,Z) = (X,-Y,Z)$. If $P$ and $Q$ are two points on $E(\mathbb{C})$, then the line through $P$ and $Q$ will intersect $E(\mathbb{C})$ at a third point $R$, and we define an addition law by setting $P \oplus Q = [-1]R$. (If $P = Q$, we take the "line through $P$ and $Q$" to be the tangent line to $E(\mathbb{C})$ at $P$.) These operations satisfy

$$P \oplus \mathcal{O} = P, \quad P \oplus [-1](P) = \mathcal{O}, \quad P \oplus Q = Q \oplus P, \quad (P \oplus Q) \oplus R = P \oplus (Q \oplus R),$$

so they make $E(\mathbb{C})$ into an abelian group. (The first three equalities are obvious, while the associativity is somewhat difficult to prove.)

An elementary calculation shows that the coordinates of $P \oplus Q$ are given by rational functions of the coordinates of $P$ and $Q$. In particular, if $a$ and $b$ are in some field $K$, and if $P$ and $Q$ have coordinates in $K$, then $P \oplus Q$ will also have coordinates in $K$. Thus $E(K) = E(\mathbb{C}) \cap \mathbb{P}^2(K)$ will be a subgroup of $E(\mathbb{C})$. Figure 1.3 illustrates the group law for points in $E(\mathbb{R})$.

Repeated addition in any abelian group gives an endomorphism of the group, so for each $d \geq 1$ we obtain a map

$$[d] : E(\mathbb{C}) \longrightarrow E(\mathbb{C}), \qquad P \longmapsto \underbrace{P \oplus P \oplus \cdots \oplus P}_{d \text{ terms}}$$

called the *multiplication-by-d map*. Setting

$$[0](P) = \mathcal{O} \qquad \text{and} \qquad [-d](P) = [-1]([d]P),$$

we obtain an (injective) homomorphism

$$\mathbb{Z} \longrightarrow \mathrm{End}(E(\mathbb{C})), \qquad d \longmapsto [d].$$

If this map is not onto, then $E$ is said to have *complex multiplication*, or CM for short. For example, the curve $E : y^2 = x^3 + x$ has complex multiplication, since it has extra endomorphisms such as $[i] : (x,y) \mapsto (-x, iy)$. Most curves do not have CM.

The multiplication map $[d]$ clearly commutes with the involution $[-1]$, so it descends to a map on the quotient space $E(\mathbb{C})/\{\pm 1\}$. The quotient space is very easy to describe,

$$E(\mathbb{C})/\{\pm 1\} \xrightarrow{\sim} \mathbb{P}^1(\mathbb{C}), \qquad (x,y) \longmapsto x,$$

so the multiplication-by-$d$ map descends to give a rational map $\phi_{E,d}$ making both squares in the following diagram commute:

$$\begin{array}{ccc} E(\mathbb{C}) & \xrightarrow{[d]} & E(\mathbb{C}) \\ \downarrow & & \downarrow \\ \dfrac{E(\mathbb{C})}{\{\pm 1\}} & \xrightarrow{[d]} & \dfrac{E(\mathbb{C})}{\{\pm 1\}} \\ {\scriptstyle (x,y) \mapsto x}\downarrow & & \downarrow{\scriptstyle (x,y) \mapsto x} \\ \mathbb{P}^1(\mathbb{C}) & \xrightarrow{\phi_{E,d}} & \mathbb{P}^1(\mathbb{C}) \end{array}$$

The map $\phi_{E,d}$ is an example of a *Lattès map*. Thus $\phi_{E,d}$ is a rational function characterized by the formula

$$\phi_{E,d}\bigl(x(P)\bigr) = x\bigl([d](P)\bigr) \qquad \text{for all } P \in E(\mathbb{C}). \tag{1.6}$$

Notice the close similarity to the defining property (1.5) of the Chebyshev polynomials. More generally, it turns out that every endomorphism $u \in \mathrm{End}(E)$ commutes with $[-1]$, even if $E$ has complex multiplication, so every endomorphism $u$ determines a rational function $\phi_{E,u}$ on $\mathbb{P}^1(\mathbb{C})$.

*Example* 1.40. Let $E : y^2 = x^3 + ax + b$ be an elliptic curve as above. Then the duplication map $[2] : E(\mathbb{C}) \to E(\mathbb{C})$ leads to the rational function

$$\phi_{E,2}(x) = \frac{x^4 - 2ax^2 - 8bx + a^2}{4x^3 + 4ax + 4b}.$$

*Example* 1.41. Let $E : y^2 = x^3 + x$ be the CM elliptic curve mentioned earlier. Then the map $[1+i] : E(\mathbb{C}) \to E(\mathbb{C})$ defined by $[1+i](P) = P \oplus [i](P)$ leads to the rational function

$$\phi_{E,1+i}(x) = \frac{1}{2i}\left(x + \frac{1}{x}\right).$$

**Proposition 1.42.** *Let $E : y^2 = x^3 + ax + b$ be an elliptic curve, let $d \geq 2$ be an integer, and let $\phi_{E,d} : \mathbb{P}^1(\mathbb{C}) \to \mathbb{P}^1(\mathbb{C})$ be the associated rational map as above. Then*

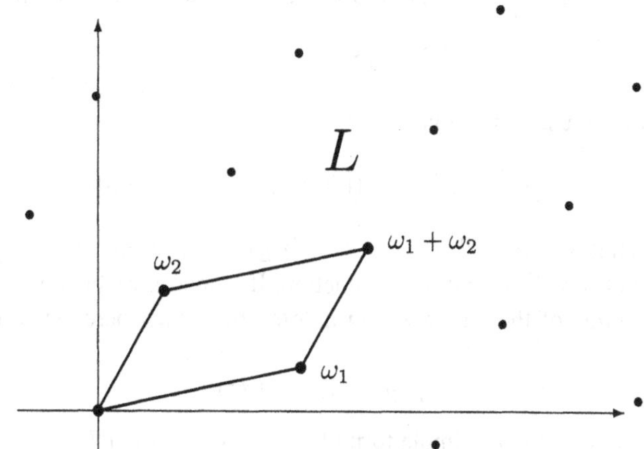


Figure 1.4: A lattice and fundamental domain associated to an elliptic curve.

$$\operatorname{PrePer}(\phi_{E,d}) = x\big(E(\mathbb{C})_{\text{tors}}\big),$$

*where we recall that the torsion subgroup* $E(\mathbb{C})_{\text{tors}}$ *of* $E(\mathbb{C})$ *is the set of elements of finite order* (*cf. Example* 0.2).

*Proof.* We leave the proof as an exercise; see Exercise 1.32. □

The dynamics of the rational function $\phi_{E,d}$ on $\mathbb{P}^1(\mathbb{C})$ can be analyzed using the group law on the larger space $E(\mathbb{C})$. The utility of this approach is further enhanced by the analytic uniformization of $E(\mathbb{C})$, which we now briefly describe.

A *lattice* $L$ in $\mathbb{C}$ is a set of all $\mathbb{Z}$-linear combinations of two $\mathbb{R}$-linearly independent complex numbers $\omega_1$ and $\omega_2$. Thus

$$L = \{m_1\omega_1 + m_2\omega_2 : m_1, m_2 \in \mathbb{Z}\}.$$

The quotient space $\mathbb{C}/L$ is a torus obtained by identifying the opposite sides of the fundamental domain

$$\{t_1\omega_1 + t_2\omega_2 : t_1, t_2 \in \mathbb{R},\ 0 \le t_1, t_2 \le 1\}.$$

A lattice $L$ and a fundamental domain are illustrated in Figure 1.4. Notice that $\mathbb{C}/L$ has a natural group structure induced by addition on $\mathbb{C}$.

Let $E(\mathbb{C})$ be an elliptic curve. Then one can prove that there exists a lattice $L$ and a complex analytic isomorphism

$$\psi_E : \mathbb{C}/L \longrightarrow E(\mathbb{C}).$$

Further, the isomorphism $\psi_E$ respects the group structure,

$$\psi_E(z_1 + z_2 \bmod L) = \psi_E(z_1) \oplus \psi_E(z_2).$$

In particular, if we set $P = \psi_E(z)$ in formula (1.6) and use the relation

$$[d](\psi_E(z)) = \psi_E(dz),$$

then we obtain the useful formula

$$\phi_{E,d}\big(x(\psi_E(z))\big) = x\big([d](\psi_E(z))\big) = x\big(\psi_E(dz)\big).$$

To ease notation, we will let $\wp(z) = x(\psi_E(z))$. The function $\wp(z)$ is a slight variant of the classical Weierstrass $\wp$-function. It is meromorphic on $\mathbb{C}$ with double poles at the points of the lattice $L$ and holomorphic elsewhere. Then our relation becomes

$$\phi_{E,d}(\wp(z)) = \wp(dz), \tag{1.7}$$

and iteration is given by the simple formula $\phi^n_{E,d}(\wp(z)) = \wp(d^n z)$.

The upshot of this discussion is that we obtain a commutative diagram

$$\begin{array}{ccc} \mathbb{C}/L & \xrightarrow{z \to dz} & \mathbb{C}/L \\ \downarrow \wp & & \downarrow \wp \\ \mathbb{P}^1(\mathbb{C}) & \xrightarrow{\phi_{E,d}} & \mathbb{P}^1(\mathbb{C}) \end{array}$$

that is of great assistance in studying the dynamics of the rational function $\phi_{E,d}$.

For example, the map $z \mapsto dz$ is clearly $d^2$-to-1 on $\mathbb{C}/L$, so we see that $\phi_{E,d}$ has degree $d^2$. Further, a point $x = \wp(z)$ will be fixed by $\phi_{E,d}$ if and only if $z$ satisfies the congruence $dz \equiv \pm z \pmod{L}$, so the fixed points of $\phi_{E,d}$ are precisely the points

$$\wp\left(\frac{a_1}{d-1}\omega_1 + \frac{a_2}{d-1}\omega_2\right) \quad \text{and} \quad \wp\left(\frac{a_1}{d+1}\omega_1 + \frac{a_2}{d+1}\omega_2\right) \quad \text{with } a_1, a_2 \in \mathbb{Z}.$$

(Note that $\wp(-z) = \wp(z)$, so there are actually only $d^2 + 1$ fixed points.)

Let $\wp(\zeta)$ be a fixed point of $\phi_{E,d}$ with $\zeta \notin \frac{1}{2}L$, which is equivalent to the assumption that $\wp'(\zeta) \neq 0$ or $\infty$. (This excludes at most four such points modulo $L$.) Differentiating (1.7) and evaluating at $z = \zeta$ yields

$$\phi'_{E,d}(\wp(\zeta))\wp'(\zeta) = d\wp'(d\zeta). \tag{1.8}$$

But we know that $\wp(z)$ is a meromorphic function on $\mathbb{C}/L$, or equivalently, it is a meromorphic function on $\mathbb{C}$ with the periodicity property

$$\wp(z + \omega) = \wp(z) \qquad \text{for all } \omega \in L.$$

Differentiation gives the same relation $\wp'(z + \omega) = \wp'(z)$ for the derivative. (Notice the analogy with the function $e^{2\pi i z}$, which is holomorphic on $\mathbb{C}$ and invariant under the translations $z \mapsto z + n$ for $n \in \mathbb{Z}$.) We evaluate at $z = d\zeta$, use the fact that $d\zeta \equiv \pm\zeta \pmod{L}$ to get $\wp'(d\zeta) = \wp'(\pm\zeta) = \pm\wp'(\zeta)$, and substitute into (1.8) to obtain the relation

$$\phi'_{E,d}(\wp(\zeta))\wp'(\zeta) = d\wp'(d\zeta) = \pm d\wp'(\zeta).$$

Finally, canceling $\wp'(\zeta)$ yields $\phi'_{E,d}(\wp(\zeta)) = \pm d$. In particular, (almost) every fixed point of $\phi_{E,d}$ is expanding and thus is in the Julia set.

But much more is true. The relation (1.7) shows immediately that

$$\phi_{E,d_1}(\phi_{E,d_2}(\wp(z))) = \phi_{E,d_1}(\wp(d_2 z)) = \wp(d_1 d_2 z) = \phi_{E,d_1 d_2}(\wp(z)),$$

so $\phi^n_{E,d} = \phi_{E,d^n}$. Thus the periodic points of $\phi_{E,d}$ of period dividing $n$ are precisely the fixed points of $\phi_{E,d^n}$, and their multipliers are (almost) all equal to $\pm d^n$. This proves that (almost) every periodic point of $\phi_{E,d}$ is expanding, hence in the Julia set. Finally, we observe that the set of points in $\mathbb{C}/L$ satisfying

$$z \equiv \pm d^n z \pmod{L} \qquad \text{for some } n = 1, 2, 3, \ldots$$

is dense in $\mathbb{C}/L$, so their image using the double cover $\wp : \mathbb{C}/L \to \mathbb{P}^1(\mathbb{C})$ is dense in $\mathbb{P}^1(\mathbb{C})$. We have thus shown that $\mathcal{J}(\phi_{E,d}) = \mathbb{P}^1(\mathbb{C})$. This proves the following theorem, which was published by Lattès in 1918 and provided the first known examples of rational maps with empty Fatou set. (See also [300, §6] for a discussion of earlier work by Schröder and Böttcher on elliptic functions and their associated rational maps, as well as an 1815 paper of Babbage in which he uses semiconjugacy to study the periodic points of certain maps.)

**Theorem 1.43.** (Lattès [260]) *Let $E$ be an elliptic curve, let $d \ge 2$ be an integer, and let $\phi_{E,d} : \mathbb{P}^1(\mathbb{C}) \to \mathbb{P}^1(\mathbb{C})$ be the rational map of degree $d^2$ characterized by* (1.6). *Then*

$$\mathcal{J}(\phi_{E,d}) = \mathbb{P}^1(\mathbb{C}) \qquad \textit{and} \qquad \mathcal{F}(\phi_{E,d}) = \emptyset.$$

# Exercises

## Section 1.1. Rational Maps and the Projective Line

**1.1.** Define a mapping from the $xy$-plane (which we identify with $\mathbb{C}$) to the unit sphere $S^2$ in $\mathbb{R}^3$ by mapping $z \in \mathbb{C}$ to the point $z^* \in S^2$ such that the line through $z$ and $z^*$ goes through the point $(0,0,1)$. Notice that $z^* \to (0,0,1)$ as $z \to \infty$, so if we set $\infty^* = (0,0,1)$, we obtain a bijection

$$\mathbb{P}^1(\mathbb{C}) = \mathbb{C} \cup \{\infty\} \longrightarrow S^2, \qquad z \longmapsto z^*,$$

as illustrated in Figure 1.1. Prove that with this identification, the chordal metric on $\mathbb{P}^1(\mathbb{C})$ is given by $\rho(z,w) = \frac{1}{2}|z^* - w^*|$.

**1.2.** (a) Prove that the inversion map $z \mapsto z^{-1}$ is an isometry of $\mathbb{P}^1(\mathbb{C})$ for the chordal metric (i) by a direct calculation using the formula for $\rho$; and (ii) by using the identification $\mathbb{P}^1(\mathbb{C}) \cong S^2$ described in Exercise 1.1.

(b) Let $a, b \in \mathbb{C}$ satisfy $|a|^2 + |b|^2 = 1$, and let $f(z) = (az - \bar{b})/(bz + \bar{a})$. Prove that $\rho\big(f(z), f(w)\big) = \rho(z,w)$ for all $z, w \in \mathbb{P}^1(\mathbb{C})$. (*Hint.* Although this can be proven directly, an alternative method is to show that $f$ corresponds to a rigid rotation of $S^2$, hence preserves chordal distances. For a nonarchimedean analogue, see Lemma 2.5.)

**1.3.** Let $\phi : \mathbb{P}^1(\mathbb{C}) \to \mathbb{P}^1(\mathbb{C})$ be a rational map.

(a) Prove that there is a constant $C(\phi)$ such that $\phi$ satisfies a Lipschitz inequality

$$\rho\big(\phi(\alpha), \phi(\beta)\big) \le C(\phi)\rho(\alpha, \beta) \qquad \text{for all } \alpha, \beta \in \mathbb{P}^1(\mathbb{C}).$$

(b) Find an explicit formula for the Lipschitz constant $C(f)$ in the case of a linear map $f(z) = (az + b)/(cz + d) \in \mathrm{PGL}_2(\mathbb{C})$.

**1.4.** Let $(\alpha, \alpha', \alpha'')$ and $(\beta, \beta', \beta'')$ be triples of distinct points in $\mathbb{P}^1$. Prove that there exists a unique automorphism $f \in \mathrm{PGL}_2(\mathbb{C})$ satisfying

$$f(\alpha) = \beta, \quad f(\alpha') = \beta', \quad f(\alpha'') = \beta''.$$

(*Hint*. First do the case that one of the triples is $(0, 1, \infty)$.)

## Section 1.2. Critical Points and the Riemann–Hurwitz Formula

**1.5.** Let $\phi(z) \in \mathbb{C}(z)$ be a rational function. In the text we defined the ramification properties of $\phi$ at a point $\alpha \in \mathbb{P}^1(\mathbb{C})$ provided that $\alpha \ne \infty$ and $\phi(\alpha) \ne \infty$. In general, let $f \in \mathrm{PGL}_2(\mathbb{C})$ be any linear fractional transformation, let $\phi^f = f^{-1} \circ \phi \circ f$, and let $\beta = f^{-1}(\alpha)$. If $\beta \ne \infty$ and $\phi^f(\beta) \ne \infty$, we say that $\phi$ is *ramified* at $\alpha$ if $(\phi^f)'(\beta) = 0$ and we define the *ramification index of $\phi$ at $\alpha$* to be

$$e_\alpha(\phi) = e_\beta(\phi^f).$$

(a) Assuming that none of $\alpha, \beta, \phi(\alpha), \phi^f(\beta)$ are equal to $\infty$, prove that

$$(\phi^f)'(\beta) = \frac{(f^{-1})'(\phi(\alpha))}{(f^{-1})'(\alpha)}\phi'(\alpha).$$

Conclude that the ramification points of a rational map $\phi : \mathbb{P}^1 \to \mathbb{P}^1$ are independent of the choice of coordinates, i.e., the vanishing of $\phi'(\alpha)$ is independent of the choice of $f$.

(b) Show that the ramification index $e_\alpha(\phi)$ is well-defined, independent of the choice of $f$.

(c) Under what conditions is it true that the derivative $\phi'(\alpha)$ is independent of the choice of $f$?

**1.6.** Let $\phi(z) \in \mathbb{C}(z)$ be a rational function of degree $d$, say given by

$$\phi(z) = \frac{F(z)}{G(z)} = \frac{a_0 + a_1 z + \cdots + a_d z^d}{b_0 + b_1 z + \cdots + b_d z^d}.$$

(a) Prove that $\phi$ is ramified at $z = 0$ if and only if $a_0 b_1 = a_1 b_0$.

(b) Prove that $\phi$ is ramified at $z = \infty$ if and only if $a_d b_{d-1} = a_{d-1} b_d$.

(c) Find a similar criterion for $e_0(\phi) \ge 3$. Same question for $e_\infty(\phi) \ge 3$.

**1.7.** Let $\phi : \mathbb{P}^1 \to \mathbb{P}^1$ be a rational function of degree $d \ge 2$. Using the standard spherical measure on $\mathbb{P}^1(\mathbb{C})$, normalized so total area is 1, prove that

$$\int_{\mathbb{P}^1(\mathbb{C})} \big|(\phi^n)'(z)\big| \, d\mu(z) \sim d^n \quad \text{as } n \to \infty.$$

Conclude that "on average," the map $\phi$ is expanding.

**1.8.** (a) Let $C$ be a compact Riemann surface with $g$ holes. Prove that if $C$ is triangulated using $V$ vertices, $E$ edges, and $F$ faces, then $V - E + F = 2 - 2g$. The number $g$ is called the *genus* of $C$.

(b) Let $\phi : C_1 \to C_2$ be a finite map of degree $d$ of compact Riemann surfaces, and let $g_i$ be the genus of $C_i$. Give a topological proof of the Riemann–Hurwitz formula,

$$2g_1 - 2 = d(2g_2 - 2) + \sum_{P \in C_1} e_P(\phi) - 1.$$

**1.9.** Let $\phi \in \mathbb{C}(z)$ be a rational map of degree $d \geq 2$.

(a) Prove that $\infty$ is a totally ramified fixed point of $\phi$ if and only if $\phi \in \mathbb{C}[z]$.

(b) Let $\alpha \in \mathbb{P}^1(\mathbb{C})$ with $\alpha \neq \infty$. Prove that $\alpha$ is a totally ramified fixed point of $\phi$ if and only if $(z-\alpha)^d/(\phi(z)-\alpha)$ is in $\mathbb{C}[z]$.

(c) Let $\alpha \in \mathbb{P}^1(\mathbb{C})$ be arbitrary. Prove that $\alpha$ is a totally ramified fixed point of $\phi$ if and only if there exists a linear factional transformation $f \in \mathrm{PGL}_2(\mathbb{C})$ such that $f^{-1}(\alpha) = \infty$ and $\phi^f(z) \in \mathbb{C}[z]$.

**1.10.** Let $K$ be a field of characteristic $p > 0$ and let $\phi(z) \in K(z)$ be a rational function.

(a) Show that the following are equivalent:

  (i) $\phi(z) \notin K(z^p)$.

  (ii) The formal derivative $\phi'(z)$ is not identically zero.

  (iii) The field extension $K(z)/K(\phi(z))$ is separable.

If these conditions hold, we say that $\phi$ is *separable*.

(b) Assume that $\phi$ is separable. Recall that the ramification index is defined by the formula $e_\alpha(\phi) = \mathrm{ord}_\alpha(\phi(z) - \phi(\alpha))$. Let $r_\alpha(\phi) = \mathrm{ord}_\alpha(\phi'(z))$. If $\alpha$ is a critical point of $\phi$, then we say that $\phi$ is *tamely ramified at* $\alpha$ if $p \nmid e_\alpha(\phi)$, and $\phi$ is *wildly ramified at* $\alpha$ if $p \mid e_\alpha(\phi)$. Prove that

$$r_\alpha(\phi) \geq e_\alpha(\phi) - 1,$$

with equality if and only if $\phi$ is either unramified or tamely ramified at $\alpha$.

(c) Prove that

$$2d - 2 \geq \sum_{\alpha \in \mathbb{P}^1(\mathbb{C})} r_\alpha(\phi) \geq \sum_{\alpha \in \mathbb{P}^1(\mathbb{C})} (e_\alpha(\phi) - 1).$$

(*Hint.* Mimic the algebraic proof of the Riemann–Hurwitz formula, Theorem 1.1.)

(d) Deduce that a separable map has at most $2d - 2$ critical points.

(e) More generally, if the rational function $\phi$ is wildly ramified at $t$ points, prove that $\phi$ has at most $2d - 2 - (p-1)t$ critical points.

**1.11.** Let $K$ be a field of characteristic $p > 0$ and let $\phi(z) \in K(z)$ be a rational function.

(a) Suppose that $\phi$ is separable and that $\phi^n \in K[z]$ for some $n \geq 1$. Prove that $\phi^2 \in K[z]$. Thus Theorem 1.7 is true in characteristic $p$ for separable maps.

(b) Let $p = 2$ and $\phi(z) = 1 + z^{-2}$. Prove that $\phi^2 \notin K[z]$ and that $\phi^3 \in K[z]$. Thus Theorem 1.7 need not be true in characteristic $p$ for inseparable maps.

(c) Suppose that $\phi^2 \notin K[z]$ and that $\phi^n \in K[z]$ for some $n \geq 3$. Prove that $\phi$ has the form $\phi = (az^q + b)/(cz^q + d)$, where $q$ is a power of $p$.

**1.12.** Let $\phi : \mathbb{P}^1 \to \mathbb{P}^1$ be a rational map of degree $d \geq 2$, and suppose that there is a point $\alpha \in \mathbb{P}^1$ and an $\epsilon > 0$ such that

$$e_\alpha(\phi^n) \geq (1+\epsilon)^n$$

for infinitely many $n \geq 1$. Prove that $\alpha$ is a preperiodic point for $\phi$, and that the eventual period of $\alpha$ is at most $(2d-2)/\epsilon$.

## Section 1.3. Periodic Points and Multipliers

**1.13.** We defined the multiplier of $\phi$ at a fixed point $\alpha$ to be the derivative $\lambda_\alpha(\phi) = \phi'(\alpha)$. This definition obviously must be modified if $\alpha = \infty$. In order to be compatible with Proposition 1.9, we must set $\lambda_\infty(\phi) = (\phi^f)'(f^{-1}(\infty))$ for all $f \in \mathrm{PGL}_2(\mathbb{C})$. Show that taking $f(z) = 1/z$ leads to the definition

$$\lambda_\infty(\phi) = \lim_{z\to 0} \frac{z^{-2}\phi'(z^{-1})}{\phi(z^{-1})^2}.$$

Prove that $\lambda_\infty(\phi)$ is finite and may be computed without taking a limit by showing that the fraction on the righthand size is a rational function in $\mathbb{C}(z)$ with no pole at $z = 0$, so it may be evaluated at $z = 0$.

**1.14.** Let $\phi : \mathbb{P}^1 \to \mathbb{P}^1$ be a rational map.

(a) Let $P \in \mathbb{P}^1$ be a point of exact period $n$ for $\phi$ and suppose that $\phi^k(P) = P$. Prove that $n \mid k$.

(b) Prove that $\mathrm{Per}_n(\phi)$ is the disjoint union of $\mathrm{Per}_m^{**}(\phi)$ over all $m \mid n$.

**1.15.** Let $\phi : \mathbb{P}^1 \to \mathbb{P}^1$ be a rational map of degree at least 2.

(a) Let $n > m \ge 0$. Prove that $\{P \in \mathbb{P}^1 : \phi^n(P) = \phi^m(P)\}$ is a finite set.

(b) Suppose that $\phi_1$ and $\phi_2$ are rational maps of degree at least 2 and suppose that $\phi_1$ and $\phi_2$ commute with one another, i.e., $\phi_1 \circ \phi_2 = \phi_2 \circ \phi_1$. Prove that

$$\mathrm{PrePer}(\phi_1) = \mathrm{PrePer}(\phi_2).$$

(*Hint.* Use (a) and apply Exercise 0.2.)

**1.16.** Suppose that $\alpha$ is a periodic point of $\phi$. Prove that the orbit of $\alpha$ contains a critical point of $\phi$ if and only if the multiplier $\lambda_\alpha(\phi)$ vanishes.

**1.17.** This exercise generalizes the multiplier sum formula described in Theorem 1.14. Let $\phi(z) \in \mathbb{C}(z)$ be a nonconstant rational map and let $\alpha \in \mathbb{C}$ be a fixed point of $\mathbb{C}$. The *residue fixed-point index* of $\phi$ at $\alpha$ is the quantity

$$\iota(\phi,\alpha) = \frac{1}{2\pi i}\int_{|z-\alpha|=\epsilon} \frac{dz}{z-\phi(z)},$$

where the integral is any sufficiently small loop around $\alpha$.

(a) If $\lambda_\alpha(\phi) \ne 1$, prove that

$$\iota(\phi,\alpha) = \frac{1}{1-\lambda_\alpha(\phi)}.$$

(b) Prove that in all cases, the index $\iota(\phi,\alpha)$ is independent of the choice of local coordinate at $\alpha$.

(c) Prove the index summation formula

$$\sum_{\alpha\in\mathrm{Fix}(\phi)} \iota(\phi,\alpha) = 1.$$

(d) Suppose that $\lambda_\alpha(\phi) = \phi'(\alpha) = 1$ and that $\phi''(\alpha) \neq 0$. Prove that

$$\iota(\phi, \alpha) = \frac{\phi'''(\alpha)/6}{\left(\phi''(\alpha)/2\right)^2}.$$

In other words, if the series expansion of $\phi(z)$ around $z = \alpha$ has the form

$$\phi(z) = \alpha + (z - \alpha) + A(z - \alpha)^2 + B(z - \alpha)^3 + \cdots \qquad \text{with } A \neq 0,$$

then $\iota(\phi, \alpha) = B/A^2$.

(e) The polynomial $\phi(z) = z + Az^2 + z^3$ has fixed points $\{0, -A, \infty\}$. The point at infinity is superattracting, so $\lambda_\infty(\phi) = 0$. Use the formula in (d) and the summation formula to compute $\lambda_{-A}(\phi)$, and then check your answer by computing $\lambda_{-A}(\phi)$ directly.

**1.18.** Let $\phi(z) \in \mathbb{C}(z)$ be a rational function of degree $d \geq 2$.
(a) Prove that $\#\operatorname{Per}_n(\phi) \leq d^n + 1$.
(b) Prove that $\#\operatorname{Per}_n(\phi) \to \infty$ as $n \to \infty$.
(c) Prove that $\operatorname{Per}_n^{**}(\phi)$ is nonempty for infinitely many values of $n$.
(d) * More precisely, prove that if $\operatorname{Per}_n^{**}(\phi)$ is empty, then $(n, d)$ is one of the pairs

$$(n, d) \in \bigl\{(2,2),\ (2,3),\ (3,2),\ (4,2)\bigr\}.$$

**1.19.** Let $\phi(z) \in \mathbb{C}[z]$ be a polynomial of degree $d \geq 2$.
(a) If $m \mid n$, prove that the polynomial $\phi^m(z) - z$ divides the polynomial $\phi^n(z) - z$.
(b) Let $\mu$ be the Möbius $\mu$ function. Prove that for every $n \geq 1$, the rational function

$$\Phi_n^*(z) = \prod_{m \mid n} \bigl(\phi_m(z) - z\bigr)^{\mu(n/m)}$$

is a polynomial. The polynomial $\Phi_n^*(z)$ is a dynamical analogue of the classical cyclotomic polynomial $\prod_{m|n}(z^m - 1)^{\mu(n/m)}$.
(c) Compute the first few polynomials $\Phi_n^*(z)$ for the quadratic polynomial $\phi_c(z) = z^2 + c$.
(d) Let $\alpha$ be a root of $\Phi_n^*(z)$. Prove that $\phi^n(\alpha) = \alpha$. Thus roots of $\Phi_n^*(z)$ are in $\operatorname{Per}_n(\phi)$.
(e) Let $\phi_c(z) = z^2 + c$. Find a value of $c$ such that $\Phi_2^*(z)$ has a root whose exact period is strictly smaller than 2?
(f) For $n = 3$ and $n = 4$, find all values of $c$ such that $\Phi_n^*(z)$ has a root whose exact period is strictly smaller than $n$.

The roots of $\Phi_n^*(z)$ are said to have *formal period* $n$. They behave in many ways as if they have exact period $n$, although their actual period is smaller than $n$.

**1.20.** Find explicit formulas for the fixed points and corresponding multipliers of the function $\phi(z) = az/(z^2 + b)$. Verify directly the formula in Theorem 1.14 (cf. Example 1.17).

## Section 1.4. The Julia Set and the Fatou Set

**1.21.** Let $(S_1, \rho_1)$ and $(S_2, \rho_2)$ be metric spaces. A collection of maps $\Phi$ from $S_1$ to $S_2$ is said to be *uniformly continuous* if for every $\epsilon > 0$ there exists a $\delta > 0$ such that

$$\rho_1(\alpha, \beta) < \delta \quad \Longrightarrow \quad \rho_2(\phi\alpha, \phi\beta) < \epsilon \quad \text{for all } \alpha, \beta \in S_1 \text{ and all } \phi \in \Phi.$$

The family is said to be *uniformly Lipschitz* if there is a constant $C = C(\Phi)$ such that

$$\rho_2\bigl(\phi(\alpha), \phi(\beta)\bigr) \leq C \cdot \rho_1(\alpha, \beta) \quad \text{for all } \alpha, \beta \in S_1 \text{ and all } \phi \in \Phi.$$

(a) If $\Phi$ is uniformly continuous, prove that $\Phi$ is equicontinuous at every point of $S_1$.
(b) If $\Phi$ is uniformly Lipschitz, prove that $\Phi$ is uniformly continuous.
(c) Let $\Phi = \{\phi_n\}_{n\geq 1}$ and suppose that for every point $\alpha \in S_1$, the limit

$$\bar{\phi}(\alpha) = \lim_{n\to\infty} \phi_n(\alpha)$$

exists. If $\Phi$ is equicontinuous at $\alpha \in S_1$, prove that $\bar{\phi}$ is continuous at $\alpha$. Give an example to show that the equicontinuity assumption is necessary.

**1.22.** Give an example to show that equicontinuity is not an open condition. Thus in order to ensure that the Fatou set is open, the definition of $\mathcal{F}(\phi)$ must include the requirement that it be open.

**1.23.** Let $\phi : S \to S$ be a surjective continuous map of a metric space with the additional property that $\phi$ maps open sets to open sets.
(a) Give a rigorous proof that the Fatou and Julia sets of $\phi$ are completely invariant, that is, $\phi^{-1}(\mathcal{F}) = \mathcal{F} = \phi(\mathcal{F})$ and $\phi^{-1}(\mathcal{J}) = \mathcal{J} = \phi(\mathcal{J})$.
(b) Is (a) true without the assumption that $\phi$ is surjective?
(c) Is (a) true without the assumption that $\phi$ maps open sets to open sets?

**1.24.** Let $\phi(z) \in \mathbb{C}[z]$ be a polynomial map, and let $\mathcal{F}_\infty$ be the connected component of $\mathcal{F}(\phi)$ containing the point $\infty$. Prove that $\phi^{-1}(\mathcal{F}_\infty) = \mathcal{F}_\infty = \phi(\mathcal{F}_\infty)$. In other words, for a polynomial map, the connected component of $\infty$ is completely invariant for $\phi$.

**1.25.** Let $\phi(z) = 1 - 2/z^2$ be the rational map from Example 1.31.
(a) Let $f(z) = 1/z$. Prove that $(\phi^f)'(0) = 0$ and conclude that $\infty$ is a critical point of $\phi$.
(b) Let $f(z) = z/(z-1)$. Prove that $(\phi^f)'(0) = 0$ and conclude that 0 is a critical point of $\phi$.

**1.26.** Let $\phi(z) \in \bar{\mathbb{Q}}(z)$ be a rational map of degree $d \geq 2$ with coefficients in $\bar{\mathbb{Q}}$, and let $\mathcal{J}(\phi;\bar{\mathbb{Q}}) = \mathcal{J}(\phi) \cap \mathbb{P}^1(\bar{\mathbb{Q}})$ be the set of points in the Julia set whose coordinates are algebraic numbers.
(a) Is $\mathcal{J}(\phi;\bar{\mathbb{Q}})$ Galois-invariant?
(b) Is $\mathcal{J}(\phi;\bar{\mathbb{Q}})$ infinite? More generally, is $\mathcal{J}(\phi;\bar{\mathbb{Q}})$ dense in $\mathcal{J}(\phi)$?
(c) ** Does $\mathcal{J}(\phi;\bar{\mathbb{Q}})$ contain points that are not preperiodic? More generally, does $\mathcal{J}(\phi;\bar{\mathbb{Q}})$ contain infinitely many nonpreperiodic points $P_i$ such that the orbits of the $P_i$ are disjoint?

## Section 1.5. Properties of Periodic Points

**1.27.** Let $\alpha$ be a periodic point of a rational function $\phi \in \mathbb{C}(z)$.
(a) If $\alpha$ is attracting, prove that $\alpha \in \mathcal{F}(\phi)$.
(b) If $\alpha$ is repelling, prove that $\alpha \in \mathcal{J}(\phi)$.

**1.28.** Let $\phi(z) = z^2 - 2$. Prove that the Julia set of $\phi$ is the closed interval on the real axis between $-2$ and 2. Find the periodic points of $\phi$, compute their multipliers, and show directly that $\mathcal{J}(\phi)$ is equal to the closure of $\operatorname{Per}(\phi)$. (*Hint.* Write $z = e^{it} + e^{-it} = 2\cos(t)$, so $\phi(z) = 2\cos(2t)$. Prove that $\phi^n(z) = 2\cos(2^n t)$ and use this formula to study the dynamics of $\phi$.)

**1.29.** Let $\phi(z) \in \mathbb{C}[z]$ be a polynomial of degree $d \geq 1$, and suppose that every $\alpha \in \operatorname{Per}_n(\phi)$ has multiplier $\lambda_\phi(\alpha) \neq 1$. Prove that the equation $\phi^n(z) = z$ has simple roots, and deduce that $\operatorname{Per}_n(\phi)$ contains exactly $d^n + 1$ distinct points (including the point $\infty$).

## Section 1.6. Dynamical Systems Associated to Algebraic Groups

**1.30.** This exercise describes algebraic properties of the Chebyshev polynomials $T_d(w)$.

(a) Prove that $T_2(w) = w^2 - 2$, $T_3(w) = w^3 - 3w$, and $T_4(w) = w^4 - 4w^2 + 2$.

(b) Prove that $T_d(T_e(w)) = T_{de}(w)$ for all $d, e \geq 0$.

(c) Prove that $T_d(-w) = (-1)^d T_d(w)$. Thus $T_d$ is an odd function if $d$ is odd and it is an even function if $d$ is even.

(d) Prove that the Chebyshev polynomials satisfy the recurrence relation

$$T_{d+1}(w) = wT_d(w) - T_{d-1}(w) \quad \text{for all } d \geq 1,$$

where we use the initial values $T_0(w) = 2$ and $T_1(w) = w$.

(e) Prove that the generating function of the Chebyshev polynomials is given by

$$\sum_{d=0}^{\infty} T_d(w)X^d = \frac{2 - wX}{1 - wX + X^2}.$$

(f) Prove that the $d^{\text{th}}$ Chebyshev polynomial is given by the explicit formula

$$T_d(w) = d \sum_{0 \leq k \leq d/2} \frac{(-1)^k}{d-k} \binom{n-k}{k} w^{d-2k}.$$

**1.31.** This exercise describes dynamical properties of the Chebyshev polynomial $T_d(w)$ for a fixed $d \geq 2$.

(a) Let $U$ be an open subinterval of $[-2, 2]$. Prove that there exists an integer $n \geq 1$ such that $T_d^n(U) = [-2, 2]$.

(b) Prove that $\lim_{n\to\infty} T_d^n(w) = \infty$ for all points $w \in \mathbb{C}$ not lying in the interval $[-2, 2]$.

(c) Prove that the Julia set $\mathcal{J}(T_d)$ is the closed interval $[-2, 2]$.

(d) Find all of the periodic points of $T_d$ and compute their multipliers. Observe that $\mathcal{J}(T_d)$ is the closure of the (repelling) periodic points.

(e) Let $F(w) \in \mathbb{C}[w]$ be a polynomial of degree $d \geq 2$ with the property that the interval $[-2, 2]$ is both forward and backward invariant for $F$. Prove that $F(w) = \pm T_d(w)$.

**1.32.** (a) Let $A$ be an abelian group, let $A_{\text{tors}} = \{P \in A : nP = 0 \text{ for some } n \geq 1\}$ be the torsion subgroup of $A$, let $d \geq 2$ be an integer, and let $\phi : A \to A$ be the map $\phi(P) = dP$. Prove that

$$\operatorname{PrePer}(\phi) = A_{\text{tors}}.$$

(This is Proposition 0.3, but try reproving it without looking back at our proof.)

(b) Let $E$ be an elliptic curve, let $d \geq 2$ be an integer, and let $\phi_{E,d} : \mathbb{P}^1 \to \mathbb{P}^1$ be the rational map associated to multiplication-by-$d$ on $E$, as described in Section 1.6.3. Prove that

$$\operatorname{PrePer}(\phi_{E,d}) = x\big(E(\mathbb{C})_{\text{tors}}\big).$$

# Chapter 2

# Dynamics over Local Fields: Good Reduction

The study of the dynamics of polynomial and rational maps over $\mathbb{R}$ and $\mathbb{C}$ has a long history and includes many deep theorems, some of which were briefly discussed in Chapter 1. A more recent development is the creation of an analogous theory over complete local fields such as the $p$-adic rational numbers $\mathbb{Q}_p$ and the completion $\mathbb{C}_p$ of an algebraic closure of $\mathbb{Q}_p$. The nonarchimedean nature of the absolute value on $\mathbb{Q}_p$ and $\mathbb{C}_p$ makes some parts of the theory easier than when working over $\mathbb{C}$ or $\mathbb{R}$. But as usual, there is a price to pay. For example, the theory of nonarchimedean dynamics must deal with the fact that $\mathbb{Q}_p$ is totally disconnected and far from being algebraically closed, while $\mathbb{C}_p$ is not locally compact.

In this chapter we begin our study of dynamics over complete local fields $K$ by concentrating on rational maps $\phi$ that have "good reduction." Roughly speaking, this means that the reduction of $\phi$ modulo the maximal ideal of the ring of integers of $K$ is a "well-behaved" rational map $\tilde{\phi}$ over the residue field $k$ of $K$. Thus studying the dynamics of $\tilde{\phi}$ over $k$ allows us to derive nontrivial information about the dynamics of $\phi$ over $K$. In Chapter 5 we take up the more difficult, but ultimately more interesting, case of rational maps with "bad reduction."

## 2.1 The Nonarchimedean Chordal Metric

We begin by quickly recalling the definition and basic properties of absolute values, especially those satisfying the ultrametric inequality.

**Definition.** An *absolute value* on a field $K$ is a map

$$|\cdot| : K \longrightarrow \mathbb{R}$$

with the following properties:

- $|\alpha| \geq 0$, and $|\alpha| = 0$ if and only if $\alpha = 0$.

- $|\alpha\beta| = |\alpha| \cdot |\beta|$ for all $\alpha, \beta \in K$.
- $|\alpha + \beta| \le |\alpha| + |\beta|$ for all $\alpha, \beta \in K$ (triangle inequality).

A *valued field* is a pair $(K, |\cdot|_K)$ consisting of a field $K$ and an absolute value on $K$, although we often omit the absolute value in the notation. A map of valued fields $i : K \to L$ is a field homomorphism that respects the absolute values,

$$|i(\alpha)|_L = |\alpha|_K \qquad \text{for all } \alpha \in K.$$

**Definition.** Let $(K, |\cdot|)$ be a valued field. If the absolute value satisfies the stronger estimate

$$|\alpha + \beta| \le \max\{|\alpha|, |\beta|\} \quad \text{for all } \alpha, \beta \in K, \tag{2.1}$$

then the absolute value is *nonarchimedean* or *ultrametric*. The associated *valuation* is the homomorphism

$$v : K^* \longrightarrow \mathbb{R}, \qquad v(\alpha) = -\log|\alpha|.$$

It satisfies

$$v(\alpha + \beta) \ge \min\{v(\alpha), v(\beta)\}.$$

The valuation $v$ is *discrete* if $v(K^*)$ is a discrete subgroup of $\mathbb{R}$, in which case the associated *normalized valuation* (sometimes denoted $\operatorname{ord}_v$) is the constant multiple of $v$ chosen to satisfy $\operatorname{ord}_v(K^*) = \mathbb{Z}$.

*Example* 2.1. The field $\mathbb{Q}$ has the usual real absolute value

$$|\alpha|_\infty = \max\{\alpha, -\alpha\}.$$

For each prime $p$ it also has a $p$-adic absolute value defined as follows. Every nonzero rational number $\alpha$ has a unique factorization of the form

$$\alpha = \pm \prod_{p \text{ prime}} p^{e_p(\alpha)} \qquad \text{with } e_p(\alpha) \in \mathbb{Z}.$$

Then

$$|\alpha|_p = p^{-e_p(\alpha)}.$$

The $p$-adic absolute values are nonarchimedean.

Two absolute values are said to be *equivalent* if there is a constant $r > 0$ such that $|\alpha|_1 = |\alpha|_2^r$ for all $\alpha \in K$. Ostrowski's theorem says that up to equivalence, the real absolute value and the $p$-adic absolute values are the only nontrivial absolute values on $\mathbb{Q}$; see [78, 1.4.2, Theorem 3] or [249, I.2, Theorem 1].

*Example* 2.2. Let $k$ be a field and let $K = k(T)$ be the field of rational functions. Then each $a \in k$ determines an absolute value on $K$ associated to the valuation $\operatorname{ord}_a$ that gives the order of vanishing of $f(T) \in k(T)$ at $T = a$. The degree map $\deg : k(T)^* \to \mathbb{Z}$ is also a valuation. If $k$ is algebraically closed, this yields the complete set of absolute values on $k(T)$ that are trivial on $k$, up to equivalence. More generally, if $k$ is not algebraically closed, there is a valuation corresponding to each monic irreducible polynomial in $k[T]$. In this book we are primarily interested in the number field scenario, i.e., the field $\mathbb{Q}$ and its extensions and completions, but the reader should be aware that it is also interesting to study the analogous case of function fields, i.e., the field $k[T]$ and its extensions and completions.

We now prove the elementary, but extremely useful, fact that if $|\alpha| \neq |\beta|$, then the ultrametric inequality (2.1) is actually an equality.

**Lemma 2.3.** *Let $K$ be a field with a nonarchimedean absolute value $|\cdot|_v$ and let $\alpha, \beta \in K$. Then*

$$|\alpha|_v \neq |\beta|_v \quad \Longrightarrow \quad |\alpha + \beta|_v = \max\{|\alpha|_v, |\beta|_v\}.$$

*Proof.* We suppose that $|\alpha|_v > |\beta|_v$. The strict inequality

$$|\beta|_v < |\alpha|_v = |(\alpha + \beta) - \beta|_v \leq \max\{|\alpha + \beta|_v, |\beta|_v\}$$

implies that the maximum on the right is $|\alpha + \beta|_v$, so we find that $|\alpha|_v \leq |\alpha + \beta|_v$. The opposite inequality is also true, since $|\alpha + \beta|_v \leq \max\{|\alpha|_v, |\beta|_v\} = |\alpha|_v$. □

Recall that the chordal metric on $\mathbb{P}^1(\mathbb{C})$, which we now denote by $\rho_\infty$, is defined by the formula

$$\rho_\infty(P_1, P_2) = \frac{|X_1 Y_2 - X_2 Y_1|}{\sqrt{|X_1|^2 + |Y_1|^2}\sqrt{|X_2|^2 + |Y_2|^2}}$$

for points $P_1 = [X_1, Y_1]$ and $P_2 = [X_2, Y_2]$ in $\mathbb{P}^1(\mathbb{C})$. In the case of a field $K$ having a nonarchimedean absolute value $|\cdot|_v$, it is convenient to use a metric given by a slightly different formula.

**Definition.** Let $K$ be a field with a nonarchimedean absolute value $|\cdot|_v$, and let $P_1 = [X_1, Y_1]$ and $P_2 = [X_2, Y_2]$ be points in $\mathbb{P}^1(K)$. The *$v$-adic chordal metric on* $\mathbb{P}^1(K)$ is

$$\rho_v(P_1, P_2) = \frac{|X_1 Y_2 - X_2 Y_1|_v}{\max\{|X_1|_v, |Y_1|_v\} \max\{|X_2|_v, |Y_2|_v\}}.$$

It is clear from the definition that $\rho_v(P_1, P_2)$ is independent of the choice of homogeneous coordinates for $P_1$ and $P_2$.

The first thing to check is that $\rho_v$ is indeed a metric. In fact, it is an ultrametric; that is, it satisfies the nonarchimedean triangle inequality.

**Proposition 2.4.** *The $v$-adic chordal metric has the following properties.*

(a) $0 \leq \rho_v(P_1, P_2) \leq 1$.

(b) $\rho_v(P_1, P_2) = 0$ *if and only if* $P_1 = P_2$.

(c) $\rho_v(P_1, P_2) = \rho_v(P_2, P_1)$.

(d) $\rho_v(P_1, P_3) \leq \max\{\rho_v(P_1, P_2), \rho_v(P_2, P_3)\}$.

*Proof.* The lower bound in (a) and parts (b) and (c) of the proposition are obvious from the definition. For the upper bound in (a), we use the nonarchimedean nature of $v$ to compute

$$\begin{aligned}|X_1Y_2 - X_2Y_1|_v &\le \max\{|X_1Y_2|_v, |X_2Y_1|_v\}\\ &\le \max\{|X_1|_v, |Y_1|_v\}\max\{|X_2|_v, |Y_2|_v\}.\end{aligned}$$

The proof of (d) requires the consideration of several cases. The following useful lemma makes the proof more transparent by allowing some freedom to change coordinates. It is the analogue of the fact that the classical chordal metric is invariant under linear fractional transformations that define rigid rotations of the Riemann sphere (cf. Exercise 1.2).

**Lemma 2.5.** *Let*

$$R = \{\alpha \in K : |\alpha|_v \le 1\}$$

*be the ring of integers of $K$, and let $f : \mathbb{P}^1 \to \mathbb{P}^1$ be a linear fractional transformation of the form*

$$f([X,Y]) = \frac{aX + bY}{cX + dY} \qquad \text{with } a,b,c,d \in R \text{ and } ad - bc \in R^*,$$

*i.e., $f \in \mathrm{PGL}_2(R)$. Then*

$$\rho_v\bigl(f(P_1), f(P_2)\bigr) = \rho_v(P_1, P_2) \qquad \text{for all } P_1, P_2 \in \mathbb{P}^1(K).$$

*(N.B. It is crucial that the quantity $ad - bc$ is a unit.)*

*Proof.* Write each point as $P_i = [X_i, Y_i]$ with $X_i, Y_i \in R$ and at least one of $X_i$ or $Y_i$ in $R^*$. Then $\max\{|X_i|_v, |Y_i|_v\} = 1$, so

$$\rho_v(P_1, P_2) = |X_1Y_2 - X_2Y_1|_v.$$

Further, the identities

$$\begin{aligned}d(aX_i + bY_i) - b(cX_i + dY_i) &= (ad - bc)X_i,\\ c(aX_i + bY_i) - a(cX_i + dY_i) &= -(ad - bc)Y_i,\end{aligned}$$

and the fact that $\max\{|X_i|_v, |Y_i|_v\} = 1$ and $|ad - bc|_v = 1$ immediately imply that

$$\max\{|aX_i + bY_i|_v, |cX_i + dY_i|_v\} = 1.$$

Thus

$$\begin{aligned}\rho_v\bigl(f(P_1), f(P_2)\bigr) &= \bigl|(aX_1 + bY_1)(cX_2 + dY_2) - (aX_2 + bY_2)(cX_1 + dY_1)\bigr|_v\\ &= \bigl|(ad - bc)(X_1Y_2 - X_2Y_1)\bigr|_v\\ &= \rho_v(P_1, P_2),\end{aligned}$$

where we have again made use of the fact that $|ad - bc|_v = 1$. □

We resume the proof of Proposition 2.4 and write each point as $P_i = [X_i, Y_i]$ with $X_i, Y_i \in R$ and at least one of $X_i$ or $Y_i$ in $R^*$. Then

$$\max\{|X_i|_v, |Y_i|_v\} = 1 \quad \text{and} \quad \rho_v(P_i, P_j) = |X_iY_j - X_jY_i|_v,$$

as usual. If $|X_2|_v > |Y_2|_v$, we apply the map $f = Y/X$ to the three points. This preserves the chordal distance (Lemma 2.5) and allows us to assume that

$$|X_2|_v \le |Y_2|_v = 1.$$

Next we apply the map $f = (Y_2 \cdot X - X_2 \cdot Y)/Y$ to the three points. Lemma 2.5 again tells us that the chordal distance is preserved (note that $|Y_2|_v = 1$), and we are reduced to the case that $P_2 = [0, 1]$. Finally, we compute

$$|X_1Y_3 - X_3Y_1|_v \le \max\{|X_1Y_3|_v, |X_3Y_1|_v\} \le \max\{|X_1|_v, |X_3|_v\},$$

which is exactly the desired inequality when $P_2$ is the point $[0, 1]$. □

## 2.2 Periodic Points and Their Properties

In Section 1.3 we described various properties of periodic points for rational maps defined over $\mathbb{C}$. Virtually all of these definitions make sense, mutatis mutandis, when we work over any field with an absolute value. In this section we briefly recall the relevant material.

Let $K$ be a field with an absolute value $|\cdot|$ and let $\phi(z) \in K(z)$ be a nonconstant rational map. The *multiplier of $\phi$ at a fixed point* $\alpha \in K$ is the derivative

$$\lambda_\alpha(\phi) = \phi'(\alpha).$$

(See Exercise 1.13 for the case $\alpha = \infty$.) The multiplier is well-defined, independent of the choice of coordinates on $\mathbb{P}^1$; see Proposition 1.9.

More generally, if $\alpha \in \mathbb{P}^1(K)$ is a point of exact period $n$ for $\phi$, then $\alpha$ is a fixed point of $\phi^n$ and we define the *multiplier of $\phi$ at* $\alpha$ to be

$$\lambda_\alpha(\phi) = \lambda_\alpha(\phi^n) = (\phi^n)'(\alpha).$$

The multiplier may be calculated using the chain rule as

$$\lambda_\alpha(\phi) = (\phi^n)'(\alpha) = \phi'(\alpha) \cdot \phi'(\phi(\alpha)) \cdot \phi'(\phi^2(\alpha)) \cdots \phi'(\phi^{n-1}(\alpha)).$$

The magnitude of the multiplier determines, to some extent, the behavior of $\phi$ in a small neighborhood of a periodic point $\alpha$. The periodic point $\alpha$ is called

*superattracting* if $\lambda_\alpha(\phi) = 0$,
*attracting* if $|\lambda_\alpha(\phi)| < 1$,
*neutral* (or *indifferent*) if $|\lambda_\alpha(\phi)| = 1$,
*repelling* if $|\lambda_\alpha(\phi)| > 1$.

The neutral periodic points are further divided into two types. The *rationally neutral periodic points* are those whose multiplier is a root of unity. The others are called *irrationally neutral*.

## 2.3 Reduction of Points and Maps Modulo $\mathfrak{p}$

One of the most important gadgets in the number theorist's toolbox is reduction modulo a prime. Thus when studying the number-theoretic properties of an object, we reduce it modulo a prime, analyze the properties of the hopefully simpler object, and then lift the information back to obtain global information. A typical example is provided by Hensel's lemma, which under certain circumstances allows us to lift solutions of a polynomial congruence $f(x) \equiv 0 \pmod{p}$ to solutions in $\mathbb{Z}_p$. Then, using information gathered from many primes, one is sometimes able to deduce results for a global field such as $\mathbb{Q}$.

Our principal objects of study are maps $\phi : \mathbb{P}^1 \to \mathbb{P}^1$. In this section we study the behavior of such maps under reduction modulo a prime. We work over a discrete valuation ring, so we set the following notation:

| | |
|---|---|
| $K$ | a field with normalized discrete valuation $v : K^* \twoheadrightarrow \mathbb{Z}$. |
| $\lvert \cdot \rvert_v$ | $= c^{-v(x)}$ for some $c > 1$, an absolute value associated to $v$. |
| $R$ | $= \{\alpha \in K : v(\alpha) \geq 0\}$, the ring of integers of $K$. |
| $\mathfrak{p}$ | $= \{\alpha \in K : v(\alpha) \geq 1\}$, the maximal ideal of $R$. |
| $R^*$ | $= \{\alpha \in K : v(\alpha) = 0\}$, the group of units of $R$. |
| $k$ | $= R/\mathfrak{p}$, the residue field of $R$. |
| $\sim$ | reduction modulo $\mathfrak{p}$, i.e., $R \to k$, $a \mapsto \tilde{a}$. |

Before studying reduction of maps, we consider the problem of reducing points modulo $\mathfrak{p}$. This is as easy in $\mathbb{P}^N$ as it is in $\mathbb{P}^1$, so we look at the general case. Let

$$P = [x_0, x_1, \dots, x_N] \in \mathbb{P}^N(K)$$

be a point defined over $K$. We cannot immediately reduce the coordinates of $P$ modulo $\mathfrak{p}$, since some of the coordinates might not be in $R$. However, since the coordinates of $P$ are homogeneous, we can replace them with

$$P = [cx_0, cx_1, \dots, cx_N]$$

for any $c \in K^*$. Choosing $c$ to be highly divisible by $\mathfrak{p}$, we can ensure that every $cx_i$ is in $R$. However, if we overdo this process and end up with every $cx_i$ in the prime ideal $\mathfrak{p}$, then when we reduce modulo $\mathfrak{p}$, we end up with $[0, 0, \dots, 0]$, which does not represent a point of projective space.

The trick is to "clear the denominators" of the $x_i$'s as efficiently as possible. To do this, we choose an element $\alpha \in K^*$ satisfying

$$v(\alpha) = \min\{v(x_0), v(x_1), \dots, v(x_N)\}. \tag{2.2}$$

For example, $\alpha$ could be the $x_j$ having minimal valuation. Then $\alpha^{-1}x_i \in R$ for every $i$, so we can reduce these quantities modulo $\mathfrak{p}$. We define the *reduction of $P$ modulo $\mathfrak{p}$* to be the point

$$\tilde{P} = \left[\widetilde{\alpha^{-1}x_0}, \widetilde{\alpha^{-1}x_1}, \dots, \widetilde{\alpha^{-1}x_N}\right] \in \mathbb{P}^1(k).$$

Note that $\tilde{P}$ has at least one nonzero coordinate, since at least one of the numbers $\alpha^{-1}x_i$ is a unit. Hence $\tilde{P}$ is a well-defined point in $\mathbb{P}^1(k)$.

We say that $P = [x_0, \ldots, x_N]$ has been written using *normalized coordinates* if

$$\min\{v(x_0), v(x_1), \ldots, v(x_N)\} = 0,$$

in which case $\tilde{P}$ is simply $[\tilde{x}_0, \ldots, \tilde{x}_N]$.

*Example* 2.6. Consider the point $P = \left[\frac{3}{14}, \frac{9}{35}, \frac{24}{49}, \frac{27}{245}\right] \in \mathbb{P}^1(\mathbb{Q})$. We can reduce $P$ modulo 11 without any modification, since every coordinate is an 11-adic integer and not all coordinates vanish modulo 11. Thus $\tilde{P} = [1, 10, 7, 9] \pmod{11}$. However, if we want to reduce $P$ modulo 3, then we first need to divide all of the coordinates by 3,

$$P = \left[\frac{3}{14}, \frac{9}{35}, \frac{24}{49}, \frac{27}{245}\right] = \left[\frac{1}{14}, \frac{3}{35}, \frac{8}{49}, \frac{9}{245}\right],$$

and then $\tilde{P} = [2, 0, 2, 0] \pmod{3}$. Similarly, in order to compute $P$ modulo 7, we first multiply the coordinates by 49,

$$P = \left[\frac{3}{14}, \frac{9}{35}, \frac{24}{49}, \frac{27}{245}\right] = \left[\frac{21}{2}, \frac{63}{5}, \frac{24}{1}, \frac{27}{5}\right],$$

and then $\tilde{P} = [0, 0, 3, 4] \pmod{7}$.

It may appear that the reduction $\tilde{P}$ depends on the choice of $\alpha$. We now check that this is not the case.

**Proposition 2.7.** *Let $P = [x_0, \ldots, x_N] \in \mathbb{P}^N(K)$. Then the reduction $\tilde{P}$ is independent of the choice of $\alpha$ satisfying* (2.2).

*Proof.* Suppose that $\alpha$ and $\beta$ both satisfy (2.2). Then $\alpha$ and $\beta$ have the same valuation, so $\alpha\beta^{-1} \in R^*$. This allows us to compute

$$\begin{aligned}\left[\widetilde{\alpha^{-1}x_0}, \widetilde{\alpha^{-1}x_1}, \ldots, \widetilde{\alpha^{-1}x_N}\right] &= \left[\widetilde{\alpha\beta^{-1}}\widetilde{\alpha^{-1}x_0}, \widetilde{\alpha\beta^{-1}}\widetilde{\alpha^{-1}x_1}, \ldots, \widetilde{\alpha\beta^{-1}}\widetilde{\alpha^{-1}x_N}\right] \\ &= \left[\widetilde{\beta^{-1}x_0}, \widetilde{\beta^{-1}x_1}, \ldots, \widetilde{\beta^{-1}x_N}\right].\end{aligned}$$

Hence the reduction of $P$ modulo $\mathfrak{p}$ is independent of the choice of $\alpha$ satisfying (2.2). □

We next prove an easy and useful lemma that relates reduction modulo $\mathfrak{p}$ to $v$-adic distance.

**Lemma 2.8.** *Let $P_1$ and $P_2$ be points in $\mathbb{P}^1(K)$. Then*

$$\tilde{P}_1 = \tilde{P}_2 \quad \textit{if and only if} \quad \rho_v(P_1, P_2) < 1.$$

*Proof.* Write $P_1 = [X_1, Y_1]$ and $P_2 = [X_2, Y_2]$ using normalized coordinates, so in particular $\rho_v(P_1, P_2) = |X_1Y_2 - X_2Y_1|_v$. Suppose first that $\tilde{P}_1 = \tilde{P}_2$. This means that there is a $\tilde{u} \in k^*$ such that $\tilde{X}_1 = \tilde{u}\tilde{X}_2$ and $\tilde{Y}_1 = \tilde{u}\tilde{Y}_2$. In other words, there is a $u \in R^*$ such that that $X_1 \equiv uX_2 \pmod{\mathfrak{p}}$ and $Y_1 \equiv uY_2 \pmod{\mathfrak{p}}$. Hence

$$X_1Y_2 - X_2Y_1 \equiv uX_2Y_2 - X_2uY_2 \equiv 0 \pmod{\mathfrak{p}},$$

which completes the proof that $\rho_v(P_1, P_2) < 1$.

Next suppose that $\rho_v(P_1, P_2) < 1$, which implies that $X_1Y_2 \equiv X_2Y_1 \pmod{\mathfrak{p}}$. If $X_1X_2 \not\equiv 0 \pmod{\mathfrak{p}}$, then $\tilde{X}_1, \tilde{X}_2 \in k^*$, so we have

$$\tilde{P}_2 = [\tilde{X}_2, \tilde{Y}_2] = [\tilde{X}_1\tilde{X}_2, \tilde{X}_1\tilde{Y}_2] = [\tilde{X}_1\tilde{X}_2, \tilde{X}_2\tilde{Y}_1] = [\tilde{X}_1, \tilde{Y}_1] = \tilde{P}_1.$$

On the other hand, if $X_1X_2 \equiv 0 \pmod{\mathfrak{p}}$, then the fact that the coordinates are normalized and the equality $X_1Y_2 \equiv X_2Y_1 \pmod{\mathfrak{p}}$ imply that $X_1 \equiv X_2 \equiv 0 \pmod{\mathfrak{p}}$, so $\tilde{P}_1 = \tilde{P}_2 = [0, 1]$. This completes the proof of the lemma. □

As a first application, we show that fractional linear transformations in $\mathrm{PGL}_2(R)$ respect reduction modulo $\mathfrak{p}$.

**Proposition 2.9.** *Let $P, Q \in \mathbb{P}^1(K)$ and $f \in \mathrm{PGL}_2(R)$. Then*

$$\tilde{P} = \tilde{Q} \quad \text{if and only if} \quad \widetilde{f(P)} = \widetilde{f(Q)}.$$

*Proof.* We combine Lemmas 2.5 and 2.8. Thus

$$\begin{aligned} \tilde{P} = \tilde{Q} \quad &\Longleftrightarrow \quad \rho_v(P, Q) < 1 && \text{from Lemma 2.8,} \\ &\Longleftrightarrow \quad \rho_v\bigl(f(P), f(Q)\bigr) < 1 && \text{from Lemma 2.5,} \\ &\Longleftrightarrow \quad \widetilde{f(P)} = \widetilde{f(Q)} && \text{from Lemma 2.8 again.} \end{aligned}$$ □

*Example* 2.10. Let $f = \left(\begin{smallmatrix} 5 & 2 \\ 5 & 8 \end{smallmatrix}\right)$, so $f \notin \mathrm{PGL}_2(\mathbb{Z}_3)$. Consider the points $P = [7, 5]$ and $Q = [4, 2]$ in $\mathbb{P}^1(\mathbb{Q}_3)$. They satisfy $\tilde{P} = \tilde{Q} = [1, 2]$ in $\mathbb{P}^1(\mathbb{F}_3)$, but

$$\begin{aligned} f(P) &= [45, 75] = [3, 5] \equiv [0, 1] \pmod 3, \\ f(Q) &= [24, 36] = [2, 3] \equiv [1, 0] \pmod 3, \end{aligned}$$

so $\widetilde{f(P)} \neq \widetilde{f(Q)}$. This shows the necessity of the condition $f \in \mathrm{PGL}_2(R)$ in Proposition 2.9.

It is easy to see that if $K$ is a field and $P_1, P_2, P_3$ are distinct points in $\mathbb{P}^1(K)$, then there is an element of $\mathrm{PGL}_2(K)$ that moves them to the points $0, 1, \infty$. (See Exercise 1.4.) The next proposition gives a stronger result for points whose reductions are distinct. It is especially useful because Lemma 2.5 says that the nonarchimedean chordal metric is invariant for maps in $\mathrm{PGL}_2(R)$, so changing coordinates via an element of $\mathrm{PGL}_2(R)$ does not change the underlying dynamics.

**Proposition 2.11.** *Let $P_1, P_2, P_3 \in \mathbb{P}^1(K)$ be points whose reductions $\tilde{P}_1, \tilde{P}_2, \tilde{P}_3$ are distinct. Then there is a linear fractional transformation $f \in \mathrm{PGL}_2(R)$ such that*

$$f(P_1) = 0, \qquad f(P_2) = 1, \qquad f(P_3) = \infty.$$

*Proof.* Write $P_i = [X_i, Y_i]$ with normalized coordinates. If $v(X_1) > v(Y_1)$, we begin by applying the map $f = Y/X \in \mathrm{PGL}_2(R)$ to each of the three points, so we may assume that $v(X_1) \le v(Y_1)$. Since the coordinates are normalized, this implies that $v(Y_1) = 0$, so $Y_1$ is a unit. We next apply the map

$$(Y_1X - X_1Y)/Y \in \mathrm{PGL}_2(R)$$

to the three points. Having done this, we see that $P_1 = [0, 1]$.

Next consider the point $P_3$. Since $\tilde{P}_3 \ne \tilde{P}_1 = [0, 1]$, we see that $v(X_3) = 0$, so we can apply the map $X/(Y_3X - X_3Y) \in \mathrm{PGL}_2(R)$ to the three points. This fixes $P_1$ and sends $P_3$ to $[1, 0]$, i.e., $P_3$ gets sent to $\infty$. Finally, since $\tilde{P}_2$ is equal to neither $\tilde{P}_1 = [0, 1]$ nor $\tilde{P}_3 = [1, 0]$, we see that $v(X_2) = v(Y_2) = 0$. Applying the map $Y_2X/X_2Y \in \mathrm{PGL}_2(R)$ to the three points then fixes $P_1$ and $P_3$ and sends $P_2$ to $[1, 1]$. □

Having looked at the reduction of a point, we next turn to the problem of reducing a rational map modulo $\mathfrak{p}$. Let $\phi : \mathbb{P}^1 \to \mathbb{P}^1$ be a rational map of degree $d$ defined over $K$, so $\phi$ is given by a pair of homogeneous polynomials of degree $d$,

$$F(X, Y), G(X, Y) \in K[X, Y].$$

Note that the map $\phi(X, Y) = [F(X, Y), G(X, Y)]$ does not change if $F$ and $G$ are each multiplied by a nonzero constant $c \in K^*$, since the coordinates are homogeneous.

**Definition.** Let $\phi : \mathbb{P}^1 \to \mathbb{P}^1$ be a rational map as above and write

$$\phi = [F(X, Y), G(X, Y)]$$

with homogeneous polynomials $F, G \in K[X, Y]$. We say that the pair $(F, G)$ is *normalized*, or that $\phi$ has been written in *normalized form*, if $F, G \in R[X, Y]$ and at least one coefficient of $F$ or $G$ is in $R^*$. Equivalently, $\phi = [F, G]$ is normalized if

$$F(X, Y) = a_0X^d + a_1X^{d-1}Y + \cdots + a_{d-1}XY^{d-1} + a_dY^d$$

and

$$G(X, Y) = b_0X^d + b_1X^{d-1}Y + \cdots + b_{d-1}XY^{d-1} + b_dY^d$$

satisfy

$$\min\{v(a_0), v(a_1), \dots, v(a_d), v(b_0), v(b_1), \dots, v(b_d)\} = 0. \qquad (2.3)$$

Given any representation $\phi = [F, G]$, it is clear that one can always find some $c \in K^*$ such that $[cF, cG]$ is a normalized representation. Further, the element $c$ is unique up to multiplication by an element of $R^*$.

Writing $\phi = [F, G]$ in normalized form, the *reduction of $\phi$ modulo $\mathfrak{p}$* is defined in the obvious way,

$$\begin{aligned}\tilde{\phi}(X,Y) &= [\tilde{F}(X,Y), \tilde{G}(X,Y)] \\ &= [\tilde{a}_0 X^d + \tilde{a}_1 X^{d-1}Y + \cdots + \tilde{a}_d Y^d, \tilde{b}_0 X^d + \tilde{b}_1 X^{d-1}Y + \cdots + \tilde{b}_d Y^d].\end{aligned}$$

In other words, $\tilde{\phi}$ is obtained by reducing the coefficients of $F$ and $G$ modulo $\mathfrak{p}$. (If the prime ideal is not clear from context, we write $\tilde{\phi}_{\mathfrak{p}}$ or $\tilde{\phi} \bmod \mathfrak{p}$.)

The fact that at least one coefficient of $F$ or $G$ is a unit ensures that at least one of $\tilde{F}$ and $\tilde{G}$ is a nonzero polynomial, so the reduction $\tilde{\phi}$ gives a well-defined map $\tilde{\phi} : \mathbb{P}^1(k) \to \mathbb{P}^1(k)$. Further, the reduced map $\tilde{\phi}$ is independent of the choice of $F$ and $G$, a fact whose proof we leave to the reader (Exercise 2.4) since it is quite similar to the proof of Proposition 2.7.

The mere existence of the reduction $\tilde{\phi}$ of a rational map $\phi$ does not imply that $\tilde{\phi}$ has good properties, as is shown by the following simple example.

*Example* 2.12. Let $a \in K^*$ and consider the rational map

$$\phi_a(X,Y) = [aX^d, Y^d].$$

If $a \in R^*$, then $\tilde{a} \neq 0$ and the reduced map $\tilde{\phi}_a(X,Y) = [\tilde{a}X^d, Y^d]$ is again a rational map of degree $d$. However, if $v(a) > 0$, then $\tilde{a} = 0$, so $\tilde{\phi}_a(X,Y) = [0, Y^d] = [0,1]$ is a constant map! Similarly, if $v(a) < 0$, then $\widetilde{a^{-1}} = 0$, so

$$\tilde{\phi}_a(X,Y) = [\widetilde{aX^d, Y^d}] = [\widetilde{X^d, a^{-1}Y^d}] = [X^d, 0] = [1,0]$$

is again a constant map, but not to the same point!! To summarize, the reduction of the map $\phi_a(X,Y) = [aX^d, Y^d]$ separates into three cases,

$$\tilde{\phi}_a = \begin{cases} [\tilde{a}X^d, Y^d] & \text{if } v(a) = 0, \\ [0,1] & \text{if } v(a) > 0, \\ [1,0] & \text{if } v(a) < 0. \end{cases}$$

Clearly it is only in the first case that the reduced map is interesting.

Keep in mind that our goal is to use the dynamics of $\tilde{\phi}$ to help us understand the dynamics of $\phi$. In the above example, if $v(a) = 0$, then it is easy to see that for any $P \in \mathbb{P}^1(K)$,

$$\widetilde{\phi(P)} = \tilde{\phi}(\tilde{P}), \quad \text{and hence by induction,} \quad \widetilde{\phi^n(P)} = \tilde{\phi}^n(\tilde{P}).$$

Thus the $\tilde{\phi}$ orbit of $\tilde{P}$ yields valuable information about the $\phi$ orbit of $P$. However, if $v(a) \neq 0$, then $\tilde{\phi}(\tilde{P})$ is constant, independent of $P$, so the $\tilde{\phi}$ orbit of $\tilde{P}$ contains no information. We formalize this notion in Section 2.5 after a preliminary discussion of the theory of resultants.

## 2.4 The Resultant of a Rational Map

A rational map $\phi : \mathbb{P}^1 \to \mathbb{P}^1$ is given by a pair of homogeneous polynomials

$$\phi = [F(X,Y), G(X,Y)]$$

having no nontrivial common roots. However, if we reduce the coefficients of $F$ and $G$ modulo some prime, they may acquire common roots in the residue field. In order to understand this phenomenon, it is useful to have a tool that characterizes the existence of common roots in terms of the coefficients of $F$ and $G$. This tool is called the resultant. Resultants and their generalizations are widely used, both theoretically and computationally, in number theory and algebraic geometry. We give in this section only a brief introduction to the theory of resultants. The reader desiring further information might consult [105, Section 3.3], [112, Chapter 3], [259, Section V.10], or [436, Sections 5.8, 5.9], while the reader interested in applications to dynamics may wish to peruse only the statements of Proposition 2.13 and Theorem 2.14 and return to the proofs at a later time.

**Proposition 2.13.** *Let*

$$A(X,Y) = a_0X^n + a_1X^{n-1}Y + \cdots + a_{n-1}XY^{n-1} + a_nY^n,$$
$$B(X,Y) = b_0X^m + b_1X^{m-1}Y + \cdots + b_{m-1}XY^{m-1} + b_mY^m$$

*be homogeneous polynomials of degrees $n$ and $m$ with coefficients in a field $K$. There exists a polynomial*

$$\operatorname{Res}(a_0,\ldots,a_n,b_0,\ldots,b_m) \in \mathbb{Z}[a_0,\ldots,a_n,b_0,\ldots,b_m],$$

*in the coefficients of $A$ and $B$, called the* resultant of $A$ and $B$, *with the following properties:*

(a) $\operatorname{Res}(A,B) = 0$ *if and only if $A$ and $B$ have a common zero in $\mathbb{P}^1(\bar{K})$.*

(b) *If $a_0b_0 \neq 0$ and if we factor $A$ and $B$ as*

$$A = a_0\prod_{i=1}^{n}(X - \alpha_iY) \qquad \textit{and} \qquad B = b_0\prod_{j=1}^{m}(X - \beta_jY),$$

*then*

$$\operatorname{Res}(A,B) = a_0^m b_0^n \prod_{i=1}^{n}\prod_{j=1}^{m}(\alpha_i - \beta_j).$$

(c) *There exist polynomials*

$$F_1, G_1, F_2, G_2 \in \mathbb{Z}[a_0,\ldots,a_n,b_0,\ldots,b_m][X,Y],$$

*homogeneous in $X$ and $Y$ of degrees $m-1$ and $n-1$, respectively, with the property that*

$$F_1(X,Y)A(X,Y) + G_1(X,Y)B(X,Y) = \operatorname{Res}(A,B)X^{m+n-1},$$
$$F_2(X,Y)A(X,Y) + G_2(X,Y)B(X,Y) = \operatorname{Res}(A,B)Y^{m+n-1}.$$

*Notice that in the first equation, the variable $Y$ has been eliminated, and similarly $X$ has been eliminated in the second equation.*

(d) *The resultant is equal to the* $(m+n)\times(m+n)$ *determinant*

$$\operatorname{Res}(A,B)=\det\left|\begin{array}{ccccccc} a_0 & a_1 & a_2 & \dots & a_n & & \\ & a_0 & a_1 & a_2 & \dots & a_n & \\ & & a_0 & a_1 & a_2 & \dots & a_n \\ & & & \ddots & & & \ddots \\ & & & a_0 & a_1 & a_2 \dots & a_n \\ b_0 & b_1 & b_2 & \dots\dots & b_m & & \\ & b_0 & b_1 & b_2 & \dots\dots & b_m & \\ & & b_0 & b_1 & b_2 & \dots\dots & b_m \\ & & & \ddots & & & \ddots \\ & & & b_0 & b_1 & b_2 \dots\dots & b_m \end{array}\right| \begin{array}{l} \left.\vphantom{\begin{array}{c}a\\a\\a\\a\\a\end{array}}\right\} m \\ \left.\vphantom{\begin{array}{c}b\\b\\b\\b\\b\end{array}}\right\} n \end{array}.$$

*In particular,* $\operatorname{Res}(A,B)$ *is homogeneous of degree* $m$ *in the variables* $a_0,\dots,a_n$ *and simultaneously homogeneous of degree* $n$ *in the variables* $b_0,\dots,b_m$.

*Proof.* We begin by showing that the following three conditions are equivalent.

(i) $A(X,Y)$ and $B(X,Y)$ have a common zero in $\mathbb{P}^1(\bar{K})$.

(ii) $A(X,Y)$ and $B(X,Y)$ have a common (nonconstant) factor in the polynomial ring $K[X,Y]$.

(iii) There are nonzero homogeneous polynomials $C,D\in K[X,Y]$ satisfying

$$A(X,Y)C(X,Y)=B(X,Y)D(X,Y) \qquad \text{with } \deg(C)\le m-1 \text{ and } \deg(D)\le n-1. \tag{2.4}$$

The equivalence of (i) and (ii) follows immediately from the fact that the greatest common divisor of $A$ and $B$ in $K[X,Y]$ vanishes at exactly the common zeros of $A$ and $B$ in $\mathbb{P}^1(\bar{K})$. (What we are really using here, of course, is the fact that the ring of homogeneous polynomials $K[X,Y]$ is a principal ideal domain.) It is also clear that (ii) implies (iii), since if $A$ and $B$ have a common factor $F$, then we simply choose $C$ and $D$ using the formulas $A=FD$ and $B=FC$. Finally, to prove that (iii) implies (i), we suppose that (2.4) is true. If we factor both sides of (2.4) into linear factors in $\bar{K}[X,Y]$, then $A(X,Y)$ has $n$ factors, while $D(X,Y)$ has at most $n-1$ factors. Therefore $A(X,Y)$ shares at least one linear factor with $B(X,Y)$ in $\bar{K}[X,Y]$, and hence they have a common zero in $\mathbb{P}^1(\bar{K})$.

We next multiply out equation (2.4), treating the coefficients of $C$ and $D$ as unknowns. This gives a system of $m+n$ homogeneous linear equations in the $m+n$ variables $c_0,\dots,c_{m-1},d_0,\dots,d_{n-1}$, and the matrix of this system is (up to changing the sign of some columns and transposing) equal to the matrix given in (d). For example, if $\deg(A)=3$ and $\deg(B)=2$, then equating coefficients in (2.4) gives the system of linear equations

$$\begin{aligned}
c_0a_0 &= d_0b_0, \\
c_0a_1 + c_1a_0 &= d_0b_1 + d_1b_0, \\
c_0a_2 + c_1a_1 &= d_0b_2 + d_1b_1 + d_2b_0, \\
c_0a_3 + c_1a_2 &= d_1b_2 + d_2b_1, \\
c_1a_3 &= d_2b_2,
\end{aligned}$$

whose associated matrix

$$\begin{pmatrix}
a_0 & & -b_0 & & \\
a_1 & a_0 & -b_1 & -b_0 & \\
a_2 & a_1 & -b_2 & -b_1 & -b_0 \\
a_3 & a_2 & & -b_2 & -b_1 \\
& a_3 & & & -b_2
\end{pmatrix}$$

becomes equal to the matrix in (d) if we change the sign of the last three columns and transpose. The general case is exactly the same.

To recapitulate, we have shown that equation (2.4) has a nontrivial solution if and only if the system of homogeneous linear equations described by the matrix in (d) has a nontrivial solution, which is equivalent to the vanishing of the determinant of the associated matrix. Hence if we take the determinant in (d) as the definition of the resultant $\operatorname{Res}(A, B)$, then the equivalence of (i) and (iii) proven above shows that $\operatorname{Res}(A, B) = 0$ if and only if $A$ and $B$ have a common zero in $\mathbb{P}^1(\bar{K})$, which proves (a).

In order to prove (c), we write

$$X^iY^{m-1-i}A \quad \text{for} \quad 0 \le i < m \qquad \text{and} \qquad X^jY^{n-1-j}B \quad \text{for} \quad 0 \le j < n$$

as a system of homogeneous equations,

$$\begin{pmatrix}
a_0 & a_1 & a_2 & \dots & a_n & & & \\
& a_0 & a_1 & a_2 & \dots & a_n & & \\
& & a_0 & a_1 & a_2 & \dots & a_n & \\
& & & \ddots & & & \ddots & \\
& & & & a_0 & a_1 & a_2 \dots & a_n \\
b_0 & b_1 & b_2 & \dots\dots & b_m & & & \\
& b_0 & b_1 & b_2 & \dots\dots & b_m & & \\
& & b_0 & b_1 & b_2 & \dots\dots & b_m & \\
& & & \ddots & & & \ddots & \\
& & & & b_0 & b_1 & b_2 \dots\dots & b_m
\end{pmatrix}
\begin{pmatrix}
X^{n+m-1} \\ X^{n+m-2}Y \\ X^{n+m-3}Y^2 \\ \\ \vdots \\ \\ XY^{n+m-2} \\ Y^{n+m-1}
\end{pmatrix}
=
\begin{pmatrix}
X^{m-1}A \\ X^{m-2}YA \\ X^{m-3}Y^2A \\ \vdots \\ Y^{m-1}A \\ X^{n-1}B \\ X^{n-2}YB \\ X^{n-3}Y^2B \\ \vdots \\ Y^{n-1}B
\end{pmatrix}.$$

Notice that the matrix $M$ appearing here is exactly the matrix in (d) whose determinant equals $\operatorname{Res}(A, B)$. We multiply on the left by the adjoint matrix $M^{\text{adj}}$ of $M$. (Recall that the entries of $M^{\text{adj}}$ are the cofactors of the matrix $M$, and that the product $M^{\text{adj}}M$ is a diagonal matrix with the quantity $\det(M)$ as its diagonal entries.) This yields the following matrix identity, where for convenience we write $\operatorname{R}(A, B)$ for $\operatorname{Res}(A, B)$:

$$\begin{pmatrix} \mathrm{R}(A,B) & 0 & 0 & \cdots & 0 \\ 0 & \mathrm{R}(A,B) & 0 & \cdots & 0 \\ 0 & 0 & \mathrm{R}(A,B) & \cdots & 0 \\ & & \ddots & & \\ 0 & 0 & 0 & \cdots & \mathrm{R}(A,B) \end{pmatrix} \begin{pmatrix} X^{n+m-1} \\ X^{n+m-2}Y \\ X^{n+m-3}Y^2 \\ \vdots \\ XY^{n+m-2} \\ Y^{n+m-1} \end{pmatrix} = M^{\mathrm{adj}} \begin{pmatrix} X^{m-1}A \\ X^{m-2}YA \\ X^{m-3}Y^2A \\ \vdots \\ Y^{m-1}A \\ X^{n-1}B \\ X^{n-2}YB \\ X^{n-3}Y^2B \\ \vdots \\ Y^{n-1}B \end{pmatrix}.$$

Examining the top entry on each side, we find that $\mathrm{Res}(A,B)X^{n+m-1}$ on the lefthand side is equal to an expression of the form

$$F_1(X,Y)A(X,Y) + G_1(X,Y)B(X,Y)$$

on the righthand side, where $F_1$ and $G_1$ are homogeneous polynomials of degrees $m-1$ and $n-1$, respectively, whose coefficients are (complicated) polynomials in the coefficients of $A$ and $B$. Similarly, the bottom entry shows that $\mathrm{Res}(A,B)Y^{n+m-1}$ is equal to an expression of the form

$$F_2(X,Y)A(X,Y) + G_2(X,Y)B(X,Y).$$

This completes the proof of part (c) of the proposition.

Finally, we leave the proof of (b) as an exercise for the reader, or see [436, Section 5.9]. □

We define the resultant of a rational map in terms of its defining pair of polynomials.

**Definition.** Let $\phi : \mathbb{P}^1 \to \mathbb{P}^1$ be a rational map defined over a field $K$ with a nonarchimedean absolute value $|\cdot|_v$. Write $\phi = [F, G]$ using a pair of normalized homogeneous polynomials $F, G \in R[X, Y]$. The *resultant of* $\phi$ is the quantity

$$\mathrm{Res}(\phi) = \mathrm{Res}(F, G).$$

Since the pair $(F, G)$ is unique up to replacement by $(uF, uG)$ for a unit $u \in R^*$, we see that $\mathrm{Res}(\phi)$ is well-defined up to multiplication by the $2d^{\text{th}}$-power of a unit. In particular, its valuation $v(\mathrm{Res}(\phi))$ depends only on the map $\phi$.

The resultant of a rational map $\phi$ provides an upper bound to the extent that $\phi$ is expanding in the chordal metric. In particular, a rational map is always Lipschitz with respect to the chordal metric, and if its resultant is a unit, then the map is nonexpanding. (See also Exercise 2.10.)

**Theorem 2.14.** *Let $\phi : \mathbb{P}^1 \to \mathbb{P}^1$ be a rational map defined over a field $K$ with a nonarchimedean absolute value $|\cdot|_v$. Then*

$$\rho_v(\phi(P_1), \phi(P_2)) \le |\mathrm{Res}(\phi)|_v^{-2} \rho_v(P_1, P_2) \qquad \textit{for all } P_1, P_2 \in \mathbb{P}^1(K).$$

*Proof.* Write $\phi = [F(X,Y), G(X,Y)]$ in normalized form. Proposition 2.13(c) says that there are homogeneous polynomials $F_1, G_1, F_2, G_2 \in R[X,Y]$ satisfying

$$F_1(X,Y)F(X,Y) + G_1(X,Y)G(X,Y) = \operatorname{Res}(\phi)X^{2d-1},$$
$$F_2(X,Y)F(X,Y) + G_2(X,Y)G(X,Y) = \operatorname{Res}(\phi)Y^{2d-1}.$$

Now let $P = [x,y] \in \mathbb{P}^1(K)$ be a point, which we assume written in normalized form. We substitute $[X,Y] = [x,y]$ into the first equation and use the nonarchimedean triangle inequality to compute

$$\begin{aligned} |\operatorname{Res}(\phi)x^{2d-1}|_v &= |F_1(x,y)F(x,y) + G_1(x,y)G(x,y)|_v \\ &\le \max\{|F_1(x,y)F(x,y)|_v, |G_1(x,y)G(x,y)|_v\} \\ &\le \max\{|F_1(x,y)|_v, |G_1(x,y)|_v\} \cdot \max\{|F(x,y)|_v, |G(x,y)|_v\} \\ &\le \max\{|F(x,y)|_v, |G(x,y)|_v\}. \end{aligned}$$

A similar calculation using the second equation gives the analogous estimate

$$|\operatorname{Res}(\phi)y^{2d-1}|_v \le \max\{|F(x,y)|_v, |G(x,y)|_v\}.$$

Since $P$ is normalized, i.e., $\max\{|x|_v, |y|_v\} = 1$, we find that

$$|\operatorname{Res}(\phi)|_v \le \max\{|F(x,y)|_v, |G(x,y)|_v\}. \tag{2.5}$$

Notice that this estimate bounds the extent to which $F(x,y)$ and $G(x,y)$ can be simultaneously divisible by high powers of $\mathfrak{p}$.

Returning to the proof of the theorem, we write $P_1 = [x_1,y_1]$, $P_2 = [x_2,y_2]$, and $\phi = [F(X,Y), G(X,Y)]$ in normalized form. Then the distance from $P_1$ to $P_2$ is

$$\rho_v(P_1,P_2) = |x_1y_2 - x_2y_1|_v,$$

while we can use the inequality (2.5) (applied to both $P_1$ and $P_2$) to estimate

$$\begin{aligned} &\rho_v(\phi(P_1),\phi(P_2)) \\ &\quad = \frac{\bigl|F(x_1,y_1)G(x_2,y_2) - F(x_2,y_2)G(x_1,y_1)\bigr|_v}{\max\{|F(x_1,y_1)|_v, |G(x_1,y_1)|_v\} \cdot \max\{|F(x_2,y_2)|_v, |G(x_2,y_2)|_v\}} \\ &\quad \le \frac{\bigl|F(x_1,y_1)G(x_2,y_2) - F(x_2,y_2)G(x_1,y_1)\bigr|_v}{|\operatorname{Res}(\phi)|_v^2}. \end{aligned}$$

To complete the proof, we observe that the polynomial

$$F(X_1,Y_1)G(X_2,Y_2) - F(X_2,Y_2)G(X_1,Y_1)$$

vanishes identically if $X_1Y_2 = X_2Y_1$. It follows that it is divisible by the polynomial $X_1Y_2 - X_2Y_1$ in the ring $R[X_1,Y_1,X_2,Y_2]$, so we can write

$$F(X_1,Y_1)G(X_2,Y_2) - F(X_2,Y_2)G(X_1,Y_1) = (X_1Y_2 - X_2Y_1)H(X_1,Y_1,X_2,Y_2)$$

for some polynomial $H \in R[X_1, Y_1, X_2, Y_2]$. Then

$$\begin{aligned} \rho_v(\phi(P_1), \phi(P_2)) &\le \frac{\left|(x_1y_2 - x_2y_1)H(x_1, y_1, x_2, y_2)\right|_v}{|\operatorname{Res}(\phi)|_v^2} \\ &\le \frac{|x_1y_2 - x_2y_1|_v}{|\operatorname{Res}(\phi)|_v^2} \\ &= \frac{\rho_v(P_1, P_2)}{|\operatorname{Res}(\phi)|_v^2}. \end{aligned}$$

□

## 2.5 Rational Maps with Good Reduction

As we saw in Example 2.12, the reduction $\tilde{\phi}$ of a rational map $\phi$ may bear little resemblance to the original map. Indeed, even the degree of the map may change. In this section we characterize maps for which $\deg(\tilde{\phi}) = \deg(\phi)$. These maps are the dynamical analogue of varieties that have good reduction, and they share many of the same properties. See [410, Chapter VII], for example, and compare the results of this section with the properties of elliptic curves that have good reduction.

**Theorem 2.15.** *Let* $\phi : \mathbb{P}^1 \to \mathbb{P}^1$ *be a rational map defined over* $K$ *and write* $\phi = [F, G]$ *in normalized form. The following are equivalent:*

(a) $\deg(\phi) = \deg(\tilde{\phi})$.

(b) *The equations* $\tilde{F}(X,Y) = \tilde{G}(X,Y) = 0$ *have no solutions* $[\alpha, \beta] \in \mathbb{P}^1(\bar{k})$.

(c) $\operatorname{Res}(\phi) \in R^*$.

(d) $\operatorname{Res}(\tilde{F}, \tilde{G}) \ne 0$.

*Proof.* The equivalence of (b), (c), and (d) is immediate from the basic properties of the resultant given in Proposition 2.13, once we observe that

$$\operatorname{Res}(\tilde{F}, \tilde{G}) = \widetilde{\operatorname{Res}(F, G)}.$$

This equality follows from the fact that the resultant is simply a polynomial in the coefficients of $F$ and $G$.

To complete the proof, we observe that the degree of $\tilde{\phi}$ is equal to the degree of $\phi$ minus any cancellation that occurs in $\tilde{F}(X,Y)/\tilde{G}(X,Y)$. In other words,

$$\deg \tilde{\phi} = \deg \phi - \begin{pmatrix} \text{Number of common roots} \\ \text{of } \tilde{F}(X,Y) = \tilde{G}(X,Y) = 0 \end{pmatrix},$$

where the roots are counted with appropriate multiplicities in $\mathbb{P}^1(\bar{k})$. In particular, $\deg \tilde{\phi} = \deg \phi$ if and only if $\tilde{F}$ and $\tilde{G}$ have no common roots, which proves the equivalence of (a) and (b). □

**Definition.** A rational map $\phi : \mathbb{P}^1 \to \mathbb{P}^1$ defined over $K$ is said to have *good reduction (modulo* $\mathfrak{p}$*)* if it satisfies any one (hence all) of the conditions of Theorem 2.15.

*Remark* 2.16. There is a fancier, but useful, characterization of good reduction in the language of schemes. The rational map $\phi$ is a morphism $\phi : \mathbb{P}^1_K \to \mathbb{P}^1_K$ over $\operatorname{Spec}(K)$, so it induces a rational map $\mathbb{P}^1_R \to \mathbb{P}^1_R$ over $\operatorname{Spec}(R)$. Then $\phi$ has good reduction if and only if this rational map over $\operatorname{Spec}(R)$ extends to a morphism. In other words, good reduction is equivalent to the existence of an $R$-morphism $\phi_R : \mathbb{P}^1_R \to \mathbb{P}^1_R$ whose restriction to the generic fiber is the original map $\phi : \mathbb{P}^1_K \to \mathbb{P}^1_K$. In this setting, the reduction $\tilde{\phi}$ is then simply the restriction of $\phi_R$ to a morphism of the special fiber $\tilde{\phi} : \mathbb{P}^1_k \to \mathbb{P}^1_k$ over $\operatorname{Spec}(k)$. See Exercise 2.15.

As a first application of the notion of good reduction, we use Theorem 2.14 to prove the somewhat surprising result that maps with good reduction have empty Julia sets. Later, in Chapter 5, we will prove that rational maps always have nonempty Fatou set. This is exactly opposite to the situation that holds over the complex numbers $\mathbb{C}$, where the Julia set is nonempty, but the Fatou set may be empty.

**Theorem 2.17.** *Let $\phi : \mathbb{P}^1 \to \mathbb{P}^1$ be a rational map that has good reduction.*
(a) *The map $\phi$ is everywhere nonexpanding,*

$$\rho_v\big(\phi(P_1), \phi(P_2)\big) \le \rho_v(P_1, P_2) \qquad \textit{for all } P_1, P_2 \in \mathbb{P}^1(K).$$

(b) *The map $\phi$ has empty Julia set.*

*Proof.* (a) This is immediate from Theorem 2.14 and the fact that good reduction is equivalent to $\operatorname{Res}(\phi) \in R^*$.
(b) It is clear from the definition of equicontinuity that a nonexpanding map is equicontinuous. Indeed, the iterates of a nonexpanding map are uniformly continuous, and indeed, even uniformly Lipschitz (cf. Section 5.4 and Exercise 5.9). □

As their name suggests, rational maps with good reduction behave well when they are reduced.

**Theorem 2.18.** *Let $\phi : \mathbb{P}^1 \to \mathbb{P}^1$ be a rational map that has good reduction.*
(a) $\tilde{\phi}(\tilde{P}) = \widetilde{\phi(P)}$ *for all $P \in \mathbb{P}^1(K)$.*
(b) *Let $\psi : \mathbb{P}^1 \to \mathbb{P}^1$ be another rational map with good reduction. Then the composition $\phi \circ \psi$ has good reduction, and*

$$\widetilde{\phi \circ \psi} = \tilde{\phi} \circ \tilde{\psi}.$$

*Proof.* (a) Write $\phi = [F(X,Y), G(X,Y)]$ in normalized form with homogeneous polynomials $F, G \in R[X,Y]$, and write $P = [\alpha, \beta]$ in normalized form with $\alpha, \beta \in R$. The good reduction assumption tells us that at least one of $F(\alpha,\beta)$ and $G(\alpha,\beta)$ is in $R^*$, so the point

$$\phi(P) = [F(\alpha,\beta), G(\alpha,\beta)]$$

is already in normalized form. Hence

$$\widetilde{\phi(P)} = [\widetilde{F(\alpha,\beta)}, \widetilde{G(\alpha,\beta)}] = [\tilde{F}(\tilde{\alpha},\tilde{\beta}), \tilde{G}(\tilde{\alpha},\tilde{\beta})] = \tilde{\phi}(\tilde{P}),$$

where the second equality simply reflects the fact that the reduction map $R \to k$ is a homomorphism.

(b) Write $\phi = [F(X,Y), G(X,Y)]$ and $\psi = [f(X,Y), g(X,Y)]$ in normalized form with homogeneous polynomials $F, G, f, g \in R[X,Y]$. Then the composition is given by

$$\begin{aligned}(\phi \circ \psi)(X,Y) &= [A(X,Y), B(X,Y)] \\ &= [F(f(X,Y), g(X,Y)), G(f(X,Y), g(X,Y))].\end{aligned}$$

Clearly $A(X,Y)$ and $B(X,Y)$ have coefficients in $R$. Suppose that their reductions $\tilde{A}$ and $\tilde{B}$ have a common root $[\alpha,\beta] \in \mathbb{P}^1(\bar{k})$. This means that

$$\tilde{F}(\tilde{f}(\alpha,\beta), \tilde{g}(\alpha,\beta)) = 0 \qquad \text{and} \qquad \tilde{G}(\tilde{f}(\alpha,\beta), \tilde{g}(\alpha,\beta)) = 0,$$

so $\tilde{F}$ and $\tilde{G}$ have the common root $[\tilde{f}(\alpha,\beta), \tilde{g}(\alpha,\beta)]$. But $\phi$ has good reduction, so $\tilde{F}$ and $\tilde{G}$ have no common root in $\mathbb{P}^1(\bar{k})$, and hence we must have

$$\tilde{f}(\alpha,\beta) = \tilde{g}(\alpha,\beta) = 0.$$

But this contradicts the assumption that $\psi$ has good reduction. This proves that the polynomials $\tilde{A}(X,Y)$ and $\tilde{B}(X,Y)$ have no common root in $\mathbb{P}^1(\bar{k})$, and therefore $\phi \circ \psi = [A,B]$ has good reduction. We have also shown that the pair $(A,B)$ is normalized, so

$$\widetilde{\phi \circ \psi} = [\tilde{A}, \tilde{B}] = [\widetilde{F(f,g)}, \widetilde{G(f,g)}] = [\tilde{F}(\tilde{f},\tilde{g}), \tilde{G}(\tilde{f},\tilde{g})] = \tilde{\phi} \circ \tilde{\psi},$$

which completes the proof of the theorem. □

*Remark* 2.19. (a) The good reduction assumption in both parts of Theorem 2.18 is essential. See Example 2.12 and Exercise 2.13.

(b) For an alternative proof of Theorem 2.18(b) that uses formal properties of resultants and provides additional information about the reduction of the composition of two maps, see Exercise 2.12.

(c) It turns out that the converse of Theorem 2.18(b) is false. In other words, a composition $\phi \circ \psi$ may have good reduction, while both $\phi$ and $\psi$ have bad reduction. For example, let $\phi([x,y]) = [x^2, py^2]$ and $\psi([x,y]) = [p^2x^2, y^2]$. Then $\tilde{\phi} = [1,0]$ and $\tilde{\psi} = [0,1]$ are constant maps, so $\phi$ and $\psi$ have bad reduction. However,

$$(\psi \circ \phi)([x,y]) = [p^2x^4, p^2y^4] = [x^4, y^4],$$

so $\psi \circ \phi$ has good reduction.

One might object to this example by noting that there is a change of variables such that $\phi(z) = z^2/p$ has good reduction, and similarly for $\psi(z) = p^2z^2$. Thus if $f(z) = pz$, then

$$\phi^f(z) = (f^{-1} \circ \phi \circ f)(z) = (f^{-1} \circ \phi)(pz) = f^{-1}(pz^2) = z^2$$

has good reduction. However, it is not difficult to modify this example so that $\phi$ and $\psi$ have bad reduction for all possible changes of variable; see Exercise 2.14.

(d) If we use the scheme-theoretic definition of good reduction as described in Remark 2.16, then both parts of Theorem 2.18 are clear. For example, if $\phi$ and $\psi$ are rational maps with good reduction, then they extend to maps $\phi_R$ and $\psi_R$ over $\operatorname{Spec}(R)$, and the commutativity of the diagram

$$\begin{array}{ccccc} \mathbb{P}^1_k & \xrightarrow{\tilde{\psi}} & \mathbb{P}^1_k & \xrightarrow{\tilde{\phi}} & \mathbb{P}^1_k \\ \downarrow & & \downarrow & & \downarrow \\ \mathbb{P}^1_R & \xrightarrow{\psi_R} & \mathbb{P}^1_R & \xrightarrow{\phi_R} & \mathbb{P}^1_R \\ \uparrow & & \uparrow & & \uparrow \\ \mathbb{P}^1_K & \xrightarrow{\psi} & \mathbb{P}^1_K & \xrightarrow{\phi} & \mathbb{P}^1_K \end{array}$$

immediately gives $\widetilde{\phi \circ \psi} = \tilde{\phi} \circ \tilde{\psi}$. Similarly, a point $P \in \mathbb{P}^1(K)$ corresponds to a unique morphism (i.e., a section) $P_R : \operatorname{Spec}(R) \to \mathbb{P}^1_R$, from which the equality $\widetilde{\phi(P)} = \tilde{\phi}(\tilde{P})$ is immediate using the fact that the composition of $R$-morphisms $\phi_R \circ P_R$ behaves well when restricted to the special fiber of $\mathbb{P}^1_R$. In other words, the following diagram commutes:

$$\begin{array}{ccccc} \operatorname{Spec}(k) & \xrightarrow{\tilde{P}} & \mathbb{P}^1_k & \xrightarrow{\tilde{\phi}} & \mathbb{P}^1_k \\ \downarrow & & \downarrow & & \downarrow \\ \operatorname{Spec}(R) & \xrightarrow{P_R} & \mathbb{P}^1_R & \xrightarrow{\phi_R} & \mathbb{P}^1_R \end{array}$$

An easy, but important, consequence of the theorem on good reduction is that periodic points behave well under reduction.

**Corollary 2.20.** *Let $\phi : \mathbb{P}^1 \to \mathbb{P}^1$ be a rational map with good reduction. Then the reduction map sends periodic points to periodic points and preperiodic points to preperiodic points:*

$$\operatorname{Per}(\phi) \longrightarrow \operatorname{Per}(\tilde{\phi}) \qquad \textit{and} \qquad \operatorname{PrePer}(\phi) \longrightarrow \operatorname{PrePer}(\tilde{\phi}).$$

*Further, if $P \in \operatorname{Per}(\phi)$ has exact period $n$ and if $\tilde{P} \in \operatorname{Per}(\tilde{\phi})$ has exact period $m$, then $m$ divides $n$.*

*Proof.* Suppose first that $P$ is periodic of exact period $n$, so $P = \phi^n(P)$. Reducing both sides modulo $\mathfrak{p}$ and using Theorem 2.18 yields

$$\tilde{P} = \widetilde{\phi^n(P)} = \tilde{\phi}^n(\tilde{P}),$$

which shows that $\tilde{P}$ is periodic. Let $m$ be the exact period of $\tilde{P}$ and write $n = mk + r$ with $0 \le r < m$. Then

$$\tilde{P} = \tilde{\phi}^n(\tilde{P}) = \tilde{\phi}^r \circ \underbrace{\tilde{\phi}^m \circ \cdots \circ \tilde{\phi}^m}_{k \text{ iterations}}(\tilde{P}) = \tilde{\phi}^r(\tilde{P}).$$

The minimality of $m$ implies that $r = 0$, and hence $m$ divides $n$. This proves the assertion about periodic points.

Similarly, if $P$ is preperiodic, say $\phi^i(P) = \phi^j(P)$, then Theorem 2.18 gives $\tilde{\phi}^i(\tilde{P}) = \tilde{\phi}^j(\tilde{P})$. Hence $\tilde{P}$ is preperiodic. □

## 2.6 Periodic Points and Good Reduction

Corollary 2.20 tells us that if $\phi$ has good reduction, then its periodic points reduce to periodic points of $\tilde{\phi}$. In this section we analyze the reduction map $\mathrm{Per}(\phi) \to \mathrm{Per}(\tilde{\phi})$ and use our results to study $\mathrm{Per}(\phi)$. We start with the following theorem, which is an amalgamation of results due to Li [266], Morton–Silverman [312, 313], Narkiewicz [325], Pezda [355], and Zieve [454].

**Theorem 2.21.** *Let $\phi : \mathbb{P}^1(K) \to \mathbb{P}^1(K)$ be a rational function of degree $d \geq 2$ defined over a local field with a nonarchimedean absolute value $|\cdot|_v$. Assume that $\phi$ has good reduction, let $P \in \mathbb{P}^1(K)$ be a periodic point of $\phi$, and define the following quantities:*

- $n$ *The exact period of $P$ for the map $\phi$.*
- $m$ *The exact period of $\tilde{P}$ for the map $\tilde{\phi}$.*
- $r$ *The order of $\lambda_{\tilde{\phi}}(\tilde{P}) = (\tilde{\phi}^m)'(\tilde{P})$ in $k^*$. (Set $r = \infty$ if $\lambda_{\tilde{\phi}}(\tilde{P})$ is not a root of unity.)*
- $p$ *The characteristic of the residue field $k$ of $K$.*

*Then $n$ has one of the following forms:*

$$n = m \quad \text{or} \quad n = mr \quad \text{or} \quad n = mrp^e.$$

*Remark* 2.22. Let $E/K$ be an elliptic curve defined over a local field, and assume that $E$ has good reduction. Then one knows [410, VII.3.1] that the reduction map $E(K) \to \tilde{E}(k)$ is injective except possibly on $p$-power torsion, where $p$ is the characteristic of the residue field $k$. This is very similar to the statement of Theorem 2.21. In the case of elliptic curves, it is also possible to bound the power of $p$ in terms of the ramification index of $p$ in $K$. We discuss below (Theorem 2.28) analogous bounds in the dynamical setting.

*Proof.* We make frequent use of Theorem 2.18, which tells us that

$$\widetilde{\phi^i(Q)} = \tilde{\phi}^i(\tilde{Q}) \quad \text{for all } Q \in \mathbb{P}^1(K) \text{ and all } i \geq 0.$$

Recall that we used this relation in Corollary 2.20 to prove that the $\phi$-period of $P$ is divisible by the $\tilde{\phi}$-period of $\tilde{P}$, which, in our current notation, says that $m$ divides $n$.

Replacing $\phi$ by $\phi^m$ and $m$ by 1, we are reduced to the case that $\tilde{P}$ is a fixed point of $\tilde{\phi}$. Having done this, we note that $\lambda_{\tilde{\phi}}(\tilde{P})$ is equal to $\widetilde{\phi'(P)}$.

If $\phi(P) = P$, then $n = m$ and we are done. We thus assume that $\phi(P) \ne P$. To simplify notation, we use Proposition 2.11 to find a transformation $f \in \mathrm{PGL}_2(R)$ with $f([0,1]) = P$. Replacing $P$ and $\phi$ with $f^{-1}(P)$ and $\phi^f = f^{-1} \circ \phi \circ f$, respectively, we may assume that $P = [0,1]$. Dehomogenizing $z = X/Y$, we write $\phi$ in the form

$$\phi(z) = \frac{a_0 z^d + a_1 z^{d-1} + \cdots + a_{d-1} z + a_d}{b_0 z^d + b_1 z^{d-1} + \cdots + b_{d-1} z + b_d}$$

with coefficients $a_0, \ldots, b_d \in R$ and at least one coefficient in $R^*$. The fact that $[0,1]$ is a fixed point of $\tilde{\phi}$ says that

$$\phi(0) = a_d/b_d \equiv 0 \pmod{\mathfrak{p}},$$

so $a_d \in \mathfrak{p}$ and $b_d \in R^*$. (We are also using the fact that $\phi$ has good reduction, of course.)

Multiplying numerator and denominator by $b_d^{-1}$, we may thus write $\phi(z)$ in the form

$$\phi(z) = \frac{a_d + a_{d-1} z + \cdots + a_1 z^{d-1} + a_0 z^d}{1 + b_{d-1} z + \cdots + b_1 z^{d-1} + b_0 z^d}.$$

The first couple of terms of the Taylor expansion of $\phi$ around $z = 0$, which in this case may be obtained by simple long division, look like

$$\phi(z) = \mu + \lambda z + \frac{A(z)}{1 + zB(z)} z^2 \tag{2.6}$$

with

$$A(z), B(z) \in R[z], \quad \lambda = \phi'(0), \quad \text{and} \quad \mu = a_d \in \mathfrak{p}.$$

A simple induction argument using (2.6) shows that

$$\phi^i(0) \equiv \mu(1 + \lambda + \lambda^2 + \cdots + \lambda^{i-1}) \pmod{\mu^2}. \tag{2.7}$$

In particular, since $\phi^n(0) = 0$ and $\mu \in \mathfrak{p}$, we find that

$$1 + \lambda + \lambda^2 + \cdots + \lambda^{n-1} \equiv 0 \pmod{\mathfrak{p}}. \tag{2.8}$$

The analysis now splits into two cases. First, suppose that $r \ge 2$, or equivalently, $\lambda \not\equiv 1 \pmod{\mathfrak{p}}$. Then formula (2.8) implies that $\lambda^n \equiv 1 \pmod{\mathfrak{p}}$, so we find that $r$ divides $n$. If $n = r$, the proof is complete. Otherwise, we replace $\phi$ by $\phi^r$ and $n$ by $n/r$. By an abuse of notation, we continue to write

$$\phi(z) = \mu + \lambda z + \frac{A(z)}{1 + zB(z)} z^2$$

with the understanding that the values of $\mu$, $\lambda$, $A(z)$, and $B(z)$ may have changed. The principal effect of replacing $\phi$ by $\phi^r$ is that we are now in the situation that

$$\lambda \equiv 1 \pmod{\mathfrak{p}},$$

i.e., the new value of $r$ is 1, which brings us to the second case that we need to consider.

To recapitulate, we have a rational function $\phi$ satisfying

$$\phi^n(0) = 0, \quad \mu = \phi(0) \equiv 0 \pmod{\mathfrak{p}}, \quad \text{and} \quad \lambda = \phi'(0) \equiv 1 \pmod{\mathfrak{p}}.$$

We are further assuming that $\phi(0) \neq 0$ (otherwise, we are done), so (2.8) becomes

$$n \equiv 1 + \lambda + \lambda^2 + \cdots + \lambda^{n-1} \equiv 0 \pmod{\mathfrak{p}}.$$

Thus $n$ is divisible by $p$. We replace $\phi$ by $\phi^p$ and $n$ by $n/p$. If now $\phi(0) = 0$, we are done. If not, the same argument shows that $n$ is again divisible by $p$. Repeating, we continue dividing $n$ by $p$ until finally we reach $n = 1$. This concludes the proof that the original period $n$ has one of the forms

$$n = m \quad \text{or} \quad n = mr \quad \text{or} \quad n = mrp^e$$

for some $e \geq 1$. □

We have seen that maps with good reduction are nonexpanding. This implies that their periodic points are nonrepelling. If the reduction $\tilde{\phi}$ is separable, we can say even more. (Recall that $\tilde{\phi}(z) \in k(z)$ is separable if it is not in $k(z^p)$. See Exercise 1.10 for details.)

**Corollary 2.23.** *Let $\phi : \mathbb{P}^1 \to \mathbb{P}^1$ be a rational map that has good reduction.*

(a) *Every periodic point of $\phi$ is nonrepelling.*

(b) *If the reduction $\tilde{\phi}$ is separable, then $\phi$ has only finitely many attracting periodic points.*

*Proof.* (a) Theorem 2.18 tells us that $\phi^n$ has good reduction. Let $P$ be a periodic point of $\phi$ of exact period $n$. Using Lemma 2.5, we can make a change of coordinates so that $P = [0, 1]$. Then we can write $\phi^n(z)$ in normalized form as

$$\phi^n(z) = \frac{F(z)}{G(z)} = \frac{a_1 z + a_2 z^2 + \cdots + a_d z^d}{b_0 + b_1 z + b_2 z^2 + \cdots + b_d z^d}.$$

The fact that $\phi^n$ has good reduction implies that $b_0 \in R^*$, since otherwise $z = 0$ would be a common root of $\tilde{F}$ and $\tilde{G}$. Hence

$$\lambda_P(\phi) = (\phi^n)'(0) = \frac{a_1}{b_0} \in R,$$

so $|\lambda_P(\phi)|_v \leq 1$, which shows that $P$ is a nonrepelling point for $\phi$.

(b) Again let $P$ be a periodic point for $\phi$ of exact period $n$, and let $m$ be the period of the reduced point $\tilde{P}$. Then we have equivalences

$$\begin{aligned}
P \text{ is attracting} &\iff |\lambda_P(\phi)|_v = |(\phi^n)'(P)|_v < 1 \\
&\iff \widetilde{(\phi^n)'(P)} = 0 \\
&\iff (\tilde{\phi}^n)'(\tilde{P}) = 0 \\
&\iff \tilde{\phi}'(\tilde{P}) \cdot \tilde{\phi}'(\tilde{\phi}\tilde{P}) \cdot \tilde{\phi}'(\tilde{\phi}^2\tilde{P}) \cdots \tilde{\phi}'(\tilde{\phi}^{n-1}\tilde{P}) = 0 \\
&\iff \left(\tilde{\phi}'(\tilde{P}) \cdot \tilde{\phi}'(\tilde{\phi}\tilde{P}) \cdot \tilde{\phi}'(\tilde{\phi}^2\tilde{P}) \cdots \tilde{\phi}'(\tilde{\phi}^{m-1}\tilde{P})\right)^{n/m} = 0 \\
&\iff (\tilde{\phi}^m)'(\tilde{P})^{n/m} = 0 \\
&\iff \mathcal{O}_{\tilde{\phi}}(\tilde{P}) \text{ contains a critical point.}
\end{aligned}$$

The fact that $\tilde{\phi}$ is separable implies that a version of the Riemann–Hurwitz formula (Theorem 1.1) is valid; see Exercise 1.10. Hence the map $\tilde{\phi}$ has finitely many (precisely, at most $2d-2$) critical points, and a fortiori, the map $\phi$ has only finitely many periodic orbits containing a critical point. In particular, there is a finite list of possible periods for $\tilde{P}$. Further, we know that the multiplier $\lambda_{\tilde{\phi}}(\tilde{P}) = (\tilde{\phi}^m)'(\tilde{P})$ of $\tilde{P}$ is 0, so Theorem 2.21 tells us that $n = m$. There are thus only finitely many possibilities for the period of $P$, and since $\phi$ has finitely many points of any given period, we conclude that $\phi$ has only finitely many attracting periodic points. □

*Remark* 2.24. The separability assumption in Proposition 2.23 is necessary, as is shown by the example $\phi(z) = z^p$, all of whose periodic points are attracting. See also Exercise 2.16.

*Example* 2.25. Let $\phi(z) \in \mathbb{Z}_2[z]$ be a polynomial of degree $d \geq 2$ whose leading coefficient is a 2-adic unit. Then $\phi$ has good reduction. Let $P \in \mathbb{P}^1(\mathbb{Q}_2)$ be a periodic point of exact period $n \geq 2$. In the notation of Theorem 2.21, $n = mr2^e$, where $m$ is the period of $\tilde{P}$ in $\mathbb{P}^1(\mathbb{F}_2)$ and $r$ is the order of $\lambda_{\tilde{\phi}}(\tilde{P})$ in $\mathbb{F}_2^*$. But $\mathbb{P}^1(\mathbb{F}_2)$ has only three points, and the fact that $\phi$ is a polynomial means that the point at infinity is not in the orbit of $\tilde{P}$, so either $m = 1$ or $m = 2$. Similarly, we note that $\mathbb{F}_2^*$ has only one element, so $r = 1$. It follows that $n = 2^s$ for some $s \geq 0$.

Similarly, if $\phi(z) \in \mathbb{Z}_3[z]$ is a polynomial of degree $d \geq 2$ with leading coefficient a 3-adic unit, and if $P \in \mathbb{P}^1(\mathbb{Q}_3)$ is a periodic point of exact period $n \geq 2$, then we find that $n = mr3^e$ with $1 \leq m \leq 3$ and $1 \leq r \leq 2$. Thus $n = 2^t \cdot 3^u$ for some $0 \leq t \leq 2$ and some $u \geq 0$.

Finally, let $\phi(z) \in \mathbb{Z}[z]$ be a polynomial of degree $d \geq 2$ whose leading coefficient is relatively prime to 6. Then $\phi$ has good reduction at 2 and 3, so the period $n$ of a periodic point $P \in \mathbb{P}^1(\mathbb{Q})$ satisfies both $n = 2^s$ and $n = 2^t \cdot 3^u$ with $t \leq 2$. This proves that $n$ is either 1, 2, or 4. The examples $\phi(z) = z^2$ with $z = 0$ and $\phi(z) = z^2 - 1$ with $z = 0$ show that $n = 1$ and $n = 2$ are possible. Can you find an example with $n = 4$? See Exercise 2.20 for a stronger version of this example.

The above example illustrates how the local result given in Theorem 2.21 can be used to derive strong bounds for the periods of periodic points defined over number fields by applying the theorem to two different primes. We now use the same argument to give a general result that, although not the strongest possible bound using these methods, is sufficient for many applications.

**Corollary 2.26.** *Let $K$ be a number field, let $\phi : \mathbb{P}^1 \to \mathbb{P}^1$ be a rational map defined over $K$, and let $\mathfrak{p}$ and $\mathfrak{q}$ be primes of $K$ such that $\phi$ has good reduction at both $\mathfrak{p}$ and $\mathfrak{q}$ and such that the residue characteristics of $\mathfrak{p}$ and $\mathfrak{q}$ are distinct. Then the period $n$ of any periodic point of $\phi$ in $\mathbb{P}^1(K)$ satisfies*

$$n \le (\mathrm{N}\mathfrak{p}^2 - 1)(\mathrm{N}\mathfrak{q}^2 - 1),$$

*where* $\mathrm{N}\mathfrak{p}$ *and* $\mathrm{N}\mathfrak{q}$ *denote the norms of* $\mathfrak{p}$ *and* $\mathfrak{q}$ *respectively.*

*In particular, the set* $\operatorname{Per}(\phi, K)$ *of $K$-rational periodic points is finite.* (*For an alternative proof of the finiteness of* $\operatorname{Per}(\phi, K)$ *using the theory of height functions, see Theorem* 3.12.)

*Proof.* Using the obvious notation, we have

$$\begin{aligned} m_{\mathfrak{p}} &= (\text{period of } \tilde{\phi}(\tilde{P}) \bmod \mathfrak{p}) \le \#\mathbb{P}^1(\mathbb{F}_{\mathfrak{p}}) = \mathrm{N}\mathfrak{p} + 1, \\ r_{\mathfrak{p}} &= (\text{period of } \lambda_{\tilde{\phi}}(\tilde{P}) \bmod \mathfrak{p}) \le \#\mathbb{F}_{\mathfrak{p}}^* = \mathrm{N}\mathfrak{p} - 1, \end{aligned}$$

and similarly for $m_{\mathfrak{q}}$ and $r_{\mathfrak{q}}$. Let $p$ and $q$ denote the residue characteristics of $\mathfrak{p}$ and $\mathfrak{q}$, respectively. Then Theorem 2.21 says that

$$n = m_{\mathfrak{p}} \cdot r_{\mathfrak{p}} \cdot p^{e_{\mathfrak{p}}} = m_{\mathfrak{q}} \cdot r_{\mathfrak{q}} \cdot q^{e_{\mathfrak{q}}}.$$

Since $p$ and $q$ are distinct primes, it follows that

$$n \le m_{\mathfrak{p}} \cdot r_{\mathfrak{p}} \cdot m_{\mathfrak{q}} \cdot r_{\mathfrak{q}} \le (\mathrm{N}\mathfrak{p} + 1)(\mathrm{N}\mathfrak{p} - 1)(\mathrm{N}\mathfrak{q} + 1)(\mathrm{N}\mathfrak{q} - 1),$$

which is the first part of the corollary.

The finiteness of $\operatorname{Per}(\phi, K)$ then follows from the fact that $\phi$ has good reduction at almost all primes of $K$ and the fact that it has only finitely many periodic points of any given period $n$. $\square$

*Remark* 2.27. The bound for rational periodic points in Corollary 2.26 depends only weakly on $\phi$ in the sense that the bound is solely in terms of the two smallest primes of good reduction for $\phi$. There are many results in the literature using local and/or global methods that describe bounds for rational periodic points that depend in various ways on the rational map. See for example [52, 87, 90, 91, 162, 171, 312, 325, 326, 328, 332, 353, 355, 358, 359, 361, 454]. However, none of these articles achieves the uniformity predicted by a conjecture that we discuss in Chapter 3 (Conjecture 3.15). This conjecture asserts that for a number field $K$ of degree $D$ and a rational map $\phi \in K(z)$ of degree $d \ge 2$, the number of $K$-rational preperiodic points of $\phi$ should be bounded solely in terms of $D$ and $d$.

If $K$ is a discrete valuation ring of characteristic 0, then it is possible to bound the exponent $e$ appearing in the formula $n = mrp^e$ in Theorem 2.21.

**Theorem 2.28.** (Zieve [454], see also Li [266] and Pezda [355]) *We continue with the notation and assumptions from Theorem 2.21. We further assume that $K$ has characteristic 0 and we let $v : K^* \twoheadrightarrow \mathbb{Z}$ be the normalized valuation on $K$. If the period $n$ of $P \in \mathbb{P}^1(K)$ has the form $n = mrp^e$, then the exponent $e$ satisfies*

$$p^{e-1} \le \frac{2v(p)}{p-1}. \tag{2.9}$$

*Further, if $p = 2$, then the upper bound may be replaced with $v(p)/(p-1)$. (Note that $v(p)$ is the ramification index of $p$ in $K$.)*

*Remark* 2.29. Let $R$ be the local ring of $K$ and let $\mathcal{F}$ be a formal group defined over $R$. (See, e.g., [410, chapter IV] for basic material on formal groups.) Theorem 2.28 is a close analogue of the fact [410, IV.6.1] that the torsion in the formal group $\mathcal{F}(R)$ consists entirely of $p$-power torsion, and that if $\alpha \in \mathcal{F}(R)$ has exact order $p^e$, then $p^{e-1} \le v(p)/(p-1)$.

*Example* 2.30. Let $q$ be a power of an odd prime $p$, let $\zeta \in \bar{\mathbb{Q}}_p$ be a primitive $q^{\text{th}}$ root of unity, let $K = \mathbb{Q}_p(\zeta)$, and let $\phi(z) = 1 + \zeta z - z^q$. The maximal ideal of $K$ is $\mathfrak{p} = (1-\zeta)$, so $\tilde{\phi} = 1 + z - z^q$ has good reduction. The point $\alpha = 1$ is a point of exact period $q$ for $\phi$, since $\phi^j(1) = \zeta^j$, while $\tilde{\alpha}$ is clearly a fixed point of $\tilde{\phi}$. Further, $\tilde{\phi}'(\tilde{\alpha}) = 1$. Hence in the notation of Theorem 2.21, we have $n = q$, $m = 1$, $r = 1$, and $e$ is determined by $q = p^e$. On the other hand, the extension $K/\mathbb{Q}_p$ is totally ramified, so the normalized valuation $v : K^* \twoheadrightarrow \mathbb{Z}$ satisfies

$$v(p) = [K : \mathbb{Q}_p] = q(1 - 1/p) = p^{e-1}(p-1).$$

Thus the inequality (2.9) in Theorem 2.28 becomes $p^{e-1} \le 2p^{e-1}$, which shows that the power of $p$ cannot be improved.

We do not give the proof of Theorem 2.28, but are content to prove the following special case, which serves to indicate some of the combinatorial issues that arise.

**Theorem 2.31.** *Let $p \ge 5$ be a prime, let $K = \mathbb{Q}_p$, or more generally, an unramified extension of $\mathbb{Q}_p$, let $\phi : \mathbb{P}^1 \to \mathbb{P}^1$ be a rational map defined over $K$ with good reduction, and let $P \in \mathbb{P}^1(K)$ be a periodic point with the property that*

$$\tilde{\phi}(\tilde{P}) = \tilde{P} \qquad \textit{and} \qquad \tilde{\phi}'(\tilde{P}) = 1.$$

*Then $\phi(P) = P$. (In the notation of Theorem 2.21, this theorem asserts that if $m = r = 1$, then $e = 0$ and $n = 1$.)*

*Proof.* As in the proof of Theorem 2.21, we begin by moving the periodic point to the origin and dehomogenizing $\phi$. Recall that during the proof of Theorem 2.21, we used a first-order Taylor expansion (2.6) for $\phi(z)$ around $z = 0$. In order to bound the exponent $e$, we need to use the second-order expansion

$$\phi(z) = \mu + \lambda z + \nu z^2 + \frac{A(z)}{1 + zB(z)} z^3, \tag{2.10}$$

where

$$A(z), B(z) \in R[z], \quad \mu \in \mathfrak{p}, \quad \lambda = \phi'(0) \equiv 1 \pmod{\mathfrak{p}}, \quad \text{and} \quad \nu \in R.$$

(In general, to prove a sharp estimate for the exponent as in Theorem 2.28, one needs to consider a longer Taylor expansion. But for the unramified case that we are considering, the second-order expansion (2.10) suffices.)

Using the results already proven, it suffices to show that if $\mu \neq 0$, i.e., if $\phi(0) \neq 0$, then $\phi^p(0) \neq 0$. This will then imply by induction that $\phi^{p^e}(0) \neq 0$ for all $e \geq 1$, contradicting the assumption that 0 is a periodic point of $\phi$.

During the proof of Theorem 2.21 we gave a simple formula (2.7) for $\phi^i(0)$ modulo $\mu^2$. In a similar manner we use (2.10) to find a formula modulo $\mu^3$. To derive this formula, we write $\phi^k(0) = \mu a_k + \mu^2 \nu b_k$, substitute into (2.10), and do some algebra to obtain

$$\begin{aligned} \mu a_{k+1} + \mu^2 \nu b_{k+1} &= \phi(\mu a_k + \mu^2 \nu b_k) \\ &\equiv \mu + \lambda(\mu a_k + \mu^2 \nu b_k) + \nu \mu^2 a_k^2 \pmod{\mu^3} \\ &\equiv \mu(1 + \lambda a_k) + \mu^2 \nu (a_k^2 + \lambda b_k) \pmod{\mu^3}. \end{aligned}$$

This yields the recurrences

$$a_{k+1} \equiv 1 + \lambda a_k \pmod{\mu^2} \qquad \text{and} \qquad b_{k+1} \equiv a_k^2 + \lambda b_k \pmod{\mu}.$$

Starting from $a_1 = 1$ and $b_1 = 0$, it is now a simple matter to find formulas for $a_k$ and $b_k$ and check them by induction. The end result is

$$\phi^k(0) \equiv \mu\left(\sum_{j=0}^{k-1} \lambda^j\right) + \mu^2 \nu \left(\sum_{i=0}^{k-2} \lambda^{k-2-i} \left(\sum_{j=0}^{i} \lambda^j\right)^2\right) \pmod{\mu^3}. \tag{2.11}$$

We are going to apply (2.11) with $k = p$. Consider the sum $\sum \lambda^j$. We know that $\lambda \equiv 1 \pmod{\mathfrak{p}}$, and we are assuming that $p$ is unramified in $K$, so $\lambda = 1 + cp$ for some $c \in R$. Assuming that $c \neq 0$, i.e., that $\lambda \neq 1$, we compute

$$\sum_{j=0}^{p-1} \lambda^j = \frac{\lambda^p - 1}{\lambda - 1} = \frac{(1+cp)^p - 1}{cp} = \sum_{i=1}^{p} \binom{p}{i} (cp)^{i-1} \equiv p \pmod{p^2}. \tag{2.12}$$

Note that the final congruence is true because every binomial coefficient $\binom{p}{i}$ with $1 \leq i < p$ is divisible by $p$ (and we are assuming that $p \neq 2$). We also observe that the congruence is true even for $\lambda = 1$, although the intermediate calculation is incorrect.

We perform a similar calculation for the more complicated sum in (2.11), but this time we are interested only in the value modulo $p$, so we can replace $\lambda$ by 1:

$$\sum_{i=0}^{p-2} \lambda^{p-2-i} \left(\sum_{j=0}^{i} \lambda^j\right)^2 \equiv \sum_{i=0}^{p-2} \left(\sum_{j=0}^{i} 1\right)^2 = \frac{(p-1)p(2p-1)}{6} \equiv 0 \pmod{p}. \tag{2.13}$$

Note that for the last step we are using the assumption that $p \geq 5$.

Substituting (2.12) and (2.13) into the iteration formula (2.11) (with $k = p$) yields

$$\phi^p(0) \equiv \mu(p + ap^2) + \mu^2 \nu bp \pmod{\mu^3},$$

where $a$ and $b$ are in $R$. The fact that $p$ is unramified in $K$ means in particular that $p$ divides $\mu$, so

$$\phi^p(0) \equiv \mu p \pmod{\mu p^2}.$$

Now using our assumption that $\mu \neq 0$, we deduce that $\phi^p(0) \neq 0$, which completes the proof of the theorem. □

## 2.7 Periodic Points and Dynamical Units

Let $\zeta$ be a primitive $p^{\text{th}}$ root of unity. Then the expression

$$\frac{\zeta^i - 1}{\zeta - 1}$$

is a unit in the cyclotomic field $\mathbb{Q}(\zeta)$, a so-called "cyclotomic unit." Similarly, if $\zeta_1$ and $\zeta_2$ are roots of unity of relatively prime orders, then the difference $\zeta_1 - \zeta_2$ is an algebraic unit. The crucial fact underlying these constructions is that distinct roots of unity remain distinct when they are reduced modulo primes. It follows that their differences are not divisible by any primes, and hence that they are units.

Theorem 2.21 can be used to deduce conditions under which distinct periodic points remain distinct when reduced modulo primes, so it can be used to construct units in a similar fashion. These constructions can be done either using different points in a single periodic orbit or using points of different periods. The following proposition provides us with the information needed to construct units of various kinds. In fact, since Lemma 2.8 tells us that the $v$-adic chordal metric satisfies

$$\rho_v(P, Q) < 1 \iff \tilde{P} = \tilde{Q},$$

the proposition actually says something stronger than the simple assertion that certain pairs of points have distinct reductions.

**Proposition 2.32.** *Let $\phi(z) \in K(z)$ be a rational function of degree $d \geq 2$ with good reduction.*

(a) *Let $P \in \mathbb{P}^1(K)$ be a point of period $n$ for $\phi$. Then*

$$\rho_v(\phi^i P, \phi^j P) = \rho_v(\phi^{i+k} P, \phi^{j+k} P) \quad \textit{for all } i, j, k \in \mathbb{Z},$$

*where for $i < 0$ we use the periodicity $\phi^n P = P$ to define $\phi^i P$.*

(b) *Let $P \in \mathbb{P}^1(K)$ be a point of exact period $n$ for $\phi$. Then*

$$\rho_v(\phi^i P, \phi^j P) = \rho_v(\phi P, P) \quad \textit{for all } i, j \in \mathbb{Z} \textit{ satisfying } \gcd(i - j, n) = 1.$$

(c) *Let $P_1, P_2 \in \mathbb{P}^1(K)$ be periodic points for $\phi$ of exact periods $n_1$ and $n_2$, respectively. Assume that $n_1 \nmid n_2$ and $n_2 \nmid n_1$. Then*

$$\rho_v(P_1, P_2) = 1.$$

*Proof.* (a) Proposition 2.14 and the good-reduction assumption imply that

$$\rho_v(Q,R) \geq \rho_v(\phi Q, \phi R) \qquad \text{for all } Q,R \in \mathbb{P}^1(K).$$

Applying this repeatedly yields

$$\begin{aligned}\rho_v(Q,R) &\geq \rho_v(\phi Q,\phi R) \geq \rho_v(\phi^2 Q,\phi^2 R)\\ &\geq \rho_v(\phi^3 Q,\phi^3 R) \geq \cdots \geq \rho_v(\phi^n Q,\phi^n R).\end{aligned}$$

If we now make the further assumption that $Q$ and $R$ are points of period $n$, then $\rho_v(\phi^n Q,\phi^n R) = \rho_v(Q,R)$, so all of the inequalities must be equalities. This proves that if $\phi^n(Q) = Q$ and $\phi^n(R) = R$, then

$$\rho_v(Q,R) = \rho_v(\phi^k Q,\phi^k R) \quad \text{for all } k \in \mathbb{Z}.$$

Substituting $Q = \phi^i P$ and $R = \phi^j P$ for the given point $P$ of period $n$ completes the proof of (a).
(b) We know from (a) that the distance $\rho_v(\phi^i P,\phi^j P)$ depends only on the difference $i-j$. We use this fact and the nonarchimedean triangle inequality (Theorem 2.4(d)) to estimate

$$\begin{aligned}\rho_v(P,\phi^k P) &\leq \max\{\rho_v(P,\phi P),\rho_v(\phi P,\phi^2 P),\ldots,\rho(\phi^{k-1}P,\phi^k P)\}\\ &= \rho_v(P,\phi P).\end{aligned} \tag{2.14}$$

In a similar fashion we obtain the estimate

$$\begin{aligned}\rho_v(P,\phi P) \leq \max\{&\rho_v(P,\phi^k P),\rho_v(\phi^k P,\phi^{2k}P),\rho_v(\phi^{2k}P,\phi^{3k}P),\\ &\ldots,\rho_v(\phi^{(m-1)k}P,\phi^{mk}P),\rho_v(\phi^{mk}P,\phi P)\}\\ = \max\{&\rho_v(P,\phi^k P),\rho_v(\phi^{mk}P,\phi P)\}.\end{aligned} \tag{2.15}$$

If we now make the further assumption that $\gcd(k,n) = 1$, then we can find an integer $m$ satisfying $mk \equiv 1 \pmod{n}$. Then $\phi^{mk}P = \phi P$, so (2.15) becomes

$$\rho_v(P,\phi P) \leq \max\{\rho_v(P,\phi^k P),\rho_v(\phi P,\phi P)\} = \rho_v(P,\phi^k P). \tag{2.16}$$

Combining (2.14) and (2.16) yields

$$\rho_v(P,\phi P) = \rho_v(P,\phi^k P) \quad \text{for all } k \text{ satisfying } \gcd(k,n) = 1.$$

In particular, if $\gcd(i-j,n) = 1$, we can use this formula and (a) to compute

$$\rho_v(\phi^i P,\phi^j P) = \rho_v(P,\phi^{i-j}P) = \rho_v(P,\phi P).$$

(c) Suppose that $\rho_v(P_1,P_2) < 1$, or equivalently from Lemma 2.8, suppose that $\tilde{P}_1 = \tilde{P}_2$. Let $m$ be the exact period of $\tilde{P}_1$ and let $r$ be the order of its multiplier $\lambda_{\tilde{\phi}}(\tilde{P}_1)$. Then Theorem 2.21 tells us that both $n_1$ and $n_2$ are in the set

$$\{m, mr, mrp, mrp^2, mrp^3, \ldots\}.$$

It follows that either $n_1$ divides $n_2$, or that $n_2$ divides $n_1$, contradicting the assumption on $n_1$ and $n_2$. Therefore $\rho_v(P_1,P_2) = 1$. □

It is a simple matter to use Proposition 2.32 to construct units from periodic points. We begin with the easier case of a polynomial mapping.

**Theorem 2.33.** (Narkiewicz [325]) *Let $\phi(z) \in R[z]$ be a polynomial of degree $d \geq 2$ whose leading coefficient is a unit in $R$. Let $\alpha \in K$ be a periodic point of $\phi$ of exact period $n$ (with $n \geq 2$), and let $i, j \in \mathbb{Z}$ be integers satisfying* $\gcd(i-j, n) = 1$. *Then*

$$\frac{\phi^i\alpha - \phi^j\alpha}{\phi\alpha - \alpha} \in R^*.$$

*Proof.* The assumption that $\phi(z)$ is a polynomial with unit leading coefficient implies that $\phi$ has good reduction, since

$$\operatorname{Res}(a_0X^d + a_1X^{d-1}Y + \cdots + a_dY^d, Y^d) = a_0^d.$$

We next observe that every periodic point of $\phi$ is integral over $R$, since it is a root of an equation $\phi^n(z) - z = 0$. Finally, we note that the distance between integral points $\alpha, \beta \in R$ is given by

$$\rho_v(\alpha, \beta) = \rho_v([\alpha, 1], [\beta, 1]) = |\alpha - \beta|_v.$$

Using this formula and Proposition 2.32(b), we compute

$$\left|\frac{\phi^i\alpha - \phi^j\alpha}{\phi\alpha - \alpha}\right|_v = \frac{|\phi^i\alpha - \phi^j\alpha|_v}{|\phi\alpha - \alpha|_v} = \frac{\rho_v(\phi^i\alpha, \phi^j\alpha)}{\rho_v(\phi\alpha, \alpha)} = 1. \qquad \square$$

We create units from periodic points of arbitrary rational maps using a cross-ratio construction.

**Definition.** Let $P_1, P_2, P_3, P_4 \in \mathbb{P}^1(K)$, and choose homogeneous coordinates $P_i = [x_i, y_i]$ for each point. The *cross-ratio* of $P_1, P_2, P_3, P_4$ is the quantity

$$\kappa(P_1, P_2, P_3, P_4) = \frac{(x_1y_3 - x_3y_1)(x_2y_4 - x_4y_2)}{(x_1y_2 - x_2y_1)(x_3y_4 - x_4y_3)}.$$

Notice that $\kappa(P_1, P_2, P_3, P_4)$ is independent of the choice of homogeneous coordinates for the points.

**Theorem 2.34.** (Morton–Silverman [313, Theorem 6.4(a)]) *Let $\phi \in K(z)$ be a rational map of degree $d \geq 2$ with good reduction. Let $P \in \mathbb{P}^1(K)$ be a periodic point for $\phi$ of exact period $n$, and let $i$ and $j$ be integers satisfying*

$$\gcd(j, n) = \gcd(i-1, n) = \gcd(i-j, n) = 1.$$

*Then*

$$\kappa(P, \phi P, \phi^i P, \phi^j P) \in R^*.$$

*Proof.* Comparing the definition of the cross-ratio to the definition of the chordal metric, we see that

$$\big|\kappa(P_1,P_2,P_3,P_4)\big|_v = \frac{\rho_v(P_1,P_3)\rho_v(P_2,P_4)}{\rho_v(P_1,P_2)\rho_v(P_3,P_4)}.$$

The assumptions on $i$ and $j$ and Proposition 2.32(b) tell us that

$$\rho_v(P,\phi P) = \rho_v(P,\phi^i P) = \rho_v(\phi P,\phi^j P) = \rho_v(\phi^i P,\phi^j P).$$

Hence

$$\big|\kappa(P,\phi P,\phi^i P,\phi^j P)\big|_v = \frac{\rho_v(P,\phi^i P)\rho_v(\phi P,\phi^j P)}{\rho_v(P,\phi P)\rho_v(\phi^i P,\phi^j P)} = 1,$$

which proves that the cross-ratio is a unit. □

We can also construct units using periodic points of different periods. This may be compared with the two different types of cyclotomic units,

$$\frac{\zeta_p^i - \zeta_p^j}{\zeta_p - 1} \qquad \text{and} \qquad \zeta_m - \zeta_n,$$

where $\zeta_n$ indicates a primitive $n^{\text{th}}$ root of unity and the condition $\gcd(m,n) = 1$ ensures that $\zeta_m - \zeta_n$ is a unit.

**Theorem 2.35.** *Let $\phi \in K(z)$ be a rational map of degree $d \ge 2$ with good reduction. Let $n_1, n_2 \in \mathbb{Z}$ be integers with $n_1 \nmid n_2$ and $n_2 \nmid n_1$, let $P_1, P_2 \in \mathbb{P}^1(K)$ be periodic points of exact periods $n_1$ and $n_2$ respectively, and write $P_i = [x_i, y_i]$ in normalized form. Then*

$$x_1y_2 - x_2y_1 \in R^*.$$

*Proof.* Since we have taken normalized homogeneous coordinates, the chordal metric is given by

$$\rho_v(P_1,P_2) = |x_1y_2 - x_2y_1|_v.$$

The assumptions on $n_1$ and $n_2$ and Proposition 2.32(c) tell us that $\rho_v(P_1,P_2) = 1$, and hence $x_1y_2 - x_2y_1$ is a unit. □

*Remark* 2.36. For further information on the geometric and arithmetic properties of periodic points and the fields that they generate, see Sections 3.9, 3.11, and 4.1–4.6. In some sense, periodic and preperiodic points play a role in arithmetic dynamics analogous to the role played by torsion points in the arithmetic of elliptic curves or abelian varieties, albeit the dynamical setting has considerably less in the way of helpful structure.

*Example* 2.37. We use the polynomial $\phi(z) = z^2 + 1$ to construct units. First we compute

$$\begin{aligned}
\phi(z) - z &= z^2 - z + 1,\\
\phi^2(z) - z &= z^4 + 2z^2 - z + 2 = (z^2 - z + 1)(z^2 + z + 2),\\
\phi^3(z) - z &= z^8 + 4z^6 + 8z^4 + 8z^2 - z + 5\\
&= (z^2 - z + 1)(z^6 + z^5 + 4z^4 + 3z^3 + 7z^2 + 4z + 5).
\end{aligned}$$

It is not surprising that $\phi(z) - z$ divides both $\phi^2(z) - z$ and $\phi^3(z) - z$, since any root of the former is clearly a root of the latter two; cf. Exercise 1.19(a). We have

$$\mathrm{Per}_1^{**}(\phi) = \left\{\frac{1 \pm \sqrt{-3}}{2}\right\} \quad \text{and} \quad \mathrm{Per}_2^{**}(\phi) = \left\{\frac{-1 \pm \sqrt{-7}}{2}\right\},$$

where recall that $\mathrm{Per}_n^{**}(\phi)$ denotes the set of points of exact period $n$ for $\phi$.

Further, the six roots of the polynomial

$$\Phi_3^*(z) := z^6 + z^5 + 4z^4 + 3z^3 + 7z^2 + 4z + 5$$

constitute a set of the form

$$\{\alpha, \phi(\alpha), \phi^2(\alpha), \beta, \phi(\beta), \phi^2(\beta)\},$$

since we know that the roots of $\Phi_3^*(z)$ are permuted by $\phi$ into two periodic cycles of period 3. In particular, the splitting field $K(\alpha, \beta)$ of $\Phi_3^*$ is a Galois extension of $K$ of degree at most 18.

Applying Theorem 2.33 with $i = 2$ and $j = 1$ gives a unit

$$u_1 = \frac{\phi^2(\alpha) - \phi(\alpha)}{\phi(\alpha) - \alpha} = \alpha^2 + \alpha + 1 \in R[\alpha].$$

We can check that $u_1$ is a unit in $R[\alpha]$ by computing

$$\mathrm{N}_{K(\alpha)/K}(\alpha^2 + \alpha + 1) = \mathrm{Res}\big(\Phi_3^*(z), z^2 + z + 1\big) = 1.$$

Similarly, we can apply Theorem 2.33 with $i = 2$ and $j = 0$ to obtain a second unit

$$u_2 = \frac{\phi^2(\alpha) - \alpha}{\phi(\alpha) - \alpha} = \alpha^2 + \alpha + 2.$$

And replacing $\alpha$ with $\phi(\alpha)$ or with $\phi^2(\alpha)$ in $u_1$ and $u_2$ gives four more units,

$$\begin{aligned}
u_1' &= \phi(\alpha)^2 + \phi(\alpha) + 1 = \alpha^4 + 3\alpha^2 + 3,\\
u_1'' &= \phi^2(\alpha)^2 + \phi^2(\alpha) + 1 = \alpha^4 + 2\alpha^2 + \alpha + 2,\\
u_2' &= \phi(\alpha)^2 + \phi(\alpha) + 2 = \alpha^4 + 3\alpha^2 + 4,\\
u_2'' &= \phi^2(\alpha)^2 + \phi^2(\alpha) + 2 = \alpha^4 + 2\alpha^2 + \alpha + 3.
\end{aligned}$$

(Note that in doing these computations, we make use of the fact that $\phi_3^*(\alpha) = 0$.)

It turns out that these six units generate a group of rank (at most) 2, since an easy computation yields

$$u_1' = -u_1^{-1}u_2, \qquad u_2' = -u_1^{-1}, \qquad u_1'' = -u_2^{-1}, \qquad u_2'' = u_1u_2^{-1}.$$

This is not surprising, since one can check that the polynomial $\Phi_3^*(x)$ is irreducible over $\mathbb{Q}$ and that all of its roots are complex. Thus the global field $\mathbb{Q}(\alpha)$ is a totally complex extension field of degree 6, so its unit group has rank 2.

We can use Theorem 2.35 to create more units. Taking $n_1 = 2$ and $n_2 = 3$ and letting $\gamma = \frac{1}{2}(-1+\sqrt{-7}\,)$, we have

$$u_{i,j} = \phi^i(\gamma) - \phi^j(\alpha) \in R[\gamma,\alpha]^* \qquad \text{for } 0 \le i \le 1 \text{ and } 0 \le j \le 2.$$

The extension $\mathbb{Q}(\alpha,\gamma)/\mathbb{Q}$ is totally complex of degree 12, so has five independent units. We leave as an exercise for the reader to compute the number of independent units in the set $\{u_{i,j}\}$.

# Exercises

## Section 2.1. The Nonarchimedean Chordal Metric

**2.1.** Let $K$ be a field that is complete with respect to an absolute value $v$ and let $\phi(z) \in K[z]$ be a polynomial. The *filled Julia set $\mathcal{K}(\phi)$ of $\phi$* is the set

$$\mathcal{K}(\phi) = \bigl\{\alpha \in K : \bigl|\phi^n(\alpha)\bigr|_v \text{ is bounded for } n = 1,2,3,\dots\bigr\}.$$

(Note that this definition applies only to polynomials, not to more general rational maps.)

(a) Prove that the filled Julia set $\mathcal{K}(\phi)$ is a closed and bounded subset of $K$. Note that if $K$ is locally compact, for example $K = \mathbb{Q}_p$ or $K = \mathbb{C}$, then this implies that $\mathcal{K}(\phi)$ is compact. However, there are cases of interest for which this is not true, e.g., the completion $\mathbb{C}_p$ of an algebraic closure of $\mathbb{Q}_p$.

(b) Prove that the complement of $\mathcal{K}(\phi)$ is equal to the set of all $P \in \mathbb{P}^1$ satisfying

$$\lim_{n\to\infty} \phi^n(P) = \infty, \tag{2.17}$$

where the limit is in the $v$-adic topology. Equivalently, prove that it is the set of $P$ satisfying $\lim \rho_v(\phi^n(P),\infty) \to 0$ or, writing $P = [\alpha,1]$, the set of points satisfying $|\phi^n(\alpha)|_v \to \infty$. The set determined by (2.17), which from (a) is an open set, is called the *attracting basin of* $\infty$.

(c) Prove that the Julia set $\mathcal{J}(\phi)$ of $\phi$ is the boundary of the filled Julia set $\mathcal{K}(\phi)$.

(d) Prove that the filled Julia set $\mathcal{K}(\phi)$ is completely invariant, i.e., prove that it satisfies

$$\phi^{-1}(\mathcal{K}(\phi)) = \mathcal{K}(\phi) = \phi(\mathcal{K}(\phi)).$$

## Section 2.3. Reduction of Maps Modulo a Prime

**2.2.** Let $f = \bigl(\begin{smallmatrix} a & b \\ c & d\end{smallmatrix}\bigr) \in \mathrm{PGL}_2(K)$ with $a,b,c,d \in R$, at least one of $a,b,c,d$ in $R^*$, and $ad \equiv bc \pmod{\mathfrak{p}}$. Prove that there exist points $P,Q \in \mathbb{P}^1(K)$ satisfying

$$\tilde{P} = \tilde{Q} \quad\text{and}\quad \widetilde{f(P)} \ne \widetilde{f(Q)}.$$

Thus Proposition 2.9 is false in general if $f \notin \mathrm{PGL}_2(R)$.

**2.3.** Generalize Proposition 2.11 to $\mathbb{P}^2$ as follows. Let

$$Q_1 = [1,0,0], \quad Q_2 = [0,1,0], \quad Q_3 = [0,0,1], \quad Q_4 = [1,1,1].$$

Suppose that $P_1, P_2, P_3, P_4 \in \mathbb{P}^2(K)$ are points whose reductions $\tilde{P}_1, \ldots, \tilde{P}_4$ are distinct and have the property that no three of them lie on a line in $\mathbb{P}^2(k)$. Prove that there is a transformation $f \in \mathrm{PGL}_3(R)$ such that $f(P_i) = Q_i$ for all $1 \le i \le 4$.

Formulate and prove an analogous statement for $\mathbb{P}^N$.

**2.4.** Let $\phi = [F, G] = [F', G']$ be two normalized representations of the rational map $\phi : \mathbb{P}^1 \to \mathbb{P}^1$. Prove that there is a constant $u \in k^*$ such that

$$u\tilde{F} = \tilde{F}' \quad \text{and} \quad u\tilde{G} = \tilde{G}'.$$

Deduce that the reduced map $\tilde{\phi} : \mathbb{P}^1(k) \to \mathbb{P}^1(k)$ does not depend on the choice of normalized representation.

## Section 2.4. The Resultant of a Rational Map

**2.5.** (a) Let $A(X,Y) = a_0X + a_1Y$ be a linear polynomial and let $B(X,Y)$ be an arbitrary homogeneous polynomial. Prove that $\mathrm{Res}(A,B) = B(-a_1, a_0)$.

(b) Let $A(X,Y) = a_0X^2 + a_1XY + a_2Y^2$ and $B(X,Y) = b_0X^2 + b_1XY + b_2Y^2$ be quadratic polynomials. Prove that

$$4\,\mathrm{Res}(A,B) = (2a_0b_2 - a_1b_1 + 2a_2b_0)^2 - (4a_0a_2 - a_1^2)(4b_0b_2 - b_1^2).$$

(Of course, in characteristic 2 one needs to cancel 4 from both sides before using this formula!)

**2.6.** With notation as in the statement of Proposition 2.13, prove that the resultant of $A$ and $B$ is related to the roots of $A$ and $B$ by

$$\mathrm{Res}(A,B) = a_0^m b_0^n \prod_{i=1}^{n}\prod_{j=1}^{m}(\alpha_i - \beta_j) = a_0^m \prod_{i=1}^{n} B(\alpha_i, 1) = (-1)^{mn} b_0^n \prod_{j=1}^{m} A(\beta_j, 1).$$

**2.7.** Let

$$A(X,Y) = a_0X^n + a_1X^{n-1}Y + \cdots + a_{n-1}XY^{n-1} + a_nY^n,$$
$$B(X,Y) = b_0X^m + b_1X^{m-1}Y + \cdots + b_{m-1}XY^{m-1} + b_mY^m,$$

be homogeneous polynomials of degrees $n$ and $m$ and let $\alpha, \beta, \gamma, \delta \in K$ be arbitrary.

(a) Prove that

$$\mathrm{Res}\bigl(A(\alpha X + \beta Y, \gamma X + \delta Y), B(\alpha X + \beta Y, \gamma X + \delta Y)\bigr) = (\alpha\delta - \beta\gamma)^{mn}\,\mathrm{Res}\bigl(A(X,Y), B(X,Y)\bigr).$$

(*Hint*. Use Proposition 2.13(b).)

(b) Suppose that $m = n = d$. Prove that

$$\mathrm{Res}\bigl(\alpha A(X,Y) + \beta B(X,Y), \gamma A(X,Y) + \delta B(X,Y)\bigr) = (\alpha\delta - \beta\gamma)^{d}\,\mathrm{Res}\bigl(A(X,Y), B(X,Y)\bigr).$$

(*Hint*. Use Proposition 2.13(d).)

(c) Continuing with the assumption that $m = n = d$, define new polynomials by the formulas

$$A^*(X,Y) = \delta A(\alpha X + \beta Y, \gamma X + \delta Y) - \beta B(\alpha X + \beta Y, \gamma X + \delta Y),$$
$$B^*(X,Y) = -\gamma A(\alpha X + \beta Y, \gamma X + \delta Y) + \alpha B(\alpha X + \beta Y, \gamma X + \delta Y).$$

Prove that
$$\operatorname{Res}(A^*, B^*) = (\alpha\delta - \beta\gamma)^{d^2+d} \operatorname{Res}(A, B).$$

(d) Suppose that $m = n = d$ and that $\operatorname{Res}(A, B) \neq 0$, so $\phi = [A, B]$ is a rational map $\phi : \mathbb{P}^1 \to \mathbb{P}^1$ of degree $d$, and suppose further that $\alpha\delta - \beta\gamma \neq 0$, so the map

$$f = [\alpha X + \beta Y, \gamma X + \delta Y]$$

is a linear fractional transformation. Let $\phi^* = [A^*, B^*]$. Prove that $\phi^* = f^{-1} \circ \phi \circ f$. Use (c) to prove that $\phi^* : \mathbb{P}^1 \to \mathbb{P}^1$ is a rational map of degree $d$.

**2.8.** Let $f \in \operatorname{PGL}_2(K)$ be a linear fractional transformation, and write

$$f(z) = \frac{az + b}{cz + d}$$

in normalized form.

(a) Prove that $\operatorname{Res}(f) = ad - bc$.

(b) Prove that

$$\rho\bigl(f(P), f(Q)\bigr) \le \bigl|\operatorname{Res}(f)\bigr|^{-1} \rho(P, Q) \qquad \text{for all } P, Q \in \mathbb{P}^1(K).$$

(Notice that this strengthens Theorem 2.14 for maps of degree 1.)

**2.9.** Let $f \in \operatorname{PGL}_2(K)$ be a linear fractional transformation, let $\phi(z) \in K(z)$ be a rational map, and let $\phi^f = f^{-1} \circ \phi \circ f$.

(a) If $\operatorname{Res}(f) \in R^*$, prove that $\operatorname{Res}(\phi^f) = \operatorname{Res}(\phi)$.

(b) If $\operatorname{Res}(f)$ is not a unit, find a formula or an inequality relating the valuations of the three quantities $\operatorname{Res}(f)$, $\operatorname{Res}(\phi)$, and $\operatorname{Res}(\phi^f)$.

**2.10.** Let $\phi : \mathbb{P}^1 \to \mathbb{P}^1$ be a rational map defined over a field $K$ with a nonarchimedean absolute value $|\cdot|_v$, and consider the statement

$$\sup_{\substack{P_1, P_2 \in \mathbb{P}^1(K) \\ P_1 \neq P_2}} \frac{\rho_v\bigl(\phi(P_1), \phi(P_2)\bigr)}{\rho_v(P_1, P_2)} = \frac{1}{\bigl|\operatorname{Res}(\phi)\bigr|_v^2}. \tag{2.18}$$

(a) Prove that (2.18) is true for the map $\phi(z) = az^d$, where $a \in R$. This example shows that Theorem 2.14 cannot be improved in general.

(b) Prove that (2.18) is not true for the map

$$\phi = [pX^2 + XY, XY + pY^2]$$

over the field $K = \mathbb{Q}_p$ by computing both sides. (You may assume that $p \neq 2$.)

**2.11.** Let $\phi : \mathbb{P}^1(\mathbb{C}) \to \mathbb{P}^1(\mathbb{C})$ be a rational map of degree $d$.

(a) Prove that there is a constant $C(d)$, depending only on $d$, such that

$$\rho_v\big(\phi(P_1),\phi(P_2)\big) \le C(d)|\operatorname{Res}(\phi)|_v^{-2}\rho_v(P_1,P_2) \qquad \text{for all } P_1,P_2 \in \mathbb{P}^1(\mathbb{C}).$$

(b) Find an explicit value for the constant $C(d)$.

Note that this exercise provides an archimedean counterpart to Theorem 2.14, whose proof may, mutatis mutandis, be helpful in doing this exercise.

### Section 2.5. Rational Maps with Good Reduction

**2.12.** Let $F(X,Y)$ and $G(X,Y)$ be homogeneous polynomials of degree $D$, let $f(X,Y)$ and $g(X,Y)$ be homogeneous polynomials of degree $d$, and let

$$A(X,Y) = F(f(X,Y),g(X,Y)) \quad \text{and} \quad B(X,Y) = G(f(X,Y),g(X,Y))$$

be their compositions.

(a) Prove that the resultants satisfy

$$\operatorname{Res}(A,B) = \operatorname{Res}(F,G)^d \cdot \operatorname{Res}(f,g)^{D^2}.$$

(b) Use (a) to give an alternative proof of Theorem 2.18(b).

(c) For any rational map $\phi : \mathbb{P}^1 \to \mathbb{P}^1$, define $\delta_v(\phi) = v(\operatorname{Res}(\phi))/\deg(\phi)$, so $\delta_v(\phi)$ is a kind of normalized resultant of $\phi$. Prove that $\delta_v$ satisfies the composition formula

$$\delta_v(\phi\circ\psi) = \delta_v(\phi) + \deg(\phi)\delta_v(\psi).$$

**2.13.** Show that the good-reduction assumptions in Theorem 2.18 are necessary by constructing the following counterexamples:

(a) Find a rational map $\phi : \mathbb{P}^1 \to \mathbb{P}^1$, which will necessarily have bad reduction, and a point $P \in \mathbb{P}^1(K)$ such that $\widetilde{\phi(P)} \ne \tilde\phi(\tilde P)$.

(b) Find rational maps $\phi : \mathbb{P}^1 \to \mathbb{P}^1$ and $\psi : \mathbb{P}^1 \to \mathbb{P}^1$ such that $\phi$ has good reduction and $\widetilde{\phi\circ\psi} \ne \tilde\phi\circ\tilde\psi$.

(c) Same as (b), except now $\psi$ is required to have good reduction and $\phi$ is allowed to have bad reduction.

**2.14.** Let $p \ge 5$ be a prime and define rational maps

$$\phi(z) = \frac{z^2+p^3z}{p^3z^2+p} \qquad \text{and} \qquad \psi(z) = \frac{p^2z^2+p^4}{p^3z+1}.$$

(a) Prove that $\phi^f$ has bad reduction modulo $p$ for all $f \in \mathrm{PGL}_2(\mathbb{Q}_p)$. (N.B. We are allowing $f$ to have coefficients in $\mathbb{Q}_p$ and/or the determinant of $f$ to be divisible by $p$.)

(b) Prove that $\psi^f$ has bad reduction modulo $p$ for all $f \in \mathrm{PGL}_2(\mathbb{Q}_p)$.

(c) Prove that the composition $\psi\circ\phi$ has good reduction at $p$.

**2.15.** Let $\phi : \mathbb{P}^1 \to \mathbb{P}^1$ be a rational map defined over $K$.

(a) Prove that the map $\phi$ has good reduction if and only if there is an $R$-morphism

$$\phi_R : \mathbb{P}^1_R \to \mathbb{P}^1_R$$

whose restriction to the generic fiber is equal to the original map $\phi : \mathbb{P}^1_K \to \mathbb{P}^1_K$.

(b) Assume that $\phi$ has good reduction. Prove that the reduction $\tilde\phi : \mathbb{P}^1_k \to \mathbb{P}^1_k$ is the restriction of $\phi_R$ to the special fiber of $\mathbb{P}^1_R$.

**2.16.** Let $K/\mathbb{Q}_p$ be a $p$-adic field.

(a) Prove that every periodic point of the map $\phi(z) = z^p$ is attracting.

(b) More generally, suppose that $\psi(z) \in K(z)$ has good reduction, and let $\phi(z) = \psi(z^p)$. Prove that every periodic point of $\phi$ is attracting.

(c) Is (b) true without the assumption that $\psi$ has good reduction?

**2.17.** Let $K/\mathbb{Q}_p$ be a $p$-adic field and let $\phi(z) \in K[z]$ be a polynomial with good reduction. Let $\alpha \in K$ be a critical point of $\phi$, i.e., $\phi'(\alpha) = 0$, and suppose that $\tilde{\alpha} \in \operatorname{Per}(\tilde{\phi})$. More precisely, suppose that the reduced point $\tilde{\alpha}$ satisfies $\tilde{\phi}^m(\tilde{\alpha}) = \tilde{\alpha}$. Prove that there is an attracting periodic point $\beta$ of $\phi$ of period $m$ such that

$$\lim_{n\to\infty} \phi^{nm}(\alpha) = \beta.$$

Generalize to the case of a rational map $\phi(z) \in K(z)$ of good reduction.

**2.18.** Let $\phi(z) \in K(z)$ be a rational map of degree $d \geq 2$ and suppose that there is a point $P \in \mathbb{P}^1(K)$ and an integer $m \geq 1$ such that the following limit exists:

$$T = \lim_{n\to\infty} \phi^{mn}(P). \tag{2.19}$$

Prove that $\phi^m(T) = T$, i.e., $T \in \operatorname{Per}_m(\phi)$. (This is true for any complete field $K$, so for example, it holds for $\mathbb{Q}_p$, $\mathbb{C}_p$, $\mathbb{R}$, $\mathbb{C}$, etc.)

Section 2.6. Periodic Points and Good Reduction

**2.19.** Let $a, b, c \in \mathbb{Z}$ and let $\phi(z) = (az^2+bz+c)/z^2$. Suppose that $P \in \mathbb{P}^1(\mathbb{Q})$ is a periodic point for $\phi$ of exact period $n$.

(a) If $\gcd(c, 6) = 1$, prove that $n = 1$, 2, or 3.

(b) Give examples to show that it is possible for $n$ to take on each of the values 1, 2, and 3.

(c) Show that if $\phi$ has a rational periodic point of exact period 3, then it has no other rational periodic points.

(d) Give a similar description of possible rational periodic points under the assumption that $\gcd(c, 10) = 1$.

(e) Same as (d), but under the assumption that $\gcd(c, 15) = 1$.

**2.20.** Let $\phi(z) \in \mathbb{Z}[z]$ be a polynomial of degree $d \geq 2$ whose leading coefficient is odd and let $P \in \mathbb{P}^1(\mathbb{Q})$ be a periodic point of exact period $n$. Prove that $n \in \{1, 2, 4\}$. This strengthens Example 2.25.

**2.21.** Let $\phi(z) \in \mathbb{Z}[z]$ be a polynomial of degree $d \geq 2$ whose leading coefficient is relatively prime to 15 and let $P \in \mathbb{P}^1(\mathbb{Q})$ be a periodic point for $\phi$ of exact period $n$. What can you say about the possible values of $n$?

**2.22.** Let $R = \mathbb{F}_p[\![T]\!]$ be the ring of formal power series in one variable with coefficients in the finite field $\mathbb{F}_p$. Prove that for every $e > 0$ there exists an element $c \in R$ with $c \equiv 1 \pmod{T}$ such that the polynomial $\phi(z) = z^p + cz$ has a periodic point $\alpha \equiv 0 \pmod{T}$ of exact order $p^e$. Thus Theorem 2.28 is false in characteristic $p$; there is no upper bound on the exponent of $p$ for the period of a periodic point.

**2.23.** Let $\phi \in K(z)$ be a rational map with good reduction and let $P \in \operatorname{Per}(\phi)$ be a periodic point. Prove that $P$ is attracting if and only if the orbit of its reduction $\mathcal{O}_{\tilde{\phi}}(\tilde{P})$ contains a critical point of $\tilde{\phi}$.

**2.24.** Let $c \in R$ and consider the quadratic map $\phi(z) = z^2 + c$. Suppose that there is an integer $m \geq 1$ such that the limit $\alpha = \lim_{n\to\infty} \phi^{mn}(0)$ exists. Prove that $\alpha$ is an attracting periodic point if and only if $\tilde{0} \in \mathrm{Per}(\tilde{\phi})$, i.e., if and only if $\tilde{0}$ is a purely periodic point of $\tilde{\phi}$.

## Section 2.7. Periodic Points and Dynamical Units

**2.25.** Theorem 2.33 gives a construction of units from periodic points of polynomial maps. Prove the following generalization.

Let $\phi \in K(z)$ be a rational map of degree $d \geq 2$. Assume that $\phi$ has good reduction and that $\infty$ is a critical fixed point of $\phi$. (That is, assume that $\phi(\infty) = \infty$ and that $e_\infty(\phi) \geq 2$.) Let $[\alpha, 1] \in \mathrm{Per}_n^{**}(\phi, K)$ be a point of exact period $n \geq 2$. Let $i, j \in \mathbb{Z}$ be integers satisfying $\gcd(i - j, n) = 1$. Prove that

$$\frac{\phi^i\alpha - \phi^j\alpha}{\phi\alpha - \alpha} \in R^*.$$

**2.26.** Prove the following version of Theorem 2.35, in which one of the periods divides the other period. Let $R$ be a local ring with residue characteristic $p$ and let $q = \#k$ be the order of the residue field. Let $\phi \in K(z)$ be a rational map of degree $d \geq 2$ with good reduction. Let $s$ be a positive integer with $p \nmid s$ and $s \nmid q - 1$. Finally, let $P_1, P_2 \in \mathbb{P}^1(K)$ be periodic points of exact periods $n$ and $sn$ respectively, and write $P_i = [x_i, y_i]$ in normalized form. Prove that

$$x_1y_2 - x_2y_1 \in R^*.$$

**2.27.** Let $P_1, P_2, P_3, P_4 \in \mathbb{P}^1$, and let $\pi$ be a permutation of the set $\{1, 2, 3, 4\}$. Prove that the cross-ratio satisfies

$$\kappa(P_1, P_2, P_3, P_4) = \kappa\left(P_{\pi(1)}, P_{\pi(2)}, P_{\pi(3)}, P_{\pi(4)}\right)^{\epsilon(\pi)},$$

where $\epsilon(\pi) = (-1)^{\mathrm{sign}(\pi)}$ equals 1 (respectively $-1$) if $\pi$ is an even (respectively odd) permutation.

**2.28.** Let $\phi(z) = z^2 + 1$, let $\gamma$ be a root of $z^2 + z + 2$, and let $\alpha$ be a root of

$$z^6 + z^5 + 4z^4 + 3z^3 + 7z^2 + 4z + 5.$$

In Example 2.37 we explained that there are units

$$u_{i,j} = \phi^i(\gamma) - \phi^j(\alpha) \in R[\gamma, \alpha]^* \qquad \text{for } 0 \leq i \leq 1 \text{ and } 0 \leq j \leq 2,$$

and that the unit group $R[\gamma, \alpha]^*$ has rank at most 5. Find generators and relations for the group of units generated by the six units $u_{i,j}$.

# Chapter 3

# Dynamics over Global Fields

Just as algebraic number theory and Diophantine geometry can be studied over local fields and over global fields, so too does arithmetic dynamics have a local theory and a global theory. In this chapter we study some of the fundamental questions in arithmetic dynamics over global fields. Many of these constructions, theorems, and conjectures have direct analogues in the theory of Diophantine equations, especially the arithmetic theory of elliptic curves. Among the topics covered in this chapter are dynamical analogues of the theory of canonical heights, finiteness of integral points, uniformity of rational torsion points, and cyclotomic and elliptic units.

We have attempted to keep this chapter self-contained, but the reader desiring further motivation and background might consult standard textbooks on arithmetic and Diophantine geometry, such as the following:

- E. Bombieri and W. Gubler [76], *Heights in Diophantine Geometry.*
- M. Hindry and J. Silverman [205], *Diophantine Geometry.*
- S. Lang [256], *Fundamentals of Diophantine Geometry.*
- J.-P. Serre [397], *Lectures on the Mordell–Weil Theorem.*
- J. Silverman [410], *The Arithmetic of Elliptic Curves.*
- J. Silverman [412], *Advanced Topics in the Arithmetic of Elliptic Curves.*

## 3.1 Height Functions

In order to study the arithmetic properties of points in projective space, it is important to have a method of measuring the size of a point. This "size" should reflect the complexity of the point in an arithmetic sense. In particular, it is good if there are only finitely many points of bounded size.

For example, suppose that we naively define the size of a rational number $\alpha$ to be its absolute value $|\alpha|$. This reflects the size of $\alpha$ as a real number, but there are infinitely many rational numbers with bounded absolute value, so it does not correctly measure $\alpha$'s size from an arithmetic perspective. If we write $\alpha = a/b$ as a fraction in lowest terms, then the integers $a$ and $b$ should each contribute to the complexity of $\alpha$, so we might define the size of $\alpha$ to be the larger of $|a|$ and $|b|$. With this definition, it is obvious that there are only finitely many rational numbers of bounded size, since there are only finitely many integers of bounded absolute value. We can easily generalize this idea to rational points in projective space.

*Example* 3.1. Let $P \in \mathbb{P}^N(\mathbb{Q})$ and write $P$ using homogeneous coordinates as

$$P = [x_0, x_1, \dots, x_N].$$

Since the coordinates are homogeneous, we can multiply through by an integer to clear the denominators, and we can also cancel any common factors, so we may assume that the homogeneous coordinates have been chosen to satisfy

$$x_0, \dots, x_N \in \mathbb{Z} \qquad \text{and} \qquad \gcd(x_0, \dots, x_N) = 1.$$

Having done this, we define the *height of* $P$ to be the quantity

$$H(P) = \max\{|x_0|, \dots, |x_N|\}.$$

It is clear that for any constant $B$, the set

$$\{P \in \mathbb{P}^N(\mathbb{Q}) : H(P) \le B\} \tag{3.1}$$

is finite. Indeed, this set clearly has fewer than $(2B+1)^{N+1}$ elements, since each coordinate $x_i$ of $P$ is an integer satisfying $|x_i| \le B$, so has at most $2B+1$ possible values. (See Exercise 3.2 for an asymptotic estimate for the size of the set (3.1).)

When trying to generalize Example 3.1 to an arbitrary number field $K$, we run into the problem that the ring of integers of $K$ may not be a principal ideal domain, so there is no uniform way to normalize the homogeneous coordinates of a point in $\mathbb{P}^N(K)$. For this reason we take a different approach based on the theory of absolute values. We now describe the absolute values on $\mathbb{Q}$, and then move on to general number fields. For further information about absolute values and completions of number fields, see any standard textbook on algebraic number theory or Diophantine geometry, for example [258, Section 2.1], [256, Chapters I, II], or [205, Section B.1].

**Definition.** The *set of standard absolute values on* $\mathbb{Q}$, denoted by $M_{\mathbb{Q}}$, consists of the following absolute values:

- $M_{\mathbb{Q}}$ contains one archimedean absolute value that is associated to its embedding $\mathbb{Q} \subset \mathbb{R}$ into the real numbers,

$$|x|_\infty = \text{usual absolute value on } \mathbb{R} = \max\{x, -x\}.$$

- Let $p$ be a prime, and for any (nonzero) integer $a$, let

$$\operatorname{ord}_p(a) = \text{exponent of highest power of } p \text{ dividing } a.$$

In other words, $p^{\operatorname{ord}_p(a)}$ is the highest power of $p$ dividing $a$. The set $M_{\mathbb{Q}}$ contains nonarchimedean ($p$-adic) absolute values, one for each prime $p$, defined by

$$\left|\frac{a}{b}\right|_p = \frac{1}{p^{\operatorname{ord}_p(a) - \operatorname{ord}_p(b)}}.$$

Now let $K/\mathbb{Q}$ be a number field. The *set of standard absolute values on* $K$ is denoted by $M_K$ and consists of all absolute values on $K$ whose restriction to $\mathbb{Q}$ is one of the absolute values in $M_\mathbb{Q}$. We write $M_K^\infty$ for the (archimedean) absolute values on $K$ lying above $|\cdot|_\infty$, and $M_K^0$ for the (nonarchimedean) absolute values of $K$ lying above the $p$-adic absolute values of $\mathbb{Q}$.

The archimedean absolute values of $K$ correspond to embeddings of $K$ into $\mathbb{R}$ or $\mathbb{C}$, while the nonarchimedean absolute values correspond to prime ideals of the ring of integers of $K$. Indeed, the *ring of integers of* $K$, denoted by $R_K$, may be characterized as the set

$$R_K = \{\alpha \in K : |\alpha|_v \le 1 \text{ for all } v \in M_K^0\}.$$

More generally, if $S \subset M_K$ is any (finite) set of absolute values containing $M_K^\infty$, then the *ring of $S$-integers of* $K$ is the set

$$R_S = \{\alpha \in K : |\alpha|_v \le 1 \text{ for all } v \notin S\}.$$

**Definition.** For any absolute value $v \in M_K$ on $K$, let $K_v$ denote the completion of $K$ at $v$. The *local degree of* $v$, denoted by $n_v$, is the quantity

$$n_v = [K_v : \mathbb{Q}_v].$$

For example, if $v$ is archimedean, then $\mathbb{Q}_v = \mathbb{R}$ and $n_v$ is 1 or 2 depending on whether $v$ corresponds to a real or complex embedding of $K$, respectively. Similarly, if $v$ is nonarchimedean, then $\mathbb{Q}_v = \mathbb{Q}_p$ is the $p$-adic rational numbers for some prime $p$, the absolute value $v$ corresponds to some prime ideal $\mathfrak{p}$ of $K$ lying over $p$, and using standard notation, $n_v = e(\mathfrak{p})f(\mathfrak{p})$ is the product of the ramification index of $\mathfrak{p}$ and the degree of the residue field modulo $\mathfrak{p}$.

The next two results, which we quote without proof, will be used to define and study height functions.

**Proposition 3.2.** (Extension Formula) *Let $L/K/\mathbb{Q}$ be a tower of number fields, and let $v \in M_K$ be an absolute value on $K$. Then*

$$\frac{1}{[L:K]} \sum_{\substack{w \in M_L \\ w|v}} n_w = n_v.$$

*Here the notation $w|v$ means that the restriction of $w$ to $K$ is equal to $v$, so the sum is over all absolute values on $L$ that extend the absolute value $v$ on $K$.*

**Proposition 3.3.** (Product Formula) *Let $K/\mathbb{Q}$ be a number field. Then*

$$\prod_{v \in M_K} |\alpha|_v^{n_v} = 1 \qquad \text{for all } \alpha \in K^*.$$

For proofs of these two formulas, see for example [258, Section II.1 and Section V.1]. The astute reader will have observed that if we take $K = \mathbb{Q}$ and $v$ nonarchimedean, then the extension formula becomes the well-known formula

$$\sum_{\mathfrak{p}|p} e(\mathfrak{p})f(\mathfrak{p}) = [L : \mathbb{Q}].$$

We now have all of the tools needed to define the height of algebraic points in projective space.

**Definition.** Let $K/\mathbb{Q}$ be a number field, and let $P \in \mathbb{P}^N(K)$ be a point with homogeneous coordinates

$$P = [x_0, \dots, x_N], \qquad x_0, \dots, x_N \in K.$$

The *height of P (relative to K)* is the quantity

$$H_K(P) = \prod_{v \in M_K} \max\{|x_0|_v, \dots, |x_N|_v\}^{n_v}.$$

Notice how the height measures the size of the coordinates of $P$ with respect to all of the absolute values on $K$. We begin by checking that the height is well-defined and proving two elementary properties.

**Proposition 3.4.** *Let $K/\mathbb{Q}$ be a number field and $P \in \mathbb{P}^N(K)$ a point.*

(a) *The height $H_K(P)$ is independent of the choice of homogeneous coordinates for P.*

(b) $H_K(P) \geq 1$.

(c) *Let $L/K$ be a finite extension. Then*

$$H_L(P) = H_K(P)^{[L:K]}.$$

*Proof.* (a) Any other choice of homogeneous coordinates for $P = [x_0, \dots, x_N]$ has the form $P = [\alpha x_0, \dots, \alpha x_N]$ for some $\alpha \in K^*$. Then the product formula (Proposition 3.3) yields

$$\prod_{v \in M_K} \max_i\{|\alpha x_i|_v\}^{n_v} = \prod_{v \in M_K} |\alpha|_v^{n_v} \max_i\{x_i|_v\}^{n_v} = \prod_{v \in M_K} \max_i\{|x_i|_v\}^{n_v}.$$

(b) Choose an index $j$ such that $x_j \neq 0$ and divide the homogeneous coordinates of $P$ by $x_j$. Then one of the homogeneous coordinates is equal to 1, so every factor in the product defining $H_K(P)$ is at least 1.

(c) We choose homogeneous coordinates for $P$ that are in $K$ and use the extension formula (Proposition 3.2) to compute

$$H_L(P) = \prod_{w \in M_L} \max_i\{|x_i|_v\}^{n_w} = \prod_{v \in M_K} \prod_{\substack{w \in M_L \\ w|v}} \max_i\{|x_i|_v\}^{n_w}$$

$$= \prod_{v \in M_K} \max_i\{|x_i|_v\}^{[L:K]n_v} = H_K(P)^{[L:K]}.$$

This completes the proof of Proposition 3.4. □

*Remark* 3.5. Now that we know that the height is well-defined, we should check that it agrees with the naive definition of the height on $\mathbb{P}^N(\mathbb{Q})$ given in Example 3.1. Thus let $P = [x_0, \ldots, x_N] \in \mathbb{P}^N(\mathbb{Q})$ with homogeneous coordinates satisfying $x_i \in \mathbb{Z}$ and $\gcd(x_i) = 1$. Then every nonarchimedean absolute value $v \in M_{\mathbb{Q}}^0$ gives $|x_i|_v \le 1$ for all indices $i$ and $|x_i|_v = 1$ for at least one index $i$. Hence in the product defining $H_{\mathbb{Q}}(P)$, only the term corresponding to the archimedean absolute value contributes, so

$$H_{\mathbb{Q}}(P) = \max\{|x_0|_\infty, \ldots, |x_N|_\infty\}.$$

We note again that the set

$$\{P \in \mathbb{P}^N(\mathbb{Q}) : H(P) \le B\}$$

is finite for any fixed bound $B$. Later in this section we prove an analogous result for arbitrary number fields.

It is sometimes easier to work with a height function that does not depend on choosing a particular number field.

**Definition.** Let $P \in \mathbb{P}^N(\bar{\mathbb{Q}})$ be a point whose coordinates are algebraic numbers. The *(absolute) height of* $P$, denoted by $H(P)$, is defined by choosing any number field $K$ such that $P \in \mathbb{P}^N(K)$ and setting

$$H(P) = H_K(P)^{1/[K:\mathbb{Q}]}.$$

The transformation formula in Proposition 3.4(c) tells us that $H(P)$ is well-defined, independent of the choice of the field $K$.

In the following, $\bar{K}$ denotes an algebraic closure of a number field $K$.

**Theorem 3.6.** *Let $K/\mathbb{Q}$ be a number field, let $P \in \mathbb{P}^N(\bar{K})$, and let $\sigma \in \mathrm{Gal}(\bar{K}/K)$. Then*

$$H(\sigma(P)) = H(P).$$

*In other words, the height is invariant under the action of the Galois group.*

*Proof.* Let $L/K$ be a finite Galois extension such that $P \in \mathbb{P}^N(L)$. For any absolute value $v \in M_L$ and any $\sigma \in \mathrm{Gal}(L/K)$, we can define a new absolute value $\sigma(v)$ on $L$ by the formula

$$|\alpha|_{\sigma(v)} = |\sigma(\alpha)|_v.$$

Using the fact that $\sigma : L \to L$ is a field automorphism, it is easy to verify that $\sigma(v)$ is an absolute value on $L$, and indeed the map $v \mapsto \sigma(v)$ is simply a permutation of the set of absolute values $M_L$. Further, the automorphism $\sigma : K \to K$ induces an isomorphism of the completions $\sigma : K_v \to K_{\sigma(v)}$, since the effect of $\sigma$ on $K$ is to transform the absolute value $v$ into the absolute value $\sigma(v)$. In particular, the local degrees

$$n_v = [K_v : \mathbb{Q}_v] \qquad \text{and} \qquad n_{\sigma(v)} = [K_{\sigma(v)} : \mathbb{Q}_{\sigma(v)}]$$

are equal. We now write $P = [x_0, \dots, x_N] \in \mathbb{P}^N(L)$ and compute

$$\begin{aligned} H_L(\sigma(P)) &= \prod_{v \in M_L} \max\{|\sigma(x_0)|_v, \dots, |\sigma(x_N)|_v\}^{n_v} \\ &= \prod_{v \in M_L} \max\left\{|x_0|_{\sigma(v)}, \dots, |x_N|_{\sigma(v)}\right\}^{n_{\sigma(v)}} \\ &= \prod_{v \in M_L} \max\{|x_0|_v, \dots, |x_N|_v\} \\ &= H_L(P). \end{aligned}$$

Taking $[L : \mathbb{Q}]^{\text{th}}$ roots completes the proof that $H(\sigma(P)) = H(P)$. □

**Definition.** Let $P = [x_0, \dots, x_N] \in \mathbb{P}^N(\bar{K})$ be a point with algebraic coordinates. The smallest field over which $P$ can be defined is called the *field of definition of* $P$ *over* $K$ and is denoted by $K(P)$. It can be constructed by choosing a nonzero coordinate $x_i$ and setting

$$K(P) = K\left(\frac{x_0}{x_i}, \frac{x_1}{x_i}, \dots, \frac{x_N}{x_i}\right).$$

It is easy to see that this field is independent of the choice of the index $i$.

Our purpose in defining the height is to have a method of measuring the arithmetic size or complexity of points in projective space, analogous to the way in which the size of a rational number is measured by taking the larger of its numerator and its denominator. The following finiteness theorem is of fundamental importance.

**Theorem 3.7.** *Let $K/\mathbb{Q}$ be a number field, and let $B$ be any constant. Then the set of points*

$$\{P \in \mathbb{P}^N(K) : H_K(P) \le B\}$$

*is finite. More generally, for any constants $B$ and $D$, the set of points*

$$\{P \in \mathbb{P}^N(\bar{\mathbb{Q}}) : H(P) \le B \quad \textit{and} \quad [\mathbb{Q}(P) : \mathbb{Q}] \le D\}$$

*is finite. In other words, there are only finitely many points in $\mathbb{P}^N(\bar{\mathbb{Q}})$ of bounded height and bounded degree.*

*Proof.* It clearly suffices to prove the second statement, i.e., that there are only finitely many points

$$P = [x_0, \dots, x_N] \in \mathbb{P}^N(\bar{\mathbb{Q}}) \quad \text{satisfying} \quad H(P) \le B \quad \text{and} \quad [\mathbb{Q}(P) : \mathbb{Q}] = D.$$

Dividing the homogeneous coordinates of $P$ by some nonzero coordinate, we may assume that some coordinate equals 1. Then

$$\mathbb{Q}(P) = \mathbb{Q}(x_0, \dots, x_N) \supset \mathbb{Q}(x_i) \qquad \text{for each } 0 \le i \le N,$$

so $[\mathbb{Q}(x_i):\mathbb{Q}] \le D$. Further, if we write $K = \mathbb{Q}(P)$ and $d = [K:\mathbb{Q}]$, then we can estimate the heights of the individual coordinates by

$$\begin{aligned}
B \ge H(P) &= \left( \prod_{v\in M_{\mathbb{Q}(P)}} \max\{|x_0|_v,\ldots,|x_N|_v\}^{n_v} \right)^{1/d} \\
&\ge \prod_{v\in M_K} \left( \left( \max\{|x_0|_v,1\}^{n_v} \cdots \max\{|x_N|_v,1\}^{n_v} \right)^{1/N} \right)^{1/d} \\
&= \left( \prod_{i=0}^{N} \left( \prod_{v\in M_K} \max\{|x_i|_v,1\}^{n_v} \right)^{1/d} \right)^{1/N} \\
&= \left( H(x_0)H(x_1)\cdots H(x_N) \right)^{1/N} \\
&\ge H(x_i)^{1/N} \qquad \text{for every } 0 \le i \le N.
\end{aligned}$$

Thus each coordinate of $P$ lies in a field of degree at most $D$ and has height at most $B^N$. Replacing $B$ by $B^N$ and taking numbers of exact degree $d$ for each $1 \le d \le D$, it suffices to prove that the set

$$\{\alpha \in \bar{\mathbb{Q}} : H(\alpha) \le B \text{ and } [\mathbb{Q}(\alpha):\mathbb{Q}] = d\} \tag{3.2}$$

is finite.

Let $\alpha$ be in the set (3.2) and write the minimal polynomial of $\alpha$ as

$$F_\alpha(X) = X^d + a_1X^{d-1} + \cdots + a_d \in \mathbb{Q}[X].$$

Also factor $F_\alpha(X)$ over $\mathbb{C}$ as

$$F_\alpha(X) = (X-\alpha_1)(X-\alpha_2)\cdots(X-\alpha_d).$$

The number $\alpha_1,\ldots,\alpha_d$ are the conjugates of $\alpha$, so Theorem 3.6 tells us that

$$H(\alpha_1) = \cdots = H(\alpha_d) = H(\alpha).$$

Further, the coefficients $a_1,\ldots,a_d$ of $F_\alpha(X)$ are the elementary symmetric polynomials of the roots (up to $\pm 1$). For example,

$$a_1 = -(\alpha_1+\alpha_2+\cdots+\alpha_d) \quad \text{and} \quad a_d = (-1)^d\alpha_1\alpha_2\cdots\alpha_d.$$

More generally, for any $1 \le k \le d$ we have

$$a_k = (-1)^k \sum_{1\le i_1<i_2<\cdots<i_k\le d} \alpha_{i_1}\alpha_{i_2}\cdots\alpha_{i_k}. \tag{3.3}$$

Note that there are $\binom{d}{k}$ terms in the sum (3.3).

We can use (3.3) and the triangle inequality to estimate the absolute values of the coefficients $a_1,\ldots,a_d$ in terms of the absolute values of the roots $\alpha_1,\ldots,\alpha_d$. Thus for any $v \in M_{\mathbb{Q}(\alpha)}$ we compute

$$\begin{aligned}|a_k|_v &= \left|\sum_{1\le i_1<i_2<\cdots<i_k\le d} \alpha_{i_1}\alpha_{i_2}\cdots\alpha_{i_k}\right|_v \\ &\le \binom{d}{k} \max_{1\le i_1<i_2<\cdots<i_k\le d} |\alpha_{i_1}\alpha_{i_2}\cdots\alpha_{i_k}|_v \\ &\le \binom{d}{k} \max\{|\alpha_1|_v,1\}\max\{|\alpha_2|_v,1\}\cdots\max\{|\alpha_d|_v,1\}.\end{aligned}$$

Further, if $v$ is nonarchimedean, then we may discard the factor of $\binom{d}{k}$. Taking the maximum over $k$ and using the fact that $\binom{d}{k} \le 2^d$ for all $k$, we find that

$$\begin{aligned}&\max\{1,|a_1|_v,|a_2|_v,\ldots,|a_d|_v\}\\ &\qquad\qquad \le 2^d \max\{|\alpha_1|_v,1\}\max\{|\alpha_2|_v,1\}\cdots\max\{|\alpha_d|_v,1\},\end{aligned}$$

where again the $2^d$ is needed only if $v$ is archimedean. Now raise to the $n_v$ power, multiply over all $v$, and take the $d^{\text{th}}$ root to obtain

$$H\big([1,a_1,a_2,\ldots,a_d]\big) \le 2^d H(\alpha_1)H(\alpha_2)\cdots H(\alpha_d).$$

Theorem 3.6 tells us that $H(\alpha_i) = H(\alpha)$ for every $i$, and we have $H(\alpha) \le B$ by assumption, so we conclude that

$$H([1,a_1,a_2,\ldots,a_d]) \le (2B)^d.$$

In other words, the height of $[1,a_1,a_2,\ldots,a_d] \in \mathbb{P}^d(\mathbb{Q})$ is bounded by $(2B)^d$. However, we already know that there are only finitely many $\mathbb{Q}$ rational points in projective space (see Example 3.1 and Remark 3.5). Hence there are only finitely many possibilities for the minimal polynomial $F_\alpha(X)$ of $\alpha$, and since each $F_\alpha(X)$ has only $d$ roots, there are only finitely many possibilities for $\alpha$. This proves that the set (3.2) is finite, which completes the proof of the theorem. □

We can use Theorem 3.7 to give a very brief proof of a famous theorem of Kronecker. It says that an algebraic integer whose conjugates all lie on the unit circle must be a root of unity.

**Theorem 3.8.** *Let $\alpha \in \bar{\mathbb{Q}}$ be a nonzero algebraic number. Then*

$$H(\alpha) = 1 \quad \textit{if and only if } \alpha \textit{ is a root of unity.}$$

*Proof.* If $\alpha$ is a root of unity, then $|\alpha|_v = 1$ for every absolute value $v$, so we clearly have $H(\alpha) = 1$. Now suppose that $H(\alpha) = 1$. Directly from the definition of the height we see that

$$H(\beta^n) = H(\beta)^n \qquad \text{for all } \beta \in \bar{\mathbb{Q}} \text{ and all } n \ge 1.$$

Thus $H(\alpha^n) = H(\alpha)^n = 1$ for all $n \ge 1$, which implies in particular that the set

$$\{\alpha, \alpha^2, \alpha^3, \ldots\}$$

is a set of bounded height. This set is also clearly contained in the number field $\mathbb{Q}(\alpha)$, so Theorem 3.7 tells us that it is a finite set. Hence there are integers $i > j > 0$ such that $\alpha^i = \alpha^j$, which shows that $\alpha$ is a root of unity. □

## 3.2 Height Functions and Geometry

The height of a point $P$ measures the arithmetic complexity of $P$. We now investigate how the height of $P$ changes when we map it to some other projective space. This will allow us to relate geometric properties of maps to the arithmetic information encapsulated by the height function.

*Remark* 3.9. For ease of exposition we restrict attention to heights on projective space, but the reader should be aware that there is a general theory of height functions on algebraic varieties due to Weil. Height functions provide a powerful tool for converting algebro-geometric relationships into number-theoretic relationships. Theorem 3.11 below provides an example; it converts the geometric information that a map $\phi$ has degree $d$ into the arithmetic information that $\phi(P)$ has height that is more-or-less the $d^{\text{th}}$ power of the height of $P$. A summary of the general theory of heights on varieties is given in Section 7.3; see [205, Part B] or [256, Chapters 3–5] for further details.

**Definition.** A *rational map of degree d* between projective spaces is a map

$$\phi : \mathbb{P}^N \longrightarrow \mathbb{P}^M$$
$$\phi(P) = [f_0(P), \dots, f_M(P)],$$

where $f_0, \dots, f_M \in \bar{K}[X_0, \dots, X_N]$ are homogeneous polynomials of degree $d$ with no common factors. (The polynomial ring $K[X_0, \dots, X_N]$ is a unique factorization domain, so it makes sense to talk about common factors.) The rational map $\phi$ is *defined at P* if at least one of the values $f_0(P), \dots, f_M(P)$ is nonzero. The rational map $\phi$ is called a *morphism* if it is defined at every point of $\mathbb{P}^N(\bar{K})$, or equivalently, if the only solution to the simultaneous equations

$$f_0(X_0, \dots, X_N) = \dots = f_M(X_0, \dots, X_N) = 0$$

is the trivial solution $X_0 = \dots = X_N = 0$. If the polynomials $f_0, \dots, f_N$ have coefficients in $K$, we say that *$\phi$ is defined over $K$*.

Our goal is to relate the height of a point $P$ to the height of its image $\phi(P)$. To do this in general, we use an important theorem from algebraic geometry called the Nullstellensatz. Here

*Null* = zero, *stellen* = places, *satz* = theorem,

so the Nullstellensatz is a theorem that relates a function to the points at which it vanishes. We give a brief overview of some basic concepts from algebra and algebraic geometry that are needed to understand the statement of the Nullstellensatz. However, we note that for rational maps $\phi(z) \in \bar{K}(z)$ in one variable (i.e., maps $\phi : \mathbb{P}^1 \to \mathbb{P}^1$), the Nullstellensatz may be replaced by a simple argument based on the fact that the ring $\bar{K}[z]$ is a principal ideal domain; see Exercise 3.8.

An ideal $I$ in $\bar{K}[X_0, \dots, X_N]$ is *homogeneous* if it is generated by homogeneous polynomials. The *radical of an ideal I* is the ideal

$$\sqrt{I} = \{f \in \bar{K}[X_0, \ldots, X_N] : f^n \in I \text{ for some } n \geq 1\}.$$

(N.B. In this definition, the quantity $f^n$ is the $n^{\text{th}}$ power of $f$, not its $n^{\text{th}}$ iterate.) The *algebraic set* attached to a homogeneous ideal $I$ is the set

$$V(I) = \{P \in \mathbb{P}^N(\bar{K}) : f(P) = 0 \text{ for all } f \in I\}.$$

The Hilbert basis theorem [259, Chapter VI, Theorem 2.1] implies that $I$ is finitely generated, so $V(I)$ is the set of simultaneous zeros of a finite collection of polynomials. If $V \subset \mathbb{P}^N(\bar{K})$ is an algebraic set, the *ideal attached to* $V$ is the ideal

$$I(V) = \begin{pmatrix} \text{ideal generated by all homogeneous} \\ \text{polynomials } f \in \bar{K}[X_0, \ldots, X_N] \\ \text{such that } f(P) = 0 \text{ for all } P \in V \end{pmatrix}.$$

It is clear that if $V = W$, then $I(V) = I(W)$. The converse is not quite true, since the algebraic set attached to the radical $\sqrt{I}$ is the same as the algebraic set attached to $I$.

**Theorem 3.10.** (Hilbert's Nullstellensatz) *Let $I$ and $J$ be homogeneous ideals properly contained in $\bar{K}[X_0, \ldots, X_N]$. Then*

$$V(I) = V(J) \quad \textit{if and only if} \quad \sqrt{I} = \sqrt{J}.$$

*Proof.* Suppose that $\sqrt{I} = \sqrt{J}$. Let $P \in V(I)$ and $f \in J$. Then $f^n \in I$ for some $n \geq 1$, so $f^n(P) = 0$, so $f(P) = 0$. This is true for every $f \in J$, so $P \in V(J)$. This proves that $V(I) \subset V(J)$, and the opposite inclusion follows by interchanging the roles and $I$ and $J$. Hence $V(I) = V(J)$. This proves the trivial direction of the theorem. For a proof of the nontrivial converse, see [198, I.1.3A] or [259, Section X.2]. □

We are now ready to prove an important result that says that up to a scalar factor, a morphism of degree $d$ causes the height to be raised to the $d^{\text{th}}$ power.

**Theorem 3.11.** *Let $\phi : \mathbb{P}^N(\bar{K}) \to \mathbb{P}^M(\bar{K})$ be a morphism of degree $d$. Then there are constants $C_1, C_2 > 0$, depending on $\phi$, such that*

$$C_1 H(P)^d \leq H(\phi(P)) \leq C_2 H(P)^d \qquad \textit{for all } P \in \mathbb{P}^N(\bar{K}).$$

*(In fact, the upper bound $H(\phi(P)) \leq C_2 H(P)^d$ is valid for rational maps provided we restrict attention to points $P$ at which $\phi$ is defined.)*

*Proof.* We begin with some notation. For any point $P = [x_0, \ldots, x_N]$ with coordinates in $K$ and any absolute value $v \in M_K$, we write

$$|P|_v = \max\{|x_0|_v, \ldots, |x_N|_v\}.$$

(This assumes that we have fixed particular homogeneous coordinates for $P$.) Similarly, we define the absolute value of a polynomial

$$f(X_0,\ldots,X_N)=\sum_{i_0,\ldots,i_N} a_{i_0\ldots i_N}X_0^{i_0}\cdots X_N^{i_N}$$

to be

$$|f|_v=\max_{i_0,\ldots,i_N}|a_{i_0\ldots i_N}|_v,$$

and if $\phi=[f_0,\ldots,f_M]$ is a collection of polynomials, we let

$$|\phi|_v=\max_j |f_j|_v.$$

Notice that the height of a point $P\in\mathbb{P}^N(K)$ may now be written in the compact form

$$H(P)=\left(\prod_{v\in M_K}|P|_v^{n_v}\right)^{1/[K:\mathbb{Q}]}.$$

We define the height of a polynomial $f$ or a collection of polynomials $\phi$ similarly,

$$H(f)=\left(\prod_{v\in M_K}|f|_v^{n_v}\right)^{1/[K:\mathbb{Q}]}\quad\text{and}\quad H(\phi)=\left(\prod_{v\in M_K}|\phi|_v^{n_v}\right)^{1/[K:\mathbb{Q}]}. \tag{3.4}$$

We set one final piece of notation that will make our computations easier. For any absolute value $v\in M_K$ and any number $m$, we set

$$\delta_v(m)=\begin{cases} m & \text{if } v\in M_K^\infty \text{ (i.e., if } v \text{ is archimedean)},\\ 1 & \text{if } v\in M_K^0 \text{ (i.e., if } v \text{ is nonarchimedean)}.\end{cases}$$

With this notation, we can write a uniform version of the triangle inequality as

$$|x_1+\cdots+x_m|_v\le\delta_v(m)\max\{|x_1|_v,\ldots,|x_m|_v\}. \tag{3.5}$$

Now let $P=[x_0,\ldots,x_N]\in\mathbb{P}^N(K)$ be a rational point and $f\in K[X_0,\ldots,X_N]$ a homogeneous polynomial of degree $d$. Then for any $v\in M_K$ we can estimate

$$\begin{aligned}|f(P)|_v&=\Biggl|\sum_{\substack{i_0,\ldots,i_N\ge 0\\ i_0+\cdots+i_N=d}} a_{i_0\ldots i_N}x_0^{i_0}\cdots x_N^{i_N}\Biggr|_v\\ &\le\delta_v(\#\text{ of terms})\max_{i_0,\ldots,i_N}\left|a_{i_0\ldots i_N}x_0^{i_0}\cdots x_N^{i_N}\right|_v.\end{aligned}$$

The number of terms in the sum is equal to at most the number of monomials of degree $d$ in $N+1$ variables, which is given by the combinatorial symbol $\binom{N+d}{d}$. For our purposes it is enough to know that it depends only on $N$ and $d$. Continuing with the computation, we find that

$$\begin{aligned}|f(P)|_v&\le\delta_v\left(\binom{N+d}{d}\right)\max_{i_0,\ldots,i_N}|a_{i_0\ldots i_N}|_v\max_{i_0,\ldots,i_N}\max_{0\le j\le N}|x_j|_v^{i_0+\cdots+i_N}\\ &=\delta_v\left(\binom{N+d}{d}\right)|f|_v|P|_v^d.\end{aligned}$$

We apply this inequality with $f = f_k$ for $k = 0, \ldots, N$ and take the maximum over $k$ to obtain

$$|\phi(P)|_v \leq \delta_v\left(\binom{N+d}{d}\right) |\phi|_v |P|_v^d. \tag{3.6}$$

Finally, we raise (3.6) to the $n_v$ power, multiply over all $v \in M_K$, and take the $[K : \mathbb{Q}]$ root to obtain

$$H(\phi(P)) \leq \binom{N+d}{d} H(\phi) H(P)^d.$$

Note that in deriving this formula, we have used the fact that

$$\prod_{v \in M_K} \delta_v(a)^{n_v} = \prod_{v \in M_K^\infty} a^{n_v} \qquad \text{and} \qquad \sum_{v \in M_K^\infty} n_v = [K : \mathbb{Q}],$$

where the latter is a special case of Theorem 3.2.

This completes the proof of the upper bound with the explicit constant

$$C_2 = \binom{N+d}{d} H(\phi).$$

And as indicated in the statement of the theorem, we never assumed that $\phi$ is a morphism, so the upper bound holds for rational maps.

In order to prove the opposite inequality, it is necessary to use the assumption that $\phi$ is a morphism, or equivalently, that $\phi = [f_0, \ldots, f_M]$ for homogeneous polynomials $f_0, \ldots, f_M$ having no common zero in $\mathbb{P}^N(\bar{K})$. This condition tells us that the ideals $(f_0, \ldots, f_M)$ and $(X_0, \ldots, X_N)$ in $\bar{K}[X_0, \ldots, X_N]$ define the same algebraic set in $\mathbb{P}^N(\bar{K})$, namely the empty set. The Nullstellensatz (Theorem 3.10) then implies that these two ideals have the same radical. In particular,

$$X_0, X_1, \ldots, X_N \in \sqrt{(f_0, f_1, \ldots, f_M)}.$$

Hence we can find an integer $e \geq 0$ such that

$$X_0^e, X_1^e, \ldots, X_N^e \in (f_0, f_1, \ldots, f_M).$$

(The Nullstellensatz says that there is an exponent $e_i$ for each $X_i$, and then we take $e$ to be the largest of the $e_i$.) Writing this out explicitly, we find homogeneous polynomials $g_{ij} \in \bar{K}[X_0, \ldots, X_N]$ such that

$$X_i^e = g_{i0} f_0 + g_{i1} f_1 + \cdots + g_{iM} f_M \qquad \text{for each } 1 \leq i \leq N.$$

We observe that each $g_{ij}$ is homogeneous of degree $e - d$, since each $f_j$ has degree $d$.

We evaluate these polynomial identities at $P = [x_0, \ldots, x_N] \in \mathbb{P}^N(K)$ and estimate $v$-adic absolute values,

$$\begin{aligned}
|P|_v^e &= \max_{0\le i\le N} |x_i|_v^e \\
&= \max_{0\le i\le N} \bigl|g_{i0}(P)f_0(P) + g_{i1}(P)f_1(P) + \cdots + g_{iM}(P)f_M(P)\bigr|_v \\
&\le \delta_v(M+1) \max_{\substack{0\le i\le N\\ 0\le j\le M}} \bigl|g_{ij}(P)f_j(P)\bigr|_v \\
&\le \delta_v(M+1) \max_{\substack{0\le i\le N\\ 0\le j\le M}} \delta_v\left(\binom{N+e-d}{e-d}\right) |g_{ij}|_v |P|_v^{e-d} |f_j(P)|_v \\
&\le \delta_v\left((M+1)\binom{N+e-d}{e-d}\right)\left(\max_{\substack{0\le i\le N\\ 0\le j\le M}} |g_{ij}|_v\right)|P|_v^{e-d}|\phi(P)|_v.
\end{aligned}$$

So if we let $|g|_v = \max_{i,j} |g_{ij}|_v$ and $a_{M,N,d,e} = (M+1)\binom{N+e-d}{e-d}$, then dividing both sides by $|P|_v^{e-d}$ yields the estimate

$$|P|_v^d \le \delta_v(a_{M,N,d,e})|g|_v|\phi(P)|_v. \tag{3.7}$$

Now raise this inequality to the $n_v$ power, multiply over all $v \in M_K$, and take the $[K:\mathbb{Q}]$ root to obtain

$$H(P)^d \le a_{M,N,d,e}H(g)H(\phi(P)),$$

where $H(g)$ is the height determined by the coefficients of all of the $g_{ij}$ polynomials. Its precise value is not important for our purposes; it is enough to note that it does not depend on the point $P$. We have thus proven that $H(P)^d \le CH(\phi(P))$ for a constant $C$ that does not depend on $P$, which completes the proof of Theorem 3.11. □

Theorem 3.11 says that the height of $\phi(P)$ is approximately equal to the $d^{\text{th}}$ power of $P$. This means that $H$ is a multiplicative kind of function, similar in some ways to an absolute value. It is often convenient notationally to work instead with an additive function, which prompts the following definition.

**Definition.** The *logarithmic height* (*relative to* $K$) is the function

$$h_K : \mathbb{P}^N(K) \longrightarrow \mathbb{R}, \qquad h_K(P) = \log H_K(P).$$

The *absolute logarithmic height* is the function

$$h : \mathbb{P}^N(\bar{\mathbb{Q}}) \longrightarrow \mathbb{R}, \qquad h(P) = \log H(P).$$

Using this logarithmic notation, Theorem 3.11 says that $h(\phi(P))$ and $dh(P)$ differ by a bounded amount. The notion of functions that differ by a bounded amount appears so frequently in mathematics that it has its own notation.

**Definition.** Let $S$ be a set and let $f, g : S \to \mathbb{R}$ be real-valued functions on $S$. The formula

$$f = g + O(1)$$

means that the quantity

$$|f(P) - g(P)|$$

is bounded as $P$ ranges over $S$. More generally, if $h : S \to \mathbb{R}$ is another function, the formula

$$f = g + O(h)$$

means that there exists a constant $C$ such that

$$|f(P) - g(P)| \le C|h(P)| \qquad \text{for all } P \in S.$$

Using the theory of heights as developed in Theorems 3.7 and 3.11, it is easy to prove that a morphism $\phi : \mathbb{P}^N \to \mathbb{P}^N$ of degree at least 2 has only finitely many preperiodic points defined over a given number field. This result was first discovered by Northcott [343] and then independently rediscovered many times in varying degrees of generality. It is instructive to compare this global proof with the local proof of finiteness of periodic points given in Corollary 2.26.

**Theorem 3.12.** (Northcott [343]) *Let $\phi : \mathbb{P}^N \to \mathbb{P}^N$ be a morphism of degree $d \ge 2$ defined over a number field $K$. Then the set of preperiodic points $\operatorname{PrePer}(\phi) \subset \mathbb{P}^N(\bar{K})$ is a set of bounded height. In particular,*

$$\operatorname{PrePer}(\phi, \mathbb{P}^N(K)) = \operatorname{PrePer}(\phi) \cap \mathbb{P}^N(K)$$

*is a finite set, and more generally, for any $D \ge 1$ the set*

$$\bigcup_{[L:K] \le D} \operatorname{PrePer}(\phi, \mathbb{P}^N(L))$$

*is finite.*

*Proof.* Theorem 3.11 tells us that there is a constant $C = C(\phi)$ such that

$$h(\phi(Q)) \ge dh(Q) - C \qquad \text{for all } Q \in \mathbb{P}^N(\bar{K}).$$

Applying this with $Q = R, \phi(R), \phi^2(R), \dots, \phi^{n-1}(R)$ yields

$$h(\phi^n(R)) \ge d^n h(R) - C(1 + d + d^2 + \cdots + d^{n-1}) \ge d^n\bigl(h(R) - C\bigr). \tag{3.8}$$

Now suppose that $P$ is preperiodic, say $\phi^{m+n}(P) = \phi^m(P)$ with $n \ge 1$ and $m \ge 0$. Setting $R = \phi^m(P)$ and using the preperiodicity yields

$$h(\phi^m(P)) = h(\phi^{m+n}(P)) = h(\phi^n(\phi^m(P))) \ge d^n\bigl(h(\phi^m(P)) - C\bigr),$$

and hence

$$h(\phi^m(P)) \le \frac{d^n}{d^n - 1} C \le 2C. \tag{3.9}$$

For the last inequality we are using the fact that $d \geq 2$ and $n \geq 1$. Next we use inequality (3.8) with $n = m$ and $R = P$,

$$d^m\big(h(P) - C\big) \leq h(\phi^m(P)). \tag{3.10}$$

Finally, combining (3.9) and (3.10) yields

$$h(P) \leq \frac{1}{d^m} h(\phi^m(P)) + C \leq \frac{1}{d^m} 2C + C \leq 3C,$$

which completes the proof that the height of $P$ is bounded by a constant depending only on the map $\phi$.

The remaining assertions of the theorem are immediate consequences of Theorem 3.7, which says that sets of bounded height have only finitely many points of bounded degree. □

*Remark* 3.13. One may also consider preperiodic points defined over infinite extensions. For example, we might fix an infinite extension $L/\mathbb{Q}$ and ask which morphisms $\phi$ have infinitely many periodic or preperiodic points in $\mathbb{P}^N(L)$. Very little is known about this question in general.

We consider a special case. Let $K/\mathbb{Q}$ be a number field and let $K^{\text{cycl}}$ denote the maximal cyclotomic extension of $K$, i.e., the field obtained by adjoining to $K$ all roots of unity. Now let $\phi(z) \in K[z]$ be a polynomial of degree $d \geq 2$ and suppose that the set $\operatorname{PrePer}\big(\phi, \mathbb{P}^1(K^{\text{cycl}})\big)$ has infinitely many points. Then Dvornicich and Zannier [146] prove that there is a Möbius transformation $f \in \mathrm{PGL}_2(\bar{K})$ such that $\phi^f(z)$ has the form $\pm z^d$ or $T_d(\pm z)$, where $T_d(z)$ is a Chebyshev polynomial (see Section 1.6.2).

Notice that an alternative dynamical definition of $K^{\text{cycl}}$ is the field obtained by adjoining to $K$ all of the points in $\operatorname{PrePer}(z^d)$ for some $d \geq 2$. This suggests a natural question.

*Question* 3.14. Let $\phi, \psi : \mathbb{P}^N \to \mathbb{P}^N$ be morphisms of degree at least 2 defined over a number field $K$. Suppose that $\phi$ has infinitely many preperiodic points defined over the field $K\big(\operatorname{PrePer}(\psi)\big)$. What can one deduce about the relationship between $\phi$ and $\psi$?

## 3.3 The Uniform Boundedness Conjecture

Northcott's Theorem 3.12 says that a morphism $\phi : \mathbb{P}^N \to \mathbb{P}^N$ has only finitely many $K$-rational preperiodic points. It is even effective in the sense that we can, in principle, find an explicit constant $C(\phi)$ in terms of the coefficients of $\phi$ such that every point $P \in \operatorname{PrePer}(\phi)$ satisfies $h(P) \leq C(\phi)$. This also allows us to compute an upper bound for $\#\operatorname{PrePer}(\phi, \mathbb{P}^N(K))$, but the bound grows extremely rapidly as the coefficients of $\phi$ become large. A better bound, at least for periodic points, may be derived from the local estimates in Chapter 2 as described in Corollary 2.26. However, even that estimate depends on the coefficients of $\phi$, since it is in terms of the two smallest primes for which $\phi$ has good reduction. The following uniformity conjecture says that there should be a bound for the size of $\mathrm{PrePer}(\phi, \mathbb{P}^N(K))$ that depends in only a minimal fashion on $\phi$ and $K$.

**Conjecture 3.15.** (Morton–Silverman [312]) *Fix integers $d \geq 2$, $N \geq 1$, and $D \geq 1$. There is a constant $C(d, N, D)$ such that for all number fields $K/\mathbb{Q}$ of degree at most $D$ and all finite morphisms $\phi : \mathbb{P}^N \to \mathbb{P}^N$ of degree $d$ defined over $K$,*

$$\#\mathrm{PrePer}\bigl(\phi, \mathbb{P}^N(K)\bigr) \leq C(d, N, D).$$

*Remark* 3.16. There are many results in the literature giving explicit bounds for the size of the sets $\mathrm{PrePer}\bigl(\phi, \mathbb{P}^N(K)\bigr)$ or $\mathrm{Per}\bigl(\phi, \mathbb{P}^N(K)\bigr)$ in terms of $\phi$, especially in the case $N = 1$. Some of these results use global methods, while others use a small prime of good (or at least not too bad) reduction for $\phi$. For example, we used local methods in Corollary 2.26 to give a weak bound for $\#\mathrm{Per}(\phi, \mathbb{P}^1(K))$. For further results, see [52, 87, 90, 91, 92, 100, 101, 137, 162, 171, 191, 192, 193, 194, 195, 227, 331, 265, 312, 325, 326, 328, 329, 330, 332, 353, 355, 358, 359, 361, 454].

*Remark* 3.17. Very little is known about Conjecture 3.15. Indeed, it is not known even in the simplest case $(d, N, D) = (2, 1, 1)$, that is, for $\mathbb{Q}$-rational points and degree-2 maps on $\mathbb{P}^1$. Specializing further, if we let $\phi_c : \mathbb{P}^1 \to \mathbb{P}^1$ denote the quadratic map $\phi_c(z) = z^2 + c$, then the conjecture implies that

$$\sup_{c \in \mathbb{Q}} \#\mathrm{Per}\bigl(\phi_c, \mathbb{P}^1(\mathbb{Q})\bigr) < \infty,$$

but the best known upper bounds for $\#\mathrm{Per}\bigl(\phi_c, \mathbb{P}^1(\mathbb{Q})\bigr)$ depend on $c$.

There are one-parameter families of $c$-values for which $\phi_c(z)$ has a $\mathbb{Q}$-rational periodic point of exact period 1, 2, or 3; see Exercise 3.9 and Example 4.9. And one can show that $\phi_c$ cannot have $\mathbb{Q}$-rational periodic points of exact period 4 or 5; see [171, 309]. Poonen has conjectured that $\phi_c$ cannot have any $\mathbb{Q}$-rational periodic points of period greater than 3. Assuming this conjecture, he gives a complete description of all possible rational preperiodic structures for $\phi_c$; see [361].

*Remark* 3.18. Another interesting collection of rational maps is the family

$$\phi_{a,b}(z) = az + \frac{b}{z}.$$

These maps have the symmetry property $\phi_{a,b}(-z) = -\phi_{a,b}(z)$, i.e., conjugation by the map $f(z) = -z$ leaves them invariant. It is known that there are one-parameter families of these maps with a $\mathbb{Q}$-rational periodic point of exact period 1 (in addition to the obvious fixed point at $\infty$), 2, or 4, and that none of the maps $\phi_{a,b}(z)$ has a $\mathbb{Q}$-rational periodic point of exact period 3. See [286] for details, and Examples 4.69 and 4.71 and Exercises 4.1, 4.40, and 4.41 for additional properties of these maps.

*Remark* 3.19. Conjecture 3.15 is an extremely strong uniformity conjecture. For example, if we consider only maps $\phi : \mathbb{P}^1 \to \mathbb{P}^1$ of degree 4 defined over $\mathbb{Q}$, then the assertion that $\#\mathrm{PrePer}\bigl(\phi, \mathbb{P}^1(\mathbb{Q})\bigr) \leq C$ for an absolute constant $C$ immediately implies Mazur's theorem [292] that the torsion subgroup of an elliptic curve $E/\mathbb{Q}$

is bounded independently of $E$. To see this, we observe that Proposition 0.3 tells us that

$$E_{\text{tors}} = \operatorname{PrePer}\bigl([2], E\bigr),$$

and hence the associated Lattès map $\phi_{E,2}$ described in Section 1.6.3 satisfies

$$x\,(E_{\text{tors}}) = \operatorname{PrePer}(\phi_{E,2}, \mathbb{P}^1).$$

Note that $\phi_{E,2}$ has degree 4.

In a similar manner, Conjecture 3.15 for maps of degree 4 on $\mathbb{P}^1$ over number fields implies Merel's theorem [297] that the size of the torsion subgroup of an elliptic curve over a number field is bounded solely in terms of the degree of the number field. Turning this argument around, Merel's theorem implies the uniform boundedness conjecture for Lattès maps, i.e., for rational maps associated to elliptic curves; see Theorem 6.65. Lattès maps are the only nontrivial family of rational maps for which the uniform boundedness conjecture is currently known.

In higher dimension, Fakhruddin [162] has shown that Conjecture 3.15 implies that there is a constant $C(N,D)$ such that if $K$ is a number field of degree at most $D$ and if $A/K$ is an abelian variety of dimension $N$, then

$$\#A(K)_{\text{tors}} \le C(N,D).$$

He also shows that if Conjecture 3.15 is true over $\mathbb{Q}$, then it is true for all number fields.

## 3.4 Canonical Heights and Dynamical Systems

It is obvious from the definition of the height that

$$h(\alpha^d) = dh(\alpha) \qquad \text{for all } \alpha \in \bar{\mathbb{Q}}. \tag{3.11}$$

Notice that Theorem 3.11 applied to the particular map $\phi(z) = z^d$ gives the less-precise statement

$$h(\phi(P)) = dh(P) + O(1). \tag{3.12}$$

Clearly the exact formula (3.11) is more attractive than the approximation (3.12). It would be nice if we could modify the height $h$ in some way so that the general formula (3.12) from Theorem 3.11 is true without the $O(1)$. It turns out that this can be done for each morphism $\phi$. To create these special heights, we follow a construction due originally to Tate.

**Theorem 3.20.** *Let $S$ be a set, let $d > 1$ be a real number, and let*

$$\phi : S \to S \qquad \text{and} \qquad h : S \to \mathbb{R}$$

*be functions satisfying*

$$h\bigl(\phi(P)\bigr) = dh(P) + O(1) \qquad \text{for all } P \in S.$$

*Then the limit*

$$\hat{h}(P) = \lim_{n\to\infty} \frac{1}{d^n} h(\phi^n(P)) \tag{3.13}$$

*exists and satisfies:*

(a) $\hat{h}(P) = h(P) + O(1)$.

(b) $\hat{h}(\phi(P)) = d\hat{h}(P)$.

*The function* $\hat{h} : S \to \mathbb{R}$ *is uniquely determined by the properties* (a) *and* (b).

*Proof.* We prove that the limit (3.13) exists by proving that the sequence is Cauchy. Let $n > m \geq 0$ be integers. We are given that there is a constant $C$ such that

$$\bigl|h(\phi(Q)) - dh(Q)\bigr| \leq C \qquad \text{for all } Q \in S. \tag{3.14}$$

We apply inequality (3.14) with $Q = \phi^{i-1}(P)$ to the telescoping sum

$$\begin{aligned} \left|\frac{1}{d^n}h(\phi^n(P)) - \frac{1}{d^m}h(\phi^m(P))\right| &= \left|\sum_{i=m+1}^{n} \frac{1}{d^i}\Bigl(h(\phi^i(P)) - dh(\phi^{i-1}(P))\Bigr)\right| \\ &\leq \sum_{i=m+1}^{n} \frac{1}{d^i}\bigl|h(\phi^i(P)) - dh(\phi^{i-1}(P))\bigr| \\ &\leq \sum_{i=m+1}^{n} \frac{C}{d^i} \leq \sum_{i=m+1}^{\infty} \frac{C}{d^i} = \frac{C}{(d-1)d^m}. \end{aligned} \tag{3.15}$$

The inequality (3.15) clearly implies that

$$\left|\frac{1}{d^n}h(\phi^n(P)) - \frac{1}{d^m}h(\phi^m(P))\right| \to 0 \qquad \text{as } m, n \to \infty,$$

so the sequence $d^{-n}h(\phi^n(P))$ is Cauchy and the limit (3.13) exists.

In order to prove (a), we take $m = 0$ in (3.15), which yields

$$\left|\frac{1}{d^n}h(\phi^n(P)) - h(P)\right| \leq \frac{C}{d-1}.$$

Next we let $n \to \infty$ to obtain

$$\bigl|\hat{h}_\phi(P) - h(P)\bigr| \leq \frac{C}{d-1},$$

which is (a) with an explicit value for the $O(1)$ constant.

The proof of (b) is a simple computation using the definition of $\hat{h}$,

$$\hat{h}_\phi(\phi(P)) = \lim_{n\to\infty} \frac{1}{d^n} h(\phi^n(\phi(P))) = \lim_{n\to\infty} \frac{d}{d^{n+1}} h(\phi^{n+1}(P))) = d\hat{h}_\phi(P).$$

Finally, to prove uniqueness, suppose that $\hat{h}' : S \to \mathbb{R}$ also has properties (a) and (b). Then the difference $g = \hat{h} - \hat{h}'$ satisfies

$$g(P) = O(1) \qquad \text{and} \qquad g(\phi(P)) = dg(P).$$

These formulas hold for all elements $P \in S$, so

$$d^n g(P) = g(\phi^n(P)) = O(1) \qquad \text{for all } n \geq 0.$$

In other words, the quantity $d^n g(P)$ is bounded as $n \to \infty$, which can happen only if $g(P) = 0$. This proves that $\hat{h}(P) = \hat{h}'(P)$, so $\hat{h}$ is unique. □

**Definition.** Let $\phi : \mathbb{P}^N \to \mathbb{P}^N$ be a morphism of degree $d \geq 2$. The *canonical height function* (*associated to* $\phi$) is the unique function

$$\hat{h}_\phi : \mathbb{P}^N(\bar{\mathbb{Q}}) \longrightarrow \mathbb{R}$$

satisfying

$$\hat{h}_\phi(P) = h(P) + O(1) \qquad \text{and} \qquad \hat{h}_\phi(\phi(P)) = d\hat{h}_\phi(P).$$

The existence and uniqueness of $\hat{h}_\phi$ follow from Theorem 3.20 applied to the maps

$$\phi : \mathbb{P}^N(\bar{\mathbb{Q}}) \longrightarrow \mathbb{P}^N(\bar{\mathbb{Q}}) \qquad \text{and} \qquad h : \mathbb{P}^N(\bar{\mathbb{Q}}) \longrightarrow \mathbb{R},$$

since Theorem 3.11 tells us that $\phi$ and $h$ satisfy

$$h(\phi(P)) = dh(P) + O(1) \qquad \text{for all } P \in \mathbb{P}^N(\bar{\mathbb{Q}}).$$

*Remark* 3.21. The definition $\hat{h}_\phi(P) = \lim_{n\to\infty} d^{-n} h(\phi^n(P))$ is not practical for accurate numerical calculations. Thus even for $P \in \mathbb{P}^1(\mathbb{Q})$, one would need to compute the exact value of $\phi^n(P)$, whose coordinates have $O(d^n)$ digits. A practical method for the numerical computation of $\hat{h}_\phi(P)$ to high accuracy uses the decomposition of $\hat{h}_\phi$ as a sum of local heights or Green functions. This decomposition is described in Sections 3.5 and 5.9. See in particular Exercise 5.29 for a detailed description of the algorithm.

The canonical height provides a useful arithmetic characterization of the preperiodic points of $\phi$.

**Theorem 3.22.** *Let $\phi : \mathbb{P}^N \to \mathbb{P}^N$ be a morphism of degree $d \geq 2$ defined over $\bar{\mathbb{Q}}$ and let $P \in \mathbb{P}^N(\bar{\mathbb{Q}})$. Then*

$$P \in \operatorname{PrePer}(\phi) \quad \textit{if and only if} \quad \hat{h}_\phi(P) = 0.$$

*Proof.* If $P$ is preperiodic, then the quantity $h(\phi^n(P))$ takes on only finitely many values, so it is clear that $d^{-n}h(\phi^n(P)) \to 0$ as $n \to \infty$.

Now suppose that $\hat{h}_\phi(P) = 0$. Let $K$ be a number field containing the coordinates of $P$ and the coefficients of $\phi$, i.e., $P \in \mathbb{P}^N(K)$ and $\phi$ is defined over $K$. Theorem 3.20 and the assumption $\hat{h}_\phi(P) = 0$ imply that

$$h(\phi^n(P)) = \hat{h}_\phi(\phi^n(P)) + O(1) = d^n \hat{h}_\phi(P) + O(1) = O(1) \quad \text{for all } n \geq 0.$$

Thus the orbit

$$\mathcal{O}_\phi(P) = \{P, \phi(P), \phi^2(P), \phi^3(P), \ldots\} \subset \mathbb{P}^N(K)$$

is a set of bounded height, so it is finite from Theorem 3.7. Therefore $P$ is a preperiodic point for $\phi$. □

*Remark* 3.23. Further material on canonical heights in dynamics may be found in Sections 3.5, 5.9, and 7.4, as well as [16, 20, 23, 36, 38, 39, 87, 88, 89, 147, 159, 231, 228, 230, 232, 234, 406, 409, 446, 453].

Theorem 3.22 is a generalization of Kronecker's theorem (Theorem 3.8), which says that $h(\alpha) = 0$ if and only if $\alpha$ is a root of unity. Kronecker's theorem follows by applying Theorem 3.22 to the $d^{\text{th}}$-power map $\phi(z) = z^d$, whose canonical height is the ordinary height $h$.

The fact that only roots of unity have height 0 leads naturally to the question of how small a nonzero height can be. If we take the relation $h(\alpha^d) = dh(\alpha)$ and substitute in $\alpha = 2^{1/d}$, we find that

$$h(2^{1/d}) = \frac{1}{d}h(2) = \frac{\log 2}{d},$$

so the height can become arbitrarily small. However, this is possible only by taking numbers lying in fields of higher and higher degree. For any algebraic number $\alpha$, let

$$D(\alpha) = [\mathbb{Q}(\alpha) : \mathbb{Q}]$$

denote the degree of its minimal polynomial over $\mathbb{Q}$.

**Conjecture 3.24.** (Lehmer's Conjecture [264]) *There is an absolute constant $\kappa > 0$ such that*

$$h(\alpha) \ge \kappa / D(\alpha)$$

*for every nonzero algebraic number $\alpha$ that is not a root of unity.*

There has been a great deal of work on Lehmer's conjecture by many mathematicians; see for example [7, 8, 75, 94, 264, 366, 421, 422, 424, 445]. (The survey [422] contains an extensive bibliography.) The best result currently known, which is due to Dobrowolski [138], says that

$$h(\alpha) \ge \frac{\kappa}{D(\alpha)} \left( \frac{\log \log D(\alpha)}{\log D(\alpha)} \right)^3.$$

The smallest known nonzero value of $D(\alpha)h(\alpha)$ is

$$D(\beta_0)h(\beta_0) = 0.1623576\ldots,$$

where $\beta_0 = 1.17628\ldots$ is a real root of

$$x^{10} + x^9 - x^7 - x^6 - x^5 - x^4 - x^3 + x + 1.$$

Theorem 3.22 tells us that $\hat{h}_\phi(P) = 0$ if and only if $P$ is a preperiodic point for $\phi$. This suggests a natural generalization of Lehmer's conjecture to the dynamical setting. (See [317] for an early version of this conjecture in a special case.)

**Conjecture 3.25.** (Dynamical Lehmer Conjecture) *Let $\phi : \mathbb{P}^N \to \mathbb{P}^N$ be a morphism defined over a number field $K$, and for any point $P \in \mathbb{P}^N(\bar{K})$, let $D(P) = [K(P) : K]$. Then there is a constant $\kappa = \kappa(K, \phi) > 0$ such that*

$$\hat{h}_\phi(P) \geq \frac{\kappa}{D(P)} \quad \textit{for all } P \in \mathbb{P}^N(\bar{K}) \textit{ with } P \notin \mathrm{PrePer}(\phi).$$

There has been considerable work on this conjecture for maps $\phi : \mathbb{P}^1 \to \mathbb{P}^1$ that are associated to groups as described in Section 1.6. For example, in the case that $\phi$ is attached to an elliptic curve $E$, it is known that

$$\hat{h}_\phi(P) \geq \begin{cases} \dfrac{\kappa}{D(P)^3 \log^2 D(P)} & \text{in general [291]}, \\ \dfrac{\kappa}{D(P)^2} & \text{if } j(E) \text{ is nonintegral [203]}, \\ \dfrac{\kappa}{D(P)} \left( \dfrac{\log\log D(P)}{\log D(P)} \right)^3 & \text{if } E \text{ has complex multiplication [263]}. \end{cases}$$

Aside from maps associated to groups, there does not appear to be a single example for which it is known that $\hat{h}_\phi(P)$ is always greater than a constant over a fixed power of $D(P)$. Using trivial estimates based on the number of points of bounded height in projective space, it is easy to prove a lower bound that decreases faster than exponentially in $D(P)$; see Exercise 3.17.

*Remark* 3.26. The Lehmer conjecture involves a single map $\phi$ and points from number fields of increasing size. Another natural question to ask about lower bounds for the canonical height involves fixing the field $K$ and letting the map $\phi$ vary. For example, consider quadratic polynomials $\phi_c(z) = z^2 + c$ as $c$ varies over $\mathbb{Q}$. Is it true that $\hat{h}_{\phi_c}(\alpha)$ is uniformly bounded away from 0 for all $c \in \mathbb{Q}$ and all nonpreperiodic $\alpha \in \mathbb{Q}$? In other words, does there exist a constant $\kappa > 0$ such that

$$\hat{h}_{\phi_c}(\alpha) \geq \kappa \quad \text{for all } c, \alpha \in \mathbb{Q} \text{ with } \alpha \notin \mathrm{PrePer}(\phi_c)?$$

We might even ask that the lower bound grow as $c$ becomes larger (in an arithmetic sense). Thus is there a constant $\kappa > 0$ such that

$$\hat{h}_{\phi_c}(\alpha) \geq \kappa h(c) \quad \text{for all } c, \alpha \in \mathbb{Q} \text{ with } \alpha \notin \mathrm{PrePer}(\phi_c)?$$

This is a dynamical analogue of a conjecture for elliptic curves that is due to Serge Lang; see [202], [254, page 92], or [410, VIII.9.9].

For the quadratic map $z^2 + c$, the height of the parameter $c$ provides a natural measure of its size, but the situation for general rational maps $\phi(z) \in K(z)$ is more complicated. We cannot simply use the height of the coefficients of $\phi$, because the canonical height is invariant under conjugation (see Exercise 3.11), while the height of the coefficients is not conjugation-invariant. We return to this question in Section 4.11 after we have developed a way to measure the size of the conjugacy class of a rational map.

## 3.5 Local Canonical Heights

The canonical height $\hat{h}_\phi$ attached to a rational map $\phi$ is a useful tool in studying the arithmetic dynamics of $\phi$. For more refined analyses, it is helpful to decompose the canonical height as a sum of local canonical heights, one for each absolute value on $K$. In this section we briefly summarize the basic properties of local canonical heights, but we defer the proofs until Section 5.9. The reader wishing to proceed more rapidly to the main arithmetic results of this chapter may safely omit this section on first reading, since the material covered is not used elsewhere in this book.

The construction of the canonical height relies on the fact that the ordinary height satisfies $h(\phi(P)) = dh(P) + O(1)$, so it is "almost canonical." The ordinary height of a point $P = [\alpha, 1]$ is equal to the sum

$$h(P) = h(\alpha) = \sum_{v \in M_K} n_v \log \max\{|\alpha|_v, 1\},$$

so for each $v \in M_K$ it is natural to define a local height function

$$\lambda_v(\alpha) = \log \max\{|\alpha|_v, 1\}.$$

We can understand $\lambda_v$ geometrically by observing that for $v \in M_K^0$,

$$\lambda_v(\alpha) = -\log \rho_v(\alpha, \infty),$$

where $\rho_v$ is the nonarchimedean chordal metric defined in Section 2.1. One says that $\lambda_v(\alpha)$ is the *logarithmic distance* from $\alpha$ to $\infty$.

Unfortunately, the function $\lambda_v$ does not transform canonically, since $\lambda_v(\phi(\alpha))$ is not equal to $d\lambda_v(\alpha) + O(1)$. To see why, note that $\lambda_v(\phi(\alpha))$ is large if $\alpha$ is close to a pole of $\phi$, while $\lambda_v(\alpha)$ is large if $\alpha$ is close to the point $\infty \in \mathbb{P}^1$. (Here the word "close" means in terms of the $v$-adic chordal metric $\rho_v$.) Thus we can hope to find a canonical local height only if we allow an appropriate correction term, as in the following theorem.

**Theorem 3.27.** *Let $\phi : \mathbb{P}^1 \to \mathbb{P}^1$ be a rational function of degree $d \geq 2$ defined over $K$, write $\phi(z) = F(z)/G(z)$ using polynomials $F, G \in K[z]$ having no common factors, and let $v$ be an absolute value on $K$. Then there is a unique function*

$$\hat{\lambda}_{\phi,v} : \mathbb{P}^1(\bar{K}_v) \smallsetminus \{\infty\} \longrightarrow \mathbb{R}$$

*with the following two properties:*

(a) *For all $\alpha \in \mathbb{P}^1(\bar{K}_v)$ with $\alpha \neq \infty$ and $\phi(\alpha) \neq \infty$, the function $\hat{\lambda}_{\phi,v}$ satisfies*

$$\hat{\lambda}_{\phi,v}(\phi(\alpha)) = d\hat{\lambda}_{\phi,v}(\alpha) - \log\bigl|G(\alpha)\bigr|_v. \tag{3.16}$$

(b) *The function*

$$\alpha \longmapsto \hat{\lambda}_{\phi,v}(\alpha) - \log \max\{|\alpha|_v, 1\}$$

*extends to a bounded continuous function on all of $\mathbb{P}^1(\bar{K}_v)$.*

*The function* $\hat{\lambda}_{\phi,v}$ *is called a* local canonical height (associated) to $\phi$.

*Proof.* We defer the proof until Section 5.9; see Theorem 5.60. □

*Remark* 3.28. The *local canonical height function* $\hat{\lambda}_{\phi,v}$ constructed in Theorem 3.27 depends on the choice of a decomposition of $\phi$ as $\phi = F/G$. If we use instead $cF/cG$ for some $c \in K^*$, so $G$ is replaced by $cG$, then it is easy to see that the new function differs from the old one by a constant,

$$\hat{\lambda}_{\phi,v,cG}(\alpha) = \hat{\lambda}_{\phi,v,G}(\alpha) + \frac{1}{d-1}\log|c|_v \qquad \text{for all } \alpha \in \mathbb{P}^1(\bar{K}_v) \smallsetminus \{\infty\}.$$

In the sequel, when we refer to $\hat{\lambda}_{\phi,v}$ without further specification, we assume that some particular $G$ has been fixed. However, we note that there are situations in which it may be convenient to normalize $\hat{\lambda}_{\phi,v}$ differently for different $v$; see Exercise 3.29.

The utility of local canonical heights is that on the one hand, they are defined on $\mathbb{P}^1(K_v)$ and have various nice metrical and analytic properties, while on the other hand, they fit together to give the global canonical height, as described in the next theorem.

**Theorem 3.29.** *Let $K$ be a number field, and for each $v \in M_K$, let $\hat{\lambda}_{\phi,v}$ be the local canonical height function constructed in Theorem* 3.27. *Then*

$$\hat{h}_\phi(\alpha) = \sum_{v \in M_K} n_v \hat{\lambda}_{\phi,v}(\alpha) \qquad \textit{for all } \alpha \in \mathbb{P}^1(K) \smallsetminus \{\infty\}. \tag{3.17}$$

*Proof.* We defer the proof until Section 5.9; see Theorem 5.61. □

*Remark* 3.30. If $\phi(z) \in K[z]$ is a polynomial, then the local height may be computed as the limit (Exercise 3.24)

$$\hat{\lambda}_{\phi,v}(\alpha) = \lim_{n\to\infty} \frac{1}{d^n}\log\max\{|\phi^n(\alpha)|_v, 1\}.$$

If $v$ is a nonarchimedean absolute value and if $\phi$ has good reduction at $v$, or more precisely, if $\phi = F/G$ with $|\operatorname{Res}(F,G)|_v = 1$, then the local height is given by the simple formula (Exercise 3.30)

$$\hat{\lambda}_{\phi,v}(\alpha) = \log\max\{|\alpha|_v, 1\}.$$

In general there is no simple limit formula to compute $\hat{\lambda}_{\phi,v}(\alpha)$. However, there is an algorithm that leads to a rapidly convergent series for $\hat{\lambda}_{\phi,v}(\alpha)$. Roughly, the $N^{\text{th}}$ partial sum of the series approximates $\hat{\lambda}_{\phi,v}(\alpha)$ to within $O(d^{-N})$. For details see Exercise 5.29, which describes an efficient algorithm to compute the Green function $\mathcal{G}_\phi$, and then use Theorem 5.60, which says that the local canonical height and the Green function are related by the formula

$$\hat{\lambda}_{\phi,v}([x,y]) = \mathcal{G}_{\phi,v}(x,y) - \log|y|_v.$$

*Remark* 3.31. Baker and Rumely [26, Chapter 7] give an interesting interpretation of the local canonical height function. They construct an invariant measure on Berkovich space and prove that the measure is the negative Laplacian of a suitable extension of the local canonical height function $\hat{\lambda}_{\phi,v}$. See Section 5.10 for a brief introduction to the geometry/topology of Berkovich space. For the more elaborate machinery required to do harmonic and functional analysis on Berkovich space, including the construction of the invariant measure, we refer the reader to [26, 29] and to the other references listed in Remark 5.77.

*Remark* 3.32. The theory of local canonical heights began with Néron's construction on abelian varieties [333], or see [256, Chapter 11]. The general theory of global and local canonical heights associated to morphisms on varieties with eigendivisor classes is described in [88].

## 3.6 Diophantine Approximation

The theory of Diophantine approximation seeks to answer the question of how closely one can approximate irrational numbers by rational numbers. The subject now includes a large and well-developed body of knowledge, while at the same time there is considerable ongoing research on many deep questions. In this section we state a famous theorem on Diophantine approximation and show how it is applied to deduce finiteness results for certain Diophantine equations. In Sections 3.7 and 3.8 we apply the theory of Diophantine approximation in a similar fashion to deduce integrality properties of points in orbits $\mathcal{O}_\phi(\alpha)$.

It is clear that any irrational number $\alpha \in \mathbb{R}$ can be approximated arbitrarily closely by rational numbers, since $\mathbb{Q}$ is dense in $\mathbb{R}$. The following elementary result quantifies this observation by relating the closeness of $x/y$ and $\alpha$ to the arithmetic complexity of the rational number $x/y$. We leave the proof as an exercise (Exercise 3.31), or see any elementary number theory text that discusses Diophantine approximation.

**Proposition 3.33.** (Dirichlet) *Let $\alpha \in \mathbb{R} \smallsetminus \mathbb{Q}$ be an irrational number. Then there are infinitely many rational numbers $x/y$ satisfying*

$$\left|\frac{x}{y} - \alpha\right| \le \frac{1}{y^2}.$$

Dirichlet's theorem says that every irrational number can be fairly well approximated by rational numbers. Some irrational numbers can be much better approximated, but it turns out that this is not true for algebraic numbers. A succession of mathematicians (Liouville, Thue, Siegel, Gel'fond, Dyson) derived ever better estimates for the approximability of algebraic numbers by rational numbers, culminating in the following deep theorem of Roth, for which he received the Fields Medal in 1958.

**Theorem 3.34.** (Roth) *Fix $\epsilon > 0$ and let $\alpha \in \bar{\mathbb{Q}}$ be an algebraic number with $\alpha \notin \mathbb{Q}$. Then there exists a constant $\kappa = \kappa(\epsilon, \alpha) > 0$ such that*

$$\left|\frac{x}{y} - \alpha\right| \geq \frac{\kappa}{|y|^{2+\epsilon}} \qquad \textit{for all } \frac{x}{y} \in \mathbb{Q}.$$

*Proof.* The proof of Roth's theorem is beyond the scope of this book. A nice exposition may be found in [393]. For more general versions, see [205, Part D] and [256, Chapter 7]. □

We apply Roth's theorem to prove an important result of Thue on the representability of integers by binary forms.

**Theorem 3.35.** (Thue) *Let $G(X,Y) \in \mathbb{Z}[X,Y]$ be a homogeneous polynomial of degree d and let $B \in \mathbb{Z}$. Assume that $G(X,Y)$ has at least three distinct roots in $\mathbb{P}^1(\mathbb{C})$. Then the equation*

$$G(X,Y) = B \tag{3.18}$$

*has only finitely many integer solutions $(X,Y) \in \mathbb{Z}^2$.*

*Proof.* We begin by factoring $G(X,Y)$ into irreducible factors in $\mathbb{Q}[X,Y]$, say

$$G(X,Y) = G_1(X,Y)^{e_1} G_2(X,Y)^{e_2} \cdots G_r(X,Y)^{e_r}.$$

Replacing $G(X,Y)$ by $cG(X,Y)$ and $B$ by $cB$ for a sufficiently large integer $c$, we may assume without loss of generality that $G_1,\ldots,G_r$ have integer coefficients.[1]

If $r \geq 2$, that is, if $G(X,Y)$ has more than one irreducible factor, then any solution $(a,b)$ to (3.18) has the property that $G_1(a,b)$ and $G_2(a,b)$ divide $B$. Since $B$ has only finitely many distinct factors, we are reduced in this case to showing that there are only finitely many solutions to the pair of simultaneous equations

$$G_1(X,Y) = B_1 \qquad \text{and} \qquad G_2(X,Y) = B_2.$$

This is clear, since the plane curves $G_1 = B_1$ and $G_2 = B_2$ have no common components, so they intersect in only finitely many points.

Next suppose that $r = 1$, so $G(X,Y) = G_1(X,Y)^{e_1}$. If $B$ is not equal to $B_1^{e_1}$ for some integer $B_1$, then $G(X,Y) = B$ has no solutions. Otherwise, we can take roots of both sides and reduce to the case that $G(X,Y)$ is irreducible in $\mathbb{Q}[X,Y]$. Then $G(X,Y)$ has only simple roots, and by assumption there are at least three such roots.

We factor $G(X,Y)$ in $\mathbb{C}[X,Y]$, say

$$G(X,Y) = a(X - \alpha_1 Y)(X - \alpha_2 Y) \cdots (X - \alpha_d Y)$$

with distinct algebraic numbers $\alpha_1,\ldots,\alpha_d$. Now divide the equation $G(X,Y) = b$ by $Y^d$ to obtain

$$\left(\frac{X}{Y} - \alpha_1\right)\left(\frac{X}{Y} - \alpha_2\right) \cdots \left(\frac{X}{Y} - \alpha_d\right) = \frac{b/a}{Y^d}. \tag{3.19}$$

[1] Or we can invoke Gauss's lemma, which implies that the factorization of $G(X,Y)$ in $\mathbb{Q}[X,Y]$ can already be achieved in $\mathbb{Z}[X,Y]$.

We observe that if $(x, y) \in \mathbb{Z}^2$ is a large solution to $G(X, Y) = b$, then the righthand side of (3.19) is small, so at least one of the factors on the lefthand side must also be small. However, since $\alpha_1, \dots, \alpha_d$ are distinct, the rational number $x/y$ can be close to at most one of $\alpha_1, \dots, \alpha_d$. Hence we can find a constant $C = C(G, b)$ such that

$$\min_{1 \le i \le d} \left| \frac{x}{y} - \alpha_i \right| \le \frac{C}{|y|^d} \qquad \text{for all } (x, y) \in \mathbb{Z}^2 \text{ satisfying } G(x, y) = b. \tag{3.20}$$

In the other direction, Roth's Theorem 3.34 applied to each of $\alpha_1, \dots, \alpha_d$ tells us that for any $\epsilon > 0$ there is a constant $\kappa = \kappa(\epsilon, \alpha_1, \dots, \alpha_d) > 0$ such that

$$\min_{1 \le i \le d} \left| \frac{x}{y} - \alpha_i \right| \ge \frac{\kappa}{|y|^{2+\epsilon}} \qquad \text{for all } \frac{x}{y} \in \mathbb{Q}. \tag{3.21}$$

Combining (3.20) and (3.21), we find that

$$|y|^{d-2-\epsilon} \le C/\kappa.$$

By assumption, $d \ge 3$, so this shows that there are only finitely many possible values for $y$. Finally, we observe that for each $y$, the equation $G(x, y) = b$ implies that there are at most $d$ possible values for $x$. Therefore the equation $G(x, y) = b$ has only finitely many integer solutions □

Siegel observed that Thue's theorem can be reformulated in terms of integral values of rational functions.

**Theorem 3.36.** (Siegel) *Let $\phi(z) \in \mathbb{Q}(z)$ be a rational function with at least three distinct poles in $\mathbb{P}^1(\mathbb{C})$. Then*

$$\{\alpha \in \mathbb{Q} : \phi(\alpha) \in \mathbb{Z}\}$$

*is a finite set.*

*Remark* 3.37. Note that Theorem 3.36 need not be true if the rational function $\phi$ has fewer than three poles. A simple example with two poles is the function

$$\phi(z) = \frac{F(z)}{(z^2 - D)^d},$$

where $D > 1$ is a squarefree integer and $F(z) \in \mathbb{Z}[z]$ is a polynomial of degree $2d$. If we now take any solution $(u, v) \in \mathbb{Z}^2$ to the Pell equation $u^2 - Dv^2 = 1$, then $\phi(u/v) = v^{2d} F(u/v) \in \mathbb{Z}$. The Pell equation has infinitely many solutions, so there are infinitely many rational numbers $u/v \in \mathbb{Q}$ with $\phi(u/v) \in \mathbb{Z}$.

*Proof.* Write

$$\phi = [F(X, Y), G(X, Y)]$$

using homogeneous polynomials $F(X, Y), G(X, Y) \in \mathbb{Z}[X, Y]$ of degree $d$ having no common factors. For any fraction $\alpha = a/b \in \mathbb{Q}$ written in lowest terms, we have

$$\phi(\alpha) = \frac{F(a,b)}{G(a,b)},$$

so $\phi(\alpha) \in \mathbb{Z}$ if and only if $G(a,b)$ divides $F(a,b)$.

Let $R = \operatorname{Res}(F,G)$ be the resultant of $F$ and $G$, which is a nonzero integer since $F$ and $G$ have no common factors. Proposition 2.13 says that there are homogeneous polynomials $f_1, g_1, f_2, g_2 \in \mathbb{Z}[X,Y]$ satisfying

$$f_1(X,Y)F(X,Y) + g_1(X,Y)G(X,Y) = RX^{2d-1},$$
$$f_2(X,Y)F(X,Y) + g_2(X,Y)G(X,Y) = RY^{2d-1}.$$

Substituting $(X,Y) = (a,b)$ into these equations, we see that if $G(a,b)$ divides $F(a,b)$, then $G(a,b)$ also divides both $Ra^{2d-1}$ and $Rb^{2d-1}$. However, $a$ and $b$ are relatively prime, so we have proven that

$$\phi(a/b) \in \mathbb{Z} \quad \text{implies that} \quad G(a,b) \text{ divides } R.$$

It is important to emphasize that the resultant $R$ depends only on $F$ and $G$ and that it is nonzero. Hence

$$\left\{\frac{a}{b} \in \mathbb{Q} : \phi\left(\frac{a}{b}\right) \in \mathbb{Z}\right\} \subset \bigcup_{D \mid R} \left\{\frac{a}{b} \in \mathbb{Q} : G(a,b) = D\right\}. \tag{3.22}$$

Thue's Theorem 3.35 tells us that each set on the righthand side of (3.22) is finite, which completes the proof of Theorem 3.36. □

*Remark* 3.38. Roth's Theorem 3.34 is not effective, in the sense that it does not provide a method for computing an allowable constant $\kappa(\epsilon, \alpha)$ in terms of $\epsilon$ and $\alpha$. Thus our proofs of Thue's and Siegel's theorems (Theorems 3.35 and 3.36) are also ineffective. Baker's theorem on linear forms in logarithms gives an effective, although weak, version of Roth's theorem, from which one can derive effective versions of Theorems 3.35 and 3.36. This means that our finiteness result on integral points in orbits (Theorem 3.43) proven in the next section can be made effective, but the stronger integrality statement (Theorem 3.48) cannot, since it relies on the full strength of Roth's theorem.

Let $C \subset \mathbb{P}^N$ be a smooth projective curve defined over $\mathbb{Q}$ and let $\phi : C \to \mathbb{P}^1$ be a nonconstant rational function on $C$. If $C$ has genus $g \geq 1$, Siegel proved that $C_\phi(\mathbb{Z}) = \{P \in C(\mathbb{Q}) : \phi(P) \in \mathbb{Z}\}$ is a finite set. For $g \geq 2$, this is superseded by Faltings' theorem that $C(\mathbb{Q})$ is finite, but for elliptic curves ($g = 1$), the set $C(\mathbb{Q})$ is often infinite. In this case, Siegel greatly strengthened the qualitative statement that $C_\phi(\mathbb{Z})$ is finite by showing that the coordinates of the points in $C(\mathbb{Q})$ have numerators and denominators of approximately the same size. The precise theorem, which we generalize to the dynamical setting in Section 3.8, is as follows.

**Theorem 3.39.** (Siegel [403, 404]) *Let $E/\mathbb{Q}$ be an elliptic curve, let $\phi \in \mathbb{Q}(E)$ be a nonconstant rational function on $E$, and for each rational point $P \in E(\mathbb{Q})$, write*

$$\phi(P) = [a(P), b(P)] \in \mathbb{P}^1(\mathbb{Q}) \quad \textit{with } a(P), b(P) \in \mathbb{Z} \textit{ and } \gcd(a(P), b(P)) = 1.$$

*Then*

$$\lim_{\substack{P \in E(\mathbb{Q}) \\ h(\phi(P)) \to \infty}} \frac{\log|a(P)|}{\log|b(P)|} = 1.$$

*Proof.* See [410, IX.3.3]. For a general version on curves of arbitrary genus $g \geq 1$, see for example [205, D.9.4], although as noted above, if $g \geq 2$, then Faltings [164, 165] proves that $C(\mathbb{Q})$ is finite. □

For the convenience of the reader, we conclude this section by stating general versions of Theorems 3.34, 3.35, and 3.36 for number fields. Proofs of these, and even more general, results may be found in [205, Part D] or [256, Chapter 7]. We set the following notation:

| | |
|---|---|
| $K$ | a number field; |
| $S$ | a finite set of absolute values on $K$; |
| $R_S$ | the ring of $S$-integers of $K$. |

**Theorem 3.40.** (Roth) *Let $\epsilon > 0$. For each $v \in S$, extend $v$ to $\bar{K}$ in some fashion, and choose an algebraic number $\alpha_v \in \bar{K}$. Then there is a constant $\kappa > 0$, depending on $K$, $S$, $\epsilon$, and $\{\alpha_v\}_{v\in S}$, such that*

$$\prod_{v\in S} \min\{|z - \alpha_v|_v, 1\}^{n_v} \geq \frac{\kappa}{H_K(z)^{2+\epsilon}} \qquad \textit{for all } z \in K.$$

**Theorem 3.41.** (Thue–Mahler) *Let $G(X,Y) \in K[X,Y]$ be homogeneous of degree $d$ with at least three distinct roots in $\mathbb{P}^1(\mathbb{C})$, and let $B \in K$. Then there are only finitely many $(X,Y) \in R_S^2$ satisfying $G(X,Y) = B$.*

**Theorem 3.42.** (Siegel) *Let $\phi(z) \in K(z)$ be a rational function with at least three distinct poles in $\bar{K}$. Then there are only finitely many $\alpha \in K$ satisfying $\phi(\alpha) \in R_S$.*

## 3.7 Integral Points in Orbits

In this section we prove that the orbit of a rational point by a rational map contains only finitely many integers, except in those cases in which that statement is clearly false. For ease of exposition, we work with $\mathbb{Q}$ and $\mathbb{Z}$, but the result holds quite generally for rings of $S$-integers in number fields; see [411] or Exercise 3.38. We begin with a definition.

**Definition.** A *wandering point* for a rational map $\phi : \mathbb{P}^1 \to \mathbb{P}^1$ is a point $P \in \mathbb{P}^1$ whose forward orbit $\mathcal{O}_\phi(P)$ is an infinite set. Thus every point is either a wandering point or a preperiodic point.[2]

[2]The reader should be aware that in topological dynamics, especially with respect to invertible maps, the standard definition says that a point is wandering if it has a wandering neighborhood. In other words, a point $P$ is *topologically wandering* if there is a neighborhood $U$ of $P$ and an integer $n_0$ such that $\phi^i(U) \cap \phi^j(U) = \emptyset$ for all $i > j \geq n_0$. Our definition coincides with the topological definition if we use the discrete topology.

**Theorem 3.43.** *Let $\phi(z) \in \mathbb{Q}(z)$ be a rational map of degree $d \geq 2$ with the property that $\phi^2(z) \notin \mathbb{Q}[z]$. Let $\alpha \in \mathbb{Q}$ be a wandering point for $\phi$. Then the orbit*

$$\mathcal{O}_\phi(\alpha) = \{\alpha, \phi(\alpha), \phi^2(\alpha), \phi^3(\alpha), \ldots\}$$

*contains only finitely many integer points.*

Theorem 3.43 is an immediate consequence of the following elementary geometric result combined with Siegel's Theorem 3.36 concerning integral values of rational functions.

**Proposition 3.44.** *Let $\phi \in \mathbb{C}(z)$ be a rational function of degree $d \geq 2$ satisfying $\phi^2(z) \notin \mathbb{C}[z]$, so Theorem 1.7 implies that no iterate of $\phi$ is a polynomial map. Then*

$$\#\phi^{-4}(\infty) \geq 3.$$

*If $d \geq 3$, then $\#\phi^{-3}(\infty) \geq 3$.*

*Proof.* We give a pictorial proof using the Riemann–Hurwitz formula. Suppose that $\phi$ is a rational map with $\#\phi^{-3}(\infty) \leq 2$. There are four possible pictures for the backward orbit of $\infty$, as illustrated in Figure 3.1.

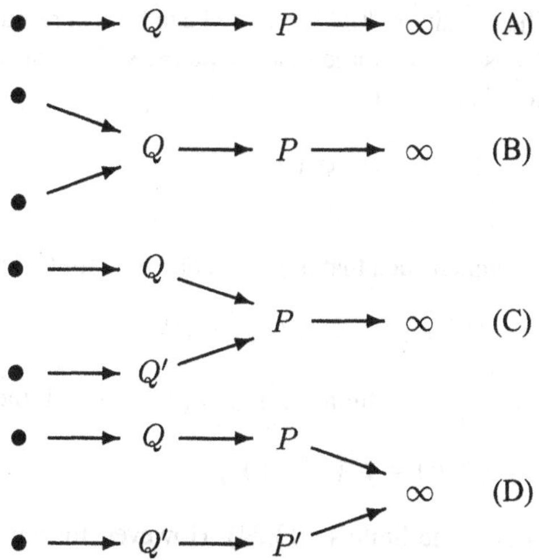


Figure 3.1: Backward orbits containing few points

The weak form of the Riemann–Hurwitz formula (Corollary 1.3) says that

$$2d - 2 = \sum_{P \in \mathbb{P}^1} \bigl(d - \#\phi^{-1}(P)\bigr). \tag{3.23}$$

We apply (3.23) to each of the four pictures in Figure 3.1. Before computing, we need to determine to what extent the points in the four pictures are distinct. For (A) and (B), the three points $P$, $Q$, and $\infty$ must be distinct, since $P = \infty$ would mean that $\phi$ is a polynomial and $Q = \infty \neq P$ would mean that $\phi^2$ is a polynomial. Similarly, in (C) we cannot have $P = \infty$, so relabeling the points in (C) and (D) if necessary, we may assume that $P$, $Q$, and $\infty$ are distinct in all four cases. We now use the Riemann–Hurwitz formula to estimate

$$2d-2 \geq \bigl(d - \#\phi^{-1}(\infty)\bigr) + \bigl(d - \#\phi^{-1}(P)\bigr) + \bigl(d - \#\phi^{-1}(Q)\bigr)$$
$$= \begin{cases} (d-1)+(d-1)+(d-1) = 3d-3 & \text{in case (A),} \\ (d-1)+(d-1)+(d-2) = 3d-4 & \text{in case (B),} \\ (d-1)+(d-2)+(d-1) = 3d-4 & \text{in case (C),} \\ (d-2)+(d-1)+(d-1) = 3d-4 & \text{in case (D).} \end{cases}$$

Thus case (A) yields $d \leq 1$, a contradiction, while the other three cases give $d \leq 2$. This proves that if $d \geq 3$, then $\#\phi^{-3}(\infty) \geq 3$. Finally, if $d = 2$, then the Riemann–Hurwitz formula tells us that $\phi$ has only two ramified points, and by inspection we see that those points already appear in the pictures for (B), (C), and (D). Since $\#\phi^{-3}(\infty) = 2$ in each case, it follows that $\#\phi^{-4}(\infty) \geq 3$. (See Exercise 3.37 for an explicit lower bound for $\#\phi^{-n}(\infty)$ in terms of $d$ and $n$.) □

*Proof of Theorem* 3.43. Proposition 3.44 tells us that $\#\phi^{-4}(\infty) \geq 3$. (If $d \geq 3$, we could even take $\phi^{-3}$.) The condition $\#\phi^{-4}(\infty) \geq 3$ can be rephrased as saying that the rational function $\phi^4$ has at least three distinct poles, so we may apply Siegel's Theorem 3.36 to conclude that the set

$$\{\beta \in \mathbb{Q} : \phi^4(\beta) \in \mathbb{Z}\} \tag{3.24}$$

is finite.

Consider the set of integers $n$ such that the $n^{\text{th}}$ iterate of $\phi$ applied to $\alpha$ is integral,

$$N_\alpha = \{n \geq 0 : \phi^n(\alpha) \in \mathbb{Z}\}.$$

It clearly suffices to prove that $N_\alpha$ is finite. If $n \in N_\alpha$ with $n \geq 4$, then

$$\phi^n(\alpha) = \phi^4\bigl(\phi^{n-4}(\alpha)\bigr) \in \mathbb{Z},$$

so we see that $\phi^{n-4}(\alpha)$ is in the finite set (3.24). However, for any fixed $\beta$ in the set (3.24), there is at most one iterate of $\alpha$ that is mapped to $\beta$ (remember that $\alpha$ is a wandering point). This completes the proof that $\mathcal{O}_\phi(\alpha) \cap \mathbb{Z}$ is finite. □

Theorem 3.43 says that if $\phi^2(z)$ is not a polynomial, then the orbit $\mathcal{O}_\phi(\alpha)$ contains only finitely many integral points. It is natural to ask how large $\mathcal{O}_\phi(\alpha) \cap \mathbb{Z}$ can be. The answer is that it may be arbitrarily large, as shown by the following construction.

*Example* 3.45. In order to create a rational function whose orbit contains a large number of integral points, we simply take a finite (reasonably random) sequence of integers $z_0, z_2, \ldots, z_{k-1}$ and treat the equations

$$\phi^{n+1}(z_n) = z_{n+1} \quad \text{for } n = 0, 1, \ldots, k-1$$

as a system of homogeneous linear equations for the $2d+2$ coefficients of $\phi$. If the degree of $\phi$ satisfies $2d+1 \geq k$, then there should be a nontrivial rational solution. We carry out this procedure with $d = 2$ to find the rational function

$$\phi(z) = \frac{899x^2 - 2002x + 275}{33x^2 - 584x + 275}$$

such that the orbit of 0 contains quite a few integer points:

$$0 \xrightarrow{\phi} 1 \xrightarrow{\phi} 3 \xrightarrow{\phi} -2 \xrightarrow{\phi} 5 \xrightarrow{\phi} -7 \xrightarrow{\phi} \cdots$$

Of course, as Theorem 3.43 predicts, the list of integer points must end, and indeed the next few iterates make it seem likely that there are no further integral points in the orbit of 0. Thus

$$-7 \xrightarrow{\phi} \frac{2917}{299} \xrightarrow{\phi} -\frac{296398306}{10198813} \xrightarrow{\phi} \frac{17011876043966969359}{938619769242091763} \xrightarrow{\phi} \cdots$$

Example 3.45 (see also Exercise 3.41) shows how to construct orbits with many integer points by using rational maps of large degree. The following trick, adapted from work of Chowla [103] and Mahler [285] on elliptic curves (see also [405]), says that $\mathcal{O}_\phi(\alpha) \cap \mathbb{Z}$ may be arbitrarily large even for rational maps of a fixed degree.

**Proposition 3.46.** *For all integers $N \geq 0$ and $d \geq 2$ there exists a rational map $\phi(z) \in \mathbb{Q}(z)$ with the following properties:*

- $\phi^2(z) \notin \mathbb{C}[z]$.
- *0 is a wandering point for $\phi$.*
- $0, \phi(0), \phi^2(0), \phi^3(0), \ldots, \phi^N(0) \in \mathbb{Z}$.

*Proof.* Let $\psi(z) \in \mathbb{Q}(z)$ be any rational map of degree $d$ for which 0 is not a preperiodic point. For each $0 \leq n \leq N$, write

$$\psi^n(0) = \frac{a_n}{b_n} \in \mathbb{Q}$$

as a fraction in lowest terms, and let $B = b_0 b_1 \cdots b_N$. Consider the rational function $\phi(z) = B\psi(z/B)$. Clearly $\phi^n(z) = B\psi^n(z/B)$, so for $0 \leq n \leq N$ we have

$$\phi^n(0) = B\psi^n(0) = b_0 b_1 \cdots b_{n-1} a_n b_{n+1} \cdots b_N \in \mathbb{Z}.$$

Hence $\phi^n(0) \in \mathbb{Z}$ for all $0 \leq n \leq N$. □

Although Proposition 3.46 shows how to make $\mathcal{O}_\phi(\alpha) \cap \mathbb{Z}$ arbitrarily large, it is in some sense a cheat. What is really happening is that we are clearing the denominators of rational points in an orbit. We can use the following notion of minimal resultant to rule out this behavior.

**Definition.** For a rational map $\phi(z) \in \mathbb{Q}(z)$, write $\phi(z)$ as $\phi(z) = F_\phi(z)/G_\phi(z)$, where $F_\phi, G_\phi \in \mathbb{Z}[z]$ have integer coefficients and the greatest common divisor of all of their coefficients is 1. Then $F_\phi$ and $G_\phi$ are uniquely determined by $\phi$ up to multiplication by $\pm 1$, and we define the *resultant of* $\phi$ to be the quantity

$$\operatorname{Res}(\phi) = \bigl|\operatorname{Res}(F_\phi, G_\phi)\bigr|.$$

(See Section 2.4 for the definition and basic properties of the resultant of two polynomials. See also Section 4.11 for a general discussion of minimal resultants of rational maps over number fields.)

We can use the resultant to rule out the denominator-clearing trick of Proposition 3.46, and having done so, we conjecture that the number of integer points in an orbit is bounded solely in terms of the degree of the map. (See Theorem 6.70 for a special case.) This is a dynamical analogue of a conjecture of Lang [254, page 140]; see also [202, 407].

**Conjecture 3.47.** *Let $\phi(z) \in \mathbb{Q}(z)$ be a rational map of degree $d \geq 2$ with $\phi^2(z) \notin \mathbb{Q}[z]$, and let $\alpha \in \mathbb{P}^1(\mathbb{Q})$ be a wandering point for $\phi$. Assume further that $\phi$ is* affine minimal *in the sense that*

$$\operatorname{Res}(\phi) = \min_{\substack{f \in \mathrm{PGL}_2(\mathbb{Q}) \\ f(z)=az+b}} \operatorname{Res}(\phi^f).$$

*(In other words, we cannot make the resultant smaller by conjugating by an affine linear transformation $az+b$.) Then there is a constant $C = C(d)$ depending only on the degree of $\phi$ such that*

$$\#\bigl(\mathcal{O}_\phi(\alpha) \cap \mathbb{Z}\bigr) \leq C.$$

## 3.8 Integrality Estimates for Points in Orbits

Theorem 3.43 says that, except for the obvious counterexamples, an orbit $\mathcal{O}_\phi(\alpha)$ contains only finitely many integer points. In this section we prove a dynamical analogue of Siegel's Theorem 3.39, which asserts that the numerators and denominators of certain rational numbers have approximately the same number of digits. Siegel's proof of Theorem 3.39 uses the existence of the multiplication-by-$m$ maps $[m] : E \to E$ on an elliptic curve. These finite unramified maps have the effect of significantly increasing the height of points while leaving distances relatively unchanged. We adapt Siegel's argument to the dynamical setting, but since our maps are on $\mathbb{P}^1$, they are always ramified. This causes some additional complications, since distances shrink significantly near ramification points.

**Theorem 3.48.** (Silverman [411]) *Let $\phi(z) \in \mathbb{Q}(z)$ be a rational map with the property that $\phi^2(z) \notin \mathbb{Q}[z]$ and $1/\phi^2(1/z) \notin \mathbb{Q}[z]$. Let $\alpha \in \mathbb{Q}$ be a wandering point for $\phi$, and write*

$$\phi^n(\alpha) = \frac{a_n}{b_n}$$

*as a fraction in lowest terms. Then*

$$\lim_{n\to\infty} \frac{\log|a_n|}{\log|b_n|} = 1. \tag{3.25}$$

*Remark* 3.49. It is clear why we must assume that $\phi^2(z) \notin \mathbb{Q}[z]$ in the statement of Theorem 3.48, but it may be less clear why we also require that $1/\phi^2(1/z) \notin \mathbb{Q}[z]$. Letting $f(z) = 1/z$, we observe that

$$1/\phi^k(1/z) = (f^{-1} \circ \phi^k \circ f)(z) = (f^{-1} \circ \phi \circ f)^k(z) = (\phi^f)^k(z).$$

So if $1/\phi^2(1/z) \in \mathbb{Z}[z]$, then $(\phi^f)^{2n}(z) \in \mathbb{Z}[z]$ for all $n \geq 1$. Now consider the orbit of $\alpha = 1/b$ for an integer $b$. We have

$$\phi^{2n}(1/b) = \phi^{2n}(f(b)) = f^{-1}\big((\phi^f)^{2n}(b)\big) = \frac{1}{(\phi^f)^{2n}(b)}.$$

The quantity $(\phi^f)^{2n}(b)$ is an integer, so $a_{2n} = 1$ and $b_{2n} = (\phi^f)^{2n}(b) \in \mathbb{Z}$. Hence the limit in (3.25) for even values of $n$ is 0. Thus the assertion of Theorem 3.48 is not true for rational maps $\phi(z)$ such that $1/\phi^2(1/z)$ is a polynomial.

*Example* 3.50. To illustrate Theorem 3.48, we take $\phi(z) = z + 1/z$ and $\alpha = 1$ and list the first few values of $\phi^n(1) = a_n/b_n$ in Table 3.1.

| $n$ | $a_n$ | $b_n$ | $\frac{\log(a_n)}{\log(b_n)}$ |
|---|---:|---:|---|
| 0 | 1 | 1 | — |
| 1 | 2 | 1 | 0.69315 |
| 2 | 5 | 2 | 1.60943 |
| 3 | 29 | 10 | 1.46239 |
| 4 | 941 | 290 | 1.20759 |
| 5 | 969581 | 272890 | 1.10128 |
| 6 | 1014556267661 | 264588959090 | 1.05110 |
| 7 | 1099331737522548368039021 | 268440386798659418988490 | 1.02613 |

Table 3.1: Orbit $\mathcal{O}_1(\phi)$ for $\phi(z) = z + 1/z$, writing $\phi^n(1) = a_n/b_n$.

Notice how both the numerator and denominator of $\phi^n(1)$ grow extremely rapidly. Of course, the elementary height estimate (Theorem 3.11) tells us that the maximum of $|a_n|$ and $|b_n|$ grows this rapidly, but the fact that they both grow at approximately an equal rate lies much deeper and is the content of Theorem 3.48. And to illustrate the speed with which the fractions grow, even for a very simple map of degree 2 such as $\phi(z) = z + 1/z$, here is the exact value of $\phi^9(1)$, which is the last value that fits on one line using very small (5 point) type:

$$\frac{1726999038066943724857508638586386504281539279376091034086485112150121338989977841573308941492781}{377908709746050392481071609609580527436122569261424131112048023467330784739529329885668846964890}.$$

This ninth iterate has logarithmic ratio

$$\frac{\log(a_n)}{\log(b_n)} = \frac{\log(17269990380\ldots492781)}{\log(37790870974\ldots964890)} \approx 1.00690.$$

*Proof.* The idea underlying the proof of Theorem 3.48 is fairly simple. Choose some $\epsilon > 0$, and suppose that

$$|a_n| \ge |b_n|^{1+\epsilon} \quad \text{for infinitely many } n \ge 0. \tag{3.26}$$

This means that $\phi^n(\alpha) = a_n/b_n$ is very large, so $\phi^n(\alpha)$ is close to $\infty$. It follows that $\alpha$ is quite close to one of the points in the inverse image $\phi^{-n}(\infty)$, say $\alpha$ is close to $\beta \in \phi^{-n}(\infty)$. But $\alpha$ is in $\mathbb{Q}$, so one can hope that if $n$ is sufficiently large, then $\alpha$ and $\beta$ are so close to one another that they contradict Roth's Theorem 3.34. Unfortunately, this naive approach does not work, because the point $\beta$ depends on $n$, so the constant in Roth's theorem changes with each new value of $n$.

A more sophisticated idea is to use a fixed (large) integer $m$ and apply Roth's theorem to the rational point $\phi^{n-m}(\alpha)$ and a nearby point in $\phi^{-m}(\infty)$. Note that

$$h(\phi^{n-m}(\alpha)) = \frac{1}{d^m} h(\phi^n(\alpha)) + O(1),$$

so the height of $\phi^{n-m}(\alpha)$ is much smaller than the height of $\phi^n(\alpha)$.

We fix an integer $m$ satisfying $d^m > 6/\epsilon$ (we will see later why this is a good choice for $m$) and we let $\beta$ be the point in $\phi^{-m}(\infty)$ that is closest to $\phi^{n-m}(\alpha)$. How close is $\beta$ to $\alpha$? It turns out that this depends on the ramification index of $\phi$ at various points. If we make the (incorrect) assumption that $\phi$ is everywhere unramified, then $\phi$ preserves distances up to a scaling factor, which allows us to make the following estimates, where the constants $C_1, C_2, \ldots$ may depend on $\phi$, $\alpha$, and $m$, but they do not depend on $n$:

$$\begin{aligned}
\frac{1}{|a_n|^\epsilon} \ge \left|\frac{b_n}{a_n}\right| &= |\phi^n(\alpha)|^{-1} && \text{since } |a_n| \ge |b_n|^{1+\epsilon} \text{ from (3.26)},\\
&\approx C_1\rho(\phi^n(\alpha), \infty), && \text{definition of } \rho, \text{ since } |\phi^n(\alpha)| > 1,\\
&= C_1\rho(\phi^n(\alpha), \phi^m(\beta)), && \text{since } \phi^m(\beta) = \infty,\\
&\approx C_2\rho(\phi^{n-m}(\alpha), \beta), && \text{assuming } \phi \text{ is unramified},\\
&\approx C_3|\phi^{n-m}(\alpha) - \beta|, && \text{definition of } \rho,\\
&\ge \frac{C_4}{H(\phi^{n-m}(\alpha))^3}, && \text{Roth's Theorem 3.34 (with exponent 3)},\\
&\approx \frac{C_5}{H(\phi^n(\alpha))^{3/d^m}}, && \text{property of heights (Theorem 3.11)},\\
&= \frac{C_5}{|a_n|^{3/d^m}}, && \text{since } |a_n| \ge |b_n|,
\end{aligned}$$

$$= \frac{C_5}{|a_n|^{\epsilon/2}}, \qquad \text{since } m \text{ satisfies } d^m > 6/\epsilon.$$

Taking logarithms, we have proven that

$$\log|a_n| \le \frac{2}{\epsilon}\log(C_5) \qquad \text{for all } n \text{ satisfying (3.26), i.e., } |a_n| \ge |b_n|^{1+\epsilon}.$$

We reiterate that the constant $C_5$ depends only on $\phi$, $\alpha$, and $m$; it does not depend on $n$. Hence if $n$ satisfies (3.26), then $a_n$, and also $b_n$, are bounded. It follows that the $\phi$-orbit of $\alpha$ contains only finitely many points satisfying (3.26), and therefore

$$\limsup_{n\to\infty} \frac{\log|a_n|}{\log|b_n|} \le 1+\epsilon. \tag{3.27}$$

Repeating the argument with the rational map $1/\phi(1/z)$ and the initial point $1/\alpha$ yields

$$\limsup_{n\to\infty} \frac{\log|b_n|}{\log|a_n|} \le 1+\epsilon. \tag{3.28}$$

Since (3.27) and (3.28) hold for all $\epsilon > 0$, we conclude that the limit in (3.25) is 1.

This would complete the proof of Theorem 3.48 except for the unfortunate fact that rational maps $\mathbb{P}^1 \to \mathbb{P}^1$ are always ramified. Further, our proof sketch made no mention of the assumption that $\phi^2(z)$ is not a polynomial, and the theorem is false for polynomials! In order to fix the proof, we begin by studying how ramification affects the distance between points.

**Lemma 3.51.** *Let $\phi : \mathbb{P}^1(\mathbb{C}) \to \mathbb{P}^1(\mathbb{C})$ be a rational map of degree $d \ge 2$, let $\rho : \mathbb{P}^1(\mathbb{C}) \times \mathbb{P}^1(\mathbb{C}) \to \mathbb{R}$ be the chordal metric as defined in Section 1.1, and for $Q \in \mathbb{P}^1(\mathbb{C})$, let*

$$\mathbf{e}_Q(\phi) = \max_{Q' \in \phi^{-1}(Q)} e_{Q'}(\phi) \tag{3.29}$$

*be the maximum of the ramification indices of the points in the inverse image of $Q$. Then there is a constant $C = C(\phi, Q)$, depending on $\phi$ and $Q$, such that*

$$\min_{Q'\in\phi^{-1}(Q)} \rho(P,Q')^{\mathbf{e}_Q(\phi)} \le C\rho(\phi(P),Q) \qquad \textit{for all } P \in \mathbb{P}^1(\mathbb{C}). \tag{3.30}$$

*In other words, if $\phi(P)$ is close to $Q$, then there is a point in the inverse image of $Q$ that is close to $P$, but ramification affects how close.*

*Proof.* We dehomogenize using a parameter $z$ such that $Q \ne \infty$ and $\infty \notin \phi^{-1}(Q)$. This means that we can write $Q = \beta$ and $P = \alpha$ and that we are looking for the $\beta' \in \phi^{-1}(\beta)$ that is closest to $\alpha$.

Writing $\phi(z) = F(z)/G(z)$ as a ratio of polynomials, the set $\phi^{-1}(\beta)$ is precisely the set of roots of the polynomial $F(z) - \beta G(z)$. If we factor this polynomial over $\mathbb{C}$ as

$$F(z) - \beta G(z) = b(z-\beta_1)^{e_1}(z-\beta_2)^{e_2}\cdots(z-\beta_r)^{e_r}$$

with distinct $\beta_1, \ldots, \beta_r$, then $\beta_i \in \phi^{-1}(\beta)$ has ramification index $e_i$. For notational convenience, we write

$$\mathbf{e} = \mathbf{e}_Q(\phi) = \max e_i.$$

We may assume that $\phi(P)$ is quite close to $Q$, i.e., that $\rho(\phi(P), Q)$ is small, since otherwise, the fact that the chordal metric satisfies $\rho \le 1$ lets us choose a $C$ for which the inequality (3.30) is true. In particular, $P \ne \infty$, and writing $\alpha = z(P)$, we may assume that $\alpha$ is quite close to at least one of the points $\beta_k$ in $\phi^{-1}(\beta)$. For example, we may require that $\alpha$ satisfy

$$|\alpha - \beta_k| < \frac{1}{2} \min_{i \ne k} |\beta_k - \beta_i| \qquad \text{and} \qquad |\alpha - \beta_k| < 1.$$

This implies in particular that

$$|\alpha - \beta_i| \ge |\beta_k - \beta_i| - |\alpha - \beta_k| \ge \frac{1}{2}|\beta_k - \beta_i| \qquad \text{for all } i \ne k.$$

Hence

$$\begin{aligned} \bigl|F(\alpha) - \beta G(\alpha)\bigr| &= |b||\alpha - \beta_1|^{e_1}|\alpha - \beta_2|^{e_2} \cdots |\alpha - \beta_r|^{e_r} \\ &= |b||\alpha - \beta_k|^{e_k} \prod_{i \ne k} |\alpha - \beta_i|^{e_i} \\ &\ge |b||\alpha - \beta_k|^{e_k} \prod_{i \ne k} \frac{1}{2}|\beta_k - \beta_i|^{e_i} \\ &= C_1|\alpha - \beta_k|^{e_k}, \end{aligned}$$

where the constant $C_1$ is positive and depends only on $\phi$ and $\beta$. Further, the exponent satisfies $e_k \le \mathbf{e}$, so we obtain the estimate

$$|F(\alpha) - \beta G(\alpha)| \ge C_1|\alpha - \beta_k|^{\mathbf{e}}.$$

The fact that $\phi(\alpha) = F(\alpha)/G(\alpha)$ is close to $\beta$ and the assumption that $G(\beta) \ne 0$ implies that $G(\alpha)$ is bounded away from 0, so dividing by $G(\alpha)$ yields

$$|\phi(\alpha) - \beta| \ge C_2|\alpha - \beta_k|^{\mathbf{e}}.$$

This in turn implies the same estimate for the chordal metric, since we can estimate $|\phi(\alpha)|$ using $|\beta|$ and we can estimate $|\alpha|$ using $|\beta_k|$. Therefore

$$\rho(\phi(\alpha), \beta) \ge C_3 \rho(\alpha, \beta_k)^{\mathbf{e}} \ge C_3 \min_{\beta' \in \phi^{-1}(\beta)} \rho(\alpha, \beta')^{\mathbf{e}}. \qquad \square$$

Next we show that if we stay away from totally ramified periodic points, then iteration of $\phi$ tends to spread out the ramification.

**Lemma 3.52.** *Let $\phi : \mathbb{P}^1 \to \mathbb{P}^1$ be a rational map of degree at least 2 and let $Q \in \mathbb{P}^1$ be a point such that $Q$ is not a totally ramified fixed point of $\phi^2$. Then*

$$\lim_{m\to\infty} \frac{\mathsf{e}_Q(\phi^m)}{(\deg \phi)^m} = 0.$$

*(See Lemma* 3.51 *for the definition of* $\mathsf{e}_Q(\phi)$.)

*Proof.* The proof of this geometric result uses the Riemann–Hurwitz Theorem 1.1 in a manner similar to the way that we have used it in the past. Let $d = \deg(\phi)$, fix an integer $m$, and let $P \in \phi^{-m}(Q)$. We use the fact that the ramification index is multiplicative,

$$e_P(\phi^m) = e_P(\phi)e_{\phi(P)}(\phi)e_{\phi^2(P)}(\phi)\cdots e_{\phi^{m-1}(P)}(\phi).$$

To ease notation, we write

$$e_i = e_{\phi^i(P)}(\phi).$$

We consider several cases.

**Case I. The points $P, \phi(P), \phi^2(P), \dots, \phi^{m-1}(P)$ are distinct.**
Note that this covers the case that $Q$ is a wandering point. The Riemann–Hurwitz formula allows us to estimate

$$\begin{aligned} e_P(\phi^m) &= e_0e_1e_2\cdots e_{m-1} \\ &\le \left(\frac{e_0+e_1+\cdots+e_{m-1}}{m}\right)^m \qquad \text{arithmetic–geometric inequality} \\ &= \left(\frac{(e_0-1)+(e_1-1)+\cdots+(e_{m-1}-1)}{m}+1\right)^m \\ &\le \left(\frac{2d-2}{m}+1\right)^m \qquad \text{Riemann–Hurwitz formula.} \end{aligned} \tag{3.31}$$

This estimate is clearly much stronger than the stated result, since the righthand side of (3.31) has a finite limit as $m \to \infty$. Indeed, it is an elementary exercise to verify that $(1 + t/m)^m \le e^t$ is valid for all $t \ge 0$, so

$$e_P(\phi^m) \le e^{2d-2} \quad \text{for all } m \ge 1 \text{ provided } P, \phi(P), \dots, \phi^{m-1}(P) \text{ are distinct.}$$

(N.B. $e_P$ is a ramification index, while the $e$ in $e^{2d-2}$ is the number 2.71828…!)

**Case II. $P$ is purely periodic of period $k \ge 3$, and $m \ge k$.**
The assumption that $\phi^k(P) = P$ implies that $e_i = e_{\phi^i(P)}(\phi)$ depends only on $i$ modulo $k$. Writing $m = qk + r$ with $0 \le r < k$, we use the multiplicativity of ramification, applied to $\phi^k$, to estimate

$$e_P(\phi^m) = \left(\prod_{j=0}^{q-1} e_{\phi^{jk}(P)}(\phi^k)\right)e_{\phi^{qk}(P)}(\phi^r) = \left(e_P(\phi^k)\right)^q e_{\phi^{qk}(P)}(\phi^r). \tag{3.32}$$

Note that $P, \phi(P), \dots, \phi^{k-1}(P)$ are distinct, so we can apply Case I to each of the ramification indices on the righthand side of (3.32). This yields

$$\begin{aligned} e_P(\phi^m) &\le \left(\frac{2d-2}{k}+1\right)^{kq}\left(\frac{2d-2}{r}+1\right)^{r} && \text{applying (3.31) to } \phi^k \text{ and } \phi^r,\\ &\le \left(\frac{2d-2}{k}+1\right)^{m-r} e^{2d-2} && \text{using } (1+t/r)^r \le e^t,\\ &\le \left(\frac{2d-2}{k}+1\right)^{m} e^{2d-2}. \end{aligned}$$

Finally, we observe that

$$\frac{1}{d}\left(\frac{2d-2}{k}+1\right) = 1-\left(1-\frac{1}{d}\right)\left(1-\frac{2}{k}\right) \le \frac{5}{6} \qquad \text{for all } d \ge 2 \text{ and } k \ge 3.$$

This proves that $e_P(\phi^m) \le e^{2d-2}\left(\frac{5}{6}d\right)^m$.

**Case III. $P$ is purely periodic of period $k \le 2$, and $m \ge k$.**
We use the assumption that $Q$ is not a totally ramified fixed point of $\phi^2$ to deduce that at least one of $e_P(\phi)$ and $e_{\phi(P)}(\phi)$ is strictly smaller than $d-1$, so

$$e_P(\phi^2) = e_P(\phi)e_{\phi(P)}(f) = e_0e_1 \le d^2 - d. \tag{3.33}$$

Now using the fact that $\phi^2(P) = P$, we see that $e_m = e_{\phi^m(P)}(\phi)$ is equal to $e_0$ if $m$ is even and equal to $e_1$ if $m$ is odd, so

$$\begin{aligned} e_P(\phi^m) &= e_0e_1e_2\cdots e_{m-1} \quad \text{multiplicativity of ramification index,}\\ &= \begin{cases} (e_0e_1)^{m/2} & \text{if } m \text{ is even,}\\ (e_0e_1)^{(m-1)/2}e_0 & \text{if } m \text{ is odd,} \end{cases}\\ &\le \begin{cases} (d^2-d)^{m/2} & \text{if } m \text{ is even,}\\ (d^2-d)^{(m-1)/2}d & \text{if } m \text{ is odd,} \end{cases}\\ &= \left(1-\frac{1}{d}\right)^{\lfloor m/2 \rfloor} d^m. \end{aligned}$$

**Case IV. $P$ is preperiodic.**
If $P$ is periodic, we are already done from earlier cases. Suppose that $P$ is not periodic and let $j \ge 1$ be the smallest integer such that $\phi^j(P)$ is periodic. Let $R = \phi^j(P)$ and let $k$ be the exact period of $R$. Then $P, \phi(P), \dots, \phi^{j-1}(P)$ are distinct and $R = \phi^j(P)$ is periodic, so from Case I we have

$$e_P(\phi^m) = e_P(\phi^j)\cdot e_{\phi^j(P)}(\phi^{m-j}) \le e^{2d-2}e_R(\phi^{m-j}) \le e^{2d-2}e_R(\phi^m).$$

We also note that $R$ is not a totally ramified fixed point of $\phi^2$, since $\phi^{-1}(R)$ contains at least the two distinct points $\phi^{k-1}(R)$ and $\phi^{j-1}(P)$. (This is where we use the assumption that $P$ is not itself periodic.) Hence we can apply Case II or III to $e_R(\phi^m)$, which gives the desired estimate for $e_P(\phi^m)$.

In all four cases we have proven estimates for $e_P(\phi^m)$ that imply that

$$e_P(\phi^m)/d^m \to 0 \quad \text{as} \quad m \to \infty.$$

This completes the proof of Lemma 3.52. More precisely, we proved that there are constants $c_1$ and $c_2$, depending only on $d$ and with $c_2 < 1$, so that

$$e_P(\phi^m) \le c_1(c_2 d)^m \qquad \text{for all } m \ge 1.$$

And if $P$ is a wandering point, we have shown that $e_P(\phi^m)$ is bounded by a constant depending only on the degree of $\phi$. □

We next prove an elementary lemma that relates the chordal metric $\rho(x,y)$ to the Euclidean distance $|x-y|$.

**Lemma 3.53.** *Let $\rho$ be the chordal metric on $\mathbb{P}^1(\mathbb{C})$. Then for all $x, y \in \mathbb{C} \subset \mathbb{P}^1(\mathbb{C})$,*

$$\rho(x,y) \le \frac{1}{2}\rho(y,\infty) \quad \Longrightarrow \quad \rho(x,y) \ge |x-y| \cdot \frac{\rho(y,\infty)^2}{2}.$$

*Proof.* Note that $\rho(z,\infty) = 1/\sqrt{|z|^2+1}$, so directly from the definition of $\rho$ we have

$$\rho(x,y) = \frac{|x-y|}{\sqrt{|x|^2+1}\sqrt{|y|^2+1}} = |x-y|\rho(x,\infty)\rho(y,\infty).$$

Then the triangle inequality for $\rho$ in the form $\rho(x,\infty)+\rho(x,y) \ge \rho(y,\infty)$ (see (1.2) in Section 1.1) yields

$$\rho(x,y) \ge |x-y|\bigl\{\rho(y,\infty) - \rho(x,y)\bigr\}\rho(y,\infty).$$

Making the further assumption that $\rho(x,y) \le \frac{1}{2}\rho(y,\infty)$, we obtain the desired inequality. □

*Proof of Theorem* 3.48. We now have the tools needed to complete the proof of Theorem 3.48. Writing $\phi^n(\alpha) = a_n/b_n$ as usual, our initial goal is to prove that the set

$$N(\phi,\alpha,\epsilon) = \bigl\{n \ge 0 : |a_n| \ge |b_n|^{1+\epsilon}\bigr\}$$

is finite. We observe that for $n \in N(\phi,\alpha,\epsilon)$, the point $\phi^n(\alpha)$ is close to $\infty$, since

$$\rho(\phi^n(\alpha),\infty) = \rho\bigl([a_n,b_n],[1,0]\bigr) = \frac{|b_n|}{\sqrt{a_n^2+b_n^2}} \le \left|\frac{b_n}{a_n}\right| \le \frac{1}{|a_n|^\epsilon}. \tag{3.34}$$

By assumption, $\phi^2(z) \notin \mathbb{C}[z]$, so $\infty$ is not a totally ramified fixed point of $\phi^2$. This allows us to apply Lemma 3.52, which gives us an integer $m_0 = m_0(d,\epsilon)$ such that

$$\mathbf{e}_\infty(\phi^m) \le \frac{\epsilon}{6}d^m \qquad \text{for all } m \ge m_0. \tag{3.35}$$

Having fixed an $m \ge m_0$, we apply Lemma 3.51 to the map $\phi^m$ and the points $P = \phi^{n-m}(\alpha)$ and $Q = \infty$ to obtain a point $\beta_n \in \phi^{-m}(\infty)$ satisfying

$$\rho\big(\phi^{n-m}(\alpha),\beta_n\big)^{e_\infty(\phi^m)} \le C_1\rho\big(\phi^n(\alpha),\infty\big), \tag{3.36}$$

where $C_1 = C_1(\phi^m,\infty)$ depends only on $m$ and $\phi$. Hence if we define

$$N(\phi,\alpha,\epsilon,\beta) = \Big\{n \in N(\phi,\alpha,\epsilon) : \rho\big(\phi^{n-m}(\alpha),\beta_n\big)^{e_\infty(\phi^m)} \le C_1\rho\big(\phi^n(\alpha),\infty\big)\Big\},$$

then $N(\phi,\alpha,\epsilon)$ is the union of $N(\phi,\alpha,\epsilon,\beta)$ for $\beta \in \phi^{-m}(\infty)$, so it suffices to prove that the set $N(\phi,\alpha,\epsilon,\beta)$ is finite for each $\beta \in \phi^{-m}(\infty)$. Note that if $N(\phi,\alpha,\epsilon,\beta)$ is infinite, then (3.34) and (3.36) imply that

$$\lim_{\substack{n\in N(\phi,\alpha,\epsilon,\beta)\\ n\to\infty}} \rho\big(\phi^{n-m}(\alpha),\beta\big) = 0. \tag{3.37}$$

In other words, $\phi^{n-m}(\alpha)$ approaches $\beta$ in $\mathbb{P}^1(\mathbb{C})$ as $n \to \infty$.

We consider first the case that $\beta \ne \infty$. The limit (3.37) tells us that

$$\rho\big(\phi^{n-m}(\alpha),\beta\big) \le \frac{1}{2}\rho(\beta,\infty) \quad \text{for all but finitely many } n \in N(\phi,\alpha,\epsilon,\beta). \tag{3.38}$$

Note that the inequality (3.38) allows us to apply Lemma 3.53 with $x = \phi^{n-m}(\alpha)$ and $y = \beta$, which yields

$$\rho\big(\phi^{n-m}(\alpha),\beta\big) \ge \big|\phi^{n-m}(\alpha) - \beta\big| \cdot \frac{\rho(\beta,\infty)^2}{2} \ge C_2\big|\phi^{n-m}(\alpha) - \beta\big|, \tag{3.39}$$

where as usual $C_2 > 0$ depends only on $\phi$ and $m$.

We are now ready to repeat the calculation from page 114, but this time done rigorously to account for the fact that $\phi$ has ramification. As usual, all constants may depend on $\phi$ and $m$, but are independent of $n$:

$$\begin{aligned}
\frac{1}{|a_n|^\epsilon} &\ge \rho\big(\phi^n(\alpha),\infty\big) && \text{from (3.34),}\\
&\ge C_3\rho\big(\phi^{n-m}(\alpha),\beta\big)^{e_\infty(\phi^m)} && \text{from (3.36),}\\
&\ge C_4\big|\phi^{n-m}(\alpha) - \beta\big|^{e_\infty(\phi^m)} && \text{from (3.39),}\\
&\ge \frac{C_5}{H\big(\phi^{n-m}(\alpha)\big)^{3e_\infty(\phi^m)}} && \text{Roth's Theorem 3.34 with exponent 3,}\\
&\ge \frac{C_6}{H\big(\phi^{n}(\alpha)\big)^{3e_\infty(\phi^m)/d^m}} && \begin{array}{l}\text{from Theorem 3.11, which says that}\\ h(P) = d^{-m}h(\phi^m(P)) + O(1),\end{array}\\
&\ge \frac{C_6}{H\big(\phi^{n}(\alpha)\big)^{\epsilon/2}} && \text{from (3.35),}\\
&= \frac{C_6}{|a_n|^{\epsilon/2}} && \text{since } H\big(\phi^n(\alpha)\big) = H(a_n/b_n) = |a_n|.
\end{aligned}$$

Hence $|a_n| \le C_7^{1/\epsilon}$ is bounded for $n \in N(\phi,\alpha,\epsilon,\beta)$, and the same is true of $|b_n|$ since $|b_n|^{1+\epsilon} \le |a_n|$, so $\phi^n(\alpha)$ takes on only finitely many values for

$n \in N(\phi, \alpha, \epsilon, \beta)$. But by assumption, $\alpha$ is a wandering point for $\phi$, so we conclude that $N(\phi, \alpha, \epsilon, \beta)$ is finite in the case that $\beta \neq \infty$.

Suppose now that $\beta = \infty$. Then a similar, but elementary, argument not requiring Roth's theorem yields

$$\begin{aligned}
\frac{1}{|a_n|^\epsilon} &\geq \rho(\phi^n(\alpha), \infty) && \text{from (3.34),} \\
&\geq C_7 \rho(\phi^{n-m}(\alpha), \beta)^{e_\infty(\phi^m)} && \text{from (3.36),} \\
&= C_7 \left( \frac{b_{n-m}}{\sqrt{a_{n-m}^2 + b_{n-m}^2}} \right)^{e_\infty(\phi^m)} && \text{definition of } \rho, \\
&\geq \frac{C_8}{H(\phi^{n-m}(\alpha))^{e_\infty(\phi^m)}} && \text{definition of height, where note that } b_{n-m} \neq 0, \text{ since } \alpha \text{ is wandering and } \infty \text{ is periodic,} \\
&\geq \frac{C_9}{H(\phi^n(\alpha))^{e_\infty(\phi^m)/d^m}} && \text{from Theorem 3.11,} \\
&\geq \frac{C_9}{H(\phi^n(\alpha))^{\epsilon/6}} && \text{from (3.35),} \\
&= \frac{C_9}{|a_n|^{\epsilon/6}} && \text{since } H(\phi^n(\alpha)) = H(a_n/b_n) = |a_n|.
\end{aligned}$$

As above, we conclude that $a_n$ and $b_n$ are bounded, and hence that $N(\phi, \alpha, \epsilon, \infty)$ is finite.

We have now proven that $N(\phi, \alpha, \epsilon, \beta)$ is finite for all $\beta \in \phi^{-m}(\infty)$, and hence that $N(\phi, \alpha, \epsilon)$ is finite. The finiteness of $N(\phi, \alpha, \epsilon)$ is equivalent to the statement that

$$\frac{\log|a_n|}{\log|b_n|} \leq 1 + \epsilon \qquad \text{for all but finitely many } n \geq 0. \tag{3.40}$$

We now apply the same argument to the rational map $\psi(z) = 1/\phi(1/z)$ and the point $1/\alpha$. It is easily verified that $\psi^n(1/\alpha) = b_n/a_n$, so we obtain the complementary inequality

$$\frac{\log|b_n|}{\log|a_n|} \leq 1 + \epsilon \qquad \text{for all but finitely many } n \geq 0. \tag{3.41}$$

Since (3.40) and (3.41) hold for all $\epsilon > 0$, it follows that

$$\lim_{n \to \infty} \frac{\log|a_n|}{\log|b_n|} = 1,$$

which completes the proof of Theorem 3.48. □

## 3.9 Periodic Points and Galois Groups

In this section we study the Galois groups of the field extensions generated by periodic points of a rational map. Much of the theory is valid for rational maps defined over an arbitrary perfect field and even for nonperfect fields, as long as $\phi$ is separable and one replaces the algebraic closure $\bar{K}$ of $K$ with the separable closure $K^{\text{sep}}$. For further material on this topic and the more general Galois theory of iterates, see [3, 179, 220, 222, 310, 311, 345, 346, 347, 425].

Let $\phi(z) \in K(z)$ be a rational function of degree $d \geq 2$ with coefficients in a perfect field $K$. The periodic points of $\phi$ have coordinates in the algebraic closure $\bar{K}$ of $K$, since they are solutions to equations of the form

$$\phi^n(z) = z.$$

More precisely, write $\phi = [F, G]$ using homogeneous polynomials $F, G \in K[X, Y]$. Then

$$\phi^n = [F_n, G_n]$$

for homogeneous polynomials $F_n, G_n \in K[X, Y]$ of degree $d^n$, and the periodic points of $\phi$ of period $n$ are the solutions in $\mathbb{P}^1(\bar{K})$ to the equation

$$YF_n(X, Y) - XG_n(X, Y) = 0.$$

Thus counted with multiplicity, there are exactly $d^n + 1$ such points.

In this section we study the field extensions of $K$ generated by the coordinates of the periodic points. Recall that if $P = [\alpha_0, \ldots, \alpha_N] \in \mathbb{P}^N(\bar{K})$, then the field of definition of $P$ over $K$ is the field $K(P)$ obtained by choosing a nonzero $\alpha_i$ and setting

$$K(P) = K\left(\frac{\alpha_0}{\alpha_i}, \frac{\alpha_1}{\alpha_i}, \ldots, \frac{\alpha_N}{\alpha_i}\right).$$

It is easy to see that this field is independent of the choice of the index $i$.

The Galois group $\operatorname{Gal}(\bar{K}/K)$ acts on the points of $\mathbb{P}^N(\bar{K})$ in the natural way,

$$\sigma(P) = [\sigma(\alpha_0), \sigma(\alpha_1), \ldots, \sigma(\alpha_N)],$$

and it is not hard to verify (Exercise 3.47) that

$$K(P) = \text{Fixed field of } \{\sigma \in \operatorname{Gal}(\bar{K}/K) : \sigma(P) = P\}.$$

Recall that the set of $n$-periodic points of $\phi$ is the set

$$\operatorname{Per}_n(\phi) = \{P \in \mathbb{P}^1(\bar{K}) : \phi^n(P) = P\}.$$

Some of the points in $\operatorname{Per}_n(\phi)$ may have period that is smaller than $n$. For example, $\operatorname{Per}_n(\phi)$ contains all of the fixed points of $\phi$. This suggests that we look at a smaller set consisting of the *primitive $n$-periodic points*,

$$\operatorname{Per}_n^{**}(\phi) = \{P \in \mathbb{P}^1(\bar{K}) : \phi^n(P) = P \text{ and } \phi^i(P) \neq P \text{ for all } 1 \leq i < n\}.$$

We begin by verifying that both $\operatorname{Per}_n(\phi)$ and $\operatorname{Per}_n^{**}(\phi)$ are Galois-invariant.

**Proposition 3.54.** *Let $\phi(z) \in K(z)$ be a rational function of degree d. The set of $n$-periodic points $\mathrm{Per}_n(\phi)$ and the set of primitive $n$-periodic points $\mathrm{Per}_n^{**}(\phi)$ are Galois-invariant sets.*

*Proof.* Let $\sigma \in \mathrm{Gal}(\bar{K}/K)$. Then

$$\phi(\sigma(P)) = \sigma(\phi(P)) \quad \text{for all points } P \in \mathbb{P}^1(\bar{K}),$$

since $\phi$ is a rational function with coefficients in $K$. Let $P \in \mathrm{Per}_n(\phi)$. Then

$$\phi^n(\sigma(P)) = \sigma(\phi^n(P)) = \sigma(P),$$

so $\sigma(P) \in \mathrm{Per}_n(\phi)$. This proves that $\mathrm{Per}_n(\phi)$ is Galois-invariant.

Similarly, we see that

$$\phi^i(\sigma(P)) = \sigma(P) \Longleftrightarrow \sigma(\phi^i(P)) = \sigma(P) \Longleftrightarrow \phi^i(P) = P,$$

which proves that $P$ is a primitive $n$-periodic point if and only if $\sigma(P)$ is a primitive $n$-periodic point. Hence $\mathrm{Per}_n^{**}(\phi)$ is also Galois-invariant. □

Proposition 3.54 tells us that the set of primitive $n$-periodic points generates a Galois extension of $K$. We denote this extension by

$$K_{n,\phi} = K\big(P : P \in \mathrm{Per}_n^{**}(\phi)\big).$$

These fields are analogues of the fields generated by roots of unity (cyclotomic fields), so we call $K_{n,\phi}$ the $n^{th}$ *dynatomic field for* $\phi$. We denote the Galois group of the $n^{\text{th}}$ dynatomic field by

$$G_{n,\phi} = \mathrm{Gal}(K_{n,\phi}/K).$$

*Example* 3.55. Consider the quadratic polynomial $\phi(z) = z^2 + 1$. The sets of primitive 2nd, 3rd, 4th, and 6th periodic points are given, respectively, by the roots of the polynomials

$$\begin{aligned}
\phi_2^*(z) &= \frac{\phi^2(z) - z}{\phi(z) - z} = z^2 + z + 2, \\
\phi_3^*(z) &= \frac{\phi^3(z) - z}{\phi(z) - z} = z^6 + z^5 + 4z^4 + 3z^3 + 7z^2 + 4z + 5, \\
\phi_4^*(z) &= \frac{\phi^4(z) - z}{\phi^2(z) - z} = z^{12} + 6z^{10} + z^9 + 18z^8 + 4z^7 + 33z^6 + 8z^5 \\
&\qquad + 40z^4 + 9z^3 + 30z^2 + 6z + 13, \\
\phi_6^*(z) &= \frac{(\phi^6(z) - z)(\phi(z) - z)}{(\phi^3(z) - z)(\phi^2(z) - z)} \\
&= z^{54} - z^{53} + 27z^{52} - 25z^{51} + \cdots - 13750z + 45833.
\end{aligned}$$

As this simple example clearly shows, it is infeasible to study dynatomic fields via explicit equations except for very small values of $n$.

The Galois group $G_{n,\phi}$ acts on the set of primitive $n$-periodic points $\mathrm{Per}_n^{**}(\phi)$, and it is clearly determined by this action, but except in trivial cases $G_{n,\phi}$ cannot be the full group of permutations of $\mathrm{Per}_n^{**}(\phi)$. The reason why it cannot be the full group of permutations is due to the relation

$$\sigma(\phi^i(P)) = \phi^i(\sigma(P)) \quad \text{for all } \sigma \in G_{n,\phi} \text{ and all } P \in \mathrm{Per}_n^{**}(\phi).$$

Thus the action of $\sigma$ on points in the orbit of $P$ is determined by its action on $P$. This suggests that we decompose the action of $G_{n,\phi}$ on $\mathrm{Per}_n^{**}(\phi)$ into its action on the set of orbits and its action within each orbit.

A nice way to visualize the action of $G_{n,\phi}$ on the points of $\mathrm{Per}_n^{**}(\phi)$ is to consider the following picture:

| $\mathcal{O}_1$ | $\mathcal{O}_2$ | $\cdots$ | $\mathcal{O}_{r-1}$ | $\mathcal{O}_r$ |
|---|---|---|---|---|
| $P_1$ | $P_2$ | $\cdots$ | $P_{r-1}$ | $P_r$ |
| $\phi(P_1)$ | $\phi(P_2)$ | $\cdots$ | $\phi(P_{r-1})$ | $\phi(P_r)$ |
| $\phi^2(P_1)$ | $\phi^2(P_2)$ | $\cdots$ | $\phi^2(P_{r-1})$ | $\phi^2(P_r)$ |
| $\vdots$ | | | | $\vdots$ |
| $\phi^{n-1}(P_1)$ | $\phi^{n-1}(P_2)$ | $\cdots$ | $\phi^{n-1}(P_{r-1})$ | $\phi^{n-1}(P_r)$ |

An element of $G_{n,\phi}$ permutes the orbits, i.e., it permutes the columns, and then within each column it performs a cyclical shift.

We can write this algebraically by observing that applying $\sigma \in G_{n,\phi}$ to a point $P_j$ gives a point $\phi^i(P_k)$ for a uniquely determined $0 \le i < n$ and $1 \le k \le r$. We indicate the dependence of $i$ and $k$ on $\sigma$ and $j$ by adopting the notation

$$\sigma(P_j) = \phi^{i_\sigma(j)}(P_{\pi_\sigma(j)}).$$

Notice that $\pi_\sigma$ is a permutation of the set $\{1, 2, \ldots, r\}$, so we can think of it as an element of the permutation group $S_r$. This gives a map $\pi : G_{n,\phi} \to S_r$.

Similarly, if we set $\mathbf{i}_\sigma = (i_\sigma(1), i_\sigma(2), \ldots, i_\sigma(r))$, then we obtain a map

$$\mathbf{i} : G_{n,\phi} \longrightarrow (\mathbb{Z}/n\mathbb{Z})^r.$$

We perform a computation to see how $\mathbf{i}$ and $\pi$ interact with one another:

$$\begin{aligned}
(\sigma\tau)(P_j) = \sigma(\tau(P_j)) &= \sigma\bigl(\phi^{i_\tau(j)}(P_{\pi_\tau(j)})\bigr) \\
&= \phi^{i_\tau(j)}\bigl(\sigma(P_{\pi_\tau(j)})\bigr) \\
&= \phi^{i_\tau(j)}\bigl(\phi^{i_\sigma(\pi_\tau(j))} P_{\pi_\sigma(\pi_\tau(j))}\bigr) \\
&= \phi^{i_\tau(j)+i_\sigma(\pi_\tau(j))}\bigl(P_{(\pi_\sigma\pi_\tau)(j)}\bigr).
\end{aligned}$$

From this equation we obtain the two formulas

$$\pi_{\sigma\tau} = \pi_\sigma\pi_\tau \qquad \text{and} \qquad \mathbf{i}_{\sigma\tau} = \mathbf{i}_\tau + \mathbf{i}_\sigma\pi_\tau. \tag{3.42}$$

Thus the map $\pi : G_{n,\phi} \to S_r$ is a homomorphism, but the map

$$\mathbf{i} : G_{n,\phi} \to (\mathbb{Z}/n\mathbb{Z})^r$$

is not a homomorphism, since it is twisted by the permutation action of $S_r$ on $(\mathbb{Z}/n\mathbb{Z})^r$. This is an example of the following general construction.

**Definition.** Let $\mathcal{S}$ and $H$ be groups, and let $A$ be an index set on which $\mathcal{S}$ acts. For simplicity, we assume that $H$ is an abelian group and we write its group law additively. The group structure on $H$ makes the collection of maps $\operatorname{Map}(A, H)$ into a group. Thus if $i_1, i_2 \in \operatorname{Map}(A, H)$, then

$$(i_1 + i_2) : A \longrightarrow H, \qquad (i_1 + i_2)(a) = i_1(a) + i_2(a).$$

There is a natural action of $\mathcal{S}$ on the group $\operatorname{Map}(A, H)$ via its action on $A$,

$$\pi : \operatorname{Map}(A, H) \longrightarrow \operatorname{Map}(A, H), \qquad \pi(i)(a) = i\big(\pi(a)\big).$$

The *wreath product of* $H$ *and* $\mathcal{S}$ (*relative to* $A$) is the twisted product of the groups $\operatorname{Map}(A, H)$ and $\mathcal{S}$ for this action. In other words, as a set the wreath product consists of the collection of ordered pairs

$$\operatorname{Wreath}(H, \mathcal{S}) = \operatorname{Map}(A, H) \times \mathcal{S},$$

and its group law is defined by twisting the natural group law on the product,

$$(i_1, \pi_1) * (i_2, \pi_2) = \big(\pi_2(i_1) + i_2, \pi_1\pi_2\big).$$

**Theorem 3.56.** *Let $\phi \in K(z)$ be a rational function of degree d. Let $\mathcal{O}_1, \ldots, \mathcal{O}_r$ be the distinct $\phi$ orbits in $\operatorname{Per}_n^{**}(\phi)$ and choose a point $P_j \in \mathcal{O}_j$ in each orbit. For each $\sigma \in G_{n,\phi}$ and each $1 \le j \le r$, define integers $0 \le i_\sigma(j) < n$ and $1 \le \pi_\sigma(j) \le r$ by the formula*

$$\sigma(P_j) = \phi^{i_\sigma(j)}(P_{\pi_\sigma(j)}).$$

*Let* $\operatorname{Wreath}(\mathbb{Z}/n\mathbb{Z}, S_r)$ *be the wreath product of the symmetric group $S_r$ and $\mathbb{Z}/n\mathbb{Z}$ relative to the natural action of $S_r$ on the set $\{1, 2, \ldots, r\}$. Then the map*

$$W : G_{n,\phi} \longrightarrow \operatorname{Wreath}(\mathbb{Z}/n\mathbb{Z}, S_r), \qquad \sigma \longmapsto (i_\sigma, \pi_\sigma),$$

*is an injective homomorphism.*

*Proof.* In order to show that $W$ is a homomorphism, we need to verify that

$$W(\sigma\tau) = W(\sigma)W(\tau).$$

Writing this out in terms of the (twisted) definition of the group law on the wreath product, we need to prove that

$$(i_{\sigma\tau}, \pi_{\sigma\tau}) = (i_\sigma, \pi_\sigma)(i_\tau, \pi_\tau) = (\pi_\tau(i_\sigma) + i_\tau, \pi_\sigma\pi_\tau) = (i_\sigma\pi_\tau + i_\tau, \pi_\sigma\pi_\tau).$$

This is precisely the formula (3.42) proven above, which shows that $W$ is a homomorphism.

Now suppose that $W(\sigma) = 1$. This means that $i_\sigma(j) = 0$ and $\pi_\sigma(j) = j$ for all $1 \le j \le r$, so $\sigma(P_j) = P_j$. Hence

$$\sigma(\phi^i(P_j)) = \phi^i(\sigma(P_j)) = \phi^i(P_j),$$

so $\sigma$ fixes every point in $\mathrm{Per}_n^{**}(\phi)$. Therefore $\sigma$ fixes $K_{n,\phi}$, so $\sigma = 1$, which proves that $W$ is injective. □

The above discussion shows that the Galois group $G_{n,\phi}$ of the dynatomic extension $K_{n,\phi}/K$ roughly splits up into two pieces, a permutation piece determined by a permutation action on the orbits, and a cyclic piece determined by the action within each orbit. The permutation piece can be quite complicated, but one might hope that the dynamics of $\phi$ can be used to study the cyclic piece. We now try to make these vague remarks more precise.

Let

$$G_{n,\phi}^0 = \mathrm{Ker}(\pi : G_{n,\phi} \longrightarrow \mathcal{S}_r) = \{\sigma \in G_{n,\phi} : \sigma(\mathcal{O}_j) = \mathcal{O}_j \text{ for all } 1 \le j \le r\},$$

and let

$$K_{n,\phi}^0 = \text{fixed field of } G_{n,\phi}^0.$$

Then

$$G_{n,\phi}^0 = \mathrm{Gal}(K_{n,\phi}/K_{n,\phi}^0) \hookrightarrow (\mathbb{Z}/n\mathbb{Z})^r, \quad \sigma \longmapsto (i_\sigma(1), i_\sigma(2), \dots, i_\sigma(r)).$$

More precisely, if we fix a particular primitive $n$-periodic point $P \in \mathrm{Per}_n^{**}(\phi)$, then the extension $K_{n,\phi}^0(P)/K_{n,\phi}^0$ is a cyclic extension of order dividing $n$ whose Galois group is determined by the homomorphism

$$i : \mathrm{Gal}(K_{n,\phi}^0(P)/K_{n,\phi}^0) \hookrightarrow \mathbb{Z}/n\mathbb{Z}, \quad \text{satisfying} \quad \sigma(P) = \phi^{i_\sigma}(P). \tag{3.43}$$

This formula gives us some control over the Galois group of the relative abelian extension $K_{n,\phi}^0(P)/K_{n,\phi}^0$.

*Remark* 3.57. If $L = K(\alpha)$ is any Galois extension and if $\sigma \in \mathrm{Gal}(L/K)$ is any automorphism, then $\sigma(\alpha)$ can always be expressed as a polynomial in $\alpha$. Thus there is nothing special, a priori, in the fact that the action of Galois in (3.43) is given by a polynomial. However, if we fix $\phi(z)$ of degree $d$ and take $n$ very large compared to $d$, then the extension $K_{n,\phi}^0(P)/K_{n,\phi}^0$ is interesting because there is an automorphism $\sigma$ described by a polynomial (or rational function) $\phi$ of small degree compared to the degree of the extension.

## 3.10 Equidistribution and Preperiodic Points

There are many theorems and conjectures concerning the distribution of torsion points and points of small height on elliptic curves and abelian varieties. In this section we describe, without proof, some dynamical analogues. For further details, see Zhang's survey article [453] and the papers listed in its references.

We begin with the Manin–Mumford conjecture, which asserts that if $X$ is an irreducible subvariety of an abelian variety $A$ such that $X \cap A_{\text{tors}}$ is Zariski dense in $X$, then $X$ is a translate by a torsion point of an abelian subvariety of $A$. The original Manin–Mumford conjecture was proven by Raynaud [367, 368]. Replacing torsion points by preperiodic points leads to a dynamical conjecture, where the dynamical analogue of a translate of an abelian subvariety of $A$ is a preperiodic subvariety of $\mathbb{P}^N$ as in the following definition.

**Definition.** Let $\phi : \mathbb{P}^N \to \mathbb{P}^N$ be a morphism. A subvariety $X \subset \mathbb{P}^N$ is a *periodic variety for* $\phi$ if there is an integer $n \geq 1$ such that $\phi^n(X) = X$. The subvariety is a *preperiodic variety for* $\phi$ if $\phi^m(X)$ is periodic for $\phi$ for some $m \geq 0$.

**Conjecture 3.58.** (Dynamical Manin–Mumford Conjecture) *Let $\phi : \mathbb{P}^N \to \mathbb{P}^N$ be a morphism defined over $\mathbb{C}$ and let $X \subset \mathbb{P}^N$ be a subvariety. Then*

$$X \cap \operatorname{PrePer}(\phi)$$

*is Zariski dense in $X$ if and only if the subvariety $X$ is preperiodic for $\phi$.*

The set of torsion points on an elliptic curve $E$, or more generally an abelian variety, is equidistributed with respect to the natural (Haar) measure on $E(\mathbb{C})$. More precisely, we identify $E(\mathbb{C}) = \mathbb{C}/L$ as described in Section 1.6.3, and then for any open set $U \subset \mathbb{C}$ lying in a fundamental domain for $E(\mathbb{C})$ we have

$$\lim_{n\to\infty} \frac{\#(E[n] \cap U)}{\#E[n]} = \operatorname{Area}(U).$$

A deeper equidistribution result of an arithmetic nature says that the Galois conjugates of torsion points are equidistributed. In order to state a dynamical analogue for morphisms $\phi : \mathbb{P}^N \to \mathbb{P}^N$, we need a $\phi$-invariant measure on $\mathbb{P}^N(\mathbb{C})$ as described in the following proposition.

**Proposition 3.59.** *Let $\phi : \mathbb{P}^N \to \mathbb{P}^N$ be a morphism of degree $d$ defined over $\mathbb{C}$. There is a unique probability measure $\mu_\phi$ on $\mathbb{P}^N(\mathbb{C})$ satisfying*

$$\phi_*\mu_\phi = \mu_\phi \qquad \textit{and} \qquad \phi^*\mu_\phi = d^N \cdot \mu_\phi.$$

*We call $\mu_\phi$ the* canonical $\phi$-invariant probability measure *on $\mathbb{P}^N(\mathbb{C})$.*

*Proof.* For a general construction that covers both archimedean and nonarchimedean base fields, see [453]. We also mention a standard result in dynamics (the Krylov–Bogolubov theorem [226, Theorem 4.1.1]), which says that any continuous map $\phi : X \to X$ on a metrizable compact topological space $X$ admits a Borel probability measure $\mu_\phi$ satisfying $\phi_*(\mu_\phi) = \mu_\phi$, i.e.,

$$\mu_\phi(\phi^{-1}(A)) = \mu_\phi(A) \quad \text{for every Borel-measurable subset } A \text{ of } X. \qquad \square$$

For the remainder of this section we fix an algebraic closure $\bar{\mathbb{Q}}$ of $\mathbb{Q}$ and an embedding $\bar{\mathbb{Q}} \hookrightarrow \mathbb{C}$. So when we speak of a number field $K$ and its algebraic closure $\bar{K}$, we assume that they come with compatible embeddings into $\mathbb{C}$.

**Definition.** Let $K/\mathbb{Q}$ be a number field. For any algebraic point $P \in \mathbb{P}^N(\bar{K})$, let $C(P/K)$ denote the set of Galois conjugates of $P$, i.e.,

$$C(P/K) = \{\sigma(P) \in \mathbb{P}^N(\bar{K}) : \sigma \in \operatorname{Gal}(\bar{K}/K)\},$$

and let $\delta_P$ denote the Dirac measure on $\mathbb{P}^N(\mathbb{C})$ supported at $P$,

$$\delta_P(U) = \begin{cases} 1 & \text{if } P \in U, \\ 0 & \text{if } P \notin U. \end{cases}$$

We associate to $P \in \mathbb{P}^N(\bar{K})$ the discrete probability measure

$$\mu_P = \frac{1}{\#C(P/K)} \sum_{Q \in C(P/K)} \delta_Q$$

supported on the Galois conjugates of $P$.

**Definition.** Let $K/\mathbb{Q}$ be a number field and let $P_1, P_2, P_3, \ldots \in \mathbb{P}^N(\bar{K})$ be a sequence of points with algebraic coordinates. Fix a probability measure $\mu$ on $\mathbb{P}^N(\mathbb{C})$. We say that the sequence $\{P_i\}_{i \ge 1}$ is *Galois equidistributed with respect to* $\mu$ if the sequence of measures $\mu_{P_i}$ converges weakly[3] to $\mu$.

We are now ready to state a dynamical equidistribution conjecture for Galois orbits of preperiodic points on $\mathbb{P}^N$, and more generally for points of small height.

**Conjecture 3.60.** (Dynamical Galois Equidistribution Conjecture) *Let $K/\mathbb{Q}$ be a number field, let $\phi : \mathbb{P}^N \to \mathbb{P}^N$ be a morphism of degree $d \ge 2$ defined over $K$, and let $P_1, P_2, P_3, \ldots \in \mathbb{P}^N(\bar{K})$ be a sequence of distinct points such that no infinite subsequence lies entirely within a preperiodic subvariety of $\mathbb{P}^N$.*

(a) *If $P_1, P_2, P_3, \ldots \in \operatorname{PrePer}(\phi)$, then the sequence $\{P_i\}_{i \ge 1}$ is Galois equidistributed in $\mathbb{P}^N(\mathbb{C})$ with respect to the canonical $\phi$-invariant probability measure $\mu_\phi$.*

(b) *If $\lim_{i \to \infty} \hat{h}_\phi(P_i) = 0$, then the sequence $\{P_i\}_{i \ge 1}$ is Galois equidistributed in $\mathbb{P}^N(\mathbb{C})$ with respect to the canonical $\phi$-invariant probability measure $\mu_\phi$.*

It is clear that Conjecture 3.60(b) implies Conjecture 3.60(a), since preperiodic points have canonical height equal to 0. A version of the conjecture is known provided that the sequence of points satisfies a somewhat stronger Zariski density condition as in the following theorem.

---

[3]Recall that a sequence of measures $\mu_i$ on a compact space $X$ *converges weakly* to $\mu$ if for every Borel-measurable set $U$, the sequence of values $\mu_i(U)$ converges to $\mu(U)$ as $i \to \infty$.

**Theorem 3.61.** (Yuan [450]) *Let $\phi : \mathbb{P}^N \to \mathbb{P}^N$ be a morphism of degree $d \ge 2$ defined over $K$ and let $P_1, P_2, P_3, \ldots \in \mathbb{P}^N(\bar{K})$ be a sequence of points satisfying the following two conditions:*

(a) *Every infinite subsequence of $\{P_i\}_{i\ge 1}$ is Zariski dense in $\mathbb{P}^N$.*

(b) *$\hat{h}_\phi(P_i) \to 0$ as $i \to \infty$.*

*(In the terminology of* [453]*, sequences with property* (a) *are called* generic *and sequences with property* (b) *are called* small.*) Then the sequence $\{P_i\}_{i\ge 1}$ is Galois equidistributed with respect to the canonical $\phi$-invariant probability measure $\mu_\phi$ on $\mathbb{P}^N(\mathbb{C})$.*

*Proof.* The proof is beyond the scope of this book. See Yuan [450] for a general version over archimedean and nonarchimedean base fields and algebraic dynamical systems on arbitrary projective varieties. Earlier results and generalizations are given by Autissier [15, 16], Baker–Ih [24], Baker–Rumely [28], Chambert-Loir [98], Chambert-Loir–Thuillier [99], Favre–Rivera-Letelier [169] and Szpiro–Ullmo–Zhang [432]. □

The classical Bogomolov conjecture, which says that sets of points of small height on abelian varieties lie on translates of abelian subvarieties, was proven by Ullmo [435] and Zhang [452]. We state a dynamical analogue.

**Conjecture 3.62.** (Dynamical Bogomolov Conjecture) *Let $\phi : \mathbb{P}^N \to \mathbb{P}^N$ be a morphism of degree $d \ge 2$ defined over a number field $K$ and let $X \subset \mathbb{P}^N$ be an irreducible subvariety that is not preperiodic. Then there is an $\epsilon > 0$ such that the set*

$$\{P \in X(\bar{K}) : \hat{h}_\phi(P) < \epsilon\}$$

*is not Zariski dense in $X$.*

Notice that Conjecture 3.62 implies Conjecture 3.58, since the set of points with $\hat{h}_\phi(P) < \epsilon$ includes all of the preperiodic points.

Finally, in closing this section, we mention that canonical invariant measures have been constructed on Berkovich spaces; see Remark 5.77 and the references listed there.

## 3.11 Ramification and Units in Dynatomic Fields

In Section 3.9 we used periodic points to construct field extensions $K_{\phi,n}$ and studied their Galois groups. We now take up the question of the arithmetic properties of these algebraic number fields. In general, the three basic questions that one would like to answer about a given number field are these: Where is it ramified? What is its ideal class group? What are its units? In addition, one wants to know how the Galois group acts on ideal classes and on units. In this section we provide partial answers to the question of ramification and units for dynatomic fields.

We first recall the classical case of cyclotomic extensions, which provide a model for the dynatomic theory. Let $\zeta$ is a primitive $n^{\text{th}}$ root of unity and $\sigma \in \mathrm{Gal}(\bar{\mathbb{Q}}/\mathbb{Q})$. Then

$$\sigma(\zeta) = \zeta^{j(\sigma)} \quad \text{for a unique } j(\sigma) \in (\mathbb{Z}/n\mathbb{Z})^*.$$

This defines an isomorphism

$$j : \mathrm{Gal}(\mathbb{Q}(\zeta)/\mathbb{Q}) \longrightarrow (\mathbb{Z}/n\mathbb{Z})^*, \quad \sigma \longmapsto j(\sigma),$$

expressing the action of $\sigma$ on $\mathbb{Q}(\zeta)$ as a polynomial action $\phi(z) = z^j$. Now let $p$ be a prime not dividing $n$, let $\mathfrak{p}$ be a prime of $\mathbb{Q}(\zeta)$ lying above $p$, and let $\sigma_p \in \mathrm{Gal}(\bar{\mathbb{Q}}/\mathbb{Q})$ be the corresponding Frobenius element. The definition of Frobenius says that

$$\sigma_p(\zeta) \equiv \zeta^p \pmod{\mathfrak{p}}. \tag{3.44}$$

However, the $n^{\text{th}}$ roots of unity remain distinct when reduced modulo $\mathfrak{p}$, so the congruence (3.44) implies an equality in $\mathbb{Q}(\zeta)$,

$$\sigma_p(\zeta) = \zeta^p.$$

This exact determination of the action of Frobenius as a polynomial map on certain generating elements of $\mathbb{Q}(\zeta)$ is of fundamental importance in the study of cyclotomic fields.

To some extent, we can carry over the analysis of $\mathbb{Q}(\zeta)$ to dynatomic fields, although the final results are not as complete as in the classical case. Let $\mathfrak{p}$ be a prime ideal of the ring of integers of $K^0_{\phi,n}$, let $p$ be the residue characteristic of $\mathfrak{p}$, and let $q$ be the norm of $\mathfrak{p}$. We assume throughout that $p \nmid n$. Choose a prime ideal $\mathfrak{P}$ in $K^0_{\phi,n}(P)$ lying above $\mathfrak{p}$. Assuming that $\mathfrak{p}$ does not ramify in $K^0_{\phi,n}(P)$, the associated Frobenius element $\sigma_{\mathfrak{p}}$ is determined by the condition

$$\sigma_{\mathfrak{p}}(\alpha) \equiv \alpha^q \pmod{\mathfrak{P}}$$

for all $\alpha$ in the ring of integers of $K^0_{\phi,n}(P)$. On the other hand, by construction the action of $\sigma_{\mathfrak{p}}$ on $P \in \mathrm{Per}^{**}_n(\phi)$ is given by by formula

$$\sigma_{\mathfrak{p}}(P) = \phi^{i_{\sigma_{\mathfrak{p}}}}(P).$$

The next proposition allows us to characterize the primes at which the extension $K^0_{\phi,n}(P)/K^0_{\phi,n}$ may be ramified.

**Proposition 3.63.** *Let $K$ be a number field, let $\phi \in K(z)$ be a rational map of degree $d \geq 2$, let $P \in \mathrm{Per}^{**}_n(\phi, K)$ be a point in $\mathbb{P}^1(K)$ of exact period $n$, and let $\mathfrak{p}$ be a prime of $K$ satisfying the following three conditions:*

$$\phi \text{ has good reduction at } \mathfrak{p}. \tag{3.45}$$

$$\mathfrak{p} \text{ does not divide } n. \tag{3.46}$$

$$\text{If } \lambda_\phi(P) \neq 1, \text{ then } \mathfrak{p} \text{ does not divide } \lambda_\phi(P) - 1. \tag{3.47}$$

*Then $\tilde{P} \bmod \mathfrak{p}$ has exact period $n$. In particular, the set*

$$\{\mathfrak{p} : \tilde{P} \bmod \mathfrak{p} \text{ has period strictly less than } n\}$$

*is a finite set of primes of $K$.*

*Proof.* Let $\mathfrak{p}$ be a prime of $K$ satisfying (3.45), (3.46), and (3.47). Let $m$ be the exact period of $\tilde{P}$ and let $r$ be the order of $\lambda_{\tilde{\phi}}(\tilde{P})$ in $\mathbb{F}_{\mathfrak{p}}^*$. Theorem 2.21 tells us that either $n = m$ or $n = mr$, since (3.46) rules out powers of $p$ appearing in $n$. If $\lambda_\phi(P) = 1$, then also $\lambda_{\tilde{\phi}}(\tilde{P}) = 1$, so $r = 1$ and $n = m$. On the other hand, if $\lambda_\phi(P) \neq 1$, then (3.47) tells us that

$$\lambda_{\tilde{\phi}}(\tilde{P})^{n/m} = (\tilde{\phi}^m)'(\tilde{P})^{n/m} = (\tilde{\phi}^n)'(\tilde{P}) = \widetilde{(\phi^n)'(P)} = \widetilde{\lambda_\phi(P)} \neq \tilde{1}.$$

But $\lambda_{\tilde{\phi}}(\tilde{P})^r = \tilde{1}$ by definition of $r$, so $n/m$ cannot equal $r$. Hence $n = m$ in all cases, which shows that $\tilde{P} \bmod \mathfrak{p}$ has exact period $n$ for all primes $\mathfrak{p}$ satisfying (3.45), (3.46), and (3.47). This proves the first part of the proposition, and since each of the three conditions is satisfied for all but finitely many primes, the second part is clear. □

**Corollary 3.64.** *Let $K$ be a number field, let $\phi(z) \in K(z)$ be a rational map of degree $d \geq 2$, and let $K_{n,\phi}$ be the $n^{th}$ dynatomic field for $\phi$. Let $S$ be the set of primes $\mathfrak{p}$ of $K$ such that either $\phi$ has bad reduction at $\mathfrak{p}$ or $\mathfrak{p}$ divides $n$ or $\mathfrak{p}$ divides the quantity*

$$\prod_{\substack{P \in \mathrm{Per}_n^{**}(\phi) \\ \lambda_\phi(P) \neq 1}} (\lambda_\phi(P) - 1). \tag{3.48}$$

*Then $K_{n,\phi}/K$ is unramified outside of $S$.*

*Proof.* The field extension $K_{n,\phi}/K$ is generated by the points of $\mathrm{Per}_n^{**}(\phi)$. Proposition 3.63 tells us that if $\mathfrak{p} \notin S$, then those points remain distinct when reduced modulo primes lying over $\mathfrak{p}$. Hence the extension $K_{n,\phi}/K$ is unramified at $\mathfrak{p}$. □

*Example* 3.65. We continue studying the quadratic polynomial $\phi(z) = z^2 + 1$ from Example 3.55 (see also Example 2.37). The polynomial $\phi(z)$ has everywhere good reduction, so if we assume for the moment that none of its periodic points have multiplier equal to 1, then the quantity (3.48) in Corollary 3.64 is equal to

$$\prod_{\phi_n^*(\alpha)=0} (\lambda_\phi(\alpha) - 1) = \prod_{\phi_n^*(\alpha)=0} ((\phi^n)'(\alpha) - 1) = \mathrm{Res}(\phi_n^*(z), (\phi^n)'(z) - 1).$$

Denoting this resultant by $\Delta_n(\phi)$, it is not hard to compute the values of $\Delta_n(\phi)$ for small values of $n$. We obtain

$$\begin{aligned}
\Delta_2(\phi) &= 7^2, \\
\Delta_3(\phi) &= (3^3 \cdot 11)^3, \\
\Delta_4(\phi) &= (3^2 \cdot 11 \cdot 13 \cdot 41)^4, \\
\Delta_5(\phi) &= (3^3 \cdot 7 \cdot 83 \cdot 331 \cdot 140869)^5, \\
\Delta_6(\phi) &= (3^4 \cdot 5 \cdot 7 \cdot 23 \cdot 73 \cdot 223 \cdot 2251 \cdot 347495839)^6.
\end{aligned}$$

The field $K_{\phi,2}$ is generated by the roots of $\phi_2^*(z) = z^2 + z + 2$, so we have explicitly $K_{\phi,2} = K(\sqrt{-7})$. For higher values of $n$, we can check the primes dividing $\Delta_n(\phi)$ by computing directly the discriminant of the polynomial $\phi_n^*(z)$, whose roots generate $K_{\phi,n}/K$. Thus for example, $\operatorname{Disc}(\phi_3^*) = -3^6 \cdot 11^3$ and $\operatorname{Disc}(\phi_4^*) = 3^4 \cdot 11^4 \cdot 13^3 \cdot 41^4$.

The local theory of units described in Section 2.7 allows us to construct units (or at least $S$-units) in dynatomic fields $K_{n,\phi}$ and their composita. For convenience, we make two definitions.

**Definition.** Let $S$ be a finite set of places of $K$ and let $L/K$ be a finite extension. We say that $u \in L$ is an *$S$-unit* if it is an $S_L$-unit, where $S_L$ is the set of places of $L$ lying over $S$.

**Definition.** Let $\phi(z) \in K(z)$ be a rational map. We write $S_\phi$ for the set

$$S_\phi = M_K^\infty \cup \{\text{primes at which } \phi \text{ has bad reduction}\}.$$

Thus $\phi$ has good reduction at all primes in the localized ring $R_S$. In particular, if

$$\phi(z) = a_0 z^d + \cdots + a_d$$

is a polynomial, then the finite primes in $S_\phi$ are the primes where some $a_i$ is nonintegral together with any primes for which $a_0$ is not a unit.

Having set this notation, we now state globalized versions of the dynamical unit theorems from Section 2.7.

**Theorem 3.66 (Global version of Theorem 2.33).** *Let $\phi(z) \in K[z]$ be a polynomial of degree $d \geq 2$, let $\alpha \in \operatorname{Per}_n^{**}(\phi)$ with $n \geq 2$, and fix integers $i$ and $j$ satisfying $\gcd(i-j,n) = 1$. Then*

$$\frac{\phi^i(\alpha) - \phi^j(\alpha)}{\phi(\alpha) - \alpha}$$

*is an $S_\phi$-unit in $K_{\phi,n}$.*

For the next result, we recall that the cross-ratio of four points $P_1, P_2, P_3, P_4$ in $\mathbb{P}^1$ is the quantity

$$\kappa(P_1, P_2, P_3, P_4) = \frac{(x_1y_3 - x_3y_1)(x_2y_4 - x_4y_2)}{(x_1y_2 - x_2y_1)(x_3y_4 - x_4y_3)}.$$

**Theorem 3.67 (Global version of Theorem 2.34).** *Let $\phi(z) \in K(z)$ be a rational function of degree $d \geq 2$, let $P \in \operatorname{Per}_n^{**}(\phi)$, and fix integers $i$ and $j$ satisfying*

$$\gcd(j,n) = \gcd(i-1,n) = \gcd(i-j,n) = 1.$$

*Then*

$$\kappa(P, \phi P, \phi^i P, \phi^j P)$$

*is an $S_\phi$-unit in $K_{\phi,n}$.*

**Theorem 3.68 (Global version of Theorem 2.35).** *Let $\phi(z) \in K(z)$ be a rational function of degree $d \geq 2$, let $m$ and $n$ be integers with $m \nmid n$ and $n \nmid m$, and let*

$$P = [x,y] \in \mathrm{Per}_m^{**}(\phi) \qquad \text{and} \qquad Q = [x',y'] \in \mathrm{Per}_n^{**}(\phi).$$

*Denote by $S_P$ the set of places*

$$S_P = \{v \in M_K^0 : v(x) > 0 \text{ and } v(y) > 0\} \cup \{v \in M_K^0 : v(x) < 0 \text{ or } v(y) < 0\},$$

*and similarly for $S_Q$. Then*

$$xy' - x'y$$

*is an $(S_\phi \cup S_P \cup S_Q)$-unit in the compositum $K_{\phi,m}K_{\phi,n}$.*

Theorems 3.66, 3.67, and 3.68 allow us to construct many $S$-units in dynatomic fields $K_{\phi,n}$ and their composita $K_{\phi,m}K_{\phi,n}$. Further, since $\mathrm{Gal}(K_{\phi,n}/K)$ frequently contains an element $\sigma_\phi$ whose action on points $P \in \mathrm{Per}_n^{**}(\phi)$ is characterized by the equation $\sigma_\phi(P) = \phi(P)$, we obtain a partial description of the action of the Galois group on the dynamical units. For example, the units in Theorem 3.67 transform via

$$\sigma_\phi\bigl(\kappa(P, \phi P, \phi^i P, \phi^j P)\bigr) = \kappa(\phi P, \phi^2 P, \phi^{i+1} P, \phi^{j+1} P).$$

We compare this to the construction of cyclotomic units in cyclotomic fields. Let $q$ be a prime power and let $\zeta$ be a primitive $q^{\text{th}}$ root of unity. Then the cyclotomic field $\mathbb{Q}(\zeta)$ contains the units

$$\frac{\zeta^i - 1}{\zeta - 1} \quad \text{for } 2 \leq i \leq \frac{q-1}{2} \text{ with } \gcd(i,q).$$

The Galois group $\mathrm{Gal}\bigl(\mathbb{Q}(\zeta)/\mathbb{Q}\bigr)$ is the set of elements $\sigma_t$ characterized by

$$\sigma_t(\zeta) = \zeta^t \quad \text{with } 0 < t < q \text{ and } p \nmid t.$$

The action of the Galois group on the cyclotomic units is given by the explicit formula

$$\sigma_t\left(\frac{\zeta^i - 1}{\zeta - 1}\right) = \frac{\zeta^{it} - 1}{\zeta^t - 1}.$$

Further, the cyclotomic units generate a subgroup of finite index in the full group of units $\mathbb{Z}[\zeta]^*$, and the index of this subgroup is related to the class number of $\mathbb{Q}(\zeta)$.

The situation for dynatomic fields is not nearly as complete. One problem is that the dynatomic fields $K_{\phi,n}$ tend to have very large degree over $K$, so the dynamical unit theorems cannot produce enough units to give a subgroup of finite index in the full unit group. Further, the Galois group $\mathrm{Gal}\bigl(K_{\phi,n}/K\bigr)$ is usually huge, and we have an explicit description only of the subgroup generated by the element $\sigma_\phi$. However, since for general number fields there is no known way to systematically produce any units with any explicit Galois action, the dynatomic construction might be said to fall under the heading of "half a loaf is better than none."

*Example* 3.69. Consider the rational map

$$\phi(z) = z^2 - 4.$$

After some algebra, we find that

$$\begin{aligned}\phi(z) - z &= z^2 - z - 4,\\ \frac{\phi^2(z) - z}{\phi(z) - z} &= z^2 + z - 3,\\ \frac{\phi^3(z) - z}{\phi(z) - z} &= (z^3 - z^2 - 6z + 7)(z^3 + 2z^2 - 3z - 5).\end{aligned}$$

It is not hard to check that $\phi(z) - z$ divides $\phi^n(z) - z$ for all $n \geq 1$ (Exercise 1.19(a)), but the further factorization of $\phi^3(z) - z$ into a product of cubics is less common; see Exercise 3.49. If we let $\alpha, \beta \in \mathbb{C}$ satisfy

$$\alpha^3 - \alpha^2 - 6\alpha + 7 = 0 \qquad \text{and} \qquad \beta^3 + 2\beta^2 - 3\beta - 5 = 0,$$

then we have

$$\operatorname{Per}_1^{**}(\phi) = \left\{\frac{1 \pm \sqrt{17}}{2}\right\}, \qquad \operatorname{Per}_2^{**}(\phi) = \left\{\frac{-1 \pm \sqrt{13}}{2}\right\},$$
$$\operatorname{Per}_3^{**}(\phi) = \left\{\alpha, \phi(\alpha), \phi^2(\alpha), \beta, \phi(\beta), \phi^2(\beta)\right\},$$

where we recall that $\operatorname{Per}_n^{**}(\phi)$ denotes the set of points of exact period $n$ for $\phi$.

Let $K = \mathbb{Q}(\alpha)$. The polynomial $z^3 - z^2 - 6z + 7$ is irreducible over $\mathbb{Q}$, but it factors completely in $K$ since its roots are $\alpha$, $\phi(\alpha)$, and $\phi^2(\alpha)$. Hence $K$ is Galois over $\mathbb{Q}$ with Galois group generated by the map $\sigma$ determined by $\sigma(\alpha) = \phi(\alpha)$. Notice that the discriminant of $z^3 - z^2 - 6z + 7$ is $19^2$, again confirming that its roots generate a cyclic cubic Galois extension. Further, $\mathbb{Z}[\alpha]$ must be the full ring of integers $R_K$ of $K$, since

$$19^2 = \operatorname{Disc}(\mathbb{Z}[\alpha]) = \left[R : \mathbb{Z}[\alpha]\right]^2 \operatorname{Disc}(K/\mathbb{Q}) > \left[R_K : \mathbb{Z}[\alpha]\right]^2.$$

(Note that $K \neq \mathbb{Q}$, so $\operatorname{Disc}(K/\mathbb{Q}) > 1$.) Then the fact that 19 is prime forces $R_K = \mathbb{Z}[\alpha]$.

We apply Theorem 3.66 with $i = 2$ and $j = 1$ to obtain the unit

$$u_1 = \frac{\phi^2(\alpha) - \phi(\alpha)}{\phi(\alpha) - \alpha} = \alpha^2 + \alpha - 4 \in R_K^*.$$

It is easy to check that $\mathrm{N}_{K/\mathbb{Q}}(u_1) = 1$, confirming that $u_1$ is a unit. Similarly, taking $i = 2$ and $j = 0$ gives the unit

$$u_2 = \frac{\phi^2(\alpha) - \alpha}{\phi(\alpha) - \alpha} = \alpha^2 + \alpha - 3 \in R_K^*.$$

In this case, $\mathrm{N}_{K/\mathbb{Q}}(u_2) = -1$.

The field $K$ is totally real of degree 3, so its unit group $R_K^*$ has rank 2. It is not hard to see that the two units $u_1$ and $u_2$ are independent, so they at least generate a subgroup of finite index. (In fact, $\{-1, u_1, u_2\}$ generates the full unit group $R_K^*$, but we leave the verification of this fact to the reader.) We can compute the action of the Galois group on $u_1$ and $u_2$,

$$\sigma(u_1) = \sigma(\alpha^2 + \alpha - 4) = \phi(\alpha)^2 + \phi(\alpha) - 4 = \alpha^4 - 7\alpha^2 + 8 = -\alpha + 1,$$
$$\sigma(u_2) = \sigma(\alpha^2 + \alpha - 3) = \phi(\alpha)^2 + \phi(\alpha) - 3 = \alpha^4 - 7\alpha^2 + 9 = -\alpha + 2.$$

These new units are related to the original units by

$$\sigma(u_1) = -u_1^{-1}u_2 \qquad \text{and} \qquad \sigma(u_2) = -u_1^{-1}.$$

Using points of period 2 and 3 for $\phi$, we can create units in larger fields. Thus let $L = K(\sqrt{13}\,)$. Then $L$ contains both $\alpha$ and the points in $\mathrm{Per}_2^{**}(\phi)$, so Theorem 3.68 with $n_1 = 2$ and $n_2 = 3$ says that

$$u_3 = \gamma - \alpha \qquad \text{with} \quad \gamma = \frac{-1+\sqrt{13}}{2}$$

is a unit in the ring of integers of $L$. If we take the norm from $L$ down to $K$, we find that $\mathrm{N}_{L/K}(u_3) = u_2$ is one of the units that we already discovered. Similarly, $\mathrm{N}_{L/K}(\gamma - \phi(\alpha)) = -\alpha + 2$.

If instead we compute the norm of $u_3$ from $L$ down to $\mathbb{Q}(\gamma)$, we find using $\gamma^2 + \gamma - 3 = 0$ that

$$\mathrm{N}_{L/\mathbb{Q}(\gamma)}(\gamma - \alpha) = \gamma^3 - \gamma^2 - 6\gamma + 7 = -\gamma + 1 = \frac{3-\sqrt{13}}{2}.$$

This unit and $-1$ generate the unit group of the ring of integers of $\mathbb{Q}(\sqrt{13})$.

For additional information about cyclic cubic extensions generated by periodic points of polynomials, see [306], and for a general analysis of units generated by 3-periodic points of quadratic polynomials, see [313, Section 8].

# Exercises

### Section 3.1. Height Functions

**3.1.** Show that the constant $C(d, N, D)$ in Conjecture 3.15 must depend on each of the quantities $d$, $N$, and $D$ by giving a counterexample if any one of them is dropped.

**3.2.** Let

$$\nu(B) = \#\{P \in \mathbb{P}^N(\mathbb{Q}) : H(P) \le B\}.$$

(a) Find positive constants $c_1$ and $c_2$ such that

$$c_1 B^{N+1} \le \nu(B) \le c_2 B^{N+1} \qquad \text{for all } B \ge 1.$$

(b) For $N = 1$, prove that
$$\lim_{B\to\infty} \frac{\nu(B)}{B^2} = \frac{12}{\pi^2}.$$

(c) More generally, prove that the limit $\lim_{B\to\infty} \nu(B)/B^{N+1}$ exists and express it in terms of a value of the Riemann $\zeta$-function.

**3.3.** Prove that
$$\#\{P \in \mathbb{P}^N(\bar{\mathbb{Q}}) : H(P) \le B \text{ and } D(P) \le D\} \le (12D)^N 2^{ND^2} B^{ND(D+1)}.$$
Aside from the constant, to what extent can you improve this estimate? In particular, can the exponent of $B$ be improved?

**3.4.** Let $F(X) = a_0X^d + a_1X^{d-1} + \cdots + a_d \in \mathbb{Q}[X]$ be a polynomial of degree $d$, and factor $F(X)$ as
$$F(X) = a_0(X-\alpha_1)(X-\alpha_2)\cdots(X-\alpha_d)$$
over the complex numbers. Prove that
$$2^{-d}H(\alpha_1)\cdots H(\alpha_d) \le H([a_0,a_1,\ldots,a_d]) \le 2^d H(\alpha_1)\cdots H(\alpha_d).$$
(*Hint.* Mimic the proof of Theorem 3.7 for the upper bound. To prove the lower bound, for each $v$ pull out the root with largest $|\alpha_i|_v$ and use induction on the degree of $F$.) Can you increase the $2^{-d}$ and/or decrease the $2^d$?

**3.5.** Prove that the number of $(N+1)$-tuples $(i_0,\ldots,i_N) \in \mathbb{Z}^{N+1}$ of integers satisfying
$$i_0,\ldots,i_N \ge 0 \qquad \text{and} \qquad i_0 + \cdots + i_N = d$$
is given by the combinatorial symbol $\binom{N+d}{d}$. Note that this is equal to the number of monomials of degree $d$ in the $N+1$ variables $x_0,\ldots,x_N$.

Section 3.2. Height Functions and Geometry

**3.6.** Let $\phi : \mathbb{P}^2 \to \mathbb{P}^2$ be the rational map $\phi(X,Y,Z) = [X^2,Y^2,XZ]$. Although $\phi$ is not defined at $[0,0,1]$, we can define $\operatorname{Per}(\phi,\mathbb{P}^2)$ to be the set of points satisfying $\phi^n(P) = P$ for some $n \ge 0$ and $\phi^i(P) \ne [0,0,1]$ for all $0 \le i < n$. Prove that Theorem 3.12 is false by showing that $\operatorname{Per}(\phi,\mathbb{P}^2(\mathbb{Q}))$ is infinite. What goes wrong with the proof? Try to find a "large" subset $S$ of $\mathbb{P}^2$ such that $\operatorname{Per}(\phi,S(K))$ is finite for every number field $K$.

**3.7.** Let $K/\mathbb{Q}$ be a number field, let $P = [x_0,\ldots,x_N] \in \mathbb{P}^N(K)$, and let $\mathfrak{b}$ be the fractional ideal generated by $x_0,\ldots,x_N$. Prove that
$$H_K(P) = \frac{1}{N_{K/\mathbb{Q}}(\mathfrak{b})} \prod_{v\in M_K^\infty} \max\{|x_0|_v,\ldots,|x_N|_v\}^{n_v}.$$

**3.8.** Let $K/\mathbb{Q}$ be a number field and let $\phi(z) \in K(z)$ be a rational map of degree $d \ge 2$. Recall that the height $H(\phi)$ is defined by writing
$$\phi = [F(X,Y),G(X,Y)] = [a_0X^d + a_1X^{d-1} + \cdots + a_dY^d, b_0X^d + \cdots + b_dY^d]$$
and setting $H(\phi) = H\bigl([a_0,\ldots,a_d,b_0,\ldots,b_d]\bigr)$. (See (3.4) on page 91.) Prove that there are positive constants $c_1(d)$ and $c_2(d)$ such that
$$c_1H(\phi)^{-(2d-1)}H(P)^d \le H(\phi(P)) \le c_2H(\phi)H(P)^d \qquad \text{for all } P \in \mathbb{P}^1(\bar{K}).$$
Find expressions for $c_1$ and $c_2$ in terms of $d$. This gives an explicit version of Theorem 3.11 for $\mathbb{P}^1$.

**3.9.** Let $\phi_c(z) = z^2 + c$.

(a) Prove that there are infinitely many $c \in \mathbb{Q}$ such that $\phi_c$ has a $\mathbb{Q}$-rational fixed point.

(b) Prove that there are infinitely many $c \in \mathbb{Q}$ such that $\phi_c$ has a $\mathbb{Q}$-rational point of exact period 2.

(c) Prove that there are infinitely many $c \in \mathbb{Q}$ such that $\phi_c$ has a $\mathbb{Q}$-rational point of exact period 3.

(d) Prove that there are no $c \in \mathbb{Q}$ such that $\phi_c$ has a $\mathbb{Q}$-rational point of exact period 4.

**3.10.** Let $d \geq 2$ and let $\phi_d(z) = z^d$. Prove that there is an absolute constant $c$ such that for all number fields $K/\mathbb{Q}$ of degree $n$ we have

$$\#\operatorname{PrePer}\big(\phi_d, \mathbb{P}^1(K)\big) \leq c[K:\mathbb{Q}] \log\log\big([K:\mathbb{Q}]\big).$$

Prove that aside from the constant, this upper bound cannot be improved. (*Hint.* You will need the fact that the Euler totient function $\varphi$ satisfies $\varphi(m) \leq cm/\log\log m$; see for example [11, Theorem 13.14].)

In particular, the uniform boundedness conjecture (Conjecture 3.15) is true for $\phi_d$. This is one of the few maps for which the conjecture is known. The others are the Chebyshev polynomials and Lattès maps; see Theorem 6.65.

## Section 3.4. Canonical Heights and Dynamical Systems

**3.11.** Let $\phi : \mathbb{P}^N \to \mathbb{P}^N$ be a morphism defined over $K$, let $f \in \mathrm{PGL}_{N+1}(K)$ be an automorphism of $\mathbb{P}^N$, and let $\phi^f = f^{-1} \circ \phi \circ f$. Prove that

$$\hat{h}_{\phi^f}(P) = \hat{h}_\phi(f(P)) \qquad \text{for all } P \in \mathbb{P}^N(\bar{K}).$$

Thus the canonical height is conjugation-invariant, i.e., it is independent of change of coordinates.

**3.12.** Let $\phi, \psi : \mathbb{P}^N \to \mathbb{P}^N$ be morphisms of degree at least two that commute with one another, i.e., $\phi\big(\psi(P)\big) = \psi\big(\phi(P)\big)$ for all points $P \in \mathbb{P}^N(\bar{\mathbb{Q}})$. Prove that

$$\hat{h}_\phi(P) = \hat{h}_\psi(P) \qquad \text{for all } P \in \mathbb{P}^N(\bar{\mathbb{Q}}).$$

**3.13.** Let $\phi : \mathbb{P}^N \to \mathbb{P}^N$ be a morphism of degree at least two defined over a number field $K$ and let $P \in \mathbb{P}^N(\bar{K})$ have the property that

$$\mathcal{O}_\phi\big(\sigma(P)\big) \cap \mathcal{O}_\phi(P) \neq \emptyset \quad \text{for all } \sigma \in \operatorname{Gal}(\bar{K}/K).$$

Prove that either $P$ is preperiodic for $\phi$ or else there is some point in the orbit of $P$ satisfying $\phi^n(P) \in \mathbb{P}^N(K)$. Is this result true if instead $K/\mathbb{Q}_p$ is a $p$-adic field?

**3.14.** In the setting of Theorem 3.20, suppose that the set $S$ is a topological space and that the maps $\phi : S \to S$ and $h : S \to \mathbb{R}$ are continuous maps. Prove that $\hat{h} : S \to \mathbb{R}$ is also a continuous map. (*Hint.* Show that the functions $h_m(P) = d^{-m} h\big(\phi^m(P)\big)$ for $m = 1, 2, 3, \ldots$ are continuous and converge uniformly to $\hat{h}$.)

**3.15.** Let $\hat{h}_\phi$ be the canonical height for $\phi(z) = z^2 + c$. For any given $c \in \mathbb{Q}$, there is a minimum nonzero value for $\hat{h}_\phi(z)$ as $z$ ranges over nonpreperiodic points in $\mathbb{Q}$. Find that minimum value for:

(a) $c = 0$.

(b) $c = -2$.
(c) $c = -1$.
(d) ** arbitrary $c \in \mathbb{Z}$.
(e) ** arbitrary $c \in \mathbb{Q}$.
(*Hint.* Try to do (a), (b), and (c) directly, but we note that it is easier to do them using the theory of local canonical heights for polynomials; see Exercises 3.24 and 3.28. For (d) and (e) the goal is to describe the minimum value of $\hat{h}_\phi$ in terms of $c$.)

**3.16.** Let $\phi(z) = z^2 - 1$. Analyze the canonical height for points in the following sets.
(a) $\mathcal{B}_\phi(\infty)$, the attracting basin of $\infty$, where we recall (Exercise 2.1) that the attracting basin of $\infty$ for a polynomial $\phi$ is the set

$$\mathcal{B}_\phi(\infty) = \Bigl\{\alpha \in \mathbb{C} : \lim_{n\to\infty} \phi^n(\alpha) = \infty\Bigr\} \cup \{\infty\}.$$

(b) $\mathcal{F}_\phi \smallsetminus \mathcal{B}_\phi(\infty)$. (Note that all of the points in this set are eventually attracted to the attracting 2-cycle $0 \overset{\phi}{\longleftrightarrow} -1$.)
(c) $\mathcal{J}_\phi$.
(See also Exercise 3.27.)

**3.17.** Let $\phi : \mathbb{P}^N \to \mathbb{P}^N$ be a morphism defined over a number field $K$, and for any point $P \in \mathbb{P}^N(\bar{K})$, let $D(P) = [K(P) : \mathbb{Q}]$. Prove that there is a constant $C = C(\phi) > 0$ such that

$$\hat{h}_\phi(P) \ge \frac{1}{d^{CD(P)^2}} \qquad \text{for all } P \in \mathbb{P}^N(\bar{K}) \text{ with } P \notin \operatorname{PrePer}(\phi).$$

(*Hint.* Use the estimate for the number of points of bounded degree and height given in Exercise 3.3.) This is a very weak version of the dynamical Lehmer Conjecture 3.25.

**3.18.** ** Let $c \in \mathbb{Q}$ with $c \ne 0, -2$ and let $\phi(z) = z^2 + c$. Prove that there exist $\kappa = \kappa(c) > 0$ and $N = N(c) > 0$ such that

$$\hat{h}_\phi(\alpha) \ge \frac{\kappa}{[\mathbb{Q}(\alpha) : \mathbb{Q}]^N} \qquad \text{for all } \alpha \in \bar{\mathbb{Q}} \text{ with } \alpha \notin \operatorname{PrePer}(\phi).$$

(This is not currently known for any value of $c$ other than $c = 0$ and $c = -2$.)

**3.19.** Let $\alpha \in \bar{\mathbb{Q}}^*$ and let

$$f(x) = a_0x^d + a_1x^{d-1} + \cdots + a_d \in \mathbb{Z}[x]$$

be the minimal polynomial of $\alpha$, normalized so that $a_0 > 0$ and $\gcd(a_0, \dots, a_d) = 1$. Factor $f(x)$ over $\mathbb{C}$ as $f(x) = a_0 \prod(x - \alpha_i)$. Prove that

$$H(\alpha)^d = a_0 \prod_{i=1}^{d} \max\bigl\{1, |\alpha_i|\bigr\} = \int_0^1 \log\bigl|f(e^{2\pi i\theta})\bigr|\, d\theta.$$

The quantity $H(\alpha)^d$, or equivalently $H_{\mathbb{Q}(\alpha)}(\alpha)$, is also known as the *Mahler measure of* $\alpha$ and denoted by $M(\alpha)$. Lehmer's conjecture (Conjecture 3.24) can be stated in terms of Mahler measure: There is an absolute constant $\kappa > 1$ such that if $\alpha \in \bar{\mathbb{Q}}^*$ is not a root of unity, then $M(\alpha) > \kappa$.

**3.20.** Let $\phi(z) \in \mathbb{Q}(z)$. Write a program to estimate $\hat{h}_\phi(\alpha)$ directly from the definition. Use your program to compute the following heights to a few decimal places. (See also Exercises 3.30 and 5.31.)
(a) $\phi(z) = z^2 - 1$ and $\alpha = \frac{1}{2}$.
(b) $\phi(z) = z^2 + 1$ and $\alpha = \frac{1}{2}$.
(c) $\phi(z) = 3z^2 - 4$ and $\alpha = 1$.
(d) $\phi(z) = z + \dfrac{1}{z}$ and $\alpha = 1$.
(e) $\phi(z) = \dfrac{3z^2 - 1}{z^2 - 1}$ and $\alpha = 1$.

**3.21.** Let $\phi(z) = z^2 - z + 1$. The $\phi$-orbit of the point 2 is called *Sylvester's sequence* [428],

$$\mathcal{O}_\phi(2) = \{2, 3, 7, 43, 1807, 3263443, 10650056950807, \\ 113423713055421844361000443, \dots\}.$$

(a) Prove that Sylvester's sequence satisfies

$$\phi^{n+1}(2) = 1 + \phi^0(2)\phi^1(2)\phi^2(2)\phi^3(2)\cdots\phi^n(2).$$

(b) A rough approximation gives $\hat{h}_\phi(2) \approx 0.468696$, so $\phi^n(2)$ is approximately equal to $e^{0.468696n^2}$. Prove a more accurate statement by showing that

$$e^{n^2\hat{h}_\phi(\alpha)} - \phi^n(2), \qquad n = 0, 1, 2, \dots,$$

is positive, strictly increasing, and converges to $\frac{1}{2}$. (*Hint*. First conjugate $\phi(z)$ to put it into the form $z^2 + c$.)

(c) Deduce that there is a real number $H$ such that $\phi^n(2)$ is the closest integer to $H^{n^2}$ for all $n \geq 0$. Note that this is far stronger than the general height estimate

$$\log \phi^n(2) = h\big(\phi^n(2)\big) = \hat{h}_\phi\big(\phi^n(2)\big) + O(1) = 2^n\hat{h}_\phi(2) + O(1).$$

**3.22.** Let $\delta > 0$, let $d \geq 2$ be an integer, and let $(x_n)_{n\geq 0}$ be a sequence of positive real numbers with the property that

$$x_0 > 1 + \delta \qquad \text{and} \qquad \left|x_{n+1} - x_n^d\right| \leq \delta x_n^{d-1} \quad \text{for all } n \geq 0.$$

(a) Prove that the sequence $x_n$ is strictly increasing and that $x_n \to \infty$ as $n \to \infty$.
(b) Prove that the limit

$$H = \lim_{n\to\infty} x_n^{1/d^n}$$

exists. (*Hint*. Take logs and use a telescoping sum to show that the sequence is Cauchy.)
(c) Prove that

$$\left|H^{n^2} - x_n\right| \leq \frac{\delta}{d-1} \quad \text{for all } n \geq 0.$$

**3.23.** Let $d \geq 2$ and let $\phi(z) \in \mathbb{Z}[z]$ be a monic polynomial of degree $d$, say

$$\phi(z) = z^d + az^{d-1} + \cdots \in \mathbb{Z}[z].$$

Prove that for every $\epsilon > 0$ there is a constant $C = C(\phi, \epsilon)$ such that for all $\alpha \in \mathbb{Z}$ satisfying $|\alpha| \geq C$ we have

$$\left|e^{\hat{h}_\phi(\alpha)n^2} + \frac{a}{d} - \phi^n(\alpha)\right| \leq \epsilon \qquad \text{for all } n \geq 0.$$

Section 3.5. Local Canonical Heights

The general theory of local canonical heights is developed in Section 5.9. However, the theory becomes much simpler if $\phi$ is a polynomial, because the local canonical height then has a simple limit definition similar to the limit used to define the global canonical height. Exercises 3.24–3.30 ask you to develop some of the theory of local canonical heights for polynomials.

**3.24.** Let $\phi(z) \in K[z]$ be a polynomial. Prove that the limit

$$\hat{\lambda}_{\phi,v}(\alpha) = \lim_{n\to\infty} \frac{1}{d^n} \log \max\bigl\{|\phi^n(\alpha)|_v, 1\bigr\} \tag{3.49}$$

exists and that the resulting function has the following two properties:

(a) For all $\alpha \in \bar{K}_v$,

$$\hat{\lambda}_{\phi,v}\bigl(\phi(\alpha)\bigr) = d\hat{\lambda}_{\phi,v}(\alpha).$$

(b) The function

$$\alpha \longmapsto \hat{\lambda}_{\phi,v}(\alpha) - \log \max\bigl\{|\alpha|_v, 1\bigr\}$$

is continuous on $\bar{K}_v$ and has a finite limit as $|\alpha|_v \to \infty$.

Hence the function defined by (3.49) is a local canonical height as described in Theorem 3.27. Prove that the (global) canonical height is equal to the sum of the local heights,

$$\hat{h}_\phi(\alpha) = \sum_{v\in M_K} n_v \hat{\lambda}_{\phi,v}(\alpha).$$

**3.25.** Let $\phi(z) \in K[z]$ be a polynomial and let $v$ be an absolute value on $K$. Prove that the local height $\hat{\lambda}_{\phi,v}$ as defined by (3.49) has the following properties.

(a) $\hat{\lambda}_{\phi,v}(\alpha) \ge 0$ for all $\alpha \in \bar{K}_v$.

(b) $\hat{\lambda}_{\phi,v}(\alpha) = 0$ if and only if $|\phi^n(\alpha)|_v$ is bounded, equivalently, if and only if $\alpha$ is in the $v$-adic filled Julia set $\mathcal{K}_v(\phi)$ of $\phi$ (see Exercise 2.1).

**3.26.** Let $\phi(z) \in \mathbb{C}[z]$ be a polynomial with complex coefficients and let $\mathcal{B}_\phi(\infty)$ be the attracting basin of $\infty$ for $\phi$. (See Exercises 2.1 and 3.16 for the definition of $\mathcal{B}_\phi(\infty)$.) Prove that the local height

$$\hat{\lambda}_\phi : \mathbb{C} \to \mathbb{R}$$

as defined by (3.49) has the following properties:

(a) $\hat{\lambda}_\phi$ is a real analytic function on $\mathcal{B}_\phi(\infty) \smallsetminus \{\infty\}$.

(b) The function $\hat{\lambda}_\phi$ is harmonic on $\mathcal{B}_\phi(\infty) \smallsetminus \{\infty\}$. In other words, writing $z = x+iy$, the function $\hat{\lambda}_\phi$ is a solution to the differential equation

$$\left(\frac{\partial^2}{\partial x^2} + \frac{\partial^2}{\partial y^2}\right)\hat{\lambda}_\phi = 0$$

on the open set $\mathcal{B}_\phi(\infty) \smallsetminus \{\infty\}$. It satisfies the boundary conditions that $\hat{\lambda}_\phi$ vanishes on the boundary of $\mathcal{B}_\phi(\infty)$ and has a logarithmic singularity at $z=\infty$, i.e.,

$$\hat{\lambda}_\phi(z) - \log|z| \quad \text{is bounded as } z \to \infty.$$

(c) $\hat{\lambda}_\phi$ is the unique function that has the properties described in (b).

In classical terminology, the function $\hat{\lambda}_\phi$ is the *Green function for the filled Julia set* $\mathcal{K}(\phi) = \mathbb{P}^1(\mathbb{C}) \smallsetminus \mathcal{B}_\phi(\infty)$.

**3.27.** Let $\phi(z) = z^2 - 1$. Analyze the local canonical height $\hat{\lambda}_{\phi,v}$ for points in:

(a) the attracting basin $\mathcal{B}_\phi(\infty)$ of $\infty$.

(b) $\mathcal{F}_\phi \smallsetminus \mathcal{B}_\phi(\infty)$.

(c) $\mathcal{J}_\phi$.

(See also Exercise 3.16.)

**3.28.** Let $\phi(z) \in \mathbb{Z}[z]$ be a monic polynomial of degree $d \ge 2$ and let $\alpha = a/b \in \mathbb{Q}$ be a rational number written in lowest terms. Prove that

$$\hat{h}_\phi(\alpha) \ge \log|b|,$$

with equality if and only if $\alpha$ is in the filled Julia set $\mathcal{K}(\phi)$ of $\phi$.

**3.29.** Let $K/\mathbb{Q}$ be a number field and let $\phi(z) \in K(z)$ be a rational map of degree $d \ge 2$. For each finite place $v \in M_K^0$, write $\phi(z) = F_v(z)/G_v(z)$ as a ratio of polynomials that are normalized for $v$, i.e., the coefficients of $F_v$ and $G_v$ are $v$-adic integers and at least one coefficient is a $v$-adic unit. Let $\hat{\lambda}^{\text{norm}}_{\phi,v}$ be the local canonical height, as described in Theorem 3.27, normalized using $G_v$ in (3.16).

(a) Prove that $\hat{\lambda}^{\text{norm}}_{\phi,v}$ is well-defined, independent of the decomposition $\phi(z)$ as a ratio of $v$-normalized polynomials. For nonarchimedean places $v$ this serves to pin down a specific function $\hat{\lambda}^{\text{norm}}_{\phi,v}$ that depends only on $\phi$ and $v$ (cf. Remark 3.28).

(b) Give an example to show that it may not be possible to write $\phi(z) = F(z)/G(z)$ such that $F$ and $G$ are simultaneously normalized for all finite places $v \in M_K^0$.

(c) More generally, prove that every $\phi \in K(z)$ can be written as $F(z)/G(z)$ with $F$ and $G$ simultaneously normalized for all finite places $v \in M_K^0$ if and only if $K$ has class number 1.

**3.30.** Suppose that $\phi(z) \in K(z)$ has good reduction at a finite place $v \in M_K$. Prove that the function

$$\hat{\lambda}_{\phi,v}(\alpha) = \log\max\bigl\{|\alpha|_v, 1\bigr\} \qquad \text{for } \alpha \in K_v$$

is a local canonical height by showing that it has the required properties.

### Section 3.6. Diophantine Approximation

**3.31.** Prove Dirichlet's Theorem 3.33, which says that for every $\alpha \in \mathbb{R} \smallsetminus \mathbb{Q}$ there are infinitely many $x/y \in \mathbb{Q}$ satisfying

$$\left|\frac{x}{y} - \alpha\right| \le \frac{1}{y^2}.$$

(*Hint.* Look at the numbers $y\alpha - \lfloor y\alpha \rfloor$ for $0 \le y \le A$. They all lie in the interval $[0, 1]$. Divide the interval into $A$ equal pieces and use the pigeonhole principle.)

**3.32.** Let $B$ be a nonzero integer.

(a) Prove that every solution $(x, y)$ in integers to the equation

$$X^3 + Y^3 = B \tag{3.50}$$

satisfies $\max\bigl\{|x|, |y|\bigr\} \le \sqrt{4B/3}$. (*Hint.* The polynomial $X^3+Y^3$ factors in $\mathbb{Q}[X, Y]$.)

(b) Try to find an explicit upper bound for $\max\{|x|, |y|\}$ for the similar-looking equation $X^3 + 2Y^3 = B$. The difficulty you face illustrates the fact, seen during the proof of Theorem 3.35, that the equation $G(X, Y) = B$ is relatively easy to solve if $G(X, Y)$ has distinct factors in $\mathbb{Q}[X, Y]$, but very difficult if it does not.

(c) Find all solutions in integers $x \geq y$ to the equation (3.50) for the following values of $B$: (i) $B = 2$. (ii) $B = 91$. (iii) $B = 728$. (iv) $B = 1729$.

## Section 3.7. Integral Points in Orbits

**3.33.** Let $\phi(z) = z + 1/z$ and write $\phi^n(1) = a_n/b_n$ in lowest terms as in Example 3.50.

(a) Prove by a direct computation that there are constants $c, c' > 0$ such that

$$c \leq \frac{\log a_n}{\log b_n} \leq c' \quad \text{for all } n \geq 2.$$

(b) Try to prove directly that $\log a_n / \log b_n \to 1$ as $n \to \infty$.

(c) Prove that $\phi^n(1) \to \infty$ as $n \to \infty$. (Notice that $\infty$ is a rationally indifferent fixed point, since $\phi'(\infty) = 1$.)

**3.34.** Let $c \in \mathbb{Z}$ be a squarefree integer and let $\phi_c(z) = z + c/z$. Further, let $\alpha \in \mathbb{Q}$ be a wandering point for $\phi_c$.

(a) If $1 - 4c$ is not a perfect square, prove that $\#\bigl(\mathcal{O}_\phi(\alpha) \cap \mathbb{Z}\bigr) \leq 2$.

(b) If $1-4c = d^2$ is a perfect square, prove that $\#\bigl(\mathcal{O}_\phi(\alpha) \cap \mathbb{Z}\bigr) \leq 2$ unless $\alpha = (-1 \pm d)/2$, in which case there are three integer points in the orbit.

**3.35.** Let $\phi(z) \in \mathbb{Q}(z)$ be a rational function of degree $d \geq 2$ such that $\phi^2$ is not a polynomial. Suppose that $\phi$ has everywhere good reduction and that $\phi(\infty) = \infty$. Let $\alpha \in \mathbb{Q}$ be a wandering point for $\phi$. Prove that

$$\#\bigl(\mathcal{O}_\phi(\alpha) \cap \mathbb{Z}\bigr) \leq 2d - 1.$$

Show that this is sharp for $d = 2$. Try to improve the bound for larger values of $d$.

**3.36.** Let $\phi(z) \in \mathbb{Q}(z)$ be a rational function of degree $d \geq 2$, let $\alpha \in \mathbb{Q}$ be a wandering point for $\phi$, and write $\phi^n(\alpha) = a_n/b_n \in \mathbb{Q}$ as a fraction in lowest terms.

(a) Suppose that there is a nonzero integer $B$ such that $b_n | B$ for infinitely many $n$. Prove that $\phi^2(z) \in \mathbb{Q}[z]$. (*Hint*. Make a change of variables and apply Theorem 3.43.)

(b) Suppose that there are infinitely many $n$ such that $b_n | b_{n+1}$. Prove that either $\infty$ is a fixed point of $\phi$ or else $\phi^2(z) \in \mathbb{Q}[z]$.

**3.37.** Let $\phi(z) \in \mathbb{C}(z)$ be a rational function of degree $d$. This exercise gives a quantitative strengthening of Proposition 3.44.

(a) Let $P$ be a nonperiodic point for $\phi$. Prove that

$$\#\phi^{-n}(P) \geq d^{n-2} \qquad \text{for all } n \geq 0.$$

(b) Let $P$ be a fixed point of $\phi$ that is not totally ramified. Prove that

$$\#\phi^{-n}(P) \geq d^{n-2} + 2 \qquad \text{for all } n \geq 2.$$

(c) Generalize (b) to the case that $P$ is a periodic point for $\phi$ of exact period $m$ under the assumption that $\phi^m$ is not totally ramified at $P$.

Use (a), (b), and (c) to deduce that if $\phi$ is not a polynomial map, then

$$\#\phi^{-n}(P) \geq 3 \quad \text{if either} \quad \begin{cases} n \geq 4 \text{ and } d = 2, \\ n \geq 3 \text{ and } d \geq 3. \end{cases}$$

**3.38.** Let $K/\mathbb{Q}$ be a number field, let $S \subset M_K$ be a finite set of absolute values on $K$, and let $R_S$ be the ring of $S$-integers of $K$. Let $\phi(z) \in K(z)$ be a rational function of degree $d \geq 2$ with $\phi^2(z) \notin K[z]$, and let $\alpha \in K$ be a wandering point for $\phi$. Prove that

$$\mathcal{O}_\phi(\alpha) \cap R_S$$

is a finite set. (This exercise generalizes Theorem 3.43.)

**3.39.** Let $K$ be a number field, let $S \subset M_K$ be a finite set of absolute values on $K$ that includes all the archimedean absolute values, and let $R_S$ be the ring of $S$-integers of $K$. Let $\phi(z) \in K(z)$ be a rational function of degree $d \geq 2$ satisfying $\phi^2(z) \notin K[z]$.

(a) Let $n = 4$ if $d = 2$ and let $n = 3$ if $d \geq 3$. Prove that the set

$$\{\alpha \in K : \phi^n(\alpha) \in R_S\}$$

is finite.

(b) Give an example with $(n, d) = (3, 2)$ and $\infty$ nonperiodic for $\phi$ such that the set in (a) is infinite.

(c) Same question as in (b) with $n = 2$ and $d$ arbitrary.

(d) Repeat (b) and (c) with $\infty$ a fixed point of $\phi$.

**3.40.** Let $\phi_1, \dots, \phi_r \in \mathbb{Q}(z)$ be rational functions of degree at least 2, and let $\Phi$ be the collection of rational functions obtained by composing an arbitrary finite number of $\phi_1, \dots, \phi_r$. Note that each $\phi_i$ may be used many times. For example, if $r = 1$, then $\Phi$ is simply the collection of iterates $\{\phi_1^n\}$. Also note that in general, composition of functions is not commutative, so if $r \geq 2$, then $\Phi$ is likely to be a very large set.

(a) A rational map $\phi \in K(z)$ is said to be of *polynomial type* if it has a totally ramified fixed point. Prove that for such a map, there is a linear fractional transformation $f \in \mathrm{PGL}_2(\bar{K})$ such that $\phi^f \in K[z]$. Further, if $\phi$ is not conjugate to $z^d$, prove that one can take $f$ to be in $\mathrm{PGL}_2(K)$.

(b) Assume that $\Phi$ contains no maps of polynomial type. Prove that there are finite subsets $\Phi_1$ and $\Phi_2$ of $\Phi$ satisfying

$$\Phi = \Phi_1 \cup \bigcup_{\phi \in \Phi_2} \phi\Phi \qquad \text{and} \qquad \#\phi^{-1}(\infty) \geq 3 \quad \text{for all } \phi \in \Phi_2.$$

(Here $\phi\Phi$ denotes the composition of $\phi$ with every map in $\Phi$.)

(c) Let $\alpha \in \mathbb{Q}$. The *$\Phi$-orbit of $\alpha$* is the set $\mathcal{O}_\Phi(\alpha) = \{\phi(\alpha) : \phi \in \Phi\}$. Continuing with the assumption that $\Phi$ contains no maps of polynomial type, prove that the $\Phi$-orbit of $\alpha$ contains only finitely many integers, i.e., prove that

$$\mathcal{O}_\Phi(\alpha) \cap \mathbb{Z} \text{ is a finite set.}$$

**3.41.** Define the (logarithmic) height of a rational map $\phi(z) \in \mathbb{Q}(z)$ by writing

$$\phi(z) = \frac{F(z)}{G(z)} = \frac{a_0 + a_1 z + \cdots + a_d z^d}{b_0 + b_1 z + \cdots + b_d z^d}$$

with $F(z), G(z) \in \mathbb{Q}[z]$ and setting

$$h(\phi) = h([a_0, a_1, \dots, a_d, b_0, b_1, \dots, b_d]).$$

Prove the following quantitative version of Proposition 3.46.

For all $d \geq 2$ there is a constant $C = C(d)$ such that for all integers $N \geq 0$ there exists a rational map $\phi(z) \in \mathbb{Q}(z)$ of degree $d$ with the following properties:

- $\phi^2(z) \notin \mathbb{C}[z]$.
- $0$ is a wandering point for $\phi$.
- $0, \phi(0), \phi^2(0), \phi^3(0), \ldots, \phi^N(0) \in \mathbb{Z}$.
- $h(\phi) \le C \cdot d^N$.

Try to do quantitatively better than this, for example, replace the height estimate with one of the form $h(\phi) \le C \cdot \delta^N$ for some $\delta < d$.

**3.42.** Recall that a rational map $\phi(z) \in \mathbb{Q}(z)$ is *affine minimal* if its resultant $\operatorname{Res}(\phi)$ cannot be made smaller via conjugation by an affine linear map $f(z) = az + b$. If $\phi$ is affine minimal and $\phi^2$ is not a polynomial, we have conjectured (Conjecture 3.47) that the size of $\mathcal{O}_\phi(\alpha) \cap \mathbb{Z}$ is bounded solely in terms of the degree of $\phi$. For each integer $d \ge 2$, let

$$C(d) = \sup\left\{ \#\big(\mathcal{O}_\phi(\alpha) \cap \mathbb{Z}\big) : \begin{array}{l} \phi(z) \in \mathbb{Q}(z),\ \phi^2(z) \notin \mathbb{Q}[z], \\ \phi \text{ is affine minimal, and} \\ \alpha \in \mathbb{P}^N(\mathbb{Q}) \smallsetminus \operatorname{PrePer}(\phi) \end{array} \right\}.$$

Thus Conjecture 3.47 says that $C(d)$ is finite.

(a) Prove that for every integer $N \ge 0$ there exists an affine minimal rational map $\phi(z) \in \mathbb{Q}(z)$ such that $\phi^2(z) \notin \mathbb{Q}[z]$, such that $0$ is not a preperiodic point for $\phi$, and such that

$$0, \phi(0), \phi^2(0), \phi^3(0), \ldots, \phi^N(0)$$

are all integers. In particular, $C(d) \to \infty$ as $d \to \infty$.

(b) Prove that the function $\phi(z)$ in (a) can be chosen to have degree $\lfloor (N-1)/2 \rfloor$. Hence $C(d) \ge 2d + 2$.

(c) Let $\phi(z)$ be the function

$$\phi(z) = \frac{481z^5 - 2565z^4 + 9385z^3 - 14955z^2 + 12094z - 3720}{z^5 - 465z^4 + 2185z^3 - 6975z^2 + 8254z - 3720}.$$

Trace the orbit of $0$. How many integers do you find?

(d) Find a rational function as in (a) of degree 2 and with the property that $\phi^i(0)$ is an integer for all $0 \le i \le 6$. Hence $C(2) \ge 7$, which improves the bound from (b). Can you find infinitely many such functions? Can you prove that $C(2) \ge 8$?

(e) Find a rational function as in (a) of degree 3 and with the property that $\phi^i(0)$ is an integer for all $0 \le i \le 8$. Hence $C(3) \ge 9$, which improves the bound from (b). Can you find infinitely many such functions? Can you prove that $C(3) \ge 10$?

## Section 3.8. Integrality Estimates for Points in Orbits

**3.43.** Let $K_v$ be a field complete with respect to the absolute value $v$, let $\phi : \mathbb{P}^1 \to \mathbb{P}^1$ be a rational map of degree $d \ge 2$ defined over $K_v$, and let $\rho_v : \mathbb{P}^1(\bar{K}_v) \times \mathbb{P}^1(\bar{K}_v) \to \mathbb{R}$ be the associated chordal metric. (See Sections 1.1 and 2.1 for the definition of the chordal metric when $v$ is archimedean and nonarchimedean, respectively.) Generalize Lemma 3.51 by showing that for every $Q \in \mathbb{P}^1(\bar{K}_v)$ there is a constant $C_v = C_v(\phi, Q)$ such that

$$\min_{Q' \in \phi^{-1}(Q)} \rho_v(P, Q')^{\mathsf{e}_Q(\phi)} \le C_v \rho_v\big(\phi(P), Q\big) \qquad \text{for all } P \in \mathbb{P}^1(\bar{K}_v).$$

Here $\mathsf{e}_Q(\phi)$ is as defined in Lemma 3.51.

**3.44.** Let $K$ be a number field and let $\phi : \mathbb{P}^1 \to \mathbb{P}^1$ be a rational map of degree $d \geq 2$ defined over $K$. Prove that there is a finite set of absolute values $S \subset M_K$ such that for all $v \notin S$ and all $Q \in \mathbb{P}^1(K)$, the constant $C_v(\phi, Q)$ in Exercise 3.43 may be taken equal to 1.

**3.45.** Let $\phi(z) \in \mathbb{Q}(z)$ be a rational map of degree $d \geq 2$ with $\phi^2(z) \notin \mathbb{Q}[z]$, let $\alpha \in \mathbb{Q}$, and write $\phi^n(\alpha) = a_n/b_n \in \mathbb{Q}$ as a fraction in lowest terms as usual. Prove that

$$\lim_{n\to\infty} \frac{\log|a_n|}{d^n} = \lim_{n\to\infty} \frac{\log|b_n|}{d^n} = \hat{h}_\phi(\alpha).$$

**3.46.** Let $K/\mathbb{Q}$ be a number field, let $v \in M_K$ be an absolute value on $K$, and let $\phi(z) \in K(z)$ be a rational function of degree $d \geq 2$. Suppose that $\alpha \in \mathbb{P}^1(K)$ is a wandering point for $\phi$ and that $\gamma \in \mathbb{P}^1(K)$ is any point that is not a totally ramified fixed point of $\phi^2$. Prove that

$$\lim_{n\to\infty} \frac{-\log \rho_v\bigl(\phi^n(\alpha), \gamma\bigr)}{h\bigl(\phi^n(\alpha)\bigr)} = 0.$$

Taking first $\gamma = \infty$ and then $\gamma = 0$, explain how to use this result to generalize Theorem 3.48 to number fields.

## Section 3.9. Periodic Points and Galois Groups

**3.47.** Let $P \in \mathbb{P}^N(\bar{K})$ be a point in projective space and let $K(P)$ be its field of definition. Prove that

$$K(P) = \text{fixed field of } \{\sigma \in \operatorname{Gal}(\bar{K}/K) : \sigma(P) = P\}.$$

In mathematical terminology, this says that the field of moduli of $P$ is a field of definition for $P$. We discuss fields of moduli and fields of definition for (equivalence classes of) rational maps in Chapter 4.

## Section 3.11. Ramification and Units in Dynatomic Fields

**3.48.** Let $\phi(z) = z^2 + c$ and let

$$\Phi_3^*(z) = \frac{\phi^3(z) - z}{\phi(z) - z}$$

be the polynomial whose roots are periodic points of period 3.

(a) Prove that $\Phi_3^*(z) \in \mathbb{Z}[c, x]$ is a polynomial in the variables $c$ and $z$ and that it has integer coefficients.

(b) Let $\alpha$ be a root of $\Phi_3^*(z)$ and assume that the field $\mathbb{Q}(c, \alpha)$ is an extension of $\mathbb{Q}(c)$ of degree 6 (i.e., $\Phi_3^*(z)$ is irreducible in $\mathbb{Q}(c)[z]$). Theorem 2.33 implies that

$$u_1 = \frac{\phi^2(\alpha) - \phi(\alpha)}{\phi(\alpha) - \alpha} \quad \text{and} \quad u_2 = \frac{\phi^2(\alpha) - \alpha}{\phi(\alpha) - \alpha}$$

are units. Compute these units explicitly as elements of $\mathbb{Q}[c, \alpha]$.

(c) Prove that there is a field automorphism $\sigma : \mathbb{Q}(c, \alpha) \to \mathbb{Q}(c, \alpha)$ characterized by the fact that it fixes $\mathbb{Q}(c)$ and satisfies $\sigma(\alpha) = \phi(\alpha) = \alpha^2 + c$. (Note that in general, $\mathbb{Q}(c, \alpha)$ will not be a Galois extension of $\mathbb{Q}(c)$.) Prove that $\sigma^3$ is the identity map.

(d) Compute the units $\sigma(u_1)$ and $\sigma(u_2)$.

(e) Compute the units $\sigma^2(u_1)$ and $\sigma^2(u_2)$.

(f) Express $\sigma(u_1)$, $\sigma(u_2)$, $\sigma^2(u_1)$, and $\sigma^2(u_2)$ in the form $\pm u_1^i u_2^j$.

**3.49.** Let $\phi(z) = z^2 + c$ and let $\Phi_3^*(z)$ be as in the previous exercise.

(a) Prove that $\Phi_3^*(z)$ factors into a product of two cubic polynomials in the ring $\mathbb{Q}(c)[z]$ if and only if $c$ has the form $c = -(e^2+7)/4$ for some $e \in \mathbb{Q}(c)$.

(b) Suppose that $c = -(e^2+7)/4$. Show that there is a polynomial $g_e(z) \in \mathbb{Q}[e,z]$ such that the factorization $\Phi_3^*(z)$ has the form $g_e(z)g_{-e}(z)$. Compute the discriminant of $g_e(z)$ and verify that it is a perfect square in $\mathbb{Q}[e]$. Conclude that if $g_e(z)$ is irreducible over $\mathbb{Q}(e)$, then its roots generate a cyclic Galois extension $K_e$ of $\mathbb{Q}(e)$.

(c) Let $\alpha$ be a root of $g_e(z)$. Use the results of the previous exercise to construct units in $\mathbb{Q}(e)(\alpha)$. Analyze these units for some small values of $e$, say $e = 1, 5, 7, 9$. (Note that we investigated the case $e = 3$ in Example 3.69.)

**3.50.** ** Let $\phi(z)$ be a generic monic polynomial of degree $d \geq 2$, i.e., $\phi(z)$ is a polynomial of the form $z^d + a_1 z^{d-1} + \cdots + a_d$, where $a_1, \ldots, a_d$ are indeterminates. Let $p$ be a prime and let $\alpha \in \text{Per}_p^{**}(\phi)$. Theorem 3.66 says that the elements

$$u_{i,j} = \frac{\phi^i(\alpha) - \phi^j(\alpha)}{\phi(\alpha) - \alpha}, \qquad 0 \leq j < i < p,$$

are units. What is the rank of the group that they generate? Is the answer the same for the polynomial $\phi(z) = z^2 + c$?

**3.51.** ** Let $\phi(z)$ be a generic rational function of degree $d \geq 2$, let $p$ be a prime, and let $P \in \text{Per}_p^{**}(\phi)$. Theorem 3.67 describes how to construct units $\kappa(P, \phi P, \phi^i P, \phi^j P)$. What is the rank of the group that they generate? Is the answer the same for the rational function $\phi(z) = (z^2 + b)/(bz^2 + 1)$?

# Chapter 4

# Families of Dynamical Systems

Most of our work in previous chapters has focused on a single rational map $\phi(z)$ and the effect its iterates have on different initial values. We now shift focus and consider the effect of varying the rational map $\phi(z)$. In order to do this, we study the set of all rational maps. This set turns out to have a natural structure as an algebraic variety, as does the set of rational maps modulo the equivalence relation defined by $\mathrm{PGL}_2$ conjugation. There are many threads to this story. The specific topics that we touch upon in this chapter include:

1. Dynatomic polynomials and fields generated by periodic points.
2. The space of quadratic polynomials (the simplest nontrivial case).
3. Rational maps that are $\mathrm{PGL}_2$-equivalent over $\bar{K}$, but not over $K$ (twists).
4. Field of defintion versus field of moduli for rational maps.
5. Minimal models for rational maps.
6. Moduli spaces of rational maps (with marked periodic points).

All of these topics have close analogues in the geometric and arithmetic theory of elliptic curves and abelian varieties. For purposes of comparison, we list the correspondences:

1. Division polynomials and fields generated by torsion points.
2. The space of elliptic curves (the simplest abelian varieties).
3. Abelian varieties isomorphic over $\bar{K}$, but not over $K$ (twists).
4. Field of definition versus field of moduli for an abelian variety.
5. Minimal Weierstrass equations and Néron models.
6. Elliptic modular curves and moduli spaces of abelian varieties.

In this chapter, we let $K$ be a perfect field and fix an algebraic closure $\bar{K}$ of $K$, although much of what we do is also valid for nonperfect fields if we use instead a separable closure of $K$. We write $\mathrm{Gal}(\bar{K}/K)$ for the absolute Galois group of $K$. It is convenient to use the profinite group $\mathrm{Gal}(\bar{K}/K)$, but we observe that it generally suffices to work with finite extensions and finite Galois groups, since the coefficients of any particular rational map $\phi(z) \in \bar{K}(z)$ lie in a finite extension of $K$.

## 4.1 Dynatomic Polynomials

Let $\phi(z) \in K[z]$ be a polynomial. Then the fixed points of $\phi$ are the roots of $\phi(z) - z$ (plus the point at $\infty$), and more generally, the points of period $n$ for $\phi$ are the roots of $\phi^n(z) - z$. However, the polynomial $\phi^n(z) - z$ may have roots of period smaller than $n$, since if $\phi^k(\alpha) = \alpha$ and $k|n$, then also $\phi^n(\alpha) = \alpha$. It is natural to try to eliminate these points of strictly smaller period and focus on the points of exact period $n$.

*Example* 4.1. We recall an analogous situation. The roots of the polynomial $z^n - 1$ include all $n^{\text{th}}$ roots of unity, not only the primitive ones. The $n^{\text{th}}$ cyclotomic polynomial is defined using an inclusion–exclusion product,

$$n^{\text{th}} \text{ cyclotomic polynomial} = \prod_{k|n} (z^k - 1)^{\mu(n/k)}. \tag{4.1}$$

It is the polynomial whose roots are the primitive $n^{\text{th}}$ roots of unity. Here $\mu$ is the Möbius function $\mu$ defined by $\mu(1) = 1$ and

$$\mu(p_1^{e_1} \cdots p_r^{e_r}) = \begin{cases} (-1)^r & \text{if } e_1 = \cdots = e_r = 1, \\ 0 & \text{if any } e_i \geq 2. \end{cases} \tag{4.2}$$

See [216, Section 2.2] or [11, Chapter 2] for basic properties of the Möbius function and the Möbius inversion formula. It is easy to check that the product (4.1) is a polynomial, using the fact that the (complex) roots of $z^n - 1$ are distinct and the following basic property of the $\mu$ function (Exercise 4.2):

$$\sum_{k|n} \mu\left(\frac{n}{k}\right) = \begin{cases} 1 & \text{if } n = 1, \\ 0 & \text{if } n > 1. \end{cases} \tag{4.3}$$

Taking our cue from the example provided by the cyclotomic polynomials, we might define the $n^{\text{th}}$ *dynatomic polynomial* by the formula

$$\Phi_n(z) = \prod_{k|n} \left(\phi^k(z) - z\right)^{\mu(n/k)}.$$

However, it is not clear that $\Phi_n(z)$ is a polynomial, since $\phi^n(z)$ may have multiple roots, as shown by the following example.

*Example* 4.2. Let $\phi(z)$ be the polynomial

$$\phi(z) = z^2 - \frac{3}{4}.$$

Then

$$\phi(z) - z = z^2 - z - \frac{3}{4} = \left(z - \frac{3}{2}\right)\left(z + \frac{1}{2}\right),$$

$$\phi^2(z) - z = z^4 - \frac{3}{2}z^2 - z - \frac{3}{16} = \left(z - \frac{3}{2}\right)\left(z + \frac{1}{2}\right)^3,$$

$$\frac{\phi^2(z) - z}{\phi(z) - z} = \left(z + \frac{1}{2}\right)^2.$$

Thus $\phi^2(z) - z$ vanishes with multiplicity 3 at the point $z = -\frac{1}{2}$, and although it is true that the ratio $\frac{\phi^2(z)-z}{\phi(z)-z}$ is a polynomial, it is somewhat distressing to observe that its root is a fixed point of $\phi$, not a point of primitive period 2.

For simplicity, the preceding discussion dealt with polynomial maps $\phi(z)$. We now turn to general rational maps $\phi(z) \in K(z)$ and develop tools that are useful for studying their periodic points. Let $\phi(z)$ be a rational function of degree $d$ and write $\phi = [F(X,Y), G(X,Y)]$ using homogeneous polynomials $F$ and $G$. Then the roots of the polynomial

$$YF(X,Y) - XG(X,Y)$$

in $\mathbb{P}^1$ are precisely the fixed points of $\phi$. If we count each fixed point according to the multiplicity of the root, then $\phi$ has exactly $d+1$ fixed points. More generally, we can apply the same reasoning to an iterate $\phi^n$ of $\phi$ and assign multiplicities to the $n$-periodic points.

**Definition.** Let $\phi(z) \in K(z)$ be a rational function of degree $d$, and for any $n \geq 0$, write

$$\phi^n = [F_n(X,Y), G_n(X,Y)]$$

with homogeneous polynomials $F_n, G_n \in K[X,Y]$ of degree $d^n$. (See Exercise 4.9 for a formal inductive definition of $F_n$ and $G_n$.) The *$n$-period polynomial of $\phi$* is the polynomial

$$\Phi_{\phi,n}(X,Y) = YF_n(X,Y) - XG_n(X,Y).$$

Notice that $\Phi_{\phi,n}(P) = 0$ if and only if $\phi^n(P) = P$, which justifies the name assigned to the polynomial $\Phi_{\phi,n}$.

The *$n^{\text{th}}$ dynatomic polynomial of $\phi$* is the polynomial[1]

$$\Phi^*_{\phi,n}(X,Y) = \prod_{k|n}\bigl(YF_k(X,Y) - XG_k(X,Y)\bigr)^{\mu(n/k)} = \prod_{k|n}\Phi_{\phi,k}(X,Y)^{\mu(n/k)},$$

where $\mu$ is the Möbius function. If $\phi$ is fixed, we write $\Phi_n$ and $\Phi^*_n$. If $\phi(z) \in K[z]$ is a polynomial, then we generally dehomogenize $[X,Y] = [z,1]$ and write $\Phi_n(z)$ and $\Phi^*_n(z)$.

All of the roots $P$ of $\Phi^*_{\phi,n}(X,Y)$ satisfy $\phi^n(P) = P$, but we saw in Example 4.2 that $\Phi^*_{\phi,n}(X,Y)$ may have roots whose periods are strictly smaller than $n$. Following Milnor [302], we make the following definitions.

[1] See Theorem 4.5 for the proof that $\Phi^*_{\phi,n}$ is indeed a polynomial.

**Definition.** Let $\phi(z) \in K(z)$ be a rational map and let $P \in \mathbb{P}^1$ be a periodic point for $\phi$.

- $P$ has *period* $n$ if $\Phi_n(P) = 0$.
- $P$ has *primitive (or exact) period* $n$ if $\Phi_n(P) = 0$ and $\Phi_m(P) \neq 0$ for all $m < n$.
- $P$ has *formal period* $n$ if $\Phi_n^*(P) = 0$.

We set the notation

$$\begin{aligned} \mathrm{Per}_n(\phi) &= \{P \in \mathbb{P}^1 : \Phi_n(P) = 0\}, \\ \mathrm{Per}_n^*(\phi) &= \{P \in \mathbb{P}^1 : \Phi_n^*(P) = 0\}, \\ \mathrm{Per}_n^{**}(\phi) &= \{P \in \mathbb{P}^1 : \Phi_n(P) = 0 \text{ and } \Phi_m(P) \neq 0 \text{ for all } 1 \leq m < n\}. \end{aligned}$$

Thus $\mathrm{Per}_n(\phi)$ is the set of points of period $n$, $\mathrm{Per}_n^*(\phi)$ is the set of points of formal period $n$, and $\mathrm{Per}_n^{**}(\phi)$ is the set of points of primitive (or exact) period $n$. Sometimes we treat these as sets of points with assigned multiplicities, e.g., if $P \in \mathrm{Per}_n^*(\phi)$, the multiplicity of $P$ is the order of vanishing of $\Phi_n^*$ at $P$.

It is clear that

$$\text{primitive period } n \quad \Longrightarrow \quad \text{formal period } n \quad \Longrightarrow \quad \text{period } n,$$

but neither of the reverse implications is true in general.

*Remark* 4.3. The polynomial $\Phi_{\phi,n}$ is homogeneous of degree $d^n + 1$, so counted with multiplicity, the map $\phi$ has exactly $d^n + 1$ points of period $n$. And if we let

$$\nu_d(n) = \deg\bigl(\Phi_{\phi,n}^*(X,Y)\bigr) = \sum_{k|n} \mu\left(\frac{n}{k}\right)(d^k + 1), \tag{4.4}$$

then counted with multiplicity, the map $\phi$ has exactly $\nu_d(n)$ points of formal period $n$. The number $\nu_d(n)$ grows very rapidly as $d$ or $n$ increases. See Exercise 4.3.

The first few period and dynatomic polynomials for $\phi(z) = z^2 + c$ are listed in Table 4.1 on page 156. Notice how complicated $\Phi_n$ and $\Phi_n^*$ are, even for small values of $n$. For example, $\Phi_6^*$ has degree 54 as a polynomial in $z$ and degree 27 as a polynomial in $c$.

*Remark* 4.4. Rather than using the homogeneous polynomials $\Phi_n$ and $\Phi_n^*$, it is sometimes more natural and convenient to consider instead the associated divisors in $\mathbb{P}^1$, especially in generalizing the theory to higher-dimensional situations. This is the approach taken in [313], where for any (nondegenerate) morphism $\phi : V \to V$ of smooth algebraic varieties, the 0-cycle $\Phi_{\phi,n}$ is defined as the pullback of the graph of $\phi^n$ by the diagonal map $\Delta : V \to V \times V$. See Exercise 4.8.

We now prove the important fact that the dynatomic polynomial $\Phi_{\phi,n}^*(X,Y)$ is indeed a polynomial.

**Theorem 4.5.** *Let $\phi(z) \in K(z)$ be a rational function of degree $d \ge 2$. For each $P \in \mathbb{P}^1(\bar{K})$, let*

$$a_P(n) = \operatorname{ord}_P\big(\Phi_{\phi,n}(X,Y)\big) \qquad \textit{and} \qquad a_P^*(n) = \operatorname{ord}_P\big(\Phi_{\phi,n}^*(X,Y)\big),$$

*or, in terms of divisors in* $\operatorname{Div}(\mathbb{P}^1)$,

$$\operatorname{div}(\Phi_{\phi,n}) = \sum_P a_P(n)(P) \qquad \textit{and} \qquad \operatorname{div}(\Phi_{\phi,n}^*) = \sum_P a_P^*(n)(P).$$

(a) $\Phi_{\phi,n}^* \in K[X,Y]$, *or equivalently,*

$$a_P^*(n) \ge 0 \quad \textit{for all } n \ge 1 \textit{ and all } P \in \mathbb{P}^1.$$

(b) *Let $P$ be a point of primitive period $m$ and let $\lambda(P) = (\phi^m)'(P)$ be the multiplier of $P$. Then $P$ has formal period $n$, i.e., $a_P^*(n) > 0$, if and only if one of the following three conditions is true:*

  (i) $n = m$.

  (ii) $n = mr$ *and $\lambda(P)$ is a primitive $r^{th}$ root of unity.*

  (iii) $n = mrp^e$, *$\lambda(P)$ is a primitive $r^{th}$ root of unity, $K$ has characteristic $p$, and $e \ge 1$.*

*In particular, if $K$ has characteristic* 0, *then $a_P^*(n)$ is nonzero for at most two values of $n$.*

*Proof.* By the definition of $\Phi_n^*$, we have the relation

$$a_P^*(n) = \sum_{m|n} \mu(n/m)a_P(m).$$

We begin with a lemma that describes the value of $a_P(n)$ for fixed points.

**Lemma 4.6.** *Let $\psi(z) \in K(z)$ be a rational function of degree $d \ge 2$, let $P \in \mathbb{P}^1(\bar{K})$ be a fixed point of $\psi$, let $\lambda = \lambda_P(\psi) = \psi'(P)$ be the multiplier of $\psi$ at $P$, and let*

$$a_P(N) = \operatorname{ord}_P\big(\Phi_{\psi,N}(X,Y)\big)$$

*be as in Theorem* 4.5. *Then for any $N \ge 2$,*

$$\lambda^N \ne 1 \implies a_P(N) = a_P(1) = 1, \tag{4.5}$$

$$\lambda \ne 1 \textit{ and } \lambda^N = 1 \implies a_P(N) > a_P(1) = 1, \tag{4.6}$$

$$\lambda = 1 \textit{ and } N \ne 0 \textit{ in } K \implies a_P(N) = a_P(1) \ge 2, \tag{4.7}$$

$$\lambda = 1 \textit{ and } N = 0 \textit{ in } K \implies a_P(N) > a_P(1) \ge 2. \tag{4.8}$$

*Proof.* Making a change of variables, we may assume that $P = 0$, and then the assumption that $P$ is a fixed point of $\psi$ means that $\psi(z)$ has the form

$$\psi(z) = \lambda z + \alpha z^e + O(z^{e+1}), \tag{4.9}$$

where $\alpha \neq 0$, $e \geq 2$, and $O(z^{e+1})$ is shorthand for a function that vanishes to order at least $e+1$ at 0. Note that

$$a_P(1) = \operatorname*{ord}_{z=0}\bigl(\psi(z) - z\bigr) = \begin{cases} 1 & \text{if } \lambda \neq 1, \\ e \geq 2 & \text{if } \lambda = 1. \end{cases}$$

We consider first the case that $\lambda \neq 1$. Using the weaker form $\psi(z) = \lambda z + O(z^2)$ of (4.9) and iterating gives

$$\psi^N(z) = \lambda^N z + O(z^2),$$

so

$$\begin{aligned} a_P(N) &= \operatorname*{ord}_{z=0}\bigl(\psi^N(z) - z\bigr) \\ &= \operatorname*{ord}_{z=0}\bigl((\lambda^N - 1)z + O(z^2)\bigr) \left\{ \begin{array}{ll} = 1 & \text{if } \lambda^N \neq 1, \\ \geq 2 & \text{if } \lambda^N = 1. \end{array} \right. \end{aligned}$$

This completes the proof of the lemma in this case.

Next we consider the case $\lambda = 1$. Then $a_P(1) = \operatorname{ord}_{z=0}\bigl(\psi(z) - z\bigr) = e$, and iteration of (4.9) yields

$$\psi^N(z) = z + N\alpha z^e + O(z^{e+1}). \tag{4.10}$$

Hence

$$\begin{aligned} a_P(N) &= \operatorname*{ord}_{z=0}\bigl(\psi^N(z) - z\bigr) \\ &= \operatorname*{ord}_{z=0}\bigl(N\alpha z^e + O(z^{e+1})\bigr) \left\{ \begin{array}{ll} = e = a_P(1) & \text{if } N \neq 0 \text{ in } K, \\ > e = a_P(1) & \text{if } N = 0 \text{ in } K. \end{array} \right. \end{aligned}$$

This completes the proof of Lemma 4.6 □

Resuming the proof of Theorem 4.5, we observe that if the primitive period $m$ does not divide $n$, then $a_P(n) = 0$, and hence also $a_P^*(n) = 0$. We may thus assume that $m|n$. Since we will deal with several different maps, we write $a_P(\phi, n)$ and $a_P^*(\phi, n)$ to indicate the dependence on the map. Let

$$\psi = \phi^m \qquad \text{and} \qquad N = \frac{n}{m}.$$

Using the fact that $a_P(\phi, k) = 0$ unless $m|k$, we find that

$$\begin{aligned} a_P^*(\phi, n) = \sum_{k|n} \mu\left(\frac{n}{k}\right) a_P(\phi, k) &= \sum_{k|N} \mu\left(\frac{Nm}{km}\right) a_P(\phi, km) \\ &= \sum_{k|N} \mu\left(\frac{N}{k}\right) a_P(\phi^m, k) = a_P^*(\psi, N). \end{aligned}$$

We further observe that $P$ is a fixed point of $\psi$, so Lemma 4.6 applies to $\psi$ and $P$. We consider several cases, the first two of which may done simultaneously.

**Case I(a). $\lambda(P)^N \neq 1$.**
**Case I(b). $\lambda(P) = 1$ and $N \neq 0$ in $K$.**
In Case I(a) we have $\lambda(P)^k \neq 1$ for all $k|N$, so we can apply (4.5) of Lemma 4.6 to conclude that $a_P(\psi, k) = a_P(\psi, 1)$ for all $k|N$. Similarly, in Case I(b) we can apply (4.7) of Lemma 4.6 to conclude that $a_P(\psi, k) = a_P(\psi, 1)$ for all $k|N$. Hence in both cases we find that

$$a_P^*(\psi, N) = \sum_{k|N} \mu\left(\frac{N}{k}\right) a_P(\psi, k) = \left(\sum_{k|N} \mu\left(\frac{N}{k}\right)\right) a_P(\psi, 1)$$
$$= \begin{cases} a_P(\psi, 1) \geq 1 & \text{if } N = 1, \\ 0 & \text{if } N > 1. \end{cases}$$

Using the equality $a_P(\psi, 1) = a_P(\phi^m, 1) = a_P(\phi, m)$, this shows in Case I(a,b) that

$$a_P^*(\phi, n) = \begin{cases} a_P(\phi, n) \geq 1 & \text{if } n = m, \\ 0 & \text{if } n > m. \end{cases}$$

**Case II. $\lambda(P) = 1$ and $N = 0$ in $K$.**
Let $p$ be the characteristic of $K$ and write $N = p^e M$ with $p \nmid M$. We observe that if $k|M$ and $0 \leq i \leq e$, then $k \neq 0$ in $K$ and $\lambda_P(\psi^{p^i}) = \lambda_P(\psi)^{p^i} = 1$, so (4.7) of Lemma 4.6 applied to $\psi^{p^i}$ tells us that

$$a_P(\psi, p^i k) = a_P(\psi^{p^i}, k) = a_P(\psi^{p^i}, 1) = a_P(\psi, p^i).$$

This allows us to compute

$$a_P^*(\psi, N) = \sum_{k|N} \mu\left(\frac{N}{k}\right) a_P(\psi, k) = \sum_{k|M} \sum_{i=0}^{e} \mu\left(\frac{p^e M}{p^i k}\right) a_P(\psi, p^i k)$$
$$= \left(\sum_{k|M} \mu\left(\frac{M}{k}\right)\right)\left(\sum_{i=0}^{e} \mu(p^{e-i}) a_P(\psi, p^i)\right)$$
$$= \begin{cases} a_P(\psi, p^e) - a_P(\psi, p^{e-1}) \geq 1 & \text{if } M = 1, \\ 0 & \text{if } M > 1. \end{cases}$$

The fact that the value is positive when $M = 1$ follows from applying condition (4.8) of Lemma 4.6 to the difference $a_P(\psi^{p^{e-1}}, p) - a_P(\psi^{p^{e-1}}, 1)$. Hence in Case II, we have shown that $a_P^*(\phi, n) \geq 0$, and further, $a_P^*(\phi, n) \geq 1$ if and only if $n = mp^e$. Since Case II includes the assumption that $\lambda(P) = 1$, this proves Theorem 4.5 in this case.

**Case III. $\lambda(P) \neq 1$ and $\lambda(P)^N = 1$.**
Let $r$ be the exact order of $\lambda(P)$ in $\bar{K}^*$, so $r|N$ and $r > 1$. Then (4.5) of Lemma 4.6

tells us that if $r \nmid k$, then $a_P(\psi,k) = a_P(\psi,1)$. This allows us to write the sum defining $a_P^*(\psi,N)$ as

$$\begin{aligned}
a_P^*(\psi,N) &= \Biggl(\sum_{\substack{k|N\\ r\nmid k}} + \sum_{\substack{k|N\\ r|k}}\Biggr)\mu\left(\frac{N}{k}\right)a_P(\psi,k) \\
&= \Biggl(\sum_{\substack{k|N\\ r\nmid k}} \mu\left(\frac{N}{k}\right)a_P(\psi,1)\Biggr) + \Biggl(\sum_{\substack{k|N\\ r|k}} \mu\left(\frac{N}{k}\right)a_P(\psi,k)\Biggr) \\
&= \Biggl(\sum_{k|N} \mu\left(\frac{N}{k}\right)a_P(\psi,1)\Biggr) + \Biggl(\sum_{\substack{k|N\\ r|k}} \mu\left(\frac{N}{k}\right)\bigl(a_P(\psi,k) - a_P(\psi,1)\bigr)\Biggr) \\
&= \sum_{k|\frac{N}{r}} \mu\left(\frac{N/r}{k}\right)\bigl(a_P(\psi,kr) - a_P(\psi,1)\bigr) \\
&= a_P^*\left(\psi^r, N/r\right) - \begin{cases} a_P(\psi,1) & \text{if } N = r, \\ 0 & \text{if } N \ne r. \end{cases}
\end{aligned} \tag{4.11}$$

If $N = r$, so $n = mr$, then we have

$$a_P^*(\phi,n) = a_P^*(\psi,N) = a_P^*(\psi^r,1) - a_P(\psi,1) = a_P(\psi,r) - a_P(\psi,1) \ge 1$$

from (4.6) of Lemma 4.6.

We may thus assume that $N > r$, so (4.11) says that

$$a_P^*(\phi,n) = a_P^*(\psi,N) = a_P^*(\psi^r, N/r).$$

We now observe that $\lambda_P(\psi^r) = \lambda_P(\psi)^r = 1$. Hence if $N \ne 0$ in $K$, then we can apply Case I(b) to $a_P^*(\psi^r, N/r)$, and if $N = 0$ in $K$, then we can apply Case II to $a_P^*(\psi^r, N/r)$. This completes the proof of Theorem 4.5. □

As an easy application of Theorem 4.5, we now prove that a rational map $\phi(z) \in K(z)$ possesses periodic points of infinitely many distinct periods, i.e., the set of primitive-$n$ periodic points $\operatorname{Per}_n^{**}(\phi)$ is nonempty for infinitely many $n$.

**Corollary 4.7.** *Let $\phi(z) \in K(z)$ be a rational map of degree $d \ge 2$. Then for all prime numbers $\ell$ except possibly for $d+2$ exceptions, the map $\phi$ has a point of primitive period $\ell$.*

*Proof.* We begin by discarding the finitely many primes $\ell$ satisfying either of the following conditions:

- $K$ has characteristic $\ell$.
- There is some $Q \in \operatorname{Fix}(\phi)$ with $\lambda(Q) \ne 1$ and $\lambda(Q)^\ell = 1$.

The set $\operatorname{Fix}(\phi)$ contains at most $d+1$ points, so these conditions eliminate at most $d+2$ primes.

For any of the remaining primes $\ell$, take a point $P \in \operatorname{Per}_\ell(\phi)$. If $P$ does not have primitive period $\ell$, then it must be a fixed point of $\phi$, since its period certainly divides $\ell$. Then our assumptions on $\ell$ imply that either $\lambda(P)=1$ or $\lambda(P)^\ell \neq 1$, and further that $\ell \neq 0$ in $K$. It follows from (4.5) and (4.7) of Lemma 4.6 that in both cases we have $a_P(\ell) = a_P(1)$. Hence

$$\sum_{P \in \operatorname{Per}_\ell(\phi)\cap\operatorname{Fix}(\phi)} a_P(\ell) = \sum_{P \in \operatorname{Per}_\ell(\phi)\cap\operatorname{Fix}(\phi)} a_P(1) \le \sum_{P \in \operatorname{Fix}(\phi)} a_P(1) = d+1.$$

Thus the total multiplicity of all of the fixed points of $\phi$ in the $\ell$-period polynomial

$$\Phi_{\phi,\ell}(X,Y) = YF_\ell(X,Y) - XG_\ell(X,Y)$$

is at most $d+1$. However, the degree of $\Phi_{\phi,\ell}$ is $d^\ell+1$, so we conclude that $\Phi_{\phi,\ell}$ has at least one root that is not fixed by $\phi$. Hence there exists at least one primitive $\ell$-periodic point. □

For many applications, Corollary 4.7 is sufficient, but a more detailed analysis yields a much stronger result. We state the full theorem in characteristic 0 and refer the reader to Pezda's two papers [354, 357] for the more complicated description required in characteristic $p$. See also [161, 172, 173, 174] for higher-dimensional results of a similar nature.

**Theorem 4.8.** (I.N. Baker [19]) *Let $\phi(z) \in K(z)$ be a rational map of degree $d \ge 2$ defined over a field $K$ of characteristic 0. Suppose that $\phi$ has no primitive $n$-periodic points. Then $(n,d)$ is one of the pairs*

$$(2,2),(2,3),(3,2),(4,2).$$

*If $\phi$ is a polynomial, then only $(2,2)$ is possible.*

*Proof.* For the proof, which is function-theoretic in nature, see [19] or [43, § 6.8]. □

## 4.2 Quadratic Polynomials and Dynatomic Modular Curves

In this section we expand on the material from the previous section in the special case of the quadratic polynomial

$$\phi(z) = \phi_c(z) = z^2 + c.$$

For further material on the iteration of quadratic polynomials, see for example [17, 18, 113, 115, 171, 220, 221, 222, 305, 309, 350, 361, 388].

$$\begin{aligned}
\Phi_1(c,z) &= z^2 - z + c \\
\Phi_2(c,z) &= z^4 + 2cz^2 - z + (c^2 + c) \\
\Phi_3(c,z) &= z^8 + 4cz^6 + (6c^2 + 2c)z^4 + (4c^3 + 4c^2)z^2 - z + (c^4 + 2c^3 + c^2 + c) \\
\Phi_4(c,z) &= z^{16} + 8cz^{14} + (28c^2 + 4c)z^{12} + (56c^3 + 24c^2)z^{10} \\
&\quad + (70c^4 + 60c^3 + 6c^2 + 2c)z^8 + (56c^5 + 80c^4 + 24c^3 + 8c^2)z^6 \\
&\quad + (28c^6 + 60c^5 + 36c^4 + 16c^3 + 4c^2)z^4 \\
&\quad + (8c^7 + 24c^6 + 24c^5 + 16c^4 + 8c^3)z^2 \\
&\quad - z + (c^8 + 4c^7 + 6c^6 + 6c^5 + 5c^4 + 2c^3 + c^2 + c)
\end{aligned}$$

$$\begin{aligned}
\Phi_1^*(c,z) &= z^2 - z + c \\
\Phi_2^*(c,z) &= z^2 + z + (c + 1) \\
\Phi_3^*(c,z) &= z^6 + z^5 + (3c + 1)z^4 + (2c + 1)z^3 + (3c^2 + 3c + 1)z^2 \\
&\quad + (c^2 + 2c + 1)z + (c^3 + 2c^2 + c + 1) \\
\Phi_4^*(c,z) &= z^{12} + 6cz^{10} + z^9 + (15c^2 + 3c)z^8 + 4cz^7 + (20c^3 + 12c^2 + 1)z^6 \\
&\quad + (6c^2 + 2c)z^5 + (15c^4 + 18c^3 + 3c^2 + 4c)z^4 + (4c^3 + 4c^2 + 1)z^3 \\
&\quad + (6c^5 + 12c^4 + 6c^3 + 5c^2 + c)z^2 + (c^4 + 2c^3 + c^2 + 2c)z \\
&\quad + (c^6 + 3c^5 + 3c^4 + 3c^3 + 2c^2 + 1) \\
\Phi_6^*(c,z) &= z^{54} - z^{53} + 27cz^{52} + (-26c + 1)z^{51} + (351c^2 + 13c - 1)z^{50} \\
&\quad + (-325c^2 + 12c)z^{49} + (2925c^3 + 325c^2 - 24c + 1)z^{48} \\
&\quad + (-2600c^3 - 12c^2 + 12c)z^{47} + \cdots \\
&\quad + (-c^{26} - 12c^{25} - 66c^{24} - 226c^{23} - \cdots + 2c^3 + c^2 + 2c - 1)z \\
&\quad + (c^{27} + 13c^{26} + 78c^{25} + 293c^{24} + 792c^{23} + \cdots + 3c^3 + c^2 - c + 1)
\end{aligned}$$

Table 4.1: Period and dynatomic polynomials for $\phi(z) = z^2 + c$.

### 4.2.1 Dynatomic Curves for Quadratic Polynomials

Note that as long as the field $K$ does not have characteristic 2, then any quadratic polynomial

$$\phi(z) = Az^2 + Bz + C$$

can be put into the form $z^2 + c$ by a simple change of variables working entirely within the field $K$. More precisely, if we let $f(z) = (2z - B)/(2A)$, then

$$\phi^f(z) = (f^{-1} \circ \phi \circ f)(z) = z^2 + \left(AC - \frac{1}{4}B^2 + \frac{1}{2}B\right). \tag{4.12}$$

The roots of the period polynomials $\Phi_n(z) = \phi_c^n(z) - z$ and associated dynatomic polynomials $\Phi_n^*(z)$ are periodic points of the map $\phi_c(z) = z^2 + c$. In order to investigate how the periodic points of $\phi_c(z)$ vary as a function of $c$, we observe that $\Phi_n^*(z)$ is a polynomial in the two variables $z$ and $c$. Thus in studying quadratic polynomial maps, it is natural to write $\Phi_n^*(c,z)$ and treat $\Phi_n^*$ as a polynomial in $\mathbb{Z}[c,z]$. (Table 4.1 lists the first few period and dynatomic polynomials for $\phi_c(z)$.) Treating $\Phi_n^*(c,z)$ as a polynomial in two variables leads to a natural association

$$\left\{\begin{array}{l}\text{polynomial } \phi(z) \in K[z] \text{ of} \\ \text{degree 2 with a point} \\ P \in K \text{ of formal period } n\end{array}\right\} \longrightarrow \left\{(c,\alpha) \in K \times K : \Phi_n^*(c,\alpha) = 0\right\}.$$

Thus given a polynomial $\phi(z) \in K[z]$ of degree 2 and a point $P$ of formal period $n$, we make a change of variables as in (4.12) so that $\phi^f(z) = z^2 + c$ and set $\alpha = f^{-1}(P)$. Note that this entire procedure takes place within the field $K$ (which we assume has characteristic different from 2). Thus the solutions to the equation $\Phi_n^*(y,z) = 0$ parameterize pairs $(\phi, P)$, where $\phi$ is a conjugacy class of quadratic polynomials and $P$ is a point of formal period $n$ for $\phi$. Further, the solution is $K$-rational if and only if $\phi$ and $P$ are $K$-rational.

**Definition.** The *dynatomic modular curve* $Y_1(n) \subset \mathbb{A}^2$ is the affine curve defined by the equation

$$\Phi_n^*(y,z) = 0.$$

The normalization of the projective closure of $Y_1(n)$ is denoted by $X_1(n)$.

*Example* 4.9. It is easy to see that $X_1(1)$ and $X_1(2)$ are rational curves. Indeed, the projective closures of $Y_1(1)$ and $Y_1(2)$ are smooth conics,

$$X_1(1) : z^2 - zw + yw = 0 \qquad \text{and} \qquad X_1(2) : z^2 + zw + yw + w^2 = 0.$$

It turns out that $X_1(3)$ is also rational, but this is less clear from the equation in Table 4.1. In order to parameterize $X_1(3)$, suppose that $\phi(z) = Az^2 + Bz + C$ is any quadratic polynomial with a periodic point of primitive period 3. Conjugating by a linear map $z \mapsto \alpha z + \beta$, we may assume that the given 3-cycle has the form $0 \to 1 \to t \to 0$ for some $t$. This gives the equations

$$\phi(0) = C = 1, \qquad \phi(1) = A + B + C = t, \qquad \phi(t) = At^2 + Bt + C = 0.$$

Solving for $A, B, C$ in terms of $t$ yields

$$\phi(z) = \frac{t^2 - t + 1}{t - t^2} z^2 - \frac{t^3 - t^2 + 1}{t - t^2} z + 1.$$

Now we apply the linear change of variables (4.12) to put $\phi$ into the form $z^2 + c$. Thus letting $f(z) = (2z - B)/(2A)$, we find that

$$\phi^f(z) = z^2 + \frac{t^6 - 4t^5 + 9t^4 - 8t^3 + 4t^2 - 2t + 1}{-4t^4 + 8t^3 - 4t^2}, \qquad f^{-1}(0) = \frac{-t^3 + t^2 - 1}{-2t^2 + 2t}.$$

Our computation shows that for every value of $t \notin \{0,1\}$, the point

$$\left(\frac{t^6 - 4t^5 + 9t^4 - 8t^3 + 4t^2 - 2t + 1}{-4t^4 + 8t^3 - 4t^2}, \frac{-t^3 + t^2 - 1}{-2t^2 + 2t}\right)$$

is a solution to the equation

$$\Phi_3^*(y,z) = z^6 + z^5 + (3y+1)z^4 + (2y+1)z^3 + (3y^2 + 3y + 1)z^2 + (y^2 + 2y + 1)z + (y^3 + 2y^2 + y + 1) = 0.$$

(You may check this directly using a computer algebra system.) We have thus constructed a nonconstant rational map

$$\mathbb{P}^1 \longrightarrow X_1(3), \qquad t \longmapsto \left(\frac{t^6 - 4t^5 + 9t^4 - 8t^3 + 4t^2 - 2t + 1}{-4t^2(t-1)^2}, \frac{t^3 - t^2 + 1}{2t(t-1)}\right). \tag{4.13}$$

General principles (Lüroth's theorem [198, IV.2.5.5]) tell us that $X_1(3)$ is birational to $\mathbb{P}^1$.

More concretely, we can prove that the map (4.13) has degree 1 by constructing its inverse. Thus let $(c,b)$ be a root of $\Phi_3^*$. We set $g(z) = (b^2 + c - b)z + b$, so $g$ sends 0 to $b$ and 1 to $\phi(b)$. Thus the 3-cycle $b \to \phi(b) \to \phi^2(b) \to 0$ becomes the following 3-cycle for $\phi^g$:

$$0 \xrightarrow{\ \phi^g\ } 1 \xrightarrow{\ \phi^g\ } b^2 + b + 1 + c \xrightarrow{\ \phi^g\ } 0.$$

This gives the map

$$X_1(3) \longrightarrow \mathbb{P}^1, \qquad (c,b) \longmapsto b^2 + b + 1 + c,$$

which is inverse to (4.13), a fact that can also be checked directly with a computer algebra system.

### 4.2.2 Dynatomic Curves as Modular Curves

The curve $Y_1(n)$ and its completion $X_1(n)$ are *modular curves* in the sense that their points are solutions to the moduli problem of describing the isomorphism classes of pairs $(\phi, \alpha)$, where $\phi$ is a polynomial of degree 2 and $\alpha \in \mathbb{A}^1$ is a point of formal period $n$ for $\phi$. Here two pairs $(\phi_1, \alpha_1)$ and $(\phi_2, \alpha_2)$ are $\mathrm{PGL}_2$-*isomorphic* if there is a linear fractional transformation $f \in \mathrm{PGL}_2$ satisfying

$$\phi_2 = \phi_1^f \qquad \text{and} \qquad \alpha_2 = f^{-1}(\alpha_1).$$

In order to state this more carefully, we define

$$\mathrm{Formal}(n) = \frac{\left\{(\phi,\alpha) : \begin{array}{c} \phi \in \bar{K}[z],\ \deg(\phi) = 2,\ \alpha \in \bar{K} \\ \alpha \text{ has formal period } n \text{ for } \phi \end{array}\right\}}{\mathrm{PGL}_2\text{-isomorphism}}.$$

We have demonstrated that the elements of $\mathrm{Formal}(n)$ are in one-to-one correspondence with the points of $Y_1(n)$. But much more is true: the correspondence is algebraic in an appropriate sense. Before stating this important result, we must define what it means for a family of maps and points to be algebraic.

**Definition.** Let $V$ be an algebraic variety. An *algebraic family of quadratic polynomials over $V$ with a marked point of formal period $n$* consists of a quadratic polynomial

$$\psi(z) = Az^2 + Bz + C, \qquad A, B, C \in \bar{K}[V],$$

whose coefficients $A, B, C$ are regular functions on $V$ and such that $A$ does not vanish on $V(\bar{K})$, and a morphism $\lambda : V \to \mathbb{A}^1$ such that for all $P \in V(\bar{K})$, the point $\lambda(P)$ is a point of formal period $n$ for the quadratic polynonomial

$$\psi_P(z) = A(P)z^2 + B(P)z + C(P) \in \bar{K}[z].$$

The family is *defined over $K$* if the variety $V$ and morphism $\lambda$ are defined over $K$ and the functions $A, B, C$ are in $K[V]$.

*Example* 4.10. The pair

$$\psi_t(z) = z^2 + (t + t^{-1} - 1)z \qquad \text{and} \qquad \lambda(t) = t - 1$$

is an algebraic family of quadratic polynomials over $\mathbb{P}^1 \smallsetminus \{0, \infty\}$ with a marked point of formal period 2.

**Theorem 4.11.** *Let $K$ be a field of characteristic different from* 2.

(a) *The map*

$$Y_1(n) \longrightarrow \operatorname{Formal}(n), \qquad (c, \alpha) \longmapsto (z^2 + c, \alpha), \tag{4.14}$$

*is a bijection of sets.*

(b) *Let $V$ be a variety and suppose that the points of $V$ algebraically parameterize a family of quadratic polynomials $\psi$ together with a marked point $\lambda$ of formal period $n$. Then there is a unique morphism of varieties*

$$\eta : V \longrightarrow Y_1(n)$$

*with the property that*

$$\eta(P) = \bigl(\psi_P(z), \lambda(P)\bigr) \in \operatorname{Formal}(n) \qquad \textit{for all } P \in V(\bar{K}), \tag{4.15}$$

*where we use* (4.14) *to identify* $\operatorname{Formal}(n)$ *with* $Y_1(n)$.

(c) *If the family is defined over the field $K$, then the morphism $\eta$ is also defined over $K$.*

*Proof.* (a) We have shown this earlier in this section.

(b) By definition $\psi$ has the form

$$\psi(z) = Az^2 + Bz + C, \qquad A, B, C \in \bar{K}[V],$$

with $A$ not vanishing on $V$ and $\lambda$ a morphism $\lambda : V \to \mathbb{A}^1$ such that for all $P \in V(\bar{K})$, the point $\lambda(P)$ is a point of formal period $n$ for the quadratic polynonomial

$$\psi_P(z) = A(P)z^2 + B(P)z + C(P) \in \bar{K}[z].$$

For any point $P \in V(\bar{K})$, let $f_P(z)$ be the linear fractional transformation

$$f_P(z) = \frac{2z - B(P)}{2A(P)}.$$

Note that $f_P$ is well-defined for every $P \in V(\bar{K})$, since we have assumed that $A$ is nonvanishing on $V$. We define a map $\eta$ from $V$ to $\mathbb{A}^2$ by the formula

$$\eta(P) = (c_P, \alpha_P) \quad \text{with} \quad \begin{cases} c_P = A(P)C(P) - \frac{1}{4}B(P)^2 + \frac{1}{2}B(P), \\ \alpha_P = A(P)\lambda(P) + \frac{1}{2}B(P). \end{cases} \tag{4.16}$$

Note that $\eta : V \to \mathbb{A}^2$ is a morphism, i.e., it is given by everywhere-defined algebraic functions on $V$. We are now going to verify that the image of $\eta$ is the curve $Y_1(n)$.

The computation that we performed in deriving formula (4.12) shows that

$$\psi_P^{f_P}(z) = z^2 + c_P.$$

To ease notation, we let $\phi_P(z) = z^2 + c_P$. We also let $\Phi_{n,P}$, $\Phi^*_{n,P}$, $\Psi_{n,P}$, and $\Psi^*_{n,P}$ be the period and dynatomic polynomials for $\phi_P$ and $\psi_P$, respectively. Note that the period polynomials are related by

$$\Psi_{n,P}^{f_P}(z) = f_P^{-1} \circ (\psi_P^n(z) - z) \circ f_P = (f_P^{-1} \circ \psi_P f_P)^n(z) - z = \phi_P^n(z) - z = \Phi_{n,P}(z).$$

Hence the dynatomic polynomials also satisfy

$$(\Psi^*_{n,P})^{f_P} = \Phi_{n,P}. \tag{4.17}$$

We are given that $\lambda(P) \in \operatorname{Per}^*_n(\psi_P)$, which is equivalent to saying that $\lambda(P)$ is a root of $\Psi^*_{n,P}$. It follows from (4.17) that

$$f_P^{-1}(\lambda(P)) = A(P)\lambda(P) + \tfrac{1}{2}B(P)$$

is a root of $\Phi^*_{n,P}$, and thus is in $\operatorname{Per}^*_n(\phi_P)$. This proves that the image of the map $\eta$ defined by (4.16) is contained in $Y_1(n)$, so $\eta$ is a morphism from $V$ to $Y_1(n)$. Further, this map $\eta$ respects the identification of $Y_1(n)$ with Formal$(n)$ from (a), since it takes $P$ to a pair $(c_P, \alpha_P)$ that is isomorphic to the pair $(\psi_P(z), \lambda(P))$ via the conjugation $f_P \in \operatorname{PGL}_2(\bar{K})$.

Finally, it is clear from the construction that the map $\eta$ is uniquely determined as a map (of sets) from $V(\bar{K})$ to Formal$(n)$, so it is the unique morphism $V \to Y_1(n)$ satisfying (4.15).

(c) The definition (4.16) of $\eta$ shows immediately that $\eta$ is defined over $K$, since all of $A$, $B$, $C$, and $\lambda$ are assumed to be defined over $K$. □

*Remark* 4.12. In the language of algebraic geometry, Theorem 4.11 says that $Y_1(n)$ is a coarse moduli space. In fact, the curve $Y_1(n)$ is actually a fine moduli space for all $n \geq 1$; see Exercise 4.18. The underlying reason is that there are no nontrivial elements of $\operatorname{PGL}_2$ that fix a quadratic polynomial and its points of formal period $n$, i.e., the moduli problem has no nontrivial automorphisms.

### 4.2.3 The Dynatomic Modular Curves $X_1(n)$ and $X_0(n)$

The curve $Y_1(n)$, and by extension $X_1(n)$, has the interesting property that the rational map $\phi$ acts on the points of $Y_1(n)$ via the map

$$\phi : Y_1(n) \longrightarrow Y_1(n), \qquad (y,z) \longmapsto (y, z^2+y) = (y, \phi_y(z)). \tag{4.18}$$

By abuse of notation, we will also use $\phi$ to denote the map (4.18). This map is well-defined, since if $\alpha$ is a point of formal period $n$ for the polynomial $\phi_y(z)$, then $\phi_y(\alpha)$ is also a point of formal period $n$ for $\phi_y$. Further, the $n^{\text{th}}$ iterate $\phi^n$ is the identity map on $Y_1(n)$ and $X_1(n)$, so $\operatorname{Aut}(Y_1(n))$ and $\operatorname{Aut}(X_1(n))$ contain subgroups of order $n$ generated by $\phi$.

In general if $V$ is any algebraic variety and if $G \subset \operatorname{Aut}(V)$ is any finite group of automorphisms of $V$, then there exist a quotient variety $W = V/G$ and a projection map $\pi : V \to W$ with the property that $\pi(y) = \pi(x)$ if and only if there is a $g \in G$ such that $y = g(x)$. (See, for example, [321, II §7].) Further, if $V$ is defined over a field $K$ and if $G$ is $K$-invariant,[2] then $V/G$ is defined over $K$.

The construction of $V/G$ when $V$ is a nonsingular curve is particularly simple. Each $g \in G$ induces an automorphism of the function field

$$g^* : \bar{K}(V) \to \bar{K}(V)$$

fixing $\bar{K}$, so we may view $G$ as a subgroup of $\operatorname{Aut}(\bar{K}(V)/\bar{K})$. Then one shows that the fixed field $\bar{K}(V)^G$ of $G$ has transcendence degree 1 over $\bar{K}$, so there exists a unique nonsingular projective curve $W/\bar{K}$ with function field $\bar{K}(W) = \bar{K}(V)^G$. The inclusion $\bar{K}(W) \subset \bar{K}(V)$ defines the projection $\pi : V \to W$ making $W$ into the quotient of $V$ by $G$. Finally, if $G$ is $K$-invariant, one shows that $W$ has a model defined over $K$.

*Remark* 4.13. We note that taking the quotient of a variety by an infinite group of automorphisms, as we will need to do in Section 4.4, is considerably more difficult than taking the quotient by a finite group of automorphisms. Indeed, in the infinite case it often happens that the quotient does not exist at all in the category of varieties.

**Definition.** With notation as above, we let $Y_0(n)$ be the quotient of $Y_1(n)$ by the finite subgroup of $\operatorname{Aut}(Y_1(n))$ generated by $\phi$. Similarly, $X_0(n)$ is the quotient of $X_1(n)$ by the finite subgroup of $\operatorname{Aut}(X_1(n))$ generated by $\phi$.

By construction, the points of $Y_0(n)$ classify isomorphism classes of pairs $(\phi, \mathcal{O})$, where $\phi$ is a quadratic polynomial and $\mathcal{O}$ is the orbit of a point of formal period $n$. The moduli-theoretic interpretation of the projection map from $Y_1(n)$ to $Y_0(n)$ is

$$Y_1(n) \longrightarrow Y_0(n), \qquad (\phi_c, \alpha) \longmapsto (\phi_c, \mathcal{O}_{\phi_c}(\alpha)).$$

---

[2]The elements in $G$ are morphisms $V \to V$ that are defined over some extension of $K$. The Galois group $\operatorname{Gal}(\bar{K}/K)$ acts on $G$ by acting on the coefficients of the polynomials defining the maps in $G$. We say that $G$ is $K$-invariant if each element of $\operatorname{Gal}(\bar{K}/K)$ maps $G$ to itself.

*Example* 4.14. We have seen in Example 4.9 that $X_1(2)$ and $X_1(3)$ are rational curves, i.e., they are isomorphic to $\mathbb{P}^1$, so their quotient curves $X_0(2)$ and $X_0(3)$ must also be rational curves. However, it is still of interest to make the quotient maps explicit. We do this for $X_1(2)$ and leave $X_1(3)$ as Exercise 4.19.

The affine curve $Y_1(2)$ has equation

$$Y_1(2) : z^2 + z + y + 1 = 0,$$

where a point $(c, \alpha) \in Y_1(2)$ corresponds to the quadratic map $\phi_c(z) = z^2 + c$ and point $\alpha \in \operatorname{Per}_2^*(\phi_c)$. The automorphism $\phi$ of $Y_1(2)$ is given by $(y, z) \mapsto (y, z^2 + y)$. From general principles we know that $\phi$ maps $Y_1(2)$ to itself, or we can see this explicitly by the formula

$$(z^2 + y)^2 + (z^2 + y) + y + 1 = (z^2 + z + y + 1)(z^2 - z + y + 1).$$

The affine coordinate ring of $Y_1(2)$ is

$$\frac{\mathbb{C}[z, y]}{(z^2 + z + y + 1)} \cong \mathbb{C}[z],$$

since we can express $y$ in terms of $z$ as $y = -z^2 - z - 1$. In terms of $z$ alone, the automorphism $\phi$ of $Y_1(2)$ corresponds to the map on $\mathbb{C}[z]$ given by

$$z \longmapsto z^2 + y = z^2 + (-z^2 - z - 1) = -z - 1.$$

So the affine coordinate ring of the quotient curve $Y_0(2)$ is the subring of $\mathbb{C}[z]$ that is fixed by the automorphism $z \mapsto -z - 1$. Since the map $Y_1(2) \to Y_0(2)$ has degree 2, we look for an invariant quadratic polynomial in $z$. Setting

$$Az^2 + Bz + C = A(-z - 1)^2 + B(-z - 1) + C$$

and equating coefficients, we find that $A = B$. Hence the invariant subring is $\mathbb{C}[z^2 + z]$, which is the affine coordinate ring of $Y_0(2)$.

Suppose now that $(\phi, \{\alpha, \beta\})$ is a quadratic polynomial and an orbit consisting of two points. How does this correspond to a point on $Y_0(2)$ whose coordinate ring is $\mathbb{C}[z^2 + z]$? First, we use a $\mathrm{PGL}_2$ conjugation as in (4.12) to put $\phi$ into the standard form $\phi_c(z) = z^2 + c$. Then $\alpha$ and $\beta$ are related by

$$\beta = \phi_c(\alpha) = \alpha^2 + c \quad \text{and} \quad \alpha = \phi_c(\beta) = \beta^2 + c.$$

Notice that this implies that

$$\alpha^2 + \alpha = (\beta - c) + (\beta^2 + c) = \beta^2 + \beta.$$

Letting $c = -\alpha^2 - \alpha - 1$ and using the isomorphisms $Y_1(2) \cong \mathbb{A}^1$ and $Y_0(2) \cong \mathbb{A}^1$ given earlier, we have the following commutative diagram:

$$\begin{array}{ccccc}
 & Y_1(2) & \xrightarrow{\ (c,\alpha)\mapsto\alpha\ } & \mathbb{A}^1 & \\
(c,\alpha) & & & & \alpha \\
\downarrow & \Big\downarrow & & \Big\downarrow & \downarrow \\
(c,\{\alpha,\alpha^2+c\}) & & & & \alpha^2+\alpha \\
 & Y_0(2) & \xrightarrow{\ (c,\{\alpha,\beta\})\mapsto\alpha^2+\alpha\ } & \mathbb{A}^1 &
\end{array}$$

*Remark* 4.15. Morton [307] describes an alternative method for finding an equation for the curve $Y_0(n)$ and the map $Y_1(n) \to Y_0(n)$. The roots of $\Phi_n^*(c,z) = 0$ can be grouped into orbits, say $\alpha_1, \ldots, \alpha_r$ are representatives for the different orbits, so

$$\operatorname{Per}_n^*(\phi) = \mathcal{O}_\phi(\alpha_1) \cup \mathcal{O}_\phi(\alpha_2) \cup \cdots \cup \mathcal{O}_\phi(\alpha_r).$$

The points in each orbit all have the same multiplier, and we define a polynomial

$$\delta_n(c,x) = \prod_{i=1}^{r} \bigl(x - \lambda_\phi(\alpha_i)\bigr).$$

Then one can show that $\delta_n(c,x) \in \mathbb{Z}[c,x]$ and that the equation

$$\delta_n(y,x) = 0$$

gives a (possibly singular) model for $Y_0(n)$. Using this model, the natural map from $Y_1(n)$ to $Y_0(n)$ is

$$Y_1(n) \longrightarrow Y_0(n), \qquad (y,z) \longmapsto \bigl(y, (\phi_y^n)'(z)\bigr).$$

See Morton's paper [307] and Exercise 4.13.

*Remark* 4.16. The dynatomic modular curves $Y_1(n)$ and $Y_0(n)$ defined in this section are analogous to the modular curves that appear in the classical theory of elliptic curves. Briefly, the elliptic modular curve $Y_1^{\text{ell}}(n)$ classifies isomorphism classes of pairs $(E,P)$, where $E$ is an elliptic curve and $P \in E$ is a point of exact order $n$. Similarly, the elliptic modular curve $Y_0^{\text{ell}}(n)$ classifies pairs $(E,C)$, where $E$ is an elliptic curve and $C \subset E$ is a cyclic subgroup of order $n$. The group $(\mathbb{Z}/n\mathbb{Z})^*$ acts on $Y_1^{\text{ell}}(n)$ via

$$m * (E,P) = \bigl(E,[m]P\bigr) \qquad \text{for} \quad m \in (\mathbb{Z}/n\mathbb{Z})^*,$$

and the quotient of $Y_1^{\text{ell}}(n)$ by this action is $Y_0^{\text{ell}}(n)$.

The reader should be aware that standard terminology is to write

$$Y_0(n),\ Y_1(n),\ X_0(n),\ X_1(n)$$

for elliptic modular curves. But since in this book we deal exclusively with dynatomic modular curves, there should be no cause for confusion. In situations in which both kinds of modular curves appear, it might be advisable to use identifying superscripts such as $Y_1^{\text{dyn}}(n)$ and $Y_1^{\text{ell}}(n)$ to distinguish them.

We showed earlier that $X_1(1)$, $X_1(2)$, and $X_1(3)$ are all (irreducible) curves of genus 0, and similar explicit computations show that $X_1(4)$ has genus 2 and $X_1(5)$ has genus 14. See [171, 305, 309] and Exercise 4.20. The geometry of $X_1(n)$ and $X_0(n)$ for general $n$ is described in the following theorem, which is an amalgamation of results due to Bousch and Morton.

**Theorem 4.17.** (a) *The affine curve $Y_1(n)$ defined by the equation*

$$Y_1(n) : \Phi^*_{n,\phi}(y,z) = 0$$

*is nonsingular.*

(b) *The dynatomic modular curves $X_1(n)$ and $X_0(n)$ are irreducible.*

(c) *The projection map*

$$X_1(n) \longrightarrow \mathbb{P}^1, \qquad (z,c) \longrightarrow c,$$

*exhibits $X_1(n)$ as a Galois cover of $\mathbb{P}^1$. The Galois group is maximal in the sense that it is the appropriate wreath product, cf. Section* 3.9.

(d) *Let $\varphi$ denote the Euler totient function (not to be confused with the rational map $\phi$), let $\mu$ be the Möbius function, and let $\kappa$ be the function*

$$\kappa(n) = \frac{1}{2}\sum_{k|n} \mu\left(\frac{n}{k}\right) 2^k.$$

*(This is essentially half the degree of $\Phi^*_{\phi,n}$; cf. Remark* 4.3.*) Then the genera of $X_1(n)$ and $X_0(n)$ are given by the formulas*

$$\text{genus}\, X_1(n) = 1 + \frac{n-3}{2}\kappa(n) - \frac{1}{2}\sum_{m|n,\ m<n} m\kappa(m)\varphi\left(\frac{n}{m}\right),$$

$$\text{genus}\, X_0(n) = 1 + \frac{n-3}{2n}\kappa(n) - \frac{1}{2}\sum_{m|n,\ m<n} \kappa(m)\varphi\left(\frac{n}{m}\right)$$

$$- \frac{1}{4n}\sum_{\substack{m|n,\ m \text{ odd} \\ n/m \text{ even}}} \mu\left(\frac{n}{m}\right) 2^{m/2}.$$

*(If $n$ is odd, then the final sum in the formula for the genus of $X_0(n)$ is empty.)*

*Proof.* The properties of $Y_1(n)$ and $X_1(n)$ were originally proven by Bousch [83], with subsequent proofs by Lau and Schleicher [261] using analytic methods and Morton [307] via algebraic arguments. The latter two papers give various generalizations, including results for the dynatomic modular curves associated to maps of the form $z^d + c$. The formula for the genus of $X_0(n)$ is due to Morton [307]. □

*Remark* 4.18. The genera of $X_1(n)$ and $X_0(n)$ grow rapidly; see Table 4.2.

| $n$ | 1 | 2 | 3 | 4 | 5 | 6 | 7 | 8 | 9 | 10 |
|---|---|---|---|---|---|---|---|---|---|---|
| genus $X_1(n)$ | 0 | 0 | 0 | 2 | 14 | 34 | 124 | 285 | 745 | 1690 |
| genus $X_0(n)$ | 0 | 0 | 0 | 0 | 2 | 4 | 16 | 32 | 79 | 162 |

Table 4.2: The genera of the dynatomic modular curves $X_1(n)$ and $X_0(n)$ for $z^2 + c$.

$$\begin{aligned}
\mathrm{Res}(\Phi_1^*, \Phi_2^*) &= 4c+3\\
\mathrm{Res}(\Phi_1^*, \Phi_3^*) &= -16c^2-4c-7\\
\mathrm{Res}(\Phi_1^*, \Phi_4^*) &= -16c^2+8c-5\\
\mathrm{Res}(\Phi_1^*, \Phi_5^*) &= -256c^4-64c^3-16c^2+36c-31\\
\mathrm{Res}(\Phi_1^*, \Phi_6^*) &= -16c^2+12c-3\\
\mathrm{Res}(\Phi_2^*, \Phi_4^*) &= -(4c+5)^2\\
\mathrm{Res}(\Phi_2^*, \Phi_6^*) &= -(16c^2+36c+21)^2\\
\mathrm{Res}(\Phi_3^*, \Phi_6^*) &= -(64c^3+128c^2+72c+81)^3
\end{aligned}$$

Table 4.3: The first few bifurcation polynomials for $z^2 + c$.

### 4.2.4 Bifurcation, Misiurewicz Points, and the Mandelbrot Set

When does the polynomial $\phi(z) = z^2 + c$ have a point of formal period $n$ whose exact period $m$ is strictly less than $n$? This will occur if and only if $\Phi_n^*(z)$ and $\Phi_m^*(z)$ have a common root, so if and only if $c$ is a root of the resultant equation

$$\mathrm{Res}(\Phi_n^*(z), \Phi_m^*(z)) = 0.$$

Note that this is a polynomial equation for the parameter $c$. We list the first few examples in Table 4.3. Thus the polynomial $\phi(z) = z^2 - \frac{3}{4}$ given in the last section is the only example with a fixed point of formal period 2. Similarly, the polynomial $\phi(z) = z^2 - \frac{5}{4}$ has a point of formal period 4 whose exact period is 2.

One pattern that is apparent from even the small list in Table 4.3 is the fact that if $m|n$, then $\mathrm{Res}(\Phi_n^*(z), \Phi_m^*(z))$ is the $m^{\text{th}}$ power of a polynomial in $\mathbb{Z}[c]$. See Exercises 4.7 and 4.12 for a description of the $m^{\text{th}}$ root of $\mathrm{Res}(\Phi_n^*(z), \Phi_m^*(z))$.

The roots of $\mathrm{Res}(\Phi_n^*(z), \Phi_m^*(z))$ are special points in the *Mandelbrot set*, which we recall is the subset of the $c$-plane given by

$$\mathcal{M} = \{c \in \mathbb{C} : \phi^n(0) \text{ is bounded as } n \to \infty\}.$$

Alternatively, the Mandelbrot set $\mathcal{M}$ is the set of $c$ for which the Julia set $\mathcal{J}(\phi_c)$ is connected. See Remark 1.34 and the picture of the Mandelbrot set (Figure 1.2) on page 27.

The solutions to the equation

$$\mathrm{Res}(\Phi_n^*(z), \Phi_m^*(z))(c) = 0$$

are called *bifurcation points*.[3] They connect the components of $\mathcal{M}$'s interior (the *bulbs* of $\mathcal{M}$). For example, the point $c = -\frac{3}{4}$ connects the main cardioid of $\mathcal{M}$ to the disk to its left. It is clear that the bifurcation points are algebraic numbers. Beyond that, little is known about their arithmetic properties, although Morton and Vivaldi conjecture that the bifurcation points of type $(m, n)$ are all Galois conjugate to one another. (See Exercise 4.12.)

An elementary property of the Mandelbrot set $\mathcal{M}$, which we now prove, is that it is contained in a disk of radius 2. Note that $\mathcal{M}$ is not contained in any smaller disk, since

$$0 \xrightarrow{\phi_{-2}} -2 \xrightarrow{\phi_{-2}} 2 \xrightarrow{\phi_{-2}} 2$$

shows that $-2 \in \mathcal{M}$.

**Proposition 4.19.** *The Mandelbrot set is contained in the disk of radius* 2,

$$\mathcal{M} \subset \{c \in \mathbb{C} : |c| \le 2\}.$$

*Proof.* Suppose that $|c| > 2$ and let $z_n = \phi_c^n(0)$. Then

$$|z_{n+1}| \ge |z_n^2| - |c| = |z_n| \cdot \bigl(|z_n| - 1\bigr) + \bigl(|z_n| - |c|\bigr). \tag{4.19}$$

We have $|z_1| = |c| > 2$ by assumption, so (4.19) and induction tell us first that $|z_n|$ is an increasing sequence, and indeed that

$$|z_{n+1}| \ge |z_1|\bigl(|z_1| - 1\bigr)^n = |c|\bigl(|c| - 1\bigr)^n.$$

Hence $|z_n| \to \infty$, so $c$ is not in $\mathcal{M}$. □

The set of quadratic maps whose critical points are preperiodic, but not periodic, defines an important subset of the Mandelbrot set.

**Definition.** A point $c \in \mathbb{C}$ is called a *Misiurewicz point* if 0 is strictly preperiodic for $\phi_c(z) = z^2 + c$. We say that $c \in \mathbb{C}$ is a *Misiurewicz point of type* $(m, n)$ if

- $m \ge 1$ is the smallest integer such that $\phi_c^m(0)$ is periodic, and
- $n$ is the primitive period of $\phi_c^m(0)$.

*Example* 4.20. Examples of Misiurewicz points include $c = -2$ and $c = i$. Thus for $c = -2$ we have $0 \to -2 \to 2 \to 2$, so $c = -2$ has type $(2, 1)$. Similarly for $c = i$ we have $0 \to i \to i - 1 \to -i \to i - 1$, so $c = i$ has type $(2, 2)$.

*Remark* 4.21. If $c$ is a Misiurewicz point, then the map $\phi_c$ is subhyperbolic without being hyperbolic, $\phi_c$ has no neutral cycles, and the orbit $\mathcal{O}_{\phi_c}(0)$ is repelling. The Misiurewicz points are all contained in the boundary $\partial\mathcal{M}$ of the Mandelbrot set, and they are dense in $\partial\mathcal{M}$. For proofs of these analytic properties of Misiurewicz points, see for example [95, §VIII.1,6] or [143].

---

[3]Bifurcation point is a general term for a point in moduli space whose polynomial $\phi$ satisfies $\operatorname{Res}(\Phi^*_{n,\phi}(z), \Phi^*_{m,\phi}(z)) = 0$. They are points (polynomials) at which periodic points of different orders merge and break apart. For the quadratic polynomial $\phi(z) = z^2 + c$, whose moduli space is the $c$-plane, the bifurcation points are also known as the *roots of the hyperbolic components of the Mandelbrot set* $\mathcal{M}$.

It is clear that every Misiurewicz point is contained in the Mandelbrot set $\mathcal{M}$, since preperiodic points certainly have bounded orbit. This suffices to prove an arithmetic result.

**Proposition 4.22.** *The set of Misiurewicz points is a set of bounded (absolute) height in $\bar{\mathbb{Q}}$. More precisely, the height of a Misiurewicz point $\gamma$ satisfies $H(\gamma) \le 2$. Hence there are only finitely many Misiurewicz points defined over any given number field.*

*Proof.* A Misiurewicz point $c = \gamma$ is the root of a polynomial of the form

$$M_{m,n}(c) = \phi_c^{n+m}(0) - \phi_c^m(0)$$

for some $m \ge 1$ and $n \ge 1$. These are monic polynomials with coefficients in $\mathbb{Z}$, so not only is $\gamma$ an algebraic number, it is an algebraic integer. Further, the minimal polynomial of $\gamma$ is a factor of $M_{m,n}(c)$. On the other hand, every root of $M_{m,n}(c)$ is in the Mandelbrot set $\mathcal{M}$, since if $c$ is a root of $M_{m,n}(c)$, then 0 is preperiodic for $\phi_c$, so it certainly has bounded orbit. It follows from Proposition 4.19 that every Galois conjugate of $\gamma$ has absolute value at most 2. Let $K = \mathbb{Q}(\gamma)$ and let $\gamma_1, \ldots, \gamma_r$ be the full set of Galois conjugates of $\gamma$. Then

$$H_K(\gamma) = \prod_{v \in M_K} \max\{1, |\gamma|_v\}^{n_v} = \prod_{i=1}^{r} \max\{1, |\gamma_i|\} \le 2^r.$$

Taking the $r^{\text{th}}$ root yields $H(\gamma) \le 2$. □

We now describe an analytic characterization of Misiurewicz points. It depends on the following deep result giving an analytic uniformization of the complement of the Mandelbrot set.

**Theorem 4.23.** (Douady–Hubbard) *There is a conformal isomorphism from the exterior of the unit disk to the complement of the Mandelbrot set,*

$$\theta : \{w \in \mathbb{C} : |w| > 1\} \xrightarrow{\ \sim\ } \mathbb{C} \smallsetminus \mathcal{M}.$$

*Proof.* It is not hard to show that for all sufficiently large $z$ (depending on $c$) there is a consistent way to choose square roots so that the limit

$$\psi_c(z) = \lim_{n \to \infty} \sqrt[2^n]{\phi_c^n(z)}$$

converges and defines a holomorphic function $\psi_c$ on some region $|z| > R_c$. (See Exercise 4.15.) If $c \notin \mathcal{M}$, then one can prove that $\psi_c$ has an analytic continuation to $\mathbb{C} \smallsetminus \mathcal{M}$. The isomorphism $\theta$ in the theorem is the inverse of the map

$$\mathbb{C} \smallsetminus \mathcal{M} \longrightarrow \{w \in \mathbb{C} : |w| > 1\}, \qquad c \longmapsto \psi_c(c).$$

See [142, 143, 141], [43, §9.10], or [95, VIII §§3,4] for details. □

*Remark* 4.24. An immediate consequence of Theorem 4.23 is the connectivity of the Mandelbrot set, since the theorem implies that $\mathbb{P}^1(\mathbb{C}) \smallsetminus \mathcal{M}$ is simply connected.

The uniformization map $\theta$ from Theorem 4.23 can be used to give an analytic description of the Misiurewicz points.

**Theorem 4.25.** *Let $\theta$ be the isomorphism $\{w \in \mathbb{C} : |w| > 1\} \to \mathbb{C} \setminus \mathcal{M}$ described in Theorem 4.23. Consider the doubling map $x \mapsto 2x$ on $\mathbb{Q}/\mathbb{Z}$. Let $t \in \mathbb{Q}/\mathbb{Z}$ be preperiodic, but not periodic, for the doubling map. Let $m$ and $n$ be the smallest positive integers for which we can write $t$ in the form*

$$t = \frac{a}{2^m(2^n - 1)}.$$

*(The fraction need not be in lowest terms. See Exercise 4.16.)*

*Then the limit*

$$c_t = \lim_{r \to 1^+} \theta(re^{2\pi i t})$$

*exists and is a Misiurewicz point of type $(m, n)$ in $\mathcal{M}$, although distinct values of $t \in \mathbb{Q}/\mathbb{Z}$ may yield the same Misiurewicz point.*

*Remark* 4.26. For a given $t \in \mathbb{Q}/\mathbb{Z}$, the "spider algorithm" [213] can be used to compute $c_t$ numerically. The spider algorithm is mainly topological and combinatorial in nature, although the limiting process that yields $c_t \in \mathbb{C}$ is analytic.

## 4.3 The Space $\operatorname{Rat}_d$ of Rational Functions

The set of quadratic polynomials $\{Az^2 + Bz + C\}$ has dimension three, since it may be identified with the set of triples $(A, B, C)$ with $A \neq 0$. In fancier language, the space of quadratic polynomials is equal to the algebraic variety

$$\{(A, B, C) \in \mathbb{A}^3 : A \neq 0\}.$$

We have seen in Section 4.2.1 that every quadratic polynomial can be conjugated to a polynomial of the form $z^2 + c$, and that polynomials with different $c$ values are not conjugate to one another. Thus the space of conjugacy classes of quadratic polynomials has dimension 1. It may be identified with the variety $\mathbb{A}^1$.

In the next few sections we study analogous parameter spaces for more general rational maps and their conjugacy classes. We begin in this section by explaining how the set $\operatorname{Rat}_d$ of rational maps of degree $d$ has a natural structure as an algebraic variety and how the natural action of the algebraic group $\operatorname{PGL}_2$ on $\operatorname{Rat}_d$ is an algebraic action. Then in Section 4.4 we discuss (mostly without proof) how to take the quotient of $\operatorname{Rat}_d$ to construct the moduli space $\mathcal{M}_d$ of conjugacy classes of rational maps of degree $d$. We continue in Section 4.5 by describing a natural collection of algebraic functions on $\mathcal{M}_d$ that are created using symmetric functions of multipliers of periodic points. These functions can be used to map $\mathcal{M}_d$ into affine space. Finally, in Section 4.6 we use these functions to prove that $\mathcal{M}_2$ is isomorphic to $\mathbb{A}^2$.

A rational map $\phi : \mathbb{P}^1 \to \mathbb{P}^1$ of degree $d$ is specified by two homogeneous polynomials

$$\phi = [F, G] = [a_0 X^d + a_1 X^{d-1} Y + \cdots + a_d Y^d, b_d X^d + b_1 X^{d-1} Y + \cdots + b_d Y^d]$$

such that $F$ and $G$ have no common factors, or equivalently from Proposition 2.13, such that the resultant $\operatorname{Res}(F, G)$ does not vanish. Thus a rational map of degree $d$ is determined by the $2d+2$ parameters $a_0, a_1, \ldots, a_d, b_0, b_1, \ldots, b_d$. However, if $u$ is any nonzero number, then $[uF, uG] = [F, G]$, so the $2d+2$ parameters that determine $\phi$ are really well-defined only up to homogeneity. This allows us to identify in a natural way the set of rational maps of degree $d$ with a subset of projective space.

To ease notation, for any $(d+1)$-tuple $\mathbf{a} = (a_0, \ldots, a_d)$, let

$$F_{\mathbf{a}}(X, Y) = a_0 X^d + a_1 X^{d-1} Y + \cdots + a_d Y^d$$

be the associated homogeneous polynomial. Similarly, if $\mathbf{a}$ and $\mathbf{b}$ are $(d+1)$-tuples, we write $[\mathbf{a}, \mathbf{b}] \in \mathbb{P}^{2d+1}$ for the point in projective space whose homogeneous coordinates are $[a_0, \ldots, a_d, b_0, \ldots, b_d]$.

**Definition.** The set of rational functions $\phi : \mathbb{P}^1 \to \mathbb{P}^1$ of degree $d$ is denoted by $\mathrm{Rat}_d$. It is naturally identified with an open subset of $\mathbb{P}^{2d+1}$ via the map

$$\begin{aligned} \{[\mathbf{a}, \mathbf{b}] \in \mathbb{P}^{2d+1} : \operatorname{Res}(F_{\mathbf{a}}, F_{\mathbf{b}}) \neq 0\} &\xrightarrow{\sim} \mathrm{Rat}_d, \\ [\mathbf{a}, \mathbf{b}] &\longmapsto [F_{\mathbf{a}}, F_{\mathbf{b}}]. \end{aligned}$$

The collection of rational maps $\mathrm{Rat}_d$, which a priori is merely a set, thus has the structure of a quasiprojective variety. In fact, $\mathrm{Rat}_d$ is an affine variety, since it is the complement of the hypersurface $\operatorname{Res}(F_{\mathbf{a}}, F_{\mathbf{b}}) = 0$ in the projective space $\mathbb{P}^{2d+1}$.

**Proposition 4.27.** *The variety* $\mathrm{Rat}_d$ *is an affine variety defined over* $\mathbb{Q}$. *The ring of regular functions* $\mathbb{Q}[\mathrm{Rat}_d]$ *of* $\mathrm{Rat}_d$ *is given explicitly by*

$$\mathbb{Q}[\mathrm{Rat}_d] = \mathbb{Q}\left[\frac{a_0^{i_0} a_1^{i_1} \cdots a_d^{i_d} b_0^{j_0} b_1^{j_1} \cdots b_d^{j_d}}{\operatorname{Res}(F_{\mathbf{a}}, F_{\mathbf{b}})} : i_0 + \cdots + i_d + j_0 + \cdots + j_d = 2d\right].$$

*Equivalently,* $\mathbb{Q}[\mathrm{Rat}_d]$ *is the ring of rational functions of degree* 0 *in the localization of* $\mathbb{Q}[a_0, a_1, \ldots, a_d, b_0, b_1, \ldots, b_d]$ *at the multiplicatively closed set consisting of the nonnegative powers of* $\operatorname{Res}(F_{\mathbf{a}}, F_{\mathbf{b}})$.

*Proof.* We remind the reader that in general, if $F \in K[X_0, \ldots, X_r]$ is a homogeneous polynomial of degree $n$, then the complement of the zero set of $F$,

$$V = \mathbb{P}^r \smallsetminus \{F = 0\},$$

is an affine variety of dimension $r$. (See [198, Exercise I.3.5].) Explicitly, each rational function

$$f_{i_0 i_1 \ldots i_r} = \frac{X_0^{i_0} X_1^{i_1} \cdots X_r^{i_r}}{F(X_0, \ldots, X_r)} \qquad \text{with } i_0 + i_1 + \cdots + i_r = n$$

is a regular (i.e., everywhere well-defined) function on $V$. There are $\binom{r+n}{r}$ such functions, and together they define an embedding

$$(f_{i_0 i_1 \dots i_r})_{i_0+i_1+\cdots+i_r=n} : V \hookrightarrow \mathbb{A}^{\binom{r+n}{r}}$$

of $V$ into affine space. The affine coordinate ring of $V$ is the ring of polynomials in these $f_{i_0\dots i_r}$,

$$K[V] = K\left[f_{i_0 i_1 \dots i_r}\right]_{i_0+i_1+\cdots+i_r=n}.$$

In the language of commutative algebra,

$$K[V] = K[X_0, \dots, X_n, 1/F]^{(0)}$$

is the set of rational functions of degree 0 in the localization of $K[X_0, \dots, X_n]$ at the multiplicatively closed set $(F^i)_{i\ge 0}$. Applying this general construction to

$$\mathrm{Rat}_d = \mathbb{P}^{2d+1} \smallsetminus \{\mathrm{Res}(F_{\mathbf{a}}, F_{\mathbf{b}}) = 0\}$$

gives the results stated in the proposition. □

*Remark* 4.28. The geometry of $\mathrm{Rat}_d(\mathbb{C})$, especially near its boundary, presents many interesting problems. See [121, 122, 369, 370].

*Example* 4.29. Let

$$\rho(\mathbf{a},\mathbf{b}) = a_2^2b_0^2 - a_1a_2b_0b_1 + a_0a_2b_1^2 + a_1^2b_0b_2 - 2a_0a_2b_0b_2 - a_0a_1b_1b_2 + a_0^2b_2^2$$

be the resultant of $a_0X^2 + a_1XY + a_2Y^2$ and $b_0X^2 + b_1XY + b_2Y^2$. Then the collection of 84 functions

$$\left(\frac{a_0^{i_0}a_1^{i_1}a_2^{i_2}b_0^{j_0}b_1^{j_1}b_2^{j_2}}{\rho(\mathbf{a},\mathbf{b})}\right)_{i_0+i_1+i_2+j_0+j_1+j_2=4}$$

gives an embedding of $\mathrm{Rat}_2$ into $\mathbb{A}^{84}$. Of course, this is not the smallest affine space into which $\mathrm{Rat}_2$ can be embedded. Projecting onto appropriately chosen hyperplanes, there is certainly an affine embedding of the 5-dimensional space $\mathrm{Rat}_2$ into $\mathbb{A}^{11}$; see [198, Exercise IV.3.11].

*Example* 4.30. The set of rational functions of degree 1 is exactly the set of linear fractional transformations,

$$\mathrm{Rat}_1 = \mathrm{PGL}_2 = \{[\alpha X + \beta Y, \gamma X + \delta Y] : \alpha\delta - \beta\gamma \ne 0\} \subset \mathbb{P}^3.$$

We note that $\mathrm{PGL}_2$ is not merely a variety, it is a group variety, which means that the maps

$$\begin{aligned} \mathrm{PGL}_2 \times \mathrm{PGL}_2 &\longrightarrow \mathrm{PGL}_2, \\ (f_1, f_2) &\longmapsto f_1 f_2, \end{aligned} \qquad \text{and} \qquad \begin{aligned} \mathrm{PGL}_2 &\longrightarrow \mathrm{PGL}_2, \\ f &\longmapsto f^{-1}, \end{aligned}$$

defining the group structure are morphisms.

Each point of $\mathrm{Rat}_d$ determines a rational map $\mathbb{P}^1 \to \mathbb{P}^1$. Further, as we vary the chosen point in $\mathrm{Rat}_d$, the rational maps "vary algebraically." We can make this vague statement precise by saying that the natural map

$$\begin{aligned} \phi : \mathbb{P}^1 \times \mathrm{Rat}_d &\longrightarrow \mathbb{P}^1 \times \mathrm{Rat}_d, \\ \bigl([X,Y],[\mathbf{a},\mathbf{b}]\bigr) &\longmapsto \bigl(\bigl[F_{\mathbf{a}}(X,Y),F_{\mathbf{b}}(X,Y)\bigr],[\mathbf{a},\mathbf{b}]\bigr), \end{aligned} \tag{4.20}$$

is a morphism of varieties. The following definition is useful for describing families that vary algebraically.

**Definition.** Let $V$ be an algebraic variety. The *projective line over* $V$ is the product

$$\mathbb{P}^1_V = \mathbb{P}^1 \times V.$$

A *morphism* $\psi : \mathbb{P}^1_V \to \mathbb{P}^1_V$ *over* $V$ is a morphism that respects the projection to $V$, i.e., the following diagram commutes, where the diagonal arrows are projection onto the second factor:

$$\begin{array}{ccc} \mathbb{P}^1_V & \xrightarrow{\ \psi\ } & \mathbb{P}^1_V \\ & \searrow \quad \swarrow & \\ & V & \end{array}$$

Then $\psi$ can be written in the form

$$\psi = \bigl[F(X,Y),G(X,Y)\bigr],$$

where $F,G \in K(V)[X,Y]$ are homogeneous polynomials with coefficients that are rational functions on $V$. The *degree of* $\psi$ is the degree of the homogeneous polynomials $F$ and $G$.[4]

Any morphism $\lambda : V \to W$ of varieties induces a natural morphism $\mathbb{P}^1_V \to \mathbb{P}^1_W$, which, by abuse of notation, we also denote by $\lambda$. Thus

$$\lambda(P,t) = \bigl(P,\lambda(t)\bigr) \qquad \text{for } (P,t) \in \mathbb{P}^1_V.$$

With this notation, the map (4.20) says that there is a natural morphism

$$\phi : \mathbb{P}^1_{\mathrm{Rat}_d} \longrightarrow \mathbb{P}^1_{\mathrm{Rat}_d}$$

over $\mathrm{Rat}_d$. The next proposition says that this $\phi$ is a universal family of rational maps of degree $d$.

**Proposition 4.31.** *Let $V$ be an algebraic variety and let*

$$\psi : \mathbb{P}^1_V \longrightarrow \mathbb{P}^1_V$$

*be a morphism over $V$ of degree $d$. Then there is a unique morphism*

[4]More generally, if $\psi : \mathbb{P}^N_S \to \mathbb{P}^N_S$ is a morphism of $S$-schemes, then $\psi^*\mathcal{O}_{\mathbb{P}^N_S}(1) \cong \mathcal{O}_{\mathbb{P}^N_S}(d)$ and $d$ is the degree of $\psi$.

$$\lambda : V \to \mathrm{Rat}_d$$

*such that the induced map* $\lambda : \mathbb{P}^1_V \to \mathbb{P}^1_{\mathrm{Rat}_d}$ *fits into the commutative diagram*

$$\begin{array}{ccc} \mathbb{P}^1_V & \xrightarrow{\psi} & \mathbb{P}^1_V \\ \big\downarrow \lambda & & \big\downarrow \lambda \\ \mathbb{P}^1_{\mathrm{Rat}_d} & \xrightarrow{\phi} & \mathbb{P}^1_{\mathrm{Rat}_d} \end{array} \tag{4.21}$$

*Remark* 4.32. The commutative diagram (4.21) in Proposition 4.31 says that any algebraic family of degree $d$ rational maps $\Psi : \mathbb{P}^1_V \to \mathbb{P}^1_V$ factors through the $\mathrm{Rat}_d$ family. Thus $\phi : \mathbb{P}^1_{\mathrm{Rat}_d} \to \mathbb{P}^1_{\mathrm{Rat}_d}$ is a universal family of rational maps of degree $d$. It is an example of a *fine moduli space*. See [322, 323, 335] for further information about moduli spaces.

*Proof.* Let $U \subset V$ be an affine open subset and write $K[U]$ for its affine coordinate ring. The fact that $\psi$ is a morphism over $V$ implies that it restricts to give a morphism $\psi : \mathbb{P}^1_U \to \mathbb{P}^1_U$ over $U$. This restriction of $\psi$ to $\mathbb{P}^1_U$ has the form

$$\psi = [\alpha_0 X^d + \alpha_1 X^{d-1}Y + \cdots + \alpha_d Y^d, \beta_0 X^d + \beta_1 X^{d-1}Y + \cdots + \beta_d Y^d],$$
$$\text{with } \alpha_0, \ldots, \alpha_d, \beta_0, \ldots, \beta_d \in K[U].$$

In other words, the coefficients of the polynomials defining $\psi$ are regular functions on the affine open set $U$. In particular, if there is a map $\lambda : V \to \mathrm{Rat}_d$ making (4.21) commute, then it must be given by $\lambda = [\alpha_0, \ldots, \alpha_d, \beta_0, \ldots, \beta_d]$ at any points of $V$ at which the $\alpha_i$ and $\beta_i$ are defined and the homogeneous polynomials in (4.21) have no common roots.

Now given any point in $V$, we can find a neighborhood $U$ of that point and $\alpha_i, \beta_i \in K[U]$ as above such that $\psi : \mathbb{P}^1_U \to \mathbb{P}^1_U$ is given by

$$\psi([x,y],t) = [\alpha_0(t)x^d + \cdots + \alpha_d(t)y^d, \beta_0(t)x^d + \cdots + \beta_d(t)y^d]$$
$$\text{for } ([x,y],t) \in \mathbb{P}^1_U. \tag{4.22}$$

Using a natural notation, we abbreviate this by writing

$$\Psi(P,t) = [F_{\boldsymbol{\alpha}(t)}(P), F_{\boldsymbol{\beta}(t)}(P)].$$

In order to make the diagram (4.21) commute on $\mathbb{P}^1_U \subset \mathbb{P}^1_V$, we are forced to define $\lambda$ on $U$ by

$$\lambda(t) = [\alpha_0(t), \alpha_1(t), \ldots, \alpha_d(t), \beta_0(t), \beta_1(t), \ldots, \beta_d(t)] \in \mathbb{P}^{2d+1}. \tag{4.23}$$

Further, this $\lambda$ will have the desired properties provided that its image lies in $\mathrm{Rat}_d$. However, the fact that $\psi$ is given by (4.22) for every point in $\mathbb{P}^1_U$ implies that for every $t \in U$, the homogeneous polynomials $F_{\boldsymbol{\alpha}(t)}$ and $F_{\boldsymbol{\beta}(t)}$ have no common root in $\mathbb{P}^1$. It follows from Proposition 2.13(a) that their resultant

$$\mathrm{Res}(F_{\boldsymbol{\alpha}}, F_{\boldsymbol{\beta}}) \in K[U]$$

is nonzero at every point in $U$. Hence the image of the map $\lambda = [\mathbf{a}, \mathbf{b}]$ defined by (4.23) is in $\mathrm{Rat}_d$.

We have now proven that every point in $V$ has a neighborhood $U$ for which there is a map $\lambda : U \to \mathrm{Rat}_d$ making (4.21) commute. Further, the maps on different $U$ must agree on the intersection by the uniqueness discussed earlier. Fitting them together, we find that there is a unique map $\lambda : V \to \mathrm{Rat}_d$ making (4.21) commute. □

An ongoing theme of this text has been the observation that a pair of rational maps $\phi_1, \phi_2 \in \mathrm{Rat}_d$ determine (arithmetically) equivalent dynamical systems if they are $\mathrm{PGL}_2$-conjugate, i.e., if there is a linear fractional transformation $f \in \mathrm{PGL}_2$ such that $\phi_2 = \phi_1^f = f^{-1}\phi f$. The conjugation action of $\mathrm{PGL}_2$ on $\mathrm{Rat}_d$ is algebraic in the following sense.

**Proposition 4.33.** *The map*

$$\mathrm{PGL}_2 \times \mathrm{Rat}_d \longrightarrow \mathrm{Rat}_d, \qquad (f, \phi) \longmapsto \phi^f = f^{-1}\phi f, \tag{4.24}$$

*is an algebraic group action of* $\mathrm{PGL}_2$ *on* $\mathrm{Rat}_d$ *and is defined over* $\mathbb{Q}$. *This means that the map* (4.24) *is both a morphism defined over* $\mathbb{Q}$ *and a group action.*

*Proof.* The proof is mostly a matter of unsorting the definitions. Let

$$f = [\alpha X + \beta Y, \gamma X + \delta Y] \in \mathrm{PGL}_2 \subset \mathbb{P}^3 \quad \text{and} \quad \phi = [F_{\mathbf{a}}, F_{\mathbf{b}}] \in \mathrm{Rat}_d \subset \mathbb{P}^{2d+1}.$$

Then

$$\begin{aligned} \phi^f([X,Y]) = \big[&\delta F_{\mathbf{a}}(\alpha X + \beta Y, \gamma X + \delta Y) - \beta F_{\mathbf{b}}(\alpha X + \beta Y, \gamma X + \delta Y), \\ &- \gamma F_{\mathbf{a}}(\alpha X + \beta Y, \gamma X + \delta Y) + \alpha F_{\mathbf{b}}(\alpha X + \beta Y, \gamma X + \delta Y)\big]. \end{aligned} \tag{4.25}$$

The homogeneity of $F_{\mathbf{a}}$ and $F_{\mathbf{b}}$ shows that (4.24) at least gives a well-defined rational map

$$\mathbb{P}^3 \times \mathbb{P}^{2d+1} \longrightarrow \mathbb{P}^{2d+1}, \qquad (f, \phi) \longmapsto \phi^f.$$

Further, an elementary resultant calculation (Exercise 2.7) shows that the resultant of the two polynomials appearing in the righthand side of (4.25) is equal to

$$(\alpha\delta - \beta\gamma)^{d^2+d}\,\mathrm{Res}(F_{\mathbf{a}}, F_{\mathbf{b}}).$$

It follows that if $f \in \mathrm{PGL}_2$ and $\phi \in \mathrm{Rat}_d$, then $\phi^f$ is a well-defined point in $\mathrm{Rat}_d$. This proves that the map (4.24) is a morphism. The fact that it is a group action is then a straightforward, albeit tedious, calculation. □

*Example* 4.34. In principle it is possible to explicitly write down the action of $\mathrm{PGL}_2$ on $\mathrm{Rat}_d$, but in practice the expressions become hopelessly unwieldy for even moderate values of $d$. As illustration, we describe the action for $d = 2$. Let

$$f = [\alpha X + \beta Y, \gamma X + \delta Y] \qquad \text{and} \qquad \phi = [F_{\mathbf{a}}, F_{\mathbf{b}}] \in \mathrm{Rat}_2,$$

and set

$$\phi^f = [F_{\mathbf{a}'}, F_{\mathbf{b}'}].$$

Multiplying out both sides of (4.25) and equating the coefficients of $X^2$, $XY$, and $Y^2$ yields the following formulas for the coefficients of $F_{\mathbf{a}'}$ and $F_{\mathbf{b}'}$:

$$\begin{aligned}
a_0' &= \alpha^2\delta a_0 + \alpha\gamma\delta a_1 + \gamma^2\delta a_2 - \alpha^2\beta b_0 - \alpha\beta\gamma b_1 - \beta\gamma^2 b_2,\\
a_1' &= 2\alpha\beta\delta a_0 + (\alpha\delta + \beta\gamma)\delta a_1 + 2\gamma\delta^2 a_2 - 2\alpha\beta^2 b_0 - (\alpha\delta + \beta\gamma)\beta b_1 - 2\beta\gamma\delta b_2,\\
a_2' &= \beta^2\delta a_0 + \beta\delta^2 a_1 + \delta^3 a_2 - \beta^3 b_0 - \beta^2\delta b_1 - \beta\delta^2 b_2,\\
b_0' &= -\alpha^2\gamma a_0 - \alpha\gamma^2 a_1 - \gamma^3 a_2 + \alpha^3 b_0 + \alpha^2\gamma b_1 + \alpha\gamma^2 b_2,\\
b_1' &= -2\alpha\beta\gamma a_0 - (\beta\gamma + \alpha\delta)\gamma a_1 - 2\gamma^2\delta a_2 + 2\alpha^2\beta b_0 + (\alpha\delta + \beta\gamma)\alpha b_1 + 2\alpha\gamma\delta b_2,\\
b_2' &= -\beta^2\gamma a_0 - \beta\gamma\delta a_1 - \gamma\delta^2 a_2 + \alpha\beta^2 b_0 + \alpha\beta\delta b_1 + \alpha\delta^2 b_2.
\end{aligned}$$

## 4.4 The Moduli Space $\mathcal{M}_d$ of Dynamical Systems

The intrinsic properties of the dynamical system associated to a rational map $\phi$ depend only on the $\mathrm{PGL}_2$-conjugacy class of $\phi$, so it is natural to take the quotient of the space $\mathrm{Rat}_d$ by the conjugation action of $\mathrm{PGL}_2$.

**Definition.** The *moduli space of rational maps of degree d on* $\mathbb{P}^1$ is the quotient space

$$\mathcal{M}_d = \mathrm{Rat}_d \,/\, \mathrm{PGL}_2,$$

where $\mathrm{PGL}_2$ acts on $\mathrm{Rat}_d$ via conjugation $\phi^f = f^{-1}\phi f$ as described in Proposition 4.33. For the moment, the quotient space $\mathcal{M}_d$ is merely a set, in the sense that for any algebraically closed field $\bar{K}$,

$$\mathcal{M}_d(\bar{K}) = \mathrm{Rat}_d(\bar{K}) / \mathrm{PGL}_2(\bar{K}).$$

We denote the natural map from $\mathrm{Rat}_d$ to $\mathcal{M}_d$ by

$$\langle\cdot\rangle : \mathrm{Rat}_d \longrightarrow \mathcal{M}_d.$$

In order to endow $\mathcal{M}_d$ with the structure of a variety, we observe that the action of $\mathrm{PGL}_2$ on $\mathrm{Rat}_d$ induces in the usual way an action of $\mathrm{PGL}_2$ on the ring of regular functions $\mathbb{Q}[\mathrm{Rat}_d]$ on $\mathrm{Rat}_d$. Thus an element $R \in \mathbb{Q}[\mathrm{Rat}_d]$ is a function whose domain is the set $\{\phi : \mathbb{P}^1 \to \mathbb{P}^1\}$ of rational functions on $\mathbb{P}^1$, and for $f \in \mathrm{PGL}_2$ we define $R^f \in \mathbb{Q}[\mathrm{Rat}_d]$ by the formula

$$R^f(\phi) = R(\phi^f).$$

In this way we obtain a map

$$\mathrm{PGL}_2 \longrightarrow \mathrm{Aut}\big(\mathbb{Q}[\mathrm{Rat}_d]\big),$$

so it makes sense to talk about functions in $\mathbb{Q}[\mathrm{Rat}_d]$ that are invariant under the action of $\mathrm{PGL}_2$. We write

$$\mathbb{Q}[\mathrm{Rat}_d]^{\mathrm{PGL}_2} = \big\{R \in \mathbb{Q}[\mathrm{Rat}_d] : R^f = R \text{ for all } f \in \mathrm{PGL}_2\big\}$$

for the ring of $\mathrm{PGL}_2$-invariant functions on $\mathrm{Rat}_d$.

*Remark* 4.35. For technical reasons, it is sometimes advantageous to replace $\mathrm{PGL}_2$ by a slightly different group. The *special linear group* is the subgroup of $\mathrm{GL}_2$ defined by

$$\mathrm{SL}_2 = \left\{ \begin{pmatrix} \alpha & \beta \\ \gamma & \delta \end{pmatrix} : \alpha\delta - \beta\gamma = 1 \right\},$$

and the *projective special linear group* $\mathrm{PSL}_2$ is the image of $\mathrm{SL}_2$ in $\mathrm{PGL}_2$. The groups $\mathrm{SL}_2$ and $\mathrm{PSL}_2$ act on $\mathrm{Rat}_d$ via conjugation in the usual way. Geometrically, the actions of $\mathrm{PSL}_2$ and $\mathrm{PGL}_2$ on $\mathrm{Rat}_d$ are identical, since it is easy to see that the natural inclusion $\mathrm{SL}_2 \subset \mathrm{GL}_2$ induces an isomorphism

$$\mathrm{PSL}_2(\bar{K}) \xrightarrow{\sim} \mathrm{PGL}_2(\bar{K}).$$

However, if $K$ is not an algebraically closed field, then the map

$$\mathrm{PSL}_2(K) \to \mathrm{PGL}_2(K)$$

need not be an isomorphism; see Exercise 4.22.

The following theorem provides the abstract quotient $\mathrm{Rat}_d / \mathrm{PSL}_2$ with the structure of an algebraic variety.

**Theorem 4.36.** *There is an algebraic variety $\mathcal{M}_d$ defined over $\mathbb{Q}$ and a morphism*

$$\langle\cdot\rangle : \mathrm{Rat}_d \longrightarrow \mathcal{M}_d \tag{4.26}$$

*defined over $\mathbb{Q}$ with the following properties:*

(a) *The map* (4.26) *is* $\mathrm{PSL}_2$*-invariant, i.e., the following diagram is commutative:*

$$\begin{array}{ccc} \mathrm{PSL}_2 \times \mathrm{Rat}_d & \xrightarrow{(f,\phi)\to\phi^f} & \mathrm{Rat}_d \\ \Big\downarrow {\scriptstyle (f,\phi)\mapsto\phi} & & \Big\downarrow {\scriptstyle\langle\cdot\rangle} \\ \mathrm{Rat}_d & \xrightarrow{\langle\cdot\rangle} & \mathcal{M}_d \end{array}$$

*In terms of elements, $\langle\phi^f\rangle = \langle\phi\rangle$ for all $\phi \in \mathrm{Rat}_d$ and all $f \in \mathrm{PGL}_2$.*

(b) *The map on complex points*

$$\langle \cdot \rangle : \mathrm{Rat}_d(\mathbb{C}) \longrightarrow \mathcal{M}_d(\mathbb{C})$$

*is surjective and each fiber is the full* $\mathrm{PSL}_2(\mathbb{C})$*-orbit of a single rational map. Thus there is a bijection of sets*

$$\langle \cdot \rangle : \mathrm{Rat}_d(\mathbb{C}) / \mathrm{PSL}_2(\mathbb{C}) \xrightarrow{\sim} \mathcal{M}_d(\mathbb{C}).$$

(c) *The variety* $\mathcal{M}_d$ *is a connected, integral (i.e., reduced and irreducible), affine variety of dimension* $2d-2$ *whose ring of regular functions is the ring of* $\mathrm{PSL}_2$*-invariant functions on* $\mathrm{Rat}_d$*,*

$$\mathbb{Q}[\mathcal{M}_d] = \mathbb{Q}[\mathrm{Rat}_d]^{\mathrm{PSL}_2}.$$

(d) *Let* $V/\mathbb{C}$ *be a variety and let* $T : \mathrm{Rat}_d \to V$ *be a morphism with the property that* $T(\phi^f) = T(\phi)$ *for all* $\phi \in \mathrm{Rat}_d(\mathbb{C})$ *and all* $f \in \mathrm{PGL}_2(\mathbb{C})$*. Then there is a unique morphism* $\bar{T} : \mathcal{M}_d \to V$ *satisfying* $\bar{T}(\langle\phi\rangle) = T(\phi)$.

*Proof Sketch.* A full proof of Theorem 4.36 (see [416]) uses the machinery of geometric invariant theory [322] and is thus unfortunately beyond the scope of this book. Geometric invariant theory tells us that there is a certain subset of $\mathbb{P}^{2d+1}$, called the *stable locus*, on which the conjugation action of $\mathrm{PSL}_2$ is well behaved. The main part of the proof is to use the $\mathrm{PSL}_2$-invariance of the resultant to verify that

$$\mathrm{Rat}_d = \bigl\{[\mathbf{a}, \mathbf{b}] \in \mathbb{P}^{2d+1} : \mathrm{Res}(F_{\mathbf{a}}, F_{\mathbf{b}}) \neq 0\bigr\}$$

is a $\mathrm{PSL}_2$-invariant subset of the stable locus of $\mathbb{P}^{2d+1}$. Then the existence of the quotient variety $\mathcal{M}_d$ with affine coordinate ring equal to the ring of invariant functions in $\mathbb{Q}[\mathrm{Rat}_d]$ follows from general theorems of geometric invariant theory [322, Chapter 1]. Further, the fact that $\mathcal{M}_d$ is connected, integral, and affine follows immediately from the corresponding property of $\mathrm{Rat}_d$ [322, Section 2, Remark 2]. The dimension of $\mathcal{M}_d$ is computed as

$$\dim \mathcal{M}_d = \dim \mathrm{Rat}_d - \dim \mathrm{PSL}_2 = (2d+1) - 3 = 2d - 2.$$

This proves (a), (b), and (c). Finally, (d) follows directly from the description of $\mathcal{M}_d$ and $\mathbb{Q}[\mathcal{M}_d]$ in (c), since the morphism $T$ induces a map $T^* : \mathcal{O}_V \to \mathbb{C}[\mathrm{Rat}_d]$, and the assumption that $T$ satisfies $T(\phi^f) = T(\phi)$ implies that the image of $T^*$ lies in the ring $\mathbb{C}[\mathrm{Rat}_d]^{\mathrm{PGL}_2(\mathbb{C})} = \mathbb{C}[\mathcal{M}_d]$. □

*Remark* 4.37. Theorem 4.36 says that the quotient $\mathrm{Rat}_d / \mathrm{PSL}_2$ is an algebraic variety. Milnor [301, 302] shows that $\mathrm{Rat}_d(\mathbb{C}) / \mathrm{PGL}_2(\mathbb{C})$ has a natural structure as a complex orbifold, which roughly means that locally it looks like the quotient of a complex manifold by the action of a finite group. Thus its singularities are of a fairly moderate type, although they can still be quite complicated. However, for rational maps of degree 2, we will see in Section 4.6 that not only is $\mathcal{M}_2$ nonsingular, it has a particularly simple structure.

*Remark* 4.38. According to Theorem 4.36(c), the affine coordinate ring of $\mathcal{M}_d$ is the ring of $\mathrm{PSL}_2$-invariant functions $\mathbb{Q}[\mathrm{Rat}_d]^{\mathrm{PSL}_2}$. However, this is the same as the ring of $\mathrm{PGL}_2$-invariant functions, since it suffices to check for invariance by the action of $\mathrm{PGL}_2(\mathbb{C}) = \mathrm{PSL}_2(\mathbb{C})$ on an element of $\mathbb{Q}[\mathrm{Rat}_d]$.

*Remark* 4.39. The moduli space $\mathcal{M}_d$ is the quotient of $\mathrm{Rat}_d$ by the conjugation action of the group $\mathrm{PSL}_2$. In particular, if $\bar{K}$ is any algebraically closed field, then the set of points $\mathcal{M}_d(\bar{K})$ is exactly the collection of cosets of $\mathrm{Rat}_d(\bar{K})$ by the conjugation action of $\mathrm{PSL}_2(\bar{K}) = \mathrm{PGL}_2(\bar{K})$. However, if $K$ is not algebraically closed, then the natural map

$$\mathrm{Rat}_d(K)/\,\mathrm{PSL}_2(K) \longrightarrow \mathcal{M}_d(K)$$

is generally neither injective nor surjective. The correct description of $\mathcal{M}_d(K)$, at least if $K$ is a perfect field, is

$$\mathcal{M}_d(K) = \left\{ \langle\phi\rangle \in \mathcal{M}_d(\bar{K}) : \begin{array}{l} \text{for every } \tau \in \mathrm{Gal}(\bar{K}/K) \text{ there is an} \\ f_\tau \in \mathrm{PGL}_2(\bar{K}) \text{ such that } \tau(\phi) = \phi^{f_\tau} \end{array} \right\}.$$

The *field of moduli* of a rational map $\phi \in \mathrm{Rat}_d(\bar{K})$ is the smallest field $L$ such that $\langle\phi\rangle \in \mathcal{M}_d(L)$. Fields of moduli and related questions are studied in detail in Section 4.10. In particular, see Example 4.85 for a map $\phi$ whose field of moduli is $\mathbb{Q}$, yet $\phi$ is not $\mathrm{PGL}_2(\mathbb{C})$-conjugate to any map in $\mathrm{Rat}_d(\mathbb{Q})$.

The moduli space $\mathcal{M}_d$ classifies rational maps up to conjugation equivalence, just as we want, but it has the defect that it is an affine variety. It is well known that if possible, it is generally preferable to work with projective varieties. How might we naturally complete $\mathcal{M}_d$ by filling in extra points "at infinity"? Note that we should not do this in an arbitrary fashion. Instead, we would like these extra points to correspond naturally to degenerate maps of degree $d$.

One possibility is simply to start with all of $\mathbb{P}^{2d+1}$ and take the quotient by the conjugation action of $\mathrm{PSL}_2$. Unfortunately, there is no natural way to give the quotient $\mathbb{P}^{2d+1}/\,\mathrm{PSL}_2$ any kind of reasonable structure. For example, it is not a variety. So $\mathbb{P}^{2d+1}$ is too large. Ideally, we would like to find a subset $S \subset \mathbb{P}^{2d+1}$ with the following properties:

1. $S$ contains $\mathrm{Rat}_d$.
2. $\mathrm{PSL}_2$ acts on $S$ via conjugation.
3. There is a variety $T$ and a morphism $S \to T$ that induces a bijection

$$S(\mathbb{C})/\,\mathrm{PSL}_2(\mathbb{C}) \longleftrightarrow T(\mathbb{C}).$$

4. The quotient variety $T$ is projective.

Geometric invariant theory gives us two candidates for $S$. The smaller candidate is the largest variety satisfying (3), but its quotient $T$ may fail to be projective. The larger candidate has a projective quotient, but the map $S(\mathbb{C})/\,\mathrm{PSL}_2(\mathbb{C}) \to T(\mathbb{C})$ in (4) may fail to be injective. The following somewhat lengthy theorem describes the application of geometric invariant theory to our situation, that is, to the conjugation action of $\mathrm{PSL}_2$ on $\mathbb{P}^{2d+1}$.

**Theorem 4.40.** *There are algebraic sets*

$$\mathrm{Rat}_d \subset \mathrm{Rat}_d^s \subset \mathrm{Rat}_d^{ss} \subset \mathbb{P}^{2d+1}$$

*with the following properties:*

(a) *The conjugation action of* $\mathrm{PSL}_2$ *on* $\mathrm{Rat}_d$ *extends to an action of* $\mathrm{PSL}_2$ *on* $\mathrm{Rat}_d^s$ *and* $\mathrm{Rat}_d^{ss}$.

(b) *There are varieties* $\mathcal{M}^s$ *and* $\mathcal{M}^{ss}$ *and morphisms*

$$\langle\cdot\rangle : \mathrm{Rat}_d^s \longrightarrow \mathcal{M}_d^s \qquad \text{and} \qquad \langle\cdot\rangle : \mathrm{Rat}_d^{ss} \longrightarrow \mathcal{M}_d^{ss} \tag{4.27}$$

*that are invariant for the action of* $\mathrm{PSL}_2$ *on* $\mathrm{Rat}_d^s$ *and* $\mathrm{Rat}_d^{ss}$. *The varieties* $\mathcal{M}^s$ *and* $\mathcal{M}^{ss}$ *and morphisms (4.27) are defined over* $\mathbb{Q}$.

(c) *Two points* $[\mathbf{a},\mathbf{b}]$ *and* $[\mathbf{a}',\mathbf{b}']$ *in* $\mathrm{Rat}_d^s(\mathbb{C})$ *have the same image in* $\mathcal{M}_d^s(\mathbb{C})$ *if and only if there is an* $f \in \mathrm{PSL}_2(\mathbb{C})$ *satisfying*

$$[\mathbf{a}',\mathbf{b}'] = [\mathbf{a},\mathbf{b}]^f.$$

*Thus as a set,* $\mathcal{M}_d^s(\mathbb{C})$ *is equal to the quotient of* $\mathrm{Rat}_d(\mathbb{C})$ *by* $\mathrm{PSL}_2(\mathbb{C})$.

(d) *Two points* $[\mathbf{a},\mathbf{b}]$ *and* $[\mathbf{a}',\mathbf{b}']$ *in* $\mathrm{Rat}_d^{ss}(\mathbb{C})$ *have the same image in* $\mathcal{M}_d^{ss}(\mathbb{C})$ *if and only if the Zariski closures of their* $\mathrm{PSL}_2(\mathbb{C})$*-orbits have a point in common,*

$$\overline{\{[\mathbf{a},\mathbf{b}]^f : f \in \mathrm{PSL}_2(\mathbb{C})\}} \cap \overline{\{[\mathbf{a}',\mathbf{b}']^f : f \in \mathrm{PSL}_2(\mathbb{C})\}} \neq \emptyset.$$

*Equivalently, they have the same image if and only if there is a holomorphic map* $f : \{t \in \mathbb{C} : 0 < |t| < 1\} \to \mathrm{SL}_2(\mathbb{C})$ *such that*

$$\lim_{t\to 0} [\mathbf{a},\mathbf{b}]^{f_t} = [\mathbf{a}',\mathbf{b}'].$$

(e) $\mathcal{M}_d^s$ *is a quasiprojective variety and* $\mathcal{M}_d^{ss}$ *is a projective variety.*

(f) *(Numerical Criterion) A point* $[\mathbf{a},\mathbf{b}] \in \mathbb{P}^{2d+1}(\mathbb{C})$ *is* not *in* $\mathrm{Rat}_d^{ss}$ *if and only if there is an* $f \in \mathrm{PSL}_2(\mathbb{C})$ *such that* $[\mathbf{a}',\mathbf{b}'] = [\mathbf{a},\mathbf{b}]^f$ *satisfies*

$$a_i' = 0 \text{ for all } i \le \frac{d-1}{2} \quad \text{and} \quad b_i' = 0 \text{ for all } i \le \frac{d+1}{2}. \tag{4.28}$$

*Similarly,* $[\mathbf{a},\mathbf{b}]$ *is* not *in* $\mathrm{Rat}_d^s$ *if and only if there is an* $f \in \mathrm{PSL}_2(\mathbb{C})$ *such that* $[\mathbf{a}',\mathbf{b}'] = [\mathbf{a},\mathbf{b}]^f$ *satisfies*

$$a_i = 0 \text{ for all } i < \frac{d-1}{2} \quad \text{and} \quad b_i = 0 \text{ for all } i < \frac{d+1}{2}. \tag{4.29}$$

(g) $\mathcal{M}_d^s$ *is isomorphic to* $\mathcal{M}_d^{ss}$ *if and only if* $d$ *is even.*

*Proof.* The proof of this theorem is beyond the scope of this book. However, we note that half of (g), namely $\mathcal{M}_d^s \cong \mathcal{M}_d^{ss}$ for even $d$, follows directly from the numerical criterion in (f), since for even $d$ the criteria (4.28) and (4.29) are the same. See [416] for a proof of a general version of Theorem 4.40 over $\mathbb{Z}$. See also [301, 302] for a similar construction over $\mathbb{C}$. □

*Remark* 4.41. In Theorem 4.40, the set denoted by $\mathrm{Rat}_d^s$ is called the set of *stable rational maps*, and the set denoted by $\mathrm{Rat}_d^{ss}$ is called the set of *semistable rational maps*. Note that points in these sets need not represent actual rational maps of degree $d$ on $\mathbb{P}^1$. The intuition is that points in $\mathrm{Rat}_d^s$ and $\mathrm{Rat}_d^{ss}$ that are not in $\mathrm{Rat}_d$ correspond to maps that want to be of degree $d$ but have degenerated in some reasonably nice way into maps of lower degree.

*Remark* 4.42. Theorem 4.40(c) says that the stable quotient $\mathcal{M}_d^s(\mathbb{C})$ has the natural quotient property, since its points correspond exactly to the $\mathrm{PSL}_2(\mathbb{C})$-orbits of points in $\mathrm{Rat}_d^s(\mathbb{C})$. Quotient varieties with this agreeable property are called *geometric quotients*. The semistable quotient $\mathcal{M}_d^{ss}(\mathbb{C})$ has a much subtler quotient property. According to Theorem 4.40(d), points in $\mathrm{Rat}_d^{ss}(\mathbb{C})$ with distinct $\mathrm{PSL}_2(\mathbb{C})$-orbits give the same point in $\mathcal{M}_d^{ss}(\mathbb{C})$ if their orbits approach one another in the limit. Quotients of this kind are called *categorical quotients*. As compensation for the less-intuitive notion of categorical quotient, Theorem 4.40(e) tells us that $\mathcal{M}_d^{ss}$ is projective, so $\mathcal{M}_d^{ss}(\mathbb{C})$ is compact. Finally, Theorem 4.40(g) says that if $d$ is even, then $\mathcal{M}_d^s$ and $\mathcal{M}_d^{ss}$ coincide, so in this case, the moduli space $\mathcal{M}_d$ has a projective closure with a natural (geometric) quotient structure.

*Remark* 4.43. Applying the full machinery of geometric invariant theory to the action of $\mathrm{PSL}_2/\mathbb{Z}$ on $\mathbb{P}_{\mathbb{Z}}^{2d+1}$, it is possible to prove versions of Theorems 4.36 and 4.40 over $\mathbb{Z}$. In other words, there is a filtration of schemes over $\mathbb{Z}$,

$$\mathrm{Rat}_d/\mathbb{Z} \subset \mathrm{Rat}_d^s/\mathbb{Z} \subset \mathrm{Rat}_d^{ss} \subset \mathbb{P}_{\mathbb{Z}}^{2d+1},$$

such that the group scheme $\mathrm{PSL}_2/\mathbb{Z}$ acts on each of these schemes and such that the quotient schemes $\mathcal{M}_d \subset \mathcal{M}_d^s \subset \mathcal{M}^{ss}$ exist in a suitable sense. In particular, Theorems 4.36 and 4.40 are true with $\mathbb{Q}$ replaced by the finite field $\mathbb{F}_p$. The proof is similar to the proof over $\mathbb{Q}$, but requires Sheshadri's theorem that reductive group schemes are geometricallly reductive. See [416] for details.

## 4.5 Periodic Points, Multipliers, and Multiplier Spectra

The moduli space $\mathcal{M}_d$ of rational functions modulo $\mathrm{PSL}_2$-equivalence is an affine variety whose ring of regular functions

$$\mathbb{Q}[\mathcal{M}_d] = \mathbb{Q}[\mathrm{Rat}_d]^{\mathrm{PSL}_2}$$

consists of all regular functions on $\mathrm{Rat}_d$ that are invariant under the action of $\mathrm{PSL}_2$, or equivalently under the action of $\mathrm{PGL}_2(\mathbb{C})$; see Remark 4.38. Abstract invariant theory, as described in the proof sketch of Theorem 4.36, says that there are many such functions. In this section we use periodic points to explicitly construct a large class of regular functions on $\mathcal{M}_d$.

*Example* 4.44. Let $\phi \in \operatorname{Rat}_d(\mathbb{C})$ be a rational function of degree $d$ defined over the complex numbers. Associated to each fixed point $P \in \operatorname{Fix}(\phi)$ is its multiplier $\lambda_P(\phi) \in \mathbb{C}^*$. A simple calculation (Proposition 1.9) shows that $\lambda_P(\phi)$ is $\mathrm{PGL}_2$-invariant in the sense that if $f \in \mathrm{PGL}_2(\mathbb{C})$, then

$$\operatorname{Fix}(\phi^f) = f^{-1}(\operatorname{Fix}(\phi)) \qquad \text{and} \qquad \lambda_{f^{-1}(P)}(\phi^f) = \lambda_P(\phi) \quad \text{for all } P \in \operatorname{Fix}(\phi).$$

Thus in some sense the function $\phi \mapsto \lambda_\phi(P)$ is a function on $\mathcal{M}_d$, where $P$ is a fixed point of $\phi$. Unfortunately, this is not quite correct, since there is no way to pick out a particular fixed point $P$ for a given map $\phi$.

Recall that a rational map $\phi$ of degree $d$ has $d+1$ fixed points counted with appropriate multiplicities, say

$$\operatorname{Fix}(\phi) = \{P_1, P_2, \dots, P_{d+1}\}.$$

The points in $\operatorname{Fix}(\phi)$ do not come in any particular order, so the set of multipliers for the fixed points,

$$\{\lambda_{P_1}(\phi), \lambda_{P_2}(\phi), \dots, \lambda_{P_{d+1}}(\phi)\},$$

is an unordered set of numbers, but as a set, it depends only on $\langle\phi\rangle$, the $\mathrm{PGL}_2$-equivalence class of $\phi$. Hence if we take any symmetric function of the elements in this set of multipliers, we get a number that depends only on $\langle\phi\rangle$.

The elementary symmetric polynomials generate the ring of all symmetric functions, so we define numbers

$$1 = \sigma_0(\phi),\ \sigma_1(\phi),\ \dots,\ \sigma_{d+1}(\phi)$$

by the formula

$$\prod_{P \in \operatorname{Fix}(\phi)} (T + \lambda_P(\phi)) = \sum_{i=0}^{d+1} \sigma_i(\phi) T^{d+1-i}.$$

In other words, the quantity $\sigma_i(\phi)$ is the $i^{\text{th}}$ elementary symmetric polynomial of the multipliers $\lambda_{P_1}(\phi), \dots, \lambda_{P_{d+1}}(\phi)$.

From this construction, it is clear that $\sigma_i(\phi^f) = \sigma_i(\phi)$ for all $f \in \mathrm{PGL}_2$. Further, if we treat the coefficients of $\phi = [F_{\mathbf{a}}, F_{\mathbf{b}}]$ as indeterminates, then the fixed points $P_i$ and the multipliers $\lambda_{P_i}(\phi)$ of $\phi$ are algebraic over $\mathbb{Q}(a_0, \dots, b_d)$ and form a complete set of Galois conjugates, from which it follows that symmetric expressions in the $\lambda_{P_i}(\phi)$, for example the functions $\sigma_i(\phi)$, are in the field $\mathbb{Q}(a_0, \dots, b_d)$. With a bit more work, which we describe in greater generality later in this section, one can show that the multipliers $\lambda_{P_i}(\phi)$ are integral over the ring

$$\mathbb{Q}\left[a_0, \dots, b_d, \operatorname{Res}(F_{\mathbf{a}}, F_{\mathbf{b}})^{-1}\right] = \mathbb{Q}[\operatorname{Rat}_d],$$

and hence that $\sigma_i(\phi) \in \mathbb{Q}[\mathcal{M}_d]$. Thus symmetric polynomials in the multipliers of the fixed points of $\phi$ are regular functions on the moduli space $\mathcal{M}_d$.

*Example* 4.45. We illustrate the construction of Example 4.44 for rational maps of degree 2. As usual, we write

$$\phi = [F_{\mathbf{a}}, F_{\mathbf{b}}] = [a_0X^2 + a_1XY + a_2Y^2, b_0X^2 + b_1XY + b_2Y^2].$$

The map $\phi$ has three fixed points $P_1, P_2, P_2$, and after much algebraic manipulation one finds that the first elementary symmetric function of the multipliers,

$$\sigma_1(\phi) = \lambda_{P_1}(\phi) + \lambda_{P_2}(\phi) + \lambda_{P_3}(\phi),$$

is given explicitly by the horrendous-looking formula

$$\sigma_1(\phi) = \frac{\begin{array}{r} a_1^3b_0-4a_0a_1a_2b_0-6a_2^2b_0^2-a_0a_1^2b_1+4a_0^2a_2b_1+4a_1a_2b_0b_1-2a_0a_2b_1^2+a_2b_1^3 \\ -2a_1^2b_0b_2+4a_0a_2b_0b_2-4a_2b_0b_1b_2-a_1b_1^2b_2+2a_0^2b_2^2+4a_1b_0b_2^2 \end{array}}{a_2^2b_0^2-a_1a_2b_0b_1+a_0a_2b_1^2+a_1^2b_0b_2-2a_0a_2b_0b_2-a_0a_1b_1b_2+a_0^2b_2^2}.$$

Notice that the denominator of $\sigma_1(\phi)$ is $\operatorname{Res}(F_{\mathbf{a}}, F_{\mathbf{b}})$, so $\sigma_1(\phi)$ is in $\mathbb{Q}[\operatorname{Rat}_d]$. It is far less obvious that this expression for $\sigma_1(\phi)$ is $\operatorname{PGL}_2$-invariant. One can verify directly that $\sigma_1(\phi) = \sigma_1(\phi^f)$ by checking that $\sigma_1(\phi)$ does not change when $a_0, \dots, b_2$ are replaced by the expressions $a_0', \dots, b_2'$ described in Example 4.34. We leave this task to the interested reader who has, we hope, access to a suitably robust computer algebra system.

We have used the set of multipliers of the fixed points of a rational map $\phi$ to create $\operatorname{PGL}_2$-invariant functions on $\operatorname{Rat}_d$. More generally, we can use the multipliers associated to periodic points of any order to create such functions.

Recall from Section 4.1 (page 149) that for any $\phi \in \operatorname{Rat}_d$ we write

$$\phi^n = [F_{\phi,n}(X,Y), G_{\phi,n}(X,Y)]$$

with homogeneous polynomials $F_{\phi,n}, G_{\phi,n} \in K[X,Y]$ of degree $d^n$. (See also Exercise 4.9.) Then the set $\operatorname{Per}_n(\phi)$ of $n$-periodic points of $\phi$ are the roots of the $n$-period polynomial

$$\Phi_{\phi,n}(X,Y) = YF_{\phi,n}(X,Y) - XG_{\phi,n}(X,Y),$$

and the set $\operatorname{Per}_n^*(\phi)$ of formal $n$-periodic points of $\phi$ are the roots of the $n^{\text{th}}$ dynatomic polynomial of $\phi$,

$$\Phi^*_{\phi,n}(X,Y) = \prod_{k|n} \bigl(YF_{\phi,k}(X,Y) - XG_{\phi,k}(X,Y)\bigr)^{\mu(n/k)} = \prod_{k|n} \Phi_{\phi,k}(X,Y)^{\mu(n/k)},$$

where we proved in Theorem 4.5 that $\Phi^*_{\phi,n}$ is a polynomial.

The polynomial $\Phi_{\phi,n}$ is homogeneous of degree $d^n+1$. For the purposes of this section, it is convenient to let $\operatorname{Per}_n(\phi)$ be a "set with multiplicity" in the sense that a point appears in $\operatorname{Per}_n(\phi)$ according to its multiplicity as a root of $\Phi_{\phi,n}$. Similarly, we denote the degree of $\Phi^*_{\phi,n}$ by $\nu_d(n)$ (see Remark 4.3) and we assume that points appear in $\operatorname{Per}_n^*(\phi)$ according to their multiplicity as roots of $\Phi^*_{\phi,n}$.

**Definition.** Let $\phi \in \operatorname{Rat}_d$. The *n-multiplier spectrum of* $\phi$ is the collection of values

$$\Lambda_n(\phi) = \{\lambda_P(\phi) : P \in \operatorname{Per}_n(\phi)\}.$$

The *formal n-multiplier spectrum of* $\phi$ is the analogous set of values

$$\Lambda_n^*(\phi) = \{\lambda_P(\phi) : P \in \operatorname{Per}_n^*(\phi)\}.$$

In both sets, the multipliers are taken with the appropriate multiplicity.

*Example* 4.46. Let $\phi(z) = z^d$ with $d \geq 2$. Then $\operatorname{Per}_n(\phi) = \{0, \infty\} \cup \boldsymbol{\mu}_{d^n-1}$ consists of the points 0, $\infty$, and the $(d^n-1)^{\text{th}}$ roots of unity. It is easy to check that $\lambda_0(\phi) = \lambda_\infty(\phi) = 0$, and for $\zeta \in \boldsymbol{\mu}_{d^n-1}$ we have

$$\lambda_\zeta(\phi) = (\phi^n)'(\zeta) = d^n \zeta^{d^n-1} = d^n.$$

Hence

$$\Lambda_n(\phi) = \{0, 0, \overbrace{d^n, d^n, \dots, d^n}^{d^n-1 \text{ copies}}\}.$$

And if $n \geq 2$, then $\Lambda_n^*(\phi)$ consists of $\varphi(d^n-1)$ copies of $d^n$, where $\varphi(m)$ is the Euler totient function (not to be confused with the rational function $\phi(z)$).

*Example* 4.47. Let $\phi(z) = z^2 + bz$. Then

$$\operatorname{Per}_1(\phi) = \{0, 1-b, \infty\} \qquad \text{and} \qquad \Lambda_1(\phi) = \{b, 2-b, 0\}.$$

Next we compute

$$\Phi_{\phi,2}^* = \frac{\phi^2(z) - z}{\phi(z) - z} = z^2 + (b+1)z + b + 1.$$

The two points of formal period 2 are the roots of $\Phi_{\phi,2}^*$,

$$\operatorname{Per}_2^*(\phi) = \left\{ \frac{-(b+1) \pm \sqrt{(b+1)^2 - 4(b+1)}}{2} \right\}. \tag{4.30}$$

Letting $\alpha$ and $\beta$ denote these two values, we substitute them into

$$(\phi^2)'(z) = 4z^3 + 6bz^2 + 2b(b+1)z + b^2$$

to compute their multipliers, which turn out to be identical,

$$\lambda_\alpha(\phi) = \lambda_\beta(\phi) = 4 + 2b - b^2.$$

Thus

$$\Lambda_2^*(\phi) = \{4 + 2b - b^2, 4 + 2b - b^2\}.$$

The multiplier spectra $\Lambda_n(\phi)$ and $\Lambda_n^*(\phi)$ depend only on the $\operatorname{PGL}_2$-equivalence class of $\phi$, so we can use them to define functions on $\mathcal{M}_d$.

**Definition.** Let $\phi \in \operatorname{Rat}_d$ and $n \geq 1$. Define quantities $\sigma_i^{(n)}(\phi)$ for $0 \leq i \leq d^n + 1$ by the relation

$$\prod_{\lambda \in \Lambda_n(\phi)} (T + \lambda) = \sum_{i=0}^{d^n+1} \sigma_i^{(n)}(\phi) T^{d^n+1-i}.$$

Similarly, define quantities $\overset{*}{\sigma}{}_i^{(n)}(\phi)$ for $0 \leq i \leq \nu_d(n)$ by

$$\prod_{\lambda \in \Lambda_n^*(\phi)} (T + \lambda) = \sum_{i=0}^{\nu_d(n)} \overset{*}{\sigma}{}_i^{(n)}(\phi) T^{d^n+1-i}.$$

*Example* 4.48. Continuing with Example 4.46, let $\phi(z) = z^d$. Then

$$\prod_{\lambda \in \Lambda_n(\phi)} (T + \lambda) = T^2 (T + d^n)^{d^n - 1} \quad \text{and} \quad \prod_{\lambda \in \Lambda_n^*(\phi)} (T + \lambda) = (T + d^n)^{\varphi(d^n)}.$$

*Example* 4.49. Continuing with Example 4.47, let $\phi(z) = z^2 + bz$. We computed $\Lambda_1(\phi) = \{b, 2 - b, 0\}$, so

$$\prod_{\lambda \in \Lambda_1(\phi)} (T + \lambda) = (T + b)(T + 2 - b)T = T^3 + 2T^2 + (2b - b^2)T,$$

which gives

$$\sigma_1^{(1)} = 2, \qquad \sigma_2^{(1)} = 2b - b^2, \qquad \sigma_3^{(1)} = 0.$$

Similarly, using the set $\Lambda_2^*(\phi)$ computed in (4.30), we find that

$$\prod_{\lambda \in \Lambda_2^*(\phi)} (T + \lambda) = (T + 4 + 2b - b^2)^2,$$

so

$$\overset{*}{\sigma}{}_1^{(2)} = 2b^2 - 4b - 8 \qquad \text{and} \qquad \overset{*}{\sigma}{}_2^{(2)} = (4 + 2b - b^2)^2.$$

**Theorem 4.50.** *For $\phi \in \operatorname{Rat}_d$, $n \geq 1$, and $i$ in the appropriate range, let $\sigma_i^{(n)}(\phi)$ and $\overset{*}{\sigma}{}_i^{(n)}(\phi)$ be the symmetric polynomials of the $n$-multiplier spectra of $\phi$.*

(a) *The functions*

$$\phi \longmapsto \sigma_i^{(n)}(\phi) \qquad \text{and} \qquad \phi \longmapsto \overset{*}{\sigma}{}_i^{(n)}(\phi) \tag{4.31}$$

*are in $\mathbb{Q}[\operatorname{Rat}_d]$, i.e., they are rational functions in the coefficients $a_0, \ldots, b_d$ of the map $\phi = [F_{\mathbf{a}}, F_{\mathbf{b}}]$ with denominators that are a power of $\operatorname{Res}(F_{\mathbf{a}}, F_{\mathbf{b}})$.*

(b) *The functions* (4.31) *are $\operatorname{PGL}_2$-invariant, and hence are in the ring of regular functions $\mathbb{Q}[\mathcal{M}_d]$ of the dynamical moduli space $\mathcal{M}_d$.*

*Proof.* We sketch the proof for $\sigma_i^{(n)}(\phi)$ and leave $\overset{*}{\sigma}{}_i^{(n)}(\phi)$ as an exercise for the reader (Exercise 4.26).

(a) We write

$$\phi^n = [F_{\mathbf{a},n}(X,Y), F_{\mathbf{b},n}(X,Y)]$$

using homogeneous polynomials of degree $d^n$. Formal properties of the resultant and an easy induction imply that

$$\operatorname{Res}(F_{\mathbf{a},n}, F_{\mathbf{b},n}) = \operatorname{Res}(F_{\mathbf{a}}, G_{\mathbf{b}})^{(2n-1)d^{n-1}}.$$

(See Exercise 4.9.) Thus the rings

$$\mathbb{Q}[a_0,\dots,b_d,\operatorname{Res}(F_{\mathbf{a},n}, F_{\mathbf{b},n})^{-1}] = \mathbb{Q}[a_0,\dots,b_d,\operatorname{Res}(F_{\mathbf{a}}, F_{\mathbf{b}})^{-1}]$$

are equal. This allows us to replace $\phi$ by $\phi^n$ and consider the quantities $\sigma_i^{(1)}(\phi)$ associated to the fixed points of $\phi$. To ease notation, for the remainder of this proof we write $\sigma_i(\phi)$ instead of $\sigma_i^{(1)}(\phi)$.

Let $L$ denote the field $\mathbb{Q}(a_0,\dots,b_d)$, where we treat $a_0,\dots,b_d$ as indeterminates. The fixed points of $\phi = [F_{\mathbf{a}}, F_{\mathbf{b}}]$ are the roots of a polynomial with coefficients in $L$, so they are defined over an algebraic extension of $L$. It follows that the fixed points $\operatorname{Fix}(\phi)$ and their set of multipliers $\Lambda_1(\phi)$ are $\operatorname{Gal}(\bar{L}/L)$-invariant sets, so the symmetric polynomials $\sigma_i(\phi)$ of the fixed points of $\phi$ are $\operatorname{Gal}(\bar{L}/L)$-invariant elements of $\bar{L}$. This proves that each $\sigma_i(\phi)$ is in $L$. Further, it is clear from the construction that $\sigma_i(\phi)$ is homogeneous in $\mathbf{a}$ and $\mathbf{b}$, in the sense that $\sigma_i(\phi)$ gives the same value for $[F_{c\mathbf{a}}, F_{c\mathbf{b}}]$ for any nonzero constant $c$. Thus the $\sigma_i(\phi)$ are in $L^{(0)}$, where the 0 denotes rational functions of $a_0,\dots,b_d$ whose numerator and denominator are homogeneous of the same degree.

Our next task is to prove that they are regular functions on $\operatorname{Rat}_d$, i.e., that they are well-defined at every point of $\operatorname{Rat}_d$. In order to do this, we must show that their only poles occur when $\operatorname{Res}(F_{\mathbf{a}}, F_{\mathbf{b}}) = 0$. Let

$$A = \mathbb{Q}[a_0, a_1, \dots, a_d, b_0, b_1, \dots, b_d]$$

be the ring of polynomials in the indeterminates $a_0,\dots,b_d$, and to simplify notation, let

$$r = \operatorname{Res}(F_{\mathbf{a}}, F_{\mathbf{b}}) \in A$$

denote the resultant of $F_{\mathbf{a}}$ and $F_{\mathbf{b}}$. With this notation, Theorem 4.27 says that the coordinate ring of $\operatorname{Rat}_d$ is $\mathbb{Q}[\operatorname{Rat}_d]$, which equals $A[r^{-1}]^{(0)}$, where again the 0 indicates that we take rational functions of degree 0. We also note that the polynomial $r$ is irreducible in $A$.[5]

We are first going to prove that $\sigma_i(\phi)$ is in the ring $A[b_d^{-1}, r^{-1}]$. We dehomogenize by setting $[X,Y] = [z,1]$ and write $\phi(z) = F_{\mathbf{a}}(z)/F_{\mathbf{b}}(z)$. The fixed points of $\phi$ are the roots of the polynomial $\phi(z) - z$, or equivalently, the roots of the polynomial

$$zF_{\mathbf{b}}(z) - F_{\mathbf{a}}(z) = b_d z^{d+1} + (b_{d-1} - a_d)z^d + (b_{d-2} - a_{d-1})z^{d-1} + \cdots + (b_0 - a_1)z + a_0. \tag{4.32}$$

[5] Indeed, the resultant polynomial is geometrically irreducible, which means that it is irreducible in $K[a_0,\dots,b_d]$ for any field $K$, see [436, §5.9].

The roots $\alpha_1, \dots, \alpha_{d+1}$ of this polynomial in $\bar{L}$ are integral over the ring $A[b_d^{-1}]$. For any such root $\alpha$, the corresponding multiplier is

$$\phi'(\alpha) = \frac{F'_{\mathbf{b}}(\alpha)F_{\mathbf{a}}(\alpha) - F'_{\mathbf{a}}(\alpha)F_{\mathbf{b}}(\alpha)}{F_{\mathbf{b}}(\alpha)^2}.$$

Let $B$ be the integral closure of $A[b_d^{-1}]$ in the field $L(\alpha)$. It is clear that the numerator of $\phi'(\alpha)$ is in $B$. We claim that its denominator is a unit in $B[r^{-1}]$. Suppose not. Then we can find a maximal ideal $\mathfrak{P} \subset B[r^{-1}]$ with $F_{\mathbf{b}}(\alpha) \in \mathfrak{P}$. But $\alpha$ is a root of (4.32), so

$$F_{\mathbf{a}}(\alpha) = \alpha F_{\mathbf{b}}(\alpha) \in \mathfrak{P}.$$

Thus $z = \alpha$ is a simultaneous root of

$$F_{\mathbf{a}}(z) \equiv F_{\mathbf{b}}(z) \equiv 0 \pmod{\mathfrak{P}},$$

so a standard property of resultants (Proposition 2.13(a)) implies that

$$r = \operatorname{Res}(F_{\mathbf{a}}, F_{\mathbf{b}}) \in \mathfrak{P}.$$

But this is a contradiction, since $r$ is a unit in $B[r^{-1}]$, so it cannot be an element of a maximal ideal. Hence $F_{\mathbf{b}}(\alpha)$ is a unit in $B[r^{-1}]$, and therefore the multiplier $\phi'(\alpha)$ is in $B[r^{-1}]$.

We have now shown that each of the multipliers $\phi'(\alpha_1), \dots, \phi'(\alpha_{d+1}) \in \bar{L}$ is integral over the ring $A[b_d^{-1}, r^{-1}]$. Hence the symmetric polynomials in these quantities, i.e., $\sigma_1(\phi), \dots, \sigma_{d+1}(\phi)$, are in $L^{(0)}$ and are integral over $A[b_d^{-1}, r^{-1}]$. It follows that they are in $A[b_d^{-1}, r^{-1}]^{(0)}$, since the ring $A[b_d^{-1}, r^{-1}]$ is integrally closed in its fraction field $L$.

A similar argument dehomogenizing $[X, Y] = [1, w]$ shows that the $\sigma_i$ are in the ring $A[a_d^{-1}, r^{-1}]^{(0)}$. Therefore

$$\sigma_1(\phi), \dots, \sigma_{d+1}(\phi) \in A[a_d^{-1}, r^{-1}]^{(0)} \cap A[b_d^{-1}, r^{-1}]^{(0)} = A[r^{-1}]^{(0)} = \mathbb{Q}[\mathrm{Rat}_d].$$

(b) Our earlier calculation (Proposition 1.9) showed that for every complex point $[\mathbf{a}, \mathbf{b}] \in \mathbb{P}^{2d+1}(\mathbb{C})$ with $\operatorname{Res}(F_{\mathbf{a}}, F_{\mathbf{b}}) \neq 0$, the set of multipliers of $\phi = [F_{\mathbf{a}}, F_{\mathbf{b}}]$ is $\mathrm{PGL}_2(\mathbb{C})$-invariant. Hence the same is true for the quantities $\sigma_i(\phi)$ for all $i$. This invariance says the following: Let $f(z) = (\alpha z + \beta)/(\gamma z + \delta)$ be a linear fractional transformation with indeterminate coefficients $\alpha, \beta, \gamma, \delta$ and consider the difference

$$\sigma_i(\phi^f) - \sigma_i(\phi). \tag{4.33}$$

Clearly (4.33) is a rational function in $\mathbb{Q}(a_0, \dots, b_d, \alpha, \beta, \gamma, \delta)$, and the $\mathrm{PGL}_2(\mathbb{C})$-invariance of $s_i(\phi)$ says that (4.33) vanishes for all choices of $\left(\begin{smallmatrix}\alpha & \beta \\ \gamma & \delta\end{smallmatrix}\right) \in \mathrm{GL}_2(\mathbb{C})$. It follows that (4.33) is identically zero.

Finally, Theorem 4.36(c) says that $\mathbb{Q}[\mathcal{M}_d]$ is the subring of $\mathrm{PGL}_2$-invariant functions in $\mathbb{Q}[\mathrm{Rat}_d]$ (see also Remark 4.38), so in particular the functions (4.31) are in $\mathbb{Q}[\mathcal{M}_d]$. □

*Remark* 4.51. As noted earlier in Remark 4.43, the moduli space $\mathcal{M}_d$ exists as a scheme over $\mathbb{Z}$. Theorem 4.50 is also valid over $\mathbb{Z}$, so in particular the affine coordinate ring $\mathbb{Z}[\mathcal{M}_d]$ is the ring of $\mathrm{PSL}_2$-invariant functions in $\mathbb{Z}[\mathrm{Rat}_d]$. See [416, Theorems 4.2 and 4.5].

The functions $\sigma_i^{(n)}$ and $\overset{*}{\sigma}{}_i^{(n)}$ constructed in Theorem 4.50 are regular functions on $\mathcal{M}_d$, so they can be used to map $\mathcal{M}_d$ to affine space. For example, in the next section we show that

$$\left(\overset{*}{\sigma}{}_1^{(1)}, \overset{*}{\sigma}{}_2^{(1)}\right) : \mathcal{M}_2 \to \mathbb{A}^2$$

is an isomorphism. In general, $\mathcal{M}_d$ is not isomorphic to $\mathbb{A}^{2d-2}$, but we might ask whether using a sufficient number of the $\sigma_i^{(n)}$ or $\overset{*}{\sigma}{}_i^{(n)}$ gives an embedding of $\mathcal{M}_d$ into affine space.

In particular, do the values of all $\sigma_i^{(n)}$ or $\overset{*}{\sigma}{}_i^{(n)}(\phi)$ determine the $\mathrm{PGL}_2(\mathbb{C})$-conjugacy class of $\phi$? The answer is no. The Lattès maps that we studied in Section 1.6.3 provide nontrivial families of rational maps whose multiplier spectra, and hence whose $\sigma_i^{(n)}$ and $\overset{*}{\sigma}{}_i^{(n)}$ values, are all the same.

*Example* 4.52. For each $t \in \mathbb{C}^*$ with $t \ne -\frac{27}{4}$, consider the rational map

$$\phi_t(x) = \frac{x^4 - 2tx^2 - 8tx + t^2}{4x^3 + 4tx + 4t}.$$

It is the Lattès map associated to multiplication-by-2 on the elliptic curve

$$E_t : y^2 = x^3 + tx + t;$$

see Section 1.6.3. Following standard notation, we write $E_t[m]$ for the points of $E_t$ of order $m$. Then the $n$-periodic points of $\phi$ are given by

$$\mathrm{Per}_n(\phi) = x\bigl(E_t[2^n - 1]\bigr) \cup x\bigl(E_t[2^n + 1]\bigr),$$

and it is not hard to compute the multipliers at these points,

$$\lambda_\alpha(\phi_t) = \begin{cases} 2^n & \text{if } \alpha \in x\bigl(E_t[2^n-1]\bigr) \text{ and } \alpha \ne \infty, \\ -2^n & \text{if } \alpha \in x\bigl(E_t[2^n+1]\bigr) \text{ and } \alpha \ne \infty, \\ 2^{2n} & \text{if } \alpha = \infty. \end{cases}$$

(For proofs of these statements, see Proposition 6.52 in Section 6.5.)

For any $m$, the set $E[m]$ of $m$-torsion points has order $m^2$, and for odd $m$ the map $x : E[m] \to \mathbb{P}^1$ is exactly 2-to-1 except at $x^{-1}(\infty)$, so we can use the listed values of $\lambda_\alpha(\phi_t)$ to compute

$$\prod_{\lambda \in \Lambda_n(\phi_t)} (T + \lambda) = (T + 2^n)^{2^{2n-1} - 2^n} (T - 2^n)^{2^{2n-1} + 2^n} (T + 2^{2n}).$$

In particular, we see that every map $\phi_t$ has the same set of multipliers, so $\sigma_i^{(n)}(\phi_t)$ does not depend on $t$. (A similar statement holds for $\overset{*}{\sigma}{}_i^{(n)}(\phi_t)$.) Hence no matter how many functions $\sigma_i^{(n)}$ we use, the resulting map $\mathcal{M}_4 \to \mathbb{A}^k$ always compresses all of the maps $\phi_t$ down to a single point. On the other hand, we will later prove that the $\phi_t$ are not $\mathrm{PGL}_2$-conjugate to one another (see Theorem 6.46). Hence the image of the map

$$\mathbb{C}^* \setminus \{-\tfrac{27}{4}\} \longrightarrow M_4, \qquad t \longmapsto \langle \phi_t \rangle,$$

is a curve in $M_4$ that is compressed to a single point in $\mathbb{A}^k$. An important theorem says that aside from Lattès examples of this kind, the map on $\mathcal{M}_d$ defined by the $\sigma_i^{(n)}$ is finite-to-one.

**Theorem 4.53.** (McMullen [294, §2]) *Fix $d \geq 2$, and for each $N \geq 1$ let*

$$\boldsymbol{\sigma}_{d,N} : \mathcal{M}_d \longrightarrow \mathbb{A}^k \tag{4.34}$$

*be the map defined using all of the functions $\sigma_i^{(n)}$ with $1 \leq n \leq N$. If $N$ is sufficiently large, then the map $\boldsymbol{\sigma}_{d,N}$ is finite-to-one on $\mathcal{M}_d(\mathbb{C})$ except for certain families of Lattès maps that it compresses down to a single point. In particular, it is finite-to-one if $d$ is not a perfect square.*

*Further, the same statement is true for the map*

$$\boldsymbol{\sigma}^*_{d,N} : \mathcal{M}_d \longrightarrow \mathbb{A}^\ell$$

*defined using all of the functions $\overset{*}{\sigma}{}_i^{(n)}$ with $1 \leq n \leq N$. (The* flexible Lattès maps *for which $\boldsymbol{\sigma}_{d,N}$ and $\boldsymbol{\sigma}_{d,N^*}$ are not finite-to-one are discussed in detail in Section* 6.5.)

McMullen's theorem says that aside from the flexible Lattès maps, the maps $\boldsymbol{\sigma}_{d,N}$ and $\boldsymbol{\sigma}_{d,N^*}$ are finite-to-one onto their image. One might hope that they are actually injective if we avoid the flexible Lattès maps, but it turns out that this is far from true.

**Theorem 4.54.** *Define the degree of $\boldsymbol{\sigma}_{d,N}$ to be the number of points in $\boldsymbol{\sigma}_{d,N}^{-1}(P)$ for a generic point $P$ in the image $\boldsymbol{\sigma}_{d,N}(\mathcal{M}_d)$. One can show that the degree of $\boldsymbol{\sigma}_{d,N}$ stabilizes as $N \to \infty$. We write* $\deg(\boldsymbol{\sigma}_d)$ *for this value. Then for every $\epsilon > 0$ there is a constant $C_\epsilon$ such that*

$$\deg(\boldsymbol{\sigma}_d) \geq C_\epsilon d^{\frac{1}{2}-\epsilon} \qquad \textit{for all } d.$$

*In particular, the multipliers of a rational function $\phi \in \mathrm{Rat}_d$ determine the conjugacy class of $\phi$ only up to $O_\epsilon(d^{\frac{1}{2}-\epsilon})$ possibilities.*

*Proof.* We will prove this in Chapter 6 using Lattès maps associated to elliptic curves with complex multiplication; see Theorem 6.62. □

**Definition.** Following Milnor [300], we define the *multiplier spectrum* of a rational map $\phi$ to be the function $\Lambda(\phi)$ that to each positive integer $n$ assigns the set $\Lambda_n(\phi)$. In other words, $\Lambda(\phi)$ is the set-valued function

$$\Lambda(\phi) : \mathbb{N} \longrightarrow \textsf{Sets with Multiplicities}, \qquad n \longmapsto \Lambda_n(\phi).$$

Two rational maps are said to be *isospectral* if they have the same multiplier spectrum. Then another way to state McMullen's Theorem 4.53 is to say that aside from the flexible Lattès maps, the multiplier spectrum determines the rational map $\phi$ up to finitely many possibilities. The fact that the flexible Lattès maps form nontrivial isospectral families is proven in Section 6.5.

*Remark* 4.55. We show in the next section that the map

$$\sigma_{2,1} : \mathcal{M}_2 \longrightarrow \mathbb{A}^3$$

is an injection, and in fact it maps $\mathcal{M}_2$ isomorphically to the plane $z = x - 2$, so $\mathcal{M}_2 \cong \mathbb{A}^2$. On the other hand, the existence of the Lattès maps described in Example 4.52 shows that $\boldsymbol{\sigma}_{4,N}$ cannot be finite-to-one on $\mathcal{M}_4$. The degree of the map

$$\boldsymbol{\sigma}_{3,N} : \mathcal{M}_3 \longrightarrow \mathbb{A}^k$$

does not appear to be known.

## 4.6 The Moduli Space $\mathcal{M}_2$ of Dynamical Systems of Degree 2

Theorem 4.50 tells us that symmetric combinations of the numbers in the multiplier spectra give well-defined functions on the moduli space $\mathcal{M}_d$ of degree $d$ dynamical systems on $\mathbb{P}^1$. In this section we describe Milnor's explicit identification of $\mathcal{M}_2$ with $\mathbb{A}^2$ using two of these functions.

**Theorem 4.56.** (Milnor [301], see also [416]) *Let $\sigma_1, \sigma_2, \sigma_3 \in \mathbb{Q}[\mathrm{Rat}_2]$ be the three functions constructed from the fixed points of a rational map of degree 2, i.e.,*

$$\prod_{P\in\mathrm{Fix}(\phi)} (T + \lambda_P(\phi)) = \sum_{i=0}^{3} \sigma_i(\phi) T^{3-i}.$$

*(In the notation from page* 183, *we have $\sigma_i = \sigma_i^{(1)}$ for $0 \le i \le 3$.)*

(a) $\sigma_1 = \sigma_3 + 2$.

(b) *The morphism*

$$\boldsymbol{\sigma} = (\sigma_1, \sigma_2) : \mathrm{Rat}_2 \longrightarrow \mathbb{A}^2$$

*has the following three properties:*

(i) $\boldsymbol{\sigma}(\phi^f) = \boldsymbol{\sigma}(\phi)$ *for all $\phi \in \mathrm{Rat}_2(\mathbb{C})$ and all $f \in \mathrm{PGL}_2(\mathbb{C})$.*

(ii) *Let $\phi_1, \phi_2 \in \mathrm{Rat}_2(\mathbb{C})$. If $\boldsymbol{\sigma}(\phi_1) = \boldsymbol{\sigma}(\phi_2)$, then there exists a Möbius transformation $f \in \mathrm{PGL}_2(\mathbb{C})$ such that $\phi_2 = \phi_1^f$.*

(iii) *Let $(s_1, s_2) \in \mathbb{A}^2(\mathbb{C})$. Then there exists a rational map $\phi \in \mathrm{Rat}_2(\mathbb{C})$ such that $\boldsymbol{\sigma}(\phi) = (s_1, s_2)$.*

*Hence $\boldsymbol{\sigma}$ induces a bijection $\boldsymbol{\sigma} : \mathcal{M}_2(\mathbb{C}) \xrightarrow{\sim} \mathbb{A}^2(\mathbb{C})$.*

(c) *The functions $\sigma_1$ and $\sigma_2$ are in $\mathbb{Q}[\mathcal{M}_2]$ and the map*

$$\boldsymbol{\sigma} = (\sigma_1, \sigma_2) : \mathcal{M}_2 \longrightarrow \mathbb{A}^2$$

*is an isomorphism of algebraic varieties defined over $\mathbb{Q}$. Equivalently, the induced map*

$$\boldsymbol{\sigma}^* : \mathbb{Q}[X,Y] \xrightarrow{\sim} \mathbb{Q}[\mathcal{M}_2], \qquad \boldsymbol{\sigma}\big(F(X,Y)\big) = F(\sigma_1, \sigma_2),$$

*is an isomorphism of rings.*

(d) *All of the functions $\sigma_i^{(n)}$ and $\overset{*}{\sigma}{}_i^{(n)}$ can be expressed as polynomials in $\sigma_1$ and $\sigma_2$ with rational coefficients.*

(e) *For any field extension $K/\mathbb{Q}$, the map $\boldsymbol{\sigma} : \mathcal{M}_2 \to \mathbb{A}^2$ in* (c) *induces a bijection $\mathcal{M}_2(K) \leftrightarrow \mathbb{A}^2(K)$. (Note that $\mathcal{M}_d(K)$ is not the same as* $\operatorname{Rat}_d(K)/\operatorname{PSL}_2(K)$*; see Remark* 4.39 *and Section* 4.10.)

*Remark* 4.57. For some applications it is useful to have an explicit description of the map

$$\boldsymbol{\sigma} = (\sigma_1, \sigma_2) : \operatorname{Rat}_2 \longrightarrow \mathbb{A}^2$$

described in Theorem 4.56(b). Let

$$\rho(\mathbf{a}, \mathbf{b}) = a_2^2 b_0^2 - a_1 a_2 b_0 b_1 + a_0 a_2 b_1^2 + a_1^2 b_0 b_2 - 2a_0 a_2 b_0 b_2 - a_0 a_1 b_1 b_2 + a_0^2 b_2^2$$

denote the resultant of $F_{\mathbf{a}}$ and $F_{\mathbf{b}}$. Then, after some algebraic manipulation, one finds that $\sigma_1$ and $\sigma_2$ are given by the expressions

$$\begin{aligned}
\rho(\mathbf{a}, \mathbf{b})\sigma_1(\phi) &= a_1^3 b_0 - 4a_0 a_1 a_2 b_0 - 6a_2^2 b_0^2 - a_0 a_1^2 b_1 + 4a_0^2 a_2 b_1 + 4a_1 a_2 b_0 b_1 \\
&\quad - 2a_0 a_2 b_1^2 + a_2 b_1^3 - 2a_1^2 b_0 b_2 + 4a_0 a_2 b_0 b_2 - 4a_2 b_0 b_1 b_2 \\
&\quad - a_1 b_1^2 b_2 + 2a_0^2 b_2^2 + 4a_1 b_0 b_2^2, \\
\rho(\mathbf{a}, \mathbf{b})\sigma_2(\phi) &= -a_0^2 a_1^2 + 4a_0^3 a_2 - 2a_1^3 b_0 + 10 a_0 a_1 a_2 b_0 + 12 a_2^2 b_0^2 - 4a_0^2 a_2 b_1 \\
&\quad - 7a_1 a_2 b_0 b_1 - a_1^2 b_1^2 + 5a_0 a_2 b_1^2 - 2a_2 b_1^3 + 2a_0^2 a_1 b_2 + 5a_1^2 b_0 b_2 \\
&\quad - 4a_0 a_2 b_0 b_2 - a_0 a_1 b_1 b_2 + 10 a_2 b_0 b_1 b_2 - 4a_1 b_0 b_2^2 + 2a_0 b_1 b_2^2 \\
&\quad - b_1^2 b_2^2 + 4b_0 b_2^3.
\end{aligned}$$

*Remark* 4.58. According to Theorem 4.56(c), every function $\overset{*}{\sigma}{}_i^{(n)} \in \mathbb{Q}[\mathcal{M}_2]$ is a polynomial in $\sigma_1$ and $\sigma_2$. In practice, it can be quite challenging to find explicit expressions. Milnor [301] gives the examples

$$\begin{aligned}
\overset{*}{\sigma}{}_1^{(2)} &= 2\sigma_1 + \sigma_2, \\
\overset{*}{\sigma}{}_1^{(3)} &= \sigma_1(2\sigma_1 + \sigma_2) + 3\sigma_1 + 3, \\
\overset{*}{\sigma}{}_2^{(3)} &= (\sigma_1 + \sigma_2)^2(2\sigma_1 + \sigma_2) - \sigma_1(\sigma_1 + 2\sigma_2) + 12\sigma_1 + 28.
\end{aligned}$$

Notice that these expressions are in $\mathbb{Z}[\sigma_1, \sigma_2]$, rather than merely in $\mathbb{Q}[\sigma_1, \sigma_2]$. This reflects the fact that the map $\boldsymbol{\sigma} : \mathcal{M}_2 \to \mathbb{A}^2$ is actually an isomorphism of schemes over $\mathbb{Z}$ as described in Remarks 4.43 and 4.51. The functions $\overset{*}{\sigma}{}_i^{(n)}$ are regular functions on the scheme $\mathcal{M}_2/\mathbb{Z}$, so they are in $\mathbb{Z}[\sigma_1, \sigma_2]$. See [416] for details.

*Proof of Theorem* 4.56. (a) Let $\phi \in \mathrm{Rat}_2(\mathbb{C})$ and let $\lambda_1, \lambda_2, \lambda_3$ be the multipliers of the fixed points of $\phi$. If we assume that none of $\lambda_1, \lambda_2, \lambda_3$ is equal to 1, then we can apply Theorem 1.14 to deduce that

$$\frac{1}{1-\lambda_1} + \frac{1}{1-\lambda_2} + \frac{1}{1-\lambda_3} = 1.$$

After some algebraic manipulation this becomes

$$\lambda_1\lambda_2\lambda_3 - \lambda_1 - \lambda_2 - \lambda_3 + 2 = 0,$$

which completes the proof that $\sigma_1(\phi) = \sigma_3(\phi) + 2$ for all $\phi \in \mathrm{Rat}_2(\mathbb{C})$ whose multipliers are not equal to 1. It is not hard to see that such $\phi$ are dense in $\mathrm{Rat}_2(\mathbb{C})$ (see Exercise 4.21), from which it follows that the function

$$\sigma_1 - \sigma_3 - 2 \in \mathbb{C}[\mathrm{Rat}_2] = \mathbb{C}\big[a_0, a_1, a_2, b_0, b_1, b_2, \mathrm{Res}(F_{\mathbf{a}}, F_{\mathbf{b}})^{-1}\big]$$

is identically zero.
(b) The first property $\boldsymbol{\sigma}(\phi^f) = \boldsymbol{\sigma}(\phi)$ is a special case of Theorem 4.50, or more directly it is an immediate consequence of our earlier calculation (Proposition 1.9) showing that the multipliers of a rational map are $\mathrm{PGL}_2$-invariant.

In order to prove the second property, it is convenient to show that every rational map of degree 2 is $\mathrm{PGL}_2$-equivalent to a map of a particular shape.

**Lemma 4.59.** (Normal Forms Lemma) *Let $\phi \in \mathrm{Rat}_2(\mathbb{C})$ be a rational map of degree 2 and let $\lambda_1, \lambda_2, \lambda_3$ be the multipliers of its fixed points.*
(a) *If $\lambda_1\lambda_2 \neq 1$, then there is an $f \in \mathrm{PGL}_2(\mathbb{C})$ such that*

$$\phi^f(z) = \frac{z^2 + \lambda_1 z}{\lambda_2 z + 1}. \tag{4.35}$$

*Further,* $\mathrm{Res}(z^2 + \lambda_1 z, \lambda_2 z + 1) = 1 - \lambda_1\lambda_2$.
(b) *If $\lambda_1\lambda_2 = 1$, then $\lambda_1 = \lambda_2 = 1$ and there is an $f \in \mathrm{PGL}_2(\mathbb{C})$ such that*

$$\phi^f(z) = z + \sqrt{1-\lambda_3} + \frac{1}{z}. \tag{4.36}$$

*Proof.* We recall that if $\alpha \in \mathrm{Fix}(\phi)$ has multiplier $\lambda_\alpha$, then the Taylor expansion of $\phi$ around $\alpha$ looks like

$$\phi(z) = \phi(\alpha) + \phi'(\alpha)(z-\alpha) + O(z-\alpha)^2 = \alpha + \lambda_\alpha(z-\alpha) + O(z-\alpha)^2.$$

Hence

$$\Phi_{\phi,1}(z) = \phi(z) - z = (\lambda_\alpha - 1)(z-\alpha) + O(z-\alpha)^2,$$

so $\phi$ has multiplicity 1 at the fixed point $\alpha$ if and only if $\lambda_\alpha \neq 1$.

We also note the formal identity

$$(X-1)^2 - (XY-1)(XZ-1) = X(X+Y+Z-2-XYZ)$$

and apply it using the relation

$$\lambda_1 + \lambda_2 + \lambda_3 = \sigma_1 = \sigma_3 + 2 = \lambda_1\lambda_2\lambda_3 + 2$$

from (a). This yields the useful formulas

$$\begin{aligned}(\lambda_1 - 1)^2 &= (\lambda_1\lambda_2 - 1)(\lambda_1\lambda_3 - 1),\\ (\lambda_2 - 1)^2 &= (\lambda_2\lambda_1 - 1)(\lambda_2\lambda_3 - 1),\\ (\lambda_3 - 1)^2 &= (\lambda_3\lambda_1 - 1)(\lambda_3\lambda_2 - 1).\end{aligned} \tag{4.37}$$

We now start with a rational map

$$\phi(z) = \frac{a_0z^2 + a_1z + a_2}{b_0z^2 + b_1z + b_2}$$

and change coordinates in order to put $\phi$ into the desired form.
(a) For this part we assume that $\lambda_1\lambda_2 \neq 1$, so (4.37) tells us that $\lambda_1 \neq 1$ and $\lambda_2 \neq 1$. Thus the fixed points associated to $\lambda_1$ and $\lambda_2$ have multiplicity 1, so in particular they are distinct and we can find an element of $\mathrm{PGL}_2(\mathbb{C})$ that moves them to 0 and $\infty$, respectively. After this change of variables has been made, the rational map $\phi$ satisfies $\phi(0) = 0$ and $\phi(\infty) = \infty$, so it has the form

$$\phi(z) = \frac{a_0z^2 + a_1z}{b_1z + b_2} \qquad \text{with} \qquad a_0b_2 \neq 0.$$

Since $a_0 \neq 0$, we can dehomogenize by setting $a_0 = 1$, and then a simple calculation yields

$$\phi'(0) = \frac{a_1}{b_2} = \lambda_1 \qquad \text{and} \qquad \phi'(\infty) = b_1 = \lambda_2.$$

Thus $f$ has the form

$$\phi(z) = \frac{z^2 + b_2\lambda_1z}{\lambda_2z + b_2} \qquad \text{with} \qquad b_2 \neq 0.$$

Finally, replacing $\phi(z)$ by $b_2^{-1}\phi(b_2z)$ yields the desired form (4.35), and we calculate

$$\operatorname{Res}(z^2 + \lambda_1z, \lambda_2z + 1) = \det\begin{vmatrix}1 & \lambda_1 & 0\\ \lambda_2 & 1 & 0\\ 0 & \lambda_2 & 1\end{vmatrix} = 1 - \lambda_1\lambda_2,$$

which completes the proof of (a).
(b) The proof of this part is similar. We begin by moving the fixed point associated to $\lambda_1$ to $\infty$, so the map $\phi$ has the form

$$\phi(z) = \frac{a_0z^2 + a_1z + a_2}{b_1z + b_2} \qquad \text{with} \qquad a_0 \neq 0 \quad \text{and} \quad \lambda_1 = \lambda_\infty(\phi) = \frac{b_1}{a_0}.$$

The assumption that $\lambda_1\lambda_2 = 1$ combined with (4.37) tells us that $\lambda_1 = \lambda_2 = 1$, so we have $b_1 = a_0$. Dehomogenizing $a_0 = 1$ puts $\phi$ into the form

$$\phi(z) = \frac{z^2 + a_1 z + a_2}{z + b_2}.$$

Next we replace $\phi(z)$ by $\phi(z - b_2) + b_2$, so now $\phi(z)$ looks like

$$\phi(z) = \frac{z^2 + a_1 z + a_2}{z} \qquad \text{with} \qquad a_2 \neq 0.$$

(Of course, the values of $a_1$ and $a_2$ have changed.) Finally, replacing $\phi(z)$ by $\phi(\sqrt{a_2}\, z)/\sqrt{a_2}$ gives $\phi(z)$ the desired form

$$\phi(z) = \frac{z^2 + a_1 z + 1}{z} = z + a_1 + \frac{1}{z}. \tag{4.38}$$

We can compute the value of $a_1$ by observing that $\phi$ has a double fixed point at $\infty$ and that its other fixed point is at $-a_1^{-1}$. (If $a_1 = 0$, then there is a triple fixed point at $\infty$.) Thus the third multiplier is

$$\lambda_3 = \phi'(-a_1^{-1}) = 1 - a_1^2,$$

so we find that $a_1 = \sqrt{1 - \lambda_3}$. (This is also correct if $a_1 = 0$.) Substituting this value of $a_1$ into (4.38) completes the proof of (b). □

We resume the proof of Theorem 4.56(b,ii). Let $\phi_1$ and $\phi_2$ be rational maps of degree 2 satisfying $\boldsymbol{\sigma}(\phi_1) = \boldsymbol{\sigma}(\phi_2)$. Thus

$$\sigma_1(\phi_1) = \sigma_1(\phi_2) \qquad \text{and} \qquad \sigma_2(\phi_1) = \sigma_2(\phi_2),$$

and then the relation $\sigma_3 = \sigma_1 - 2$ from (a) implies that also $\sigma_3(\phi_1) = \sigma_3(\phi_2)$. It follows that the set of multipliers of the fixed points of $\phi_1$ and $\phi_2$ are the same, say $\{\lambda_1, \lambda_2, \lambda_3\}$, since they are the three roots (counted with multiplicity) of the polynomial

$$T^3 - \sigma_1 T^2 + \sigma_2 T - \sigma_3.$$

We consider two cases. Suppose first that $\lambda_1\lambda_2 \neq 1$. Then Lemma 4.59(a) says that $\phi_1$ and $\phi_2$ are 2-equivalent to the function $(z^2 + \lambda_1 z)/(\lambda_2 z + 1)$, so they are $\mathrm{PGL}_2$-equivalent to each other, and similarly if $\lambda_1\lambda_3 \neq 1$ or if $\lambda_2\lambda_3 \neq 1$.

We are left to consider the case $\lambda_1\lambda_2 = \lambda_1\lambda_3 = \lambda_2\lambda_3 = 1$. Then Lemma 4.59(b) tells us that $\lambda_1 = \lambda_2 = \lambda_3 = 1$ and that $\phi_1$ and $\phi_2$ are both $\mathrm{PGL}_2$-equivalent to the rational map $z + z^{-1}$.

This completes the proof of (b,ii), so we turn to (b,iii), the surjectivity of $\boldsymbol{\sigma}$. Given $(s_1, s_2) \in \mathbb{A}^2(\mathbb{C})$, we set $s_3 = s_1 - 2$ and factor the polynomial

$$T^3 - s_1 T^2 + s_2 T - s_3 = (T - \lambda_1)(T - \lambda_2)(T - \lambda_3) \in \mathbb{C}[T].$$

Notice that the condition $s_3 = s_1 - 2$ gives the familiar relation

$$\lambda_1 + \lambda_2 + \lambda_3 = s_1 = s_3 + 2 = \lambda_1\lambda_2\lambda_3 + 2, \tag{4.39}$$

which in turn implies the usual formal identities (4.37) for $\lambda_1, \lambda_2, \lambda_3$.

Suppose first that some $\lambda_i$ is not equal to 1, say $\lambda_1 \neq 1$. Then $\lambda_1\lambda_2 \neq 1$ from (4.37), so we can set

$$\phi(z) = \frac{z^2 + \lambda_1 z}{\lambda_2 z + 1},$$

since $\operatorname{Res}(z^2 + \lambda_1 z, \lambda_2 z + 1) = 1 - \lambda_1\lambda_2$ from Lemma 4.59(a). The fixed points of $\phi$ are $\alpha_1 = 0$, $\alpha_2 = \infty$, and $\alpha_3 = \frac{1-\lambda_1}{1-\lambda_2}$. Its multipliers at $\alpha_1$ and $\alpha_2$ are easily computed,

$$\lambda_{\alpha_1}(\phi) = \lambda_1 \qquad \text{and} \qquad \lambda_{\alpha_2}(\phi) = \lambda_2,$$

and the multiplier at $\alpha_3$ is

$$\lambda_{\alpha_3}(\phi) = \frac{2 - \lambda_1 - \lambda_2}{1 - \lambda_1\lambda_2} = \frac{\lambda_3 - \lambda_1\lambda_2\lambda_3}{1 - \lambda_1\lambda_2} = \lambda_3,$$

where we use (4.39) for the middle equality. Hence $\sigma_1(\phi) = s_1$ and $\sigma_2(\phi) = s_2$.

We are left with the case $\lambda_1 = \lambda_2 = \lambda_3 = 1$, which corresponds to the values $s_1 = s_2 = 3$. It is easy to check that the rational map $\phi(z) = z + z^{-1}$ has a triple-order fixed point at $\infty$ and that all three multipliers are equal to 1, so $\boldsymbol{\sigma}(\phi) = (3, 3)$.

(c) We briefly sketch the proof, which uses basic methods from algebraic geometry. We refer the reader to [416, §5] for further details.

The first step is to verify that the map $\boldsymbol{\sigma} : \mathcal{M}_2 \to \mathbb{A}^2$ is proper. This follows easily from the valuative criterion [198, II.4.7] using the fact proven in (b) that every fiber $\boldsymbol{\sigma}^{-1}(t)$ consists of a single point. (Roughly speaking, a morphism of varieties over $\mathbb{C}$ is proper if its fibers are complete.) Next one checks that $\boldsymbol{\sigma}$ is finite, which can be proven using the fact that both $\mathcal{M}_2$ and $\mathbb{A}^2$ are affine varieties (cf. [416, Lemma 5.6]). Alternatively, one can show in general that a proper quasifinite map is finite ([299, I, Proposition 1.10]). Then one uses the fact that $\mathbb{A}^2$ is nonsingular and the bijectivity of $\boldsymbol{\sigma}$ on complex points to prove that $\boldsymbol{\sigma}$ is an isomorphism (cf. [416, Lemma 5.7]). Finally, it is clear from the explicit formulas for $\sigma_1$ and $\sigma_2$ in Remark 4.57 that $\boldsymbol{\sigma}$ is defined over $\mathbb{Q}$.

(d) This is immediate from (c), which says that $\mathbb{Q}[\mathcal{M}_2] = \mathbb{Q}[\sigma_1, \sigma_2]$, since we already know from Theorem 4.50(b) that $\sigma_i^{(n)}$ and $\overset{*}{\sigma}{}_i^{(n)}$ are in $\mathbb{Q}[\mathcal{M}_2]$.

(e) This is a consequence of the fact that the isomorphism $\boldsymbol{\sigma} : \mathcal{M}_2 \to \mathbb{A}^2$ in (c) is defined over $\mathbb{Q}$. □

*Remark* 4.60. Regarding the proof sketch of Theorem 4.56(c), we observe that a morphism of irreducible varieties $V \to W$ may be bijective on complex points, yet not be an isomorphism. A simple example of this phenomenon is the cuspidal cubic map

$$F : \mathbb{A}^1 \to \{(x, y) : y^2 = x^3\}, \qquad F(t) = (t^2, t^3).$$

Thus in the proof that $\boldsymbol{\sigma} : \mathcal{M}_2 \to \mathbb{A}^2$ is an isomorphism, it is crucial that $\boldsymbol{\sigma}$ maps $\mathcal{M}_2$ onto the *nonsingular* variety $\mathbb{A}^2$. It does not appear to be known in general whether $\mathcal{M}_d$ is nonsingular.

The content of Theorem 4.56 is that the moduli space $\mathcal{M}_2$ of dynamical systems of degree 2 is isomorphic to $\mathbb{A}^2$ and an explicit isomorphism is provided by the pair of functions $(\sigma_1, \sigma_2)$ created from the multipliers of the fixed points of a rational map. From our general theory (Theorem 4.40), the space $\mathcal{M}_2$ sits naturally inside two larger spaces $\mathcal{M}_2^s$ and $\mathcal{M}_2^{ss}$, but since $d = 2$ is even, these two spaces coincide and will be denoted by $\overline{\mathcal{M}}_2$. We conclude this section with a description of $\overline{\mathcal{M}}_2$.

**Theorem 4.61.** *Let $\overline{\mathcal{M}}_2 = \mathcal{M}_2^s = \mathcal{M}_2^{ss}$ be the completion of $\mathcal{M}_2$ constructed using geometric invariant theory in Theorem* 4.40. *Then the isomorphism*

$$\sigma = (\sigma_1, \sigma_2) : \mathcal{M}_2 \to \mathbb{A}^2$$

*extends to an isomorphism $\bar{\sigma} : \overline{\mathcal{M}}_2 \to \mathbb{P}^2$ such that the following diagram commutes:*

$$\begin{array}{ccc} \mathcal{M}_2 & \xrightarrow[\sim]{\sigma} & \mathbb{A}^2 \\ \big\downarrow & & \big\downarrow {\scriptstyle (x,y) \mapsto [x,y,1]} \\ \overline{\mathcal{M}}_2 & \xrightarrow[\sim]{\bar{\sigma}} & \mathbb{P}^2 \end{array}$$

*The points in $\overline{\mathcal{M}}_2(\mathbb{C})$ that are not in $\mathcal{M}_2(\mathbb{C})$ correspond to degenerate maps of degree* 2 *that may be informally described as maps of the form*

$$\phi_{A,B}(X,Y) = [AXY, XY + BY^2], \qquad [A,B] \in \mathbb{P}^1(\mathbb{C}). \tag{4.40}$$

*The point $[A,B]$ is uniquely determined up to reversing $A$ and $B$.*

*Proof.* See [302] and [416, Theorem 6.1 and Lemmas 6.2 and 6.3]. □

*Remark* 4.62. We expand briefly on what it means to say that the points in the set $(\overline{\mathcal{M}}_2 \smallsetminus \mathcal{M}_2)(\mathbb{C})$ correspond to the maps $\phi_{A,B}$ given by (4.40). Let $U \subset \mathbb{A}^1$ be a neighborhood of 0 and let

$$\phi : U \longrightarrow \mathrm{Rat}_2$$

be a rational map that is a morphism away from 0. We denote the image of $t \in U$ by $\phi_t$ to help remind the reader that $\phi_t$ is itself a map, i.e., $\phi_t : \mathbb{P}^1 \to \mathbb{P}^1$.

Consider the composition

$$U \xrightarrow{\phi} \mathrm{Rat}_2 \longrightarrow \mathcal{M}_2 \hookrightarrow \overline{\mathcal{M}}_2,$$

which by abuse of notation we again denote by $\phi$. It is a rational map, and since $U \subset \mathbb{A}^1$ and $\overline{\mathcal{M}}_2$ is complete, we see that it is a morphism from $U$ to $\overline{\mathcal{M}}_2$. In particular, the point $\phi_0$ is a well-defined point in $\overline{\mathcal{M}}_2(\mathbb{C})$. If $\phi_0 \in \mathcal{M}_2(\mathbb{C})$, there is nothing to say, so we suppose that $\phi_0 \notin \mathcal{M}_2(\mathbb{C})$.

Then the second part of Theorem 4.61 means that possibly after choosing a smaller neighborhood of 0, there exists a morphism

$$f : U \longrightarrow \mathrm{PGL}_2$$

such that the conjugate map $\phi^f$ has the form

$$\phi^f(X,Y) = \left[a_0X^2 + a_1XY + a_2Y^2, b_0X^2 + b_1XY + b_2Y^2\right]$$

with $a_0, \dots, b_2$ regular functions on $U$ and satisfying

$$a_0(0) = a_2(0) = b_0(0) = 0, \qquad b_1(0) = 1, \qquad \text{and} \qquad a_1(0), b_2(0) \text{ not both } 0.$$

In other words, the degeneration of $\phi^f$ at $t = 0$ has the form $[AXY, XY + BY^2]$. Further, the map $\phi$ determines the point $[A, B] = [a_1(0), b_2(0)] \in \mathbb{P}^1(\mathbb{C})$ up to switching the two coordinates. In this way we identify

$$\mathbb{P}^1(\mathbb{C})/\mathcal{S}_2 \longleftrightarrow (\overline{\mathcal{M}}_2 \smallsetminus \mathcal{M}_2)(\mathbb{C}), \qquad [A, B] \longmapsto [AXY, XY + BY^2],$$

where the symmetric group on two letters $\mathcal{S}_2$ acts on $\mathbb{P}^1$ by interchanging the coordinates. (It is an exercise to show that $\mathbb{P}^1/\mathcal{S}_2$ is isomorphic to $\mathbb{P}^1$.)

## 4.7 Automorphisms and Twists

As we have repeatedly seen, from a dynamical perspective the geometric properties of a rational map $\phi(z)$ and its conjugates $\phi^f = f^{-1}\phi f$ are the same, since conjugation by $f \in \mathrm{PGL}_2(\mathbb{C})$ is simply an invertible change of variables. However, matters become more complicated if we restrict the coefficients of $f$ to lie in a field that is not algebraically closed. This leads to a notion of conjugation equivalence relative to a particular field, as in the following definitions.

**Definition.** Two rational maps $\phi(z), \psi(z) \in K(z)$ are $\mathrm{PGL}_2$*-equivalent* if there is a linear fractional transformation $f \in \mathrm{PGL}_2(\bar{K})$ such that $\psi = \phi^f$. More generally, we say that $\phi$ and $\psi$ are $\mathrm{PGL}_2(K)$*-equivalent* if there is a linear fractional transformation $f \in \mathrm{PGL}_2(K)$ such that $\psi = \phi^f$. When $\mathrm{PGL}_2$ is clear from context, we refer more simply to $\bar{K}$*-equivalence* and $K$*-equivalence*. We leave as an exercise (Exercise 4.31) the proof that these are equivalence relations.

We denote the set of rational maps that are $\bar{K}$-equivalent to $\phi$ by

$$[\phi] = \{\phi^f : f \in \mathrm{PGL}_2(\bar{K})\},$$

and similarly the set of rational maps that are $K$-equivalent by

$$[\phi]_K = \{\phi^f : f \in \mathrm{PGL}_2(K)\}.$$

*Remark* 4.63. In some sense we already have a notation $\langle\phi\rangle$ for the $\mathrm{PGL}_2(\bar{K})$ equivalence class of $\phi$. However, we generally view $\langle\,\cdot\,\rangle$ as a map $\mathrm{Rat}_d \to \mathcal{M}_d$ and $\langle\phi\rangle$ as a point in the moduli space $\mathcal{M}_d$, while we think of $[\phi]$ as a set of rational maps. This is the reason for the notational distinction.

Not all conjugates $\phi^f$ of $\phi$ need be distinct. For example, the rational map

$$\phi(z) = az + \frac{b}{z} \qquad \text{satisfies} \qquad -\phi(-z) = \phi(z),$$

so $\phi^f = \phi$ for the linear fractional transformation $f(z) = -z$. The set of transformations $f \in \mathrm{PGL}_2(\bar{K})$ that fix $\phi$ is an interesting group whose properties play an important role both geometrically and arithmetically.

**Definition.** Let $\phi(z) \in \bar{K}(z)$ be a rational map. The *automorphism group of* $\phi$ is the group

$$\mathrm{Aut}(\phi) = \{f \in \mathrm{PGL}_2(\bar{K}) : \phi^f(z) = \phi(z)\}.$$

(Another common name for $\mathrm{Aut}(\phi)$ is the *group of self similarities of* $\phi$.)

*Remark* 4.64. It is easy to check that $\mathrm{Aut}(\phi)$ is a subgroup of $\mathrm{PGL}_2(\bar{K})$ and that for any $h \in \mathrm{PGL}_2(\bar{K})$ the map

$$\mathrm{Aut}(\phi) \longrightarrow \mathrm{Aut}(\phi^h), \qquad f \longmapsto h^{-1}fh,$$

is an isomorphism (see Exercise 4.32). Thus as an abstract group, $\mathrm{Aut}(\phi)$ depends only on the $\mathrm{PGL}_2$-conjugacy equivalence class of $\phi$; more precisely, the $\bar{K}$-equivalence class of $\phi$ determines $\mathrm{Aut}(\phi)$ in $\mathrm{PGL}_2$ up to conjugation.

**Proposition 4.65.** *Let $\phi(z) \in \bar{K}(z)$ be a rational map of degree $d \geq 2$. Then $\mathrm{Aut}(\phi)$ is a finite subgroup of $\mathrm{PGL}_2(\bar{K})$, and its order is bounded by a function of $d$.*

*Proof.* Let $f \in \mathrm{Aut}(\phi)$. Then for any point $P \in \mathbb{P}^1(\bar{K})$ and any $n \geq 1$ we have

$$\phi^n(P) = (\phi^f)^n(P) = (f^{-1}\phi^n f)(P),$$

and hence

$$f(\phi^n(P)) = \phi^n(f(P)).$$

In particular, if $P$ is a periodic point of (primitive) period $n$, then $f(P)$ is also a periodic point of (primitive) period $n$. Thus each $f \in \mathrm{Aut}(\phi)$ induces a permutation of each of the sets $\mathrm{Per}_n(\phi)$ and $\mathrm{Per}_n^{**}(\phi)$.

Choose three distinct integers $n_1, n_2, n_3$ such that $\phi$ contains primitive $n$ periodic points for each value of $n$. Corollary 4.7 says that we can find such values, and further that their magnitude can be bounded solely in terms of $d$. More precisely, they may be chosen from among the first $d + 5$ primes. Letting

$$N_i = \#\,\mathrm{Per}_{n_i}^{**}(\phi) \geq 1 \qquad \text{for } i = 1, 2, 3,$$

the action of $\phi$ on the sets of primitive periodic points yields a homomorphism

$$\mathrm{Aut}(\phi) \longrightarrow S_{N_1} \times S_{N_2} \times S_{N_3} \tag{4.41}$$

from $\mathrm{Aut}(\phi)$ into a product of three symmetric groups.

We claim that the homomorphism (4.41) is injective. To see this, we observe that any $f$ in the kernel of (4.41) fixes each $\mathrm{Per}_{n_i}^{**}(\phi)$; hence $f$ fixes at least three points in $\mathbb{P}^1(\bar{K})$; hence $f$ is the identity map. Thus the homomorphism (4.41) is injective, which clearly implies that $\mathrm{Aut}(\phi)$ is finite. □

*Remark* 4.66. Proposition 4.65 tells us that the automorphism group $\operatorname{Aut}(\phi)$ of a rational map is a finite subgroup of $\mathrm{PGL}_2(\bar{K})$. For $\bar{K} = \mathbb{C}$, or more generally for any field of characteristic 0, the classical description of finite subgroups of $\mathrm{PGL}_2(\mathbb{C})$ says that every such subgroup is conjugate to either a cyclic group, a dihedral group, or the symmetry group of a regular three-dimensional solid (i.e., the tetrahedral, octahedral, and icosohedral groups). See, e.g., [414].

*Example* 4.67. Let $\zeta$ be an $n^{\text{th}}$ root of unity, let $\psi(z) \in \bar{K}(z)$ be any rational map, and let $\phi(z) = z\psi(z^n)$. Then $\operatorname{Aut}(\phi)$ contains a cyclic subgroup of order $n$ generated by the map $f(z) = \zeta z$.

*Example* 4.68. The map $\phi(z) = (z^2 - 2z)/(-2z + 1)$ has an automorphism group $\operatorname{Aut}(\phi)$ that is isomorphic to the symmetric group $\mathcal{S}_3$ on three letters. More precisely (see Exercise 4.36), the automorphism group of $\phi$ consists of the following six linear fractional transformations:

$$\operatorname{Aut}(\phi) = \left\{ z, \frac{1}{z}, \frac{z-1}{z}, \frac{1}{1-z}, \frac{z}{z-1}, 1-z \right\} \cong \mathcal{S}_3.$$

*Example* 4.69. Consider the rational map

$$\phi_b(z) = z + \frac{b}{z},$$

whose automorphism group $\operatorname{Aut}(\phi_b) = \{z, -z\}$ has order 2. These maps are all $\mathrm{PGL}_2$-equivalent, since the linear fractional transformation $f(z) = z\sqrt{b/c}$ gives $\phi_c = \phi_b^f$. Thus the geometric dynamical properties of $\phi_b$ are the same for all $b$. However, the arithmetic properties of $\phi_b$ may change quite substantially depending on the value of $b$, since the change of variable involves a square root. So although $\phi_b$ and $\phi_c$ are always $\mathrm{PGL}_2(\bar{K})$-equivalent, they are not $\mathrm{PGL}_2(K)$-equivalent unless $b/c$ is a perfect square in $K$. The underlying reason for the existence of these "twists" is the fact that $\operatorname{Aut}(\phi_1)$ is nontrivial.

**Definition.** Let $\phi(z) \in K(z)$ be a rational map. The *set of twists of* $\phi$ *over* $K$ is the set

$$\operatorname{Twist}(\phi/K) = \left\{ \begin{matrix} K\text{-equivalence classes of maps } \psi \text{ such} \\ \text{that } \psi \text{ is } \bar{K}\text{-equivalent to } \phi \end{matrix} \right\}.$$

*Remark* 4.70. As noted earlier, the geometric dynamical properties of the maps in $\operatorname{Twist}(\phi/K)$ are identical, but their arithmetic properties may be significantly different. For example, if two rational maps are $K$-isomorphic, then the field extensions of $K$ generated by their periodic points are the same. This is clear from the fact that

$$\operatorname{Per}_n(\phi^f) = \{f^{-1}(P) : P \in \operatorname{Per}_n(\phi)\},$$

so if $f \in \mathrm{PGL}_2(K)$, the fields generated by $\operatorname{Per}_n(\phi)$ and $\operatorname{Per}_n(\phi^f)$ are identical. However, if $\phi$ and $\psi$ are only $\bar{K}$-isomorphic, these fields may well be different. This often provides a convenient method for proving that two maps are not $K$-isomorphic, as in the following example.

*Example* 4.71. Continuing with Example 4.69, for each $b \in K^*$ we let

$$\phi_b(z) = z + \frac{b}{z}.$$

We saw that these maps are all $\bar{K}$-isomorphic, which gives a map of sets

$$K^* \longrightarrow \operatorname{Twist}(\phi_1/K), \qquad b \longmapsto [\phi_b]_K. \tag{4.42}$$

Note that if $b/c$ is a square in $K$, say $b/c = a^2$, then $\phi_b$ and $\phi_c$ are $K$-isomorphic, since $\phi_c = \phi_b^f$ for the map $f(z) = az \in \mathrm{PGL}_2(K)$. Thus (4.42) induces a well-defined map

$$K^*/K^{*2} \longrightarrow \operatorname{Twist}(\phi_1/K), \qquad b \longmapsto [\phi_b]_K. \tag{4.43}$$

We can use Remark 4.70 to prove that the map (4.43) is injective. (We assume that $K$ does not have characteristic 2.) A quick computation shows that $\Phi^*_{\phi_b,2}(X,Y) = 2X^2 + bY^2$, so the primitive 2-periodic points of $\phi_b$ are $\pm\sqrt{-b/2}$. Hence if $\phi_b$ and $\phi_c$ are $K$-isomorphic, then the fields $K(\sqrt{-b/2})$ and $K(\sqrt{-c/2})$ are the same, so $b/c$ must be a square[6] in $K$. Notice that we could not use $\operatorname{Per}_1(\phi_b)$ to prove this result, since the only fixed point of $\phi_b$ is $\infty$.

These quadratic twists of $\phi_1(z)$ are analogous to quadratic twists of elliptic curves (cf. [410, §X.5]). Thus fix $A, B \in K$ and, for each $D \in K^*$, let $E_D$ be the elliptic curve $E_D : DY^2 = X^3 + AX + B$. Then all of the $E_D$ are isomorphic over $\bar{K}$, but $E_{D_1}$ and $E_{D_2}$ are isomorphic over $K$ if and only if the ratio $D_1/D_2$ is a square in $K$. This gives a natural map $K^*/(K^*)^2 \to \operatorname{Twist}(E_1/K)$.

*Remark* 4.72. How does the theory of twists fit in with the moduli spaces $\mathcal{M}_d$ constructed in Section 4.4? The answer is that if $\phi$ and $\psi$ are twists of one another, then their corresponding points $\langle\phi\rangle$ and $\langle\psi\rangle$ in the moduli space $\mathcal{M}_d$ are equal. This is true because points in $\mathcal{M}_d(\bar{K})$ classify rational maps of degree $d$ modulo $\mathrm{PGL}_2(\bar{K})$ conjugation, so points in $\mathcal{M}_d(\bar{K})$ do not detect whether the conjugation is defined over $K$. In other words, there is a natural map

$$\mathrm{Rat}_d(K)/\mathrm{PGL}_2(K) \longrightarrow \mathcal{M}_d(K), \tag{4.44}$$

but this map is not one-to-one. It fails to be injective precisely for those maps in $\mathrm{Rat}_d(K)$ that have nontrivial twists; cf. Remark 4.39. On the other hand, the next proposition and Exercise 4.38 imply that the map (4.44) is injective on a Zariski open subset of $\mathrm{Rat}_d$.

We now prove that a rational map with no automorphisms has no nontrivial twists. Later, in Theorem 4.79, we prove a much stronger result.

**Proposition 4.73.** *Let $\phi(z) \in K(z)$ be a rational map of degree $d \geq 2$ and assume that its automorphism group* $\operatorname{Aut}(\phi)$ *is trivial. Then $\phi$ has no nontrivial twists, i.e.,* $\operatorname{Twist}(\phi/K)$ *has only one element, the $K$-equivalence class of $\phi$ itself.*

[6] It is an easy exercise to prove that $K(\sqrt{A})$ and $K(\sqrt{B})$ are isomorphic if and only if $A/B$ is a square in $K$, assuming that $K$ does not have characteristic 2.

*Proof.* Suppose that $\psi \in K(z)$ is a twist of $\phi$, so there is an $f \in \mathrm{PGL}_2(\bar{K})$ such that $\psi = \phi^f$. We let an element $\sigma \in \mathrm{Gal}(\bar{K}/K)$ act on $f(z)$ and $\phi(z)$ in the natural way by applying $\sigma$ to each of the coefficients. Notice that $\sigma(\phi) = \phi$ and $\sigma(\psi) = \psi$, since the coefficients of $\phi$ and $\psi$ are in $K$. Hence

$$\begin{aligned}\phi^f &= \psi = \sigma(\psi) = \sigma(\phi^f)\\ &= \sigma(f\phi f^{-1}) = \sigma(f)\sigma(\phi)\sigma(f^{-1}) = \sigma(f)\phi\sigma(f)^{-1} = \phi^{\sigma(f)}.\end{aligned}$$

Hence $\sigma(f)f^{-1} \in \mathrm{Aut}(\phi)$. But $\mathrm{Aut}(\phi)$ is trivial by assumption, so $\sigma(f) = f$. This is true for all $\sigma \in \mathrm{Gal}(\bar{K}/K)$, so we conclude[7] that $f \in \mathrm{PGL}_2(K)$. Hence $\psi = \phi^f$ is $K$-isomorphic to $\phi$, so it represents the trivial twist. □

*Remark* 4.74. As we have seen in Section 4.3, the set $\mathrm{Rat}_d$ of all rational maps of degree $d$ is a Zariski open subset of $\mathbb{P}^{2d+1}$. It is the complement of the hypersurface described by the vanishing of the resultant $\mathrm{Res}(F_{\mathbf{a}}, F_{\mathbf{b}}) = 0$. One can show that the set of rational maps $\phi \in \mathrm{Rat}_d$ having nontrivial automorphism group forms a proper Zariski closed subset of $\mathrm{Rat}_d$; see Exercise 4.38. Thus most rational maps have no nontrivial automorphisms, and those that do, fall into finitely many irreducible algebraic families. Further, since for all $f \in \mathrm{PGL}_2$ the automorphism groups of $\phi$ and $\phi^f$ are isomorphic as abstract groups, there is a proper Zariski closed subset of $\mathcal{M}_d$ determined by the conjugacy classes of rational maps with nontrivial automorphisms. It is an interesting geometric problem to describe the irreducible subvarieties making up this set and an interesting arithmetic question to describe their rational and integral points. See Exercises 4.28 and 4.41.

## 4.8 General Theory of Twists

In this section we develop the basic theory of Galois twists in an abstract setting. Although we apply this material only to rational maps, in Section 4.9, and to $\mathbb{P}^1$, in Section 4.10, we develop the theory in some generality in order to clarify the underlying principles.

Let $X$ and $Y$ be objects defined over the field $K$. For example, $X$ and $Y$ might be curves or algebraic varieties or rational maps.[8] We consider $X$ and $Y$ to be equivalent if they are isomorphic over $K$. However, it may happen that they are isomorphic over $\bar{K}$, but not over $K$. This leads us to consider the following set.

**Definition.** Let $X$ be an object defined over the field $K$.The set of *twists of* $X/K$ is the set

[7]The proof that $\mathrm{PGL}_2(K)$ is the subgroup of $\mathrm{PGL}_2(\bar{K})$ fixed by $\mathrm{Gal}(\bar{K}/K)$ is not hard, although it does use Hilbert's Theorem 90.

[8]Formally, $X$ and $Y$ should be objects in a category on which the Galois group $\mathrm{Gal}(\bar{K}/K)$ acts in an appropriate way. We do not concern ourselves with such formalism and leave it to the reader either to formulate the correct abstract concepts or to restrict attention to those situations, namely algebraic varieties and rational maps, in which the Galois action is clear.

$$\operatorname{Twist}(X/K) = \left\{ \begin{matrix} K\text{-isomorphism classes of objects } Y \\ \text{such that } Y \text{ is defined over } K \text{ and} \\ Y \text{ is isomorphic to } X \text{ over } \bar{K} \end{matrix} \right\}.$$

In other words, an element of $\operatorname{Twist}(X/K)$ is an object $Y$ that is defined over $K$ such that there is an isomorphism $i : X \to Y$, but the isomorphism $i$ might be defined only over an extension of $K$. Two elements $Y$ and $Y'$ in $\operatorname{Twist}(X/K)$ are considered to be equivalent if there is an isomorphism $j : Y \to Y'$ that is defined over $K$. The following examples should help to clarify this definition.

*Example* 4.75. Our first example deals with twists of rational maps. Let $\phi(z) \in K(z)$ be an odd rational map, i.e., a rational map satisfying $\phi(-z) = -\phi(z)$. Then for each $b \in K^*$ we can define a new rational map $\phi_b$ by the formula

$$\phi_b(z) = \frac{1}{\sqrt{b}}\phi(\sqrt{b}\,z).$$

The odd parity of $\phi(z)$ implies that $\phi$ has the form $\phi(z) = z\psi(z^2)$ for some rational map $\psi(z) \in K(z)$, so $\phi_b(z) = z\psi(bz^2)$ is in $K(z)$. Further, if $b \in K^{*2}$, say $b = c^2$, then $\phi_b$ is $\mathrm{PGL}_2(K)$-conjugate to $\phi$ via $\phi_b = \phi^f$ with $f(z) = cz$. In this way we obtain a map

$$K^*/K^{*2} \longrightarrow \operatorname{Twist}(\phi/K), \qquad b \longmapsto [\phi_b]_K. \tag{4.45}$$

If $\operatorname{Aut}(\phi) = \{z, -z\}$, then it is not hard to show that (4.45) is an isomorphism. Notice how this example generalizes Example 4.71.

*Example* 4.76. Our second example deals with twists of the variety $\mathbb{P}^1$. For any nonzero $a \in K^*$, let $C_a$ be the plane curve

$$C_a : x^2 + y^2 = a.$$

All of these curves are isomorphic over $\bar{K}$ via the explicit isomorphism

$$i : C_a \longrightarrow C_b, \qquad i(x, y) = \left(x\sqrt{b/a}, y\sqrt{b/a}\right). \tag{4.46}$$

Further, if $b/a$ is a square in $K$, say $b/a = c^2$, then $C_a$ and $C_b$ are isomorphic over $K$ via the isomorphism $i(x, y) = (cx, cy)$. Thus exactly as in Example 4.75, we obtain a natural map

$$K^*/K^{*2} \longrightarrow \operatorname{Twist}(C_1/K), \qquad a \longmapsto [C_a]_K,$$

where $[C_a]_K$ denotes the $K$-isomorphism class of $C_a$.

Notice that if two curves $C$ and $C'$ are $K$-isomorphic, then the $K$-isomorphism $i : C \to C'$ identifies their $K$-rational points $i : C(K) \to C'(K)$. This suffices to prove that the curves $C_1$ and $C_{-1}$ are not isomorphic over $\mathbb{Q}$, since

$$C_1(\mathbb{Q}) \neq \emptyset \qquad \text{and} \qquad C_{-1}(\mathbb{Q}) = \emptyset.$$

Hence $C_{-1}$ represents a nontrivial element of $\operatorname{Twist}(C_1/\mathbb{Q})$, and indeed a nontrivial element of $\operatorname{Twist}(C_1/\mathbb{R})$. More generally, a famous theorem of Fermat says that if $p$ is an odd prime number, then

$$C_p(\mathbb{Q}) \neq \emptyset \quad \text{if and only if} \quad p \equiv 1 \pmod 4.$$

Hence if $p \equiv 3 \pmod 4$, then $C_p$ represents a nontrivial element of $\operatorname{Twist}(C_1/\mathbb{Q})$. It is not hard to show that different primes $p \equiv 3 \pmod 4$ yield distinct elements of $\operatorname{Twist}(C_1/\mathbb{Q})$; see Exercise 4.44.

Returning to the general situation, let $X$ be an object defined over $K$ and let $Y$ represent an element of $\operatorname{Twist}(X/K)$. This means that there is a $\bar{K}$-isomorphism

$$i : Y \longrightarrow X.$$

We wish to determine whether $X$ and $Y$ are $K$-isomorphic. If $i$ is itself a $K$-isomorphism, then we are done; but even if $i$ is not a $K$-isomorphism, it may be possible to modify $i$ and turn it into a $K$-isomorphism.

In order to measure the extent to which $i$ fails to be a $K$-isomorphism, it is natural to make use of Galois theory, since $i$ is defined over $K$ if and only $\sigma(i) = i$ for every $\sigma \in \operatorname{Gal}(\bar{K}/K)$. Thus for each element $\sigma \in \operatorname{Gal}(\bar{K}/K)$, we consider the composition of maps

$$g_\sigma : X \xrightarrow{\sigma(i^{-1})} Y \xrightarrow{\ i\ } X.$$

The map $g_\sigma = i\sigma(i^{-1})$ is a $\bar{K}$-automorphism of $X$, i.e., it is an isomorphism from $X$ to itself defined over $\bar{K}$. If $i$ is already defined over $K$, then $g_\sigma(x) = x$ is the identity map, but in general $g_\sigma$ will be a nontrivial automorphism.

**Proposition 4.77.** *Let $X$ be an object defined over $K$, let $Y$ be a twist of $X/K$, choose a $\bar{K}$-isomorphism $i : Y \to X$, and define a map*

$$g : \operatorname{Gal}(\bar{K}/K) \longrightarrow \operatorname{Aut}(X), \qquad g_\sigma(x) = (i\sigma(i^{-1}))(x).$$

(a) *The map*

$$\operatorname{Gal}(\bar{K}/K) \longrightarrow \operatorname{Aut}(X), \qquad \sigma \longmapsto g_\sigma,$$

*satisfies*

$$g_{\sigma\tau} = g_\sigma \sigma(g_\tau) \qquad \text{for all } \sigma, \tau \in \operatorname{Gal}(\bar{K}/K).$$

*Maps satisfying $g_{\sigma\tau} = g_\sigma \sigma(g_\tau)$ are called* 1-cocycles.

(b) *$Y$ is the trivial twist of $X$ if and only if there is an $f \in \operatorname{Aut}(X)$ satisfying*

$$g_\sigma = f\sigma(f^{-1}) \qquad \text{for all } \sigma \in \operatorname{Gal}(\bar{K}/K).$$

*Maps of the form $f\sigma(f^{-1})$ are called* 1-coboundaries.

*Proof.* (a) We have $g_{\sigma\tau} = i \circ (\sigma\tau)(i^{-1}) = i \circ \sigma(\tau(i)^{-1})$. Consider the following commutative diagram of maps:

$$\begin{array}{ccccc} X & \xrightarrow{\sigma(\tau(i)^{-1})} & Y & \xrightarrow{\ i\ } & X \\ & & \sigma(i)\big\downarrow & & \big\uparrow i \\ & & X & \xrightarrow{\sigma(i^{-1})} & Y \end{array}$$

The top row is $g_{\sigma\tau}$, but if we travel around the diagram the long way, we get

$$i \circ \sigma(i^{-1}) \circ \sigma(i) \circ \sigma(\tau(i)^{-1}) = g_\sigma \circ \sigma(g_\tau).$$

(b) Suppose first that $g$ is a coboundary, say $g_\sigma = f\sigma(f^{-1})$ for some $f \in \mathrm{Aut}(X)$. We verify that the isomorphism $f^{-1}i : Y \to X$ is defined over $K$. For any $\sigma \in \mathrm{Gal}(\bar{K}/K)$ we have

$$\sigma(f^{-1}i) = \sigma(f^{-1})\sigma(i) = f^{-1}g_\sigma\sigma(i) = f^{-1}i\sigma(i^{-1})\sigma(i) = f^{-1}i.$$

This proves that $f^{-1}i$ is defined over $K$, so $Y$ is $K$-isomorphic to $X$.

Next suppose that $Y$ is $K$-isomorphic to $X$, say $j : Y \to X$ is a $K$-isomorphism. Then $\sigma(j) = j$ for every $\sigma \in \mathrm{Gal}(\bar{K}/K)$, so we have

$$g_\sigma = i \circ \sigma(i^{-1}) = i \circ j^{-1} \circ \sigma(j) \circ \sigma(i^{-1}) = (i \circ j^{-1}) \circ \sigma((i \circ j^{-1})^{-1}).$$

Thus if we let $f = ij^{-1} \in \mathrm{Aut}(X)$, then $g_\sigma = f\sigma(f^{-1})$, which proves that $g$ is a coboundary and completes the proof of the proposition. □

As the terminology suggests, there is a cohomology theory underlying Proposition 4.77.

**Definition.** Let $\Gamma$ be a group that acts on another group $A$. A 1-*cocycle* (*from* $\Gamma$ *to* $A$) is a map

$$g : \Gamma \longrightarrow A \quad \text{satisfying} \quad g_{\sigma\tau} = g_\sigma\sigma(g_\tau) \quad \text{for all } \sigma, \tau \in \Gamma,$$

and a 1-*coboundary* (*from* $\Gamma$ *to* $A$) is a map of the form

$$\Gamma \longrightarrow A, \quad \sigma \longmapsto f\sigma(f^{-1}) \quad \text{for some } f \in A.$$

The *cohomology set* $H^1(\Gamma, A)$ is defined to be the collection of 1-cocycles $\Gamma \to A$ modulo the equivalence relation that two 1-cocycles $g_1$ and $g_2$ are *cohomologous* if the map $g_1^{-1}g_2$ is a 1-coboundary.

*Remark* 4.78. If the group $A$ is abelian, then $H^1(\Gamma, A)$ is itself an abelian group, and in this situation it is possible to define cohomology groups $H^n(\Gamma, A)$ for all $n \geq 0$. For the general theory of group (and Galois) cohomology, with many important applications to class field theory, Diophantine equations, and arithmetic geometry, see for example [97, 396, 410].

A slight elaboration of the proof of Proposition 4.77 shows that if $\mathrm{Aut}(X)$ is abelian, then there is a well-defined one-to-one map

$$\mathrm{Twist}(X/K) \longrightarrow H^1(\mathrm{Gal}(\bar{K}/K), \mathrm{Aut}(X)), \qquad [Y]_K \longmapsto (\sigma \mapsto i_Y\sigma(i_Y^{-1})), \tag{4.47}$$

where $i_Y : Y \to X$ is a $\bar{K}$-isomorphism; see Exercise 4.42. In some cases, for example when $X$ is an algebraic variety, the map (4.47) is an isomorphism.

## 4.9 Twists of Rational Maps

According to Proposition 4.77, every twist corresponds to a 1-cocycle, and a twist is trivial if and only if its 1-cocycle is a 1-coboundary. The natural question that arises is whether every 1-cocycle comes from a twist. It turns out that the answer depends on the category from which the objects are being chosen. For example, in the category of algebraic varieties, every 1-cocycle does come from a twist. We will not prove this general result, but in the next section we treat the case of twists of $\mathbb{P}^1$. In this section we describe what happens for rational maps.

**Definition.** We define an action of the Galois group $\operatorname{Gal}(\bar{K}/K)$ on the group of linear fractional transformations $\operatorname{PGL}_2(\bar{K})$ in the natural way, thus

$$\sigma(f) = \sigma\left(\frac{az+b}{cz+d}\right) = \frac{\sigma(a)z+\sigma(b)}{\sigma(c)z+\sigma(d)} \qquad \text{for } f = \frac{az+b}{cz+d} \in \operatorname{PGL}_2(\bar{K}) \text{ and } \sigma \in \operatorname{Gal}(\bar{K}/K).$$

The automorphism group $\operatorname{Aut}(\phi)$ of a rational map $\phi$ is a subgroup of the full group $\operatorname{PGL}_2$ of linear fractional transformations. Thus if we have a 1-cocycle

$$g : \operatorname{Gal}(\bar{K}/K) \longrightarrow \operatorname{Aut}(\phi)$$

with values in $\operatorname{Aut}(\phi)$, we can compose it with the inclusion of $\operatorname{Aut}(\phi)$ into $\operatorname{PGL}_2$ to obtain a 1-cocycle

$$\operatorname{Gal}(\bar{K}/K) \xrightarrow{\;g\;} \operatorname{Aut}(\phi) \hookrightarrow \operatorname{PGL}_2(\bar{K}) \tag{4.48}$$

with values in $\operatorname{PGL}_2(\bar{K})$. This way of extending $g$ provides the key to describing which 1-cocycles correspond to actual twists of $\phi$.

**Theorem 4.79.** *Let $\phi(z) \in K(z)$ be a nonzero rational map and let*

$$g : \operatorname{Gal}(\bar{K}/K) \to \operatorname{Aut}(\phi)$$

*be a 1-cocycle with values in $\operatorname{Aut}(\phi)$. Then the following are equivalent:*
(a) *There is twist of $\phi/K$ whose 1-cocycle is $g$.*
(b) *The 1-cocycle $g$ becomes a 1-coboundary when it is extended to a 1-cocycle with values in $\operatorname{PGL}_2(\bar{K})$.*

*Hence in cohomological terms, the set of twists of $\phi/K$ is given by*

$$\operatorname{Twist}(\phi/K) = \Bigl\{\xi \in H^1\bigl(\operatorname{Gal}(\bar{K}/K), \operatorname{Aut}(\phi)\bigr) : \\ \xi \text{ becomes trivial in } H^1\bigl(\operatorname{Gal}(\bar{K}/K), \operatorname{PGL}_2(\bar{K})\bigr)\Bigr\}.$$

*Proof.* Suppose first that there is a twist $\phi^f$ of $\phi/K$ whose 1-cocycle is $g$. Then by the definition given in Proposition 4.77, the 1-cocycle $g$ is given by $g_\sigma = f\sigma(f^{-1})$.

Hence $g$ is the $\mathrm{PGL}_2(\bar{K})$ 1-coboundary associated to the element $f \in \mathrm{PGL}_2(\bar{K})$. This proves that (a) implies (b).

Next suppose that $g$ is an $\mathrm{Aut}(\phi)$ 1-cocycle that is a $\mathrm{PGL}_2(\bar{K})$ 1-coboundary. This means that there is an $f \in \mathrm{PGL}_2(\bar{K})$ with the property that $g_\sigma = f\sigma(f^{-1})$ for all $\sigma \in \mathrm{Gal}(\bar{K}/K)$. We claim that $\phi^f$ is a twist of $\phi/K$ with 1-cocycle $g$. What we need to check is that $\phi^f$ is defined over $K$, i.e., that $\phi^f \in K(z)$, since once we know that, it is clear from the definitions that $g$ is its associated 1-cocycle.

Let $\sigma \in \mathrm{Gal}(\bar{K}/K)$. Then

$$\begin{aligned}
\phi^{g_\sigma} &= \phi && \text{since we are given that } g_\sigma \in \mathrm{Aut}(\phi),\\
\phi g_\sigma &= g_\sigma \phi && \text{by definition of } \phi^{g_\sigma},\\
\phi f\sigma(f^{-1}) &= f\sigma(f^{-1})\phi && \text{since } g_\sigma = f\sigma(f^{-1}) \text{ by assumption},\\
f^{-1}\phi f &= \sigma(f^{-1})\phi\sigma(f) && \text{multiplying by } f^{-1} \text{ and by } \sigma(f),\\
\phi^f &= \phi^{\sigma(f)} && \text{since } \sigma(f^{-1}) = \sigma(f)^{-1},\\
\phi^f &= \sigma(\phi)^{\sigma(f)} = \sigma(\phi^f) && \text{since } \phi(z) \in K(z), \text{ so } \sigma(\phi) = \phi.
\end{aligned}$$

We have proven that $\phi^f = \sigma(\phi^f)$ for all $\sigma \in \mathrm{Gal}(\bar{K}/K)$, which shows that $\phi^f(z) \in K(z)$ and thus completes the proof that (b) implies (a).

In order to prove the cohomological description of $\mathrm{Twist}(\phi/K)$, we first use Proposition 4.77, which says that there is a natural embedding of $\mathrm{Twist}(\phi/K)$ into the cohomology set $H^1\big(\mathrm{Gal}(\bar{K}/K), \mathrm{Aut}(\phi)\big)$ (cf. also Remark 4.78). Then the equivalence of (a) and (b) tells us that an element of $H^1\big(\mathrm{Gal}(\bar{K}/K), \mathrm{Aut}(\phi)\big)$ comes from an element of $\mathrm{Twist}(\phi/K)$ if and only if it becomes trivial when mapped to $H^1\big(\mathrm{Gal}(\bar{K}/K), \mathrm{PGL}_2(\bar{K})\big)$. □

*Remark* 4.80. A formal argument with commutative diagrams shows that

$$H^1\big(\mathrm{Gal}(\bar{K}/K), \mathrm{PGL}_2(\bar{K})\big) \hookrightarrow H^2\big(\mathrm{Gal}(\bar{K}/K), \bar{K}^*\big)[2] \cong \mathrm{Br}(K)[2],$$

where $\mathrm{Br}(K)$ is the *Brauer group* of $K$. The Brauer group plays an important role in class field theory and many other areas of number theory and arithmetic geometry. For example, $\mathrm{Br}(\mathbb{Q}_p) = \mathbb{Q}/\mathbb{Z}$, so $\mathrm{Br}(\mathbb{Q}_p)[2]$ has only two elements, and the same is true for finite extensions of $\mathbb{Q}_p$. The Brauer group of a number field is more complicated.

*Example* 4.81. Let $K$ be a field of characteristic not dividing $n$, and let $\phi(z) \in K(z)$ be a rational map whose automorphism group is

$$\mathrm{Aut}(\phi) = \{\zeta z : \zeta \in \boldsymbol{\mu}_n\},$$

where we recall that $\boldsymbol{\mu}_n \subset \bar{K}$ denotes the set of $n^{\text{th}}$ roots of unity. Then there is an isomorphism

$$K^*/K^{*n} \xrightarrow{\ \sim\ } \mathrm{Twist}(\phi/K), \qquad b \longmapsto \left[\frac{1}{\sqrt[n]{b}}\phi\big(\sqrt[n]{b}\,z\big)\right]_K. \tag{4.49}$$

Note that by assumption we have $\phi(z) = \zeta^{-1}\phi(\zeta z)$ for all $\zeta \in \boldsymbol{\mu}_n$. In particular, the function $z^{-1}\phi(z)$ is invariant under the substitution $z \to \zeta z$, so it has the form $\phi(z) = z\psi(z^n)$ for some $\psi(z) \in K(z)$ (Exercise 4.37). Thus the twist of $\phi(z)$ given in (4.49) is equal to $z\psi(bz^n)$, so it is indeed in $K(z)$.

Since we are assuming that the automorphism group of $\phi$ is isomorphic to $\boldsymbol{\mu}_n$, not merely as an abstract group, but also in terms of the way in which $\operatorname{Gal}(\bar{K}/K)$ acts on $\operatorname{Aut}(\phi)$ and on $\boldsymbol{\mu}_n$, a standard result in Galois cohomology[9] says that there is an isomorphism

$$K^*/K^{*n} \xrightarrow{\sim} H^1(\operatorname{Gal}(\bar{K}/K), \boldsymbol{\mu}_n), \qquad b \longmapsto \left(\sigma \mapsto \frac{\sigma(\sqrt[n]{b})}{\sqrt[n]{b}}\right). \tag{4.50}$$

This allows us to identify $\operatorname{Twist}(\phi/K)$ with a subset of $K^*/K^{*n}$. In order to show that they are isomorphic, Theorem 4.79 says that we must show that every cocycle in (4.50) becomes a coboundary when we consider it as a cocycle with values in $\operatorname{PGL}_2(\bar{K})$. But this is clear, since with our identification of $\boldsymbol{\mu}_n$ with $\operatorname{Aut}(\phi)$, the cocycle in (4.50) is equal to

$$\sigma \longmapsto \frac{\sigma(\sqrt[n]{b})}{\sqrt[n]{b}} z = f\sigma(f^{-1})(z) \qquad \text{with } f(z) = \frac{z}{\sqrt[n]{b}} \in \operatorname{PGL}_2(\bar{K}).$$

Note that this example generalizes Example 4.75, which dealt with the case $n = 2$.

*Example* 4.82. Let $\phi(z) \in K(z)$ be a rational map with automorphism group $\operatorname{Aut}(\phi) = \{z, z^{-1}\}$. (See Exercise 4.35.) The Galois group $\operatorname{Gal}(\bar{K}/K)$ acts trivially on $\operatorname{Aut}(\phi)$, so we have

$$H^1(\operatorname{Gal}(\bar{K}/K), \operatorname{Aut}(\phi)) = H^1(\operatorname{Gal}(\bar{K}/K), \boldsymbol{\mu}_2) \cong K^*/K^{*2}.$$

The isomorphism is given explicitly by associating to any $b \in K^*/K^{*2}$ the cocycle

$$\sigma \longmapsto \begin{cases} z & \text{if } \sigma(\sqrt{b}) = \sqrt{b}, \\ z^{-1} & \text{if } \sigma(\sqrt{b}) = -\sqrt{b}. \end{cases} \tag{4.51}$$

To ease notation, we let $\beta = \sqrt{b}$ and let $g$ be the cocycle described by (4.51). Then Theorem 4.79 says that $g$ comes from a twist of $\phi$ if and only if there is some $f \in \operatorname{PGL}_2(\bar{K})$ satisfying $g_\sigma = f\sigma(f^{-1})$. Using the fact that $g_\sigma(z) \in \{z, z^{-1}\}$, we are looking for an $f \in \operatorname{PGL}_2(\bar{K})$ satisfying

[9] This statement is equivalent to $H^1\big(\operatorname{Gal}(\bar{K}/K), \bar{K}^*\big) = 0$, which is a version of Hilbert's Theorem 90. Using this and taking Galois invariants of the Kummer sequence

$$1 \longrightarrow \boldsymbol{\mu}_n \longrightarrow \bar{K}^* \xrightarrow{\alpha \mapsto \alpha^n} \bar{K}^* \longrightarrow 1$$

yields the cohomology long exact sequence

$$K^* \xrightarrow{\alpha \mapsto \alpha^n} K^* \longrightarrow H^1\big(\operatorname{Gal}(\bar{K}/K), \boldsymbol{\mu}_n\big) \longrightarrow H^1\big(\operatorname{Gal}(\bar{K}/K), \bar{K}^*\big) = 0.$$

$$\sigma(f)(z) = \begin{cases} f(z) & \text{if } \sigma(\beta) = \beta, \\ 1/f(z) & \text{if } \sigma(\beta) = -\beta. \end{cases}$$

It is not hard to construct such an $f$, for example

$$f(z) = \frac{\beta z + 1}{-\beta z + 1}.$$

Note that $\det\left(\begin{smallmatrix} \beta & 1 \\ -\beta & 1 \end{smallmatrix}\right) = 2\beta \neq 0$, so $f$ is in $\mathrm{PGL}_2(\bar{K})$. Hence every $g$ gives a twist of $\phi$, and indeed we can write the twist associated to $b = \beta^2 \in K^*/K^{*2}$ explicitly using $f$,

$$\phi^f(z) = \frac{\beta\phi\left(\frac{\beta z+1}{-\beta z+1}\right) + \beta}{-\phi\left(\frac{\beta z+1}{-\beta z+1}\right) + 1}.$$

For example, let $M_d(z) = z^d$ be the $d^{\text{th}}$-power map. Then a judicious use of the binomial theorem yields a formula for the $b$-twist $M_d^{(b)}$ of $M_d$,

$$M_d^{(b)}(z) = \sum_k \binom{d}{2k} b^k z^{2k} \Big/ \sum_k \binom{d}{2k+1} b^k z^{2k+1}.$$

In particular, the first few $b$-twists of $M_d$ are

$$M_2^{(b)}(z) = \frac{1 + bz^2}{2z}, \qquad M_3^{(b)}(z) = \frac{1 + 3bz^2}{3z + bz^3}, \qquad M_4^{(b)}(z) = \frac{1 + 6bz^2 + b^2z^4}{4z + 4bz^3}.$$

## 4.10 Fields of Definition and the Field of Moduli

If $\phi(z)$ is a rational map whose coefficients lie in a field $K$, then it is interesting to study the orbits of points whose coordinates lie in $K$ or in finite extensions of $K$. The smallest field $K$ containing the coefficients of $\phi$ can generally be determined by inspection. For example, the map $\phi(z) = z^3 + 1$ is clearly in $\mathbb{Q}(z)$ and the map $\phi(z) = z^3 + i$ is just as clearly in $\mathbb{Q}(i)(z)$. However, if we make a change of variables $f(z) = iz$ in the latter map $\phi(z) = z^3 + i$, we find that

$$\phi^f(z) = f^{-1}(\phi(f(z))) = -i\phi(iz) = -i(-iz^3 + i) = -z^3 + 1 \in \mathbb{Q}(z).$$

Thus the map $\phi(z) = z^3 + i$ is really a $\mathbb{Q}(z)$ map that has been altered by an injudicious change of variables.

**Definition.** Let $K$ be a field of characteristic 0, let $\phi(z) \in \bar{K}(z)$ be a rational map with coefficients in an algebraic closure of $K$, and let $K'/K$ be an extension field. We say that $K'$ is a *field of definition for* $\phi$ if there is a linear fractional transformation $f(z) \in \mathrm{PGL}_2(\bar{K})$ such that

$$\phi^f = f^{-1}\phi f \in K'(z).$$

In other words, $K'$ is a field of definition for $\phi$ (or an FOD for short) if, after a change of variables, $\phi$ has coefficients in $K'$.

As in Section 4.9, we can use Galois theory to investigate the possible fields of definition of a given rational function $\phi(z) \in \bar{K}(z)$. We let $\operatorname{Gal}(\bar{K}/K)$ act on $\bar{K}(z)$ in the natural way by applying $\sigma \in \operatorname{Gal}(\bar{K}/K)$ to the coefficients of $\phi \in \bar{K}(z)$,

$$\sigma(\phi) = \sigma\left(\frac{a_0 + a_1 z + \cdots + a_d z^d}{b_0 + b_1 z + \cdots + b_d z^d}\right) = \frac{\sigma(a_0) + \sigma(a_1) z + \cdots + \sigma(a_d) z^d}{\sigma(b_0) + \sigma(b_1) z + \cdots + \sigma(b_d) z^d}.$$

Then Galois theory (and Hilbert's Theorem 90) tell us that

$$\phi \in K'(z) \quad \text{for the field} \quad K' = \Bigl(\text{fixed field of } \bigl\{\sigma \in \operatorname{Gal}(\bar{K}/K) : \sigma(\phi) = \phi\bigr\}\Bigr).$$

Suppose instead that $\phi$ is merely equivalent to a map whose coefficients are in some field $K'$. That is, suppose that there is a linear fractional transformation $f \in \operatorname{PGL}_2(\bar{K})$ such that $\phi^f \in K'(z)$. Then $\sigma(\phi^f) = \phi^f$ for all $\sigma \in \operatorname{Gal}(\bar{K}/K')$, so

$$f^{-1}\phi f = \phi^f = \sigma(\phi^f) = \sigma(f^{-1}\phi f) = \sigma(f^{-1})\sigma(\phi)\sigma(f) = \sigma(f)^{-1}\sigma(\phi)\sigma(f).$$

Solving for $\sigma(\phi)$ yields

$$\sigma(\phi) = \sigma(f) f^{-1}\phi f \sigma(f^{-1}) = \bigl(f\sigma(f^{-1})\bigr)^{-1}\phi\bigl(f\sigma(f^{-1})\bigr) = \phi^{f\sigma(f^{-1})}.$$

Thus $\sigma(\phi)$ is equal to $\phi$ conjugated by the map $f\sigma(f^{-1}) \in \operatorname{PGL}_2(\bar{K})$, so in particular $\sigma(\phi)$ and $\phi$ are equivalent. We have proven that

$$\begin{pmatrix} K' \text{ is a field of} \\ \text{definition for } \phi \end{pmatrix} \Longrightarrow \begin{pmatrix} \text{for every } \sigma \in \operatorname{Gal}(\bar{K}/K') \text{ there exists} \\ \text{a } g_\sigma \in \operatorname{PGL}_2(\bar{K}) \text{ such that } \sigma(\phi) = \phi^{g_\sigma} \end{pmatrix}. \tag{4.52}$$

Turning this around, we start with an arbitrary rational map $\phi(z) \in \bar{K}(z)$ and study the automorphisms $\sigma \in \operatorname{Gal}(\bar{K}/K)$ for which $\sigma(\phi)$ is equivalent to $\phi$.

**Definition.** Let $\phi(z) \in \bar{K}(z)$ be a rational map. We associate to $\phi$ a subgroup $G_\phi$ of $\operatorname{Gal}(\bar{K}/K)$ and a field $K_\phi$ defined by

$$G_\phi = \{\sigma \in \operatorname{Gal}(\bar{K}/K) : \sigma(\phi) = \phi^{g_\sigma} \text{ for some } g_\sigma \in \operatorname{PGL}_2(\bar{K})\},$$
$$K_\phi = \text{fixed field of } G_\phi = \{\alpha \in \bar{K} : \sigma(\alpha) = \alpha \text{ for all } \sigma \in G_\phi\}.$$

The field $K_\phi$ is called the *field of moduli* (FOM) *of* $\phi$.

*Remark* 4.83. The group $\operatorname{PGL}_2(\bar{K})$ acts on the space of rational maps via the usual conjugation action, $\phi^f = f^{-1}\phi f$. If we consider as usual the equivalence class

$$[\phi] = \{\phi^f : f \in \operatorname{PGL}_2(\bar{K})\}$$

consisting of all maps that are conjugate to a particular map $\phi$, then $G_\phi$ is the subgroup of $\operatorname{Gal}(\bar{K}/K)$ consisting of elements that map the set $[\phi]$ to itself. Equivalently, if $\phi$ has degree $d$, let $\langle\phi\rangle$ be the image of $\phi$ in the moduli space $\mathcal{M}_d = \operatorname{Rat}_d/\operatorname{PSL}_2$ that we defined and studied in Section 4.4. The space $\mathcal{M}_d$ is defined over $\mathbb{Q}$ (Theorem 4.36) and the field of moduli of $\phi$ is exactly equal to the field generated by the coordinates of the point $\langle\phi\rangle \in \mathcal{M}_d$.

We begin by proving some elementary properties of FOM and FOD, in particular, the important fact that the field of moduli is contained in every field of definition.

**Proposition 4.84.** *Let $\phi(z) \in \bar{K}(z)$.*
(a) *The set $G_\phi$ is a subgroup of* $\mathrm{Gal}(\bar{K}/K)$.
(b) *Let $K'$ be a field of definition for $\phi$. Then $K_\phi \subseteq K'$. Informally, we say that* FOM $\subseteq$ FOD.

*Proof.* The proof of this proposition is simply a matter of unsorting definitions. Thus let $\sigma, \tau \in G_\phi$. Then

$$(\sigma\tau)(\phi) = \sigma(\tau(\phi)) = \sigma(\phi^{g_\tau}) = \sigma(\phi)^{\sigma(g_\tau)} = (\phi^{g_\sigma})^{\sigma(g_\tau)} = \phi^{g_\sigma \sigma(g_\tau)}.$$

Thus $(\sigma\tau)(\phi)$ is equivalent to $\phi$, so $\sigma\tau \in G_\phi$. Similarly,

$$\sigma(\phi) = \phi^g \quad \Longrightarrow \quad \phi = (\sigma^{-1}\phi)^{\sigma^{-1}(g)} \quad \Longrightarrow \quad \sigma^{-1}(\phi) = \phi^{\sigma^{-1}(g^{-1})},$$

which shows that $\sigma^{-1}(\phi)$ is equivalent to $\phi$, and hence that $\sigma^{-1} \in G_\phi$. This proves that $G_\phi$ is a subgroup of $\mathrm{Gal}(\bar{K}/K)$.

Next let $K'$ be a field of definition for $\phi$. Under this assumption, we proved earlier (4.52) that for every $\sigma \in \mathrm{Gal}(\bar{K}/K')$ there is a $g_\sigma \in \mathrm{PGL}_2(\bar{K})$ such that $\sigma(\phi) = \phi^{g_\sigma}$. In other words, $\sigma(\phi)$ is equivalent to $\phi$, so $\sigma \in G_\phi = \mathrm{Gal}(\bar{K}/K_\phi)$ by definition. This proves that $\mathrm{Gal}(\bar{K}/K') \subset \mathrm{Gal}(\bar{K}/K_\phi)$, and hence by Galois theory, that $K_\phi \subset K'$. (We are also using the fact that $K_\phi$ is a finite extension of $K$.) □

It follows from Proposition 4.84 that the smallest possible field of definition for $\phi$ is the field of moduli of $\phi$, but it is not clear whether the field of moduli is always a field of definition. The following example shows that we can have FOM $\ne$ FOD.

*Example* 4.85. Let

$$\phi(z) = i\left(\frac{z-1}{z+1}\right)^3.$$

Clearly $\mathbb{Q}(i)$ is a field of definition for $\phi$. Let $\sigma$ be complex conjugation, so $\mathrm{Gal}(\mathbb{Q}(i)/\mathbb{Q}) = \{1, \sigma\}$, and let $g(z) = -1/z$. Then we obtain

$$\begin{aligned}\phi^g(z) = (g^{-1}\phi g)(z) &= -\frac{1}{\phi(-1/z)} = -\frac{1}{i\left(\dfrac{-1/z-1}{-1/z+1}\right)^3} \\ &= i\left(\frac{-1/z+1}{-1/z-1}\right)^3 = -i\left(\frac{z-1}{z+1}\right)^3 = \sigma(\phi)(z).\end{aligned}$$

This shows that $\sigma \in G_\phi$, so $G_\phi = \{1, \sigma\}$ and $K_\phi = \mathbb{Q}$. In other words, $\mathbb{Q}$ is the field of moduli of $\phi$.

Now suppose that $\mathbb{Q}$ is actually a field of definition for $\phi$. This means that we can find some $f \in \mathrm{PGL}_2(\bar{\mathbb{Q}})$ such that $\phi^f \in \mathbb{Q}(z)$. In particular, letting $\sigma$ denote complex conjugation, we have

$$\phi^f = \sigma(\phi^f) = \sigma(\phi)^{\sigma(f)} = \phi^{g\sigma(f)}.$$

It is not hard to verify (see Exercise 4.39) that $\operatorname{Aut}(\phi) = 1$, so we deduce that $f = g\sigma(f)$. Using the fact that $g(z) = -1/z$, this equation says that

$$\frac{az+b}{cz+d} = -\frac{\bar{c}z+\bar{d}}{\bar{a}z+\bar{b}},$$

where we use an overscore to denote complex conjugation. The fact that these two linear fractional transformations are equal means that there is a $\lambda \in \bar{\mathbb{Q}}^*$ such that

$$a = -\lambda\bar{c}, \quad b = -\lambda\bar{d}, \quad c = \lambda\bar{a}, \quad d = \lambda\bar{b}.$$

Multiplying the first and third equations gives

$$\lambda|a|^2 = -\lambda|c|^2.$$

This is a contradiction, since $\lambda \neq 0$ and we cannot have $a = c = 0$. Hence $\mathbb{Q}$ is not a field of definition for $\phi$, so we have an example of a map with FOM $\neq$ FOD.

In order to investigate more closely the question of when the field of moduli of a rational map $\phi$ is a field of definition, we make the simplifying assumption that

$$\operatorname{Aut}(\phi) = 1. \tag{4.53}$$

Replacing $K$ by $K_\phi$, we may assume that $K$ is the field of moduli of $\phi$. This means that for every $\sigma \in \operatorname{Gal}(\bar{K}/K)$ there exists a $g_\sigma \in \operatorname{PGL}_2(\bar{K})$ satisfying

$$\sigma(\phi) = \phi^{g_\sigma}.$$

Note that the assumption (4.53) that $\operatorname{Aut}(\phi) = 1$ implies that $g_\sigma$ is uniquely determined by $\sigma$. The next proposition describes some of the properties of the map $\sigma \mapsto g_\sigma$.

**Proposition 4.86.** *Let $\phi \in \bar{K}(z)$ be a rational map of degree $d \geq 2$ with field of moduli $K$ and satisfying $\operatorname{Aut}(\phi) = 1$, and for each $\sigma \in \operatorname{Gal}(\bar{K}/K)$ write $\sigma(\phi) = \phi^{g_\sigma}$ as above.*

(a) *The map*

$$\operatorname{Gal}(\bar{K}/K) \longrightarrow \operatorname{PGL}_2(\bar{K}), \qquad \sigma \longmapsto g_\sigma,$$

*is a 1-cocycle, i.e., it satisfies*

$$g_{\sigma\tau} = g_\sigma\sigma(g_\tau) \qquad \text{for all } \sigma,\tau \in \operatorname{Gal}(\bar{K}/K).$$

(b) *$K$ is a field of definition for $\phi$ if and only if $g$ is a 1-coboundary, i.e., if and only if there is an $f \in \operatorname{PGL}_2(\bar{K})$ such that*

$$g_\sigma = f\sigma(f^{-1}) \qquad \text{for all } \sigma \in \operatorname{Gal}(\bar{K}/K).$$

*Proof.* (a) Let $\sigma, \tau \in \operatorname{Gal}(\bar{K}/K)$. Then

$$\phi^{g_{\sigma\tau}} = \sigma(\tau(\phi)) = \sigma(\phi^{g_\tau}) = \sigma(\phi)^{\sigma(g_\tau)} = \phi^{g_\sigma \sigma(g_\tau)}.$$

Since $\operatorname{Aut}(\phi) = 1$, we conclude that $g_{\sigma\tau} = g_\sigma \sigma(g_\tau)$.
(b) Suppose first that $K$ is a field of definition for $\phi$. This means that there is an $f \in \mathrm{PGL}_2(\bar{K})$ such that $\phi^f \in K(z)$. Hence for any $\sigma \in \operatorname{Gal}(\bar{K}/K)$ we have

$$\phi^f = \sigma(\phi^f) = \sigma(\phi)^{\sigma(f)} = \phi^{g_\sigma \sigma(f)}.$$

Again using the fact that $\operatorname{Aut}(\phi) = 1$, we find that $f = g_\sigma \sigma(f)$, and hence $g_\sigma = f\sigma(f^{-1})$.

Conversely, suppose that there is an $f \in \mathrm{PGL}_2(\bar{K})$ such that $g_\sigma = f\sigma(f^{-1})$. Then

$$\sigma(\phi^f) = \sigma(\phi)^{\sigma(f)} = \phi^{g_\sigma \sigma(f)} = \phi^f.$$

This is true for every $\sigma \in \operatorname{Gal}(\bar{K}/K)$, so $\phi^f \in K(z)$. Hence $K$ is a field of definition for $\phi$. □

The criterion for FOM = FOD given in Proposition 4.86 may seem complicated, but it represents a tremendous simplification. If we try to use the definition of FOD directly, we need to search for an $f \in \mathrm{PGL}_2(\bar{K})$ that makes the coefficients of $\phi^f$ lie in $K$. The substitution

$$\phi^f(z) = \frac{d\phi\left(\frac{az+b}{cz+d}\right) - b}{-c\phi\left(\frac{az+b}{cz+d}\right) + a},$$

even for a rational function $\phi(z)$ of small degree, has coefficients that are very complicated expressions in the quantities $a, b, c, d$. It is thus difficult to determine whether there is some choice of $a, b, c, d$ that makes the coefficients lie in $K$. On the other hand, the cocycle–coboundary criterion

$$\text{FOM} = \text{FOD} \quad \Longleftrightarrow \quad g_\sigma \text{ has the form } f\sigma(f^{-1})$$

in Proposition 4.86 involves only linear functions, i.e., elements of $\mathrm{PGL}_2(\bar{K})$, so it is often considerably easier to apply.

The 1-cocycles that arise in the FOD = FOM question have the form

$$g : \operatorname{Gal}(\bar{K}/K) \longrightarrow \mathrm{PGL}_2(\bar{K}).$$

The group $\mathrm{PGL}_2(\bar{K})$ is the automorphism group of the projective line $\mathbb{P}^1$, so these cocycles should be associated to twists of $\mathbb{P}^1$. We saw in Section 4.8 that every twist gives rise to a 1-cocycle, but in general it is a delicate question to determine whether every 1-cocycle comes from a twist. It turns out that this is true for algebraic varieties, but since we do not need the most general result, we are content to construct the twists of $\mathbb{P}^1$ that are needed to answer our question about FOD = FOM. (See also Example 4.87.)

**Proposition 4.87.** *Let*

$$g : \mathrm{Gal}(\bar{K}/K) \longrightarrow \mathrm{PGL}_2(\bar{K})$$

*be a* 1*-cocycle, and assume that $g$ has the property that there is a finite extension $L/K$ such that $g_\sigma = 1$ for all $\sigma \in \mathrm{Gal}(\bar{K}/L)$.*[10] *Then there is an algebraic curve $C$ defined over $K$ and an isomorphism $i : C \to \mathbb{P}^1$ defined over $\bar{K}$ such that the* 1*-cocycle*

$$\mathrm{Gal}(\bar{K}/K) \longrightarrow \mathrm{PGL}_2(\bar{K}), \qquad \sigma \longmapsto i\sigma(i^{-1}),$$

*is equal to the* 1*-cocycle $g$. Hence $C$ is a twist of $\mathbb{P}^1/K$, and*

$$C \text{ is the trivial twist of } \mathbb{P}^1/K \quad \text{if and only if} \quad g \text{ is a 1-coboundary.}$$

*Proof.* We construct the curve $C$ by describing its field of rational functions. Note that the field of rational functions for the curve $\mathbb{P}^1$ is the field $\bar{K}(z)$ and that the Galois group $\mathrm{Gal}(\bar{K}/K)$ acts naturally on $\bar{K}(z)$ by acting on $\bar{K}$ and leaving $z$ fixed.

Now consider another field of rational functions in one variable $\bar{K}(w)$, but this time with a "twisted action" of $\mathrm{Gal}(\bar{K}/K)$. We define this twisted action by letting $\mathrm{Gal}(\bar{K}/K)$ act on $\bar{K}$ as usual, but setting

$$\sigma(w) = g_\sigma^{-1}(w).$$

In other words, if $g_\sigma(z) = (az+b)/(cz+d)$, then

$$\sigma(w) = g_\sigma^{-1}(w) = \frac{dw - b}{-cw + a},$$

and for any rational function

$$F(w) = \frac{a_0 + a_1 w + a_2 w^2 + \cdots + a_d w^d}{b_0 + b_1 w + b_2 w^2 + \cdots + b_d w^d} \in \bar{K}(w),$$

we have

$$\sigma(F)(w) = \frac{\sigma(a_0) + \sigma(a_1) g_\sigma^{-1}(w) + \sigma(a_2) g_\sigma^{-1}(w)^2 + \cdots + \sigma(a_d) g_\sigma^{-1}(w)^d}{\sigma(b_0) + \sigma(b_1) g_\sigma^{-1}(w) + \sigma(b_2) g_\sigma^{-1}(w)^2 + \cdots + \sigma(b_d) g_\sigma^{-1}(w)^d}.$$

It is clear that $\sigma(F+G) = \sigma(F) + \sigma(G)$ and that $\sigma(FG) = \sigma(F)\sigma(G)$. However, to be an action, we must also have $(\sigma\tau)(F) = \sigma(\tau(F))$. We use the cocycle relation to verify this condition,

$$\sigma(\tau(w)) = \sigma(g_\tau^{-1} w) = \sigma(g_\tau^{-1})\sigma(w) = \sigma(g_\tau^{-1}) g_\sigma^{-1}(w) = g_{\sigma\tau}^{-1}(w) = (\sigma\tau)(w).$$

We now look at the subfield of $\bar{K}(w)$ that is fixed by the twisted action,

$$\mathcal{K} = \{F \in \bar{K}(w) : \sigma(F) = F \text{ for all } \sigma \in \mathrm{Gal}(\bar{K}/K)\}.$$

We prove that the field $\mathcal{K}$ has the following properties:

[10]One says that $g$ is a continuous 1-cocycle for the profinite topology on $\mathrm{Gal}(\bar{K}/K)$ and the discrete topology on $\mathrm{PGL}_2(\bar{K})$.

(i) $\mathcal{K} \cap \bar{K} = K$.

(ii) $\mathcal{K}$ has transcendence degree 1 over $K$.

(iii) $\bar{K}\mathcal{K} = \bar{K}(z)$.

To verify (i), let $a \in \mathcal{K} \cap \bar{K}$. Then $a \in \bar{K}$, so the action of $\mathrm{Gal}(\bar{K}/K)$ on $a$ is the usual Galois action. On the other hand, the fact that $a \in \mathcal{K}$ means that the Galois action is trivial. Hence $a \in K$. This shows that $\mathcal{K} \cap \bar{K} \subset K$, and the opposite inclusion is obvious.

Next let $L/K$ be a finite Galois extension with the property that $g_\tau = 1$ for every $\tau \in \mathrm{Gal}(\bar{K}/L)$. This implies that the action of an element $\sigma \in \mathrm{Gal}(\bar{K}/K)$ on $w$ depends only on the image of $\sigma$ in the quotient group

$$\mathrm{Gal}(L/K) = \mathrm{Gal}(\bar{K}/K)/\,\mathrm{Gal}(\bar{K}/L),$$

since if $\tau \in \mathrm{Gal}(\bar{K}/L)$, then

$$(\sigma\tau)(w) = \sigma(g_\tau(w)) = \sigma(w).$$

It follows that the polynomial

$$P(T) = \prod_{\lambda \in \mathrm{Gal}(L/K)} (T - \lambda(w)) = \prod_{\lambda \in \mathrm{Gal}(L/K)} (T - g_\lambda^{-1}(w)) \in \bar{K}(w)[T] \tag{4.54}$$

is well-defined. Further, we claim that the coefficients of $P(T)$ are in $\mathcal{K}$. To see this, for any $\sigma \in \mathrm{Gal}(\bar{K}/K)$ we note that

$$\sigma(P)(T) = \prod_{\lambda \in \mathrm{Gal}(L/K)} (T - \sigma\lambda(w)) = P(T),$$

since the effect of replacing $\lambda$ with $\sigma\lambda$ is simply to permute the order of the factors in the product.

We have now constructed a polynomial $P(T) \in \mathcal{K}[T]$, and we observe that $w$ is a root of $P(T)$. Hence by definition, the extension $\mathcal{K}(w)/\mathcal{K}$ is an algebraic extension. Since the element $w$ is transcendental over $K$, this proves that $\mathcal{K}$ has transcendence degree at least 1 over $K$. On the other hand, $\mathcal{K}$ is contained in the field $\bar{K}(w)$, and it is clear that $\bar{K}(w)/\bar{K}$ has transcendence degree 1 over $K$. Therefore $\mathcal{K}$ has transcendence degree exactly 1 over $K$, which proves (ii).

Finally, consider the splitting field $\mathcal{L}$ over $\mathcal{K}$ of the polynomial $P(T)$. As already noted, we have $w \in \mathcal{L}$. The definition of $P(T)$ shows that there is a natural surjection $\mathrm{Gal}(L/K) \to \mathrm{Gal}(\mathcal{L}/\mathcal{K})$, which implies in particular that $\mathcal{L} = L\mathcal{K}$. Hence $w \in L\mathcal{K}$, which proves that $\bar{K}(z) \subset \bar{K}\mathcal{K}$. This gives (iii), since the other inclusion is true from the definition of $\mathcal{K}$.

We now have the tools needed to complete the proof of Proposition 4.87, but we pause briefly for an example illustrating the general construction.

*Example* 4.88. Define a map $g$ by the rule

$$g : \operatorname{Gal}(\bar{\mathbb{Q}}/\mathbb{Q}) \longrightarrow \operatorname{PGL}_2(\bar{\mathbb{Q}}), \qquad g_\sigma(z) = \begin{cases} z & \text{if } \sigma(i) = i, \\ -1/z & \text{if } \sigma(i) = -i. \end{cases}$$

It is easy to check that $g$ is a 1-cocycle, and indeed it is the 1-cocycle described in Example 4.85. Fix an embedding of $\bar{\mathbb{Q}}$ into $\mathbb{C}$ and let $\rho \in \operatorname{Gal}(\bar{\mathbb{Q}}/\mathbb{Q})$ denote complex conjugation. The twisted action on $\bar{\mathbb{Q}}(w)$ is given by

$$\rho(a) = \bar{a} \quad \text{for } a \in \bar{\mathbb{Q}} \qquad \text{and} \qquad \rho(w) = -1/w.$$

The field $\mathcal{K}$ is the subfield of $\bar{\mathbb{Q}}(w)$ consisting of elements that are fixed by the twisted action. The coefficients of the polynomial $P(T)$ defined by equation (4.54) give us elements in $\mathcal{K}$,

$$P(T) = (T - w)(T - \rho(w)) = (T - w)\left(T + \frac{1}{w}\right) = T^2 - \left(w - \frac{1}{w}\right)T - 1.$$

This yields one interesting element in $\mathcal{K}$, namely $u = w - w^{-1}$, and we observe that the quantity $v = i(w + w^{-1})$ gives another element in $\mathcal{K}$. It is not hard to show that $u$ and $v$ generate $\mathcal{K}$,

$$\mathcal{K} = K(u, v) = K\left(w - \frac{1}{w}, i\left(w + \frac{1}{w}\right)\right).$$

(See Exercise 4.45.) Of course, $u$ and $v$ are not independent, since

$$u^2 + v^2 = \left(w - \frac{1}{w}\right)^2 + \left(i\left(w + \frac{1}{w}\right)\right)^2 = -4.$$

The field $\mathcal{K}$ is the function field of the curve $C : u^2 + v^2 = -4$. Notice that $C$ is defined over $\mathbb{Q}$, and $C$ is $\bar{\mathbb{Q}}$-isomorphic to $\mathbb{P}^1$, but $C$ is not $\mathbb{Q}$-isomorphic to $\mathbb{P}^1$, since $C(\mathbb{Q}) = \emptyset$. From our general theory, the fact that $C$ is a nontrivial twist of $\mathbb{P}^1/\mathbb{Q}$ is equivalent to the fact (proven by a direct calculation in Example 4.85) that the 1-cocycle $g$ is not a 1-coboundary.

Resuming the proof of Proposition 4.87, we have constructed a field $\mathcal{K}$ that is the function field of a curve $C/K$, and we have an isomorphism

$$\bar{K}\mathcal{K} = \bar{K}(w) \cong \bar{K}(z), \qquad w \longleftrightarrow z,$$

that induces a $\bar{K}$-isomorphism $i : C \to \mathbb{P}^1$. In other words, the functions $w$ on $C$ and $z$ on $\mathbb{P}^1$ are related by the formula $w = z \circ i$. The curve $C$ is a twist of $\mathbb{P}^1/K$, and its associated cocycle is given by $\sigma \to i\sigma(i^{-1})$. We compute

$$\begin{aligned} z\circ\sigma(i) &= \sigma(z\circ i) && \text{since } \sigma(z)=z,\\ &= \sigma(w) && \text{since } w = z\circ i,\\ &= g_\sigma^{-1}(w) && \text{by definition of the twisted action on } \bar K(w),\\ &= \frac{dw-b}{-cw+a} && \text{where } g_\sigma(z) = \frac{az+b}{cz+d} \in \mathrm{PGL}_2(\bar K),\\ &= \frac{dz\circ i - b}{-cz\circ i + a} && \text{since } w = z\circ i,\\ &= z\circ g_\sigma^{-1}\circ i. \end{aligned}$$

Thus $\sigma(i) = g_\sigma^{-1}\circ i$, which proves that $g_\sigma = i\sigma(i^{-1})$ is the 1-cocycle associated to the $\bar K$-isomorphism $i : C \to \mathbb{P}^1$.

This completes the proof that the algebraic curve $C$ is a twist of $\mathbb{P}^1/K$ whose associated 1-cocycle is $g$. Finally, Proposition 4.77 tells us that $C$ is the trivial twist if and only if its associated 1-cocycle is a 1-coboundary. □

Returning to the question of FOD = FOM, let $\phi(z) \in \bar K(z)$ be a rational function with field of moduli $K$ and trivial automorphism group. We have constructed a 1-cocycle

$$g : \mathrm{Gal}(\bar K/K) \longrightarrow \mathrm{PGL}_2(\bar K)$$

that is associated to $\phi$ (Proposition 4.86) and a twist $C_\phi$ of $\mathbb{P}^1/K$ that is associated to the 1-cocycle $g$ (Proposition 4.87). We have also proven the following chain of equivalences:

$$\begin{aligned} &K \text{ is a field of definition for } \phi\\ &\qquad\iff g \text{ is a 1-coboundary} && \text{(Proposition 4.86)},\\ &\qquad\iff C_\phi \text{ is } K\text{-isomorphic to } \mathbb{P}^1 && \text{(Proposition 4.87)}. \end{aligned}$$

It remains to find a way of determining whether a twist of $\mathbb{P}^1$ is $K$-isomorphic to $\mathbb{P}^1$. We will use the Riemann–Roch theorem to provide two sufficient conditions for resolving this problem. For the convenience of the reader, we recall the general statement of the Riemann–Roch theorem, although we will need it only for curves of genus 0.

**Theorem 4.89.** (Riemann–Roch Theorem) *Let $C/K$ be a smooth projective curve of genus $g$ defined over $K$.*

(a) *There is a divisor on $C$ of degree $2g-2$ that is defined over $K$.*

(b) *Let $D$ be a divisor on $C$ that is defined over $K$ and assume that the degree of $D$ satisfies* $\deg(D) \ge 2g-1$. *Then there is a function $f \in K(C)$ satisfying*

$$\mathrm{div}(f) + D \ge 0.$$

**Corollary 4.90.** (Riemann–Roch in Genus 0) *Let $C/K$ be a smooth projective curve of genus 0 defined over $K$.*

(a) *There is a $K$-rational divisor of degree 2 on $C$.*

(b) *Let $D \in \operatorname{Div}(C)$ be a divisor defined over $K$ and satisfying* $\deg(D) \geq 1$. *Then there is a function $f \in K(C)$ with* $\operatorname{div}(f) + D \geq 0$.

*Proof.* The proof of the Riemann–Roch theorem over an algebraically closed field is given in most introductory texts on algebraic geometry, such as [198, IV.1.3], or see [255, Chapter I] for an elementary proof due to Weil and [410, II, §5] for an overview. The divisor of degree $2g-2$ in (a) is a canonical divisor, i.e., the divisor of any $K$-rational differential form such as $df$ for any nonzero $f \in K(C)$. In particular, for curves of genus 0 we get a $K$-rational divisor $D$ of degree $-2$, so $-D$ is a $K$-rational divisor of degree 2. □

**Proposition 4.91.** *Let $C$ be a twist of $\mathbb{P}^1/K$. The following are equivalent:*

(a) *$C$ is the trivial twist of $\mathbb{P}^1/K$, i.e., $C$ is $K$-isomorphic to $\mathbb{P}^1$.*
(b) *$C(K)$ is nonempty, i.e., $C$ has a point with coordinates in $K$.*
(c) *There is a divisor $D = \sum n_i(P_i)$ on $\mathbb{P}^1(\bar{K})$ of odd degree such that $D$ is defined over $K$, i.e., for all $\sigma \in \operatorname{Gal}(\bar{K}/K)$ we have $\sum n_i(\sigma(P_i)) = \sum n_i(P_i)$ as a formal sum of points.*

*Proof.* If $C$ is the trivial twist of $\mathbb{P}^1/K$, then there is a $K$-isomorphism $j : C \to \mathbb{P}^1$. In particular, $j : C(K) \to \mathbb{P}^1(K)$ is a bijection, so $C(K)$ is certainly nonempty. This proves that (a) implies (b).

Next, it is clear that (b) implies (c), since if $P \in C(K)$, then the divisor $D = (P)$ has odd degree (one is an odd number!) and is clearly defined over $K$.

Finally, suppose that the degree $n = \deg(D) = \sum n_i$ of $D = \sum n_i(P_i)$ is odd and that $D$ is defined over $K$. We also use the fact that $C$ and $\mathbb{P}^1$ are $\bar{K}$-isomorphic, so in particular $C$ is a curve of genus zero. It follows from the Riemann–Roch theorem (Corollary 4.90(a)) that there is a $K$-rational divisor on $C$ having degree $-2$, say

$$D' = (Q_1) + (Q_2).$$

Consider the divisor

$$E = D + \frac{n-1}{2}D' = (P_1) + (P_2) + \cdots + (P_n) - \frac{n-1}{2}\big((Q_1) + (Q_2)\big).$$

The divisor $E$ is defined over $K$ and has degree 1, so the Riemann–Roch theorem (Corollary 4.90(b)) tells us that there is a rational function $\psi$ on $C$ such that $\psi$ is defined over $K$ and $\psi$ has degree 1. In other words, $\psi$ is a map $\psi : C \to \mathbb{P}^1$ of degree 1 defined over $K$, and hence $C$ is $K$-isomorphic to $\mathbb{P}^1$. This shows that (c) implies (a) and completes the proof of the theorem. □

We have now assembled all of the tools that we need to prove the main theorem of this section. We state the theorem in full generality, but give the proof only for rational maps with trivial automorphism groups. The general case is proven similarly, but there are many additional technical complications.

**Theorem 4.92.** *Let $\phi(z) \in \bar{K}(z)$ be a rational map of degree $d \geq 2$. Then the field of moduli of $\phi$ is a field of definition for $\phi$ in the following two situations:*

(a) $\phi(z)$ *has even degree.*
(b) $\phi(z)$ *is a polynomial.*

*Proof.* We prove the theorem under the assumption that $\operatorname{Aut}(\phi) = 1$. See [414] for a proof in the general case.

Without loss of generality, we may assume that $K$ is the field of moduli of $\phi$. Let $g : \operatorname{Gal}(\bar{K}/K) \to \operatorname{PGL}_2(\bar{K})$ be the 1-cocycle associated to $\phi$ (Proposition 4.86) and let $C_\phi$ be the twist of $\mathbb{P}^1/K$ associated to the 1-cocycle $g$ (Proposition 4.87). This means that there is a $\bar{K}$-isomorphism

$$i : C_\phi \longrightarrow \mathbb{P}^1 \qquad \text{such that} \qquad g_\sigma = i\sigma(i^{-1}) \quad \text{for all } \sigma \in \operatorname{Gal}(\bar{K}/K).$$

Using many of the results proven in this chapter, we have the following chain of equivalences:

$$\begin{aligned}
&K \text{ is a field of definition for } \phi \\
&\quad \Longleftrightarrow g \text{ is a 1-coboundary} && \text{(Proposition 4.86),} \\
&\quad \Longleftrightarrow C_\phi \text{ is } K\text{-isomorphic to } \mathbb{P}^1 && \text{(Proposition 4.87),} \\
&\quad \Longleftrightarrow C_\phi(K) \text{ is not empty} && \text{(Proposition 4.91),} \\
&\quad \Longleftrightarrow \begin{pmatrix} \text{There is a divisor on } C_\phi(\bar{K}) \text{ of} \\ \text{odd degree and defined over } K \end{pmatrix} && \text{(Proposition 4.91).}
\end{aligned}$$

We are going to produce divisors and points on the curve $C_\phi$ using the map $\psi : C_\phi \to C_\phi$ defined by the composition

$$\psi : C_\phi \xrightarrow{\ i\ } \mathbb{P}^1 \xrightarrow{\ \phi\ } \mathbb{P}^1 \xrightarrow{\ i^{-1}\ } C_\phi.$$

We begin by checking that the map $\psi = i^{-1}\phi i$ is defined over $K$. To verify this, we let $\sigma \in \operatorname{Gal}(\bar{K}/K)$ and compute

$$\begin{aligned}
\sigma(\psi) &= \sigma(i^{-1}\phi i) = \sigma(i^{-1})\sigma(\phi)\sigma(i) = \sigma(i^{-1})\phi^{g_\sigma}\sigma(i) \\
&= \sigma(i^{-1})g_\sigma^{-1}\phi g_\sigma \sigma(i) = \sigma(i^{-1})(i\sigma(i^{-1}))^{-1}\phi i\sigma(i^{-1})\sigma(i) = i^{-1}\phi i = \psi.
\end{aligned}$$

Hence $\psi : C_\phi \to C_\phi$ is defined over $K$. We also note that $\phi$ and $\psi$ have the same degree, since $i$ is an isomorphism.

(a) Let $D_\psi$ be the divisor of fixed points of $\psi$, that is, the collection of fixed points of $\psi$ counted with appropriate multiplicities. In the language of algebraic geometry, $D_\psi$ is the pullback by the diagonal map

$$C_\phi \longrightarrow C_\phi \times C_\phi, \qquad P \longmapsto (P, P),$$

of the graph $\{(x, \psi(x)) : x \in C_\phi\}$ of $\psi$.

A map of degree $d$ has exactly $d + 1$ fixed points (counted with multiplicities), and if the map is defined over $K$, then the divisor of fixed points is defined over $K$, i.e., it is fixed by the Galois group. Thus $D_\psi$ is a divisor of degree $d + 1$ on $C_\phi$

and $D_\psi$ is defined over $K$. By assumption, $d$ is even, so $D_\psi$ has odd degree. Hence by the chain of equivalences derived earlier, the field $K$ is a field of definition for $\phi$.
(b) The map $\phi : \mathbb{P}^1 \to \mathbb{P}^1$ is a polynomial by assumption, so it has a totally ramified fixed point $P$. In other words, $\phi(P) = P$ and the ramification index of $\phi$ at $P$ satisfies $e_P(\phi) = d$. The map $i : C_\phi \to \mathbb{P}^1$ is an isomorphism, so the point

$$Q = i^{-1}(P) \in C_\phi(\bar{K})$$

is a totally ramified fixed point of $\psi$. Suppose first that $P$ is the only fixed point of $\phi$ in $\mathbb{P}^1(\bar{K})$ with ramification index $d$. Then $Q$ is the only fixed point of $\psi$ in $C_\phi(\bar{K})$ with ramification index $d$. However, for any $\sigma \in \operatorname{Gal}(\bar{K}/K)$ and any point $R \in C_\phi(\bar{K})$ we have

$$\psi(\sigma(R)) = \sigma(\psi)(\sigma(R)) = \sigma(\psi(R)) \quad \text{and} \quad e_R(\psi) = e_{\sigma(R)}(\sigma(\psi)) = e_{\sigma(R)}(\psi),$$

since $\psi$ is defined over $K$. In particular, taking $R = Q$ to be the given fixed point of ramification index $d$, we see that $\psi(\sigma(Q)) = \sigma(Q)$ and $d = e_Q(\psi) = e_{\sigma(Q)}(\psi)$, so $\sigma(Q)$ is also a fixed point of $\psi$ of ramification index $d$. Hence $\sigma(Q) = Q$, and since this holds for every $\sigma \in \operatorname{Gal}(\bar{K}/K)$, we conclude that $Q \in C_\phi(K)$. Hence $C_\phi(K)$ is not empty, so we again conclude that $K$ is a field of definition for $\phi$.

We are left to consider the case that $\phi$ has a second fixed point of ramification index $d$, say $\phi(P') = P'$ and $e_{P'}(\phi) = d$. (Note that the Riemann–Hurwitz formula, Theorem 1.1, precludes more than two such points.) Choose some linear fractional transformation $h \in \mathrm{PGL}_2(\bar{K})$ satisfying $h(\infty) = P$ and $h(0) = P'$. Then $\phi^h$ satisfies

$$\phi^h(\infty) = \infty, \quad \phi^h(0) = 0, \quad \text{and} \quad e_0(\phi^h) = e_\infty(\phi^h) = d,$$

so $\phi^h$ must have the form

$$\phi^h(z) = cz^d \quad \text{for some } c \in \bar{K}^*.$$

But any rational map of this form has a nontrivial automorphism group. Indeed, its automorphism group is a dihedral group of order $2(d-1)$ generated by the maps

$$z \longmapsto \zeta z \qquad \text{and} \qquad z \longmapsto a/z,$$

where $\zeta^{d-1} = 1$ and $a^{d-1} = 1/c$. In any case, we have ruled out this case by our assumption that $\operatorname{Aut}(\phi) = 1$, which completes the proof of Theorem 4.92. □

*Remark* 4.93. The distinction between the field of moduli and fields of definition is important in the study of abelian varieties; see for example the work of Shimura [398]. More recently, the FOM = FOD question has been investigated for the collection of of covering maps $\phi : X \to B$ up to automorphism of the base $B$. In particular, this is much studied for $B = \mathbb{P}^1$ (cf. Grothendieck's "dessins d'enfant"). If also $X = B = \mathbb{P}^1$, then one studies the set of rational maps $\phi(z) \in \bar{K}(z)$ under the left composition equivalence relation $\phi \sim f\phi$ for $f \in \mathrm{PGL}_2(\bar{K})$. This bears considerable resemblance to the material in this section, where we use instead the equivalence relation $\phi \sim f^{-1}\phi f$, but there are significant differences. For example, Couveignes [111] shows that using the relation equivalence $\phi \sim f\phi$, there are polynomials in $\mathbb{Q}[z]$ with FOM $\neq$ FOD, in direct contrast to Theorem 4.92. For further results, see for example work of Débes and Douai [117, 118, 119] and Débes and Harbater [120].

## 4.11 Minimal Resultants and Minimal Models

Let $P(X,Y) = \sum a_{ij}X^iY^j \in K[X,Y]$ be a polynomial and let $\mathfrak{p}$ be a prime of $K$. We define the order of $P$ at $\mathfrak{p}$ to be

$$\operatorname{ord}_{\mathfrak{p}}(P) = \min_{i,j} \operatorname{ord}_{\mathfrak{p}}(a_{ij}).$$

Notice that $\operatorname{ord}_{\mathfrak{p}}(P) = 0$ if and only if all of the coefficients of $\phi$ are $\mathfrak{p}$-integral and at least one coefficient is a $\mathfrak{p}$-unit.

In Section 2.4 we defined the resultant of a rational map $\phi$ (with respect to $\mathfrak{p}$) by writing $\phi = [F, G]$ using homogeneous polynomials $F, G \in K[X,Y]$ satisfying $\min\{\operatorname{ord}_{\mathfrak{p}}(F), \operatorname{ord}_{\mathfrak{p}}(G)\} = 0$ and setting

$$\operatorname{Res}_{\mathfrak{p}}(\phi) = \operatorname{Res}(F, G).$$

The resultant $\operatorname{Res}_{\mathfrak{p}}(\phi)$ is well-defined up to multiplication by the $2d^{\text{th}}$ power of a $\mathfrak{p}$-adic unit, so in particular $\operatorname{ord}_{\mathfrak{p}}(\operatorname{Res}_{\mathfrak{p}}(\phi))$ depends only on $\phi$ and $\mathfrak{p}$. We also recall that $\phi$ has good reduction at $\mathfrak{p}$ if and only if its resultant is a $\mathfrak{p}$-adic unit.

*Example* 4.94. Even if $\phi$ has bad reduction at $\mathfrak{p}$, it may be possible to change coordinates and achieve good reduction. In other words, there may be some $f \in \operatorname{PGL}_2(K)$ such that $\operatorname{Res}(\phi^f)$ is a $\mathfrak{p}$-unit. For example, the map $\phi(z) = z + p^2z^{-1}$ has bad reduction at $p$, since

$$\phi = [X^2 + p^2Y^2, XY], \quad \text{so} \quad \operatorname{Res}_p(\phi) = \operatorname{Res}(X^2 + p^2Y^2, XY) = p^2.$$

However, if we let $f(z) = pz$, then $\phi^f(z) = z + z^{-1}$, which has good reduction at $p$.

Our aim in this section is to study this phenomenon. In particular, we study the extent to which we can eliminate, or at least ameliorate, bad reduction in $\phi$ by conjugating $\phi$ with a linear fractional transformation in $\operatorname{PGL}_2(K)$.

**Definition.** Let $\phi = [F, G]$ be a rational map given by homogeneous polynomials $F, G \in K[X,Y]$, let $f \in \operatorname{PGL}_2(K)$, and choose a matrix

$$A = \begin{pmatrix} \alpha & \beta \\ \gamma & \delta \end{pmatrix} \in \operatorname{GL}_2(K)$$

representing $f$. We define polynomials $F_A, G_A \in K[X,Y]$ by the formulas

$$F_A(X,Y) = \delta F(\alpha X + \beta Y, \gamma X + \delta Y) - \beta G(\alpha X + \beta Y, \gamma X + \delta Y), \tag{4.55}$$
$$G_A(X,Y) = -\gamma F(\alpha X + \beta Y, \gamma X + \delta Y) + \alpha G(\alpha X + \beta Y, \gamma X + \delta Y). \tag{4.56}$$

It is often convenient to write this in matrix notation as

$$\begin{bmatrix} F_A \\ G_A \end{bmatrix} = A^{\mathrm{adj}} \begin{bmatrix} F \circ A \\ G \circ A \end{bmatrix},$$

where $A^{\mathrm{adj}} = \left(\begin{smallmatrix} \delta & -\beta \\ -\gamma & \alpha \end{smallmatrix}\right)$ is the adjoint matrix to $A$. Note that the conjugate $\phi^f = f^{-1} \circ \phi \circ f$ of $\phi$ by $f$ is equal to

$$\phi^f(X,Y) = \big[F_A(X,Y), G_A(X,Y)\big].$$

**Proposition 4.95.** *Let $\phi = [F,G]$ be a rational map of degree $d$ described by homogeneous polynomials $F,G \in K[X,Y]$ and let $\mathfrak{p}$ be a prime ideal.*

(a) *The valuation of the minimal resultant of $\phi$ is given by the formula*

$$\operatorname{ord}_{\mathfrak{p}}\big(\operatorname{Res}_{\mathfrak{p}}(\phi)\big) = \operatorname{ord}_{\mathfrak{p}}(\operatorname{Res}(F,G)) - 2d \min\big\{\operatorname{ord}_{\mathfrak{p}}(F), \operatorname{ord}_{\mathfrak{p}}(G)\big\}. \tag{4.57}$$

*Note that there is no requirement that the coefficients of $F$ and $G$ be $\mathfrak{p}$-integral or that some coefficient be a $\mathfrak{p}$-adic unit.*

(b) *Let $A \in \mathrm{GL}_2(K)$. Then with $F_A$ and $G_A$ defined by* (4.55) *and* (4.56),

$$\operatorname{ord}_{\mathfrak{p}}\big(\operatorname{Res}(F_A,G_A)\big) = \operatorname{ord}_{\mathfrak{p}}\big(\operatorname{Res}(F,G)\big) + (d^2+d)\operatorname{ord}_{\mathfrak{p}}(\det A),$$
$$\min\big\{\operatorname{ord}_{\mathfrak{p}}(F_A), \operatorname{ord}_{\mathfrak{p}}(G_A)\big\} \ge \min\big\{\operatorname{ord}_{\mathfrak{p}}(F), \operatorname{ord}_{\mathfrak{p}}(G)\big\} + (d+1)\operatorname{ord}_{\mathfrak{p}}(A),$$

*where $\operatorname{ord}_{\mathfrak{p}}(A)$ denotes the minimum of the order of the coordinates of the matrix $A$.*

(c) *In particular, if $U \in \mathrm{GL}_2(R_{\mathfrak{p}})$, then*

$$\operatorname{ord}_{\mathfrak{p}}\big(\operatorname{Res}(F_U,G_U)\big) = \operatorname{ord}_{\mathfrak{p}}\big(\operatorname{Res}(F,G)\big),$$
$$\min\big\{\operatorname{ord}_{\mathfrak{p}}(F_U), \operatorname{ord}_{\mathfrak{p}}(G_U)\big\} = \min\big\{\operatorname{ord}_{\mathfrak{p}}(F), \operatorname{ord}_{\mathfrak{p}}(G)\big\}.$$

*Proof.* (a) Choose a constant $c \in K^*$ satisfying

$$\operatorname{ord}_{\mathfrak{p}}(c) = \min\big\{\operatorname{ord}_{\mathfrak{p}}(F), \operatorname{ord}_{\mathfrak{p}}(G)\big\}.$$

Then

$$\operatorname{Res}_{\mathfrak{p}}(\phi) = \operatorname{Res}(c^{-1}F, c^{-1}G) = c^{-2d}\operatorname{Res}(F,G),$$

where we have made use of the homogeneity property of the resultant (Proposition 2.13(d)). This gives the desired result (4.57).

(b) An elementary calculation (see Exercise 2.7(c)) shows that

$$\operatorname{Res}(F_A,G_A) = (\det A)^{d^2+d}\operatorname{Res}(F,G),$$

so taking $\operatorname{ord}_{\mathfrak{p}}$ gives the first part of (b). For the second part, we observe that every coefficient of $F_A$ and $G_A$ is a sum of terms of the form

$$\left(\begin{array}{c}\text{coefficient}\\ \text{of } F \text{ or } G\end{array}\right) \times \left(\begin{array}{c}\text{homogeneous polynomial of}\\ \text{degree } d+1 \text{ in } \mathbb{Z}[\alpha,\beta,\gamma,\delta]\end{array}\right).$$

Hence

$$\min\{\operatorname{ord}_\mathfrak{p}(F_A), \operatorname{ord}_\mathfrak{p}(G_A)\} \ge \min\{\operatorname{ord}_\mathfrak{p}(F), \operatorname{ord}_\mathfrak{p}(G)\} + (d+1)\operatorname{ord}_\mathfrak{p}(A).$$

(c) The assumption that $U \in \mathrm{GL}_2(R_\mathfrak{p})$ is equivalent to the two conditions

$$\operatorname{ord}_\mathfrak{p}(U) \ge 0 \qquad \text{and} \qquad \operatorname{ord}_\mathfrak{p}(\det U) = 0,$$

i.e., the coefficients of $U$ are $\mathfrak{p}$-adic integers and the determinant is a $\mathfrak{p}$-adic unit. Applying (b) with $A = U$ gives

$$\begin{gathered}\operatorname{ord}_\mathfrak{p}\bigl(\operatorname{Res}(F_U, G_U)\bigr) = \operatorname{ord}_\mathfrak{p}\bigl(\operatorname{Res}(F,G)\bigr),\\ \min\{\operatorname{ord}_\mathfrak{p}(F_U), \operatorname{ord}_\mathfrak{p}(G_U)\} \ge \min\{\operatorname{ord}_\mathfrak{p}(F), \operatorname{ord}_\mathfrak{p}(G)\}.\end{gathered} \tag{4.58}$$

This almost completes (c). We next apply (b) to the polynomials $F_U$ and $G_U$ and to the matrix $A = U^{-1} \in \mathrm{GL}_2(R_\mathfrak{p})$. Using the fact that $(F_U)_{U^{-1}} = F$ and similarly for $G$, we find from (b) that

$$\begin{aligned}\min\{\operatorname{ord}_\mathfrak{p}(F), \operatorname{ord}_\mathfrak{p}(G)\} &= \min\{\operatorname{ord}_\mathfrak{p}((F_U)_{U^{-1}}), \operatorname{ord}_\mathfrak{p}((G_U)_{U^{-1}})\}\\ &\ge \min\{\operatorname{ord}_\mathfrak{p}(F_U), \operatorname{ord}_\mathfrak{p}(G_U)\}.\end{aligned}$$

This gives the opposite inequality to (4.58), which completes the proof of (c). □

**Definition.** Let $K$ be a number field and let $\phi(z) \in K(z)$ be a rational map. For each prime $\mathfrak{p}$ of $K$, define a nonnegative integer by

$$\epsilon_\mathfrak{p}(\phi) = \min_{f \in \mathrm{PGL}_2(K)} \operatorname{ord}_\mathfrak{p} \operatorname{Res}_\mathfrak{p}(\phi^f).$$

In other words, $\epsilon_\mathfrak{p}(\phi)$ is the exponent of the power of $\mathfrak{p}$ dividing the resultant of the conjugate $\phi^f$ that is closest to having good reduction at $\mathfrak{p}$. Then the *(global) minimal resultant of* $\phi$ is the integral ideal

$$\mathfrak{R}_\phi = \prod_\mathfrak{p} \mathfrak{p}^{\epsilon_\mathfrak{p}(\phi)}. \tag{4.59}$$

We say that $(F,G)$ is a *minimal model for* $\phi$ *at* $\mathfrak{p}$ if $\phi = [F,G]$, the coefficients of $F$ and $G$ are $\mathfrak{p}$-integral and

$$\operatorname{ord}_\mathfrak{p} \operatorname{Res}(F,G) = \operatorname{ord}_\mathfrak{p} \mathfrak{R}_\phi.$$

(See Section 6.3.5 for the analogous definition of mimimal models of an elliptic curve.)

*Remark* 4.96. The product (4.59) defining $\mathfrak{R}_\phi$ makes sense, since $\epsilon_\mathfrak{p}(\phi) = 0$ for all but finitely primes $\mathfrak{p}$. To see this, write $\phi = [F,G]$ for any $F$ and $G$ having coefficients in the ring of integers of $K$. Then there are only finitely many primes $\mathfrak{p}$ with $\operatorname{ord}_\mathfrak{p} \operatorname{Res}(F,G) > 0$, and it is clear that $\epsilon_\mathfrak{p}(\phi) = 0$ for all other primes.

The minimal resultant of a rational map is clearly invariant under $\mathrm{PGL}_2(K)$-conjugation. It measures the extent to which the conjugates of $\phi$ have bad reduction, so provides a convenient measure of the arithmetic complexity of the conjugacy class of $\phi$. A coarser way to measure arithmetic complexity is simply to take the product of the primes with bad reduction, which we denote by

$$\mathfrak{N}_\phi = \prod_{\mathfrak{p} \mid \mathfrak{R}_\phi} \mathfrak{p}. \tag{4.60}$$

It is clear that $\mathfrak{N}_\phi$ divides $\mathfrak{R}_\phi$, and it is tempting to conjecture an inequality in the opposite direction that would be a dynamical analogue of Szpiro's conjecture [205, F.3.2] for elliptic curves.

**Conjecture 4.97.** *Let $K/\mathbb{Q}$ be a number field and $d \geq 2$. There is a constant $c = c(K,d)$ such that for all rational maps $\phi(z) \in K(z)$ of degree $d$,*

$$\mathrm{N}_{K/\mathbb{Q}} \mathfrak{R}_\phi \leq \left(\mathrm{N}_{K/\mathbb{Q}} \mathfrak{N}_\phi\right)^c.$$

The minimal resultant gives one way to measure the arithmetic complexity of a rational map, but note that there are infinitely many $\mathrm{PGL}_2(K)$-inequivalent rational maps of a given degree whose minimal resultants are the same. For example, the minimal resultants of the polynomials

$$\phi_{uc}(z) = z^2 + uc$$

are the same as $u$ ranges over all units in the ring of integers of $K$.

The moduli space $\mathcal{M}_d$ provides an alternative way to measure the arithmetic complexity of the conjugacy class of a rational map. If we fix a projective embedding $\mathcal{M}_d \hookrightarrow \mathbb{P}^N$, then $\phi \in K(z)$ determines a point $\langle\phi\rangle \in \mathcal{M}_d(K)$ and we can take the height of the corresponding point in $\mathbb{P}^N(K)$. However, this way of measuring arithmetic complexity is also not entirely satisfactory, since twists of a rational map give the same point in $\mathcal{M}_d$, yet are arithmetically quite different. Note that the same situation arises in the theory of elliptic curves, where curves with the same $j$-invariant need not be arithmetically identical.

This suggests combining the primes of bad reduction with the height coming from moduli space. We do this and formulate a dynamical version of a conjecture of Lang (cf. [202], [254, page 92], or [410, VIII.9.9]). Recall that the canonical height of a point $P$ satisfies $\hat{h}_\phi(P) = 0$ if and only if $P$ is preperiodic for $\phi$ (Theorem 3.22). The following conjecture says that the height of nonpreperiodic points grows as $\phi$ becomes more arithmetically complicated.

**Conjecture 4.98.** *Fix an embedding of the moduli space $\mathcal{M}_d$ in projective space and let $h_{\mathcal{M}_d}$ denote the associated height function. Let $K$ be a number field and $d \geq 2$ an integer. Then there is a positive constant $c = c(K,d)$ such that for all rational maps $\phi \in K(z)$ of degree $d$ and all wandering (i.e., non-preperiodic) points $P \in \mathbb{P}^1(K)$,*

$$\hat{h}_\phi(P) \geq c \max\left\{\log \mathrm{N}_{K/\mathbb{Q}} \mathfrak{R}_\phi,\ h_{\mathcal{M}_d}(\langle\phi\rangle)\right\}.$$

In the theory of elliptic curves, the notion of global miminal Weierstrass equation is extremely useful; see the discussion in Section 6.3.5 and [410, VIII, §8] for further details. We briefly discuss a dynamical analogue.

**Definition.** Let $K$ be a number field and let $\phi(z) \in K(z)$ be a rational map. Then $\phi$ has a *global minimal model* if there is a linear fractional transformation $f \in \mathrm{PGL}_2(K)$ and homogeneous polynomials $F$ and $G$ satisfying $\phi^f = [F, G]$ with the property that the coefficients of $F$ and $G$ are in the ring of integers of $K$ and

$$\operatorname{ord}_{\mathfrak{p}}\bigl(\operatorname{Res}(F,G)\bigr) = \operatorname{ord}_{\mathfrak{p}}(\mathfrak{R}_\phi) \qquad \text{for every prime } \mathfrak{p}.$$

In other words, the pair $(F, G)$ is simultaneously a minimal model for $\phi$ at every prime $\mathfrak{p}$ of $K$.

In the remainder of this section we develop some tools that help to determine whether a given rational map has a global minimal model.

**Proposition 4.99.** *Let $K$ be a number field, let $\phi(z) \in K(z)$ be a rational map of degree d, and write $\phi = [F, G]$ with polynomials F and G as usual.*
(a) *There is a (fractional) ideal $\mathfrak{a}_{F,G}$ of K satisfying*

$$\mathfrak{R}_\phi = \begin{cases} \operatorname{Res}(F,G)\mathfrak{a}_{F,G}^{2d} & \text{if } d \text{ is odd,} \\ \operatorname{Res}(F,G)\mathfrak{a}_{F,G}^{d} & \text{if } d \text{ is even.} \end{cases} \tag{4.61}$$

(b) *If d is odd, then the ideal class of $\mathfrak{a}_{F,G}$ depends only on $\phi$, independent of the choice of F and G.*
(c) *If d is even, then $\mathfrak{a}_{F,G}$ depends only on $\phi$ up to multiplication by the square of a principal ideal.*

*Proof.* For any $f \in \mathrm{PGL}_2(K)$ with corresponding $A \in \mathrm{GL}_2(K)$, we use Proposition 4.95(a,b) to compute

$$\begin{aligned} \operatorname{ord}_{\mathfrak{p}}\bigl(\operatorname{Res}_{\mathfrak{p}}(\phi^f)\bigr) &= \operatorname{ord}_{\mathfrak{p}}(\operatorname{Res}(F_A, G_A)) - 2d\min\bigl\{\operatorname{ord}_{\mathfrak{p}}(F_A), \operatorname{ord}_{\mathfrak{p}}(G_A)\bigr\} \\ &= \operatorname{ord}_{\mathfrak{p}}(\operatorname{Res}(F,G)) + (d^2+d)\operatorname{ord}_{\mathfrak{p}}(\det A) \\ &\quad - 2d\min\bigl\{\operatorname{ord}_{\mathfrak{p}}(F_A), \operatorname{ord}_{\mathfrak{p}}(G_A)\bigr\}. \end{aligned} \tag{4.62}$$

For each prime ideal $\mathfrak{p}$ we choose a linear fractional transformation $f_{\mathfrak{p}} \in \mathrm{PGL}_2(K)$ and corresponding matrix $A_{\mathfrak{p}}$ so as to minimize the resultant of $\phi^{f_{\mathfrak{p}}}$. In other words,

$$\operatorname{ord}_{\mathfrak{p}}\bigl(\operatorname{Res}_{\mathfrak{p}}(\phi^{f_{\mathfrak{p}}})\bigr) = \operatorname{ord}_{\mathfrak{p}}(\mathfrak{R}_\phi). \tag{4.63}$$

Combining (4.62) and (4.63) yields

$$\begin{aligned} &\operatorname{ord}_{\mathfrak{p}}(\mathfrak{R}_\phi) - \operatorname{ord}_{\mathfrak{p}}\bigl(\operatorname{Res}(F,G)\bigr) \\ &\qquad = d\bigl[(d+1)\operatorname{ord}_{\mathfrak{p}}(\det A_{\mathfrak{p}}) - 2\min\bigl\{\operatorname{ord}_{\mathfrak{p}}(F_{A_{\mathfrak{p}}}), \operatorname{ord}_{\mathfrak{p}}(G_{A_{\mathfrak{p}}})\bigr\}\bigr]. \end{aligned} \tag{4.64}$$

Hence if we define an ideal $\mathfrak{a}_{F,G}$ by the rule

$$\operatorname{ord}_\mathfrak{p}(\mathfrak{a}_{F,G}) = \begin{cases} \frac{1}{2}(d+1)\operatorname{ord}_\mathfrak{p}(\det A_\mathfrak{p}) - \min\{\operatorname{ord}_\mathfrak{p}(F_{A_\mathfrak{p}}), \operatorname{ord}_\mathfrak{p}(G_{A_\mathfrak{p}})\}, & d \text{ odd,} \\ (d+1)\operatorname{ord}_\mathfrak{p}(\det A_\mathfrak{p}) - 2\min\{\operatorname{ord}_\mathfrak{p}(F_{A_\mathfrak{p}}), \operatorname{ord}_\mathfrak{p}(G_{A_\mathfrak{p}})\}, & d \text{ even,} \end{cases}$$

then (4.64) says that $\mathfrak{a}_{F,G}$ satisfies the desired formula (4.61). It remains to determine the extent to which the ideal $\mathfrak{a}_{F,G}$ depends on the choice of $F$ and $G$.

Let $\phi = [F, G] = [F', G']$ be two lifts of $\phi$. Then there is a constant $c \in K^*$ satisfying $F' = cF$ and $G' = cG$. Suppose first that $d$ is odd. We use (4.61) to compute

$$c^{2d}\operatorname{Res}(F,G)\mathfrak{a}_{cF,cG}^{2d} = \operatorname{Res}(cF,cG)\mathfrak{a}_{cF,cG}^{2d} = \mathfrak{R}_\phi = \operatorname{Res}(F,G)\mathfrak{a}_{F,G}^{2d}.$$

This is an equality of ideals, so by unique factorization of ideals we conclude that $c\mathfrak{a}_{cF,cG} = \mathfrak{a}_{F,G}$. Hence the ideals $\mathfrak{a}_{cF,cG}$ and $\mathfrak{a}_{F,G}$ differ by a principal ideal.

Next suppose that $d$ is even. Then by a similar calculation we find that

$$c^{2d}\operatorname{Res}(F,G)\mathfrak{a}_{cF,cG}^{d} = \operatorname{Res}(cF,cG)\mathfrak{a}_{cF,cG}^{d} = \mathfrak{R}_\phi = \operatorname{Res}(F,G)\mathfrak{a}_{F,G}^{d}.$$

Hence $(c^2)\mathfrak{a}_{cF,cG} = \mathfrak{a}_{F,G}$, so $\mathfrak{a}_{cF,cG}$ and $\mathfrak{a}_{F,G}$ differ by the square of a principal ideal. □

**Definition.** Let $K$ be a number field, let $\phi(z) \in K(z)$ be a rational map of degree $d$, and write $\phi = [F, G]$ with polynomials $F$ and $G$ as usual. If $d$ is odd, we write $\bar{\mathfrak{a}}_{\phi/K}$ for the ideal class of $\mathfrak{a}_{F,G}$ in the ideal class group of $K$. If $d$ is even, we write $\bar{\mathfrak{a}}_{\phi/K}$ for the image of $\mathfrak{a}_{F,G}$ in the group of fractional ideals modulo squares of principal ideals. In both cases, by analogy with the theory of minimal equations of elliptic curves (cf. [410, VIII §8]), we call $\bar{\mathfrak{a}}_{\phi/K}$ the *Weierstrass class of $\phi$ over the field $K$*. By Proposition 4.99 the Weierstrass class $\bar{\mathfrak{a}}_{\phi/K}$ depends only on $\phi$, independent of the chosen lift $[F, G]$.

The triviality of the Weierstrass class gives a necessary condition for the existence of a global minimal model.

**Proposition 4.100.** *Let $K$ be a number field and $\phi(z) \in K(z)$ a rational map of degree $d \geq 2$. If $\phi$ has a global minimal model over $K$, then its Weierstrass class $\bar{\mathfrak{a}}_{\phi/K}$ is trivial.*

*Proof.* Replacing $\phi$ by $\phi^f$ for an appropriate choice of $f \in \mathrm{PGL}_2(K)$, we may assume that $\phi = [F, G]$ with polynomials $F$ and $G$ having coefficients in the ring of integers of $K$ and satisfying

$$\operatorname{ord}_\mathfrak{p}\bigl(\operatorname{Res}(F,G)\bigr) = \operatorname{ord}_\mathfrak{p}(\mathfrak{R}_\phi) \qquad \text{for every prime } \mathfrak{p}.$$

It follows from the defining equation (4.61) of $\mathfrak{a}_{F,G}$ that

$$\operatorname{ord}_\mathfrak{p}(\mathfrak{a}_{F,G}) = 0 \qquad \text{for every prime } \mathfrak{p}.$$

Hence $\mathfrak{a}_{F,G} = (1)$, so its image in the ideal class group (if $d$ is odd) or in the group of ideals modulo squares of principal ideals (if $d$ is even) is also trivial. □

# Exercises

## Section 4.1. Dynatomic Polynomials

**4.1.** Let $\phi(z) = z + 1/z$, or in homogeneous form, $\phi([X,Y]) = [X^2 + Y^2, XY]$.

(a) Compute the first few dynatomic polynomials $\Phi^*_{\phi,n}(X,Y)$ for $\phi$, say for $n = 1,2,3,4$.

(b) Prove that for all $n \ge 2$, the dynatomic polynomial $\Phi^*_{\phi,n}$ satisfies $\Phi^*_{\phi,n}(\pm X, \pm Y) = \Phi^*_{\phi,n}(X,Y)$. Deduce that $\Phi^*_{\phi,n}(X,Y) \in \mathbb{Z}[X^2, Y^2]$.

(c) Prove that the field $\mathbb{Q}_{2,\phi}$ generated by the points of exact period 2 is the field $\mathbb{Q}(\sqrt{-2})$.

(d) Prove that the field $\mathbb{Q}_{3,\phi}$ generated by the points of exact period 3 is an $S_3$ extension of $\mathbb{Q}$. (*Hint.* Show that the roots of $\Phi_{\phi,3}(1,\sqrt{w},) \in \mathbb{Z}[w]$ generate a cyclic cubic extension of $\mathbb{Q}$.)

(e) Prove that $\Phi^*_{\phi,4}(X,Y)$ factors into a polynomial of degree 4 and a polynomial of degree 8. (See Exercise 4.40 for a more general result.) Describe the fields generated by the roots of each factor.

**4.2.** Prove the following elementary properties of the Möbius function. (See (4.2) on page 148 for the definition of the Möbius function.)

(a) Let $n \ge 1$ be an integer. Then

$$\sum_{d|n} \mu(d) = \begin{cases} 1 & \text{if } n = 1, \\ 0 & \text{if } n \ge 2. \end{cases}$$

(b) Let $g(n)$ be a function whose domain is the positive integers, and define a new function $f(n)$ by $f(n) = \sum_{d|n} g(d)$. Prove that

$$g(n) = \sum_{d|n} f(d)\mu\left(\frac{n}{d}\right).$$

This is called the *Möbius inversion formula.*

(c) Prove that the $n^{\text{th}}$ cyclotomic polynomial

$$\prod_{k|n} (z^k - 1)^{\mu(n/k)}$$

is indeed a polynomial. (*Hint.* Use (b) and the fact $z^k - 1$ has distinct complex roots.)

**4.3.** Let $\nu_d(n) = \deg(\Phi^*_{\phi,n})$ be the number of formal $n$-periodic points of a rational map of degree $d$, counted with multiplicity (see Remark 4.3).

(a) Prove that $\nu_d(1) = d + 1$ and

$$\nu_d(n) = \sum_{k|n} \mu\left(\frac{n}{k}\right) d^k \qquad \text{for } n \ge 2.$$

(b) Make a table of values of $\nu_d(n)$ for some small values of $n$ and $d$.

(c) Prove that

$$\sum_{n\ge1} \nu_d(n)\frac{x^n}{n} = x - \sum_{n\ge1} \frac{\mu(n)}{n} \log(1 - dx^n)$$

formally as power series. For what range of $x > 0$ do the series converge?

| Type of Point | Multiplier | Char $K$ | $a_P^*(n)$ values |
|---|---|---|---|
| Wandering | | | $a_P^*(n) = 0$ for all $n$ |
| Periodic | not root of unity | | $a_P^*(n) > 0$ for exactly one $n$ |
| Periodic | root of unity | 0 | $a_P^*(n) > 0$ for exactly two $n$'s |
| Periodic | root of unity | $p > 0$ | $a_P^*(n) > 0$ for $n = tp^k$, exactly two $t$'s and all $k \geq 0$ |

Table 4.4: Values of $a_P^*(n)$.

**4.4.** Let $\phi(z) \in K(z)$ be a rational function of degree $d \geq 2$ and suppose that $z = 0$ is a fixed point of $\phi$ with multiplier $\lambda = 1$. Write $\phi(z) = z + z^e\psi(z)$ with $e \geq 2$ and $\psi(0) \neq 0$, where $e = a_0(\phi, 1)$ in the notation of Theorem 4.5.

(a) Prove that

$$\phi^n(z) = z + nz^e\psi(z) + O(z^{2e-1}).$$

(b) Assume further that $K$ has characteristic $p > 0$. Prove that

$$a_0^*(\phi, p^k) \geq 2^{k-1}(e-1) \qquad \text{for all } k \geq 1.$$

**4.5.** Verify that the description of the values of $a_P^*(n)$ given in Table 4.4 is correct.

**4.6.** Let $\phi(z)$ be a rational function, let $n \geq 1$, let $p$ be a prime with $p \nmid n$, and let $P \in \mathbb{P}^1$.

(a) Prove that

$$a_P^*(n, \phi^p) = a_P^*(n, \phi) + a_P^*(np, \phi).$$

(b) Deduce that the $n^{\text{th}}$ dynatomic polynomial for $\phi^p$ factors as

$$\Phi_{n,\phi^p}^* = \Phi_{n,\phi}^*(z)\Phi_{np,\phi}^*.$$

(c) More generally, if $\gcd(n, k) = 1$, prove that

$$\Phi_{nk,\phi}^* = \prod_{j|k} (\Phi_{n,\phi^j}^*)^{\mu(k/j)}.$$

**4.7.** * Let

$$\phi(z) = z^d + a_{d-1}z^{d-1} + \cdots + a_2z^2 + a_1z + a_0$$

be a monic polynomial of degree $d$ and let $\Phi_{n,\phi}^*(z)$ be its $n^{\text{th}}$ dynatomic polynomial. Each root $\alpha$ of $\Phi_{n,\phi}^*(z)$ has an associated multiplier $\lambda_\phi(\alpha) = (\phi^n)'(\alpha)$.

(a) Prove that there is a (unique) monic polynomial

$$\delta_{n,\phi}(x) \in \mathbb{Z}[a_0, a_1, \ldots, a_{d-1}][x]$$

whose $n^{\text{th}}$ power satisfies

$$\delta_{n,\phi}(x)^n = \prod_{\Phi_{n,\phi}^*(\alpha)=0} \bigl(x - \lambda_\phi(\alpha)\bigr) = \operatorname{Res}_z\bigl(\Phi_{n,\phi}^*(z), x - (\phi^n)'(z)\bigr),$$

where $\operatorname{Res}_z$ means to take the resultant with respect to the $z$ variable. (*Hint.* All $n$ of the points in the orbit $\mathcal{O}_\phi(\alpha)$ have the same multiplier.)

(b) Let $C_k(x) \in \mathbb{Z}[x]$ denote the $k^{\text{th}}$ cyclotomic polynomial, and for integers $m|n$ with $m < n$, define

$$\Delta_{n,m} = \operatorname{Res}\bigl(C_{n/m}, \delta_{m,\phi}\bigr) \in \mathbb{Z}[a_0, a_1, \ldots, a_{d-1}].$$

Prove that the discriminant of $\Phi^*_{n,\phi}(x)$ is given by the formula

$$\operatorname{Disc} \Phi^*_{n,\phi} = \pm\delta_{n,\phi}(1)^n \prod_{\substack{m|n \\ m<n}} \Delta^{-m}_{n,m}.$$

(c) Let $m|n$ with $m < n$. Prove that the resultant of $\Phi^*_{n,\phi}$ and $\Phi^*_{m,\phi}$ is given by the formula

$$\operatorname{Res}(\Phi^*_{n,\phi}, \Phi^*_{m,\phi}) = \pm\Delta^m_{n,m}.$$

Conclude that $\phi$ has a point of formal period $n$ whose exact period $m$ is strictly smaller than $n$ if and only if there is a point of formal period $m$ whose multiplier is a primitive $(n/m)^{\text{th}}$ root of unity.

**4.8.** This exercise describes an analogue of Theorem 4.5 for automorphisms of projective space. Let $\phi : \mathbb{P}^N \to \mathbb{P}^N$ be an automorphism defined over a field $K$, i.e., $\phi \in \mathrm{PGL}_{N+1}(K)$. We say that $\phi$ is *nondegenerate* if the equation $\phi(P) = P$ has only finitely many solutions in $\mathbb{P}^N(\bar{K})$.

(a) Let $A \in \mathrm{GL}_{N+1}(K)$ be an invertible matrix with coefficients in $K$ representing the map $\phi$. Prove that $\phi$ is nondegenerate if and only if every eigenspace of $A$ has dimension 1.

(b) Assume that $\phi^n$ is nondegenerate, let $\Gamma_{\phi^n} \subset \mathbb{P}^N \times \mathbb{P}^N$ be the graph of $\phi^n$, and denote the diagonal map by $\Delta : \mathbb{P}^N \to \mathbb{P}^N \times \mathbb{P}^N$. Following Remark 4.4, we define the 0-cycle of $n$-periodic points of $\phi$ to be the pullback

$$\Phi_{\phi,n} = \Delta^*(\Gamma_\phi) = \sum_{P\in\mathbb{P}^N} a_P(\phi, n)(P)$$

and the 0-cycle of formal $n$-periodic points to be

$$\Phi^*_{\phi,n} = \sum_{k|n} \mu\left(\frac{n}{k}\right) \Phi_{\phi,n} = \sum_{P\in\mathbb{P}^N} a^*_P(\phi, n)(P).$$

Let $P \in \mathbb{P}^N$ be a point of primitive period $m$ for $\phi$. Prove that

$$a_P(n) = \begin{cases} a_P(m) & \text{if } m|n, \\ 0 & \text{if } m \nmid n, \end{cases} \qquad \text{and} \qquad a^*_P(n) = \begin{cases} a_P(m) & \text{if } m = n, \\ 0 & \text{if } m \neq n. \end{cases}$$

In particular, $\Phi^*_{\phi,n}$ is an effective (i.e., positive) cycle.

**4.9.** Let $F, G \in K[X, Y]$ be homogeneous polynomials of degree $d \geq 2$ with no common factors and let $\phi = [F, G] : \mathbb{P}^1 \to \mathbb{P}^1$ be the associated rational map. Define two sequences of polynomials inductively starting with $F_0(X, Y) = X$ and $G_0(X, Y) = Y$ and then for $n \geq 0$,

$$F_{n+1} = F_n\bigl(F(X,Y), G(X,Y)\bigr) \qquad \text{and} \qquad G_{n+1} = G_n\bigl(F(X,Y), G(X,Y)\bigr).$$

(a) Prove that $F_n$ and $G_n$ have no common factors.

(b) More precisely, prove that the resultant of $F_n$ and $G_n$ is given by

$$\operatorname{Res}(F_n, G_n) = \operatorname{Res}(F, G)^{(2n-1)d^{n-1}}.$$

(*Hint.* Use Exercise 2.12.)

(c) Prove that $\phi^n = [F_n, G_n]$.

(d) For all $n, m \ge 0$, prove that

$$F_{n+m}(X,Y) = F_n\big(F_m(X,Y), G_m(X,Y)\big),$$
$$G_{n+m}(X,Y) = G_n\big(F_m(X,Y), G_m(X,Y)\big).$$

**4.10.** With notation as in Exercise 4.9, we define the (*generalized*) $(m,n)$-*period polynomial of* $\phi$ to be the polynomial

$$\Phi_{m,n}(X,Y) = \Phi_n\big(F_m(X,Y), G_m(X,Y)\big),$$

where $\Phi_n(X,Y) = YF_n - XG_n$ is the usual $n$-period polynomial of $\phi$.

(a) Prove that

$$\Phi_{m,n}(X,Y) = G_m(X,Y)F_{n+m}(X,Y) - F_m(X,Y)G_{n+m}(X,Y).$$

(b) Let $P \in \mathbb{P}^1(\bar{K})$. Prove that

$$\Phi_{m,n}(P) = 0 \quad \text{if and only if} \quad \phi^{m+n}(P) = \phi^m(P).$$

Thus $P$ is a root of $\Phi_{m,n}$ if and only if $P$ is a preperiodic point with "tail" of length at most $m$ and with period dividing $n$.

(c) Prove that for all $m, n \ge 1$, the quotient

$$\frac{\Phi_{m,n}}{\Phi_{m-1,n}}$$

is a polynomial.

**4.11.** We continue with the notation from Exercises 4.9 and 4.10. Let $\Phi_n^*(X,Y)$ be the $n^{\text{th}}$ dynatomic polynomial of $\phi$. Then for $m, n \ge 1$ we define the (*generalized*) $(m,n)$-*dynatomic "polynomial" of* $\phi$ to be

$$\Phi_{m,n}^*(X,Y) = \frac{\Phi_n^*\big(F_m(X,Y), G_m(X,Y)\big)}{\Phi_n^*\big(F_{m-1}(X,Y), G_{m-1}(X,Y)\big)}.$$

(a) Prove that $\Phi_{m,n}^*(P) = 0$ if and only if $\phi^m(P)$ has formal period $n$. Points satisfying $\Phi_{m,n}^*(P) = 0$ are called *preperiodic points with formal preperiod* $(m,n)$.

(b) ** Prove (or disprove) that $\Phi_{m,n}^*(X,Y)$ is a polynomial.

## Section 4.2. Quadratic Polynomials and Dynatomic Modular Curves

**4.12.** We continue with the notation from Exercise 4.7. Thus for any monic polynomial $\phi(z)$, we define a monic polynomial $\delta_{n,\phi}(x)$ by

$$\delta_{n,\phi}(x)^n = \prod_{\Phi_{n,\phi}(\alpha)=0} \big(x - \lambda_\phi(\alpha)\big) = \prod_{\Phi_{n,\phi}(\alpha)=0} \big(x - (\phi^n)'(\alpha)\big), \tag{4.65}$$

and for $m|n$ with $m < n$ we set

$$\Delta_{n,m} = \operatorname{Res}\big(C_{n/m}(x), \delta_{m,\phi}(x)\big),$$

where $C_k(x)$ is the $k^{\text{th}}$ cyclotomic polynomial. We now specialize to $\phi_c(z) = z^2 + c$ and write $\delta_n(c,x) = \delta_{n,\phi_c}(x)$ and $\Delta_{n,m}(c)$ to indicate the dependence on $c$.

(a) Prove that $\delta_n(c,x) \in \mathbb{Z}[c,x]$ and that $\Delta_{n,m}(c) \in \mathbb{Z}[c]$. Then use Exercise 4.7(c) to explain the powers that appear in Table 4.3.

(b) Prove that

$$\tilde{\delta}_n(c,x) = 2^{-\deg_z \Phi_n^*}\delta_n(c, 2^n x) \quad \text{is in} \quad \mathbb{Z}[c,x].$$

(This eliminates many powers of 2 from the coefficients of $\delta_n(c,x)$.) Compute $\tilde{\delta}_n(c,x)$ for $n = 1,2,3,4$.

(c) ** Prove that there is a *monic* polynomial $\Psi_{n,m}(x) \in \mathbb{Z}[x]$ such that

$$\Delta_{n,m}(c) = \Psi_{n,m}(4c),$$

and deduce that the set

$$\left\{c \in \bar{\mathbb{Q}} : \Delta_{n,m}(c) = 0 \text{ for some } m < n \text{ with } m|n\right\}$$

is a set of bounded height. Conclude that the set of bifurcation points in the Mandelbrot set $\mathcal{M}$ is a set of bounded height. (*Hint.* The formula $\Delta_{n,m}(c) = \Psi_{n,m}(4c)$ bounds the denominator of $c$, then mimic the proof of Proposition 4.22.)

(d) Prove that $\Delta_{n,1}(c)$ and $\Delta_{2n,2}(c)$ are irreducible in $\mathbb{Q}[c]$ for all $n \geq 1$.

(e) ** Prove the following conjecture of Morton–Vivaldi [314]: The polynomial $\Delta_{n,m}(c)$ is irreducible in $\mathbb{Q}[c]$ for all $m|n$ with $m < n$.

**4.13.** Let $\phi_c(z) = z^2 + c$ and write $\Phi_n^*(c,z)$ for the $n^{\text{th}}$ dynatomic polynomial of $\phi_c$. Continuing with the notation from Exercise 4.12, we let $\delta_n(c,x) \in \mathbb{Z}[c,x]$ be the polynomial defined by (4.65) and $\tilde{\delta}_n(c,x) = 2^{-\deg \Phi_n^*}\delta_n(c,2^n x)$ the normalized polynomial in Exercise 4.12(b).

(a) Recall that the dynatomic curve $Y_1(n) \subset \mathbb{A}^2$ attached to the polynomial $\phi_c(z)$ is defined by the equation $\Phi_n^*(y,z) = 0$. Let $Z(n) \subset \mathbb{A}^2$ be similarly defined by the equation

$$Z(n) : \tilde{\delta}_n(y,x) = 0.$$

Prove that there is a morphism

$$F : Y_1(n) \longrightarrow Z(n), \qquad (y,z) \longmapsto \left(y, 2^{-n}(\phi_y^n)'(z)\right).$$

(b) Define an action of $i \in \mathbb{Z}/n\mathbb{Z}$ on $Y_1(n)$ by

$$(y,z) = \left(y, \phi_y^i(z)\right).$$

(See Section 4.2.3.) Prove that the map $F$ in (b) is invariant for this action, i.e., prove that

$$F\left(y, \phi_y^i(z)\right) = F(y,z) \quad \text{for all } i \in \mathbb{Z}/n\mathbb{Z}.$$

Deduce that there is a unique morphism $Y_0(n) \to Z(n)$ such that the composition

$$Y_1(n) \to Y_0(n) \to Z(n)$$

is the map $F$.

(c) Prove that the map $Y_0(n) \to Z(n)$ in (b) has degree 1, so the equation $\tilde{\delta}_n(y,x) = 0$ gives a (possibly singular) model for $Y_0(n)$.

(d) Prove that $Z(1)$ and $Z(2)$ are nonsingular and that $Z(3)$ and $Z(4)$ are singular. Resolve the singularities of $Z(3)$ and $Z(4)$ and check directly that $Y_0(3)$ and $Y_0(4)$ are curves of genus 0.

**4.14.** Let $\phi_c(z) = z^2 + c$ and let $\Phi^*_{m,n}$ be the generalized dynatomic polynomial defined in Exercise 4.11.

(a) Use a computer algebra system to compute the first few generalized dynatomic polynomials $\Phi^*_{m,n}$ for $\phi_c$, say for $(m,n)$ in the list

$$\{(1,1),(2,1),(3,1),(4,1),(1,2),(2,2),(3,2),(4,2),(1,3),(2,3),(1,4)\}.$$

(b) From your list, it should be apparent that many of the leading terms of $\Phi^*_{m,1}$ and $\Phi^*_{m,2}$ coincide. Prove that

$$\deg(\Phi^*_{m,2} - \Phi^*_{m,1}) = \frac{1}{2}\deg(\Phi^*_{m,2}) = \frac{1}{2}\deg(\Phi^*_{m,1}) \qquad \text{for all } m \geq 1.$$

(Note that this is a special property of the generalized dynatomic polynomials for $\phi_c$.)

**4.15.** Let $c \in \mathbb{C}$ and let $\phi_c(z) = z^2 + c$. Prove that there is a number $R_c > 0$ and a holomorphic function

$$\psi_c : \{z \in \mathbb{C} : |z| > R_c\} \longrightarrow \mathbb{C}$$

satisfying

$$\psi_c(\phi_c(z)) = \psi_c(z)^2 \qquad \text{and} \qquad \log|\psi_c(z)| \sim \log|z| \quad \text{as } |z| \to \infty.$$

(*Hint.* Show that there is a consistent way to choose square roots so that $\psi_c$ can be defined as $\lim_{n\to\infty} \sqrt[2^n]{\phi_c^n(z)}$ .)

**4.16.** Consider the doubling map

$$D : \mathbb{Q}/\mathbb{Z} \longrightarrow \mathbb{Q}/\mathbb{Z}, \qquad D(t) = 2t \bmod \mathbb{Z}.$$

Fix $t \in \mathbb{Q}/\mathbb{Z}$ and let $m \geq 0$ and $n \geq 1$ be the smallest integers such that the denominator of $t$ divides $2^m(2^n - 1)$.

(a) Prove that $t$ is periodic for $D$ if $m = 0$ and that $t$ is strictly preperiodic for $D$ if $m \geq 1$.

(b) Prove that $m \geq 0$ and $n \geq 1$ are the smallest positive integers such that $t$ can be written as a (not necessarily reduced) fraction of the form

$$t = \frac{a}{2^m(2^n - 1)}.$$

(c) If $t$ is preperiodic for $D$, prove that the preperiod of $t$ is equal to $m$ and that $D^m(t)$ is periodic with exact period $n$.

(d) For $t$ as above, we say that $t$ is of type $(m,n)$ for the doubling map $D$. For a given pair of positive integers $(m,n)$, how many $t \in \mathbb{Q}/\mathbb{Z}$ are of type $(m,n)$?

The map $D$ is used in the analytic characterization of Misiurewicz points; see Theorem 4.25. However, note that (d) does not count the exact number of Misiurewicz points of type $(m,n)$, because distinct $t$ may give the same Misiurewicz point.

**4.17.** ** Let $\phi_c(z) = z^2 + c$ as usual, and for integers $m,n \geq 1$, define a rational function (of $c$) by

$$F_{m,n}(c) = \prod_{k\mid n}\left(\frac{\phi_c^{m+k}(0) - \phi_c^m(0)}{\phi_c^{m-1+k}(0) - \phi_c^{m-1}(0)}\right)^{\mu(n/k)} = \prod_{k\mid n}\left(\phi_c^{m-1+k}(0) + \phi_c^{m-1}(0)\right)^{\mu(n/k)}.$$

Then set

$$G_{m,n}(c) = \begin{cases} F_{m,n}(c) & \text{if } m \not\equiv 1 \pmod{n} \text{ or } m = 1, \\ F_{m,n}(c)/F_{1,n}(c) & \text{if } m \equiv 1 \pmod{n} \text{ and } m \neq 1. \end{cases}$$

(a) Prove that $G_{m,n}(c)$ is a polynomial in $c$. More precisely, prove that $G_{m,n}(c)$ is in $\mathbb{Z}[c]$.
(b) If $n \ge 2$, prove that the roots of $G_{m,n}$ are the Misiurewicz points of type $(m,n)$.
(c) Prove that $G_{m,n}(c)$ is irreducible in $\mathbb{Q}[c]$.

**4.18.** Let $V$ be a variety and suppose that the points of $V$ algebraically parameterize a family of quadratic polynomials $\psi$ together with a marked point $\lambda$ of formal period $n$. Theorem 4.11 says that there is a unique morphism $\eta : V \to Y_1(n)$ satisfying

$$\eta(P) = \big(\psi_P(z), \lambda(P)\big) \in \text{Formal}(n).$$

Note that by construction, each point $\gamma = (c,\alpha) \in Y_1(n)$ is identified with a quadratic polynomial $\phi_\gamma$ and point $\mu(\gamma)$ of formal period $n$ via

$$\phi_\gamma(z) = z^2 + c \quad \text{and} \quad \mu(\gamma) = \alpha.$$

Prove that there is a unique morphism of varieties

$$f : V \longrightarrow \text{PGL}_2$$

such that the following identities are true for all $P \in V(\bar{K})$:
(a) $(f_P^{-1} \circ \psi_P \circ f_P)(z) = \phi_{\eta(P)}(z)$.
(b) $(\eta \circ f^{-1} \circ \lambda)(P) = (\mu \circ \eta)(P)$.
In other words, prove that the following two diagrams commute:

$$\begin{array}{ccc} \mathbb{P}^1_V & \xrightarrow{\psi^f} & \mathbb{P}^1_V \\ \downarrow{\scriptstyle\eta} & & \downarrow{\scriptstyle\eta} \\ \mathbb{P}^1_{Y_1(n)} & \xrightarrow{\phi} & \mathbb{P}^1_{Y_1(n)} \end{array} \qquad\qquad \begin{array}{ccc} V & \xrightarrow{f^{-1}\circ\lambda} & \mathbb{P}^1_V \\ \downarrow{\scriptstyle\eta} & & \downarrow{\scriptstyle\eta} \\ Y_1(n) & \xrightarrow{\mu} & \mathbb{P}^1_{Y_1(n)} \end{array}$$

(*Hint*. During the proof of Theorem 4.11 we defined a linear fractional transformation $f_P(z)$. Prove that $f_P$ is uniquely determined by $P$. Deduce that the map $P \mapsto f_P$ is a morphism.)

**4.19.** In Example 4.14 we gave an explicit description of the quotient map $Y_1(2) \to Y_0(2)$. Perform a similar analysis and describe the map $Y_1(3) \to Y_0(3)$.

**4.20.** We proved that the dynatomic modular curves $X_1(1)$, $X_1(2)$, and $X_1(3)$ are isomorphic to $\mathbb{P}^1$. This exercise asks you to investigate $X_1(n)$ for other small values of $n$.
(a) Prove that $X_1(4)$ is a curve of genus 2 and $X_0(4)$ is a curve of genus 1.
(b) Prove that $X_0(4)(\mathbb{Q})$ is finite, and $Y_0(4)(\mathbb{Q})$ is the empty set.
(c) Prove that $X_1(5)$ has genus 14 and $X_0(5)$ has genus 2.
(d) * Prove that $X_0(5)(\mathbb{Q})$ is finite, and $Y_0(5)(\mathbb{Q})$ is the empty set.
(e) Compute the genera of $X_1(6)$ and $X_0(6)$.
(f) ** Find all rational points in $Y_0(6)(\mathbb{Q})$.

## Section 4.3. The Space $\text{Rat}_d$ of Rational Functions

**4.21.** Let $\phi = [F_{\mathfrak{a}}, F_{\mathfrak{b}}] \in \text{Rat}_d$ be a rational map of degree $d$. Prove that the following are equivalent:
(a) At least one of the fixed points of $\phi$ has multiplier equal to 1.
(b) The polynomial $YF_{\mathfrak{b}} - XF_{\mathfrak{a}}$ has a multiple root.

Deduce that there is a nonzero homogeneous polynomial $D(\mathbf{a},\mathbf{b}) \in \mathbb{Q}[a_0,\ldots,b_d]$ such that

$$\begin{pmatrix} \phi \text{ has a fixed point whose} \\ \text{multiplier equals } 1 \end{pmatrix} \iff D(\mathbf{a},\mathbf{b}) = 0.$$

Hence

$$\bigl\{\phi \in \mathrm{Rat}_d : \phi \text{ has a fixed point whose multiplier equals } 1\bigr\}$$

is a proper Zariski closed subset of $\mathrm{Rat}_d$.

## Section 4.4. The Moduli Space $\mathcal{M}_d$ of Dynamical Systems

**4.22.** Let $\mathrm{GL}_n$ be the group of $n \times n$ matrices with nonzero determinant, let $\mathrm{SL}_n$ be its subgroup of matrices with determinant 1, let $\mathrm{PGL}_n$ be the quotient of $\mathrm{GL}_n$ by its subgroup of diagonal matrices, and similarly let $\mathrm{PSL}_n$ be the quotient of $\mathrm{SL}_n$ by its subgroup of diagonal matrices.

(a) Let $\bar{K}$ be an algebraically closed field. Prove that the natural map from $\mathrm{PSL}_n(\bar{K})$ to $\mathrm{PGL}_n(\bar{K})$ is an isomorphism.

(b) More generally, prove that for any field $K$ there is an exact sequence

$$1 \longrightarrow \mathrm{PSL}_n(K) \longrightarrow \mathrm{PGL}_n(K) \xrightarrow{\ \det\ } K^*/K^{*n} \longrightarrow 1.$$

**4.23.** We say that a separable rational map $\phi(z) \in K(z)$ is *very highly ramified* if there is a point $P \in \mathbb{P}^1$ such that the ramification index of $\phi$ satisfies $e_P(\phi) \ge 3$. Let

$$V = \bigl\{\phi \in \mathrm{Rat}_d : \phi \text{ is very highly ramified}\bigr\}.$$

(a) Prove that $V$ is a proper Zariski closed subset of $\mathrm{Rat}_d$. Hence "most" rational maps are not very highly ramified. (One says that a *generic map of degree d* is not very highly ramified.)

(b) Prove that $V$ is invariant under the conjugation action of $\mathrm{PGL}_2$.

(c) Prove that the quotient $V/\mathrm{PSL}_2$ is a Zariski closed subset of $\mathcal{M}_d = \mathrm{Rat}_d/\mathrm{PSL}_2$.

**4.24.** Let $\phi(z) \in K(z)$ be a nonconstant rational function. The *Schwarzian derivative* $\mathcal{S}\phi$ of $\phi$ is the function

$$(\mathcal{S}\phi)(z) = \frac{\phi'''(z)}{\phi'(z)} - \frac{3}{2}\left(\frac{\phi''(z)}{\phi'(z)}\right)^2.$$

It measures the difference between $\phi$ and the best approximation to $\phi$ by linear fractional transformations.

(a) If $f(z) = (az+b)/(cz+d)$ is a linear fractional transformation, prove that $(\mathcal{S}f)(z) = 0$.

(b) Let $f$ be a linear fractional transformation. Prove that $\phi$ and $f \circ \phi$ have the same Schwarzian derivative.

(c) Let $\phi(z), \psi(z) \in K(z)$ be nonconstant rational functions. Prove that

$$\bigl(\mathcal{S}(\phi\circ\psi)\bigr)(z) = (\mathcal{S}\phi)\bigl(\psi(z)\bigr)\cdot\bigl(\psi'(z)\bigr)^2 + (\mathcal{S}\psi)(z).$$

(d) In particular, if $f$ is a linear fractional transformation, prove that

$$(\mathcal{S}\phi^f)(z) = f'(z)^2 \cdot (\mathcal{S}\phi)\bigl(f(z)\bigr),$$

and deduce that the quadratic differential form

$$\omega_\phi(z) = (\mathcal{S}\phi)(z)\,(dz)^2$$

is invariant under the substitution $(\phi, z) \mapsto (\phi^f, f^{-1}z)$. Thus the map $\phi \mapsto \omega_\phi$ induces a natural map

$$\mathcal{M}_d \longrightarrow (\text{quadratic differential forms on } \mathbb{P}^1).$$

(e) Suppose that $\phi(z)$ has a multiple zero or pole at $z = \alpha$, say

$$\phi(z) = a(z-\alpha)^m + \cdots \quad \text{with } a \neq 0 \text{ and } |m| \geq 2.$$

Prove that $\mathcal{S}\phi$ has a double pole at $\alpha$. More precisely, prove that the Laurent series expansion of $\mathcal{S}\phi$ around $\alpha$ looks like

$$(\mathcal{S}\phi)(z) = \frac{1-m^2}{2}(z-\alpha)^{-2} + \cdots.$$

(f) Prove that the map

$$\operatorname{Rat}_d \times \mathbb{P}^1 \longrightarrow \mathbb{P}^1, \qquad (\phi, P) \longmapsto (\mathcal{S}\phi)(P),$$

is a morphism.

**4.25.** ** An algebraic variety $V$ is called *unirational* if there is a rational map $\mathbb{P}^N \to V$ whose image is Zariski dense in $V$, and the variety $V$ is called *rational* if there is a rational map $\mathbb{P}^N \to V$ that is an isomorphism from an open subset of $\mathbb{P}^N$ to an open subset of $V$. It is clear that every moduli space $\mathcal{M}_d$ is unirational, since $\operatorname{Rat}_d$ is an open subset of $\mathbb{P}^{2d+1}$ and the map $\operatorname{Rat}_d \to \mathcal{M}_d$ is surjective. We also know that $\mathcal{M}_2$ is rational, since Theorem 4.56 says that $\mathcal{M}_2 \cong \mathbb{A}^2$. For which values of $d$ is $\mathcal{M}_d$ a rational variety? In particular, is $\mathcal{M}_3$ rational?

### Section 4.5. Periodic Points, Multipliers, and the Multiplier Spectrum

**4.26.** This exercise asks you to prove the part of Theorem 4.50 that was left undone in the text. Prove that the map

$$\operatorname{Rat}_d \longrightarrow \mathbb{A}^1, \qquad \phi \longmapsto \overset{*}{\sigma}{}_i^{(n)}(\phi),$$

defines a function in $\mathbb{Q}[\operatorname{Rat}_d]$. Prove that it is $\operatorname{PGL}_2$-invariant and deduce that it defines a function in $\mathbb{Q}[\mathcal{M}_d]$.

**4.27.** ** What is the degree of the map

$$\sigma_{3,N} : \mathcal{M}_3 \longrightarrow \mathbb{A}^k$$

for sufficiently large $N$? Similarly, what is the degree of $\sigma_{4,N}$ on $\mathcal{M}_4$ away from the Lattès locus on which it is constant?

### Section 4.6. The Moduli Space $\mathcal{M}_2$ of Dynamical Systems of Degree 2

**4.28.** Prove that $\phi \in \operatorname{Rat}_2$ is conjugate to a polynomial if and only $\sigma_1(\phi) = 2$. Thus the polynomial maps in $\mathcal{M}_2 \cong \mathbb{A}^2$ trace out the line $x = 2$.

**4.29.** Let $\phi \in \operatorname{Rat}_2(\mathbb{C})$ be a rational map of degree 2 and suppose that one of its fixed points has a nonzero multiplier $\lambda$.

(a) Prove that there are an $f \in \mathrm{PGL}_2(\mathbb{C})$ and a $c \in \mathbb{C}$ such that

$$\phi^f(z) = \frac{1}{\lambda}\left(z + c + \frac{1}{z}\right). \tag{4.66}$$

This generalizes the normal form given in Lemma 4.59.

(b) Verify that the multiplier of $\phi^f$ at the fixed point $z = \infty$ is $\lambda_\infty(\phi^f) = \lambda$.

(c) Let $\lambda_1, \lambda_2, \lambda_3$ be the multipliers of the fixed points of $\phi$ with $\lambda_1 = \lambda$. Prove that the number $c$ in (4.66) satisfies $c^2 = 4 - \lambda_1(2 + \lambda_2 + \lambda_3)$.

**4.30.** Fix $d \geq 2$ and consider the subset of $\mathrm{Rat}_d$ defined by

$$\mathrm{BiCrit}_d = \{\phi \in \mathrm{Rat}_d : \phi \text{ has exactly two critical points}\}.$$

For $d = 2$ we have $\mathrm{BiCrit}_2 = \mathrm{Rat}_2$, but for larger $d$ the set $\mathrm{BiCrit}_d$ is a proper subset of $\mathrm{Rat}_d$.

(a) Prove that $\mathrm{BiCrit}_d$ is an algebraic variety of dimension 5.

(b) Prove that conjugation induces a natural action of $\mathrm{PGL}_2$ on $\mathrm{BiCrit}_d$, i.e., there is a morphism of varieties

$$\mathrm{PGL}_2 \times \mathrm{BiCrit}_d \longrightarrow \mathrm{BiCrit}_d, \qquad (f, \phi) \longmapsto \phi^f.$$

(c) Suppose that $\phi \in \mathrm{BiCrit}_d$ has critical points at 0 and $\infty$. Prove that $\phi(z)$ has the form

$$\phi(z) = \frac{\alpha z^d + \beta}{\gamma z^d + \delta}. \tag{4.67}$$

(d) Let $\phi \in \mathrm{BiCrit}_d$ and apply a conjugation to move the critical points of $\phi$ to 0 and $\infty$, so $\phi$ has the form (4.67) described in (c). Prove that the following quantities depend only on the conjugacy class of $\phi$:

$$\omega(\phi) = \frac{\beta\gamma}{\alpha\delta - \beta\gamma}, \qquad \sigma(\phi) = \frac{\alpha^{d+1}\beta^{d-1}}{(\alpha\delta - \beta\gamma)^d}, \qquad \tau(\phi) = \frac{\gamma^{d-1}\delta^{d+1}}{(\alpha\delta - \beta\gamma)^d}. \tag{4.68}$$

(e) Let $\phi, \psi \in \mathrm{BiCrit}_d$ and suppose that

$$\omega(\phi) = \omega(\psi), \qquad \sigma(\phi) = \sigma(\psi), \qquad \tau(\phi) = \tau(\psi).$$

Prove that $\phi$ and $\psi$ are $\mathrm{PGL}_2$ conjugate (working over an algebraically closed field).

(f) Prove that the three quantities (4.68) described in (d) satisfy the relation

$$\sigma\tau = \omega^{d-1}(\omega + 1)^{d+1}$$

and no other relations. Conclude that $\mathcal{M}_d^{\mathrm{BiCrit}}$ is isomorphic to $\mathbb{A}^2$. (This generalizes Theorem 4.56, since $\mathcal{M}_2^{\mathrm{BiCrit}} = \mathcal{M}_2$.)

(g) ** Describe the stable and semistable completions of $\mathcal{M}_d^{\mathrm{BiCrit}}$ coming from geometric invariant theory.

## Section 4.7. Automorphisms and Twists

**4.31.** In Section 4.7 we defined two rational maps $\phi$ and $\psi$ to be equivalent if there is an $f \in \mathrm{PGL}_2(\bar{K})$ such that $\psi = \phi^f$, and similarly to be $K$-equivalent if there is an $f \in \mathrm{PGL}_2(K)$ such that $\psi = \phi^f$.

(a) Prove that these definitions do indeed define equivalence relations on the set of rational maps.
(b) More generally, suppose that a group $G$ acts on a set $X$. Define a relation on $X$ by setting $y \sim x$ if there exists a $\sigma \in G$ such that $y = \sigma(x)$. Prove that this is an equivalence relation.

**4.32.** Let $\phi(z) \in \bar{K}(z)$ be a rational map.
(a) Prove that $\operatorname{Aut}(\phi)$ is a subgroup of $\operatorname{PGL}_2(\bar{K})$.
(b) Let $h \in \operatorname{PGL}_2(\bar{K})$. Prove that $\operatorname{Aut}(\phi^h) = h^{-1}\operatorname{Aut}(\phi)h$, so $\operatorname{Aut}(\phi)$ and $\operatorname{Aut}(\phi^h)$ are conjugate subgroups of $\operatorname{PGL}_2(\bar{K})$.

**4.33.** Let $\phi(z) \in \bar{K}(z)$, let $f \in \operatorname{Aut}(\phi)$, and let $\alpha$ be a critical point of $\phi$ (i.e., $\phi'(\alpha) = 0$). Prove that $f(\alpha)$ is also a critical point of $\phi$. More generally, prove that $\alpha$ and $f(\alpha)$ have the same ramification index.

**4.34.** Describe all polynomials $\phi(z) \in \bar{K}[z]$ whose automorphism group $\operatorname{Aut}(\phi)$ is nontrivial. (*Hint.* If $f \in \operatorname{Aut}(\phi)$, what does $f$ do to the totally ramified fixed point(s) of $\phi$?)

**4.35.** Let $\phi(z) \in K(z)$ be a rational map of degree $d$ and write $\phi(z)$ as a quotient of polynomials

$$\phi(z) = \frac{a_0 + a_1 z + \cdots + a_d z^d}{b_0 + b_1 z + \cdots + b_d z^d}.$$

Prove that $f(z) = z^{-1}$ is in $\operatorname{Aut}(\phi)$ if and only if $b_i = a_{d-i}$ for all $i$.

**4.36.** Let $\phi(z) = (z^2 - 2z)/(-2z + 1)$.
(a) Prove that $\operatorname{Aut}(\phi)$ contains the maps

$$\left\{ z, \frac{1}{z}, \frac{z-1}{z}, \frac{1}{1-z}, \frac{z}{z-1}, 1 - z \right\},$$

and that they form a group isomorphic to $\mathcal{S}_3$. Prove that $\#\operatorname{Aut}(\phi) = 6$, so in fact $\operatorname{Aut}(\phi) \cong \mathcal{S}_3$. (*Hint.* Find the fixed points of $\phi$.)
(b) Compute the values of $\sigma_1(\phi)$, $\sigma_2(\phi)$, and $\sigma_3(\phi)$.

**4.37.** Let $\zeta \in K$ be a primitive $n^{\text{th}}$ root of unity and let $\phi(z) \in K(z)$ be a nonconstant rational map.
(a) Prove that $\operatorname{Aut}(\phi)$ contains the map $f(z) = \zeta z$ if and only if there is a rational map $\psi(z) \in K(z)$ such that $\phi(z) = z\psi(z^n)$.
(b) Prove that $\operatorname{Aut}(\phi)$ contains both of the maps $f(z) = \zeta z$ and $g(z) = 1/z$ if and only if there is a polynomial $F(z) \in K[z]$ such that the rational map $\psi(z) \in K(z)$ in (a) has the form $\psi(z) = F(z)/(z^d F(z^{-1}))$, where $d = \deg(F)$. Verify that the group generated by $f$ and $g$ is the dihedral group of order $2n$.
(c) Let $\phi(z) = z^d$. Prove that $\operatorname{Aut}(\phi)$ is a dihedral group of order $2d - 2$.

**4.38.** We identify the set of rational functions $\operatorname{Rat}_d$ of degree $d$ with an open subset of $\mathbb{P}^{2d+1}$ as described in Section 4.3.
(a) Let $f \in \operatorname{PGL}_2$. Prove that the set $\{\phi \in \operatorname{Rat}_d : \phi^f = \phi\}$ is a (possibly empty) Zariski closed subset of $\operatorname{Rat}_d$. If $f(z) \neq z$, prove that it is a proper subset of $\operatorname{Rat}_d$.
(b) Let $A \subset \operatorname{PGL}_2(\bar{K})$ be a nontrivial finite subgroup. Prove that the set

$$\{\phi \in \operatorname{Rat}_d : \phi^f = \phi \text{ for all } f \in A\}$$

is a proper Zariski closed subset of $\operatorname{Rat}_d$.

(c) Prove that up to conjugation, $\mathrm{PGL}_2(\bar{K})$ has only finitely many distinct finite subgroups of any given order.

(d) Prove that the set $\{\phi \in \mathrm{Rat}_d : \mathrm{Aut}(\phi) \neq 1\}$ is a proper Zariski closed subset of $\mathrm{Rat}_d$. (*Hint.* Note that the order of $\mathrm{Aut}(\phi)$ is bounded by Proposition 4.65.)

(e) Prove that the set $\{\phi \in \mathrm{Rat}_d : \mathrm{Aut}(\phi) \neq 1\}$ is $\mathrm{PGL}_2$-invariant and defines a proper Zariski closed subset of $\mathcal{M}_d$. (*Hint.* The groups $\mathrm{Aut}(\phi)$ and $\mathrm{Aut}(\phi^f)$ are conjugate subgroups of $\mathrm{PGL}_2$; see Remark 4.64 and Exercise 4.32.)

**4.39.** Let $\alpha \in \mathbb{C}^*$ and $d \geq 1$ and consider the rational function

$$\phi(z) = \alpha\left(\frac{z+1}{z-1}\right)^d.$$

Prove that $\mathrm{Aut}(\phi) = \{f \in \mathrm{PGL}_2(\mathbb{C}) : \phi^f = \phi\}$ is trivial.

**4.40.** Let $\phi(z) \in K(z)$ be a rational map of degree $d \geq 2$ and assume that $\mathrm{Aut}(\phi)$ contains the element $h(z) = -z$ of order 2.

(a) Suppose that $d$ is even. Prove that at least one of the fixed points of $\phi(z)$ is defined over $K$.

(b) Suppose further that $d = 2$. Prove that there is an $f \in \mathrm{PGL}_2(K)$ such that $h^f = h$ and such that $\phi^f$ has the form

$$\phi^f(z) = az + \frac{b}{z}.$$

(c) Write $\phi_{a,b}(z) = az + bz^{-1}$. Prove that for all $c \in K^*$, the maps $\phi_{a,bc^2}$ and $\phi_{a,b}$ are $\mathrm{PGL}_2(K)$-equivalent.

(d) In homogeneous form we have $\phi_{a,b}([X,Y]) = [aX^2 + bY^2, XY]$. Let $\Phi_n^*(a,b;X,Y)$ be the associated dynatomic polynomial. Prove that $\Phi_n^*(a,b;X,Y) \in \mathbb{Z}[a,b,X,Y]$.

(e) * If $n \geq 4$ is even, prove that $\Phi_n^*(a,b;X,Y)$ is reducible. More precisely, prove that there are nonconstant polynomials $\Psi_n, \Lambda_n \in \mathbb{Z}[a,b,X,Y]$ such that

$$\Phi_n^*(a,b;X,Y) = \Psi_n(a,b;X,Y)\Lambda_n(a,b;X,Y).$$

(*Hint.* Divide the set of points $P$ of formal period $n$ into two subsets depending on whether the involution $h$ permutes the orbit $\mathcal{O}_\phi(P)$ or sends it to a different orbit.) This result is due to Manes [286].

**4.41.** Let $\phi = az + bz^{-1} \in \mathrm{Rat}_2$.

(a) Prove that $(\sigma_1(\phi), \sigma_2(\phi))$ depends only on $a$.

(b) We saw in Exercise 4.40 that $\mathrm{Aut}(\phi)$ contains $\{\pm z\}$, so it has order at least 2. Find all values of $a$ for which $\mathrm{Aut}(\phi)$ is strictly larger than $\{\pm z\}$.

(c) As $a$ varies, prove that the set of points

$$(\sigma_1(\phi), \sigma_2(\phi))$$

describes the curve

$$C : 2x^3 - 8xy + x^2y - 4y^2 - x^2 + 12x + 12y - 36 = 0$$

in $\mathbb{A}^2 \cong \mathcal{M}_2$.

(d) Prove that the curve $C$ in (c) is singular at the point $(-6, 12)$.

(e) Move the singular point to the origin and perform a further change of variables to prove that the curve $C$ is isomorphic to the curve

$$y^2 = 4x^3 + x^4.$$

Thus the singularity of $C$ is a cubic cusp.

(f) The point $(-6, 12)$ is the unique singular point of the curve $C$, so the rational map $\phi$ satisfying $\bigl(\sigma_1(\phi), \sigma_2(\phi)\bigr) = (-6, 12)$ should be special in some way. How is it special? (*Hint.* Compare with the answer to (a). See also Exercise 4.36.)

## Section 4.8. General Theory of Twists

**4.42.** Let $X$ be an object defined over $K$, and for each twist $Y$ of $X$, fix a $\bar{K}$-isomorphism $i_Y : Y \to X$. Assume that $\operatorname{Aut}(X)$ is abelian. Prove that

$$\operatorname{Twist}(X/K) \longrightarrow H^1\bigl(\operatorname{Gal}(\bar{K}/K), \operatorname{Aut}(X)\bigr), \qquad [Y]_K \longmapsto \bigl(\sigma \mapsto i_Y \sigma(i_Y^{-1})\bigr),$$

is a well-defined one-to-one map of sets. (See Remark 4.78.)

## Section 4.10. Fields of Definition and the Field of Moduli

**4.43.** Let $\phi \in \bar{K}(z)$ be a rational function. We know from Proposition 4.84 that the field of moduli $K_\phi$ is contained in every field of definition of $\phi$. Prove that $K_\phi$ is equal to the intersection over all fields of definition of $\phi$.

**4.44.** Let $C_a : x^2 + y^2 = a$ be the family of curves studied in Example 4.76.

(a) Let $p$ be a prime number satisfying $p \equiv 1 \pmod 4$. Prove that $C_p$ is $\mathbb{Q}$-isomorphic to $C_1$.

(b) Let $p$ be a prime number satisfying $p \equiv 3 \pmod 4$. Prove that $C_p(\mathbb{Q}) = \emptyset$. Deduce that $C_p$ is not $\mathbb{Q}$-isomorphic to $C_1$, and hence that $[C_p]_K$ represents a nontrivial element of $\operatorname{Twist}(C_1/\mathbb{Q})$.

(c) Let $p$ and $q$ be distinct prime numbers that are congruent to 3 modulo 4. Prove that $C_p$ and $C_q$ are not $\mathbb{Q}$-isomorphic. Deduce that $\operatorname{Twist}(C_1/\mathbb{Q})$ is an infinite set.

**4.45.** Let $g$ be the cocycle

$$g : \operatorname{Gal}(\bar{\mathbb{Q}}/\mathbb{Q}) \longrightarrow \operatorname{PGL}_2(\bar{\mathbb{Q}}), \qquad g_\sigma(z) = \begin{cases} z & \text{if } \sigma(i) = i, \\ -1/z & \text{if } \sigma(i) = -i, \end{cases}$$

described in Example 4.88, and let $\mathcal{K}$ be the associated fixed field in $\bar{\mathbb{Q}}(w)$.

(a) Prove that $\mathcal{K} = \bar{\mathbb{Q}}(u, v)$, where $u = w - 1/w$ and $v = i(w + 1/w)$. Deduce that $C : u^2 + v^2 = -4$ is the twist of $\mathbb{P}^1/\mathbb{Q}$ associated to the cocycle $g$.

(b) Let $\phi(z) = i((z-1)/(z+1))^3$ be the rational map from Example 4.85 satisfying $\sigma(f) = \phi^{g_\sigma}$. Our general theory says that $K$ is a field of definition for $\phi$ if and only if $C(K) \neq \emptyset$, where $C$ is the curve in (a). For example, $C(\mathbb{Q}(\sqrt{-2})) \neq \emptyset$, so $\mathbb{Q}(\sqrt{-2})$ must be a field of definition for $\phi(z)$. Find an explicit linear fractional transformation $f \in \operatorname{PGL}_2(\bar{\mathbb{Q}})$ such that $\phi^f(z) \in \mathbb{Q}(\sqrt{-2})(z)$.

## Section 4.11. Minimal Resultants and Minimal Models

**4.46.** ** This exercise raises some natural questions concerning global minimal models.

(a) Is it true that every rational map $\phi(z) \in \mathbb{Q}(z)$ of (odd) degree $d \geq 2$ has a global minimal model over $\mathbb{Q}$?

(b) Let $K$ be a number field, let $R$ be the ring of integers of $K$, let $S$ be a finite set of primes, and let $\phi(z) \in K(z)$ be a rational map of degree $d \geq 2$. We say that $\phi$ has a *global $S$-minimal model* if there is a linear fractional transformation $f \in \mathrm{PGL}_2(K)$ and homogeneous polynomials $F$ and $G$ satisfying $\phi^f = [F, G]$ with the property that the coefficients of $F$ and $G$ are in the ring of $S$-integers $R_S$ and

$$\mathrm{ord}_{\mathfrak{p}}\bigl(\mathrm{Res}(F,G)\bigr) = \mathrm{ord}_{\mathfrak{p}}(\mathfrak{R}_\phi) \qquad \text{for every prime } \mathfrak{p} \notin S.$$

Suppose that $R_S$ is a principal ideal domain. Is it true that every $\phi$ of (odd) degree has a global $S$-minimal model?

(c) Let $K$ be a number field and $\phi(z) \in K(z)$ a rational map of (odd) degree $d \geq 2$. Suppose that the Weierstrass class $\bar{\mathfrak{a}}_{\phi/K}$ is trivial. Is it then necessarily true that $\phi$ has a global minimal model? (This is true in an analogous situation for elliptic curves; see [410, VIII.8.2].)

**4.47.** ** Let $K$ be a number field and let $\phi \in K(z)$ be a rational map of degree $d \geq 2$. Let $S$ be a finite set of primes of $K$. Prove that

$$\bigl\{\psi \in \mathrm{Twist}(\phi/K) : \psi \text{ has good reduction at all } \mathfrak{p} \notin S\bigr\}$$

is a finite set.

**4.48.** ** Fix an embedding of the moduli space $\mathcal{M}_d$ in projective space and let $h_{\mathcal{M}_d}$ denote the associated height function. Let $K/\mathbb{Q}$ be a number field, let $d \geq 2$ be an integer, and let $B \geq 1$ be a number. Prove that the set

$$\bigl\{\phi \in \mathrm{Rat}_d(K) : \mathrm{N}_{K/\mathbb{Q}}\mathfrak{R}_\phi \leq B \text{ and } h_{\mathcal{M}_d}\bigl(\langle\phi\rangle\bigr) \leq B\bigr\}$$

contains only finitely many $\mathrm{PGL}_2(K)$-conjugacy classes of rational maps.

**4.49.** * Let $K$ be a number field, let $R$ be the ring of integers of $K$, let $S$ be a finite set of primes of $R$, and let $d \geq 2$ be an integer. Prove that there is a finite set of rational maps $\mathcal{B}_{K,S,d} \subset \mathrm{Rat}_d(K)$ such that if $\phi \in \mathrm{Rat}_d(K)$ is a rational map satisfying

(1) $\phi$ has three or more critical points,

(2) the critical points of $\phi$ remain distinct modulo all primes not in $S$,

(3) the critical values of $\phi$ remain distinct modulo all primes not in $S$, where we recall that a *critical value of* $\phi$ is the image of a critical point of $\phi$,

then there are automorphisms $f, g \in \mathrm{PGL}_2(R_S)$ such that

$$f \circ \phi \circ g \in \mathcal{B}_{K,S,d}.$$

(Note that when we say that the critical points or values are distinct modulo a prime $\mathfrak{p}$, we really mean that they are distinct modulo $\mathfrak{P}$ for all primes $\mathfrak{P}$ lying above $\mathfrak{p}$ in a suitable finite extension of $K$.)

# Chapter 5

# Dynamics over Local Fields: Bad Reduction

In this chapter we return to the study of dynamical systems over complete local fields such as $\mathbb{Q}_p$. We saw in Chapter 2 that if a rational map $\phi(z) \in \mathbb{Q}_p(z)$ has good reduction, then its Julia set is empty, in which case considerable information about the dynamics of $\phi$ on $\mathbb{P}^1(\mathbb{Q}_p)$ may be deduced by studying the dynamics of the reduction $\tilde{\phi}$ on $\mathbb{P}^1(\mathbb{F}_p)$. But if $\phi(z)$ has bad reduction, then the situation is far more complicated. Indeed, since "interesting" unpredictable dynamics occurs only in the Julia set, we might say that the good reduction scenario studied in Chapter 2 is the uninteresting situation. This chapter is devoted to the interesting case!

The field $\mathbb{Q}_p$ and its finite extensions have the agreeable property that they are complete, but they are not algebraically closed, so they are more analogous to $\mathbb{R}$ than they are to $\mathbb{C}$. This suggests that we work instead with an algebraic closure $\bar{\mathbb{Q}}_p$ of $\mathbb{Q}_p$, but unfortunately we then lose the completeness property! Going one step further, we take the completion of $\bar{\mathbb{Q}}_p$, and it turns out that this field, denoted by

$$\mathbb{C}_p = \text{the completion of the algebraic closure of } \mathbb{Q}_p,$$

is both complete and algebraically closed. For proofs of the basic properties of $\bar{\mathbb{Q}}_p$ and $\mathbb{C}_p$, see for example [81, 175, 183, 249, 382].

The fields $\mathbb{C}$ and $\mathbb{C}_p$ share many common properties, but they also differ in crucial ways. In particular, the field of $p$-adic complex numbers $\mathbb{C}_p$ is not locally compact! However, it is often essential to work in $\mathbb{C}_p$, rather than in a finite extension of $\mathbb{Q}_p$, for example if we want to guarantee the existence of large numbers of periodic points. Table 5.1 compares some of the properties of the complete fields $\mathbb{Q}_p$, $\bar{\mathbb{Q}}_p$, $\mathbb{C}_p$, $\mathbb{R}$, and $\mathbb{C}$.

The subject of $p$-adic and more general nonarchimedean dynamics is relatively new. After some early articles [50, 201, 278, 433, 438] in the 1980s, there was an explosion of interest that put the subject on a firm footing with a body of significant theorems and, just as importantly, an array of fascinating conjectures. In this brief

| | $\mathbb{Q}_p$ | $\bar{\mathbb{Q}}_p$ | $\mathbb{C}_p$ | $\mathbb{R}$ | $\mathbb{C}$ |
|---|---|---|---|---|---|
| Nonarchimedean metric | ✓ | ✓ | ✓ | | |
| Algebraically closed | | ✓ | ✓ | | ✓ |
| Complete | ✓ | | ✓ | ✓ | ✓ |
| Locally compact | ✓ | | | ✓ | ✓ |
| Totally disconnected | ✓ | ✓ | ✓ | | |

Table 5.1: Comparison of complete fields.

chapter we can provide only a glimpse into this active area of current research, with many important topics omitted entirely in order to keep the chapter at a manageable length. For example, we do not touch on the important concept of local conjugacy, nor do we describe Rivera-Letelier's classification of Fatou domains [373, 375]. For the reader desiring further information, we mention the following articles on $p$-adic dynamics that are listed in the references: [4, 5, 13, 14, 22, 23, 26, 30, 29, 50, 63, 53, 54, 56, 57, 58, 59, 60, 62, 70, 71, 72, 73, 82, 84, 104, 116, 170, 168, 169, 185, 188, 189, 201, 206, 208, 220, 222, 236, 237, 238, 239, 240, 241, 242, 243, 244, 245, 262, 266, 267, 268, 271, 272, 273, 274, 278, 280, 282, 283, 320, 318, 319, 334, 337, 338, 339, 340, 344, 348, 352, 355, 372, 373, 374, 375, 376, 378, 389, 427, 433, 438, 449].

## 5.1 Absolute Values and Completions

We recall from Section 2.1 that a *valued field* is a pair $(K, |\cdot|)$ consisting of a field $K$ and an absolute value $|\cdot|$ on $K$. In this section we briefly remind the reader of the construction and basic properties of the completion of a valued field.

**Definition.** Let $(K, |\cdot|)$ be a valued field. A sequence $\alpha_1, \alpha_2, \alpha_3, \ldots \in K$ is *Cauchy* if for every $\epsilon > 0$ there exists an $N = N(\epsilon)$ such that

$$|\alpha_n - \alpha_m| \le \epsilon \qquad \text{for all } m, n \ge N.$$

In other words, the sequence $\{\alpha_n\}$ is Cauchy if $|\alpha_n - \alpha_m| \to 0$ as $m, n \to \infty$.

It is clear that every convergent sequence is Cauchy. A valued field $K$ is said to be *complete* if every Cauchy sequence in $K$ converges.

*Example* 5.1. The real numbers $\mathbb{R}$ and the complex numbers $\mathbb{C}$ are complete with respect to their usual absolute values.

**Theorem 5.2.** *Let $(K, |\cdot|_K)$ be a valued field. Then there exists a valued field $(\hat{K}, |\cdot|_{\hat{K}})$, unique up to isomorphism of valued fields, with the following properties:*

(a) *There is an inclusion $K \subset \hat{K}$ as valued fields.*

(b) *$\hat{K}$ is complete.*

(c) *$K$ is a dense subset of $\hat{K}$.*

*Further, $\hat{K}$ is the smallest valued field satisfying* (a) *and* (b) *in the following sense: Let $(L, |\cdot|_L)$ be any complete valued field containing $K$. Then there is a unique inclusion $\hat{K} \hookrightarrow L$ of valued fields that respects the inclusions of $K$ into $\hat{K}$ and $L$.*

*The field $\hat{K}$ is called the* completion of $K$ with respect to the absolute value $|\cdot|_K$.

*Proof.* (Sketch) The field $\hat{K}$ may be constructed as follows. Let $C$ be the set of all Cauchy sequences in $K$ and make $C$ into a ring by setting

$$\{\alpha_n\} + \{\beta_n\} = \{\alpha_n + \beta_n\} \qquad \text{and} \qquad \{\alpha_n\} \cdot \{\beta_n\} = \{\alpha_n \cdot \beta_n\}.$$

We define an absolute value on $C$ by setting

$$\bigl|\{\alpha_n\}\bigr| = \lim_{n\to\infty} |\alpha_n|,$$

and we define an equivalence relation $\sim$ on $C$ by

$$\{\alpha_n\} \sim \{\beta_n\} \quad \text{if} \quad \lim_{n\to\infty} |\alpha_n - \beta_n| = 0.$$

Then one checks that these definitions are consistent and that the quotient

$$\hat{K} = C/\sim$$

is a complete field with the desired properties. Note that $K$ is identified with the subfield of $\hat{K}$ consisting of constant sequences. For further details on completions, see for example [190, 4 §7] or [217, IV §1]. □

**Theorem 5.3.** *Let $(K, |\cdot|_K)$ be a complete field and let $L/K$ be a finite extension. Then there is a unique absolute value $|\cdot|_L$ on $L$ extending the absolute value on $K$. The field $L$ is complete with respect to $|\cdot|_L$.*

*Proof.* (Sketch) One checks that $|\cdot|_L$ is given by

$$|\alpha|_L = \bigl|\mathrm{N}_{L/K}(\alpha)\bigr|_K^{1/[L:K]}.$$

See [259, Proposition XII.2.6] or [382, II §4]. □

*Example* 5.4. The following table gives some examples of completions.

| Field | Absolute Value | Completion |
|---|---|---|
| $\mathbb{Q}$ | $\lvert\alpha\rvert_\infty = \max\{\alpha, -\alpha\}$ | $\mathbb{R}$ |
| $\mathbb{Q}(i)$ | $\lvert a+bi\rvert = \sqrt{a^2+b^2}$ | $\mathbb{C}$ |
| $\mathbb{Q}$ | $\lvert\alpha\rvert_p = p^{-\operatorname{ord}_p(\alpha)}$ | $\mathbb{Q}_p$ |

*Remark* 5.5. Theorem 5.3 says that for a finite extension $L/K$, there is a unique absolute value on $L$ extending the absolute value on $K$, and further that $L$ is complete with respect to this extended absolute value. If we go to the algebraic closure $\bar{K}$ of $K$, then we still get a unique absolute value, since $\bar{K}$ is the compositum of the finite extensions of $K$, but unfortunately $\bar{K}$ may not be complete. (See, e.g., [249, III.4, Theorem 12] for a proof that $\bar{\mathbb{Q}}_p$ is not complete.) So we use Theorem 5.2 to form the completion of the valued field $\bar{K}$ as in the next theorem.

**Theorem 5.6.** *Let*

$$\mathbb{C}_p = \hat{\bar{\mathbb{Q}}}_p = \textit{the completion of the algebraic closure of } \mathbb{Q}_p.$$

*Then $\mathbb{C}_p$ is both complete and algebraically closed. It is the smallest complete algebraically closed field containing $\mathbb{Q}_p$.*

*Proof.* More generally, the completion of an algebraically closed nonarchimedean field is algebraically closed. We briefly sketch the proof. Let $\alpha$ be algebraic over $\mathbb{C}_p$, say a root of $f(X) = \sum a_i X^i \in \mathbb{C}_p[X]$. Choose $b_i \in \bar{\mathbb{Q}}_p$ that are very close to $a_i$, which can be done since $\bar{\mathbb{Q}}_p$ is dense in $\mathbb{C}_p$. Then the roots of $g(X) = \sum b_i X^i$ can be matched with the roots of $f(X)$, so we can find a root $\beta$ of $g(X)$ that is very close to $\alpha$. Note that $\beta \in \bar{\mathbb{Q}}_p$, since $\bar{\mathbb{Q}}_p$ is algebraically closed. Thus we can find elements in $\bar{\mathbb{Q}}_p$ (so a fortiori in $\mathbb{C}_p$) that are arbitrarily close to $\alpha$. This gives a sequence of elements of $\mathbb{C}_p$ that converges to $\alpha$, so the fact that $\mathbb{C}_p$ is complete implies that $\alpha \in \mathbb{C}_p$. For details of the proof, see [249, III.4, Theorem 13] or [382, Lemma II.4.2]. □

## 5.2 A Primer on Nonarchimedean Analysis

Throughout this section we take $K$ to be a field that is complete with respect to an absolute value $|\cdot|$ satisfying the nonarchimedean (ultrametric) triangle inequality

$$|\alpha + \beta| \le \max\{|\alpha|, |\beta|\} \quad \text{for all } \alpha, \beta \in K. \tag{5.1}$$

We recall from Lemma 2.3 that if $|\alpha| \ne |\beta|$, then (5.1) is an equality.

We set the following notation for the ring of integers of $K$, its unit group, and its maximal ideal:

$$\begin{aligned} R = R_K &= \{\alpha \in K : |\alpha| \le 1\}, \\ R^* = R_K^* &= \{\alpha \in K : |\alpha| = 1\}, \\ \mathfrak{M} = \mathfrak{M}_K &= \{\alpha \in K : |\alpha| < 1\}. \end{aligned}$$

We define "open disks," "closed disks," and "circles" in $K$ by the usual formulas:

$$\begin{aligned} D(a,r) &= \{u \in K : |u - a| < r\} = \text{open disk of radius } r \text{ at } a, \\ \bar{D}(a,r) &= \{u \in K : |u - a| \le r\} = \text{closed disk of radius } r \text{ at } a, \\ S(a,r) &= \{u \in K : |u - a| = r\} = \text{circle of radius } r \text{ at } a. \end{aligned}$$

However, it is important to note that in the nonarchimedean setting, the disks $D(a,r)$ and $\bar{D}(a,r)$ and the circle $S(a,r)$ are simultaneously open and closed sets! To see this, let $x \in S(a,r)$. Then for any $s < r$ we have $D(x,s) \subset S(a,r)$, since any $y \in D(x,s)$ satisfies

$$|y-x| < s < r, \quad \text{so} \quad |y-a| = \max\{|y-x|, |x-a|\} = r.$$

This shows that $S(a,r)$ is open, and it is clearly closed by definition. It follows that $\bar{D}(a,r) = D(a,r) \cup S(a,r)$ is open and that $D(a,r) = \bar{D}(a,r) \setminus S(a,r)$ is closed.

Notice that $\bar{D}(0,r)$ and $D(0,r)$ are groups under addition, i.e., they are closed under addition and negation, with a nontopological meaning of the word "closed." More precisely, the closed unit disk $\bar{D}(0,1)$ is the ring of integers $R$ of $K$ and the open unit disk $D(0,1)$ is the maximal ideal $\mathfrak{M}$ of $R$.

Recall that the chordal distance between two points $P_1 = [X_1, Y_1]$ and $P_2 = [X_2, Y_2]$ in $\mathbb{P}^1(K)$ is the quantity (cf. Section 2.1)

$$\rho(P_1, P_2) = \frac{|X_1Y_2 - X_2Y_1|}{\max\{|X_1|, |Y_1|\}\max\{|X_2|, |Y_2|\}}.$$

The *chordal metric* $\rho$ satisfies $0 \le \rho(P_1, P_2) \le 1$, so it is bounded. In particular, if $K$ is a finite extension of $\mathbb{Q}_p$, so $K$ is locally compact, then $\mathbb{P}^1(K)$ is compact, since it is a locally compact bounded metric space. On the other hand, the field $\mathbb{C}_p$ is not locally compact, so although $\mathbb{P}^1(\mathbb{C}_p)$ is a bounded space, it is not compact, nor even locally compact.

We can define open and closed disks in $\mathbb{P}^1(K)$ using the chordal metric. We will use the same notation as above:

$$D(Q,r) = \{P \in \mathbb{P}^1(K) : \rho(P,Q) < r\} = \text{open disk of radius } r \text{ in } \mathbb{P}^1(K),$$
$$\bar{D}(Q,r) = \{P \in \mathbb{P}^1(K) : \rho(P,Q) \le r\} = \text{closed disk of radius } r \text{ in } \mathbb{P}^1(K).$$

It should be clear from context whether we are working in $K$ or in $\mathbb{P}^1(K)$. We observe that $\mathbb{P}^1(K)$ is equal to the union of the two unit disks

$$\mathbb{P}^1(K) = \{[x,1] : |x| \le 1\} \cup \{[1,y] : |y| \le 1\},$$

each of which is metrically isomorphic to $R_K$.

*Remark* 5.7. It is sometimes convenient to make the assumption that the radius $r$ of a disk is equal to the absolute value of some element of $K^*$. In this case we say that the disk has *rational radius*. We note that if $K/\mathbb{Q}_p$ is a finite extension of ramification degree $e$, then the value set $|K^*|$ has the form $\{p^{n/e} : n \in \mathbb{Z}\}$. Similarly, the value set $|\mathbb{C}_p^*|$ is equal to $\{p^r : r \in \mathbb{Q}\}$. Thus rational radius really means that the radius is a rational power of $p$.

The nonarchimedean nature of the absolute value implies that a sequence $\{a_i\}_{i\ge 0}$ in $K$ is Cauchy if and only if

$$\lim_{i\to\infty} |a_{i+1} - a_i| = 0.$$

This follows immediately from the inequality

$$|a_n - a_m| \le \max_{m \le i < n} |a_{i+1} - a_i|.$$

This observation and the completeness of $K$ tell us that a power series

$$\phi(z) = \sum_{i=0}^{\infty} a_i (z-a)^i \in K[\![z]\!]$$

converges if and only if

$$\lim_{i\to\infty} \bigl|a_i(z-a)^i\bigr| = 0.$$

In other words, the "$n^{\text{th}}$ term test" from elementary calculus becomes both necessary and sufficient in the nonarchimedean setting.

**Definition.** A function $\phi : \bar{D}(a,r) \to K$ is *holomorphic* (or *analytic*) if it is represented by a power series

$$\phi(z) = \sum_{i=0}^{\infty} a_i (z-a)^i \in K[\![z-a]\!] \tag{5.2}$$

that converges for all $z \in \bar{D}(a,r)$. The *order of $\phi$ at $a$*, denoted by $\operatorname{ord}_a(\phi)$, is the smallest index $i$ such that $a_i \ne 0$.

A *meromorphic function on $\bar{D}(a,r)$* is a quotient $\phi = \phi_1/\phi_2$ of functions $\phi_1$ and $\phi_2 \ne 0$ that are holomorphic functions on $\bar{D}(r,a)$.[1] A meromorphic function $\phi = \phi_1/\phi_2$ induces a well-defined map

$$\phi : \bar{D}(a,r) \longrightarrow \mathbb{P}^1(K), \qquad z \longmapsto \bigl[\phi_1(z), \phi_2(z)\bigr].$$

The *order of $\phi$ at $a$* is the difference

$$\operatorname{ord}_a(\phi) = \operatorname{ord}_a(\phi_1) - \operatorname{ord}_a(\phi_2).$$

We say that $\phi$ has a *zero* (respectively a *pole*) at $z = a$ if $\operatorname{ord}_\alpha(\phi) > 0$ (respectively if $\operatorname{ord}_\alpha(\phi) < 0$).

The next proposition describes some elementary properties of nonarchimedean holomorphic and meromorphic functions.

**Proposition 5.8.** (a) *Let $\phi(z)$ be a holomorphic function on the closed disk $\bar{D}(a,r)$ and let $b \in \bar{D}(a,r)$. Then $\phi(z)$ is a holomorphic function on $\bar{D}(b,r)$, i.e., $\phi(z)$ is given by a convergent power series in $K[\![z-b]\!]$.*

(b) *Let $\phi(z)$ be a nonzero holomorphic function on $\bar{D}(a,r)$. Then the zeros of $\phi(z)$ in $\bar{D}(a,r)$ are isolated. This means that if $\phi(b) = 0$, then there is a disk $\bar{D}(b,\epsilon)$ such that $\phi(z) \ne 0$ for all $z \in \bar{D}(b,\epsilon) \smallsetminus \{b\}$.*

---

[1] More precisely, a meromorphic function is an equivalence class of pairs $(\phi_1, \phi_2)$ with $\phi_2 \ne 0$, with two pairs $(\phi_1, \phi_2)$ and $(\psi_1, \psi_2)$ being equivalent if $\phi_1\psi_2 = \phi_2\psi_1$.

(c) *Let $\phi(z)$ be a meromorphic function on $\bar{D}(a,r)$ and suppose that the only pole of $\phi(z)$ in $\bar{D}(a,r)$, if any, is $z = a$. Then $\phi(z)$ is represented by a convergent Laurent series,*

$$\phi(z) = \sum_{i=-m}^{\infty} a_i(z-a)^i \in K[\![z]\!] \qquad \text{for all } z \in \bar{D}(a,r) \smallsetminus \{a\}. \tag{5.3}$$

(d) *Let $\phi(z)$ be a meromorphic function on $\bar{D}(a,r)$. Then for every $b \in \bar{D}(a,r)$ there is an $s$ such that $\phi(z)$ is represented by a convergent Laurent series on $\bar{D}(b,s)$.*

*Proof.* (a) Let $\phi(z) = \sum a_i(z-a)^i$, and for $k \ge 0$, define coefficients $b_k$ by the formula

$$b_k = \sum_{i=k}^{\infty} \binom{i}{k} a_i(b-a)^{i-k}.$$

(These values are not mysterious. If $K$ has characteristic 0 then they are the usual Taylor coefficients $b_k = (1/k!)(d^k\phi/dz^k)(b)$.) The series defining $b_k$ converges since the convergence of $\phi(z)$ on $\bar{D}(a,r)$ implies that

$$\left|a_i(b-a)^i\right| \le |a_i|r^i \to 0 \quad \text{as } i \to \infty.$$

Further, we have the estimate

$$|b_k| \le \max_{i \ge k} \left| \binom{i}{k} a_i(b-a)^{i-k} \right| \le \max_{i \ge k} |a_i|r^{i-k}, \qquad \text{so} \qquad |b_k|r^k \le \max_{i \ge k} |a_i|r^i.$$

Hence $|b_k|r^k \to 0$ as $k \to \infty$, so the power series $\sum_{k=0}^{\infty} b_k(z-b)^k$ converges on $\bar{D}(b,r)$. Finally, we check that the series represents $\phi$ by computing

$$\begin{aligned}\sum_{k=0}^{\infty} b_k(z-b)^k &= \sum_{k=0}^{\infty}\sum_{i=k}^{\infty} \binom{i}{k} a_i(b-a)^{i-k}(z-b)^k \\ &= \sum_{i=0}^{\infty} a_i \sum_{k=0}^{i} \binom{i}{k} (b-a)^{i-k}(z-b)^k \\ &= \sum_{i=0}^{\infty} a_i(z-a)^i.\end{aligned}$$

(b) Let $b \in \bar{D}(a,r)$ be a point with $\phi(b) = 0$. We will find a deleted neighborhood of $b$ on which $\phi(z)$ is nonvanishing. We use (a) to expand $\phi$ as a power series $\phi(z) = \sum b_i(z-b)^i$ centered at $b$ and converging on $\bar{D}(b,r)$. The assumption that $\phi \ne 0$ means that some coefficient is nonzero; we let $j$ be the smallest index such that $b_j \ne 0$. The fact that $\phi$ converges on $\bar{D}(b,r)$ implies that $|b_i|r^i \to 0$ as $i \to \infty$, so there is a constant $C$ such that $|b_i|r^i < C$ for all $i$.

Let $\epsilon = r^{j+1}|b_j|/2C$. We claim that the only zero of $\phi(z)$ on $\bar{D}(b,\epsilon)$ is $z = b$. To see this, we take any $z \in \bar{D}(b,\epsilon)$ and estimate

$$\max_{i\ge j+1}|b_i|\cdot|z-b|^i \le \max_{i\ge j+1} C\left(\frac{|z-b|}{r}\right)^i \le C\left(\frac{|z-b|}{r}\right)^{j+1}$$
$$= \frac{C|z-b|}{r^{j+1}}\cdot\left|(z-b)^j\right| \le \frac{C\epsilon}{r^{j+1}}\cdot\left|(z-b)^j\right| = \frac{1}{2}\left|b_j(z-b)^j\right|.$$

Hence for all $z \in \bar{D}(b,\epsilon)$ with $z \ne b$, the first term in the series

$$\phi(z) = b_j(z-b)^j + \sum_{i=j+1}^{\infty} b_i(z-b)^i$$

has absolute value strictly larger than any of the other terms, so the nonarchimedean triangle inequality (Lemma 2.3) implies that

$$\left|\phi(z)\right| = \left|b_j(z-b)^j\right| \qquad \text{for all } z \in \bar{D}(b,\epsilon).$$

In particular, $\phi(z) \ne 0$ for all $z \in \bar{D}(b,\epsilon)$ with $z \ne b$.

(c) Write $\phi(z) = \phi_1(z)/\phi_2(z)$ as a quotient of functions that are holomorphic on $\bar{D}(a,r)$. We give the proof of (c) in the case that $\phi_2(z)$ is a polynomial, which is the only case that we will need. For the general case, see [81, 175] or Exercise 5.7.

Let $b \in \bar{D}(a,r)$ with $b \ne a$ and $\phi_2(b) = 0$. Taking the Taylor series of $\phi_1$ and $\phi_2$ around $b$, we can write

$$\phi_1(z) = (z-b)^{\operatorname{ord}_b(\phi_1)}\psi_1(z) \qquad \text{and} \qquad \phi_2(z) = (z-b)^{\operatorname{ord}_b(\phi_2)}\psi_2(z),$$

where $\psi_1$ and $\psi_2$ are nonvanishing holomorphic functions on $\bar{D}(b,r)$. Then the assumption that $\operatorname{ord}_b(\phi) \ge 0$ implies that we can write $\phi(z)$ as a quotient

$$\phi(z) = \frac{(z-b)^{\operatorname{ord}_b(\phi)}\psi_1(z)}{\psi_2(z)}$$

of holomorphic functions on $\bar{D}(a,r)$ such that the denominator does not vanish at $b$. Repeating this process for each of the zeros of $\phi_2$ in $\bar{D}(a,r)$ other than $z=a$, we find that $\phi(z)$ is a quotient of holomorphic functions on $\bar{D}(a,r)$ such that the denominator does not vanish except possibly at $a$. Further, this cancellation process must stop, since we have assumed that $\phi_2(z)$ is a polynomial, so it has only finitely many zeros.

By abuse of notation, we again write $\phi(z) = \phi_1(z)/\phi_2(z)$, where we may now assume that the polynomial $\phi_2(z)$ has no zeros in $\bar{D}(a,r)$ other than $z=a$. We assume for the moment that $K$ is algebraically closed and factor $\phi_2(z)$ as

$$\phi_2(z) = c(z-a)^e\prod_{i=1}^{n}(z-\alpha_i)^{e_i}.$$

By assumption, the roots satisfy $\alpha_i \notin \bar{D}(a,r)$, or equivalently $|\alpha_i - a| > r$, for all $1 \le i \le n$.

The reciprocal $1/\phi_2(z)$ has a partial fraction expansion

$$\frac{1}{\phi_2(z)} = \sum_{j=1}^{e} \frac{A_j}{(z-a)^j} + \sum_{i=1}^{n}\sum_{j=1}^{e_i} \frac{B_{ij}}{(z-\alpha_i)^j}$$

for certain coefficients $A_j, B_{ij} \in K$. (See [259, IV §5] or [436, §5.10].) The terms with negative powers of $z-a$ form part of the desired Laurent series. We claim that all of the other terms are holomorphic on $\bar{D}(a,r)$.

To verify this, we let $\alpha \in K$ with $|\alpha - a| > r$, and for each $k \geq 0$ we consider the function $1/(z-\alpha)^{k+1}$. If $k=0$ we get a geometric series

$$\frac{1}{z-\alpha} = \frac{1}{(z-a)-(\alpha-a)} = \frac{-(\alpha-a)^{-1}}{1-(\alpha-a)^{-1}(z-a)} = \frac{-1}{\alpha-a}\sum_{n=0}^{\infty}\left(\frac{z-a}{\alpha-a}\right)^n. \tag{5.4}$$

This series converges for all $z$ satisfying $|z-a| < |\alpha - a|$, so in particular for all $z \in \bar{D}(a,r)$, since $|\alpha - a| > r$. Hence $1/(z-\alpha)$ is holomorphic on $\bar{D}(a,r)$.

More generally, the same argument works using the identity[2]

$$\frac{1}{(z-\alpha)^{k+1}} = \frac{1}{((z-a)-(\alpha-a))^{k+1}} = \frac{(-1)^{k+1}}{(\alpha-a)^{k+1}}\sum_{n=0}^{\infty}\binom{n+k}{k}\left(\frac{z-a}{\alpha-a}\right)^n.$$

This completes the proof that $1/\psi_2(z)$ is holomorphic on $\bar{D}(r,a)$ in the case that $K$ is algebraically closed.

If $K$ is not algebraically closed, we let $L$ be the completion of the algebraic closure of $K$ and use the above argument to write $\phi(z)$ as a power series in $L[\![z-a]\!]$ that converges on $\bar{D}(a,r)$ for $L$. We now observe that when we add up the series for the different terms in the partial fraction expansion, the coefficients of each power of $z-a$ are symmetric expressions in the roots of $\psi_2(z)$. Thus the resulting power series actually lives in $K[\![z-a]\!]$.

We have shown that $1/\phi_2(z)$ is represented by a Laurent series on $\bar{D}(a,r)$. Multiplying that Laurent series by the holomorphic function $\phi_1(z)$ gives a convergent Laurent series (cf. Exercise 5.3) representing $\phi(z)$ on $\bar{D}(a,r)$.

(d) Write $\phi(z) = \phi_1(z)/\phi_2(z)$. Then (b) says that there is a neighborhood $\bar{D}(b,s)$ of $b$ such that $\phi_2(z) \neq 0$ for $z \in \bar{D}(b,s)$ with $z \neq b$. Hence $\phi$ has no poles other than $b$ on $\bar{D}(b,s)$, so (c) says that it is represented by a Laurent series on $\bar{D}(b,s)$. □

**Definition.** Let $\phi : \bar{D}(a,r) \to K$ be a holomorphic function represented by a power series as in (5.2). The *norm of* $\phi$*, relative to the disk* $\bar{D}(a,r)$, is the quantity

$$\|\phi\| = \|\phi\|_{r,a} = \sup_{i\geq 0} |a_i| r^i.$$

*Remark* 5.9. The assumption that $\phi$ converges on $\bar{D}(a,r)$ implies that $|a_i|r^i \to 0$, so there will be at least one, and at most finitely many, indices $i$ satisfying $\|\phi\| = |a_i|r^i$.

[2] Over a field a characteristic 0, this identity can be derived by taking the $k^{\text{th}}$ derivative of (5.4) and dividing both sides by $k!$. This proof fails in characteristic $p$ if $k \geq p$, but one can either prove the formula directly by algebraic manipulations, or else one can clear denominators, observe that this yields an identity in the two-variable polynomial ring $\mathbb{Z}[\alpha, z]$, and then reduce modulo $p$.

Further, if $D(a,r)$ has rational radius, then $r \in |K^*|$, so there is a $b \in K^*$ satisfying $\|\phi\| = |b|$.

The nonarchimedean nature of the absolute value gives immediately the inequality

$$|\phi(z)| \le \sup_{i\ge 0} |a_i| \cdot |z-a|^i \le \|\phi\| \quad \text{for all } z \in \bar{D}(a,r).$$

We now show that $\|\phi\|$ is more or less a Lipschitz constant for a holomorphic map $\phi$.

**Proposition 5.10.** *Let $\phi(z) \in K[\![z]\!]$ be a power series converging on $\bar{D}(a,r)$. Then*

$$|\phi(z) - \phi(w)| \le \frac{\|\phi\|}{r}|z-w| \quad \textit{for all } z, w \in \bar{D}(a,r).$$

*Proof.* We take $z, w \in \bar{D}(a,r)$ and compute

$$\begin{aligned}
|\phi(z) - \phi(w)| &= \left|\sum_{i=0}^{\infty} a_i\big((z-a)^i - (w-a)^i\big)\right| \\
&= |z-w| \left|\sum_{i=0}^{\infty} a_i \sum_{j=0}^{i-1} (z-a)^{i-1-j}(w-a)^j\right| \\
&\le |z-w| \sup_{i\ge 0} \sup_{0\le j<i} |a_i| \cdot |z-a|^{i-1-j}|w-a|^j \\
&\le |z-w| \sup_{i\ge 0} |a_i| r^{i-1}, \qquad \text{since } |z-a|, |w-a| \le r, \\
&= |z-w| \frac{\|\phi\|}{r}.
\end{aligned}$$

□

## 5.3 Newton Polygons and the Maximum Modulus Principle

A powerful tool in complex analysis is the maximum modulus principle,[3] which asserts that a holomorphic function $\phi(z)$ on an open set $U \subset \mathbb{C}$ has no maximum on $U$. Equivalently, if $\bar{D} \subset U$ is any closed disk in $U$, then $|\phi(z)|$ attains its maximum value on the boundary $\partial\bar{D}$ of $\bar{D}$. In this section we prove a nonarchimedean analogue of the maximum modulus principle that is of similar fundamental importance in the theory of nonarchimedean analysis. However, in the nonarchimedean setting we cannot prove the maximum modulus principle using path integrals and Cauchy's residue theorem. In their place we substitute the powerful method of the Newton polygon.

[3] Indeed, Ahlfors [1] says that "because of its simple and explicit formulation it is one of the most useful general theorems in the theory of functions. As a rule, all proofs based on the maximum principle are very straightforward, and preference is quite justly given to proofs of this kind."

The Newton polygon of a nonarchimedean power series $\phi(z)$ is very easy to describe, and it can be used to give a precise description of the distribution of zeros of the power series. To ease notation, we let

$$v(z) = -\log_p |z| \quad \text{for } z \in \mathbb{C}_p.$$

Notice that the valuation $v$ is a surjective homomorphism $v : \mathbb{C}_p^* \twoheadrightarrow \mathbb{Q}$, since we have normalized $|\cdot|$ so that $|p| = p^{-1}$.

**Definition.** Let $\phi(z) = \sum a_n z^n \in \mathbb{C}_p[\![z]\!]$ be a power series. The *Newton polygon of* $\phi$ is the convex hull of the set of points

$$\{(n, v(a_n)) : n = 0, 1, 2, \dots\},$$

where by convention we set $v(0) = \infty$. Informally, the Newton polygon is created as follows: take a vertical ray starting at the point $(0, v(a_0))$ and aiming down the $y$-axis. Then rotate the ray counterclockwise, keeping the point $(0, v(a_0))$ fixed, until it bends around all of the points $(n, v(a_n))$.

The Newton polygon consists of a set of *line segments* that connect the dots required to create the convex hull. A typical Newton polygon is illustrated in Figure 5.1(a). It has a segment from $(0, 5)$ to $(2, 1)$, a segment from $(2, 1)$ to $(4, -1)$, a segment from $(4, -1)$ to $(7, -1)$, etc. A Newton polygon may have infinitely many line segments, or it may terminate with an infinite ray.

A fundamental theorem says that the Newton polygon of an analytic function contains a tremendous amount of information about the roots of the function. It provides a very powerful tool for studying nonarchimedean power series.

**Theorem 5.11.** *Let $\phi(z) \in \mathbb{C}_p[\![z]\!]$ be a power series. Suppose that the Newton polygon of $\phi$ includes a line segment of slope $m$ whose horizontal length is $N$, i.e., the Newton polygon has a line segment running from*

$$(n, v(a_n)) \quad \text{to} \quad (n+N, v(a_{n+N}))$$

*whose slope is*

$$m = \frac{v(a_{n+N}) - v(a_n)}{N}.$$

*Suppose further that $\phi$ converges on the closed disk of radius $p^m$. Then $\phi(z)$ has exactly $N$ roots $\alpha$, counted with multiplicity, satisfying $|\alpha| = p^m$.*

*Proof.* See [249, IV.4, Corollary to Theorem 14]. We observe that the proof of this result for polynomials or rational functions is quite easy. For power series, one first proves a $p$-adic version of the Weierstrass preparation theorem saying, roughly, that $\phi(z)$ factors into the product of a polynomial $g(z)$ and a nonvanishing power series $\psi(z)$ such that the initial parts of the Newton polygons of $\phi(z)$ and $g(z)$ coincide. Then the theorem for power series follows immediately by applying the elementary result for polynomials to $g(z)$. □

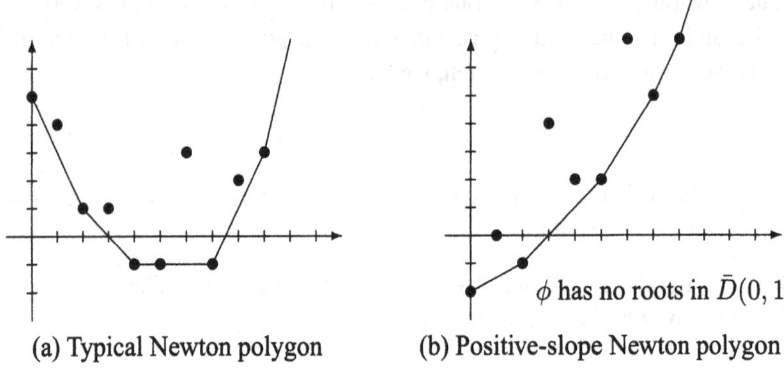


(a) Typical Newton polygon

(b) Positive-slope Newton polygon

Figure 5.1: Examples of Newton polygons.

*Example* 5.12. The Newton polygon of the power series

$$\phi(z) = p^5 + p^4 z + p z^2 + p z^3 + p^{-1} z^4 + p^{-1} z^5 + p^3 z^6 + p^{-1} z^7 + p^2 z^8 + p^3 z^9 + \cdots$$

is illustrated in Figure 5.1(a). The leftmost line segment has slope $-2$ and width 2, so $\phi(z)$ has exactly 2 roots $\alpha$ satisfying $|\alpha| = p^{-2}$ (assuming that $\phi(z)$ converges on the appropriate disk). Similarly, $\phi(z)$ has exactly 2 roots satisfying $|\alpha| = p^{-1}$, exactly 3 roots satisfying $|\alpha| = 1$, and exactly 2 roots satisfying $|\alpha| = p^2$.

Our first application of the Newton polygon is a nonarchimedean version of the classical maximum modulus principle from complex analysis.

**Theorem 5.13.** (Maximum Modulus Principle) *Let $\phi(z) \in \mathbb{C}_p[\![z]\!]$ be a power series that converges on a disk $\bar{D}(a,r)$ of rational radius.*

(a) *There is a point $\beta \in \bar{D}(a,r)$ satisfying*

$$|\phi(\beta)| = \sup_{z \in \bar{D}(a,r)} |\phi(z)| = \|\phi\|.$$

(b) *If $\phi$ does not vanish on $\bar{D}(a,r)$, then*

$$|\phi(z)| = \|\phi\| \quad \textit{for all } z \in \bar{D}(a,r).$$

*In other words, either $\phi$ has a zero in $\bar{D}(a,r)$, or else it has constant magnitude on $\bar{D}(a,r)$.*

*Remark* 5.14. For many applications, it suffices to know that the maximum modulus principle is true for rational functions $\phi(z) \in \mathbb{C}_p(z)$. In this case, both (a) and (b) are quite easy to prove. We do (b) and leave (a) for the reader. Replacing $z$ by $(z-a)/c$ for some $c$ with $|c| = r$, we may assume that $\phi(z)$ is well-defined and nonvanishing on the disk $\bar{D}(0,1)$. We factor $\phi(z)$ as

$$\phi(z) = z^k \frac{(1-\alpha_1 z)(1-\alpha_2 z)\cdots(1-\alpha_r z)}{(1-\beta_1 z)(1-\beta_2 z)\cdots(1-\beta_s z)} \qquad \text{with } \alpha_i, \beta_i \neq 0.$$

The assumption that $\phi$ has no zero or pole at $z = 0$ implies that $k = 0$. Further, we have $|\alpha_i| < 1$ and $|\beta_i| < 1$ for all $i$, since otherwise $\phi$ would have a zero at $\alpha_i^{-1} \in \bar{D}(0,1)$ or a pole at $\beta_i^{-1} \in \bar{D}(0,1)$. It follows that

$$|1 - \alpha_i z| = |1 - \beta_i z| = 1 \quad \text{for all } z \in \bar{D}(0,1),$$

and hence $\bigl|\phi(z)\bigr| = 1$ for all $z \in \bar{D}(0,1)$.

*Proof of Theorem* 5.13. Write $\phi(z) = \sum a_i(z-a)^i$ and choose constants $b, c \in K^*$ with $|c| = r$ and $|b| = \|\phi\|$ (cf. Remark 5.9). Consider the series

$$\psi(z) = b^{-1}\phi(cz+a) = b^{-1}\sum_{i=0}^{\infty} a_i(cz)^i = \sum_{i=0}^{\infty} \frac{a_i c^i}{b} z^i.$$

The convergence of $\phi$ on $\bar{D}(a,r)$ clearly implies the convergence of $\psi$ on $\bar{D}(0,1)$, and we have

$$\|\psi\| = \sup_{i\ge 0}\left|\frac{a_i c^i}{b}\right| = \sup_{i\ge 0}\frac{|a_i| r^i}{|b|} = \frac{\|\phi\|}{|b|} = 1.$$

Replacing $\phi$ by $\psi$, we are reduced to proving the theorem under the assumptions that $\phi$ converges on the unit disk $\bar{D}(0,1)$ and that $\|\phi\| = 1$.

(a) The condition $\|\phi\| = 1$ says that every coefficient of $\phi$ lies in $R$, and the fact that $\phi$ converges on $\bar{D}(0,1)$ implies that all but finitely many coefficients lie in the maximal ideal $\mathfrak{M}$ of $\mathbb{C}_p$. So when we reduce $\phi(z)$ modulo $\mathfrak{M}$, we get a nonzero polynomial

$$\phi(z) \equiv A_0 + A_1 z + \cdots + A_d z^d \pmod{\mathfrak{M}}.$$

Let $\alpha_1, \ldots, \alpha_r \in R$ be representatives for the distinct roots of $\phi(z)$ mod $\mathfrak{M}$ in the residue field $R/\mathfrak{M}$. The residue field is infinite, since $K$ is algebraically closed, so we can find a $\beta \in R$ satisfying

$$\beta \not\equiv \alpha_i \pmod{\mathfrak{M}} \qquad \text{for all } 1 \le i \le r.$$

Then $\phi(\beta) \not\equiv 0 \pmod{\mathfrak{M}}$, so $\bigl|\phi(\beta)\bigr| = 1 = \|\phi\|$.

(b) Write $\phi(z) = \sum a_i z^i$ as usual and consider the Newton polygon of $\phi(z)$. Theorem 5.11 asserts that if some line segment of the Newton polygon of $\phi(z)$ were to have slope $m \le 0$, then $\phi(z)$ would have at least one root $\alpha$ satisfying $|\alpha| = p^m \le 1$, and hence $\phi(z)$ would have a root in $\bar{D}(0,1)$. Thus our assumption that $\phi(z)$ has no roots in $\bar{D}(0,1)$ implies that every line segment of the Newton polygon has strictly positive slope. (An illustrative Newton polygon is given in Figure 5.1(b).) In particular, directly from the definition of the Newton polygon, this implies that

$$v(a_i) > v(a_0) \quad \text{for all } i \ge 1.$$

Equivalently, we have $|a_i| < |a_0|$ for all $i \ge 1$. We first observe that this gives

$$|a_0| = \sup |a_i| = \|\phi\|.$$

Second, we note that it implies that the constant term $a_0$ in the series

$$\phi(z) = a_0 + a_1 z + a_2 z^2 + a_3 z^3 + \cdots \qquad \text{with } z \in \bar{D}(0,1)$$

has absolute value strictly larger than any of the other terms, so the ultrametric inequality (Lemma 2.3) says that

$$|\phi(z)| = |a_0| = \|\phi\| \qquad \text{for all } z \in \bar{D}(0,1).$$ □

*Remark* 5.15. The maximum modulus principle (Theorem 5.13(a)), which we stated over $\mathbb{C}_p$, is true more generally as long as $K$ has infinite residue field, but otherwise it need not be true. For example, let

$$K = \mathbb{Q}_p \quad \text{and} \quad \phi(z) = z - z^p \quad \text{on } \bar{D}(0,1).$$

Clearly $\|\phi\| = 1$, but for every $\beta \in \mathbb{Z}_p = \bar{D}(0,1)$, Fermat's little theorem tells us that $\beta \equiv \beta^p \pmod{p}$. Hence

$$|\phi(\beta)| \le p^{-1} < \|\phi\| \quad \text{for all } \beta \in \bar{D}(0,1).$$

We conclude this section with some useful consequences of the maximum modulus principle, including the fundamental fact that $p$-adic holomorphic and rational functions send closed disks to closed disks.

**Proposition 5.16.** *Let $\phi(z) \in \mathbb{C}_p[\![z]\!]$ be a nonconstant power series that converges on a disk $\bar{D}(a,r)$ of rational radius.*
(a) *$\phi(\bar{D}(a,r))$ is a closed disk.*
(b) *Write $\phi(\bar{D}(a,r)) = \bar{D}(\phi(a), s)$. Then $|\phi'(a)| \le s/r$.*
(c) *If $\phi'(a) \ne 0$, then there exists a radius $t > 0$ such that*

$$|\phi(z) - \phi(w)| = |\phi'(a)| \cdot |z - w| \qquad \textit{for all } z, w \in \bar{D}(a,t).$$

*Proof.* Replacing $\phi(z)$ with $\phi(cz + a) - \phi(a)$ for some $c \in \mathbb{C}_p$ satisfying $|c| = r$, we may assume that $\phi(z) \in \mathbb{C}_p[\![z]\!]$ converges on $\bar{D}(0,1)$ and that $\phi(0) = 0$. Write $\phi(z) = \sum_{n \ge 1} a_n z^n$ as usual (note that $a_0 = 0$ since $\phi(0) = 0$), and let

$$s = \sup_{z \in \bar{D}(0,1)} |\phi(z)|.$$

(a) The maximum modulus principle (Theorem 5.13(a)) says that

$$s = \|\phi\| = \sup_{n \ge 0} |a_n|.$$

Let $j \ge 1$ be the smallest index such that $s = |a_j|$.

The definition of $s$ clearly implies that $\phi(\bar{D}(0,1)) \subseteq \bar{D}(0,s)$, so we are reduced to proving the opposite inclusion. Let $\beta \in \bar{D}(0,s)$. Consider the Newton polygon of the power series

$$\phi(z) - \beta = -\beta + a_1 z + a_2 z^2 + a_3 z^3 + \cdots.$$

The fact that $|\beta| \le s = |a_j|$ and $|a_n| \le |a_j|$ for all $n \ge 1$ implies that

$$v(\beta) \ge v(a_j) \qquad \text{and} \qquad v(a_n) \ge v(a_j) \quad \text{for all } n \ge 1,$$

so the Newton polygon of $\phi(z) - \beta$ includes one or more line segments connecting $(0, v(\beta))$ to $(j, v(a_j))$. Further, since the point $(0, v(\beta))$ is no lower than the point $(j, v(a_j))$, at least one of those line segments has slope $m \le 0$. The fundamental theorem on Newton polygons (Theorem 5.11) then tells us that the power series $\phi(z) - \beta$ has a least one root $\alpha$ satisfying $|\alpha| = p^m \le 1$. This proves that there is a point $\alpha \in \bar{D}(0,1)$ satisfying $\phi(\alpha) = \beta$, and since $\beta \in \bar{D}(0,s)$ was arbitrary, this completes the proof that $\phi(\bar{D}(0,1)) = \bar{D}(0,s)$.

(b) As in (a), the maximum modulus principle (Theorem 5.13(a)) gives

$$s = \sup_{z \in \bar{D}(0,1)} |\phi(z)| = \|\phi\| = \sup_{n \ge 0} |a_n| \ge |a_1|.$$

This is the desired result, since $r = 1$ and $a_1 = \phi'(0)$.

(c) Continuing with the notation from (b), we compute

$$\begin{aligned}
|\phi(z) - \phi(w)| &= \left| \sum_{n=1}^{\infty} a_n (z^n - w^n) \right| \\
&= |z - w| \cdot \left| \sum_{n=1}^{\infty} a_n \sum_{i=0}^{n-1} z^i w^{n-1-i} \right| \\
&= |z - w| \cdot \left| \phi'(0) + \sum_{n=2}^{\infty} \sum_{i=0}^{n-1} a_n z^i w^{n-1-i} \right|. \qquad (5.5)
\end{aligned}$$

For all $n \ge 2$, all $0 \le i < n$, all $t \le 1$, and all $z, w \in \bar{D}(0,t)$ we have

$$|a_n z^i w^{n-1-i}| \le |a_n| t^{n-1} \le \|\phi\| \cdot t^{n-1} \le \|\phi\| \cdot t.$$

Hence if we choose a value of $t$ satisfying $0 < t < |\phi'(0)|/\|\phi\|$, then the double sum in the righthand side of (5.5) has absolute value strictly smaller than $|\phi'(0)|$, and (5.5) reduces to the desired inequality

$$|\phi(z) - \phi(w)| \le |z - w| \cdot |\phi'(0)|.$$

This completes the proof of (c) and provides an explicit value for $t$. □

**Corollary 5.17.** *In each of the following situations, the indicated map $\phi$ is both open and continuous*:

(a) $\phi : \bar{D}(a,r) \to \mathbb{C}_p$ *is a nonconstant analytic map.*

(b) $\phi : \bar{D}(a,r) \to \mathbb{P}^1(\mathbb{C}_p)$ *is a nonconstant meromorphic map.*

(c) $\phi : \mathbb{P}^1(\mathbb{C}_p) \to \mathbb{P}^1(\mathbb{C}_p)$ *is a nonconstant rational function.*

*Proof.* (a) The continuity is a consequence of Proposition 5.10, which gives the stronger assertion that $\phi$ is Lipschitz. The openness of $\phi$ follows easily from Proposition 5.16, since the collection of "closed" disks (which is also a collection of open sets)

$$\{\bar{D}(b,t) : t > 0 \text{ and } b \in \mathbb{C}_p\}$$

forms a base for the topology of $\mathbb{C}_p$.
(b) Let $b \in \bar{D}(a,r)$. Proposition 5.8(d) says that we can write $\phi$ as a Laurent series in some neighborhood $\bar{D}(b,s)$ of $b$. If $\phi$ does not have a pole at $b$, then (a) completes the proof. If $\phi$ does have a pole at $b$, we consider instead the meromorphic function $1/\phi(z)$. It too can be written as a Laurent series in some neighborhood $\bar{D}(b,s)$, and it has no pole at $b$, so again we are done using (a).
(c) A rational function is clearly everywhere meromorphic, since it is the ratio of two power series (i.e., polynomials) that converge on all of $\mathbb{C}_p$. (For the point at $\infty$, change coordinates and use $1/\phi(z)$.) Hence we are done from (b). □

## 5.4 The Nonarchimedean Julia and Fatou Sets

In this section we recall some basic notions of convergence for collections of functions, define the Fatou and Julia sets of a rational map $\phi(z) \in K(z)$ over a field $K$ with an absolute value, and use a formula from Chapter 1 to show that in the nonarchimedean setting, the Fatou set is always nonempty.

We begin with three definitions.

**Definition.** Let $U$ be an (open) subset of $\mathbb{P}^1(K)$ and let $\Phi$ be a collection of functions $\phi : U \to \mathbb{P}^1(K)$.

(a) $\Phi$ is *equicontinuous* on $U$ if for every $P \in U$ and every $\epsilon > 0$ there exists a $\delta > 0$ such that

$$\phi\big(D(P,\delta)\big) \subset D\big(\phi(P),\epsilon\big) \qquad \text{for every } \phi \in \Phi.$$

(b) $\Phi$ is *uniformly continuous* on $U$ if for every $\epsilon > 0$ there exists a $\delta > 0$ such that

$$\phi\big(D(P,\delta) \cap U\big) \subset D\big(\phi(P),\epsilon\big) \qquad \text{for every } \phi \in \Phi \text{ and every } P \in U.$$

(c) $\Phi$ is *uniformly Lipschitz* on $U$ if there is a constant $C$ such that

$$\rho\big(\phi(P),\phi(Q)\big) \le C \cdot \rho(P,Q) \qquad \text{for every } \phi \in \Phi \text{ and every } P, Q \in U.$$

In the case that $\Phi = \{\phi^n\}$ is the collection of iterates of a single function, we say simply that $\phi$ is equicontinuous, uniformly continuous, or uniformly Lipschitz.

It is important to understand that equicontinuity is weaker than uniform continuity, because equicontinuity is relative to a particular point, while uniform continuity is uniform with respect to all points in $U$. In particular, uniform continuity is an open condition, whereas equicontinuity is not. Similarly, the uniform Lipschitz property is an open condition, and indeed it is even stronger than uniform continuity. The following implications are easy consequences of the definitions:

| uniformly Lipschitz | $\Longrightarrow$ | uniformly continuous | $\Longrightarrow$ | equicontinuous at every point |
|---|---|---|---|---|

As we will discover throughout this chapter, in a nonarchimedean setting it is often just as easy to prove that a family of maps is uniformly Lipschitz as it is to prove that it is equicontinuous.

**Definition.** Assume first that $K$ is algebraically closed. Then the *Fatou set* $\mathcal{F}(\phi)$ is the union of all open subsets of $\mathbb{P}^1(K)$ on which $\phi$ is equicontinuous, i.e., $\mathcal{F}(\phi)$ is the largest open set on which $\phi$ is equicontinuous. The *Julia set* $\mathcal{J}(\phi)$ is the complement of the Fatou set. In general, the Fatou set of $\phi$ over $K$, which we denote by $\mathcal{F}(\phi, K)$, is the intersection of $\mathcal{F}(\phi, \bar{K})$ with $\mathbb{P}^1(K)$. Similarly, the Julia set $\mathcal{J}(\phi, K)$ is the complement of $\mathcal{F}(\phi, K)$ in $\mathbb{P}^1(K)$.

**Proposition 5.18.** *For every integer $n \geq 1$,*

$$\mathcal{F}(\phi^n) = \mathcal{F}(\phi) \qquad \textit{and} \qquad \mathcal{J}(\phi^n) = \mathcal{J}(\phi).$$

*Proof.* We proved this over $\mathbb{C}$ in Proposition 1.25, and the same proof works for nonarchimedean fields using the nonarchimedean Lipschitz property (Theorem 2.14) in place of the archimedean version cited in Chapter 1. □

We recall Theorem 1.14, which says that the multipliers of the fixed points of $\phi$ satisfy

$$\sum_{P \in \operatorname{Fix}(\phi)} \frac{1}{1 - \lambda_P(\phi)} = 1 \qquad \text{provided that } \lambda_P(\phi) \neq 1 \text{ for all } P \in \operatorname{Fix}(\phi).$$

The following corollary of this formula has the useful consequence that nonarchimedean Fatou sets are never empty, a fact that is false in the archimedean setting (cf. Theorem 1.30, Example 1.31, and Theorem 1.43).

**Corollary 5.19.** *Let $K$ be an algebraically closed field of characteristic 0 that is complete with respect to a nonarchimedean absolute value. Let $\phi(z) \in K(z)$ be a rational function of degree $d \geq 2$. Then $\phi$ has a nonrepelling fixed point.*

*Proof.* If some fixed point $P$ has multiplier $\lambda_P(\phi) = 1$, then $P$ is nonrepelling and we are done. Otherwise, we can use Theorem 1.14 to estimate

$$1 = \left| \sum_{P \in \operatorname{Fix}(\phi)} \frac{1}{1 - \lambda_P(\phi)} \right| \leq \max_{P \in \operatorname{Fix}(\phi)} \left| \frac{1}{1 - \lambda_P(\phi)} \right| = \frac{1}{\min\limits_{P \in \operatorname{Fix}(\phi)} \left|1 - \lambda_P(\phi)\right|}.$$

Hence there is at least one fixed point $Q$ satisfying

$$\left|1 - \lambda_Q(\phi)\right| \leq 1.$$

It follows that

$$|\lambda_Q(\phi)| \le \max\{|\lambda_Q(\phi) - 1|, |1|\} \le 1,$$

so $Q$ is nonrepelling. □

**Proposition 5.20.** *Let $K$ be an algebraically closed field of characteristic 0 that is complete with respect to a nonarchimedean absolute value. Let $\phi(z) \in K(z)$ be a rational function of degree $d \ge 2$.*

(a) *Let $P \in \mathbb{P}^1(K)$ be a nonrepelling periodic point for $\phi$. Then $P$ is in the Fatou set $\mathcal{F}(\phi)$.*
(b) *Let $P \in \mathbb{P}^1(K)$ be a repelling periodic point for $\phi$. Then $P$ is in the Julia set $\mathcal{J}(\phi)$.*
(c) *The Fatou set $\mathcal{F}(\phi)$ of $\phi$ is nonempty.*

*Proof.* Making a change of variables, we can move $P$ to 0, and then Proposition 5.18 lets us replace $\phi$ by $\phi^n$, so we may assume that 0 is a fixed point. This puts $\phi$ into the form

$$\phi(z) = \lambda z + \frac{z^2F(z)}{G(z)}$$

with $G(0) \ne 0$ and $\lambda = \lambda_0(\phi) \in K$. Thus $|G(z)|$ is bounded away from 0 if we stay away from the roots of $G$, and clearly $F(z)$ is bounded on any disk around 0, so there are a disk $D(0,r)$ and a constant $C$ such that $|F(z)/G(z)| \le C$ for all $z \in D(0,r)$. Let $\delta = \min\{r, 1/C\}$.
(a) The assumption that 0 is a nonrepelling fixed point means that $|\lambda| \le 1$. Hence for $z \in D(0,\delta)$ we have

$$|\phi(z)| = \left|\lambda z + \frac{z^2F(z)}{G(z)}\right| \le |z| \max\left\{|\lambda|, \left|\frac{zF(z)}{G(z)}\right|\right\} \le |z|.$$

This proves that $\phi$ is nonexpanding on the ball $D(0,\delta)$, so its iterates are uniformly Lipschitz on that disk (with Lipschitz constant 1). Hence 0 is in the Fatou set.
(b) For this part the assumption that 0 is a repelling fixed point means that $|\lambda| > 1$. Hence for $z \in D(0,\delta)$ with $z \ne 0$ we have

$$\left|\frac{z^2F(z)}{G(z)}\right| \le |z| < |\lambda z|.$$

The strict inequality allows us to conclude that

$$|\phi(z)| = \left|\lambda z + \frac{z^2F(z)}{G(z)}\right| = |\lambda z| \quad \text{for all } z \in D(0,\delta).$$

Suppose now that $0 \in \mathcal{F}(\phi)$. This implies that there is some $\epsilon > 0$ such that if $\alpha \in D(0,\epsilon)$, then every iterate satisfies $\phi^n(\alpha) \in D(0,\delta)$. But then

$$|\phi^n(\alpha)| = |\lambda|^n|\alpha| \longrightarrow \infty \quad \text{as } n \to \infty,$$

contradicting $\phi^n(\alpha) \in D(0,\delta)$. Hence $0 \in \mathcal{J}(\phi)$.
(c) Corollary 5.19 says that $\phi$ has a nonrepelling fixed point and (a) tells us that nonrepelling periodic points are in the Fatou set. Hence $\mathcal{F}(\phi) \ne \emptyset$. □

*Remark* 5.21. Corollary 5.19 is also true in characteristic $p$, since it follows directly from the fixed point multiplier formula (Theorem 1.14), which can be proven in characteristic $p$ either by reduction modulo $p$ from the characteristic-0 case or by using the abstract theory of residues (see Exercise 5.10). It follows that Proposition 5.20 is also true in characteristic $p$.

## 5.5 The Dynamics of $(z^2 - z)/p$

In this section we illustrate the general theory by studying the $p$-adic dynamics of the map

$$\phi(z) = \frac{z^2 - z}{p} \qquad \text{for a prime } p \geq 3.$$

Note that $\phi$ has bad reduction at $p$. An important observation is that the Julia set sits inside two disjoint disks.

**Proposition 5.22.** *Let* $\phi(z) = (z^2 - z)/p$.

(a) $\quad |z| > \dfrac{1}{p}$ *and* $|z - 1| > \dfrac{1}{p} \quad \Longrightarrow \quad \lim_{n\to\infty} \bigl|\phi^n(z)\bigr| = \infty.$

(b) $\quad \mathcal{J}(\phi) \subset \bar{D}(0, 1/p) \cup \bar{D}(1, 1/p).$

*Proof.* (a) We consider two cases. First, if $|z| > 1$, then $|z| = |z - 1|$, so we find that

$$\bigl|\phi(z)\bigr| = \left|\frac{z(z-1)}{p}\right| = p \cdot |z| \cdot |z - 1| = p \cdot |z|^2.$$

In particular, $\bigl|\phi(z)\bigr| > 1$, so we can apply this inequality again. Repeating the process shows that

$$\bigl|\phi^n(z)\bigr| = p^{2^n - 1} \cdot |z|^{2^n} \xrightarrow[n\to\infty]{} \infty.$$

Next suppose $|z| \leq 1$. Then $\max\bigl\{|z|, |z-1|\bigr\} = 1$, so using the assumption that $|z| > 1/p$ and $|z - 1| > 1/p$, we compute

$$\bigl|\phi(z)\bigr| = \left|\frac{z(z-1)}{p}\right| = p \cdot \max\bigl\{|z|, |z-1|\bigr\} \cdot \min\bigl\{|z|, |z-1|\bigr\} > 1.$$

Now we can apply the earlier result to $\phi(z)$ to conclude that $\bigl|\phi^n(z)\bigr| \to \infty$.
(b) From (a) we see that every point outside of the two disks $\bar{D}(0, 1/p) \cup \bar{D}(1, 1/p)$ is attracted to the superattracting fixed point at $\infty$, so such points are in the Fatou set. □

The proposition tells us that the Julia set is contained in $\bar{D}(0, 1/p) \cup \bar{D}(1, 1/p)$, but not every point in these disks is in the Julia set. For example,

$$\phi(p\alpha) = p\alpha^2 - \alpha \equiv -\alpha \pmod{p},$$

so if $-\alpha$ is not in $\bar{D}(0,1/p) \cup \bar{D}(1,1/p)$, then $\phi^n(p\alpha)$ is attracted to $\infty$, hence it is in the Fatou set. It is not easy to predict which points in the two disks are attracted to $\infty$. However, since only points with bounded orbit can be in $\mathcal{J}(\phi)$, we let

$$\Lambda = \Lambda(\phi) = \{z \in \mathbb{C}_p : |\phi^n(z)| \text{ is bounded for } n \geq 0\}.$$

It is clear that $\Lambda$ is a completely invariant set, and Proposition 5.22 tells us that $\Lambda$ is contained in two disjoint disks, which for notational convenience we henceforth denote by $I_0$ and $I_1$:

$$\Lambda \subset I_0 \cup I_1, \qquad \text{where} \quad I_0 = \bar{D}(0,1/p) \quad \text{and} \quad I_1 = \bar{D}(1,1/p).$$

**Definition.** The orbit of a point $z \in \Lambda$ is contained within $\Lambda$, so each iterate $\phi^n(z)$ is contained in one of the two disjoint disks $I_0$, $I_1$. The *itinerary of* $z$ is the sequence $[\beta_0\beta_1\beta_2\ldots]$ of numbers $\beta_n \in \{0,1\}$ determined by the condition

$$\phi^n(z) \in I_{\beta_n} \qquad \text{for } n = 0,1,2,\ldots.$$

In other words, the itinerary $[\beta_n]$ specifies how the points $\phi^n(z)$ jump back and forth between the disks $I_0$ and $I_1$.

Our goal is to show that the dynamics of the itineraries of the points in $\Lambda$ accurately reflects the actual dynamics of $\phi$ on the set $\Lambda$. In particular, we will prove that $\Lambda = \mathcal{J}(\phi)$. In order to do this, we make a brief digression to discuss symbolic dynamics.

### 5.5.1 Symbolic Dynamics

Symbolic dynamics is a tool for modeling seemingly more complicated dynamical systems. It has a long history and is used in many areas of mathematics. See [276] for a thorough introduction to the subject. In this section we briefly develop enough of the theory to complete our analysis of the dynamics of $(z^2 - z)/p$.

Let $S = \{\sigma_1, \sigma_2, \ldots, \sigma_s\}$ be a finite set of symbols and let

$$S^{\mathbb{N}} = \left\{ \begin{matrix} \text{sequences of} \\ \text{elements in } S \end{matrix} \right\}$$

be the set of sequences $[\beta_0\beta_1\beta_2\ldots]$, where each $\beta_n \in S$. (Here $\mathbb{N} = \{0,1,2,\ldots\}$ denotes the set of natural numbers.) We put a metric on $S^{\mathbb{N}}$ by fixing a number $p > 1$ and setting

$$\rho(\alpha,\beta) = p^{-(\text{smallest } n \text{ with } \alpha_n \neq \beta_n)}.$$

Thus the more that the initial terms of $\alpha$ and $\beta$ agree, the closer they are to one another. It is easy to check that $\rho$ is a metric, and indeed a nonarchimedean metric:

- $1 \geq \rho(\alpha,\beta) \geq 0$.
- $\rho(\alpha,\beta) = 0$ if and only if $\alpha = \beta$.

- $\rho(\alpha,\gamma) \le \max\{\rho(\alpha,\beta),\rho(\beta,\gamma)\}$.

Symbolic dynamics is the study of the dynamics of continuous maps $S^{\mathbb{N}} \to S^{\mathbb{N}}$.

An important map on $S^{\mathbb{N}}$ is the *left shift map* $L : S^{\mathbb{N}} \to S^{\mathbb{N}}$, defined by

$$L([\beta_0\beta_1\beta_2\ldots]) = [\beta_1\beta_2\beta_3\ldots].$$

In other words, $L$ simply discards the first term in the sequence and shifts each of the remaining terms to the left. More formally, the sequence $L(\beta)$ is defined by $L(\beta)_n = \beta_{n+1}$. The next proposition describes some elementary properties of the map $L$.

**Proposition 5.23.** *Let $S^{\mathbb{N}}$ be the space of $S$-sequences with associated metric as above and let $L : S^{\mathbb{N}} \to S^{\mathbb{N}}$ be the left shift map.*

(a) *If $\rho(\alpha,\beta) < 1$, then $\rho\bigl(L(\alpha),L(\beta)\bigr) = p\cdot\rho(\alpha,\beta)$.*

(b) *$L$ is continuous (indeed Lipschitz), and it is uniformly expanding on each of the "disks"*

$$\{\alpha \in S^{\mathbb{N}} : \alpha_0 = \sigma_i\}, \quad i = 1,2,\ldots,s.$$

(c) *The set $S^{\mathbb{N}}$ contains exactly $s^n$ points satisfying $L^n(\alpha) = \alpha$. In other words, $\mathrm{Per}_n(S^{\mathbb{N}},L)$ has $s^n$ elements.*

(d) *The periodic points of $L$ are dense in $S^{\mathbb{N}}$.*

(e) *There exists a point $\gamma \in S^{\mathbb{N}}$ whose orbit $\mathcal{O}_L(\gamma) = \{L^n(\gamma) : n \ge 0\}$ is dense in $S^{\mathbb{N}}$. (Maps with this property are called* topologically transitive.*)*

*Proof.* (a) The condition $\rho(\alpha,\beta) < 1$ is equivalent to $\alpha_0 = \beta_0$. It is then clear from the definition that $\rho\bigl(L(\alpha),L(\beta)\bigr) = p\cdot\rho(\alpha,\beta)$, since $L(\alpha)_n \ne L(\beta)_n$ if and only if $\alpha_{n+1} \ne \beta_{n+1}$.

(b) The metric $\rho$ satisfies $0 \le \rho \le 1$, so (a) implies that $\rho\bigl(L(\alpha),L(\beta)\bigr) \le p\cdot\rho(\alpha,\beta)$ for all $\alpha,\beta$. Hence $L$ is Lipschitz. Further, (a) says that $L$ is expanding by a factor of $p$ on each of the disks.

(c) A sequence $\alpha \in S^{\mathbb{N}}$ satisfies $L^n(\alpha) = \alpha$ if and only if its first $n$ terms repeat, i.e., it has the form

$$\alpha = [\underbrace{\alpha_0\alpha_1\ldots\alpha_{n-1}}_{\text{initial } n \text{ terms}}\underbrace{\alpha_0\alpha_1\ldots\alpha_{n-1}}_{\text{same } n \text{ terms}}\underbrace{\alpha_0\alpha_1\ldots\alpha_{n-1}}_{\text{same } n \text{ terms}}\ldots]$$

There are $s$ choices for each of $\alpha_0,\alpha_1,\ldots,\alpha_{n-1}$, so $s^n$ possible elements.

(d,e) We leave the proof of these elementary results as exercises for the reader (Exercise 5.12). $\square$

## 5.5.2 The Dynamics of $(z^2 - z)/p$

Recall that we are studying the $p$-adic dynamics of the map $\phi(z) = (z^2 - z)/p$ for a prime $p \ge 3$, and that we have defined

$$I_0 = \bar{D}(0,1/p) \qquad \text{and} \qquad I_1 = \bar{D}(1,1/p),$$
$$\Lambda = \{z \in \mathbb{C}_p : \phi^n(z) \text{ is bounded for all } n \ge 0\} \subset I_0 \cup I_1,$$
$$\beta(z) = [\beta_0\beta_1\beta_2\ldots] = (\text{Itinerary of } z \in \Lambda), \quad \text{where } \beta_n \text{ is defined by } \phi^n(z) \in I_{\beta_n}.$$

The sequence $\beta(z)$ is an element of the space of binary sequences $\{0,1\}^{\mathbb{N}}$, so we obtain a map (of sets)

$$\beta : \Lambda \to \{0,1\}^{\mathbb{N}}, \qquad \beta(z) = \text{itinerary of } z.$$

**Proposition 5.24.** *With notation as above, the itinerary map $\beta : \Lambda \to \{0,1\}^{\mathbb{N}}$ has the following properties:*
(a) *$\beta$ is injective.*
(b) *$\beta(\Lambda \cap \mathbb{Q}_p) = \{0,1\}^{\mathbb{N}}$, i.e., $\beta$ restricted to $\Lambda \cap \mathbb{Q}_p$ is surjective.*
(c) *$\beta$ respects the metrics on $\Lambda$ and $\{0,1\}^{\mathbb{N}}$, i.e.,*

$$|z - w| = \rho\bigl(\beta(z), \beta(w)\bigr) \qquad \text{for all } z, w \in \Lambda.$$

(d) *Let $L : \{0,1\}^{\mathbb{N}} \to \{0,1\}^{\mathbb{N}}$ be the left shift map on $\{0,1\}^{\mathbb{N}}$. Then $\beta \circ \phi = L \circ \beta$. In other words, the following diagram is commutative:*

$$\begin{array}{ccc} \Lambda & \xrightarrow{\ \phi\ } & \Lambda \\ {\scriptstyle\beta}\big\downarrow & & {\scriptstyle\beta}\big\downarrow \\ \{0,1\}^{\mathbb{N}} & \xrightarrow{\ L\ } & \{0,1\}^{\mathbb{N}} \end{array}$$

*Proof.* (a) We begin with the following observation. Let $u$ be 0 or 1.

$$\text{If } z, w \in I_u, \text{ then } \bigl|\phi(z) - \phi(w)\bigr| = p \cdot |z - w|. \tag{5.6}$$

To verify (5.6), we use the assumption that $z, w \in I_u$ to write $z = u + px$ and $w = u + py$ with $|x| \le 1$ and $|y| \le 1$. Then

$$\phi(z) - \phi(w) = (x - y)\bigl((2u - 1) + p(x + y)\bigr).$$

The quantity $2u - 1$ is $\pm 1$, so the final factor is a unit and we have

$$\bigl|\phi(z) - \phi(w)\bigr| = |x - y| = \left|\frac{z - w}{p}\right| = p \cdot |z - w|.$$

Now suppose that $z, w \in \Lambda$ have the same itinerary $\beta$. Then $\phi^n(z)$ and $\phi^n(w)$ are in the same $I_{\beta_n}$ for every $n \ge 0$, so applying (5.6) to these two points, we find that

$$\bigl|\phi^{n+1}(z) - \phi^{n+1}(w)\bigr| = p \cdot \bigl|\phi^n(z) - \phi^n(w)\bigr| \qquad \text{for all } n \ge 0.$$

Hence by induction,

$$\bigl|\phi^n(z) - \phi^n(w)\bigr| = p^n \cdot |z - w| \qquad \text{for all } n \ge 0. \tag{5.7}$$

The lefthand side of (5.7) is bounded as $n \to \infty$, since $z$ and $w$ are in $\Lambda$. Hence we must have $|z - w| = 0$.
(b) Let $w \in I_0 \cup I_1$. We claim that $\phi^{-1}(w)$ consists of two points $\{z_0, z_1\}$, one of which is in $I_0$ and one of which is in $I_1$, and further, if $w \in \mathbb{Q}_p$, then $z_0$ and $z_1$ are in $\mathbb{Q}_p$. To see why this is true, we fix $w$ and solve

$$\phi(z) = w, \qquad \text{or equivalently, solve} \qquad z^2 - z - pw = 0.$$

The quadratic formula, the binomial theorem, and some elementary algebra gives the two solutions explicitly as

$$\frac{1 \pm (1+4pw)^{1/2}}{2} = \frac{1}{2} \pm \frac{1}{2}\sum_{k=0}^{\infty} \binom{1/2}{k} (4pw)^k = \frac{1}{2} \pm \frac{1}{2}\sum_{k=0}^{\infty} \frac{(-1)^{k-1}}{2k-1}\binom{2k}{k}(pw)^k.$$

Note that the series converge for all $|w| \le 1$. The plus sign gives a solution satisfying $z_1 \equiv 1 \pmod{p}$ and the minus sign gives a solution satisfying $z_2 \equiv 0 \pmod{p}$, so one solution is in $I_0$ and the other is in $I_1$. Further, if $w \in \mathbb{Q}_p$, then $z_1$ and $z_2$ are also in $\mathbb{Q}_p$.

We are going to apply this observation to (nonempty) sets $U \subset I_0 \cup I_1$. The inverse image $\phi^{-1}(U)$ of such a set thus consists of two nonempty disjoint pieces, one in $I_0$ and one in $I_1$. And since the inverse images of $\mathbb{Q}_p$ points in $I_0 \cup I_1$ are again in $\mathbb{Q}_p$, we have

$$\phi^{-1}(U \cap \mathbb{Q}_p) = \phi^{-1}(U) \cap \mathbb{Q}_p.$$

In particular, if $U$ contains a point in $\mathbb{Q}_p$, then each of the pieces of $\phi^{-1}(U)$ contains a point in $\mathbb{Q}_p$.

Let $J_0 = I_0 \cap \mathbb{Q}_p$ and $J_1 = I_1 \cap \mathbb{Q}_p$, or equivalently, $J_0$ and $J_1$ are the open unit disks in $\mathbb{Z}_p$ centered at 0 and 1, respectively. For any binary sequence of $\alpha_0\alpha_1\alpha_2\ldots\alpha_n$ with $n \ge 1$, define a set

$$\begin{aligned} J_{\alpha_0\alpha_1\ldots\alpha_n} &= \bigl\{z \in \mathbb{Q}_p : z \in J_{\alpha_0} \text{ and } \phi(z) \in J_{\alpha_1} \text{ and } \phi^2(z) \in J_{\alpha_2} \\ &\qquad\qquad \ldots \text{ and } \phi^n(z) \in J_{\alpha_n}\bigr\} \\ &= \mathbb{Q}_p \cap J_{\alpha_0} \cap \phi^{-1}(J_{\alpha_1}) \cap \phi^{-2}(J_{\alpha_2}) \cap \cdots \cap \phi^{-n}(J_{\alpha_n}). \end{aligned}$$

Notice that if $z \in J_{\alpha_0\alpha_1\ldots\alpha_n} \cap \Lambda$, then the initial $n$ terms in the itinerary of $z$ are $\alpha_0\alpha_1\ldots\alpha_n$.

The sets $J_{\alpha_0\ldots\alpha_n}$ are closed in $\mathbb{Q}_p$, since $\phi$ is continuous and $J_0$ and $J_1$ are closed, and they are nested in the sense that

$$J_{\alpha_0\alpha_1\ldots\alpha_n} = J_{\alpha_0\alpha_1\ldots\alpha_{n-1}} \cap \phi^{-n}(J_{\alpha_n}) \subset J_{\alpha_0\alpha_1\ldots\alpha_{n-1}}.$$

We claim that they are also nonempty. To prove this by induction, suppose that we know that $J_{\beta_0\beta_1\ldots\beta_{n-1}}$ is nonempty for all sequences $\beta_0\beta_1\ldots\beta_{n-1}$ of length $n$. We use the equality

$$J_{\alpha_0\alpha_1\ldots\alpha_n} = J_{\alpha_0} \cap \phi^{-1}\bigl(J_{\alpha_1\alpha_2\ldots\alpha_n}\bigr)$$

and apply the inductive hypothesis to see that $J_{\alpha_1\alpha_2\ldots\alpha_n}$ is nonempty. Then from our earlier remarks we know that $\phi^{-1}(J_{\alpha_1\alpha_2\ldots\alpha_n})$ consists of two pieces, one in $J_0$ and one in $J_1$, so its intersection with $J_{\alpha_0}$ is nonempty.

Let $\alpha = [\alpha_0\alpha_1\alpha_2\ldots]$ be any binary sequence. Then

$$J_\alpha \stackrel{\text{def}}{=} \bigcap_{n\geq 0} J_{\alpha_0\alpha_1\ldots\alpha_n}$$

is the intersection of a nested sequence of nonempty closed bounded subsets of $\mathbb{Q}_p$. By compactness, the set $J_\alpha$ is nonempty, and by construction, any point in $J_\alpha$ has bounded orbit and itinerary $\alpha$. This proves that the itinerary map

$$\beta : \Lambda \cap \mathbb{Q}_p \to \{0,1\}^{\mathbb{N}}$$

is surjective.

We also note that since every point in $J_\alpha$ has itinerary $\alpha$ and since we know from (a) that $\beta$ is injective, it follows that $J_\alpha$ consists of a single point.

(c) If $z = w$, there is nothing to prove. Assume that $z \neq w$ and let $k$ be the first index at which their itineraries diverge, so by definition, $\rho\big(\beta(z), \beta(w)\big) = p^{-k}$. This means that for each $0 \leq n < k$, the points $\phi^n(z)$ and $\phi^n(w)$ are in the same disk (either $J_0$ or $J_1$), but $\phi^k(z)$ and $\phi^k(w)$ are in different disks. Repeated application of (5.6) tells us that

$$\big|\phi^n(z) - \phi^n(w)\big| = p^n \cdot |z - w| \qquad \text{for all } 0 \leq n \leq k. \tag{5.8}$$

On the other hand, the assumption that $\phi^k(z)$ and $\phi^k(w)$ are in different disks (one in $J_0$ and one in $J_1$) implies that

$$\big|\phi^k(z) - \phi^k(w)\big| = 1. \tag{5.9}$$

Combining (5.8) and (5.9) yields

$$|z - w| = \frac{\big|\phi^k(z) - \phi^k(w)\big|}{p^k} = \frac{1}{p^k} = \rho\big(\beta(z), \beta(w)\big).$$

(d) It is clear that the itinerary of $\phi(z)$ is the left shift of the itinerary of $z$. □

Proposition 5.24 allows us to identify the dynamics of the polynomial map $\phi(z) = (z^2 - z)/p$ with the dynamics of the shift map on the space of binary sequences. It then becomes an easy matter to read off a great deal of information about the dynamics of $\phi(z)$ from elementary properties of the shift map.

**Corollary 5.25.** *Let $p \geq 3$ be a prime and let*

$$\phi(z) = \frac{z^2 - z}{p} \qquad \textit{and} \qquad \Lambda = \{z \in \mathbb{C}_p : \phi^n(z) \textit{ is bounded for all } n \geq 0\}.$$

(a) $\mathcal{J}(\phi) = \Lambda \subset \mathbb{Q}_p$.

(b) $\#\mathrm{Per}_n(\phi) = 2^n + 1$ *for all* $n \geq 1$, *and aside from the fixed point at* $\infty$, *every periodic point of* $\phi$ *is repelling.*

(c) *The repelling periodic points are dense in* $\mathcal{J}(\phi)$.

(d) *There exists a point* $w \in \mathcal{J}(\phi)$ *such that the orbit* $\mathcal{O}_\phi(w)$ *is dense in* $\mathcal{J}(\phi)$. (*Thus* $\phi$ *is* topologically transitive *on* $\mathcal{J}(\phi)$.)

*Proof.* Let $\beta : \Lambda \to \{0,1\}^{\mathbb{N}}$ be the itinerary map. Proposition 5.24 tells us that $\beta$ is injective, and further that it is surjective even when restricted to $\Lambda \cap \mathbb{Q}_p$. It follows that $\Lambda \subset \mathbb{Q}_p$, which proves one part of (a).

For the other part, we note that Proposition 5.24 says that $\beta$ is an isomorphism of metric spaces

$$\beta : \Lambda \to \{0,1\}^{\mathbb{N}}.$$

(The proposition says that $\beta$ is bijective and respects the metrics, so its inverse also respects the metrics.) The proposition also tells us that this isomorphism transforms the map $\phi$ into the left shift map $L$, i.e., $\beta \circ \phi = L \circ \beta$. Hence the dynamical properties of $\phi$ acting on $\Lambda$ are identical to the dynamical properties of $L$ acting on the space of binary sequences.

We proved earlier that $\mathcal{J}(\phi) \subset \Lambda$, so

$$\beta : \mathcal{J}(\phi) \xrightarrow{\sim} \mathcal{J}(L).$$

However, the shift map $L$ is uniformly expanding on each of the disks $\beta(I_0)$ and $\beta(I_1)$ (Proposition 5.23(b)). A uniformly expanding map is nowhere equicontinuous, so $\mathcal{J}(L) = L$, from which we deduce that $\mathcal{J}(\phi) = \Lambda$.

(b) The points not in $\Lambda$ are attracted to $\infty$, so the only periodic point in $\mathbb{P}^1(\mathbb{C}_p) \smallsetminus \Lambda$ is the fixed point at $\infty$. The periodic points in $\Lambda$ are determined by the isomorphism

$$\beta : \mathrm{Per}_n(\phi) \cap \Lambda \xrightarrow{\sim} \mathrm{Per}_n(L),$$

and we proved (Proposition 5.23(c)) that $\#\mathrm{Per}_n(L) = 2^n$.

Further, we know that the shift map $L$ is uniformly expanding (by a constant factor $p$) on each of $\beta(I_0)$ and $\beta(I_1)$, so via the metric isomorphism $\beta$ and the identification of $\phi$ with $L$, we find that $\left|(\phi^n)'(z)\right| = p^n$ for every $z \in \mathrm{Per}_n(\phi)$ except $z = \infty$. In particular, every periodic point other than $\infty$ is repelling.

(c,d) The density of the (repelling) periodic points in $\mathcal{J}(\phi)$ and the topological transitivity of $\phi$ on $\mathcal{J}(\phi) = \Lambda$ follow from the corresponding facts for $L$ acting on $\{0,1\}^{\mathbb{N}}$ (Proposition 5.23(d,e)). □

## 5.6 A Nonarchimedean Montel Theorem

In this section we give an important characterization of the Fatou and Julia sets and use it to draw a number of conclusions. The analogous results over $\mathbb{C}$ are classical and due mainly to Fatou and Julia. The nonarchimedean results are due to Hsia, whose paper [208] we follow. We assume throughout that $K$ is complete and algebraically closed.

**Theorem 5.26.** (Hsia [208]) *Let $\Phi$ be a collection of power series that converge on $\bar{D}(a,r)$, and suppose that there is a point $\alpha \in K$ such that*

$$\alpha \notin \bigcup_{\phi\in\Phi} \phi(\bar{D}(a,r)).$$

*Then $\Phi$ is uniformly Lipschitz on $\bar{D}(a,r)$ with respect to the chordal metric $\rho$, so in particular, it is equicontinuous on $\bar{D}(a,r)$.*

*Proof.* We choose some $c \in K$ with $|c| = r$ (see Remark 5.7) and replace each function $\phi(z) \in \Phi$ by $\phi(cz+a)$. This changes the chordal metric by only a bounded amount (Theorem 2.14), so we are reduced to the case of power series converging on the unit disk $\bar{D}(0,1)$. By assumption, the function $\phi(z) - \alpha$ does not vanish on $\bar{D}(0,1)$, so Theorem 5.13(b) tells us that

$$|\phi(z) - \alpha| = \|\phi - \alpha\| \quad \text{is constant for all } z \in \bar{D}(0,1). \tag{5.10}$$

Suppose that $\alpha \ne 0$ (the case $\alpha = 0$ is similar) and consider the following decomposition of the set $\Phi$:

$$\Phi_1 = \{\phi \in \Phi : \|\phi - \alpha\| \le |\alpha|\},$$
$$\Phi_2 = \{\phi \in \Phi : \|\phi - \alpha\| > |\alpha|\}.$$

If $\phi \in \Phi_1$ and $z, w \in \bar{D}(0,1)$, then

$$\begin{aligned} \rho\bigl(\phi(z),\phi(w)\bigr) &\le \bigl|\phi(z) - \phi(w)\bigr| \\ &= \bigl|(\phi(z)-\alpha) - (\phi(w)-\alpha)\bigr| \\ &\le \|\phi - \alpha\| \cdot |z - w| && \text{from Proposition 5.10,} \\ &\le |\alpha| \cdot |z - w| && \text{since } \phi \in \Phi_1, \\ &= |\alpha|\rho(z,w) && \text{since } z,w \in \bar{D}(0,1). \end{aligned}$$

This shows that the functions in $\Phi_1$ satisfy a uniform Lipschitz inequality.

Next let $\phi \in \Phi_2$. Note that the definition of $\Phi_2$ and (5.10) imply that

$$|\phi(z) - \alpha| = \|\phi - \alpha\| > |\alpha| \quad \text{for all } z \in \bar{D}(0,1). \tag{5.11}$$

Using (5.11) and Lemma 2.3, and then applying (5.10) again, we find that

$$\bigl|\phi(z)\bigr| = \bigl|(\phi(z)-\alpha)+\alpha\bigr| = \bigl|\phi(z)-\alpha\bigr| = \|\phi - \alpha\| \quad \text{for all } z \in \bar{D}(0,1). \tag{5.12}$$

We use this formula to compute, for $z, w \in \bar{D}(0,1)$,

$$\begin{aligned} \rho\bigl(\phi(z),\phi(w)\bigr) &= \frac{\bigl|\phi(z)-\phi(w)\bigr|}{\max\bigl\{1,|\phi(z)|\bigr\}\cdot\max\bigl\{1,|\phi(w)|\bigr\}} \\ &= \frac{\bigl|(\phi(z)-\alpha)-(\phi(w)-\alpha)\bigr|}{\max\bigl\{1,|\phi(z)|\bigr\}\cdot\max\bigl\{1,|\phi(w)|\bigr\}} \end{aligned}$$

$$
\begin{aligned}
&\le \frac{\|\phi-\alpha\|\cdot|z-w|}{\max\{1,\|\phi-\alpha\|\}^2} && \text{from (5.12) and Proposition 5.10,}\\
&\le |z-w| && \text{regardless of the value of } \|\phi-\alpha\|,\\
&= \rho(z,w) && \text{since } z,w\in\bar{D}(0,1).
\end{aligned}
$$

Thus functions in the set $\Phi_2$ are nonexpanding, so in particular they satisfy a uniform Lipschitz condition.

Finally, we note that if $\alpha = 0$, then the same argument works using the sets $\Phi_1 = \{\phi \in \Phi : \|\phi\| \le 1\}$ and $\Phi_2 = \{\phi \in \Phi : \|\phi\| > 1\}$. We leave the details to the reader. □

It is now a simple matter to use Theorem 5.26 to prove a nonarchimedean version of Montel's theorem for rational functions.

**Theorem 5.27.** (Nonarchimedean Montel Theorem, Hsia [208]) *Let $\Phi$ be a collection of rational, or more generally meromorphic, functions $\bar{D}(a,r) \to \mathbb{P}^1(K)$, and suppose that the union*

$$\bigcup_{\phi\in\Phi} \phi(\bar{D}(a,r)) \tag{5.13}$$

*omits two or more points of $\mathbb{P}^1(K)$. Then $\Phi$ satisfies a uniform Lipschitz inequality on $\bar{D}(a,r)$, so in particular $\Phi$ is an equicontinuous family of functions on $\bar{D}(a,r)$.*

*Proof.* Let $\alpha = [\alpha_1,\alpha_2]$ and $\beta = [\beta_1,\beta_2]$ be two points of $\mathbb{P}^1(K)$ that are not in the union (5.13). Consider the family of rational (or meromorphic) functions

$$\Psi = \left\{ \frac{\alpha_2 f(z) - \alpha_1 g(z)}{\beta_2 f(z) - \beta_1 g(z)} \;:\; \frac{f(z)}{g(z)} \in \Phi \right\}.$$

By construction, we have

$$\psi(z) \ne \infty \quad \text{for all } \psi\in\Psi \text{ and all } z\in D(a,r),$$

so Proposition 5.8(c) tells us that the functions in $\Psi$ are holomorphic on $D(a,r)$ (i.e., they are given by convergent power series). We also know that $\psi(z) \ne 0$ for all $z \in \bar{D}(a,r)$, so the functions in $\Psi$ omit at least one point in $K$. It follows from Theorem 5.26 that there is a constant $C_1$ such that

$$\rho(\psi(z),\psi(w)) \le C_1\rho(z,w) \quad \text{for all } \psi\in\Psi \text{ and all } z,w\in\bar{D}(a,r). \tag{5.14}$$

Finally, let $A(z)$ be the linear fractional transformation

$$A(z) = \frac{\alpha_2 z - \alpha_1}{\beta_2 z - \beta_1}.$$

The assumption that $\alpha$ and $\beta$ are distinct implies that $A$ is invertible. Then a very special case of Theorem 2.14 (for the rational map $A^{-1}$) says that there is a constant $C_2 = C_2(A) = C_2(\alpha,\beta) > 0$ such that

$$\rho\big(A^{-1}(z), A^{-1}(w)\big) \le C_2\rho(z,w) \qquad \text{for all } z,w \in \mathbb{P}^1(K). \tag{5.15}$$

Further, by construction we have $\phi \in \Phi$ if and only if $A \circ \phi \in \Psi$, so for any $z, w \in \bar{D}(a,r)$,

$$\begin{aligned} C_1\rho(z,w) &\ge \rho\big(A(\phi(z)), A(\phi(w))\big) && \text{from (5.14)},\\ &\ge C_2^{-1}\rho\big(\phi(z),\phi(w)\big) && \text{from (5.15)}. \end{aligned}$$

This completes the proof of the nonarchimedean version of Montel theorem. □

*Remark* 5.28. The proof of Theorem 5.27 is fairly straightforward. Later, in Section 5.10.3.4, we describe a deeper $p$-adic version of Montel's theorem on Berkovich space; see Theorem 5.80.

In the classical setting, there are a number of important properties of the Julia set that follow more or less formally from Montel's theorem. We conclude this section with a few instances.

**Proposition 5.29.** *Let $\phi : \mathbb{P}^1(K) \to \mathbb{P}^1(K)$ be a rational map of degree $d \ge 2$ and let $U \subset \mathbb{P}^1(K)$ be an open set such that $U \cap \mathcal{J}(\phi) \ne \emptyset$. In particular, we are assuming that the Julia set of $\phi$ is nonempty.*
(a) *The set $\bigcup_{n\ge 0} \phi^n(U)$ omits at most one point of $\mathbb{P}^1(K)$.*
(b) *Suppose that the set in* (a) *does omit a point. Then $\phi$ is a polynomial function and the omitted point is the totally ramified fixed point.* (*In other words, there is a change of variables $f \in \mathrm{PGL}_2(K)$ such that $\phi^f(z) \in K[z]$ and the omitted point has been moved to $\infty$.*)

*Proof.* The set $U$ is covered by disks $\bar{D}(a,r)$, which are both open and closed, so it suffices to prove the proposition under the assumption that $U = \bar{D}(a,r)$. If the union omits two or more points of $\mathbb{P}^1(K)$, then Montel's theorem (Theorem 5.27) implies that $\phi$ is equicontinuous on $U$, contradicting the assumption that $U$ contains a Julia point. This proves (a).

If the union omits $\alpha$, then $\phi^{-1}(\alpha) = \{\alpha\}$, so $\alpha$ is a totally ramified fixed point of $\phi$. Hence $\phi$ is a polynomial map by definition (page 17), and after a change of variables it becomes a polynomial in $K[z]$ (Exercise 1.9(c)). □

Let $\phi : \mathbb{P}^1(K) \to \mathbb{P}^1(K)$ be a rational map of degree $d \ge 2$. Recall that a subset $E$ of $\mathbb{P}^1(K)$ is said to be *completely invariant for* $\phi$ if it is both forward and backward invariant,

$$\phi^{-1}(E) = E = \phi(E).$$

Proposition 1.24 says that the Fatou set $\mathcal{F}(\phi)$ and the Julia set $\mathcal{J}(\phi)$ are completely invariant. (The proof in Chapter 1 is over $\mathbb{C}$, but the proof works, mutatis mutandis, for any complete field.)

In Chapter 1 we used the Riemann–Hurwitz formula to characterize all finite completely invariant sets (see Theorem 1.6). More precisely, we showed that a finite completely invariant set $E$ has at most two elements. Further, if $\#E = 1$, then after a change of variables, $\phi(z) \in K[z]$ is a polynomial and $E = \{\infty\}$, and if $\#E = 2$, then again after a change of variables, $\phi(z) = z^d$ or $\phi(z) = z^{-d}$ and $E = \{0, \infty\}$. We now show that except for these trivial cases, the Julia set is the smallest closed completely invariant subset of $\mathbb{P}^1(K)$.

**Proposition 5.30.** *Let $\phi : \mathbb{P}^1(K) \to \mathbb{P}^1(K)$ be a rational map of degree $d \ge 2$, and let $E \subseteq \mathbb{P}^1(K)$ be a closed completely invariant subset for $\phi$ containing at least three points. Then $E$ is an infinite set and $E \supseteq \mathcal{J}(\phi)$.*

*Proof.* Theorem 1.6 tells us that a finite completely invariant subset contains at most two points, so our assumption that $\#E \ge 3$ implies that $E$ is infinite. Notice that the complete invariance of the closed set $E$ implies the complete invariance of its complement $U$, which is an open set. It follows that the union $\bigcup_{n\ge 0} \phi^n(U)$ omits at least two points, since it in fact omits the infinite set $E$. Montel's theorem (Theorem 5.27) tells us that $U \subseteq \mathcal{F}(\phi)$. Hence $E \supseteq \mathcal{J}(\phi)$. □

*Remark* 5.31. Proposition 5.30 tells us that if the Julia set $\mathcal{J}(\phi)$ is nonempty, then it is the smallest closed completely invariant set containing at least two points. (Notice that the case of exactly two points is ruled out by the fact that if $\phi$ has a completely invariant subset containing exactly two points, then $\phi$ is conjugate to either $z^d$ or $z^{-d}$, in which case its Julia set is empty.)

**Corollary 5.32.** *Let $\phi : \mathbb{P}^1(K) \to \mathbb{P}^1(K)$ be a rational map of degree $d \ge 2$, and assume that $\mathcal{J}(\phi) \ne \emptyset$.*

(a) *$\mathcal{J}(\phi)$ has empty interior.*

(b) *Let $P \in \mathcal{J}(\phi)$ and let*

$$\mathcal{O}_\phi^-(P) = \bigcup_{n\ge 0} \phi^{-n}(P)$$

*be the backward orbit of $P$. The Julia set $\mathcal{J}(\phi)$ is equal to the closure of $\mathcal{O}_\phi^-(P)$ in $\mathbb{P}^1(K)$.*

(c) *$\mathcal{J}(\phi)$ is a perfect set, i.e., for every point $P \in \mathcal{J}(\phi)$, the closure of $\mathcal{J}(\phi) \smallsetminus \{P\}$ contains $P$.*

(d) *$\mathcal{J}(\phi)$ is an uncountable set.*

*Proof.* (a) Let $\partial\mathcal{J}(\phi)$ denote the boundary of the Julia set $\mathcal{J}(\phi)$. Theorem 1.24 tells us that $\mathcal{F}(\phi)$ and $\partial\mathcal{J}(\phi)$ are completely invariant, so the same is true of their union $\partial\mathcal{J}(\phi) \cup \mathcal{F}(\phi)$. This union is also closed, since its complement is the interior of $\mathcal{J}(\phi)$. Proposition 5.20 says that the Fatou set $\mathcal{F}(\phi)$ is always nonempty, and since it is open, it must contain infinitely many points. Hence the union $\partial\mathcal{J}(\phi) \cup \mathcal{F}(\phi)$ is an infinite, closed, completely invariant set, so Proposition 5.30 tells us that

$$\mathcal{J}(\phi) \subseteq \partial\mathcal{J}(\phi) \cup \mathcal{F}(\phi).$$

But $\mathcal{J}(\phi)$ and $\mathcal{F}(\phi)$ are disjoint by definition, which proves that $\mathcal{J}(\phi) = \partial\mathcal{J}(\phi)$, i.e., the Julia set has empty interior.

(b) We know that $\mathcal{J}(\phi)$ is completely invariant, so in particular $\mathcal{O}_\phi^-(P) \subset \mathcal{J}(\phi)$ for any point $P \in \mathcal{J}(\phi)$.

Next let $U$ be any open set with $U \cap \mathcal{J}(\phi) \neq \emptyset$. Then Proposition 5.29(a) tells us that $\bigcup_{n\geq 0} \phi^n(U)$ omits at most one point, and Proposition 5.29(a) says that if it does omit a point, that point is a totally ramified fixed point, hence is in the Fatou set. In particular, the possible omitted point cannot be $P$, since $P \in \mathcal{J}(\phi)$ by assumption. This proves that $P \in \bigcup_{n\geq 0} \phi^n(U)$, or equivalently, that there is some $n \geq 0$ such that

$$U \cap \phi^{-n}(P) \neq \emptyset.$$

This proves that every open set $U$ that intersects $\mathcal{J}(\phi)$ nontrivially also intersects the backward orbit $\mathcal{O}_\phi^-(P)$ nontrivially. Hence $\mathcal{J}(\phi)$ is contained in the closure of $\mathcal{O}_\phi^-(P)$.

(c) Let $P_0 \in \mathcal{J}(\phi)$. We claim that the backward orbit $\mathcal{O}_\phi^-(P_0)$ must contain a nonperiodic point. To see this, suppose instead that $\mathcal{O}_\phi^-(P_0)$ consists entirely of periodic points. Then $\phi^{-1}(P_0)$ consists of a single point, so $P_0$ is a totally ramified periodic point and hence in the Fatou set, contrary to assumption. Therefore we can find a nonperiodic point $P_1 \in \mathcal{O}_\phi^-(P_0)$.

The point $P_1$ is in $\mathcal{J}(\phi)$, since $\mathcal{J}(\phi)$ is completely invariant, so (b) tells us that

$$P_0 \in \text{closure of } \mathcal{O}_\phi^-(P_1).$$

On the other hand, $P_0$ is not in $\mathcal{O}_\phi^-(P_1)$, since otherwise $P_1$ would be periodic. Hence

$$P_0 \in \text{closure of } \bigl(\mathcal{J}(\phi) \smallsetminus \{P\}\bigr).$$

(d) The Baire category theorem [387, §5.1,5.2] implies that a nonempty perfect subset of $\mathbb{P}^1(K)$ is uncountable. □

## 5.7 Periodic Points and the Julia Set

Our goal in this section is to show that the Julia set $\mathcal{J}(\phi)$ of a rational map $\phi$ is contained in the closure of the periodic points of $\phi$. We begin with an elementary lemma that is obvious in the classical setting by a compactness argument, but which requires a different proof over a non-locally compact field such as $\mathbb{C}_p$.

**Lemma 5.33.** *Let $\phi_1(z)$ and $\phi_2(z)$ be power series that converge on $\bar{D}(a,r)$, and suppose that $\phi_1\bigl(\bar{D}(a,r)\bigr) \cap \phi_2\bigl(\bar{D}(a,r)\bigr) = \emptyset$. Then*

$$\inf_{z\in\bar{D}(a,r)} \rho\bigl(\phi_1(z), \phi_2(z)\bigr) > 0.$$

*Proof.* Let

$$M_1 = \sup_{z\in\bar{D}(a,r)} \bigl|\phi_1(z)\bigr| \qquad \text{and} \qquad M_2 = \sup_{z\in\bar{D}(a,r)} \bigl|\phi_2(z)\bigr|.$$

The maximum modulus principle (Theorem 5.13(a)) says that there are points $z_1, z_2 \in \bar{D}(a,r)$ such that $\phi_1(z_1) = M_1$ and $\phi_2(z_2) = M_2$. In particular, $M_1$ and $M_2$ are finite, since $\phi_1$ and $\phi_2$ are power series that converge on $\bar{D}(a,r)$.

Let $M = \max\{M_1, M_2, 1\}$. Then for any $z \in \bar{D}(a,r)$ we have

$$\rho\big(\phi_1(z), \phi_2(z)\big) = \frac{\big|\phi_1(z) - \phi_2(z)\big|}{\max\{|\phi_1(z)|, 1\} \cdot \max\{|\phi_2(z)|, 1\}} \ge \frac{\big|\phi_1(z) - \phi_2(z)\big|}{M^2}.$$

On the other hand, the function $\phi_1 - \phi_2$ does not vanish on $\bar{D}(a,r)$ by assumption, so Theorem 5.13(b) tells us that

$$\big|\phi_1(z) - \phi_2(z)\big| = \|\phi_1 - \phi_2\| \qquad \text{for all } z \in \bar{D}(a,r).$$

Hence

$$\inf_{z \in \bar{D}(a,r)} \rho\big(\phi_1(z), \phi_2(z)\big) \ge \frac{\|\phi_1 - \phi_2\|}{M^2} > 0. \qquad \square$$

The next lemma is used in conjunction with Lemma 5.33 to move a varying set of pairs of points $\{\alpha, \beta\}$ to the specific pair $\{0, 1\}$.

**Lemma 5.34.** *Let $A, B \subset \mathbb{C}_p$ be bounded sets that are at a positive distance from one another. In other words, there are constants $\Delta, \delta > 0$ such that*

$$\sup_{\alpha \in A} |\alpha| \le \Delta, \qquad \sup_{\beta \in B} |\beta| \le \Delta, \qquad \text{and} \qquad \inf_{\alpha \in A,\, \beta \in B} \rho(\alpha, \beta) = \delta > 0. \quad (5.16)$$

*For each $(\alpha, \beta) \in A \times B$, define a linear fractional transformation*

$$L_{\alpha,\beta}(z) = (\beta - \alpha)z + \alpha.$$

*Then there is a constant $C > 0$, depending only on $\Delta$ and $\delta$, such that*

$$\rho\big(L_{\alpha,\beta}(z), L_{\alpha',\beta'}(z')\big) \le C \cdot \max\big\{\rho(\alpha, \alpha'), \rho(\beta, \beta'), \rho(z, z')\big\}$$
$$\textit{for all } \alpha, \alpha' \in A, \textit{ all } \beta, \beta' \in B, \textit{ and all } z, z' \in \mathbb{P}^1(\mathbb{C}_p).$$

*Remark* 5.35. Although Lemma 5.34 appears somewhat technical, it is not saying anything mysterious. The linear fractional transformation $L_{\alpha,\beta}$ is determined by the three conditions

$$L_{\alpha,\beta}(0) = \alpha, \qquad L_{\alpha,\beta}(1) = \beta, \qquad L_{\alpha,\beta}(\infty) = \infty.$$

The lemma is asserting, roughly, that if we take two nearby $(\alpha, \beta)$ values, then the associated transformations are close to one another, where we use the *chordal sup norm*

$$\rho(L, L') = \sup_{P \in \mathbb{P}^1(\mathbb{C}_p)} \{\rho(L(P), L'(P))\}$$

to measure the closeness of two maps. Thus Lemma 5.34 is equivalent to the assertion that the map

$$A \times B \times \mathbb{P}^1(\mathbb{C}_p) \longrightarrow \mathbb{P}^1(\mathbb{C}_p), \qquad (\alpha, \beta, z) \longmapsto L_{\alpha,\beta}(z) = (\beta - \alpha)z + \alpha$$

is Lipschitz.

*Proof of Lemma* 5.34. To ease notation, for $x, y \in \mathbb{C}_p$ we write

$$|x, y| = \max\{|x|, |y|\}.$$

We also assume (without loss of generality) that $\Delta \geq 1$ and $\delta \leq 1$. Then for any $\alpha, \alpha' \in A$ and $\beta, \beta' \in B$ we have

$$\begin{aligned} |\alpha - \alpha'| &= \rho(\alpha, \alpha') \cdot |\alpha, 1| \cdot |\alpha', 1| \leq \Delta^2 \rho(\alpha, \alpha'), \\ |\beta - \beta'| &= \rho(\beta, \beta') \cdot |\beta, 1| \cdot |\beta', 1| \leq \Delta^2 \rho(\beta, \beta'). \end{aligned} \tag{5.17}$$

Let $\alpha, \alpha' \in A$, let $\beta, \beta' \in B$, and let $z, z' \in \mathbb{P}^1(\mathbb{C}_p)$. Directly from the definitions of $L_{\alpha,\beta}$ and the chordal metric, we have

$$\rho\bigl(L_{\alpha,\beta}(z), L_{\alpha',\beta'}(z')\bigr) = \frac{\bigl|((\beta' - \alpha')z' + \alpha') - ((\beta - \alpha)z + \alpha)\bigr|}{\max\{|(\beta - \alpha)z + \alpha|, 1\} \cdot \max\{|(\beta' - \alpha')z' + \alpha'|, 1\}}. \tag{5.18}$$

Assuming for the moment that $z \neq \infty$ and $z' \neq \infty$, we multiply out the numerator and estimate it using the triangle inequality:

$$\begin{aligned} &\bigl|((\beta' - \alpha')z' + \alpha') - ((\beta - \alpha)z + \alpha)\bigr| \\ &= \bigl|\beta' z' - \beta z - \alpha' z' + \alpha z + \alpha' - \alpha\bigr| \\ &= \bigl|\beta'(z' - z) + (\beta' - \beta)z - \alpha'(z' - z) + (\alpha - \alpha')z + (\alpha' - \alpha)\bigr| \\ &\leq \max\{|\beta' - \beta| \cdot |z|,\ |\alpha' - \alpha| \cdot |z, 1|,\ |z' - z| \cdot |\alpha', \beta'|\} \\ &\leq \max\{\Delta^2 \rho(\beta, \beta') \cdot |z|,\ \Delta^2 \rho(\alpha, \alpha') \cdot |z, 1|,\ |z' - z| \cdot |\alpha', \beta'|\} \\ &\qquad\qquad\qquad\qquad\qquad\qquad\qquad\qquad\text{from (5.17)}, \\ &\leq \max\{\Delta^2 \rho(\beta, \beta') \cdot |z|,\ \Delta^2 \rho(\alpha, \alpha') \cdot |z, 1|,\ \Delta \cdot |z' - z|\} \\ &\qquad\qquad\qquad\qquad\qquad\qquad\qquad\qquad\text{from (5.16)}, \\ &\leq \max\{\Delta^2 \rho(\beta, \beta') \cdot |z|,\ \Delta^2 \rho(\alpha, \alpha') \cdot |z, 1|,\ \Delta \rho(z, z') \cdot |z, 1| \cdot |z', 1|\} \\ &\qquad\qquad\qquad\qquad\qquad\qquad\qquad\qquad\text{definition of } \rho, \\ &\leq \Delta^2 \cdot \max\{\rho(\beta, \beta'), \rho(\alpha, \alpha'), \rho(z, z')\} \cdot |z, 1| \cdot |z', 1|. \end{aligned}$$

Substituting this into (5.18) and doing a little bit of algebra yields

$$\begin{aligned} &\frac{\rho\bigl(L_{\alpha,\beta}(z), L_{\alpha',\beta'}(z')\bigr)}{\max\{\rho(\beta, \beta'), \rho(\alpha, \alpha'), \rho(z, z')\}} \\ &\qquad \leq \Delta^2 \cdot \frac{|z, 1|}{\max\{|(\beta - \alpha)z + \alpha|, 1\}} \cdot \frac{|z', 1|}{\max\{|(\beta' - \alpha')z' + \alpha'|, 1\}}. \end{aligned} \tag{5.19}$$

We are left to show that the righthand side is bounded in terms of $\delta$ and $\Delta$. By symmetry, it suffices to bound the first fraction.

We consider two cases. First, if $|z| \leq \Delta/\delta$, then we have the trivial estimate

$$\frac{|z, 1|}{\max\{|(\beta - \alpha)z + \alpha|, 1\}} \leq |z, 1| \leq \frac{\Delta}{\delta}.$$

Second, suppose that $|z| > \Delta/\delta$. Then the fact (5.16) that $|\beta - \alpha| \ge \delta$ implies that

$$\bigl|(\beta-\alpha)z\bigr| > \Delta \ge |\alpha|, \qquad \text{so} \qquad |(\beta-\alpha)z+\alpha| = \bigl|(\beta-\alpha)z\bigr| \ge \delta|z|.$$

Hence

$$\frac{|z,1|}{\max\bigl\{|(\beta-\alpha)z+\alpha|,1\bigr\}} = \frac{|z,1|}{\max\{\delta|z|,1\}} \le \frac{1}{\delta} \le \frac{\Delta}{\delta}.$$

Thus $\Delta/\delta$ serves as an upper bound in both cases, and substituting this bound into (5.19) yields the estimate

$$\rho\bigl(L_{\alpha,\beta}(z), L_{\alpha',\beta'}(z')\bigr) \le \frac{\Delta^4}{\delta^2} \max\bigl\{\rho(\beta,\beta'), \rho(\alpha,\alpha'), \rho(z,z')\bigr\}.$$

This completes the proof of Lemma 5.34 with explicit dependence on $\delta$ and $\Delta$ in the case that $z \ne \infty$ and $z' \ne \infty$. The remaining cases are similar and are left to the reader. □

We next prove a version of Montel's theorem in which the two omitted points are allowed to move. Lemma 5.34 is the key technical tool that allows us to uniformly replace the two moving points with two particular points, thereby reducing the proof to our earlier result.

**Theorem 5.36.** (Montel Theorem with Moving Targets, Hsia [208]) *Let $\phi_1, \phi_2$ be power series that converge on $\bar{D}(a,r)$, and suppose that*

$$\phi_1\bigl(\bar{D}(a,r)\bigr) \cap \phi_2\bigl(\bar{D}(a,r)\bigr) = \emptyset.$$

*Further let $\Phi$ be a collection of rational, or more generally meromorphic, functions on $\bar{D}(a,r)$ such that*

$$\phi(z) \ne \phi_1(z) \quad \textit{and} \quad \phi(z) \ne \phi_2(z) \qquad \textit{for all } \phi \in \Phi \textit{ and all } z \in \bar{D}(a,r).$$

*Then $\Phi$ satisfies a uniform Lipschitz inequality on $\bar{D}(a,r)$, so in particular $\Phi$ is an equicontinuous family of functions on $\bar{D}(a,r)$.*

*Proof.* The proof is very similar to the proof of Theorem 5.26, but somewhat more elaborate.

First we note that since $\phi_1, \phi_2 : \bar{D}(a,r) \to K$ have disjoint images, there is at least one point $\alpha$ omitted by both of them. Making a linear change of variables $z \mapsto z - \alpha$, we may assume that $\phi_1$ and $\phi_2$ omit the value 0. Then Theorem 5.13(b) tells us that

$$\bigl|\phi_1(z)\bigr| = \|\phi_1\| \quad \text{and} \quad \bigl|\phi_2(z)\bigr| = \|\phi_2\| \qquad \text{for all } z \in \bar{D}(a,r).$$

We are going to apply Lemma 5.34 to the disjoint bounded sets $\phi_1\bigl(\bar{D}(a,r)\bigr)$ and $\phi_2\bigl(\bar{D}(a,r)\bigr)$. Thus for each point $w \in \bar{D}(a,r)$, if we let

$$L_w(z) = \bigl(\phi_2(w) - \phi_1(w)\bigr)z + \phi_1(w) \in \mathrm{PGL}_2(\mathbb{C}_p),$$

then Lemma 5.34 says that there is a constant $C$ such that

$$\rho\bigl(L_w(z), L_u(z')\bigr) \le C \max\bigl\{\rho\bigl(\phi_1(w), \phi_1(u)\bigr), \rho\bigl(\phi_2(w), \phi_2(u)\bigr), \rho(z, z')\bigr\}$$
$$\text{for all } w, u \in \bar{D}(a,r) \text{ and all } z, z' \in \mathbb{P}^1(\mathbb{C}_p). \quad (5.20)$$

(In the notation of Lemma 5.34, we have set $\alpha = \phi_1(w)$, $\beta = \phi_2(w)$, $\alpha' = \phi_1(u)$, $\beta' = \phi_2(u)$.)

We next use the fact that $\phi_1$ and $\phi_2$ themselves satisfy a Lipschitz condition. More precisely, Proposition 5.10 says that

$$\bigl|\phi_i(w) - \phi_i(u)\bigr| \le \frac{\|\phi_i\|}{r}|w-u| \qquad \text{for all } w, u \in \bar{D}(a,r) \text{ and } i = 1, 2.$$

Since $w$ and $u$ are bounded, this implies that there is a constant $C' \ge 1$ such that

$$\rho\bigl(\phi_i(w), \phi_i(u)\bigr) \le C'\rho(w,u) \qquad \text{for all } w, u \in \bar{D}(a,r) \text{ and } i = 1, 2.$$

Substituting this into the inequality (5.20) yields

$$\rho\bigl(L_w(z), L_u(z')\bigr) \le CC' \max\bigl\{\rho(w,u), \rho(z,z')\bigr\}$$
$$\text{for all } w, u \in \bar{D}(a,r) \text{ and all } z, z' \in \mathbb{P}^1(\mathbb{C}_p). \quad (5.21)$$

We are now ready to prove Theorem 5.36. The idea is that we know that each $\phi \in \Phi$ omits at least two values, but the omitted values may vary with $\phi$, so we use the linear transformations $L_w$ to move the omitted values to two specific points. Thus for each $\phi \in \Phi$ we define a new function

$$\psi_\phi(w) = L_w^{-1}\bigl(\phi(w)\bigr),$$

and we consider the family of functions

$$\Psi = \{\psi_\phi : \phi \in \Phi\}.$$

Each $\psi_\phi$ can be expressed as a rational function of $\phi_1$, $\phi_2$, and $\phi$, so it is a meromorphic function on $\bar{D}(a,r)$. Further, since we are given that $\phi(z) \ne \phi_1(z)$ and $\phi(z) \ne \phi_2(z)$ for all $z \in \bar{D}(a,r)$, it follows that

$$\psi_\phi(w) \ne 0 \qquad \text{and} \qquad \psi_\phi(w) \ne 1 \qquad \text{for all } w \in \bar{D}(a,r).$$

To see why this is true, note that

$$\psi_\phi(w) = 0 \iff L_w^{-1}\bigl(\phi(w)\bigr) = 0 \iff \phi(w) = L_w(0) = \phi_1(w),$$

contradicting the assumption on $\phi$, and similarly for $\psi_\phi(w) = 1$. Hence the family of maps $\Psi$ on $\bar{D}(a,r)$ omits at least the two values 0 and 1, so Montel's theorem (Theorem 5.27) tells us that $\Psi$ satisfies a uniform Lipschitz inequality,

$$\rho\bigl(\psi_\phi(u), \psi_\phi(w)\bigr) \le C''\rho(u,w) \qquad \text{for all } \phi \in \Phi \text{ and all } u, w \in \bar{D}(a,r). \quad (5.22)$$

Using this and our earlier estimates, we compute, for $\phi \in \Phi$ and $u, w \in \bar{D}(a,r)$,

$$\begin{aligned}
\rho(\phi(u),\phi(w)) &= \rho(L_u(\psi_\phi(u)), L_w(\psi_\phi(w))) && \text{by definition of } \psi_\phi,\\
&\le CC' \max\{\rho(u,w), \rho(\psi_\phi(u),\psi_\phi(w))\} && \text{from (5.21)},\\
&\le CC'C''\rho(u,w) && \text{from (5.22)}.
\end{aligned}$$

This completes the proof that the family of maps $\Phi$ is uniformly Lipschitz. □

We now have the tools to prove the main theorem of this section.

**Theorem 5.37.** (Hsia [208]) *Let $\phi(z) \in K(z)$ be a rational function of degree d with $d \ge 2$. Then*

$$\mathcal{J}(\phi) \subset \overline{\mathrm{Per}(\phi)},$$

*i.e., the closure of the periodic points of $\phi$ contains the Julia set of $\phi$.*

*Proof.* We may clearly assume that $\mathcal{J}(\phi)$ is not empty. Take any open set $U$ having nontrivial intersection with $\mathcal{J}(\phi)$. We must show that $U$ contains a periodic point.

The Julia set is a perfect set (Corollary 5.32), so the open set $U$ actually intersects $\mathcal{J}(\phi)$ in infinitely many points. In particular, there is a point $P \in U \cap \mathcal{J}(\phi)$ that is not the image of a ramification point of $\phi$, since $\phi$ has at most $2d-2$ ramification points. This implies that there is a neighborhood $\bar{D}(P,r) \subset U$ of $P$ such that

$$\phi^{-1}(\bar{D}(P,r)) = V_1 \cup V_2 \cup \cdots \cup V_d$$

consists of $d$ disjoint open sets with the property that the maps

$$\phi : V_i \to \bar{D}(P,r) \qquad \text{for } 1 \le i \le d$$

are bijective. In particular, they have inverses

$$\phi_i : \bar{D}(P,r) \xrightarrow{\sim} V_i$$

given by convergent power series. (This is a $p$-adic version of the one variable inverse function theorem. See Exercise 5.5.) We take $i = 1$ and $i = 2$ and consider the maps $\phi_1$ and $\phi_2$ and the disjoint sets $V_1$ and $V_2$ as illustrated in Figure 5.2.

We now examine the effect of applying the iterates $\phi^n$ of $\phi$ to the disk $\bar{D}(P,r)$. The assumption that $\bar{D}(P,r)$ contains a point of the Julia set of $\phi$ means that $\phi$ is not equicontinuous on $\bar{D}(P,r)$, so Theorem 5.36 tells us that the iterates of $\phi$ cannot avoid both of the moving targets described by the power series $\phi_1$ and $\phi_2$. Hence there exists an iterate $\phi^n$ of $\phi$ and a point $Q \in \bar{D}(P,r)$ such that either

$$\phi^n(Q) = \phi_1(Q) \qquad \text{or} \qquad \phi^n(Q) = \phi_2(Q).$$

Applying $\phi$ to both sides and using the fact that $\phi\circ\phi_1$ and $\phi\circ\phi_2$ are both the identity map on $\bar{D}(P,r)$ yields

$$\phi^{n+1}(Q) = Q,$$

so $Q \in \bar{D}(P,r) \subset U$ is a periodic point. □

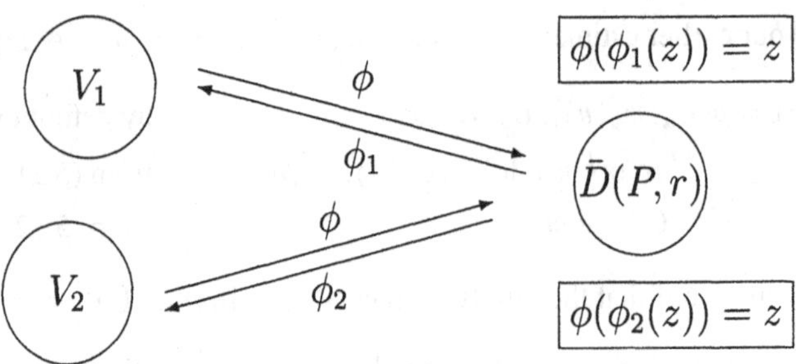


Figure 5.2: Inversion of $\phi$ over a critical-point-free neighborhood.

*Remark* 5.38. In the classical setting over $\mathbb{C}$, one can further show that the Julia set is *equal* to the closure of the repelling periodic points. This follows from the complex analogue of Theorem 5.37 combined with the fact that a rational function over $\mathbb{C}$ has only finitely many nonrepelling periodic points. Unfortunately, a rational function over $\mathbb{C}_p$ may well have nonempty Julia set and infinitely many nonrepelling periodic points. However, one still hopes that the classical result is true in the nonarchimedean setting.

**Conjecture 5.39.** (Hsia [208]) *Let $\phi(z) \in K(z)$ be a rational function of degree $d \geq 2$. Then the Julia set $\mathcal{J}(\phi)$ is equal to the closure of the repelling periodic points of $\phi$.*

Some evidence for Conjecture 5.39 is provided by the following result of Bézivin. It says that if the conjecture is false, then there are maps with nonempty Julia set containing no periodic points.

**Theorem 5.40.** (Bézivin) *If a rational function $\phi(z) \in \mathbb{C}_p(z)$ has at least one repelling periodic point, then $\mathcal{J}(\phi)$ is the closure of the repelling periodic points of $\phi$. In particular, one repelling periodic point implies infinitely many repelling periodic points.*

*Proof.* See [71] for the first assertion. The second then follows immediately from Corollary 5.32(d), since an uncountable set cannot be the closure of a finite set. □

However, some evidence against Conjecture 5.39 is provided by Benedetto [58, Example 9], who shows that it is possible for a rational function to have a sequence of attracting periodic points whose limit is a repelling periodic point! Further, a slight variation of [57, Example 3] shows that for every $n > 0$ there is a polynomial $\phi(z) \in \mathbb{C}_p[z]$ of degree $p + 2$ that has no repelling periodic points of period smaller than $n$, yet $\phi(z)$ does have repelling periodic points of higher periods.

*Example* 5.41. Consider the polynomial map

$$\phi(z) = \frac{z^p - z}{p}.$$

It is clear that the Julia set of $\phi$ is contained in $\bar{D}(0,1)$, since if $|\alpha| > 1$, then $|\alpha^p| > |\alpha|$, so

$$|\phi(\alpha)| = \left|\frac{\alpha^p - \alpha}{p}\right| = p \cdot |\alpha^p - \alpha| = p \cdot |\alpha^p| > p|\alpha|.$$

Hence $\phi^n(\alpha) \to \infty$, so $\alpha \in \mathcal{F}(\phi)$, since $\alpha$ is attracted to the attracting fixed point at infinity.

We also observe that if $\alpha \in \bar{D}(0,1) \cap \mathbb{Q}_p = \mathbb{Z}_p$, then Fermat's little theorem tells us that $\alpha^p \equiv \alpha \pmod{p}$, so $\phi(\alpha) \in \mathbb{Z}_p$. Thus $\mathbb{Z}_p$ is a completely invariant subset of $\phi$. Hsia [206, Example 4.11] (see also [449]) explains how to identify the dynamics of $\phi$ on $\mathbb{Z}_p$ with a shift map on $p$ symbols, similar to the example studied in Section 5.5, from which one deduces the following facts:

- $\mathcal{J}(\phi, \mathbb{Q}_p) = \mathbb{Z}_p$.
- $\mathcal{J}(\phi, \mathbb{Q}_p)$ contains all of the periodic points of $\phi$ (other than $\infty$), so in particular all of the periodic points of $\phi$ are defined over $\mathbb{Q}_p$, and all except $\infty$ are repelling.
- $\mathcal{J}(\phi, \mathbb{Q}_p) = \mathcal{J}(\phi, \mathbb{C}_p)$, since Theorem 5.37 tells us that $\mathcal{J}(\phi, \mathbb{C}_p)$ is contained in the closure of the periodic points of $\phi$. Thus $\mathcal{J}(\phi, \mathbb{C}_p)$ is compact. (See also [73].)

*Example* 5.42. Let $p \geq 5$ be a prime, and let $\phi(z) = pz^3 + az^2 + b$ with $a, b \in \mathbb{Z}_p^*$. We first consider the fixed points of $\phi$, which are the roots of the equation

$$pz^3 + az^2 - z + b = 0.$$

The assumption that $a, b \in \mathbb{Z}_p^*$ implies that the roots satisfy $|\alpha_1| = p$ and $|\alpha_2| = |\alpha_3| = 1$. (Look at the Newton polygon!) We also observe that $p\alpha_1^3$ and $a\alpha_1^2$ have norm $p^2$, while $\alpha_1 - b$ has norm $p$, so $\alpha_1$ must have the form

$$\alpha_1 = -\frac{a}{p} + c \qquad \text{for some } c \text{ with } |c| = 1.$$

This allows us to compute

$$\begin{aligned} |\phi'(\alpha_1)| &= |3p\alpha_1^2 + 2a\alpha_1 - 1| \\ &= |\alpha_1(3p\alpha_1 + 2a) - 1| = |\alpha_1(-a + 3pc) - 1| = |\alpha_1| = p. \end{aligned}$$

Thus $\alpha_1$ is a repelling fixed point of $\phi$.

A similar, but more involved, calculation can be used to show that there are repelling periodic points of higher orders. Alternatively, we can invoke Bézivin's Theorem 5.40, which says that the existence of the single repelling fixed point $\alpha_1$ implies that $\phi$ has infinitely many repelling points whose closure is $\mathcal{J}(\phi)$.

In order to study the periodic points in the Fatou set, we observe that $\phi$ is non-expanding on $\bar{D}(0,1)$. To see this, note that $\phi$ maps the disk $\bar{D}(0,1)$ to itself, so in particular $\|\phi\| \leq 1$. Applying Proposition 5.10 yields

$$|\phi(z) - \phi(w)| \leq |z - w| \qquad \text{for all } z, w \in \bar{D}(0,1).$$

Hence $\phi$ is uniformly Lipschitz on $\bar{D}(0,1)$, so $\bar{D}(0,1) \subset \mathcal{F}(\phi)$. Notice that the $n^{\text{th}}$ iterate of $\phi$ has the following form (think about the reduction of $\phi^n$ modulo $p$, which is the same as the $n^{\text{th}}$ iterate of the reduction $\tilde{\phi}(z) = \tilde{a}z^2 + \tilde{b}$):

$$\phi^n(z) = \underbrace{Az^{3^n} + \cdots + Bz^{2^n+1}}_{\text{coefficients in } p\mathbb{Z}_p} + \underset{\substack{\uparrow \\ C \in \mathbb{Z}_p^*}}{Cz^{2^n}} + \underbrace{Dz^{2^n-1} + \cdots + Ez + F}_{\text{coefficients in } \mathbb{Z}_p}.$$

Again using the Newton polygon, we see that the polynomial $\phi^n(z) - z$ has exactly $2^n$ roots $\alpha$ (counted with multiplicity) satisfying $|\alpha| \le 1$. These roots are points of period dividing $n$ for $\phi$ and they are in the Fatou set, since we showed that $\bar{D}(0,1) \subset \mathcal{F}(\phi)$. One can prove that this gives infinitely many distinct periodic points in $\bar{D}(0,1)$. See [206, Example 4.11].

## 5.8 Nonarchimedean Wandering Domains

We first recall a famous theorem from complex dynamics (see Theorem 1.36).

**Theorem 5.43.** (Sullivan's No Wandering Domains Theorem [426])
*Let $\phi(z) \in \mathbb{C}(z)$ be a rational function of degree $d \ge 2$ over the complex numbers and let $U \subset \mathcal{F}(\phi)$ be a connected component of the Fatou set of $\phi$. Then $U$ is preperiodic in the sense that there are integers $n > m > 0$ such that*

$$\phi^n(U) = \phi^m(U).$$

*In other words, the connected component $U$ does not wander, whence the name of the theorem.*

The first obstacle to translating Sullivan's theorem to the nonarchimedean setting is the fact that $\mathbb{Q}_p$ and $\mathbb{C}_p$ are totally disconnected. Thus with the classical definition of connectivity, there is no good way to break up the Fatou set into "connected" components. Various alternatives have been studied, including disk connectivity, rigid analytic connectivity, and the use of Berkovich spaces.

In this section we consider disk connectivity, which is the simplest of the three to describe. We state a theorem of Benedetto to the effect that a large class of rational maps in $\mathbb{Q}_p(z)$ have no wandering "disk domains" and illustrate the result by proving it under the somewhat stronger hypothesis that there are no critical points in the Julia set of $\phi$. We also note that Benedetto has shown that the statement is false over $\mathbb{C}_p$, that is, there are rational maps over $\mathbb{C}_p$ that do have wandering disk domains.

Much work has been done on this problem, especially in a series of papers by Benedetto [54, 56, 57, 58, 59, 63, 62] and Rivera-Letelier [372, 373, 375, 376, 378], but there are still many open questions. The material that we cover in this section is due to Benedetto.

### 5.8.1 Disk Domains and Disk Components

The ordinary definition of connectivity is not useful for studying the totally disconnected spaces $\mathbb{Q}_p$ and $\mathbb{C}_p$, nor is the notion of path connectedness helpful. We begin with an abstract notion of connectivity that uses a chain of "disks" in place of a path.

**Definition.** Let $X$ be a topological space and let $\mathcal{D}$ be a collection of open subsets of $X$. (In practice, $\mathcal{D}$ will be a base for the toplogy of $X$.) For convenience, we refer to the sets in $\mathcal{D}$ as *disks*.

Let $U \subset X$ be an open subset and let $P \in U$. The *disk component of $U$ containing $P$* (relative to $\mathcal{D}$) is the set of $Q \in U$ with the property that there is a finite sequence of disks $D_1, D_2, \ldots, D_n \subset U$ such that

$$P \in D_1, \qquad Q \in D_n, \qquad D_i \cap D_{i+1} \neq \emptyset \quad \text{for all } 1 \le i < n.$$

In other words, $Q$ is connected to $P$ by a path of disks, as illustrated in Figure 5.3, although as we shall see, in the nonarchimedean world, Figure 5.3 is somewhat misleading. Note that we define disk components only for *open* subsets of $X$. It is easy to see that $U$ breaks up into a disjoint union of disk components (Exercise 5.22).

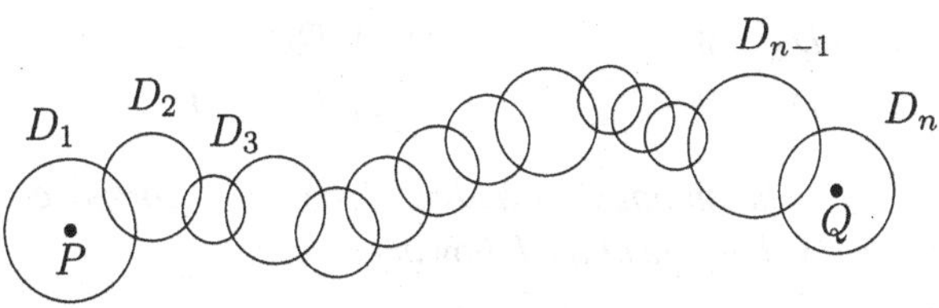


Figure 5.3: A path of disks from $P$ to $Q$.

*Example* 5.44. Let $X = \mathbb{C}$ and let $\mathcal{D}$ be the usual collection of open disks in $\mathbb{C}$. Then the disk components of an open set $U \subset X$ are the same as the usual path-connected components. This is clear, since if $\Gamma$ is a path from $P$ to $Q$, then $\Gamma$ can be covered by open disks contained in $U$, and the compactness of $\Gamma$ shows that it suffices to take a finite subcover. Thus the definition of disk components and the related notion of disk connectivity (see Exercise 5.23) are reasonable. For example, Sullivan's theorem may equally well be stated in terms of the disk components of the open set $\mathcal{F}(\phi)$.

For the purposes of this section, we modify slightly our definition of disks in $\mathbb{P}^1(\mathbb{C}_p)$. The resulting topology is the same, and indeed the disks contained within the unit disk $\bar{D}(0,1)$ are the same, but the alternative definition is more convenient for working with disks that may contain the point at infinity.

**Definition.** The *standard collection of closed disks in* $\mathbb{P}^1(\mathbb{C}_p)$, denoted by $\mathcal{D}_{\text{closed}}$, is given by

$$\mathcal{D}_{\text{closed}} = \begin{cases} \text{all closed disks } \bar{D}(a,r) \text{ and} \\ \text{the complement } \mathbb{P}^1(\mathbb{C}_p) \smallsetminus D(a,r) \text{ of all open disks } D(a,r). \end{cases}$$

Of course, despite the name, all of the disks in $\mathcal{D}_{\text{closed}}$ are both open and closed sets. Note that $\mathcal{D}_{\text{closed}}$ includes all such disks, not only the disks of rational radius (cf. Remark 5.7). One can show (Exercise 5.24) that the disks of rational radius in $\mathcal{D}_{\text{closed}}$ are exactly the images of the unit disk $\bar{D}(0,1)$ via elements of $\text{PGL}_2(\mathbb{C}_p)$.

Similarly, the *standard collection of open disks in* $\mathbb{P}^1(\mathbb{C}_p)$, denoted by $\mathcal{D}_{\text{open}}$, is given by

$$\mathcal{D}_{\text{open}} = \begin{cases} \text{all open disks } D(a,r) \text{ and} \\ \text{the complement } \mathbb{P}^1(\mathbb{C}_p) \smallsetminus \bar{D}(a,r) \text{ of all closed disks } \bar{D}(a,r). \end{cases}$$

In the nonarchimedean world, every disk component has a very simple form. More precisely, it is either a disk, the complement of a single point, or all of $\mathbb{P}^1(\mathbb{C}_p)$.

**Proposition 5.45.** *Let $\mathcal{D}_{\text{open}}$ and $\mathcal{D}_{\text{closed}}$ be, respectively, the collections of standard open and closed disks in $\mathbb{P}^1(\mathbb{C}_p)$ as defined above.*

(a) *Let $D_1, D_2 \in \mathcal{D}_{\text{closed}}$. Then one of the following is true:*

(i) $D_1 \cap D_2 = \emptyset$. (iii) $D_1 \subseteq D_2$.

(ii) $D_1 \cup D_2 = \mathbb{P}^1(\mathbb{C}_p)$. (iv) $D_2 \subseteq D_1$.

(b) *Let $U \subset \mathbb{P}^1(\mathbb{C}_p)$ be an open set and let $V$ be a disk component of $U$ relative to $\mathcal{D}_{\text{closed}}$. Then $V$ has one of the following forms:*

(i) $V = \mathbb{P}^1(\mathbb{C}_p)$. (ii) $V = \mathbb{P}^1(\mathbb{C}_p) \smallsetminus \{P\}$. (iii) $V \in \mathcal{D}_{\text{closed}} \cup \mathcal{D}_{\text{open}}$.

*Proof.* (a) If $D_1 \cup D_2 = \mathbb{P}^1(\mathbb{C}_p)$, we are done. Otherwise, choose any point in the complement of $D_1 \cup D_2$ and use a linear fractional transformation to move that point to $\infty$. This reduces us to the case that neither $D_1$ nor $D_2$ contains $\infty$, so they have the form

$$D_1 = \bar{D}(a_1, r_1) \quad \text{and} \quad D_2 = \bar{D}(a_2, r_2).$$

If $D_1$ and $D_2$ are disjoint, we are done, so we may assume that there is a point

$$\alpha \in D_1 \cap D_2,$$

and switching $D_1$ and $D_2$ if necessary, we may also assume that $r_1 \le r_2$. Let $\beta \in D_1$. Then

$$|\beta - a_2| \le \max\{|\beta - a_1|, |a_1 - \alpha|, |\alpha - a_2|\} \le \max\{r_1, r_1, r_2\} = r_2,$$

so $\beta \in D_2$. Hence $D_1 \subseteq D_2$.

(b) If $V = \mathbb{P}^1(\mathbb{C}_p)$, we are done, so we assume that $V \ne \mathbb{P}^1(\mathbb{C}_p)$. Using a linear fractional transformation to move a point not in $V$ to $\infty$, we are reduced to the case that $\infty \notin V$.

Let $D_1, \ldots, D_n \in \mathcal{D}_{\text{closed}}$ be any path of disks contained in $V$. Each pair of adjacent disks $(D_i, D_{i+1})$ has nonempty intersection, so (a) tells us that one of them is contained within the other. Applying this reasoning to each pair, we see that the union $\bigcup_{i=1}^{n} D_i$ is itself a closed disk, i.e., the union is in $\mathcal{D}_{\text{closed}}$. This shows that every disk path in $V$ actually consists of a single disk.

Fix some point $a \in V$ and let

$$R = \sup\{r \geq 0 : \bar{D}(a, r) \subset V\}.$$

Note that $R > 0$, since $a \in V$ and $V$ is open. If $R = \infty$, then $V = \mathbb{C}_p$ and we are done, so we assume that $R < \infty$. We claim that

$$D(a, R) \subseteq V \subseteq \bar{D}(a, R). \tag{5.23}$$

The lefthand inclusion is clear from the definition of $R$. For the righthand inclusion, suppose that $b \in V$ and consider a disk path from $a$ to $b$ lying within $V$. From our previous remarks, this disk path consists of a single disk $\bar{D}(a, s)$. The definition of $R$ and the fact that $\bar{D}(a, s) \subseteq V$ tell us that $s \leq R$, and then the fact that $b \in \bar{D}(a, s)$ tells us that $b \in \bar{D}(a, R)$. This gives the other inclusion.

But we get even more. If there is even a single $b \in V$ satisfying $|b - a| = R$, then $s = R$ and $\bar{D}(a, R) = \bar{D}(b, s) \subseteq V$, so we find that $V = \bar{D}(a, R) \in \mathcal{D}_{\text{closed}}$. On the other hand, if $|b - a| < R$ for every $b \in V$, then $V \subset D(a, R)$, so (5.23) tells us that $V = D(a, R) \in \mathcal{D}_{\text{open}}$. This completes the proof of Proposition 5.45. □

### 5.8.2 Hyperbolic Maps over Nonarchimedean Fields

In this section we prove that the Julia set of a rational map $\phi$ contains no critical points if and only if $\phi$ is strictly expanding on its Julia set. This result is used later to prove that such maps, which we call *(p-adically) hyperbolic*, satisfy a $p$-adic version of Sullivan's no wandering domains theorem. On first reading, the reader may wish to omit the somewhat technical proof of the main theorem in this section and proceed directly to the application of the theorem in proving Theorem 5.55 in Section 5.8.3.

**Theorem 5.46.** (Benedetto [56]) *Let $K/\mathbb{Q}_p$ be a finite extension of $p$-adic fields and let $\phi(z) \in K(z)$ be a rational function of degree $d \geq 2$. Proposition* 5.20(c) *tells us that $\mathcal{F}(\phi) \neq \emptyset$, so changing variables if necessary, we may assume that $\infty \in \mathcal{F}(\phi)$. Then the following are equivalent*:

(a) *There are no critical points in $\mathcal{J}(\phi)$.*

(b) *For every finite extension $L/K$ there exists an integer $m \geq 1$ such that*

$$\left|(\phi^m)'(\alpha)\right| > 1 \qquad \text{for all } \alpha \in \mathcal{J}(\phi) \cap L.$$

*In other words, $\phi^m$ is strictly expanding on $\mathcal{J}(\phi) \cap L$.*

**Definition.** Let $K/\mathbb{Q}_p$ be a finite extension. A rational function $\phi \in K(z)$ will be called *p-adically hyperbolic* if it satisfies the equivalent conditions of Theorem 5.46. See Exercise 5.25 for the relationship with the classical definition of hyperbolicity.

*Remark* 5.47. The classical analogue of Theorem 5.46 over $\mathbb{C}$ is much weaker. It says that some iterate of $\phi$ is strictly expanding on the Julia set if and only if the closure of the *postcritical set* is disjoint from the Julia set. (The postcritical set is the union of the forward orbits of the critical points.)

*Proof of Theorem* 5.46. The implication (b) $\Rightarrow$ (a) is clear, since if $\alpha$ is a critical point in $\mathcal{J}(\phi)$, we can take $L = K(\alpha)$ and observe that

$$(\phi^m)'(\alpha) = \prod_{i=0}^{m-1} \phi'(\phi^i(\alpha)) = 0 \qquad \text{(since } \phi'(\alpha) = 0\text{)}.$$

Thus the existence of a critical point in $\mathcal{J}(\phi)$ implies that $(\phi^m)'$ has a root in $\mathcal{J}(\phi)$ for every $m$, so (b) cannot hold.

The other implication is more difficult. Using the assumption that $\phi'$ does not vanish on $\mathcal{J}(\phi)$, we can apply Proposition 5.16(c) to every point $\alpha$ in $\mathcal{J}(\phi) \cap L$ to find a disk $\bar{D}(\alpha, r_\alpha)$ with the property that

$$\bigl|\phi(z) - \phi(w)\bigr| = \bigl|\phi'(\alpha)\bigr| \cdot |z - w| \qquad \text{for all } z, w \in \bar{D}(\alpha, r_\alpha).$$

These disks form an open covering of the compact set $\mathcal{J}(\phi) \cap L$, so we can find a finite subcover. Let $\epsilon$ be the smallest radius of the disks in this finite subcover. We then consider a finite covering of $\mathcal{J}(\phi) \cap L$ by disks of radius $\epsilon$, say

$$\mathcal{J}(\phi) \cap L \subset \bar{D}(\alpha_1, \epsilon) \cup \bar{D}(\alpha_2, \epsilon) \cup \cdots \cup \bar{D}(\alpha_n, \epsilon).$$

The (nonarchimedean) triangle inequality implies that each disk in this covering is contained in one of the disks of the previous covering, so we conclude that $\phi$ stretches by a constant factor on each $\bar{D}(\alpha_i, \epsilon)$. In other words, there are constants $S_i$ such that for each $1 \le i \le n$,

$$\bigl|\phi(z) - \phi(w)\bigr| = S_i |z - w| \qquad \text{for all } z, w \in \bar{D}(\alpha_i, \epsilon).$$

The next step is to show that for any fixed $\alpha \in \mathcal{J}(\phi) \cap L$, the set of derivatives

$$\bigl|(\phi^n)'(\alpha)\bigr|, \qquad n = 1, 2, 3, \ldots, \tag{5.24}$$

is unbounded. We prove this by contradiction, using the following claim:

**Claim 5.48.** *Let $\alpha \in \mathcal{J}(\phi) \cap L$ and suppose that there is an $R > 1$ such that*

$$\bigl|(\phi^n)'(\alpha)\bigr| \le R \quad \textit{for all } n \ge 1.$$

*Then*

$$\phi^n\bigl(\bar{D}(\alpha, \epsilon/R)\bigr) \subset \bar{D}\bigl(\phi^n(\alpha), \epsilon\bigr) \quad \textit{for all } n \ge 1. \tag{5.25}$$

*Proof of Claim* 5.48. We verify the claim by induction on $n$. The inclusion (5.25) is clearly true for $n = 0$, since $R > 1$. Suppose the inclusion (5.25) is known for all $0 \le n < N$. Our choice of $\epsilon$ ensures that for any $\beta \in \mathcal{J}(\phi) \cap L$, the map $\phi$ stretches by a constant factor on the disk $\bar{D}(\beta, \epsilon)$, so in particular each of the maps

$$\phi : \phi^n\bigl(\bar{D}(\alpha,\epsilon/R)\bigr) \longrightarrow \phi^{n+1}\bigl(\bar{D}(\alpha,\epsilon/R)\bigr), \qquad 0 \le n < N,$$

stretches by a constant factor. It follows that

$$\phi^N : \bar{D}(\alpha,\epsilon/R) \longrightarrow \phi^N\bigl(\bar{D}(\alpha,\epsilon/R)\bigr)$$

stretches by a constant factor. In other words, there is a constant $S$ such that

$$\bigl|\phi^N(z) - \phi^N(w)\bigr| = S|z-w| \qquad \text{for all } z, w \in \bar{D}(\alpha,\epsilon/R).$$

Taking $w = \alpha$ and letting $z \to \alpha$, we see that $S = \bigl|(\phi^N)'(\alpha)\bigr|$, and then taking $w = \alpha$ and $z \in \bar{D}(\alpha,\epsilon/R)$ arbitrary, we conclude that

$$\begin{aligned} \bigl|\phi^N(z) - \phi^N(\alpha)\bigr| &= \bigl|(\phi^N)'(\alpha)\bigr| \cdot |z-\alpha| \\ &\le R|z-\alpha| \qquad \text{since } \bigl|(\phi^n)'(\alpha)\bigr| \le R \text{ for all } n \text{ by assumption,} \\ &\le \epsilon \qquad \text{since } z \in \bar{D}(\alpha,\epsilon/R). \end{aligned}$$

This shows that the inclusion (5.25) is true for $n = N$, hence for all $n$ by induction. □

Recall that we are in the midst of a proof by contradiction that the derivatives (5.24) are unbounded. Under the assumption that the derivatives are bounded, we have proven the inclusion (5.25), which in turn certainly implies that

$$\bigcup_{n\ge1} \phi^n\bigl(\bar{D}(\alpha),\epsilon/R\bigr) \subset \bigcup_{n\ge1} \bar{D}\bigl(\phi^n(\alpha,\epsilon)\bigr) \subset \bigcup_{\beta\in\mathcal{J}(\phi)\cap L} \bar{D}(\beta,\epsilon).$$

However, the Julia set is bounded (since we assumed that $\infty \in \mathcal{F}(\phi)$), so an $\epsilon$-neighborhood of the Julia set is also bounded. In particular, the above union omits at least two points, so by the nonarchimedean Montel theorem (Theorem 5.27), the map $\phi$ is equicontinuous on $\bar{D}(\alpha,\epsilon/R)$. This is a contradiction, since $\alpha \in \mathcal{J}(\phi)$ by assumption, which completes the proof that the derivatives (5.24) are unbounded.

We now know that for each point $\alpha \in \mathcal{J}(\phi) \cap L$ there is some integer $m_\alpha$ such that $\bigl|(\phi^{m_\alpha})'(\alpha)\bigr| \ge 2$. (There is no significance to the number 2; any number strictly larger than 1 would suffice.) By continuity, there is a disk $\bar{D}(\alpha, s_\alpha)$ such that $\bigl|(\phi^{m_\alpha})'\bigr| \ge 2$ at every point in the disk. Taking a finite subcovering, we can cover $\mathcal{J}(\phi) \cap L$ by disks

$$\mathcal{J}(\phi) \cap L \subset \bar{D}_1 \cup \bar{D}_2 \cup \cdots \cup \bar{D}_t$$

with the property that there are integers $m_1, \ldots, m_t \ge 1$ such that

$$\bigl|(\phi^{m_i})'(z)\bigr| \ge 2 \qquad \text{for all } 1 \le i \le t \text{ and all } z \in \bar{D}_i.$$

For convenience, we may assume that the disks $\bar{D}_1, \ldots, \bar{D}_t$ are pairwise disjoint, since if two closed disks have a point in common, then one is contained in the other and may discarded. It remains to show that there is a single iterate $\phi^m$ that is expanding on all of the disks. In fact, we show that this is true for all sufficiently large $m$.

To ease notation, we let $\psi_i = \phi^{m_i}$. Given any point $\alpha \in \mathcal{J}(\phi) \cap L$, we map out an orbit and an itinerary for $\alpha$ by applying $\psi_i$ whenever we land in $\bar{D}_i$. More formally, define inductively an orbit $\alpha_0, \alpha_1, \ldots$ and a sequence of indices $i_0, i_1, \ldots$ by the following procedure:

$$\begin{aligned}
\alpha_0 &= \alpha && \text{and} && i_0 = (\text{index so that } \alpha_0 \in \bar{D}_{i_0}),\\
\alpha_1 &= \psi_{i_0}(\alpha_0) && \text{and} && i_1 = (\text{index so that } \alpha_1 \in \bar{D}_{i_1}),\\
\alpha_2 &= \psi_{i_1}(\alpha_1) && \text{and} && i_2 = (\text{index so that } \alpha_2 \in \bar{D}_{i_2}),\\
&\;\;\vdots && && \vdots\\
\alpha_k &= \psi_{i_{k-1}}(\alpha_{k-1}) && \text{and} && i_k = (\text{index so that } \alpha_k \in \bar{D}_{i_k}).
\end{aligned}$$

In other words, if $\alpha_n$ is in $\bar{D}_i$, then $\alpha_{n+1}$ is obtained by applying $\psi_i = \phi^{m_i}$ to $\alpha_n$. We note that

$$\bigl|(\psi_{i_n})'(\alpha_{i_n})\bigr| \ge 2 \qquad \text{for all } n = 0, 1, 2, \ldots, \tag{5.26}$$

since by construction, the point $\alpha_{i_n}$ is in $D_{i_n}$ and $\psi_{i_n} = \phi^{m_{i_n}}$.

We next observe that the derivative $\phi'(z)$ is continuous and nonvanishing on the compact set $\mathcal{J}(\phi) \cap L$. (Continuity follows from the assumption that $\infty \notin \mathcal{J}(\phi)$ and nonvanishing is our assumption (a) that $\mathcal{J}(\phi)$ contains no critical points.) It follows that there is a constant $\mu > 0$ such that

$$\bigl|\phi'(\beta)\bigr| \ge \mu \qquad \text{for all } \beta \in \mathcal{J}(\phi) \cap L.$$

(We may assume that $\mu \le 1$.) Using the chain rule and applying this repeatedly yields

$$\bigl|(\phi^n)'(\beta)\bigr| = \prod_{i=0}^{n-1} \bigl|\phi'(\phi^i\beta)\bigr| \ge \mu^n \qquad \text{for all } \beta \in \mathcal{J}(\phi) \cap L \text{ and all } n \ge 0. \tag{5.27}$$

Let $M = \max\{m_1, \ldots, m_t\}$ and let $N$ be any integer satisfying

$$2^N > 4^M \mu^{-M^2}. \tag{5.28}$$

We claim that $\bigl|(\phi^N)'\bigr| > 1$ on $\mathcal{J}(\phi) \cap L$, which will complete the proof of the theorem.

To verify this claim, let $\alpha \in \mathcal{J}(\phi) \cap L$ and let $\alpha_0, \alpha_1, \ldots$ and $i_0, i_1, \ldots$ be the orbit and itinerary of $\alpha$ as described earlier. We use the chain rule to compute

$$(\phi^N)'(\alpha) = \psi'_{i_0}(\alpha_0)\psi'_{i_1}(\alpha_1)\psi'_{i_2}(\alpha_2)\cdots\psi'_{i_k}(\alpha_k) \cdot (\phi^{N - m_{i_0} - m_{i_1} - \cdots - m_{i_k}})'(\alpha_{k+1}).$$

We continue this process until the exponent $N - m_{i_0} - m_{i_1} - \cdots - m_{i_k}$ first becomes smaller than $M$. Notice that this implies that

$$M > N - m_{i_0} - m_{i_1} - \cdots - m_{i_k} \ge N - (k+1)M,$$

so $k + 2 > N/M$. We have thus found integers $k$ and $r$ satisfying

$$(\phi^N)'(\alpha) = \psi'_{i_0}(\alpha_0)\psi'_{i_1}(\alpha_1)\psi'_{i_2}(\alpha_2)\cdots\psi'_{i_k}(\alpha_k)\cdot(\phi^r)'(\alpha_{k+1})$$
$$\text{with } k > N/M - 2 \text{ and } r < M.$$

Hence

$$\begin{aligned} \left|(\phi^N)'(\alpha)\right| &= \underbrace{\left|\psi'_{i_0}(\alpha_0)\psi'_{i_1}(\alpha_1)\psi'_{i_2}(\alpha_2)\cdots\psi'_{i_k}(\alpha_k)\right|}_{\geq 2^k \text{ from (5.26)}} \cdot \underbrace{\left|(\phi^r)'(\alpha_{k+1})\right|}_{\geq \mu^r \text{ from (5.27)}} \\ &\geq 2^{N/M-2}\cdot\mu^M \qquad \text{since } k > N/M - 2 \text{ and } r < M, \\ &> 1 \qquad \text{from the choice (5.28) of } N. \end{aligned}$$

This shows that $|(\phi^N)'(\alpha)| > 1$ for all $N > M^2\log_2(\mu^{-1})$ and all $\alpha \in \mathcal{J}(\phi) \cap L$, which completes the proof of Theorem 5.46. □

### 5.8.3 Wandering Disk Domains

If $U$ is a disk component of the Fatou set $\mathcal{F}(\phi)$ of a rational map $\phi \in \mathbb{C}_p(z)$, then $\phi(U)$ may not be a full disk component of $\mathcal{F}(\phi)$. This situation, which does not occur in the complex case, prompts the following definition.

**Definition.** Let $\phi(z) \in \mathbb{C}_p(z)$ be a rational map defined over a nonarchimedean field, and let

$$\mathcal{DC}(\phi) = \{\text{disk components of the Fatou set } \mathcal{F}(\phi)\}$$

be the collection of disk components of the Fatou set of $\phi$. Then $\phi$ induces a map of the set $\mathcal{DC}(\phi)$ to itself according to the rule

$$\mathcal{DC}(\phi) \longrightarrow \mathcal{DC}(\phi), \qquad U \longmapsto (\text{disk component of } \mathcal{F}(\phi) \text{ containing } \phi(U)). \tag{5.29}$$

We say that $U \in \mathcal{DC}(\phi)$ is *periodic*, *preperiodic*, or *wandering* according to the behavior of $U$ under iteration of the map (5.29).

*Example* 5.49. Let $p$ be an odd prime and let

$$\phi(z) = \frac{z^2 - z}{p}$$

be the function that we studied in Section 5.5. We proved (Proposition 5.22) that the Julia set of $\phi$ is contained in the union of two open disks,

$$\mathcal{J}(\phi) \subset D(0,1) \cup D(1,1),$$

and that $\mathcal{J}(\phi)$ contains the repelling fixed points 0 and 1.

For $\alpha \in \mathbb{C}_p$, let $B(\alpha)$ denote the disk component of $\alpha$ in $\mathcal{F}(\phi)$. We claim that $B(-1) = D(-1,1)$. To see this, we observe that $B(-1)$ cannot contain any larger disk, since it does not contain 0. On the other hand, $D(-1,1)$ is in $\mathcal{F}(\phi)$, since it is disjoint from $D(0,1) \cup D(1,1)$. Hence $B(-1) = D(-1,1)$.

Now consider the image point $\phi(-1) = 2p^{-1}$ and the image of the associated disk component

$$\phi\big(B(-1)\big) = \phi\big(D(-1,1)\big) = D(2p^{-1},p).$$

The disk $D(2p^{-1},p)$ is contained in $\mathcal{F}(\phi)$, but it is certainly not the largest disk around $2p^{-1}$ contained in $\mathcal{F}(\phi)$. Indeed,

$$2p^{-1} \in \mathbb{P}^1(\mathbb{C}_p) \smallsetminus \bar{D}(0,1) \subset \mathcal{F}(\phi).$$

It is not hard to check that $\mathbb{P}^1(\mathbb{C}_p) \smallsetminus \bar{D}(0,1)$ is the full disk component of $\mathcal{F}(\phi)$ containing $\phi\big(B(-1)\big)$.

**Conjecture 5.50.** (No Wandering Disk Domains Conjecture) *Let $K/\mathbb{Q}_p$ be a finite extension and let $\phi(z) \in K(z)$ be a rational map of degree $d \geq 2$. Then the Fatou set $\mathcal{F}(\phi)$ has no wandering disk components.*

Benedetto proves Conjecture 5.50 for a large class of rational maps. In order to state his result, we give four definitions (some of which we already know):

**Definition.** Let $\phi \in \mathbb{C}_p(z)$ be a rational map of degree $d \geq 2$ and let $P \in \mathbb{P}^1(\mathbb{C}_p)$. We say that $P$ is:

(i) *Julia* if $P \in \mathcal{J}(\phi)$.

(ii) *critical* if $e_P(\phi) \geq 2$.

(iii) *wildly critical* if $e_P(\phi) \equiv 0 \pmod{p}$.

(iv) *recurrent* if there is a sequence of integers $n_i \to \infty$ such that $\phi^{n_i}(P) \to P$, i.e., if $P$ is in the closure of the set $\{\phi^n(P) : n \geq 1\}$.

**Theorem 5.51.** (Benedetto [54]) *Let $K/\mathbb{Q}_p$ be a finite extension of p-adic fields and let $\phi(z) \in K(z)$ be a rational map of degree $d \geq 2$. Assume that $\phi$ has no wildly critical recurrent Julia points (defined over $\bar{K}$). Then the Fatou set $\mathcal{F}(\phi)$ has no wandering disk components.*

We make three short observations concerning Theorem 5.51.

*Remark* 5.52. If $p$ is odd, then Theorem 5.51 is true for "almost all" rational maps in $\mathbb{C}_p(z)$. This is true because if $\phi(z)$ has a wild critical point $P$, then in particular it has a point whose ramification index satisfies $e_P(\phi) \geq p \geq 3$. It is not hard to show that all of the critical points of a "generic" rational map of degree $d$ have ramification index equal to 2. (See Exercise 4.23 for a more precise statement.)

*Remark* 5.53. For a fixed degree $d$, Theorem 5.51 is true for all primes $p > d$, since $p > d$ rules out the existence of wild critical points.

*Remark* 5.54. If a recurrent critical point $P$ is in the Fatou set $\mathcal{F}(\phi)$, then one can show that $P$ is in fact periodic (Exercise 5.26). On the other hand, if $P$ is critical, then locally around $P$ the map $\phi$ looks like $\phi(z) = \phi(P) + a(z - z(P))^k + \cdots$ for some $a \in \mathbb{C}_p$ and some $k \geq 2$. Thus if $Q$ is sufficiently close to $P$, then

$\rho(\phi(Q), \phi(P)) = \rho(P, Q)^k$, so $\phi$ is highly contractive near $P$. And if $P$ is also recurrent, then $\phi^n(P)$ gets very close to $P$ infinitely often, so it receives the highly contractive effect of $\phi$ infinitely often. This should cause points near $P$ to remain near $P$, and thus force $P$ into the Fatou set. The critical recurrent condition and the Julia condition are thus in opposition to one another, which means that nonperiodic recurrent critical points should be quite rare. On the other hand, Rivera-Letelier [378] has shown that there are maps having wildly critical recurrent points (which are then necessarily in the Julia set) in $\mathbb{P}^1(\mathbb{Q}_p)$. It is not known whether Rivera-Letelier's examples have wandering disk domains.

In the other direction, it is known that there are rational maps defined over $\mathbb{C}_p$ that have wandering disk domains [63, 59, 62, 378]. In these examples the critical points are in the Fatou set, so Theorem 5.55 implies that the maps cannot be defined over a finite extension of $\mathbb{Q}_p$.

We now use Theorem 5.46 and a simple compactness argument to prove that the Fatou sets of $p$-adically hyperbolic maps over finite extensions of $\mathbb{Q}_p$ have no wandering disk domains. This is not as strong as Theorem 5.51, but still covers an important class of maps. The proof of Theorem 5.51 uses similar ideas, but is more complicated; see [54].

**Theorem 5.55.** (Benedetto [56]) *Let $K/\mathbb{Q}_p$ be a finite extension of p-adic fields, let $\phi(z) \in K(z)$ be a rational map of degree $d \geq 2$, and assume that the Julia set $\mathcal{J}(\phi)$ contains no critical points of $\phi$, i.e., assume that $\phi$ is p-adically hyperbolic. Then the Fatou set $\mathcal{F}(\phi)$ has no wandering disk components.*

*Proof.* Proposition 5.20 assures us that $\mathcal{F}(\phi)$ is nonempty, and it is open, so it contains an algebraic point. (Note that $\bar{\mathbb{Q}}_p$ is dense in $\mathbb{C}_p$.) Replacing $K$ by a finite extension and changing coordinates, we may assume that $\infty \in \mathcal{F}(\phi)$, and indeed we may even assume that $\mathcal{J}(\phi) \subset \bar{D}(0,1)$. Equivalently, we may assume that the disk component of $\infty$ contains $\mathbb{P}^1(\mathbb{C}_p) \setminus \bar{D}(0,1)$.

We suppose that $U \subset \mathcal{F}(\phi)$ is a wandering disk component of $\mathcal{F}(\phi)$ and derive a contradiction. Replacing $U$ with the disk component containing $\phi^n(U)$ for a sufficiently large $n$, we may assume that the orbit of $U$ does not include the disk component at $\infty$. In particular, $\phi^n(U) \subset \bar{D}(0,1)$ for all $n \geq 0$.

Again taking a finite extension of $K$ if necessary, we can find a point $\alpha_0 \in K$ and a radius $r_0$ such that

$$\bar{D}(\alpha_0, r_0) \subset U.$$

At this stage we fix the field $K$ and we use Theorem 5.46 to find an integer $m$ such that $|(\phi^m)'| > 1$ on $\mathcal{J}(\phi) \cap K$. Replacing $\phi$ by $\phi^m$, it suffices to consider the case that $|\phi'| > 1$ on $\mathcal{J}(\phi) \cap K$.

For each $n \geq 0$, the image $\phi^n(\bar{D}(\alpha_0, r_0))$ is a (closed) disk centered at the point $\alpha_n = \phi^n(\alpha_0)$, say

$$\phi^n\big(\bar{D}(\alpha_0, r_0)\big) = \bar{D}(\alpha_n, r_n).$$

In particular, $\phi(\bar{D}(\alpha_n, r_n)) = \bar{D}(\alpha_{n+1}, r_{n+1})$, so applying Proposition 5.16(b) yields

$$r_n|\phi'(\alpha_n)| \le r_{n+1}. \tag{5.30}$$

We also know that the disks $\bar{D}(\alpha_n, r_n)$ are disjoint, since $\bar{D}(\alpha_n, r_n)$ is contained in $\phi^n(U)$, and further, each disk $\bar{D}(\alpha_n, r_n)$ contains a point of $\bar{D}(0,1) \cap K$. It follows that the radii must satisfy

$$\lim_{n\to\infty} r_n = 0, \tag{5.31}$$

since the set $\bar{D}(0,1) \cap K$ has finite volume, so can contain only finitely many nonempty disjoint disks of any fixed radius.

It follows from (5.30) and (5.31) that there are infinitely many $n$ with the property that

$$|\phi'(\alpha_n)| < 1,$$

i.e., since $r_n \to 0$, we must have $r_{n+1} < r_n$ infinitely often. Consider the infinite set of points

$$\{\alpha_n : |\phi'(\alpha_n)| < 1, \quad n = 1, 2, 3, \ldots\}.$$

It is contained in the compact set $\bar{D}(0,1) \cap K$, so it contains an accumulation point $\beta \in K$. The continuity of $\phi'$ implies that $|\phi'(\beta)| \le 1$, which shows that $\beta \in \mathcal{F}(\phi)$, since we used Theorem 5.46 to ensure that $|\phi'| > 1$ on $\mathcal{J}(\phi)$.

Let $V$ be the disk component of $\mathcal{F}(\phi)$ containing $\beta$. Then by construction $V$ contains infinitely many of the iterates $\alpha_n = \phi^n(\alpha_0)$. Since the radii of $\phi^n(U)$ and $\phi^n(V)$ shrink to 0 as $n \to \infty$, it follows that some iterate $\phi^n(U)$ is contained in $V$ and that some (nontrivial) iterate of $\phi^n(V)$ is contained in $V$. Therefore $U$ is not wandering, contradicting our original assumption. □

### 5.8.4 Wandering Disk Domains Exist in $\mathbb{C}_p$

We have proven that $p$-adically hyperbolic maps defined over $\bar{\mathbb{Q}}_p$ have no wandering disk domains. More generally, Theorem 5.51 shows that rational maps $\phi(z) \in \bar{\mathbb{Q}}_p(z)$ with wandering disk domains are very rare, if they exist at all. And of course, Sullivan's theorem 5.43 says that rational maps $\phi(z) \in \mathbb{C}(z)$ defined over the complex numbers never have wandering domains. It is thus somewhat surprising to discover that there are very simple polynomial maps defined over $\mathbb{C}_p$ that have wandering disk domains.

**Theorem 5.56.** (Benedetto [59]) *For $c \in \mathbb{C}_p$, let $\phi_c(z)$ be the polynomial*

$$\phi_c(z) = (1-c)z^{p+1} + cz^p.$$

*Then there exists a value $a \in \mathbb{C}_p$ such that:*
(1) *$\mathcal{J}(\phi_a) \ne \emptyset$,*
(2) *$\mathcal{F}(\phi_a)$ has a wandering disk domain,*
(3) *$\mathcal{F}(\phi_a)$ contains every critical point of $\phi_a$.*

*Proof.* See [59] for a proof of this specific theorem, and see [63, 62, 373, 378, 380] for generalizations and related results. □

## 5.9 Green Functions and Local Heights

The canonical height $\hat{h}_\phi$ associated to a morphism $\phi : \mathbb{P}^1 \to \mathbb{P}^1$ is defined as the limit of $d^{-n}h(\phi^n(P))$ as $n \to \infty$. The utility of $\hat{h}_\phi$ lies in the two formulas

$$\hat{h}_\phi(P) = h(P) + O(1) \qquad \text{and} \qquad \hat{h}_\phi(\phi(P)) = d\hat{h}_\phi(P),$$

where the first says that $\hat{h}_\phi$ contains arithmetic information and the second says that $\hat{h}_\phi$ transforms canonically.

It is tempting to try a similar construction locally and define (say)

$$(v\text{-adic local height of } \alpha) = \lim_{n\to\infty} \frac{1}{d^n} \log \max\{|\phi^n(\alpha)|_v, 1\}. \tag{5.32}$$

It is clear that if the limit (5.32) exists, then it transforms canonically, and indeed if $\phi(z)$ is a polynomial, then the limit does exist and everything works quite well (see Exercise 3.24). Unfortunately, for general rational maps the limit (5.32) may not exist.

Rather than working directly with a rational map $\phi : \mathbb{P}^1 \to \mathbb{P}^1$, it turns out to be easier to develop a theory of local heights by first lifting $\phi$ to a map $\Phi : \mathbb{A}^2 \to \mathbb{A}^2$ and then constructing a real-valued function $\mathcal{G}$ on $\mathbb{A}^2$ that satisfies the canonical transformation formula $\mathcal{G}(\Phi(x,y)) = d\mathcal{G}(x,y)$. In this section we construct the Green function $\mathcal{G}$, prove some of its basic properties, and then use $\mathcal{G}$ to construct local canonical height functions on $\mathbb{P}^1$ as described in Theorem 3.27.

A point in projective space $[x,y] \in \mathbb{P}^1$ is given by homogeneous coordinates, so it is really an equivalence class of pairs $(x,y)$. We make explicit the natural projection map

$$\pi : (\mathbb{A}^2 \smallsetminus \{0,0\}) \longrightarrow \mathbb{P}^1, \qquad (x,y) \longmapsto [x,y],$$

that sends a point $(x,y) \in \mathbb{A}^2$ to its equivalence class $[x,y] \in \mathbb{P}^1$. To ease notation, we write

$$\mathbb{A}^2_* = \mathbb{A}^2 \smallsetminus \{0,0\}$$

for the affine plane with the origin removed.

Let $\phi : \mathbb{P}^1 \to \mathbb{P}^1$ be a rational map of degree $d$. Then $\phi$ can be written as usual in the form $\phi = [F,G]$ with homogeneous polynomials $F$ and $G$ of degree $d$ having no common factors. The polynomials $F$ and $G$ then define a map

$$\Phi : \mathbb{A}^2 \to \mathbb{A}^2, \qquad \Phi(x,y) = (F(x,y), G(x,y)),$$

that fits into the commutative diagram

$$\begin{array}{ccc} \mathbb{A}^2_* & \xrightarrow{\ \Phi\ } & \mathbb{A}^2_* \\ {\scriptstyle\pi}\big\downarrow & & {\scriptstyle\pi}\big\downarrow \\ \mathbb{P}^1 & \xrightarrow{\ \phi\ } & \mathbb{P}^1 \end{array}$$

We call $\Phi$ a *lift of* $\phi$. By homogeneity, if $\Phi = [F,G]$ is one lift of $\phi$, then every other lift of $\phi$ has the form $c\Phi = [cF, cG]$ for some constant $c \in K^*$.

**Definition.** Let $K$ be a field and let $v$ be an absolute value on $K$. We denote the *absolute value* (or *sup norm*) *of a point* $(x, y) \in \mathbb{A}^2(K)$ by

$$\|(x,y)\|_v = \max\{|x|_v, |y|_v\}.$$

Similarly, the absolute value (or sup norm) of one or more polynomials is the maximum of the absolute values of their coefficients. (We have already made use of this convention in the proof of Theorem 3.11.)

We begin by recalling how a map $\Phi : \mathbb{A}^2(K) \to \mathbb{A}^2(K)$ affects the $v$-adic norm of a point.

**Proposition 5.57.** *Let $K$ be a field with an absolute value $v$. Let*

$$\Phi = (F, G) : \mathbb{A}^2 \to \mathbb{A}^2$$

*be given by homogeneous polynomials $F, G \in K[x, y]$ of degree $d \geq 1$, and assume that $F$ and $G$ have no common factors in $K[x, y]$.*
(a) *There are constants $c_1, c_2 > 0$, depending only on $\Phi$ and $v$, such that*

$$c_1 \leq \frac{\|\Phi(x,y)\|_v}{\|(x,y)\|_v^d} \leq c_2 \qquad \textit{for all } (x,y) \in \mathbb{A}^2_*(K). \tag{5.33}$$

(b) *If $v$ is nonarchimedean and $\Phi$ satisfies $\|\Phi\|_v = 1$, then* (a) *is true with the explicit constants*

$$|\mathrm{Res}(F,G)|_v \leq \frac{\|\Phi(x,y)\|_v}{\|(x,y)\|_v^d} \leq 1 \qquad \textit{for all } (x,y) \in \mathbb{A}^2_*(K). \tag{5.34}$$

*Proof.* (a) We proved inequality (5.33) for general morphisms $\mathbb{P}^N \to \mathbb{P}^M$ during the course of proving Theorem 3.11. More precisely, see (3.6) on page 92 for the upper bound with an explicit value for $c_2(\Phi, v)$, and see (3.7) on page 93 for the lower bound.
(b) By homogeneity, it suffices to prove (5.34) for points satisfying $\|(x,y)\|_v = 1$. Then the upper bound is obvious from the triangle inequality, and the lower bound was proven during the course of proving Theorem 2.14, see (2.5) on page 57. □

The next result describes a kind of $v$-adic canonical height associated to a map $\Phi : \mathbb{A}^2 \to \mathbb{A}^2$. The construction is the same as the one that we used to construct canonical heights in Section 3.4.

**Proposition 5.58.** *Let $K$ be a field with an absolute value $v$, let $\phi : \mathbb{P}^1 \to \mathbb{P}^1$ be a morphism of degree $d \geq 2$, and let $\Phi = (F, G) : \mathbb{A}^2 \to \mathbb{A}^2$ be a lift of $\phi$.*
(a) *For all $(x, y) \in \mathbb{A}^2_*(K)$ the following limit exists:*

$$\mathcal{G}_\Phi(x,y) = \lim_{n\to\infty} \frac{1}{d^n} \log\|\Phi^n(x,y)\|_v. \tag{5.35}$$

*We call $\mathcal{G}_\Phi$ the* Green function *of $\Phi$.*

(b) *The Green function is the unique function* $\mathbb{A}^2_*(K) \to \mathbb{R}$ *having the following two properties*:

$$\mathcal{G}_\Phi\big(\Phi(x,y)\big) = d\mathcal{G}_\Phi(x,y) \qquad \textit{for all } (x,y) \in \mathbb{A}^2_*(K). \tag{5.36}$$

$$\mathcal{G}_\Phi(x,y) = \log\big\|(x,y)\big\|_v + O(1) \qquad \textit{for all } (x,y) \in \mathbb{A}^2_*(K). \tag{5.37}$$

(c) *If* $v$ *is nonarchimedean and* $\Phi$ *satisfies* $\|\Phi\|_v = 1$ *and* $\big|\mathrm{Res}(F,G)\big|_v = 1$*, i.e., if the map* $\phi = [F,G] : \mathbb{P}^1 \to \mathbb{P}^1$ *has good reduction at* $v$*, then*

$$\mathcal{G}_\Phi(x,y) = \log\big\|(x,y)\big\|_v \qquad \textit{for all } (x,y) \in \mathbb{A}^2_*(K).$$

*(The converse is also true. See Exercise 5.27.)*

(d) *For all* $(x,y) \in \mathbb{A}^2_*(K)$ *and all* $c \in K^*$*, the Green function* $\mathcal{G}_\Phi$ *has the following homogeneity properties*:

$$\mathcal{G}_\Phi(cx,cy) = \mathcal{G}_\Phi(x,y) + \log|c|_v. \tag{5.38}$$

$$\mathcal{G}_{c\Phi}(x,y) = \mathcal{G}_\Phi(x,y) + \frac{1}{d-1}\log|c|_v. \tag{5.39}$$

(e) *The Green function* $\mathcal{G}_\Phi : \mathbb{A}^2_*(K) \to \mathbb{R}$ *is continuous. (In fact,* $\mathcal{G}_\Phi$ *is Hölder continuous, but this is more difficult to prove. See Exercise 5.28.)*

*Proof.* We consider the two functions

$$\Phi : \mathbb{A}^2_*(K) \longrightarrow \mathbb{A}^2_*(K) \qquad \text{and} \qquad \log\|\cdot\| : \mathbb{A}^2_*(K) \longrightarrow \mathbb{R}.$$

Proposition 5.57(a) tells us that they satisfy

$$\log\big\|\Phi(x,y)\big\|_v = d\log\big\|(x,y)\big\|_v + O(1) \qquad \text{for all } (x,y) \in \mathbb{A}^2_*(K). \tag{5.40}$$

This is exactly the situation needed to apply Theorem 3.20, from which we conclude that the limit (5.35) exists and satisfies (5.36) and (5.37). Further, Theorem 3.20 says that $\mathcal{G}_\Phi$ is the unique function satisfying (5.36) and (5.37). This completes the proof of (a) and (b).

(c) The assumptions $\|\Phi\|_v = 1$ and $\big|\mathrm{Res}(F,G)\big|_v = 1$ combine with Proposition 5.57(b) to tell us that $\big\|\Phi(x,y)\big\|_v = \big\|(x,y)\big\|_v^d$ for all points $(x,y) \in \mathbb{A}^2_*(K)$. Hence by induction we obtain

$$\big\|\Phi^n(x,y)\big\|_v = \big\|(x,y)\big\|_v^{d^n}.$$

Then the definition of $\mathcal{G}_\Phi$ immediately gives $\mathcal{G}_\Phi(x,y) = \log\big\|(x,y)\big\|_v$, which proves (c).

(d) The map $\Phi^n$ is homogeneous of degree $d^n$, so $\Phi^n(cx,cy) = c^{d^n}\Phi^n(x,y)$. Substituting this into the definition of $\mathcal{G}_\Phi$ gives (5.38).

Similarly, if we let $\Phi_c(x,y) = c\Phi(x,y)$, then homogeneity and an easy induction argument show that

$$\Phi_c^n(x,y) = c^{1+d+d^2+\cdots+d^{n-1}}\Phi(x,y).$$

Hence

$$\begin{aligned}\mathcal{G}_{c\Phi}(x,y) &= \lim_{n\to\infty}\frac{1}{d^n}\log\bigl\|\Phi_c^n(x,y)\bigr\|_v\\ &= \lim_{n\to\infty}\frac{1}{d^n}\left(\log\bigl\|\Phi^n(x,y)\bigr\|_v + \frac{d^n-1}{d-1}\log|c|_v\right)\\ &= \mathcal{G}_\Phi(x,y) + \frac{1}{d-1}\log|c|_v.\end{aligned}$$

(e) Let $\mathcal{G}_n(x,y) = d^{-n}\log\bigl\|\Phi^n(x,y)\bigr\|_v$. Then for any fixed value of $n$, the function $\mathcal{G}_n : \mathbb{A}_*^2(K) \to \mathbb{R}$ is continuous, since it is the composition of the continuous maps $\Phi^n : \mathbb{A}_*^2(K) \to \mathbb{A}_*^2(K)$ and $\log\|\cdot\|_v : \mathbb{A}_*^2(K) \to \mathbb{R}$. Further, the telescoping sum argument used in Theorem 3.20 to prove the existence of $\lim_{n\to\infty}\mathcal{G}_n(x,y)$ implies that

$$\bigl|\mathcal{G}_\Phi(x,y) - \mathcal{G}_n(x,y)\bigr| \le \frac{C}{(d-1)d^n} \qquad \text{for all } n \ge 1 \text{ and all } (x,y) \in \mathbb{A}_*^2(K),$$

where $C = C(\Phi, v)$ depends only on the $O(1)$ constant in (5.40). (In the inequality (3.15) on page 98, switch the roles of $m$ and $n$ and then let $m \to \infty$.) Hence the sequence of continuous functions $\mathcal{G}_n(x,y)$ converges uniformly to $\mathcal{G}_\Phi(x,y)$, so the limiting function $\mathcal{G}_\Phi$ is also continuous. (See also Exercise 3.14.) □

The Green function allows us to decompose the canonical height $\hat{h}_\phi$ into a sum of local terms.

**Theorem 5.59.** *Let $K$ be a number field, let $\phi : \mathbb{P}^1 \to \mathbb{P}^1$ be a rational function of degree $d \ge 2$ defined over $K$, and let $\Phi$ be a fixed lift of $\phi$. For each absolute value $v \in M_K$, let $\mathcal{G}_{\Phi,v}$ be the associated Green function as described in Proposition 5.58, where we now include $v$ in the notation so as to distinguish between different absolute values. Then the (global) canonical height decomposes as a sum*

$$\hat{h}_\phi(P) = \sum_{v\in M_K} n_v\mathcal{G}_{\Phi,v}(x,y) \qquad \textit{for all } P = [x,y] \in \mathbb{P}^1(K).$$

*Proof.* Let

$$\eta(x,y) = \sum_{v\in M_K} n_v\mathcal{G}_{\Phi,v}(x,y) \qquad \text{for } (x,y) \in \mathbb{A}_*^2(K),$$

so a priori the function $\eta$ is a function on $\mathbb{A}_*^2(K)$. However, applying the Green function homogeneity property (5.38) from Proposition 5.58(d) and the product formula (Proposition 3.3), we see that

$$\eta(cx,cy) = \sum_{v\in M_K} n_v\bigl(\mathcal{G}_{\Phi,v}(x,y) + \log|c|_v\bigr) = \sum_{v\in M_K} n_v\mathcal{G}_{\Phi,v}(x,y) = \eta(x,y),$$

so $\eta$ gives a well-defined function on $\mathbb{P}^1(K)$.

Next we use the transformation property (5.36) from Proposition 5.58(b) to compute

$$\eta(\phi(P)) = \sum_{v\in M_K} n_v \mathcal{G}_{\Phi,v}(\Phi(P)) = \sum_{v\in M_K} n_v d\mathcal{G}_{\Phi,v}(P) = d\eta(P).$$

Similarly, we use the normalization property (5.37) from Proposition 5.58(b) to estimate

$$\begin{aligned}\eta(P) &= \sum_{v\in M_K} n_v \mathcal{G}_{\Phi,v}(P) \\ &= \sum_{v\in M_K} n_v\big(\log\|P\|_v + O_v(1)\big) = h(P) + \sum_{v\in M_K} n_v O_v(1),\end{aligned} \tag{5.41}$$

where $h(P)$ is the usual height of the point $P \in \mathbb{P}^1(K)$ and the $O_v(1)$ are the bounded functions appearing in (5.37). We further observe that Proposition 5.58(c) says that we may take $O_v(1) = 0$ for all but finitely many $v \in M_K$. More precisely, we may take $O_v(1) = 0$ for all $v$ satisfying

$$\text{(i) } v \in M_K^0, \qquad \text{(ii) } \|\Phi\|_v = 1, \qquad \text{and} \qquad \text{(iii) } \big|\operatorname{Res}(\Phi)\big|_v = 1.$$

Hence the final sum in (5.41) is a bounded function.

We have now proven that $\eta : \mathbb{P}^1(K) \to \mathbb{R}$ satisfies

$$\eta(\phi(P)) = d\eta(P) \qquad \text{and} \qquad \eta(P) = h(P) + O(1).$$

It follows from the uniqueness of the canonical height (Theorem 3.20) that $\eta$ is equal to $\hat{h}_\phi$. □

In Section 3.5 we described a decomposition of the canonical height $\hat{h}_\phi$ as a sum of local canonical height functions $\hat{\lambda}_{\phi,v}$, but we deferred the proof. The intuition in Section 3.5 was that the local canonical height should measure

$$\hat{\lambda}_{\phi,v}(P) = -\log(v\text{-adic distance from } P \text{ to } \infty).$$

More generally, it is convenient to define a local canonical height that measures the $v$-adic distance from $P$ to a collection of points. In the following theorem we describe a set of points, possibly with multiplicities greater than 1, by specifying the homogeneous polynomial $E \in K[x,y]$ at which they vanish. In slightly fancier terminology, we are identifying positive divisors in $\operatorname{Div}(\mathbb{P}^1)$ with homogeneous polynomials in $K[x,y]$.

**Theorem 5.60.** *Let $K$ be a field with an absolute value $v$ and let $\phi : \mathbb{P}^1 \to \mathbb{P}^1$ be a rational function of degree $d \ge 2$ defined over $K$. Fix a lift $\Phi = (F,G)$ of $\phi$ and let $\mathcal{G}_\Phi$ be the associated Green function.*

(a) *For any homogeneous polynomial* $E(x,y) \in K[x,y]$ *of degree* $e \geq 1$ *we define*

$$\hat{\lambda}_{\phi,E}([x,y]) = e\mathcal{G}_\Phi(x,y) - \log\bigl|E(x,y)\bigr|_v$$
$$\textit{for } [x,y] \in \mathbb{P}^1(K) \textit{ with } E(x,y) \neq 0.$$

*Then* $\hat{\lambda}_{\phi,E}$ *is a well-defined function on* $\mathbb{P}^1$*, i.e., the value of* $\hat{\lambda}_{\phi,E}(P)$ *does not depend on the choice of homogeneous coordinates* $[x,y]$ *for* $P$.

*The function* $\hat{\lambda}_{\phi,E}$ *is called a* local canonical height associated to $\phi$ and $E$. *In the special case that* $E(x,y) = y$*, we drop* $E$ *from the notation and refer simply to a local canonical height associated to* $\phi$.

(b) *For all* $P \in \mathbb{P}^1(K)$ *with* $E(P) \neq 0$ *and* $E(\Phi(P)) \neq 0$ *we have*

$$\hat{\lambda}_{\phi,E}(\phi(P)) = d\hat{\lambda}_{\phi,E}(P) - \log\left|\frac{(E\circ\Phi)(P)}{E(P)^d}\right|_v.$$

*(Note that the homoegeneity of* $E$ *and* $\Phi$ *ensures that the ratio* $(E\circ\Phi)/E^d$ *is a well-defined function on* $\mathbb{P}^1$*.)*

(c) *The function*

$$P \longmapsto \hat{\lambda}_{\phi,E}(P) + \log\frac{|E(P)|_v}{\|P\|_v^e}, \tag{5.42}$$

*which a priori is defined only at points* $P$ *satisfying* $E(P) \neq 0$*, extends to a bounded continuous function on all of* $\mathbb{P}^1(K)$.

(d) *Given the particular choice of lift* $\Phi$*, the function* $\hat{\lambda}_{\phi,E}$ *defined in* (a) *is the unique real-valued function*

$$\mathbb{P}^1(K) \smallsetminus \{E=0\} \longrightarrow \mathbb{R}$$

*satisfying* (b) *and* (c). *If* $c\Phi$ *is a different lift, then with the obvious notation,*

$$\hat{\lambda}_{c\Phi,E}(P) = \hat{\lambda}_{\Phi,E}(P) + \frac{e}{d-1}\log|c|_v.$$

*In particular, any two local canonical heights differ by a constant.*

*Proof.* The Green function satisfies $\mathcal{G}_\Phi(cx,cy) = \mathcal{G}_\Phi(x,y) + \log|c|_v$, while the polynomial $E$ satisfies $E(cx,cy) = c^eE(x,y)$, so the difference

$$e\mathcal{G}_\Phi(x,y) - \log|E(x,y)|_v$$

defining $\hat{\lambda}_{\phi,E}$ does not change if we replace $(x,y)$ by $(cx,cy)$. This proves (a).

(b) We compute

$$\begin{aligned}
\hat{\lambda}_{\phi,E}(\phi(P)) &= e\mathcal{G}_\Phi(\Phi(x,y)) - \log\bigl|E(\Phi(x,y))\bigr|_v \\
&= ed\mathcal{G}_\Phi(x,y) - \log\bigl|E(\Phi(x,y))\bigr|_v \qquad \text{from Proposition 5.58(b),} \\
&= d\hat{\lambda}_{\phi,E}(P) + d\log|E(x,y)|_v - \log\bigl|E(\Phi(x,y))\bigr|_v.
\end{aligned}$$

(c) Directly from the definition of $\hat{\lambda}_{\phi,E}$ we see that the righthand side of (5.42) is equal to

$$e\bigl(\mathcal{G}_\Phi(P) - \log\|P\|_v\bigr). \tag{5.43}$$

The boundedness of (5.43) is exactly (5.37) in Proposition 5.58(b). Further, we know from Proposition 5.58(e) that $\mathcal{G}_\Phi(x,y)$ is a continuous function on $\mathbb{A}^2_*(K)$, and it is clear that $\log\bigl\|(x,y)\bigr\|_v$ is also a continuous function on $\mathbb{A}^2_*(K)$, so the difference (5.43) is a continuous function on $\mathbb{A}^2_*(K)$. Further, the difference is invariant under $(x,y)\mapsto(cx,cy)$, so it descends to a continuous function on $\mathbb{P}^1(K)$.

(d) Suppose that $\hat{\lambda}$ and $\hat{\lambda}'$ both satisfy (b) and (c), and let $\hat{\lambda}'' = \hat{\lambda} - \hat{\lambda}'$. Writing

$$\hat{\lambda}''(P) = \left(\hat{\lambda}(P) + \log\frac{|E(P)|_v}{\|P\|_v^e}\right) - \left(\hat{\lambda}'(P) + \log\frac{|E(P)|_v}{\|P\|_v^e}\right),$$

we see from (c) that $\hat{\lambda}''$ extends to a continuous bounded function on all of $\mathbb{P}^1(K)$. Let $C$ be an upper bound for $|\hat{\lambda}''|$.

From (b) we find that

$$\hat{\lambda}''\bigl(\phi(P)\bigr) = d\hat{\lambda}''(P) \quad \text{for all } P\in\mathbb{P}^1(K) \text{ with } E(P)\ne 0 \text{ and } E\bigl(\phi(P)\bigr)\ne 0,$$

and iterating this relation yields

$$\hat{\lambda}''\bigl(\phi^n(P)\bigr) = d^n\hat{\lambda}''(P) \quad \text{provided } E\bigl(\phi^i(P)\bigr)\ne 0 \text{ for all } 0\le i\le n.$$

Hence

$$\bigl|\hat{\lambda}''(P)\bigr| = \frac{1}{d^n}\bigl|\hat{\lambda}''\bigl(\phi^n(P)\bigr)\bigr| \le \frac{C}{d^n} \tag{5.44}$$

for all points $P\in\mathbb{P}^1(K)$ satisfying $E\bigl(\phi^i(P)\bigr)\ne 0$ for all $0\le i\le n$.

But each equation $E\bigl(\phi^i(P)\bigr) = 0$ eliminates only finitely many points, so the inequality (5.44) is true for all but finitely many points of $\mathbb{P}^1(K)$. Then the continuity of $\hat{\lambda}''$ tells us that $\bigl|\hat{\lambda}''(P)\bigr| \le Cd^{-n}$ is true for all $P\in\mathbb{P}^1(K)$. Since $n$ is arbitrary, this proves that $\hat{\lambda}''(P)=0$, so $\hat{\lambda}=\hat{\lambda}'$.

Finally, the effect of replacing $\Phi$ by $c\Phi$ follows immediately from the definition of $\hat{\lambda}_{\phi,E}$ in terms of the Green function $\mathcal{G}_\Phi$ and from the corresponding transformation formula for $\mathcal{G}_\Phi$ given in Proposition 5.58(d). □

Finally, as promised in Section 3.5, we prove that the global canonical height is equal to the sum of the local canonical heights.

**Theorem 5.61.** *Let $K$ be a number field, let $\phi:\mathbb{P}^1\to\mathbb{P}^1$ be a rational function of degree $d\ge 2$ defined over $K$, and fix a lift $\Phi=(F,G)$ of $\phi$. Choose a homogeneous polynomial $E(x,y)\in K[x,y]$, and for each absolute value $v\in M_K$, let $\hat{\lambda}_{\phi,E,v}$ be the associated local canonical height described in Theorem 5.60, where we now include $v$ in the notation so as to distinguish between different absolute values. Then the (global) canonical height has a decomposition as a sum of local canonical heights,*

$$\hat{h}_\phi(P) = \frac{1}{\deg E}\sum_{v\in M_K} n_v\hat{\lambda}_{\phi,E,v}(P) \qquad \textit{for all } P\in\mathbb{P}^1(K) \textit{ with } E(P)\ne 0.$$

*Proof.* We use the definition of $\hat{\lambda}_{\phi,E,v}$ in terms of the associated Green function $\mathcal{G}_{\Phi,v}$ from Theorem 5.60(a) to compute

$$\begin{aligned}\frac{1}{\deg E}\sum_{v\in M_K} n_v\hat{\lambda}_{\phi,E,v}(P) &= \frac{1}{\deg E}\sum_{v\in M_K} n_v\left((\deg E)\mathcal{G}_{\Phi,v}(P) - \log\bigl|E(P)\bigr|_v\right) \\ &= \sum_{v\in m_K} n_v\mathcal{G}_{\Phi,v}(P) - \frac{1}{\deg E}\sum_{v\in M_K} n_v\log\bigl|E(P)\bigr|_v.\end{aligned}$$

Theorem 5.59 says that the first sum is equal to $\hat{h}_\phi(P)$, while the product formula (Proposition 3.3) tells us that the second sum is 0. (Note that this is where we use the assumption that $E(P)\neq 0$.) □

*Remark* 5.62. If $v\in M_K^0$ is nonarchimedean and $\phi$ has good reduction at $v$, then the Green function and the local canonical height are given by the simple formulas

$$\mathcal{G}_{\Phi,v}(x,y) = \log\max\bigl\{|x|_v,|y|_v\bigr\} \text{ and } \hat{\lambda}_{\phi,E,v}\bigl([x,y]\bigr) = \log\left(\frac{\max\bigl\{|x|_v,|y|_v\bigr\}}{\bigl|E(x,y)\bigr|_v}\right).$$

Thus it is only for maps with bad reduction that the Green and local canonical height functions are interesting. This should not come as a surprise to the reader, since bad reduction is the situation in which dynamics itself becomes truly interesting. Of course, this is said with the understanding that every rational map over $\mathbb{C}$ has "bad reduction," so the dynamics of holomorphic maps on $\mathbb{P}^1(\mathbb{C})$ are always interesting and complicated.

*Remark* 5.63. For additional material on dynamical Green functions and dynamical local heights, see [21, 88, 234, 233].

## 5.10 Dynamics on Berkovich Space

We have seen that $\mathbb{C}_p$ is a natural space over which to study nonarchimedean dynamics, since it is both complete and algebraically closed. However, the field $\mathbb{C}_p$ has various unpleasant properties:

- $\mathbb{C}_p$ is totally disconnected.
- $\mathbb{C}_p$ is not locally compact, so the unit disk $\{z\in\mathbb{C}_p : |z|\le 1\}$ and projective line $\mathbb{P}^1(\mathbb{C}_p)$ are not compact.
- The value group $|\mathbb{C}_p^*| = p^{\mathbb{Q}}$ consists of the rational powers of $p$, so it is not discrete in $\mathbb{R}_{>0}$, yet neither is it all of $\mathbb{R}_{>0}$.

This list suggests that it might be better to work in some larger space. There is a general construction, due to Berkovich [64, 67], that solves these problems for $\mathbb{C}_p$ and other more complicated spaces. The study of dynamics on Berkovich spaces started during the 1990s and is an area of much current research. In this section we briefly describe the Berkovich disk and associated affine and projective lines and discuss some very basic dynamical results. In a final subsection we state without proof some recent results. For further reading, see [26, 51] for an introduction to dynamics on Berkovich space and see [373, 375, 376, 379, 377, 380, 381] for Rivera-Letelier's fundamental work in this area.

## 5.10.1 The Berkovich Disk over $\mathbb{C}_p$

The unit Berkovich disk $\bar{D}^{\mathcal{B}}$ is a compact connected metric space that contains the totally disconnected non-locally compact unit disk in $\mathbb{C}_p$. We describe two constructions of $\bar{D}^{\mathcal{B}}$, the first an explicit description as the union of four types of points and the second as a set of bounded seminorms on $\mathbb{C}_p[z]$.

### 5.10.1.1 The Four Types of Berkovich Points

The most concrete description of $\bar{D}^{\mathcal{B}}$ is as the union of the following four sets of points:

**Type-I Berkovich Points.** Each point $a$ in the standard unit disk $\bar{D}(0,1)$ of $\mathbb{C}_p$ is associated to a point of the Berkovich disk, which we denote by

$$\xi_{a,0} \in \bar{D}^{\mathcal{B}}.$$

**Type-II Berkovich Points.** Each closed disk $\bar{D}(a,r)$ contained in $\bar{D}(0,1)$ with radius $r \in |\mathbb{C}_p^*| = p^{\mathbb{Q}}$ is associated to a point of the Berkovich disk, which we denote by

$$\xi_{a,r} \in \bar{D}^{\mathcal{B}}.$$

**Type-III Berkovich Points.** Similarly, each closed disk $\bar{D}(a,r) \subseteq \bar{D}(0,1)$ with positive radius $r \notin |\mathbb{C}_p^*| = p^{\mathbb{Q}}$ is associated to a point of the Berkovich disk, which we naturally also denote by $\xi_{a,r}$.

**Type-IV Berkovich Points.** These are the trickiest points in $\bar{D}^{\mathcal{B}}$. They are associated to nested sequences of closed disks

$$\bar{D}(0,1) \supset \bar{D}(a_1,r_1) \supset \bar{D}(a_2,r_2) \supset \bar{D}(a_3,r_3) \supset \cdots$$

with the property that

$$\bigcap_{n\ge 1} \bar{D}(a_n,r_n) = \emptyset.$$

We denote these points by $\xi_{\mathbf{a},\mathbf{r}}$, where, as the notation suggests, the vectors $\mathbf{a}$ and $\mathbf{r}$ are $\mathbf{a} = (a_1, a_2, \ldots)$ and $\mathbf{r} = (r_1, r_2, \ldots)$.

*Remark* 5.64. Note that Berkovich points $\xi_{a,r}$ of Types II and III are disks $\bar{D}(a,r)$, so different values of $a$ may yield the same Berkovich point. Indeed, we have

$$\xi_{a,r} = \xi_{b,s} \qquad \text{if and only if} \qquad r = s \ \text{ and } \ |a-b| \le r,$$

since these are the conditions for the disks $\bar{D}(a,r)$ and $\bar{D}(b,s)$ to coincide. Similarly, two Berkovich points of type IV are the same if their sequences of disks can be suitably intertwined. See Exercise 5.40 for details.

*Remark* 5.65. A point $\xi_{a,r}$ of Type-I, II, or III corresponds to a disk (possibly of radius 0), so we define the *radius of* $\xi_{a,r}$ to be $r$. The radii $r_0, r_1, r_2, \ldots$ of a Type-IV point $\xi_{\mathbf{a},\mathbf{r}}$ are nonincreasing, so the limit $r = \lim_{i\to\infty} r_i$ exists and is called the *radius of* $\xi_{\mathbf{a},\mathbf{r}}$.

We claim that the radius of a Type-IV point is strictly positive. To see this, suppose that $\xi_{\mathbf{a},\mathbf{r}}$ has radius 0. Then the sequence of points $a_1, a_2, \ldots$ is a Cauchy sequence in $\mathbb{C}_p$, so it converges to a point $a \in \mathbb{C}_p$. Let

$$\delta_i = \inf_{z\in\bar{D}(a_i,r_i)} |z-a|$$

be the distance from $a$ to the $i^{\text{th}}$ disk. Notice that $0 \le \delta_1 \le \delta_2 \le \delta_3 \le \cdots$, since the disks form a decreasing sequence. On the other hand, $\delta_i \le |a - a_i| \to 0$ as $i \to \infty$. Hence $\delta_i = 0$ for all $i$, so $a \in \bar{D}(a_i, r_i)$ for all $i$ and the intersection is nonempty, contradicting the assumption that $\xi_{\mathbf{a},\mathbf{r}}$ is a Type-IV point.

#### 5.10.1.2 The Berkovich Disk as a Set of Seminorms

The description of the Berkovich disk $\bar{D}^{\mathcal{B}}$ as a union of points of Types I, II, III, and IV is very concrete, but it can be awkward to apply, since one must deal with four different kinds of points.[4] There is an alternative description of $\bar{D}^{\mathcal{B}}$ as a collection of seminorms on the ring $\mathbb{C}_p[z]$ that is sometimes easier to use and that also naturally generalizes to other rings and other spaces.

**Definition.** Fix $R > 0$. The *Gauss norm* $\|\cdot\|_R$ on $\mathbb{C}_p[z]$ is defined by

$$\|f\|_R = \max_{n\ge0} |c_n| R^n \qquad \text{for} \quad f(z) = \sum c_n z^n \in \mathbb{C}_p[z].$$

The maximum modulus principle (Theorem 5.13(a)) tells us that the Gauss norm is equal to the sup norm,

$$\|f\|_R = \sup_{z\in\bar{D}(0,R)} |f(z)|.$$

**Definition.** A (nontrivial) $\|\cdot\|_R$*-bounded seminorm on* $\mathbb{C}_p[z]$ is a nonconstant map

$$|\cdot| : \mathbb{C}_p[z] \longrightarrow \mathbb{R}$$

with the following properties:

1. $|f| \ge 0$ for all $f \in \mathbb{C}_p[z]$.
2. $|fg| = |f|\cdot|g|$ for all $f, g \in \mathbb{C}_p[z]$.

[4]One can unify the four types of points by defining all of them as equivalence classes of nested sequences of closed disks. See Exercise 5.40.

3. $|f+g| \le \max\{|f|, |g|\}$ for all $f, g \in \mathbb{C}_p[z]$.
4. $|f| \le \|f\|_R$ for all $f \in \mathbb{C}_p[z]$. (This is the boundedness condition.)

Thus a seminorm has all of the properties of an absolute value except that there may be nonzero elements $f \in \mathbb{C}_p[z]$ with $|f| = 0$.

**Definition.** The (*closed unit*) *Berkovich disk* $\bar{D}^{\mathcal{B}}$ is the set

$$\bar{D}^{\mathcal{B}} = \bigl\{\| \cdot \|_1\text{-bounded seminorms on } \mathbb{C}_p[z]\bigr\}.$$

More generally, the *closed Berkovich disk of radius* $R$ is the set

$$\bar{D}^{\mathcal{B}}_R = \bigl\{\| \cdot \|_R\text{-bounded seminorms on } \mathbb{C}_p[z]\bigr\}.$$

The definition of $\bar{D}^{\mathcal{B}}$ as a set of bounded seminorms is quite unintuitive, but its utility becomes clear when one examines Table 5.2 and sees how each of the four types of Berkovich points naturally defines a seminorm on $\mathbb{C}_p[z]$. A fundamental theorem of Berkovich, whose proof we omit, says that every bounded seminorm on $\mathbb{C}_p[z]$, and more generally on certain power series rings containing $\mathbb{C}_p[z]$, comes from one of the four types of Berkovich points. (See [64, page 18] or [26, Proposition 1.1].) Notice that the seminorm $|\cdot|_{0,1}$ corresponding to the Berkovich point $\xi_{0,1}$ is exactly the Gauss norm $|f|_{0,1} = \|f\|_1$, so following Baker and Rumely, we call $\xi_{0,1}$ the *Gauss point*.

| | Point | Seminorm |
|---|---|---|
| Type-I | $\xi_{a,0}$ | $\lvert f\rvert_{a,0} = \lvert f(a)\rvert$ |
| Types-II & III | $\xi_{a,r}$ | $\lvert f\rvert_{a,r} = \sup\limits_{z\in\bar{D}(a,r)} \lvert f(z)\rvert$ |
| Type-IV | $\xi_{\mathbf{a},\mathbf{r}}$ | $\lvert f\rvert_{\mathbf{a},\mathbf{r}} = \lim\limits_{n\to\infty} \lvert f\rvert_{a_n,r_n}$ |

Table 5.2: Seminorms on $\mathbb{C}_p[z]$ associated to Berkovich points.

*Remark* 5.66. It is easy to see that the seminorms associated to points of Type-II, III, and IV are actually norms. (See Exercise 5.35.) However, the seminorm associated to a point $\xi_{a,0}$ of Type-I is not a norm, since $|f|_{a,0} = 0$ if and only if $f$ vanishes at $a$. For example, $|z-a|_{a,0} = 0$. This explains why the Berkovich disk is defined using seminorms, rather than norms.

*Remark* 5.67. In order to properly develop function theory on the Berkovich disk and to glue disks together to create larger spaces, it is advantageous to use sets of seminorms on more general rings, such as power series rings. For example, let $\mathbb{T}_R$ be the ring of power series in $\mathbb{C}_p[\![z]\!]$ that converge on the closed disk $\bar{D}(0,R)$ of radius $R$, i.e.,

$$\mathbb{T}_R = \Bigl\{f(z) = \sum_{n\ge 0} c_n z^n \in \mathbb{C}_p[\![z]\!] : \lim_{n\to\infty} |c_n| R^n = 0\Bigr\}.$$

The ring $\mathbb{T}_R$ is a Banach algebra over $\mathbb{C}_p$ using the *Gauss norm*,

$$\|f\|_R = \max_{n\geq 0} |c_n| R^n,$$

which the maximum modulus principle tells us is the same as the sup norm. The Berkovich disk $\bar{D}_R^{\mathcal{B}}$ is often defined to be the set of bounded seminorms on the ring $\mathbb{T}_R$. One can show, using the Weierstrass preparation theorem, that this leads to the same set of points and the same topology as taking the $\|\cdot\|_R$-bounded seminorms on the polynomial ring $\mathbb{C}_p[z]$; see [26]. The ring $\mathbb{T}_R$ is an example of a *Tate algebra.* Applying the same construction to more general Tate algebras allows one to construct Berkovich spaces for the associated rigid analytic spaces.

#### 5.10.1.3 Visualizing the Berkovich Disk

In order to visualize the Berkovich disk, we place the Gauss point $\xi_{0,1}$ at the top of the page and observe that there is a line segment running from any point $\xi \neq \xi_{0,1}$ up to the Gauss point. If $\xi = \xi_{a,r}$ is of Type-I, II, or III, then this line segment is the set of points

$$L_{a,r} = \{\xi_{a,t} : r \leq t \leq 1\}.$$

Notice that any two line segments $L_{a,r}$ and $L_{b,s}$ merge with one another at the point $\xi_{a,t} = \xi_{b,t}$ determined by

$$t = \max\{r, s, |b-a|\}.$$

This is the smallest allowable value of $t$ for which the disks $\bar{D}(a,t)$ and $\bar{D}(b,t)$ acquire a common point, hence for which they coincide. Thus one can imagine the various line segments continually merging as they run upward toward the Gauss point at the top of the tree.

The line segment running up from a Type-IV point $\xi_{\mathbf{a},\mathbf{r}}$ is slightly more complicated. It is the set of points

$$L_{\mathbf{a},\mathbf{r}} = \{\xi_{\mathbf{a},\mathbf{r}}\} \cup \bigcup_{i=1}^{\infty} \{\xi_{a_i,t} : r_i \leq t \leq r_{i-1}\} = \{\xi_{\mathbf{a},\mathbf{r}}\} \cup \bigcup_{i=1}^{\infty} L_{a_i,r_i}. \tag{5.45}$$

(Note that $r_0 = 1$ by definition.) The Type-IV point $\xi_{\mathbf{a},\mathbf{r}}$ is included in the Berkovich disk precisely to provide an endpoint for the union of line segments $\bigcup L_{a_i,r_i}$.

Now imagine starting at the Gauss point and moving downward through the tree. We claim that at any instant, there are three possible scenarios:

1. If you have reached a point $\xi_{a,0}$ of Type I or a point $\xi_{\mathbf{a},\mathbf{r}}$ of Type IV, then you have hit the end of a segment and cannot proceed further.
2. If you have reached a point $\xi_{a,r}$ of Type II, then $r > 0$ is in the value group of $\mathbb{C}_p$ and there are countably infinitely many branches along which you can move down the tree.
3. If you have reached a point $\xi_{a,r}$ of Type III, then $r > 0$ is not in the value group of $\mathbb{C}_p$ and there is only one direction to move down the tree.

The picture for points of Type I is clear. In order to understand Types II and III, suppose that we fix a point $\xi_{a,r}$ of Type II or III. Each $b \in \bar{D}(a,r)$ gives a line segment $L_{b,0}$ that runs up from the Type-I point $\xi_{b,0}$ and through the point $\xi_{r,a}$. Two such line segments $L_{b,0}$ and $L_{b',0}$ merge before reaching $\xi_{a,r}$ if and only if $|b-b'|<r$, so it is really each open disk $D(b,r)$ inside the closed disk $\bar{D}(a,r)$ that gives a line segment running up to $\xi_{a,r}$.

If $\xi_{a,r}$ is of Type III, then $D(b,r)=\bar{D}(a,r)$ for any $b \in \bar{D}(a,r)$, so there is only one segment running downward from $\xi_{a,r}$.

The situation is much more interesting, and complicated, if $\xi_{a,r}$ is of Type II. In this case $\bar{D}(a,r)$ is covered by a countable union of open disks $D(b,r)$, so there is a countable set of branches downward from $\xi_{a,r}$. A convenient, although noncanonical, way to describe the branches is as follows. Let $\mathfrak{P}=\{z \in \mathbb{C}_p : |z|<1\}$ denote the maximal ideal in the ring of integers of $\mathbb{C}_p$ and fix some $c \in \mathbb{C}_p$ with $|c|=r$. Then the open disks of radius $r$ are in one-to-one correspondence with the residue field $\bar{\mathbb{F}}_p$ via the map

$$\{\text{disks } D(b,r) \text{ inside } \bar{D}(a,r)\} \longrightarrow \bar{\mathbb{F}}_p, \qquad D(b,r) \longmapsto (b/c) \bmod \mathfrak{P}.$$

The surjectivity of this map is clear, and the injectivity follows from the fact that $D(b,r)=D(b',r)$ if and only if $|b-b'|<r=|c|$.

In order to fit the Type-IV points into the picture, let

$$\bar{D}(a_1,r_1) \supset \bar{D}(a_2,r_2) \supset \bar{D}(a_3,r_3) \supset \cdots$$

be any sequence of nested disks and consider what happens as we move down the line segments $L_{a_i,r_i}$. The fact that the disks $\bar{D}(a_i,r_i)$ are nested implies that the line segments $L_{a_i,r_i}$ extend one another downward as $i$ increases. If the intersection $\bigcap \bar{D}(a_i,r_i)$ is nonempty, then it is equal to $\bar{D}(a,r)$ for some $a \in \mathbb{C}_p$ and some $r \geq 0$, so the intersection corresponds to a point $\xi_{a,r}$ of Type I, II, or III and the union $\bigcup L_{a_i,r_i}$ together with the endpoint $\xi_{a,r}$ forms a closed line segment.

However, if the intersection $\bigcap \bar{D}(a_i,r_i)$ is empty, then there is no actual disk $\bar{D}(a,r)$ sitting at the bottom of the union of the line segments $\bigcup L_{a_i,r_i}$. Thus as already noted, the points of Type IV exist precisely to remedy this situation and to ensure that every downward path has a termination point. Further, this explains why we defined $L_{\mathbf{a},\mathbf{r}}$ by (5.45) to be the line segment running from $\xi_{\mathbf{a},\mathbf{r}}$ to $\xi_{0,1}$. (See Exercise 5.41.)

The Berkovich disk $\bar{D}^{\mathcal{B}}$ is "illustrated" in Figure 5.4. Of course, Figure 5.4 is at best a pale imitation of the true glory of the Berkovich disk, since in a complete picture of $\bar{D}^{\mathcal{B}}$, every line segment contains a countable number of Type-II points, off of each of which there is a countable number of downward branches. Unfortunately, despite advances in modern technology, there are still no computer packages capable of fully rendering a (countably) infinitely branched broomstick.

#### 5.10.1.4 The Gel'fond Topology on the Berkovich Disk

The next step is to put a topology on the Berkovich disk.

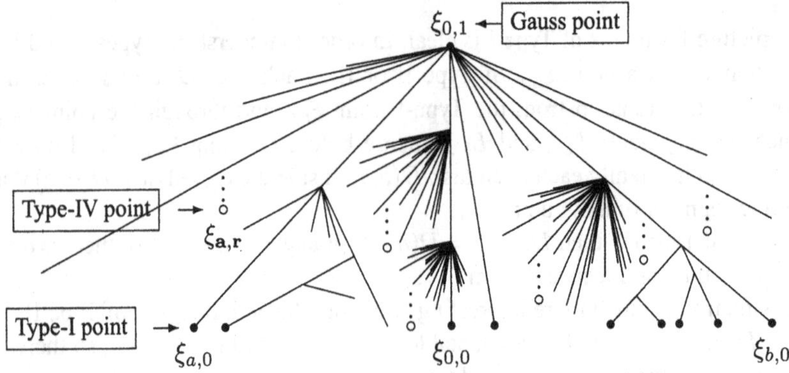


Figure 5.4: The Berkovich disk $\bar{D}^{\mathcal{B}}$.

**Definition.** The Gel'fond topology on $\bar{D}^{\mathcal{B}}$ is the weakest topology such that for every $f \in \mathbb{C}_p[z]$ and every $B > 0$, the following sets are open:

$$U(f,B) = \{x \in \bar{D}^{\mathcal{B}} : |f|_x < B\} \qquad \text{and} \qquad V(f,B) = \{x \in \bar{D}^{\mathcal{B}} : |f|_x > B\}.$$

**Theorem 5.68.** (Berkovich) *The Berkovich disk $\bar{D}^{\mathcal{B}}$ with the Gel'fond topology is a compact path-connected Hausdorff space.*

*Proof.* See [64, Theorem 1.2.1] or [29, Appendix D] for the proof that $\bar{D}^{\mathcal{B}}$ is compact and Hausdorff and [64, Corollary 3.2.3] for the proof that it is path connected. □

Basic open sets for the Gel'fond topology on $\bar{D}^{\mathcal{B}}$, viewed as a tree, can be described by taking (and deleting) branches of the tree as described in the following definition.

**Definition.** The *closed branch of $\bar{D}^{\mathcal{B}}$ rooted at* $(a,r)$, denoted by $\bigwedge_{a,r}$, consists of all points $\xi_{b,s}$ such that $\xi_{a,r}$ is on the line segment $L_{b,s}$ running from $\xi_{b,s}$ to $\xi_{0,1}$, together with whatever Type-IV points are needed to finish off the bottom of any open line segments. Thus

$$\bigwedge_{a,r} = \{\xi_{b,s} : s \le r \text{ and } |b-a| \le r\} \cup \{\text{appropriate Type-IV points}\}.$$

The *open branch rooted at* $(a,r)$, denoted by $\bigwedge^{\circ}_{a,r}$ is obtained by starting with the closed branch $\bigwedge_{a,r}$, removing the point $\xi_{a,r}$, and then taking the connected component that contains $\xi_{a,0}$. If $\xi_{a,r}$ is of Type III, this is simply the closed branch at $\xi_{a,r}$ with the single point $\xi_{a,r}$ removed; but if $\xi_{a,r}$ is of Type II, then there are countably many branches at $\xi_{a,r}$ and we select the one that includes the point $\xi_{a,0}$. It is not hard to see that

$$\bigwedge^{\circ}_{a,r} = \{\xi_{b,s} : s < r \text{ and } |b-a| < r\} \cup \{\text{appropriate Type-IV points}\}.$$

Then a base of open sets on $\bar{D}^{\mathcal{B}}$ for the Gel'fond topology consists of all sets of the following three types:

- Open branches.
- Open branches with a finite number of closed subbranches removed.
- The entire tree with a finite number of closed branches removed.

*Remark* 5.69. There is a natural inclusion

$$\bar{D}(0,1) \hookrightarrow \bar{D}^{\mathcal{B}}, \qquad a \longmapsto \xi_{a,0},$$

that identifies the unit disk $\bar{D}(0,1)$ as the set of Type-I points in the Berkovich disk $\bar{D}^{\mathcal{B}}$. It is not hard to check that the Gel'fond toplogy on $\bar{D}^{\mathcal{B}}$, restricted to $\bar{D}(0,1)$, is the usual topology induced by the metric on $\mathbb{C}_p$. See Exercise 5.44.

## 5.10.2 The Berkovich Affine and Projective Lines

It is relatively easy to construct the Berkovich affine line $\mathbb{A}^{\mathcal{B}}$ and the Berkovich projective line $\mathbb{P}^{\mathcal{B}}$ as topological spaces, which we do in this section. It is more difficult to construct them as ringed spaces with sheaves of functions appropriate for doing analysis. See Remark 5.74 for a brief discussion and references.

### 5.10.2.1 The Berkovich Affine Line $\mathbb{A}^{\mathcal{B}}$

Recall that the Berkovich disk $\bar{D}^{\mathcal{B}}$ consists of four types of points. Each may be described in terms of disks that are contained in the closed unit disk $\bar{D}(0,1)$. Equivalently, $\bar{D}^{\mathcal{B}}$ is the collection of $\|\cdot\|_1$-bounded seminorms on $\mathbb{C}_p[z]$.

More generally, we define the Berkovich disk $\bar{D}^{\mathcal{B}}_R$ of radius $R$ to be the collection of $\|\cdot\|_R$-bounded seminorms on $\mathbb{C}_p[z]$. It is given the Gel'fond topology and has its own Gauss point $\xi_{0,R}$ corresponding to the seminorm

$$|f|_{0,R} = \max_{n\ge 0} |c_n|R^n = \sup_{z\in\bar{D}(0,R)} |f(z)| \quad \text{for} \quad f(z) = \sum c_n z^n \in \mathbb{C}_p[z].$$

It is clear how to define points of Type I, II, III, and IV in $\bar{D}^{\mathcal{B}}_R$ using closed disks contained in $\bar{D}(0,R)$, just as we did for $\bar{D}^{\mathcal{B}}$. Further, there is an inclusion

$$\bar{D}^{\mathcal{B}}_{R_1} \subseteq \bar{D}^{\mathcal{B}}_{R_2} \qquad \text{for } R_1 \le R_2.$$

This is clear from the definition of $\bar{D}^{\mathcal{B}}_R$ as a set of seminorms. In terms of the picture of $\bar{D}^{\mathcal{B}}_R$ as a branched tree, we see that $\bar{D}^{\mathcal{B}}_{R_1}$ is the closed branch of $\bar{D}^{\mathcal{B}}_{R_2}$ lying below the Gauss point $\xi_{0,R_1}$ of $\bar{D}^{\mathcal{B}}_{R_1}$.

**Definition.** The *Berkovich affine line* $\mathbb{A}^{\mathcal{B}}$ is the union of the increasing collection of Berkovich disks,

$$\mathbb{A}^{\mathcal{B}} = \bigcup_{R>0} \bar{D}^{\mathcal{B}}_R,$$

with the topology induced by the direct limit topology on the individual Berkovich disks. It suffices, of course, to take the union over any increasing sequence of radii, for example, over $R = p^k$ with $k \to \infty$.

Thus every point $a \in \mathbb{C}_p$, every disk $\bar{D}(a,r) \subset \mathbb{C}_p$, and every nested sequence of disks with empty intersection gives a point in the Berkovich affine line $\mathbb{A}^{\mathcal{B}}$, and $\mathbb{A}^{\mathcal{B}}$ is composed of exactly this collection of points. In particular, there is a natural inclusion of $\mathbb{C}_p$ as the set of Type-I points in $\mathbb{A}^{\mathcal{B}}$,

$$\mathbb{C}_p = \mathbb{A}^1(\mathbb{C}_p) \subset \mathbb{A}^{\mathcal{B}}.$$

We also note that $\mathbb{A}^{\mathcal{B}}$ inherits a tree structure from the natural tree structure of the Berkovich disks $\bar{D}^{\mathcal{B}}_R$. However, the tree $\mathbb{A}^{\mathcal{B}}$ extends infinitely far upward; there is no Gauss point sitting at the top of $\mathbb{A}^{\mathcal{B}}$.

#### 5.10.2.2 The Berkovich Projective Line $\mathbb{P}^{\mathcal{B}}$

The easiest way to construct the Berkovich projective line $\mathbb{P}^{\mathcal{B}}$ as a topological space is to glue together two copies of the Berkovich disk $\bar{D}^{\mathcal{B}}$ along their annuli

$$\mathrm{Ann}^{\mathcal{B}} = \{\xi_{a,r} \in \bar{D}^{\mathcal{B}} : |a| = 1\}$$

using the map

$$f : \mathrm{Ann}^{\mathcal{B}} \longrightarrow \mathrm{Ann}^{\mathcal{B}}, \qquad f(\xi_{a,r}) = \xi_{a^{-1},r}.$$

We note that the map $f$ is induced from the inversion map $f(z) = 1/z$, since it is easy to check (Exercise 5.37) that if $0 \notin \bar{D}(a,r)$, then

$$f\big(\bar{D}(a,r)\big) = \{z^{-1} : |z-a| \le r\} = \bar{D}(a^{-1}, r/|a|^2).$$

In particular, if $|a| = 1$, then $f\big(\bar{D}(a,r)\big) = \bar{D}(a^{-1},r)$, so $f(\xi_{a,r}) = \xi_{a^{-1},r}$.

The full Berkovich disk is the disjoint union of the annulus and the open branch containing $\xi_{0,0}$,

$$\bar{D}^{\mathcal{B}} = \mathrm{Ann}^{\mathcal{B}} \cup \mathring{\Lambda}_{0,1}.$$

Thus when we glue two copies of $\bar{D}^{\mathcal{B}}$ along their annuli, the only parts of the two disks that are not identified are the two open branches $\mathring{\Lambda}_{0,1}$. Hence another way to construct $\mathbb{P}^{\mathcal{B}}$ is to take one copy of $\bar{D}^{\mathcal{B}}$ and attach one extra copy of $\mathring{\Lambda}_{0,0}$ running vertically upward from the Gauss point $\xi_{0,1}$. The result is illustrated in Figure 5.5.

It is natural to denote the extra vertical branch by $\mathring{\Lambda}_{\infty}$ and to label its points using the reciprocals of the points in $D(0,1)$,

$$\mathring{\Lambda}_{\infty} = \{\xi_{a,r} : |a| > 1 \quad\text{and}\quad r < 1\} \cup \{\xi_{\infty,0}\}.$$

The open and closed branches in $\mathring{\Lambda}_{\infty}$ are defined using the natural identification

$$\mathring{\Lambda}_{0,1} \xleftrightarrow{\ \sim\ } \mathring{\Lambda}_{\infty}, \qquad \xi_{a,r} \longleftrightarrow \xi_{a^{-1},r}.$$

For example, a basic open neighborhood of the Gauss point $\xi_{0,1}$ is obtained by removing from $\mathbb{P}^{\mathcal{B}}$ a finite number of closed branches, some of which may be in the vertical branch $\mathring{\Lambda}_{\infty}$ at infinity.

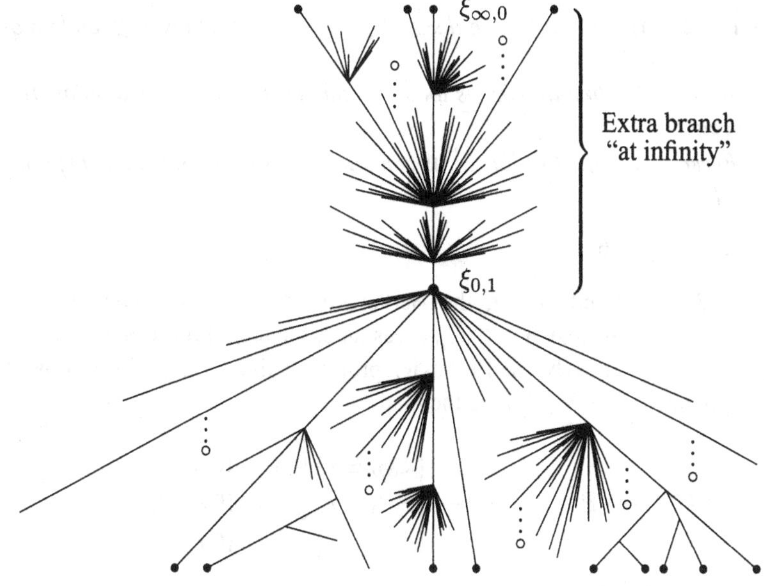


Figure 5.5: The Berkovich projective line $\mathbb{P}^{\mathcal{B}}$.

*Remark* 5.70. There is a natural embedding of $\mathbb{A}^{\mathcal{B}}$ into $\mathbb{P}^{\mathcal{B}}$. However, the inversion map $f(z) = 1/z$ used to glue together the two pieces of $\mathbb{P}^{\mathcal{B}}$ can cause notational confusion regarding the "radius" of points of Types II and III. For example, the point $\xi_{p^{-2},p^{-3}}$ in the branch $\mathring{\wedge}_{\infty}$ of $\mathbb{P}^{\mathcal{B}}$ would be denoted by $\xi_{p^{-1},p}$ when viewed as a point in $\mathbb{A}^{\mathcal{B}}$. Thus it might be wiser to denote the points in $\mathring{\wedge}_{\infty}$ using some alternative notation, for example $\hat{\xi}_{a,r}$, but we will not do so.

*Remark* 5.71. A useful alternative construction of $\mathbb{P}^{\mathcal{B}}$ mimics the construction of the scheme $\mathbb{P}^1_{\mathbb{Z}}$ as the set of homogeneous prime ideals. It starts with the set of bounded seminorms on the two-variable polynomial ring $\mathbb{C}_p[x,y]$ that extend the usual absolute value on $\mathbb{C}_p$ and that do not vanish on the maximal ideal $(x,y)$. Two seminorms $|\cdot|_1$ and $|\cdot|_2$ are considered equivalent if there is a constant $c > 0$ such that

$$|f|_1 = c^{\deg f}|f|_2 \qquad \text{for all homogeneous } f \in \mathbb{C}_p[x,y].$$

Then $\mathbb{P}^{\mathcal{B}}$ is the set of equivalence classes of such seminorms. For details of this construction, see [66].

Let $P \in \mathbb{P}^1(\mathbb{C}_p)$. To create a seminorm from $P$, choose homogeneous coordinates $P = [a,b]$ and set $|f| = |f(a,b)|$. Notice that a different choice of homogeneous coordinates for $P$ gives an equivalent seminorm. This embeds $\mathbb{P}^1(\mathbb{C}_p)$ into $\mathbb{P}^{\mathcal{B}}$.

#### 5.10.2.3 Properties of $\mathbb{A}^{\mathcal{B}}$ and $\mathbb{P}^{\mathcal{B}}$

As repayment for the effort required to construct them, Berkovich spaces have many nice properties that $\mathbb{C}_p$ lacks.

**Theorem 5.72.** (a) *The Berkovich disks $\bar{D}_R^{\mathcal{B}}$ are compact, Hausdorff, and uniquely path connected.*[5]

(b) *The Berkovich affine line $\mathbb{A}^{\mathcal{B}}$ is locally compact, Hausdorff, and uniquely path connected.*

(c) *The Berkovich projective line $\mathbb{P}^{\mathcal{B}}$ is compact, Hausdorff, and uniquely path connected.*

*Proof.* See [26] and [64]. □

*Remark* 5.73. As noted earlier, the Berkovich affine line $\mathbb{A}^{\mathcal{B}}$ contains a copy of $\mathbb{A}^1(\mathbb{C}_p) = \mathbb{C}_p$, since each $a \in \mathbb{C}_p$ gives an associated Type-I point $\xi_{a,0}$ in $\bar{D}_R^{\mathcal{B}}$ provided $R \ge |a|$. Similarly, the Berkovich projective line $\mathbb{P}^{\mathcal{B}}$ contains a copy of the classical projective line $\mathbb{P}^1(\mathbb{C}_p)$ via the map

$$\mathbb{P}^1(\mathbb{C}_p) \longrightarrow \mathbb{P}^{\mathcal{B}}, \qquad a \longmapsto \begin{cases} \xi_{a,0} \in \bar{D}(0,1) & \text{if } |a| \le 1, \\ \xi_{a,0} \in \mathring{\Lambda}_\infty & \text{if } 1 < |a| < \infty, \\ \xi_{\infty,0} \in \mathring{\Lambda}_\infty & \text{if } a = \infty. \end{cases}$$

One can show that the restriction of the Gel'fond topology on $\mathbb{A}^{\mathcal{B}}$ and $\mathbb{P}^{\mathcal{B}}$ to $\mathbb{A}^1(\mathbb{C}_p)$ and $\mathbb{P}^1(\mathbb{C}_p)$, respectively, gives the topology induced by the usual metric on $\mathbb{A}^1(\mathbb{C}_p)$ and the chordal metric on $\mathbb{P}^1(\mathbb{C}_p)$. See Exercise 5.44. This explains why the Gel'fond topology is the "right" topology to use on Berkovich spaces.

*Remark* 5.74. We have constructed $\mathbb{A}^{\mathcal{B}}$ and $\mathbb{P}^{\mathcal{B}}$ purely as topological spaces. It is more difficult, but very important, to refine this construction and make $\mathbb{A}^{\mathcal{B}}$ and $\mathbb{P}^{\mathcal{B}}$ into ringed spaces with structure sheaves built up naturally from rings of functions. There are two approaches, both due to Berkovich. The first takes unions of open Berkovich disks, which have a natural structure as analytic spaces, and glues them along open annuli; see [26, 64]. The second glues affinoids (which are closed) using nets; see [65]. This second construction is less intuitive, but it allows one to functorially attach a Berkovich analytic space to any reasonable rigid analytic space.

### 5.10.3 Dynamics on Berkovich Space

Having constructed the Berkovich spaces $\bar{D}^{\mathcal{B}}$, $\mathbb{A}^{\mathcal{B}}$, and $\mathbb{P}^{\mathcal{B}}$, we are finally ready to study iteration of maps on these spaces.

#### 5.10.3.1 Polynomial and Rational Maps on Berkovich Space

Let $\phi(z) \in \mathbb{C}_p[z]$ be a polynomial. There is a natural way to extend the map $\phi : \mathbb{A}^1(\mathbb{C}_p) \to \mathbb{A}^1(\mathbb{C}_p)$ to a map on Berkovich affine space $\phi : \mathbb{A}^{\mathcal{B}} \to \mathbb{A}^{\mathcal{B}}$. In terms of seminorms, the action of $\phi$ is simply given by composition,

$$\phi : \mathbb{A}^{\mathcal{B}} \longrightarrow \mathbb{A}^{\mathcal{B}}, \qquad |f|_{\phi(\xi)} = |f \circ \phi|_\xi.$$

[5] Recall that a topological space $X$ is *path connected* if given any two points $x_0, x_1 \in X$ there is a continuous map $f : [0,1] \to X$ with $f(0) = x_0$ and $f(1) = x_1$. Then $X$ is *uniquely path connected* if any two such paths $f_1$ and $f_2$ are homotopic to one another, i.e., $f_1$ an be continuously deformed to $f_2$.

However, it is perhaps easier to understand the map $\phi : \mathbb{A}^{\mathcal{B}} \to \mathbb{A}^{\mathcal{B}}$ by looking at the action of $\phi$ on points of Types I–IV. Recall (Proposition 5.16) that $\phi$ maps disks to disks, say

$$\phi\bigl(\bar{D}(a,r)\bigr) = \bar{D}\bigl(\phi(a), R\bigr) \qquad \text{for some } R = R(\phi, a, r).$$

Then for points of Types I, II, and III we define

$$\phi(\xi_{a,r}) = \xi_{\phi(a), R(\phi,a,r)},$$

and for points of Type IV we take the usual limit

$$\phi(\xi_{\mathbf{a},\mathbf{r}}) = \lim_{i\to\infty} \phi(\xi_{a_i,r_i}) = \lim_{i\to\infty} \xi_{\phi(a_i), R(\phi,a_i,r_i)}.$$

*Remark* 5.75. The maximum modulus principle (Theorem 5.13) allows us to explicitly describe the radius $R(\phi, a, r)$ of the image $\phi(\bar{D}(a,r))$. First expand $\phi(z)$ as a polynomial in powers of $z - a$, say

$$\phi(z) = \sum_{i=0}^{d} c_i(\phi, a)(z-a)^i.$$

Then

$$R(\phi, a, r) = \max_{1\le i\le d} \bigl|c_i(\phi,a)\bigr| r^i = \sup_{z\in\bar{D}(a,r)} \bigl|\phi(z) - \phi(a)\bigr|.$$

A rational function $\phi(z) \in \mathbb{C}_p(z)$ similarly induces a map on the Berkovich projective line $\mathbb{P}^{\mathcal{B}}$ extending the usual map on $\mathbb{P}^1(\mathbb{C}_p)$. If $\phi$ has no zeros or poles on the disk $\bar{D}(a,r)$, then it is relatively easy to describe the value of $\phi(\xi_{a,r})$. We know in this situation that $\phi\bigl(\bar{D}(a,r)\bigr)$ is a disk, say equal to $\bar{D}\bigl(\phi(a), s\bigr)$. Then

$$\phi(\xi_{a,r}) = \begin{cases} \xi_{\phi(a),s} & \text{if } |\phi(a)| \le 1, \\ \xi_{\phi(a)^{-1}, s/|\phi(a)|^2} & \text{if } |\phi(a)| > 1. \end{cases}$$

Note that the assumption that $\phi$ does not vanish on $\bar{D}(a,r)$ is equivalent to the inequality $|\phi(a)| > s$, so the indicated points are in $\bar{D}^{\mathcal{B}}$.

The description of $\phi(\xi_{a,r})$ when $\phi$ has zeros and/or poles on $\bar{D}(a,r)$ is more complicated. An explicit description in terms of open annuli is given by Rivera-Letelier [373, 375, 376]. (See also [26, Section 2].) A succinct, but less explicit, way to specify the induced map $\phi : \mathbb{P}^{\mathcal{B}} \to \mathbb{P}^{\mathcal{B}}$ is to use the construction of $\mathbb{P}^{\mathcal{B}}$ as a space of homogeneous seminorms as described in Remark 5.71. Then for a given seminorm $\xi \in \mathbb{P}^{\mathcal{B}}$, the seminorm $\phi(\xi)$ is determined by writing $\phi = [F, G]$ using homogeneous polynomials $F$ and $G$ and setting

$$\bigl|f(x,y)\bigr|_{\phi(\xi)} = \bigl|f(F(x,y), G(x,y))\bigr| \qquad \text{for all homogeneous } f \in \mathbb{C}_p[x,y].$$

#### 5.10.3.2 The Julia and Fatou Sets in Berkovich Space

A natural way to put a metric on the Berkovich spaces $\bar{D}^{\mathcal{B}}$, $\mathbb{A}^{\mathcal{B}}$, and $\mathbb{P}^{\mathcal{B}}$ is to use the underlying tree structure and measure distances along line segments. Unfortunately, this path-length metric does not give the Gel'fond topology, and as we have observed, it is the Gel'fond topology that extends the natural metric topologies on $\bar{D}(0,1)$, $\mathbb{A}^1(\mathbb{C}_p)$, and $\mathbb{P}^1(\mathbb{C}_p)$ . (See Exercises 5.42 and 5.44.) It is possible to define a metric that does yield the Gel'fond topology, but the definition of the "Gel'fond" metric is quite indirect. See [26, Corollary 1.3].

So rather than using a metric, we instead characterize the Fatou and Julia sets in Berkovich space using an abstract topological version of equicontinuity.

**Definition.** Let $X$ and $Y$ be topological spaces and let $\Phi$ be a collection of continuous maps $X \to Y$. The set $\Phi$ is *(topologically) equicontinuous at* $x$ if for every point $y \in Y$ and every neighborhood $V \subset Y$ of $y$ there are neighborhoods $U \subset X$ of $x$ and $W \subset Y$ of $y$ such that for every $\phi \in \Phi$, the following implication is true:

$$\phi(U) \cap W \neq \emptyset \quad \Longrightarrow \quad \phi(U) \subset V.$$

*Intuition*: $\Phi$ is equicontinuous at $x$ if for each $y \in Y$, whenever $\phi \in \Phi$ sends some point close to $x$ to a point that is close to $y$, then $\phi$ sends every point close to $x$ to a point that is close to $y$.

One can show that if $Y$ is a compact metric space, then topological equicontinuity agrees with the usual metric definition of equicontinuity. (See [26, Proposition 7.17].) We say that $\Phi$ is *(topologically) equicontinuous on* $X$ if it is topologically equicontinuous at every point of $X$.

**Definition.** Let $\phi(z) \in \mathbb{C}_p(z)$ be a rational map. The *(Berkovich) Fatou set of* $\phi$ is the largest open subset of $\mathbb{P}^{\mathcal{B}}$ on which $\phi$ is equicontinuous, or more precisely, on which the set of iterates $\{\phi^n\}$ is equicontinuous. The *(Berkovich) Julia set of* $\phi$ is the complement of the Berkovich Fatou set. We denote these sets by $\mathcal{F}^{\mathcal{B}}(\phi)$ and $\mathcal{J}^{\mathcal{B}}(\phi)$, respectively.

*Remark* 5.76. Recall that the classical points in $\mathbb{P}^{\mathcal{B}}$, i.e., the points of Type-I, form a copy of $\mathbb{P}^1(\mathbb{C}_p)$ sitting inside $\mathbb{P}^{\mathcal{B}}$. As noted earlier in Remark 5.73, the restriction of the Gel'fond topology on $\mathbb{P}^{\mathcal{B}}$ to the classical points gives the same topology on $\mathbb{P}^1(\mathbb{C}_p)$ as that induced by the chordal metric. Using this one can show that equicontinuity at a classical point of $\mathbb{P}^{\mathcal{B}}$ using the Gel'fond topology is equivalent to equicontinuity using the chordal metric. Hence the classical Fatou and Julia sets sit within their Berkovich counterparts:

$$\mathcal{F}(\phi) = \mathcal{F}^{\mathcal{B}}(\phi) \cap \mathbb{P}^1(\mathbb{C}_p) \qquad \text{and} \qquad \mathcal{J}(\phi) = \mathcal{J}^{\mathcal{B}}(\phi) \cap \mathbb{P}^1(\mathbb{C}_p).$$

*Remark* 5.77. Let $\phi(z) \in \mathbb{C}_p(z)$ be a rational map of degree at least 2. Various authors have shown that there is a unique probability measure $\mu_\phi$ on $\mathbb{P}^{\mathcal{B}}$ satisfying

$$\phi^* \mu_\phi = d \cdot \mu_\phi \qquad \text{and} \qquad \phi_* \mu_\phi = \mu_\phi.$$

(Recall that a *probability measure* is a nonnegative measure of total mass 1.) We call $\mu_\phi$ the *canonical measure associated to* $\phi$, since the property $\phi^*\mu_\phi = d \cdot \mu_\phi$ resembles the analogous property $\hat{h}_\phi(\phi(P)) = d \cdot \hat{h}_\phi(P)$ of the canonical height. The reader should be aware that other common names for $\mu_\phi$ in the literature include Brolin measure, Lyubich measure, and invariant measure. For the construction and applications of $\mu_\phi$, see Baker and Rumely [26, Theorem 7.14], [29], Chambert-Loir [98], Thuillier [434], and Favre and Rivera-Letelier [168], as well as [360].

**Theorem 5.78.** *Let $\phi(z) \in \mathbb{C}_p(z)$ be a rational map of degree at least 2. The support of the canonical measure $\mu_\phi$ is equal to the Julia set $\mathcal{J}^{\mathcal{B}}(\phi)$. In particular, the Berkovich Julia set $\mathcal{J}^{\mathcal{B}}(\phi)$ is not empty.*

*Proof.* This theorem is an amalgamation of results due to Baker, Rumely, and Rivera-Letelier. We refer the reader to [26, Section 7.5] for the construction of the canonical measure and to [26, Theorem 7.18], [27, Theorems 8.9 and A.7], and [381] for the proof that $\mu_\phi$ is supported exactly on $\mathcal{J}^{\mathcal{B}}(\phi)$. The last part of the theorem is then clear, since the empty set cannot provide the support for a nontrivial measure. □

*Example* 5.79. Let $\phi(z) \in \mathbb{C}_p(z)$ be a rational map of degree at least 2 and suppose that $\phi$ has good reduction. We know (Theorem 2.17) that the classical Julia set $\mathcal{J}(\phi) \subset \mathbb{P}^1(\mathbb{C}_p)$ is empty. Using the construction of the canonical measure, it is not hard to show [26, Example 7.2] that for a map of good reduction, the canonical measure is entirely supported at the Gauss point, i.e.,

$$\mu_\phi(U) = 1 \quad \text{if} \quad \xi_{0,1} \in U \qquad \text{and} \qquad \mu_\phi(U) = 0 \quad \text{if} \quad \xi_{0,1} \notin U.$$

Thus $\mathcal{J}^{\mathcal{B}}(\phi) = \{\xi_{0,1}\}$, so the nonempty Julia set guaranteed by Theorem 5.78 is not very interesting, since it consists of a single point. Hence even in Berkovich space, the most interesting dynamical behavior occurs for maps of bad reduction. On the other hand, if the conjugates $\phi^f$ of $\phi$ have bad reduction for every $f \in \mathrm{PGL}_2(\mathbb{C}_p)$, then $\mathcal{J}(\phi)$ is a perfect set, and hence uncountable. (See Theorem 5.82.)

#### 5.10.3.3 The Map $\phi(z) = z^2$ on Berkovich Space

To conclude our brief foray into Berkovich space, we illustrate Berkovich dynamics by studying the simplest possible map, namely $\phi(z) = z^2$. For any $a \in \mathbb{C}_p$, we expand

$$\phi(z) - \phi(a) = z^2 - a^2 = 2a(z-a) + (z-a)^2.$$

Assuming henceforth that $p \geq 3$ and using our convention that all points in $\mathbb{P}^{\mathcal{B}}$ have radius satisfying $r \leq 1$, we find that

$$\phi(\xi_{a,r}) = \xi_{\phi(a),s} \qquad \text{with} \quad s = \max\{|2a|r, r^2\} = r \cdot \max\{|a|, r\}. \tag{5.46}$$

This explicitly gives the action of $\phi$ on points of Types I, II, and III in $\bar{D}^{\mathcal{B}}$, and the action of $\phi$ on Type-IV points is given by the appropriate limit.

The formula (5.46) allows us to compute many orbits $\mathcal{O}(\xi_{a,r})$. For example, suppose that $|a|<1$ and $r<1$. Then $\bigl|\phi^n(a)\bigr|\to 0$, so the radius of $\phi^n(\xi_{a,r})$ also goes to 0. Thus

$$|a|<1 \quad\text{and}\quad r<1 \quad\Longrightarrow\quad \lim_{n\to\infty}\phi^n(\xi_{a,r})=\xi_{0,0}.$$

In other words, the open branch $\mathring{\Lambda}_{0,1}$ is in the attracting basin of the fixed point $\xi_{0,0}$. (We leave it to the reader to make this informal argument rigorous using the Gel'fond topology on $\bar{D}^{\mathcal{B}}$.)

What are the fixed points of $\phi$? For a point $\xi_{a,r}\in\bar{D}^{\mathcal{B}}$ of Type I, II, or III in the Berkovich disk, we have

$$\begin{aligned}\phi(\xi_{a,r})=\xi_{a,r} \quad&\Longleftrightarrow\quad r=r\cdot\max\bigl\{|a|,r\bigr\} \quad\text{and}\quad \bar{D}(a,r)=\bar{D}(a^2,r)\\ &\Longleftrightarrow\quad \begin{cases} r=0 \quad\text{and}\quad a=a^2, \quad\text{or}\\ \max\bigl\{|a|,r\bigr\}=1 \quad\text{and}\quad |a-a^2|\le r.\end{cases}\end{aligned}$$

There are three cases to consider. First, if $r=0$, then $a=a^2$, so we see that $\xi_{0,0}$ and $\xi_{1,0}$ are the only fixed points of Type I in $\bar{D}^{\mathcal{B}}$. Second, if $r=1$, then $\xi_{a,r}$ is equal to the Gauss point $\xi_{0,1}$, which is clearly fixed by $\phi(z)$. Finally, suppose that $0<r<1$. Then $\xi_{a,r}$ is fixed if and only if

$$|a|=1 \qquad\text{and}\qquad |a-1|=|a-a^2|\le r<1.$$

Thus $\xi_{a,r}$ is fixed if and only if $a\in\bar{D}(1,r)$, in which case $\xi_{a,r}=\xi_{1,r}$. This exhausts the Type-I,-II, and-III fixed points in $\bar{D}^{\mathcal{B}}$. A similar analysis on the branch leading up to $\xi_{\infty,0}$ yields one more fixed point, namely the Type-I point $\xi_{\infty,0}$ at the top of the tree. Hence aside from Type-IV points, the fixed-point set of $\phi$ on $\mathbb{P}^{\mathcal{B}}$ consists of two (attracting) Type-I fixed points $\xi_{0,0}$ and $\xi_{\infty,0}$ and the line segment running from the Gauss point $\xi_{0,1}$ down to the (neutral) Type-I fixed point $\xi_{1,0}$,

$$\operatorname{Fix}(\phi,\mathbb{P}^{\mathcal{B}})=\{\xi_{0,0},\xi_{\infty,0}\}\cup\{\xi_{1,r}:0\le r\le 1\}.$$

We leave it as an exercise to show that $\phi(z)=z^2$ has no fixed points of Type IV (Exercise 5.36).

We have seen that every point in $\mathring{\Lambda}_{0,0}$ is attracted to $\xi_{0,0}$, and the Gauss point $\xi_{0,1}$ is fixed. We next show that every other point in $\bar{D}^{\mathcal{B}}$ of Type II or III is preperiodic. Let $\xi_{a,r}\in\bar{D}^{\mathcal{B}}$ be such a point, which means that

$$0<r<1 \qquad\text{and}\qquad |a|=1.$$

Then the disk $\bar{D}(a,r)$ has positive radius, so it contains points that are algebraic over $\mathbb{Q}_p$ (note that $\bar{\mathbb{Q}}_p$ is dense in $\mathbb{C}_p$). Replacing $a$ with such a point, we may assume that $a$ is algebraic over $\mathbb{Q}_p$. Next we observe that (5.46) combined with the assumption that $|a|=1$ implies

$$\phi(\xi_{a,r})=\xi_{\phi(a),r}, \qquad\text{so by iteration}\qquad \phi^n(\xi_{a,r})=\xi_{\phi^n(a),r}.$$

Hence

$$\phi^m(\xi_{a,r}) = \phi^n(\xi_{a,r}) \quad \Longleftrightarrow \quad \bar{D}(\phi^m(a),r) \cap \bar{D}(\phi^n(a),r) \neq \emptyset.$$

Let $K = \mathbb{Q}_p(a)$ and notice that

$$\bigcup_{n \geq 0} \bar{D}(\phi^n(a),r) \cap K \subset \bar{D}(0,1) \cap K.$$

The disk $\bar{D}(0,1) \cap K$ cannot contain infinitely many disjoint disks of radius $r$, so there must exist $m > n$ such that

$$\bar{D}(\phi^m(a),r) \cap \bar{D}(\phi^n(a),r) \cap K \neq \emptyset.$$

Then $\phi^m(\xi_{a,r}) = \phi^n(\xi_{a,r})$, so $\xi_{a,r}$ is preperiodic.

We conclude this section by using the definition of topological equicontinuity to directly demonstrate that the Gauss point $\xi_{0,1}$ is in the Julia set of $\phi$. The intuition is that any neighborhood of $\xi_{0,1}$ contains points $\xi_{0,r}$ on the line segment connecting $\xi_{0,1}$ to $\xi_{0,0}$. If $r < 1$, then the iterates $\phi^n(\xi_{0,r})$ approach $\xi_{0,0}$, but the Gauss point $\xi_{0,1}$ is fixed. Hence $\phi^n(\xi_{0,1})$ does not remain close to $\phi^n(\xi_{0,r})$. We now make this argument rigorous.

We suppose that $x = \xi_{0,1} \in \mathcal{F}^{\mathcal{B}}(\phi)$ and derive a contradiction. Let $y = \xi_{0,0}$, and let $V = \mathring{\bigwedge}_{0,1/2}$ be our chosen neighborhood of $\xi_{0,0}$. The definition of equicontinuity says that there are neighborhoods $\xi_{0,1} \in U$ and $\xi_{0,0} \in W$ such that for all $n \geq 0$,

$$\phi^n(U) \cap W \neq \emptyset \quad \Longrightarrow \quad \phi^n(U) \subset \mathring{\bigwedge}_{0,1/2}. \tag{5.47}$$

The implication (5.47) remains true if we replace $U$ and $W$ by smaller neighborhoods, so we may assume that

$$U = \mathbb{P}^{\mathcal{B}} \smallsetminus \bigcup_{i=1}^{k} \bigwedge_{a_i,r_i} \qquad \text{with } 0 < r_i < 1 \text{ for all } i, \text{ and}$$

$$W = \mathring{\bigwedge}_{0,r} \qquad \text{with } 0 < r < 1.$$

Choose a value of $s$ satisfying

$$\max_{1 \leq i \leq k} r_i < s < 1.$$

Then $\xi_{0,s} \in U$, since $\xi_{0,s}$ is not on any of the closed branches $\bigwedge_{a_i,r_i}$. On the other hand,

$$\phi^n(\xi_{0,s}) = \xi_{0,s^n} \in W = \mathring{\bigwedge}_{0,r} \quad \text{for sufficiently large } n,$$

since we just need to ensure that $s^n < r$. This proves that

$$\xi_{0,s} \in \phi^n(U) \cap W,$$

so the assumption that $\phi$ is equicontinuous at $\xi_{0,1}$ implies that

$$\phi^n(U) \subset \mathring{\Lambda}_{0,1/2}.$$

But this inclusion is clearly not true, since, for example, $\xi_{0,1} \in U$ is fixed by $\phi$, but $\xi_{0,1} \notin \mathring{\Lambda}_{0,1/2}$. Therefore $\phi$ is not equicontinuous at $\xi_{0,1}$, so $\xi_{0,1} \in \mathcal{J}^{\mathcal{B}}(\phi)$.

A similar case-by-case argument using the explicit description (5.46) of the action of $\phi$ shows that every other point in $\mathbb{P}^{\mathcal{B}}$ is in the Fatou set. We leave the details for the reader.

#### 5.10.3.4 Further Results

We briefly describe, without proof, some deeper results on Berkovich dynamics. Our exposition follows [27], and the author is grateful to Baker and Rumely for making their preprint available.

**Theorem 5.80.** (Strong Montel Theorem on $\mathbb{P}^{\mathcal{B}}$) *Let $\phi \in \mathbb{C}_p(z)$ be a rational map of degree at least 2, let $\xi \in \mathbb{P}^{\mathcal{B}}$, let $U \subset \mathbb{P}^{\mathcal{B}}$ be an open neighborhood of $\xi$, and let $V$ be the union of $\phi^n(U)$ for all $n \geq 1$.*
(a) *If $\mathbb{P}^1(\mathbb{C}_p) \smallsetminus V$ contains at least 3 points, then $\xi \in \mathcal{F}^{\mathcal{B}}(\phi)$.*
(b) *If $\mathbb{P}^{\mathcal{B}} \smallsetminus \big(V \cup \mathbb{P}^1(\mathbb{C}_p)\big)$ is nonempty, then $\xi \in \mathcal{F}^{\mathcal{B}}(\phi)$.*

*Proof.* This theorem is due to Baker and Rumely [27, Theorem 7.1] for maps $\phi$ defined over a finite extension of $\mathbb{Q}_p$, and to Rivera-Letelier in the general case; see [27, Theorem A.1] and [381]. □

The proof is based on Rivera-Letelier's classification of periodic components in the Fatou set. In order to describe this classification and its applications, we need to define what it means for a periodic point in Berkovich space to be attracting or repelling. For Type-I points, i.e., for points in $\mathbb{P}^1(\mathbb{C}_p)$, we use the usual definition. It turns out that all attracting periodic points in $\mathbb{P}^{\mathcal{B}}$ are Type-I points. (See Exercise 5.45 for an explanation of why this is reasonable.) The definition of repelling periodic points is more complicated. Repelling periodic points are all of Type I or II [376, Proposition 5.5], where we use Rivera-Letelier's definition that a Type-II periodic point is repelling if its residual degree (see [376, Section 5]) is at least 2.

**Definition.** Let $\phi \in \mathbb{C}_p(z)$ be a rational map of degree at least 2 and let $\xi \in \mathbb{P}^{\mathcal{B}}$ be an attracting periodic point of period $n$. The *basin of attraction of* $\xi$ is the set

$$\big\{\eta \in \mathbb{P}^{\mathcal{B}} : \phi^{nk}(\eta) \to \xi \text{ as } k \to \infty\big\}.$$

The connected component of this set is called the *immediate basin of attraction of* $\xi$.

**Definition.** Recall that a point $P$ is called *recurrent for* $\phi$ if it is in the closure of $\big\{\phi^n(P) : n \geq 1\big\}$. The *domain of quasiperiodicity of* $\phi$ is the interior of the set of points in $\mathbb{P}^{\mathcal{B}}$ that are recurrent for $\phi$.

We are now ready to state Rivera-Letelier's strong Montel theorem for rational maps on the Berkovich projective line.

**Theorem 5.81.** (Rivera-Letelier) *Let $\mathcal{U}_\phi$ be the set of points $\xi \in \mathbb{P}^{\mathcal{B}}$ with the property that there is a neighborhood $U$ of $\xi$ such that*

$$\mathbb{P}^1(\mathbb{C}_p) \smallsetminus \bigcup_{n\geq 1} \phi^n(U)$$

*contains at least 3 points.*

(a) *Every periodic connected component of $\mathcal{U}_\phi$ is either an immediate basin of attraction of $\phi$ or a connected component of the domain of quasiperiodicity of $\phi$. All such components are in $\mathcal{F}^{\mathcal{B}}(\phi)$.*

(b) *Every wandering component of $\mathcal{U}_\phi$ is contained in $\mathcal{F}^{\mathcal{B}}(\phi)$.*

*Proof.* The proof of (a) is given in [27, Theorem A.2] and the proof of (b) is in [27, Corollary A.5]. □

As in the classical setting, Montel's theorem has a large number of important consequences, some of which we state here.

**Theorem 5.82.** *Let $\phi \in \mathbb{C}_p(z)$ be a rational map of degree at least 2.*

(a) *If there is some $f \in \mathrm{PGL}_2(\mathbb{C}_p)$ such that the conjugate $\phi^f$ has good reduction, then the Julia set $\mathcal{J}^{\mathcal{B}}(\phi)$ consists of a single point.*

(b) *If every conjugate $\phi^f$ of $\phi$ has bad reduction, then the Julia set $\mathcal{J}^{\mathcal{B}}(\phi)$ is a perfect set, and hence in particular it is uncountable.*

(c) *Let $\xi \in \mathcal{J}^{\mathcal{B}}(\phi)$, let $U \in \mathbb{P}^{\mathcal{B}}$ be an open neighborhood of $\xi$, and let $V$ be the union of $\phi^n(U)$ for all $n \geq 1$. Then* (i) *$V$ contains $\mathbb{P}^{\mathcal{B}} \smallsetminus \mathbb{P}^1(\mathbb{C}_p)$;* (ii) *$V$ contains $\mathcal{J}^{\mathcal{B}}(\phi)$; and* (iii) *$\mathbb{P}^1(\mathbb{C}_p) \smallsetminus V$ consists of at most two points.*

(d) *The Julia set $\mathcal{J}^{\mathcal{B}}(\phi)$ is either connected or else it has infinitely many connected components. Further, it has empty interior.*

(e) *Let $\xi \in \mathcal{J}^{\mathcal{B}}(\phi)$. Then the backward orbit of $\xi$ is dense in $\mathcal{J}^{\mathcal{B}}(\phi)$.*

(f) *Let $V \subset \mathbb{P}^{\mathcal{B}}$ be a closed completely $\phi$-invariant set containing at least 3 points. Then $V \supseteq \mathcal{J}^{\mathcal{B}}(\phi)$.*

(g) *The Julia set $\mathcal{J}^{\mathcal{B}}(\phi)$ is exactly equal to the closure of the repelling periodic points in $\mathbb{P}^{\mathcal{B}}$.*

*Proof.* (a) See [27, Lemma 8.1].

(b) See [27, Corollary 8.6].

(c) See [27, Theorem 8.2].

(d) See [27, Corollary 8.3 and Corollary 8.7].

(e) See [27, Corollary 8.5].

(f) See [27, Corollary 8.8].

(g) See [27, Theorem A.7]. □

# Exercises

## Section 5.1. Absolute Values and Completions

**5.1.** Prove that up to equivalence, the only nontrivial absolute values on $\mathbb{Q}$ are the usual archimedean absolute value and the $p$-adic absolute values. (This result is known as Ostrowski's theorem.)

## Section 5.2. A Primer on Nonarchimedean Analysis

**5.2.** (a) Let $\phi(z)$ be a holomorphic function on $\bar{D}(a,r)$. Prove that the Taylor series coefficients of $\phi(z)$ are uniquely determined by $\phi$.

(b) Let $\phi(z)$ be a function that is represented by a convergent Laurent series on the punctured disk $\bar{D}(a,r) \smallsetminus \{a\}$. Prove that the coefficients of its Laurent series are uniquely determined by $\phi$.

**5.3.** (a) Let $\phi(z)$ and $\psi(z)$ be holomorphic functions on $\bar{D}(a,r)$. Prove that the product $\phi(z)\psi(z)$ is a holomorphic function on $\bar{D}(a,r)$.

(b) Let $\phi(z)$ and $\psi(z)$ be meromorphic functions represented by Laurent series on $\bar{D}(a,r)$. Prove that the product $\phi(z)\psi(z)$ is a meromorphic function and is represented by a Laurent series on $\bar{D}(a,r)$.

**5.4.** Let $\phi(z) \in \mathbb{C}_p[z]$ be a nonconstant polynomial.

(a) Prove directly (i.e., without using Newton polygons) that $\phi$ sends a closed disk $\bar{D}(a,r)$ to a closed disk $\bar{D}\bigl(\phi(a),s\bigr)$.

(b) Prove that $\phi$ maps $\bar{D}(a,r)$ bijectively to $\bar{D}\bigl(\phi(a),s\bigr)$ if and only if

$$\bigl|\phi(z)-\phi(a)\bigr| = \frac{s}{r}|z-a| \qquad \text{for all } z \in \bar{D}(a,r).$$

**5.5.** This exercise outlines a prove of a $p$-adic inverse function theorem. Let $\phi \in \mathbb{C}_p(z)$ be a rational function of degree $d \geq 1$ and let $P \in \mathbb{P}^1(\mathbb{C}_p)$ be a point that is not a critical value of $\phi$, i.e., $P$ is not the image of a critical point of $\phi$.

(a) Prove that there is a disk $\bar{D}(P,r)$ centered at $P$ such that $\phi^{-1}\bigl(\bar{D}(P,r)\bigr)$ consists of $d$ *disjoint* open sets

$$\phi^{-1}\bigl(\bar{D}(P,r)\bigr) = V_1 \cup V_2 \cup \cdots \cup V_d.$$

Show that the $V_i$ are disks.

(b) Let $\phi_i : V_i \to \bar{D}(P,r)$ be the restriction of $\phi$ to $V_i$ for each $1 \leq i \leq d$. Prove that $\phi_i$ is bijective.

(c) Possibly after reducing $r$, prove that the inverse maps $\phi_i^{-1} : \bar{D}(P,r) \to V_i$ are given by convergent power series. (If $\infty \in \phi^{-1}(P)$, either change coordinates or instead use a convergent Laurent series.)

**5.6.** (a) Let $K$ be an algebraically closed field that is complete with respect to a (nonarchimedean) absolute value and let $\phi : \mathbb{P}^1(K) \to \mathbb{P}^1(K)$ be a rational map. Prove that $\phi$ is an open map, i.e., the image of an open set is an open set.

(b) Consider the map $\phi : \mathbb{P}^1(\mathbb{Q}_3) \to \mathbb{P}^1(\mathbb{Q}_3)$ defined by $\phi(z) = z^2$. Prove that $\phi$ is not an open map by showing that the image of the open disk $D(0,1)$ in $\mathbb{P}^1(\mathbb{Q}_3)$ is not open. Thus rational maps over complete, but not algebraically closed, fields need not be open.

**5.7.** Let $\phi(z) \in \mathbb{C}_p[\![z]\!]$ be analytic on $\bar{D}(a,r)$ and assume that $\phi(a) \neq 0$. Prove that $1/\phi(z)$ is analytic on some closed disk centered at $a$. More precisely, prove that $1/\phi(z)$ is analytic on $\bar{D}(a,t)$ for any $t < r|\phi(a)|/\|\phi\|$.

## Section 5.3. Newton Polygons and the Maximum Modulus Principle

**5.8.** This exercise generalizes Proposition 5.16(b) by asking you to prove a general Cauchy estimate for $p$-adic analytic functions. Let $\phi(z) \in \mathbb{C}_p[\![z]\!]$ be a power series that converges on $\bar{D}(a,r)$ and let $\phi(\bar{D}(a,r)) = \bar{D}(\phi(a),s)$. Prove that

$$\left|\frac{d^n\phi}{dz^n}(a)\right| \le \frac{s}{|n!|r^n} \qquad \text{for all } n \ge 1.$$

(Note that as $n$ increases, the estimate becomes worse, since $|n!| \to 0$ as $n \to \infty$.)

## Section 5.4. The Nonarchimedean Julia and Fatou Sets

**5.9.** Prove the implications between uniform Lipschitz, uniform continuity, and equicontinuity stated in Section 5.4. More generally, give a definition of what it means for a family of functions to be *equi-Lipschitz* and prove the following implications:

| uniformly Lipschitz | $\Longrightarrow$ | equi-Lipschitz at every point |
|---|---|---|
| $\Downarrow$ | | $\Downarrow$ |
| uniformly continuous | $\Longrightarrow$ | equicontinuous at every point |

**5.10.** This exercise develops the abstract theory of residues (on $\mathbb{P}^1$) in order to prove Theorem 1.14 in arbitrary characteristic. Let $\phi(z) \in K(z)$ and $\alpha \in K$. The function $\phi(z)$ can be written as a partial Laurent series

$$\phi(z) = \frac{a_{-N}}{(z-\alpha)^N} + \frac{a_{-N+1}}{(z-\alpha)^{N-1}} + \cdots + \frac{a_{-1}}{(z-\alpha)} + \psi(z) \quad \text{with } \psi(z) \in K[\![z-\alpha]\!],$$

and we define the *residue of $\phi$ at $\alpha$* to be

$$\operatorname*{Res}_{z=\alpha}\bigl(\phi(z)\,dz\bigr) = a_{-1}.$$

The residue of $\phi$ at $\infty$ is defined by using the substitution $z = 1/w$; thus

$$\operatorname*{Res}_{z=\infty}\bigl(\phi(z)\,dz\bigr) = \operatorname*{Res}_{w=0}\bigl(-\phi(w^{-1})w^{-2}\,dw\bigr).$$

(a) Compute all of the nonzero residues of each of the following functions. In each case, check that the sum of the residues is 0.

(i) $\phi_1(z) = \dfrac{z^2+2}{z^2-3z+2}$. (ii) $\phi_2(z) = \dfrac{z^2+2}{z^3-4z^2+5z-2}$.

(b) If $\phi(z)$ has a simple pole at $\alpha$, prove that the residue of $\phi$ at $\alpha$ is the value of the function $(z-\alpha)\phi(z)$ evaluated at $z = \alpha$.

(c) More generally, for any integer $n \ge 0$, let $\psi_n(z)$ be the function

$$\psi_n(z) = \frac{1}{n!}\frac{d^n}{dz^n}\bigl((z-\alpha)^{n+1}\phi(z)\bigr).$$

Here we are taking formal derivatives in $K(z)$. Note that in characteristic $p$, one must be careful to "cancel" the $n!$ before setting $p = 0$. (If you have not seen this kind of computation before, rewrite the expression for $\psi_n(z)$ to make it clear that it makes sense in any characteristic.) Suppose that $\phi(z)$ has a pole of exact order $n \geq 1$ at $\alpha$, i.e., writing $\phi(z) = F(z)/G(z)$ as a ratio of polynomials with no common factors, we have $n = \operatorname{ord}_{z=\alpha}\bigl(G(z)\bigr)$. Prove that

$$\operatorname*{Res}_{z=\alpha}\bigl(\phi(z)\,dz\bigr) = \psi_{n-1}(\alpha).$$

(d) Prove that

$$\bigl\{P \in \mathbb{P}^1(K) : \operatorname*{Res}_{P}\bigl(\phi(z)\,dz\bigr) \neq 0\bigr\}$$

is a finite set.

(e) Assume that $K$ is algebraically closed and prove the Cauchy residue formula

$$\sum_{P\in\mathbb{P}^1(K)} \operatorname*{Res}_{P}\bigl(\phi(z)\,dz\bigr) = 0.$$

(This is a hard exercise. Try it first under the assumption that $\phi(z)$ has simple poles and $\phi(\infty) \neq \infty$, which suffices for (f).)

(f) Use (e) to prove that Theorem 1.14 is true in arbitrary characteristic, and use it to deduce that Corollary 5.19 and Proposition 5.20(c) are true. Hence Fatou sets are nonempty for all algebraically closed nonarchimedean fields.

**5.11.** (a) Let $p \geq 3$ be a prime and let $c \in \mathbb{C}_p$ satisfy $|c| > 1$. Prove that the function $\phi(z) = z^2 + c$ has exactly one nonrepelling periodic point, namely the totally ramified fixed point at infinity.

(b) * Suppose that $\phi(z) \in \mathbb{C}_p(z)$ has an indifferent periodic point, i.e., a periodic point $P$ whose multiplier satisfies $|\lambda_P(\phi)| = 1$. Prove that $\phi(z)$ has infinitely many indifferent periodic points. (This result is due to Rivera-Letelier [372].)

(c) ** Do there exist rational maps $\phi(z) \in \mathbb{C}_p(z)$ that have at least two, but only finitely many, nonrepelling periodic points? (From (b), the nonrepelling periodic points would have to be attracting.)

## Section 5.5. The Dynamics of $(z^2 - z)/p$

**5.12.** Let $S$ be a finite set, let $S^{\mathbb{N}}$ be the space of sequences on $S$ with the symbolic dynamics metric, and let $L : S^{\mathbb{N}} \to S^{\mathbb{N}}$ be the left shift map. (See Section 5.5.1.)

(a) Let $\alpha, \beta \in S^{\mathbb{N}}$ and suppose that $\lim_{n\to\infty} \rho\bigl(L^n(\alpha), L^n(\beta)\bigr) = 0$. Prove that there exists an $m$ such that $L^m(\alpha) = L^m(\beta)$.

(b) Prove that the periodic points of $L$ are dense in $S^{\mathbb{N}}$.

(c) Prove that there is an element $\gamma \in S^{\mathbb{N}}$ with the property that the orbit

$$\mathcal{O}_L(\gamma) = \{L^n(\gamma) : n \in \mathbb{N}\}$$

is dense in $S^{\mathbb{N}}$. One says that $L$ is *topologically transitive on* $S^{\mathbb{N}}$. (*Hint*. Create $\gamma$ by first listing all possible blocks of length 1, then all possible blocks of length 2, and so on.)

(d) Let $\alpha \in S^{\mathbb{N}}$. Prove that the backward orbit $\mathcal{O}_L^-(\alpha) = \bigcup_{n\geq 0} L^{-n}(\alpha)$ is dense in $S^{\mathbb{N}}$.

(e) More precisely, prove that the backward orbit of any $\alpha \in S^{\mathbb{N}}$ is equidistributed in the following sense: For all $\beta \in S^{\mathbb{N}}$ and all $k \ge 0$,

$$\lim_{n\to\infty} \frac{\#\{\gamma \in L^{-n}(\alpha) : \rho(\gamma,\beta) \le p^{-k}\}}{\#L^{-n}(\alpha)} = \frac{1}{(\#S)^k}.$$

(In fact, prove that the limit stabilizes as soon as $n > k$.)

**5.13.** Let $\phi(z) = (z^2 - z)/p$ with $p \ge 3$. Use Exercise 5.12 and the identification provided by Proposition 5.24 to prove that the backward orbit of a point $a \in \mathcal{J}(\phi)$ is equidistributed in the following sense: For all $b \in \mathcal{J}(\phi)$ and all radii $r = p^{-k}$,

$$\lim_{n\to\infty} \frac{\#\bigl(\phi^{-n}(a) \cap \bar{D}(b,r)\bigr)}{2^n} = \frac{1}{2^k}.$$

**5.14.** Let $p \ge 3$ and for any $c \in \mathbb{C}_p$, let $\phi_c(z) = (z^2 - cz)/p$.
(a) If $c \in \mathbb{Z}_p^*$, prove that the statement of Corollary 5.25 is true for $\phi_c$.
(b) More generally, prove an analogous statement if $c$ is a unit in a finite extension of $\mathbb{Z}_p$.
(c) What happens if $c \in p\mathbb{Z}_p$?

**5.15.** Let $d \ge 2$ be an integer, let $p > d$ be a prime, and consider the dynamics of the map

$$\phi(z) = \frac{z(z-1)(z-2)\cdots(z-d+1)}{p}$$

over $\mathbb{C}_p$. Prove that $\mathcal{J}(\phi)$ can be described using symbolic dynamics and use this identification to prove the following generalization of Corollary 5.25: $\mathcal{J}(\phi) \subset \mathbb{Q}_p$, the set $\operatorname{Per}_n(\phi)$ consists of the fixed point at $\infty$ and $d^n$ repelling points, the Julia set $\mathcal{J}(\phi)$ is the closure of the repelling periodic points, and $\phi$ is topologically transitive on $\mathcal{J}(\phi)$.

## Section 5.6. A Nonarchimedean Version of Montel's Theorem

**5.16.** Complete the proof of Theorem 5.26 (see page 265) by writing down the details in the case that $\alpha = 0$.

**5.17.** Let $\Phi$ be a collection of power series that converge on $\bar{D}(a,r)$, and suppose that there is a point $\alpha \in K$ such that $\alpha \notin \phi\bigl(\bar{D}(a,r)\bigr)$ for all $\phi \in \Phi$. We proved in Theorem 5.26 that there is a constant $C = C(\alpha, a, r)$ such that

$$\rho(\phi(z),\phi(w)) \le C\rho(z,w) \quad \text{for all } \phi \in \Phi \text{ and all } z,w \in \bar{D}(a,r).$$

Find an explicit value for the Lipschitz constant $C$ in terms of $\alpha$, $a$, and $r$.

**5.18.** Let $\Phi$ be a collection of rational, or more generally meromorphic, functions

$$\bar{D}(a,r) \longrightarrow \mathbb{P}^1(K),$$

and suppose that $\alpha, \beta \in \mathbb{P}^1(K)$ do not lie in $\phi\bigl(\bar{D}(a,r)\bigr)$ for all $\phi \in \Phi$. Our proof of the nonarchimedean Montel theorem (Theorem 5.27) shows that there is a constant $C = C(\alpha,\beta,a,r)$ such that

$$\rho(\phi(z),\phi(w)) \le C\rho(z,w) \quad \text{for all } \phi \in \Phi \text{ and all } z,w \in \bar{D}(a,r).$$

Find an explicit value of $C$ in terms of $\alpha$, $\beta$, $a$, and $r$.

## Section 5.7. Periodic Points and the Julia Set

**5.19.** As in the statement of Lemma 5.34, let $A, B \subset \mathbb{C}_p$ be bounded sets for which there are constants $0 < \delta \le 1 \le \Delta$ such that

$$\sup_{\alpha\in A} |\alpha| \le \Delta, \quad \sup_{\beta\in B} |\beta| \le \Delta, \quad \text{and} \quad \inf_{\alpha\in A,\,\beta\in B} \rho(\alpha,\beta) = \delta > 0.$$

For each $(\alpha,\beta) \in A \times B$, let $L_{\alpha,\beta}(z) = (\beta-\alpha)z + \alpha$. During the proof of Lemma 5.34 we showed that

$$\rho\bigl(L_{\alpha,\beta}(z), L_{\alpha',\beta'}(z')\bigr) \le \frac{\Delta^4}{\delta^2} \cdot \max\bigl\{\rho(\alpha,\alpha'), \rho(\beta,\beta'), \rho(z,z')\bigr\}$$
$$\text{for all } \alpha,\alpha' \in A, \text{ all } \beta,\beta' \in B, \text{ and all } z,z' \in \mathbb{P}^1(\mathbb{C}_p).$$

Improve this result by reducing the exponent of $\Delta$ and/or $\delta$ in the constant. Try to find the best-possible exponents.

**5.20.** Let $A, B, C \subset \mathbb{P}^1(\mathbb{C}_p)$ be three sets that are at a positive distance from each other, i.e., there is a constant $\delta > 0$ such that

$$\inf_{\alpha\in A,\,\beta\in B} \rho(\alpha,\beta) \ge \delta, \qquad \inf_{\alpha\in A,\,\gamma\in C} \rho(\alpha,\gamma) \ge \delta, \qquad \inf_{\beta\in B,\,\gamma\in C} \rho(\beta,\gamma) \ge \delta.$$

For any triple of points $(\alpha,\beta,\gamma) \in A \times B \times C$, define

$$L_{\alpha,\beta,\gamma}(z) \in \mathrm{PGL}_2(\mathbb{C}_p)$$

to be the unique linear fractional transformation satisfying

$$L_{\alpha,\beta,\gamma}(0) = \alpha, \qquad L_{\alpha,\beta,\gamma}(1) = \beta, \qquad L_{\alpha,\beta,\gamma}(\infty) = \gamma.$$

Prove that the map

$$A \times B \times C \times \mathbb{P}^1(\mathbb{C}_p) \longrightarrow \mathbb{P}^1(\mathbb{C}_p), \qquad (\alpha,\beta,\gamma,z) \longmapsto L_{\alpha,\beta,\gamma}(z),$$

is Lipschitz and find an explicit Lipschitz constant depending only on $\delta$. (This exercise generalizes Lemma 5.34, which is the case that $C = \{\infty\}$ is the single point at infinity.)

**5.21.** This exercise can be used in place of Lemma 5.34 to prove Montel's theorem with moving targets (Theorem 5.36).

(a) Let $P, Q \in \mathbb{P}^1(K)$ be distinct points. Prove that there is a linear fractional transformation $\Lambda = \Lambda_{P,Q} \in \mathrm{PGL}_2(K)$ satisfying

$$\Lambda(0) = P, \qquad \Lambda(\infty) = Q, \qquad \bigl|\mathrm{Res}(\Lambda)\bigr| = \rho(P,Q).$$

(*Hint.* If $\Lambda = \left(\begin{smallmatrix} a & b \\ c & d\end{smallmatrix}\right)$ is normalized to satisfy $\max\bigl\{|a|,|b|,|c|,|d|\bigr\} = 1$, then Exercise 2.8 says that the resultant of $\Lambda$ is equal to the determinant $ad-bc$.)

(b) Let $A, B \subset \mathbb{C}_p$ be bounded sets that are at a positive distance from each other. Prove that the map

$$A \times B \times \mathbb{P}^1(K) \longrightarrow \mathbb{P}^1(K), \qquad (P,Q,R) \longmapsto \Lambda_{P,Q}(R),$$

is Lipschitz.

## Section 5.8. Nonarchimedean Wandering Domains

**5.22.** Let $X$ be a topological space and let $\mathcal{D}$ be a collection of open subsets of $X$ that form a base for the toplogy of $X$. We use $\mathcal{D}$ to define disk components as described on page 277. (Note that the sets in $\mathcal{D}$ need not be actual disks, since the space $X$ is merely assumed to be a topological space.)

(a) Let $U \subset X$ be an open set, let $P_1, P_2 \in U$ be points, and let $V_1$ and $V_2$ be the disk components of $U$ containing $P_1$ and $P_2$, respectively. Prove that either $V_1 = V_2$ or $V_1 \cap V_2 = \emptyset$.

(b) Prove that $U$ is a disjoint union of disk components.

(c) Prove that the disk components of $U$ are open.

**5.23.** Let $X$ be a topological space and let $\mathcal{D}$ be a collection of open subsets (disks) of $X$ that form a base for the topology of $X$. An open set $U \subset X$ is defined to be *disk-connected* (relative to $\mathcal{D}$) if for every pair of points $P, Q \in U$ there is a finite sequence of disks $D_1, D_2, \ldots, D_n \subset U$ such that

$$P \in D_1, \qquad Q \in D_2, \qquad D_i \cap D_{i+1} \neq \emptyset \quad \text{for all } 1 \le i < n.$$

(a) Let $U_1$ and $U_2$ be disk-connected subsets of $X$. Prove that either

$$U_1 = U_2 \quad \text{or} \quad U_1 \cap U_2 = \emptyset.$$

(b) Let $U \subset X$ be an open set and let $P \in U$. Prove that there is a maximal disk-connected open subset of $U$ containing $P$.

(c) Prove that the maximal disk-connected open subset of $U$ containing $P$ described in (b) is in fact the disk component of $U$ containing $P$. This gives an alternative definition of disk component.

**5.24.** Prove that the standard closed disks with rational radius (cf. Remark 5.7) in $\mathbb{P}^1(\mathbb{C}_p)$ are exactly the images via elements of $\mathrm{PGL}_2(\mathbb{C}_p)$ of the unit disk $\bar{D}(0,1)$. Similarly, the standard open disks of rational radius are the images of $D(0,1)$.

**5.25.** Let $K/\mathbb{Q}_p$ be a finite extension and let $\phi(z) \in K(z)$ be a rational function of degree $d \ge 2$ with $\infty \in \mathcal{F}(\phi)$. Prove that $\phi$ is hyperbolic, i.e., $\phi$ satisfies the conditions of Theorem 5.46, if and only if $\phi$ has the following property:

> For every finite extension $L/K$ there are a set $U \subset L$ containing $\mathcal{J}(\phi) \cap L$, positive constants $b > a > 0$, and a continuous function $\sigma : U \to [a,b]$ such that
>
> $$\sigma\bigl(\phi(\alpha)\bigr)\bigl|\phi'(\alpha)\bigr| \ge \sigma(\alpha) \qquad \text{for all } \alpha \in U.$$

This shows that $p$-adic hyperbolicity and classical (complex) hyperbolicity can both be characterized as saying that a map is everywhere expanding on the Julia set with respect to some reasonable metric. For further information about hyperbolic maps in the classical setting; see, for example [95] or [302].

**5.26.** * Let $\phi : \mathbb{P}^1 \to \mathbb{P}^1$ be a rational map of degree at least 2 and let $P \in \mathbb{P}^1(\mathbb{C}_p)$ be a recurrent critical point that is in the Fatou set $\mathcal{F}(\phi)$. Prove that $P$ is a periodic point.

## Section 5.9. Green Functions and Local Heights

**5.27.** Let $K$ be a field with a nonarchimedean absolute value $v$, let $\phi : \mathbb{P}^1 \to \mathbb{P}^1$ be a morphism of degree $d \ge 2$, let $\Phi = (F, G) : \mathbb{A}^2 \to \mathbb{A}^2$ be a lift of $\phi$ satisfying $\|\Phi\|_v = 1$, and let $\mathcal{G}_\Phi$ be the associated Green function. If $\bigl|\operatorname{Res}(F,G)\bigr|_v \ne 1$, or equivalently if $\phi$ has bad reduction at $\phi$, prove that there exists a point $(x, y) \in \mathbb{A}^2_*(K)$ such that

$$\mathcal{G}_\Phi(x, y) \ne \log\bigl\|(x, y)\bigr\|_v.$$

This is the converse to Proposition 5.58(c).

**5.28.** Let $K$ be a field with an absolute value $v$, let $\phi : \mathbb{P}^1 \to \mathbb{P}^1$ be a morphism of degree $d \ge 2$, and let $\Phi : \mathbb{A}^2 \to \mathbb{A}^2$ be a lift of $\phi$.

(a) Prove that the map

$$g(x, y) = \log\bigl\|\Phi(x, y)\bigr\|_v - d\log\bigl\|(x, y)\bigr\|_v$$

induces a well-defined function $g : \mathbb{P}^1(K) \to \mathbb{R}$.

(b) Prove that $g$ is Lipschitz, i.e., prove that there is a constant $C = C(\phi)$ such that

$$\bigl|g(P) - g(Q)\bigr| \le C\rho_v(P, Q) \quad \text{for all } P, Q \in \mathbb{P}^1(K).$$

(c) If $v$ is nonarchimedean, prove that $g$ is locally constant. More precisely, if the lift $\Phi$ is chosen to satisfy $\|\Phi\|_v = 1$, prove that

$$g(P) = g(Q) \quad \text{for all } P, Q \in \mathbb{P}^1(K) \text{ with } \rho_v(P, Q) < \bigl|\operatorname{Res}(\Phi)\bigr|_v.$$

(d) Define a modified Green function $\hat{\mathcal{G}}_\Phi$ by

$$\hat{\mathcal{G}}_\Phi(x, y) = \mathcal{G}_\Phi(x, y) - \log\bigl\|(x, y)\bigr\|_v.$$

Prove that $\hat{\mathcal{G}}_\Phi$ is a Hölder continuous function on $\mathbb{P}^1(K)$. In other words, prove that $\hat{\mathcal{G}}_\Phi$ is well-defined on $\mathbb{P}^1(K)$ and that there are positive constants $C$ and $\delta$ such that

$$\bigl|\hat{\mathcal{G}}_\Phi(P) - \hat{\mathcal{G}}_\Phi(Q)\bigr| \le C\rho_v(P, Q)^\delta \qquad \text{for all } P, Q \in \mathbb{P}^1(K).$$

(*Hint.* Show that $\hat{\mathcal{G}}_\Phi(P)$ is given by the telescoping series $\sum_{n=0}^{\infty} d^{-n} g\bigl(\phi^n(P)\bigr)$. For small $n$, say $n \le N$, estimate the difference $\bigl|g\bigl(\phi^n(P)\bigr) - g\bigl(\phi^n(Q)\bigr)\bigr|$ using (b) and the fact that $\phi$ is Lipschitz, and for large $n$ use an elementary bound. Then make an appropriate choice for $N$.)

**5.29.** The definition of the canonical height $\hat{h}_\phi(P)$ as the limit of $d^{-n}h\bigl(\phi^n(P)\bigr)$ is not practical for numerical calculations, even for $P \in \mathbb{P}^1(\mathbb{Q})$, because one would need to compute the exact value of points $\phi^n(P)$ whose coordinates have $O(d^n)$ digits. A better method to compute $\hat{h}_\phi(P)$ is as the sum of the Green functions $\mathcal{G}_{\Phi,v}(P)$ as described in Theorem 5.59.

Let $\phi : \mathbb{P}^1 \to \mathbb{P}^1$ be a morphism of degree $d \ge 2$ and fix a lift $\Phi$ of $\phi$. If $v \in M_K$ is a nonarchimedean absolute value such that $\|\Phi\|_v = 1$ and $|\operatorname{Res}(\Phi)|_v = 1$, then Proposition 5.58(c) says that

$$\mathcal{G}_{\Phi,v}(x, y) = \log\bigl\|(x, y)\|_v.$$

This covers all but finitely many absolute values, so it remains to devise an efficient method to compute $\mathcal{G}_{\Phi,v}(x, y)$ in the remaining cases.

Let $P = [x, y] \in \mathbb{P}^1(K)$ and use the algorithm described in Figure 5.6 to define a sequence of triples

$$(u_i, x_i, y_i)_{i\geq 0} \qquad \text{with } u_i, x_i, y_i \in K.$$

This exercise asks you to prove that $N$ iterations of the algorithm gives the value of $\mathcal{G}_{\Phi,v}(x,y)$ to within $O(d^{-N})$.

(a) Prove that $|u_i|_v \leq 1$ for all $i \geq 0$. Hence the computation of quantities such as $F(u_i, 1)$ in the algorithm do not involve excessively large numbers.

(b) Prove that for all $i \geq 1$ we have

$$\log\bigl\|(x_i, y_i)\bigr\|_v = \log\bigl\|\Phi^i(x,y)\bigr\|_v - d\log\bigl\|\Phi^{i-1}(x,y)\bigr\|_v.$$

(c) As in Exercise 5.28, we let

$$g(X,Y) = \log\bigl\|\Phi(X,Y)\bigr\|_v - d\log\bigl\|(X,Y)\bigr\|_v.$$

The homogeneity of $\Phi$ shows that $g$ is a well-defined function on $\mathbb{P}^1(K)$, and Proposition 5.57 says that $g$ is bounded. Prove that

$$\mathcal{G}_\Phi(x,y) = \log\bigl\|(x,y)\bigr\|_v + \sum_{i=1}^{\infty} \frac{1}{d^i} g\bigl(\Phi^{i-1}(x,y)\bigr).$$

(d) Prove that

$$\mathcal{G}_\Phi(x,y) = \log\bigl\|(x,y)\bigr\|_v + \sum_{i=1}^{N} \frac{1}{d^i} \log\max\bigl\{|x_i|_v, |y_i|_v\bigr\} + O(d^{-N}), \tag{5.48}$$

where the big-$O$ constant depends only on $\Phi$. Deduce that the algorithm described in Figure 5.6 computes $\mathcal{G}_\Phi(x,y)$ to within $O(d^{-N})$.

(e) Find an explicit value for the big-$O$ constant in (5.48) in terms of the quantities $d$, $\|\Phi\|_v$, and $\bigl|\operatorname{Res}(\Phi)\bigr|_v$.

**5.30.** Implement the algorithm described in Figure 5.6 to compute the Green function $\mathcal{G}_{\phi,\infty}$ for the archimedean absolute value on $\mathbb{P}^1(\mathbb{R})$.

(a) Let $\phi(z) = z + 1/z$ and compute $\mathcal{G}_{\phi,\infty}(x,y)$ to (say) 8 decimal places for each of the points $(1,1)$, $(2,1)$, and $(5,2)$. Check your program by verifying that your values satisfy

$$\mathcal{G}_{\phi,\infty}(5,2) \approx 2\mathcal{G}_{\phi,\infty}(2,1) \approx 4\mathcal{G}_{\phi,\infty}(1,1).$$

Compute the canonical height $\hat{h}_\phi(1)$ and compare it with the value that you obtained in Exercise 3.20. (*Hint.* The map $\phi$ has good reduction at all primes.)

(b) Let $\phi(z) = (3z^2-1)/(z^2-1)$ and note that $\phi^2(1) = 3$ and $\phi^3(1) = \frac{13}{4}$. Compute $\mathcal{G}_{\phi,\infty}(x,y)$ for each of the points $(1,1)$, $(3,1)$, and $(13,4)$. Why does $\mathcal{G}_{\phi,\infty}(3,1)$ not equal $4\mathcal{G}_{\phi,\infty}(1,1)$? What is the difference between these two values?

**5.31.** Let $\phi \in \mathbb{Q}(z)$ be a rational map of degree at least 2 and let $P \in \mathbb{P}^1(\mathbb{Q})$. Write an efficient computer program to compute $\hat{h}_\phi(P)$ as a sum of Green functions. (For primes $p$ not dividing $\operatorname{Res}(\Phi)$, use Proposition 5.58 to compute $\mathcal{G}_{\phi,p}(x,y)$. For primes dividing $\operatorname{Res}(\Phi)$ and for the archimedean place, use the Green function algorithm in Figure 5.6.) Compute $\hat{h}_\phi(\alpha)$ to 8 decimal places for each of the following maps and points.

(a) $\phi(z) = z^2 - 1$, $\alpha = \frac{1}{2}$.

```
INITIALIZATION
    Write Φ as Φ = (F, G) with F, G ∈ K[X, Y]
    Set N = Desired number of iterations
    Set x_0 = x and y_0 = y and Green = 0
MAIN LOOP:  i = 0, 1, 2, ..., N
    Increment Green By d^{-i} log max{|x_i|_v, |y_i|_v}
    If |x_i|_v ≤ |y_i|_v
        Set u_i = x_i/y_i
        Compute x_{i+1} = F(u_i, 1) and y_{i+1} = G(u_i, 1)
    Else |y_i|_v < |x_i|_v
        Set u_i = y_i/x_i
        Compute x_{i+1} = F(1, u_i) and y_{i+1} = G(1, u_i)
END MAIN LOOP
Return the Value Green
```

Figure 5.6: An algorithm to compute the Green function $\mathcal{G}_\Phi(x,y)$.

(b) $\phi(z) = z^2 + 1, \quad \alpha = \frac{1}{2}$.
(c) $\phi(z) = 3z^2 - 4, \quad \alpha = 1$.
(d) $\phi(z) = z + \dfrac{1}{z}, \quad \alpha = 1$.
(e) $\phi(z) = \dfrac{3z^2 - 1}{z^2 - 1}, \quad \alpha = 1$ and $\alpha = 3$. Check that $\hat{h}_\phi(3) \approx 4\hat{h}_\phi(1)$.

**5.32.** Let $K$ be an algebraically closed field with an absolute value $v$, let $\phi : \mathbb{P}^1 \to \mathbb{P}^1$ be a morphism of degree $d \ge 2$, fix a lift $\Phi$ of $\phi$, and for each homogeneous polynomial $E \in K[x,y]$, let $\hat{\lambda}_{\phi,E}$ be the associated local canonical height function. (See Theorem 5.60.)

(a) If $E_1, E_2 \in K[x,y]$ are homogeneous polynomials, prove that

$$\hat{\lambda}_{\phi,E_1E_2}(P) = \hat{\lambda}_{\phi,E_1}(P) + \hat{\lambda}_{\phi,E_2}(P)$$

at all points such that $E_1(P) \neq 0$ and $E_2(P) \neq 0$.

(b) Let $D = n_1(Q_1) + n_2(Q_2) + \cdots + n_r(Q_r)$ be a divisor on $\mathbb{P}^1$, i.e., $D$ is a formal sum of points with $n_1, \ldots, n_r \in \mathbb{Z}$. Use (a) to associate to $D$ a local height function

$$\hat{\lambda}_{\phi,D} : \mathbb{P}^1(K) \smallsetminus \{Q_1, \ldots, Q_r\} \longrightarrow \mathbb{R}.$$

(c) Prove that there is a rational function $f_D \in K(z)$ with the property that

$$\hat{\lambda}_{\phi,D}\bigl(\phi(P)\bigr) = d\hat{\lambda}_{\phi,D}(P) - \log\bigl|f_D(P)\bigr|_v$$

at all points where $\hat{\lambda}_{\phi,D}\bigl(\phi(P)\bigr)$ and $\hat{\lambda}_{\phi,D}(P)$ are defined.

(d) Let $g(z) \in K(z)$ be a rational function and let $D_g$ be the *divisor of* $g$, which by definition is the formal sum

$$D_g = \sum_{Q\in\mathbb{P}^1(K)} \operatorname{ord}_Q(g)(Q).$$

(Here $\operatorname{ord}_Q(g)$ is the order of vanishing of $g$ at the point $Q$; see Example 2.2.) Prove that the function

$$P \longmapsto \hat{\lambda}_{\phi,D_g}(P) + \log\bigl|g(P)\bigr|_v$$

extends to a bounded continuous function on all of $\mathbb{P}^1(K)$.

## Section 5.10. Berkovich Space and Dynamics

**5.33.** Let $\phi(z) \in \mathbb{C}_p[z]$ be a polynomial and let $a \in \bar{D}(0,1)$. Write $\phi(z)$ as a polynomial in $z-a$, say

$$\phi(z) = c_0 + c_1(z-a) + c_2(z-a)^2 + \cdots + c_d(z-a)^d.$$

Prove that

$$\phi\bigl(\bar{D}(a,r)\bigr) = \bar{D}\bigl(\phi(a), s\bigr),$$

where the radius $s$ of the image disk is given by

$$s = r\cdot\max\bigl\{|c_1|,\ |c_2|r,\ |c_3|r^2,\ \ldots,\ |c_d|r^d\bigr\}.$$

(*Hint*. We already proved that $\phi\bigl(\bar{D}(a,r)\bigr)$ is a disk; see Proposition 5.16. Now use the maximum modulus principle to find the radius.)

**5.34.** This exercise develops some elementary properties of bounded seminorms as defined on page 296.

(a) Let $|\cdot|$ be a nonconstant seminorm. Prove that $|0| = 0$ and $|1| = 1$.

(b) Suppose that $|\cdot|$ is an $\|\cdot\|_R$-bounded seminorm. Let $f(z) = c$ be a constant polynomial. Prove that $|f| = |c|$ is the usual absolute value on $\mathbb{C}_p$.

(c) Suppose that we replace property (3) of bounded seminorm with the usual triangle inequality $|f+g| \le |f| + |g|$. Prove that $|\cdot|$ satisfies (3).

(d) Suppose that we replace property (4) of bounded seminorm with the weaker statement that there is a constant $\kappa$ such that $|f| \le \kappa\|f\|_R$ for all $f \in \mathbb{C}_p[z]$. Prove that (4) is true, i.e., we can take $\kappa = 1$.

**5.35.** Prove that the seminorms associated to points of Types II, III, and IV are actually norms. (*Hint*. For Type IV, use the fact that the limiting radius is positive. See Remark 5.65.)

**5.36.** Let $d \ge 2$. Prove that the map $\phi(z) = z^d$ has no Type-IV fixed points in $\mathbb{A}^{\mathcal{B}}$. Does $\phi(z)$ have any Type-IV periodic or preperiodic points?

**5.37.** Let $\bar{D}(a,r)$ be a disk with $0 \notin \bar{D}(a,r)$ and let $f(z) = 1/z$. Prove that

$$f\bigl(\bar{D}(a,r)\bigr) = \bigl\{z^{-1} : |z-a| \le r\bigr\} = \bar{D}\bigl(a^{-1}, r/|a|^2\bigr).$$

**5.38.** Let

$$\phi(z) = \sum_{i=0}^{d} c_i z^i \in \mathbb{C}_p[z] \quad \text{be a polynomial with } |c_i| \le 1 \text{ and } |c_d| = 1. \tag{5.49}$$

The polynomial $\phi$ induces a map $\phi : \bar{D}^{\mathcal{B}} \to \bar{D}^{\mathcal{B}}$ on the Berkovich disk.

(a) Prove that the Gauss point $\xi_{0,1}$ is fixed by $\phi$.

(b) Let $a \in \mathbb{C}_p$ be a fixed point of $\phi : \mathbb{C}_p \to \mathbb{C}_p$. Prove that $\xi_{a,0}$ is fixed by $\phi$.

(c) Let $a \in \mathbb{C}_p$ be a neutral fixed point, i.e., $|\phi'(a)| = 1$, and let $0 \le r \le 1$. Prove that $\xi_{a,r}$ is fixed by $\phi$. (In other words, if $a$ is a neutral fixed point, then the entire line segment $L_{a,0} \subset \bar{D}^{\mathcal{B}}$ is fixed by $\phi$.)

(d) Prove that every Berkovich fixed point of $\phi$ of Type I, II, or III is one of the three types listed in (a), (b), (c). (*Hint.* Use the explicit description of $\phi(\xi_{a,r})$ given in Exercise 5.33.)

(e) Can $\phi$ have fixed points of Type IV?

**5.39.** Let $\phi(z) = \sum c_i z^i \in \mathbb{C}_p[z]$ be a polynomial as in (5.49). Prove directly that the Julia set of $\phi$ is $\mathcal{J}^{\mathcal{B}}(\phi) = \{\xi_{1,0}\}$. (This is a special case of the result described in Example 5.79, since the conditions on $\phi$ imply that it has good reduction.)

**5.40.** For any nested sequence of closed disks

$$\bar{D}(0,1) \supset \bar{D}(a_1,r_1) \supset \bar{D}(a_2,r_2) \supset \cdots,$$

regardless of whether the intersection is empty, we define a seminorm in the usual way,

$$|f|_{\mathbf{a},\mathbf{r}} = \lim_{i\to\infty} \sup_{z\in\bar{D}(a_i,r_i)} |f(z)|.$$

(a) Let $r = \lim r_i$. If the intersection $\bigcap \bar{D}(a_i,r_i)$ is nonempty and $a$ is a point in the intersection, prove that $|f|_{\mathbf{a},\mathbf{r}} = |f|_{a,r}$. Thus every point in $\bar{D}^{\mathcal{B}}$, regardless of whether it is of Type I, II, III, or IV, can be represented by a nested sequence of closed disks.

(b) Two nested sequences of closed disks

$$\bar{D}(0,1) \supset \bar{D}(a_1,r_1) \supset \bar{D}(a_2,r_2) \supset \cdots,$$
$$\bar{D}(0,1) \supset \bar{D}(b_1,s_1) \supset \bar{D}(b_2,s_2) \supset \cdots,$$

are defined to be *equivalent* if the following two conditions are true:

- For every $i \ge 1$ there exists a $j \ge 1$ such that $\bar{D}(b_j,s_j) \subset \bar{D}(a_i,r_i)$.
- For every $i \ge 1$ there exists a $j \ge 1$ such that $\bar{D}(a_j,r_j) \subset \bar{D}(b_i,s_i)$.

Prove that two sequences of disks are equivalent if and only if their seminorms $|\cdot|_{\mathbf{a},\mathbf{r}}$ and $|\cdot|_{\mathbf{b},\mathbf{s}}$ are equal.

**5.41.** Let $\xi_{\mathbf{a},\mathbf{r}} \in \bar{D}^{\mathcal{B}}$ be a Type-IV point. We defined the "line segment" running from $\xi_{\mathbf{a},\mathbf{r}}$ to the Gauss point $\xi_{0,1}$ to be the set

$$L_{\mathbf{a},\mathbf{r}} = \{\xi_{\mathbf{a},\mathbf{r}}\} \cup \bigcup_{i=1}^{\infty} \{\xi_{a_i,t} : r_i \le t \le r_{i-1}\}.$$

(Note that $r_0 = 1$ by definition.) Let $r = \lim_{i\to\infty} r_i$. Prove that the map

$$h : [r,1] \longrightarrow L_{\mathbf{a},\mathbf{r}}, \qquad h(t) = \begin{cases} \xi_{\mathbf{a},\mathbf{r}} & \text{if } t = r, \\ \xi_{a_i,t} & \text{if } r_i \le t \le r_{i-1}, \end{cases}$$

is a homeomorphism, where $[r,1] \subset \mathbb{R}$ has the usual topology and $\bar{D}^{\mathcal{B}}$ has the Gel'fond topology. Thus $L_{\mathbf{a},\mathbf{r}}$ is indeed a line segment.

**5.42.** There is a natural metric on the Berkovich disk $\bar{D}^{\mathcal{B}}$ coming from the tree structure. We first identify the interval $[0,1]$ with each of the line segments

$$L_{a,0} = \{\xi_{a,r} : 0 \le r \le 1\} \in \bar{D}^{\mathcal{B}},$$

so each $L_{a,0}$ has length 1. We then define the distance between two points $\xi_{a,r}$ and $\xi_{b,s}$ to be the length of the shortest path in the tree connecting them. Denote this distance by $\kappa(\xi_{a,r}, \xi_{b,s})$. (For Type-IV points, take the limit.)

(a) If $\xi_{a,r}$ and $\xi_{b,s}$ both lie on some line segment $L_{c,0}$, prove that $\kappa(\xi_{a,r}, \xi_{b,s}) = |r-s|$ is simply the distance between them on that line segment.

(b) In general, prove that

$$\kappa(\xi_{a,r}, \xi_{b,s}) = \max\bigl\{|r-s|, 2|b-a| - r - s\bigr\}.$$

(*Hint*. How far above $\xi_{a,r}$ and $\xi_{b,s}$ do the line segments $L_{a,r}$ and $L_{b,s}$ merge? Go up one line segment and then down the other.)

(c) Prove that the set

$$\left\{\xi \in \bar{D}^{\mathcal{B}} : \kappa(\xi, \xi_{0,1}) < \frac{1}{2}\right\}$$

contains no Type-I points. Prove that every neighborhood of $\xi_{0,1}$ in the Gel'fond topology contains infinitely many Type-I points. Deduce that the path metric $\kappa$ does not define the Gel'fond topology, and indeed that $\kappa$ is not even continuous in the Gel'fond topology!

In Baker and Rumely's terminology, the set $\bar{D}^{\mathcal{B}}$ with the metric $\kappa$ is called the *small model*. There is also a *big model*, in which the edges are reparameterized so that if $\bar{D}(a,r) \subset \bar{D}(b,s)$, then the distance from $\xi_{a,r}$ to $\xi_{b,s}$ is $\bigl|\log(r/s)\bigr|$. In particular, points of Type I are at infinite distance from each other and from all points of Types II, III, and IV. It is the big model that is best adapted for doing potential theory on Berkovich space; see [26] for details.

**5.43.** For points of Type I, II, and III in $\bar{D}^{\mathcal{B}}$, the *Hsia kernel* is defined to be

$$\delta(\xi_{a,r}, \xi_{b,s}) = \max\bigl(r, s, |a-b|\bigr).$$

(For points of Type IV, one takes the appropriate limit.) The Hsia kernel is used in studying potential theory on $\bar{D}^{\mathcal{B}}$; see [26, 209].

(a) Prove that $\delta(\xi_{a,r}, \xi_{a,r}) = r$.

(b) Prove that $\delta(\xi_{a,0}, \xi_{b,0}) = |a-b|$. Thus $\delta$ extends the usual norm on $\mathbb{C}_p$, where we identify the unit disk in $\mathbb{C}_p$ with the Type-I points in $\bar{D}^{\mathcal{B}}$.

(c) Let $\xi_{c,t}$ be the point where the line segments $L_{a,r}$ and $L_{b,s}$ first meet. Prove that $\delta(\xi_{a,r}, \xi_{b,s}) = t$.

(d) Prove that

$$\delta(\xi, \xi'') \le \max\bigl\{\delta(\xi, \xi'), \delta(\xi', \xi'')\bigr\} \qquad \text{for all } \xi, \xi', \xi'' \in \bar{D}^{\mathcal{B}}.$$

**5.44.** There are natural inclusions

$$\bar{D}(0,1) \hookrightarrow \bar{D}^{\mathcal{B}}, \qquad \mathbb{A}^1(\mathbb{C}_p) \hookrightarrow \mathbb{A}^{\mathcal{B}}, \qquad \mathbb{P}^1(\mathbb{C}_p) \hookrightarrow \mathbb{P}^{\mathcal{B}}.$$

In each case, prove that the restriction of the Gel'fond topology on the Berkovich space yields the usual metric topology on the smaller space, where we use the usual $p$-adic metric on $\bar{D}(0,1)$ and $\mathbb{A}^1(\mathbb{C}_p)$ and the $p$-adic chordal metric on $\mathbb{P}^1(\mathbb{C}_p)$.

**5.45.** Let $\phi$ be a rational map of degree at least 2, let $P \in \mathbb{P}^{\mathcal{B}}$, and let $U \subset \mathbb{P}^{\mathcal{B}}$ be a neighborhood of $P$ with the property that

$$\phi(U) \subseteq U \qquad \text{and} \qquad \bigcap_{n \geq 0} \phi^n(U) = \{P\}.$$

Prove that $P$ is a Type-I point and that, considered as a point in $\mathbb{P}^1(\mathbb{C}_p)$, the point $P$ is an attracting fixed point for $\phi$.

# Chapter 6

# Dynamics Associated to Algebraic Groups

In the forest of untamed rational maps live a select few whose additional structure allows them to be more easily domesticated. They are the power maps, Chebyshev polynomials, and Lattès maps, whose complex dynamics were briefly discussed in Section 1.6. The underlying structure that they possess comes from an algebraic group, namely the multiplicative group for the power maps and Chebyshev polynomials and elliptic curves for the Lattès maps. Although such maps are special in many ways, they yet provide important examples, testing grounds, and boundary conditions for general results in dynamics. In this chapter we investigate some of the algebraic and arithmetic properties of the rational maps associated to algebraic groups.

## 6.1 Power Maps and the Multiplicative Group

The simplest rational maps are the power maps given by monic monomials,

$$M_d : \mathbb{P}^1 \longrightarrow \mathbb{P}^1, \qquad M_d(z) = z^d,$$

where the integer $d$ may be positive or negative. These maps obviously commute with one another under composition,

$$M_d\bigl(M_e(z)\bigr) = z^{de} = M_e\bigl(M_d(z)\bigr).$$

A more intrinsic description of the maps associated to the monic monomials $M_d$ is that they are endomorphisms of the multiplicative group,

$$M_d : \mathbb{G}_m \longrightarrow \mathbb{G}_m, \qquad \alpha \longmapsto M_d(\alpha) = \alpha^d.$$

More precisely, there is an isomorphism of rings,

$$\mathbb{Z} \xrightarrow{\sim} \operatorname{End}(\mathbb{G}_m), \qquad d \longmapsto M_d.$$

The fact that the $M_d$ are endomorphisms of the multiplicative group $\mathbb{G}_m$ makes it quite easy to describe their preperiodic points.

**Proposition 6.1.** *Let $d \in \mathbb{Z}$ with $|d| \ge 2$ and let $M_d : \mathbb{P}^1 \to \mathbb{P}^1$ be the power map $M_d(z) = z^d$. Then*

$$\operatorname{PrePer}(M_d) = (\mathbb{G}_m)_{\text{tors}} = \{\zeta \in \mathbb{G}_m : \zeta^n = 1 \textit{ for some } n \ge 1\} = \bigcup_{n\ge1} \boldsymbol{\mu}_n,$$

*where recall that $\boldsymbol{\mu}_n$ denotes the group of $n^{th}$ roots of unity.*

*Proof.* We proved this long ago for any abelian group $G$ and homomorphism $z \mapsto z^d$ with $d \ge 2$; see Proposition 0.3. The proof for $d \le -2$ is similar and left to the reader. □

The iterates of $M_d$ are given by

$$M_d^n(z) = z^{d^n} = M_{d^n}(z),$$

so the periodic points of $M_d$ are also easy to characterize,

$$\operatorname{Per}_n(M_d) = \operatorname{Fix}(M_d^n) = \operatorname{Fix}(M_{d^n}) = \{\zeta \in \mathbb{G}_m : \zeta^{d^n} = \zeta\} = \boldsymbol{\mu}_{d^n-1} \cup \{0\}.$$

**Proposition 6.2.** *Let $|d| \ge 2$ and let $\zeta \in \operatorname{Per}_n^{**}(M_d)$ be a point of exact period $n \ge 2$. Then the multiplier of $M_d$ at $\zeta$ is given by*

$$\lambda_\zeta(M_d) = d^n.$$

*Proof.* Using $M_d^n(z) = z^{d^n}$, we can directly compute

$$\lambda_\zeta(M_d) = \frac{dM_d^n(z)}{dz}\bigg|_{z=\zeta} = \frac{dz^{d^n}}{dz}\bigg|_{z=\zeta} = d^n\zeta^{d^n-1} = d^n. \qquad \square$$

In particular, if we are working over $\mathbb{C}$, then every periodic point of $M_d$ in $\mathbb{C}^*$ is repelling. On the other hand, over a $p$-adic field with $p \nmid d$, the multiplier $d^n$ is a unit, so the periodic points are indifferent. And if $p \mid d$, then all of the periodic points are attracting. Of course, there are also the two superattracting fixed points $0$ and $\infty$.

It is not hard to find all rational maps that commute with the power maps. In particular, we can compute the automorphism group $\operatorname{Aut}(M_d)$.

**Proposition 6.3.** *Let $K$ be a field and let $M_d(z) = z^d$ be the $d^{th}$-power map for some $|d| \ge 2$. Further, if $K$ has finite characteristic $p$, assume that $p \nmid d$.*

(a) *The set of rational maps that commute with $M_d(z)$ is given by*

$$\{\phi(z) \in K(z) : \phi \circ M_d = M_d \circ \phi\} = \{cz^e : c \in \boldsymbol{\mu}_{d-1} \textit{ and } e \in \mathbb{Z}\}.$$

(b) *The automorphism group of* $M_d$ *is*

$$\operatorname{Aut}(M_d) = \{az : a \in \boldsymbol{\mu}_{d-1}\} \cup \{bz^{-1} : b \in \boldsymbol{\mu}_{d-1}\},$$

*where we recall that the automorphism group* $\operatorname{Aut}(\phi)$ *of any rational map* $\phi$ *is the set of* $f \in \operatorname{PGL}_2(\bar{K})$ *satisfying* $\phi^f = \phi$. *In particular,* $\operatorname{Aut}(M_d)$ *is a dihedral group of order* $2n$, *where* $n$ *is the number of* $(d-1)^{st}$ *roots of unity in* $\bar{K}^*$.

*Proof.* (a) It is clear by a direct computation that the indicated maps $cz^e$ commute with $M_d(z)$, so it suffices to prove that they are the only commuting maps. Suppose that $\phi(z) \in K(z)$ commutes with $M_d(z)$, so

$$\phi(z^d) = \phi(z)^d.$$

Let $\zeta \in \bar{K}$ be a primitive $d^{\text{th}}$ root of unity. (This is where we use the assumption that $p \nmid d$ if $K$ has positive characteristic $p$.) Then $\phi(\zeta z)^d = \phi(z)^d$, so

$$\phi(\zeta z) = \zeta^k \phi(z) \quad \text{for some } d^{\text{th}} \text{ root of unity } \zeta^k.$$

Consider the function $\psi(z) = z^{-k}\phi(z)$. It satisfies

$$\psi(\zeta z) = (\zeta z)^{-k}\phi(\zeta z) = (\zeta z)^{-k}\zeta^k\phi(z) = z^{-k}\phi(z) = \psi(z).$$

Hence $\psi$ is a function of $z^d$. In other words, there is a rational function $\psi_1(z) \in \bar{K}(z)$ such that $\psi(z) = \psi_1(z^d)$, and thus $\phi(z) = z^k\psi_1(z^d)$.

More generally, for any $n \geq 1$ we have $\phi \circ M_d^n = M_d^n \circ \phi$, so $\phi(z^{d^n}) = \phi(z)^{d^n}$. The above argument then yields an integer $0 \leq k_n < d^n$ and a rational function $\psi_n(z) \in \bar{K}(z)$ such that

$$\phi(z) = z^{k_n}\psi_n(z^{d^n}).$$

We write $\psi_n(z) = z^j\lambda_n(z)$ for some integer $j$ and some $\lambda_n(z) \in \bar{K}(z)$ satisfying $\lambda_n(0) \neq 0$ and $\lambda_n(0) \neq \infty$. Then

$$\begin{aligned}\deg(\phi(z)) &= \deg(z^{k_n}\psi_n(z^{d^n}))\\ &= \deg(z^{k_n+jd^n}\lambda_n(z^{d^n})) \geq \deg(\lambda_n(z^{d^n})) = d^n\deg(\lambda_n(z)).\end{aligned}$$

Hence $\deg(\lambda_n) = 0$ for sufficiently large $n$, which proves that $\phi(z)$ has the form $\phi(z) = cz^e$ for some $c \in \bar{K}^*$ and some $e \in \mathbb{Z}$. With this information in hand, it remains only to observe that $\phi(z^d) = \phi(z)^d$ if and only if $c = c^d$, i.e., if and only if either $c = 0$ or $c \in \boldsymbol{\mu}_{d-1}$.

(b) By definition, $\operatorname{Aut}(\phi)$ is the set of rational maps of degree 1 that commute with $\phi$. It follows from (a) that $\operatorname{Aut}(\phi)$ is the set of all $cz^{\pm 1}$ with $c \in \boldsymbol{\mu}_{d-1}$. In particular, if we let $f_a(z) = az$ and $g(z) = z^{-1}$, then

$$\operatorname{Aut}(M_d) = \{f_a : a \in \boldsymbol{\mu}_{d-1}\} \cup \{f_a \circ g : a \in \boldsymbol{\mu}_{d-1}\},$$

and the dihedral nature of the group law is evident from the identities

$$g^2 = f_1, \qquad f_a \circ f_b = f_{ab}, \qquad f_a \circ g = g \circ f_{a^{-1}}. \qquad \square$$

*Example* 6.4. We can use the map $M_d(z) = z^d$ to illustrate the construction of dynamical units in Section 3.11. First we use Theorem 3.66, which says that if $\alpha$ has exact order $n$ and $\gcd(i-j,n)=1$, then

$$\frac{\alpha^{d^i}-\alpha^{d^j}}{\alpha^d-\alpha} \text{ is a unit.}$$

Taking $j=0$ and $\gcd(i,n)=1$, this implies that

$$\frac{\alpha^{d^i-1}-1}{\alpha^{d-1}-1} \text{ is a unit for all primitive } (d^n-1)^{\text{st}} \text{ roots of unity } \alpha.$$

These are examples of classical cyclotomic units.

Similarly, let $m$ and $n$ be positive integers with $m \nmid n$ and $n \nmid m$. Then Theorem 3.68 says that if $\alpha$ is a primitive $(d^m-1)^{\text{st}}$ root of unity and if $\beta$ is a primitive $(d^n-1)^{\text{st}}$ root of unity, then $\alpha-\beta$ is a unit. These are again classical examples of cyclotomic units.

*Example* 6.5. Recall that a nontrivial twist of a rational map $\phi(z) \in K(z)$ is a rational map $\psi(z) \in K(z)$ such that $\psi(z)$ is $\mathrm{PGL}_2(\bar{K})$-conjugate to $\phi(z)$, but $\psi(z)$ is not $\mathrm{PGL}_2(K)$-conjugate to $\phi(z)$. See Section 4.9 for the general theory. Since the power maps $M_d(z) = z^d$ have a large automorphism group, they tend to have many twists. For example, the twists associated to the subgroup $\{cz : c \in \boldsymbol{\mu}_{d-1}\}$ of $\mathrm{Aut}(\phi)$ are given by (cf. Example 4.81)

$$K^*/(K^*)^{d-1} \longrightarrow \mathrm{Twist}(M_d), \qquad a \longmapsto az^d.$$

There are also some rather complicated-looking twists associated to the subgroup

$$\{z, z^{-1}\} \subset \mathrm{Aut}(M_d).$$

Each $b \in K^*/(K^*)^2$ leads to a twist

$$\sum_{0\le k\le d/2} \binom{d}{2k} b^k z^{2k} \Big/ \sum_{0\le k\le (d-1)/2} \binom{d}{2k+1} b^k z^{2k+1}. \tag{6.1}$$

See Example 4.82 for the derivation of this formula.

## 6.2 Chebyshev Polynomials

The multiplicative group $\mathbb{G}_m$ has a nontrivial automorphism given by inversion $z \mapsto z^{-1}$, and the quotient of $\mathbb{G}_m$ by this automorphism is isomorphic to the affine line $\mathbb{A}^1$ via the map

$$\mathbb{G}_m/\{z=z^{-1}\} \xrightarrow{\ \sim\ } \mathbb{A}^1, \qquad z \longmapsto z+z^{-1}.$$

The inversion automorphism commutes with the $d^{\text{th}}$-power map $M_d(z) = z^d$, so when we take the quotient of $\mathbb{G}_m$, we find that $M_d(z)$ descends to give a map on the quotient space $\mathbb{A}^1$. This leads to the following definition (cf. Section 1.6.2).

**Definition.** The $d^{\text{th}}$ *Chebyshev polynomial* is the polynomial $T_d(w) \in \mathbb{Z}[w]$ satisfying the identity

$$T_d(z+z^{-1}) = z^d + z^{-d} \qquad \text{in the field } \mathbb{Q}(z). \tag{6.2}$$

Of course, we need to show that that $T_d$ exists. In the next proposition we prove the existence of the Chebyshev polynomials and describe some of their algebraic properties.

**Proposition 6.6.** *For each integer $d \ge 0$ there exists a unique polynomial $T_d(w) \in \mathbb{Q}[w]$ satisfying*

$$T_d(z+z^{-1}) = z^d + z^{-d} \qquad \textit{in the field } \mathbb{Q}(z). \tag{6.3}$$

*We call $T_d$ the $d^{th}$* Chebyshev polynomial.

(a) *$T_d(w)$ is a monic polynomial of degree $d$ in $\mathbb{Z}[w]$.*
(b) *$T_d(T_e(w)) = T_{de}(w)$ for all $d, e \ge 0$.*
(c) *$T_d(-w) = (-1)^d T_d(w)$. Thus $T_d$ is an odd function if $d$ is odd and it is an even function if $d$ is even.*
(d) *The Chebyshev polynomials satisfy the recurrence relation*

$$T_{d+2}(w) = wT_{d+1}(w) - T_d(w) \quad \textit{for all } d \ge 0. \tag{6.4}$$

(e) *For all $d \ge 1$, the $d^{th}$ Chebyshev polynomial is given by the explicit formula*

$$T_d(w) = \sum_{0 \le k \le d/2} (-1)^k \frac{d}{d-k} \binom{d-k}{k} w^{d-2k}. \tag{6.5}$$

*Proof.* Suppose first that there do exist polynomials $T_d(w)$ satisfying (6.3). Then

$$\begin{aligned} T_0(z+z^{-1}) &= z^0 + z^{-0} = 2, \\ T_1(z+z^{-1}) &= z + z^{-1}, \\ T_2(z+z^{-1}) &= z^2 + z^{-2} = (z+z^{-1})^2 - 2, \end{aligned}$$

so we see that $T_0(w) = 2$, $T_1(w) = w$, and $T_2(w) = w^2 - 2$ are uniquely determined monic polynomials with integer coefficients. Still assuming that the set of Chebyshev polynomials exists, we next observe that they satisfy

$$\begin{aligned} (z+z^{-1})&T_{d+1}(z+z^{-1}) - T_d(z+z^{-1}) \\ &= (z+z^{-1})(z^{d+1} + z^{-d-1}) - (z^d + z^{-d}) \\ &= z^{d+1} + z^{-d-2} \\ &= T_{d+2}(z+z^{-1}). \end{aligned}$$

Putting $z + z^{-1} = w$, this means that $T_{d+2}(w) = wT_{d+1}(w) - T_d(w)$. Hence if the Chebyshev polynomials exist, they are unique, because they are completely determined by the recurrence

$$T_0(w) = 2, \quad T_1(w) = w, \quad \text{and} \quad T_{d+2}(w) = wT_{d+1}(w) - T_d(w) \quad \text{for } d \geq 0. \tag{6.6}$$

And from this recurrence we see immediately by induction that $T_d(w)$ is a monic polynomial of degree $d$ in $\mathbb{Z}[w]$.

We now turn this argument around and use the recurrence (6.6) to *define* a sequence of polynomials $T_d(w)$. We claim that $T_d(w)$ then satisfies (6.3). This is clear for $T_0$ and $T_1$, so we assume that (6.3) is true for $T_0, T_1, \ldots, T_{d+1}$ and use (6.6) with $w = z + z^{-1}$ to compute

$$\begin{aligned} T_{d+2}(z+z^{-1}) &= (z+z^{-1})T_{d+1}(z+z^{-1}) - T_d(z+z^{-1}) \\ &= (z+z^{-1})(z^{d+1}+z^{-d-1}) - (z^d+z^{-d}) = z^{d+2}+z^{-d-2}. \end{aligned}$$

Hence (6.3) is true for $T_{d+2}$, so by induction we have

$$T_d(z+z^{-1}) = z^d + z^{-d} \qquad \text{for all } d \geq 0.$$

This proves that the recurrence defines polynomials satisfying (6.3), so Chebyshev polynomials of every degree exist. We have now shown that Chebyshev polynomials exist, are unique, and have the properties in (a) and (d).

Next we make repeated use of (6.3) to compute

$$T_d\big(T_e(z+z^{-1})\big) = T_d(z^e+z^{-e}) = (z^e)^d + (z^e)^{-d} = z^{de}+z^{-de} = T_{de}(z+z^{-1}).$$

Hence $T_d\big(T_e(w)\big) = T_{de}(w)$, which proves (b).

To prove (c), we replace $z$ by $-z$ in (6.3) to obtain

$$\begin{aligned} T_d(-(z+z^{-1})) &= T_d(-z+(-z)^{-1}) \\ &= (-z)^d + (-z)^{-d} = (-1)^d(z^d+z^{-d}) = (-1)^d T_d(z+z^{-1}). \end{aligned}$$

Therefore $T_d(-w) = (-1)^d T_d(w)$, which is (c).

Next we prove the explicit summation formula (6.5) given in (e). Substituting $d = 1$ and $d = 2$ into the formula yields the correct values $T_1(w) = w$ and $T_2(w) = w^2 - 2$. We now assume that the formula is correct up to $T_{d+1}$ and use the recurrence (6.4) to check it for $d+2$. Thus

$$\begin{aligned} T_{d+2}(w) &= wT_{d+1}(w) - T_d(w) \\ &= w \sum_{0 \leq k \leq (d+1)/2} (-1)^k \frac{d+1}{d+1-k}\binom{d+1-k}{k} w^{d+1-2k} \\ &\qquad - \sum_{0 \leq k \leq d/2} (-1)^k \frac{d}{d-k}\binom{d-k}{k} w^{d-2k} \\ &= \sum_{0 \leq k \leq (d+1)/2} (-1)^k \frac{d+1}{d+1-k}\binom{d+1-k}{k} w^{d+2-2k} \\ &\qquad - \sum_{1 \leq k \leq d/2+1} (-1)^{k-1} \frac{d}{d-k+1}\binom{d-k+1}{k-1} w^{d-2k+2} \end{aligned}$$

$$
\begin{aligned}
T_2 &= w^2 - 2\\
T_3 &= w^3 - 3w\\
T_4 &= w^4 - 4w^2 + 2\\
T_5 &= w^5 - 5w^3 + 5w\\
T_6 &= w^6 - 6w^4 + 9w^2 - 2\\
T_7 &= w^7 - 7w^5 + 14w^3 - 7w\\
T_8 &= w^8 - 8w^6 + 20w^4 - 16w^2 + 2\\
T_9 &= w^9 - 9w^7 + 27w^5 - 30w^3 + 9w\\
T_{10} &= w^{10} - 10w^8 + 35w^6 - 50w^4 + 25w^2 - 2\\
T_{11} &= w^{11} - 11w^9 + 44w^7 - 77w^5 + 55w^3 - 11w\\
T_{12} &= w^{12} - 12w^{10} + 54w^8 - 112w^6 + 105w^4 - 36w^2 + 2
\end{aligned}
$$

Table 6.1: The first few Chebyshev polynomials.

$$= \sum_{0\le k\le d/2+1} (-1)^k \frac{(d+1)\binom{d+1-k}{k} + d\binom{d+1-k}{k-1}}{d+1-k} w^{d+2-2k},$$

where we use the standard convention that $\binom{n}{m} = 0$ if $n < m$ or if $m < 0$. A simple algebraic calculation that we leave for the reader (Exercise 6.4) shows that

$$\frac{(d+1)\binom{d+1-k}{k} + d\binom{d+1-k}{k-1}}{d+1-k} = \frac{d+2}{d+2-k}\binom{d+2-k}{k}.$$

Hence

$$T_{d+2}(w) = \sum_{0\le k\le (d+2)/2} (-1)^k \frac{d+2}{d+2-k}\binom{d+2-k}{k} w^{d+2-2k},$$

which completes the proof of (e). □

*Remark* 6.7. As mentioned in Section 1.6.2, the classical normalization for the Chebyshev polynomials is

$$\tilde{T}_d\left(\frac{z+z^{-1}}{2}\right) = \frac{z^d + z^{-d}}{2},$$

or equivalently,

$$\tilde{T}_d(\cos\theta) = \cos(d\theta) \qquad \text{for all } \theta \in \mathbb{R}.$$

The two normalizations are related by the simple formula $\tilde{T}_d(w) = \frac{1}{2}T_d(2w)$. We have chosen the alternative normalization because it has better arithmetic properties. In particular, the map $T_d : \mathbb{P}^1 \to \mathbb{P}^1$ has good reduction at all primes. The classical normalization $\tilde{T}_d$ has bad reduction at 2.

As with the power maps, it is not difficult to describe the periodic points of the Chebyshev polynomials and to compute their multipliers. We state the result and leave the computation as an exercise.

**Proposition 6.8.** *Let $T_d(w)$ be the $d^{th}$ Chebyshev polynomial for some $d \geq 2$.*
(a) *The fixed points of $T_d$ in $\mathbb{A}^1(\mathbb{C})$ are*

$$\left\{2\cos\left(\frac{2\pi j}{d+1}\right) : 0 \leq j \leq \frac{d+1}{2}\right\} \cup \left\{2\cos\left(\frac{2\pi j}{d-1}\right) : 0 < j < \frac{d-1}{2}\right\}.$$

(b) *The multipliers of $T_d$ at its fixed points are given by*

$$\lambda_{T_d}\left(2\cos\left(\frac{2\pi j}{d+1}\right)\right) = -d \qquad \text{for } 0 < j < \frac{d+1}{2},$$
$$\lambda_{T_d}\left(2\cos\left(\frac{2\pi j}{d-1}\right)\right) = d \qquad \text{for } 0 < j < \frac{d-1}{2},$$
$$\lambda_{T_d}(\pm 2) = d^2.$$

*(Note that $-2 \in \operatorname{Fix}(T_d)$ if and only if $d$ is odd.)*
*In general, the periodic points and multipliers of $T_d$ can be derived from the above formulas using $T_d^n = T_{d^n}$ and $\operatorname{Per}_n(T_d) = \operatorname{Fix}(T_d^n)$.*

*Proof.* See Exercise 6.5. □

We now prove an analogue of Proposition 6.3 for Chebyshev polynomials.

**Theorem 6.9.** *Let $K$ be a field and let $T_d(w)$ be the $d^{th}$ Chebyshev polynomial for some $d \geq 2$. Further, if $K$ has finite characteristic $p$, assume that $p \nmid d$.*
(a) *The automorphism group of $T_d$ is given by*

$$\operatorname{Aut}(T_d) = \begin{cases} 1 & \text{if } d \text{ is even}, \\ \mu_2 & \text{if } d \text{ is odd}. \end{cases}$$

(b) *Assume that $K$ does not have characteristic 2. Let $\phi(w) \in K(w)$ be a rational map that commutes with $T_d(w)$, i.e., $\phi(T_d(w)) = T_d(\phi(w))$. Then*

$$\phi(w) = \pm T_e(w) \qquad \text{for some } e \geq 1.$$

*The minus sign is allowed if and only if $d$ is odd. (See Theorem 6.79 for a stronger result.)*

*Proof.* (a) The assertion that $\mathrm{Aut}(T_d) \subset \mu_2$ is an immediate consequence of (b), since (b) implies that any $f \in \mathrm{Aut}(T_d)$ satisfies $f(w) = \pm T_1(w) = \pm w$. However, since the proof of (b) is somewhat intricate, we give a direct and elementary proof of (a).

Suppose that $f \in \mathrm{PGL}_2(\bar{K})$ satisfies $T_d^f = T_d$. The polynomial $T_d$ has a unique totally ramified fixed point at $\infty$ (cf. Exercise 6.8), and $T_d^f$ similarly has a unique totally ramified fixed point at $f^{-1}(\infty)$, so the equality $T_d^f = T_d$ tells us that $f^{-1}(\infty) = \infty$. Hence $f(w) = aw + b$ is an affine transformation. (The same argument applies to any polynomial not of the form $aw^d$.)

Proposition 6.6(c) says that $T_d(w)$ satisfies $T_d(-w) = (-1)^d T_d(w)$, so in particular,

$$T_d(w) = w^d + (\text{terms of degree at most } d-2). \tag{6.7}$$

The identity $T_d^f(w) = T_d(w)$ with $f(w) = aw + b$ can be written as

$$T_d(aw + b) = aT_d(w) + b.$$

We evaluate both sides using (6.7) and look at the top degree terms. This gives

$$\begin{aligned} a^d w^d + da^{d-1} b w^{d-1} &+ (\text{terms of degree at most } d-2) \\ &= aw^d + (\text{terms of degree at most } d-2). \end{aligned}$$

Hence

$$a^d = a \qquad \text{and} \qquad da^{d-1} b = 0.$$

By assumption, $d \neq 0$ in the field $K$, so we conclude that $a^{d-1} = 1$ and $b = 0$.

In order to pin down the value of $a$, we use the explicit formula for $T_d(w)$ given in Proposition 6.6(e). In fact, we need only the top two terms,

$$T_d(w) = w^d - dw^{d-2} + (\text{terms of degree at most } d-4).$$

By assumption we have $T_d(aw) = aT_d(w)$, so

$$a^d w^d + da^{d-2} w^{d-2} + \cdots = aw^d - adw^{d-2} + \cdots.$$

Hence $a^d = a$ and $a^{d-2} = a$, where we again use the assumption that $d \neq 0$ in the field $K$. It follows that $a^2 = 1$. Further, $a = -1$ is possible only if $(-1)^d = -1$, so when $d$ is odd. This completes the proof that $\mathrm{Aut}(T_d)$ is trivial if $d$ is even and is equal to $\mu_2$ if $d$ is odd.

(b) It is easy to verify that the Chebyshev polynomial $T_d(w)$ cannot be conjugated to a polynomial of the form $cw^d$ (Exercise 6.8). It follows that any rational map commuting with $T_d(w)$ is necessarily a polynomial, a fact whose proof we defer until later in this chapter; see Theorem 6.80. We now describe a proof due to Bertram [69] that the only polynomials commuting with $T_d$ are $\pm T_e$. We begin with two lemmas. The first characterizes the Chebyshev polynomials as the solutions of a nonlinear differential equation, and the second explains how to exploit such equations.

**Lemma 6.10.** *Assume that $K$ does not have characteristic 2. Let $d \ge 1$ and let $F(w)$ be a polynomial solution to the differential equation*

$$(4-w^2)F'(w)^2 = d^2\bigl(4-F(w)^2\bigr). \tag{6.8}$$

*Then $F(w) = \pm T_d(w)$.*

*Proof.* We first check that $\pm T_d(w)$ are solutions. We differentiate the functional equation (6.2) defining the Chebyshev polynomials to obtain the identity

$$T_d'(z+z^{-1})(1-z^{-2}) = dz^{d-1} - dz^{-d-1},$$

and then solve for $T_d'$,

$$T_d'(z+z^{-1}) = d\frac{z^d - z^{-d}}{z-z^{-1}}.$$

Putting $w = z + z^{-1}$ as usual and noting that $w^2 - 4 = (z-z^{-1})^2$, we compute

$$\begin{aligned}(4-w^2)T_d'(w)^2 &= \bigl(4-(z+z^{-1})^2\bigr)d^2\left(\frac{z^d-z^{-d}}{z-z^{-1}}\right)^2 \\ &= -d^2(z^d - z^{-d})^2 \\ &= d^2\bigl(4-(z^d+z^{-d})^2\bigr) \\ &= d^2\bigl(4-T_d(w)^2\bigr).\end{aligned}$$

This proves that $\pm T_d(w)$ are solutions to (6.8).

Next suppose that $F(w)$ is any polynomial solution to (6.8). If $F'(w)$ is identically 0, then (6.8) implies that $F(w) = \pm 2 = \pm T_0(w)$, so we are done. We may thus assume that $F'(w) \ne 0$. We differentiate both sides of (6.8) and divide by $2F'(w)$ to obtain

$$(4-w^2)F''(w) - wF'(w) + d^2F(w) = 0. \tag{6.9}$$

In particular, $T_d(w)$ is a solution to (6.9). Suppose now that $F$ is any polynomial of degree $k$ that is a solution to (6.9). We write $F(w) = aw^k + bw^{k-1} + \cdots$ with $a \ne 0$ and substitute into (6.9). The leading term is

$$a(-k(k-1) - k + d^2)w^k = a(d^2 - k^2)w^k,$$

so we must have $k = d$. In other words, we have shown that every nonzero polynomial solution of (6.9) has degree $d$. But $F(w) - aT_d(w)$ is a polynomial of degree strictly less than $d$ that satisfies (6.9); hence $F(w) = aT_d(w)$. Finally, substituting $w = 2$ into (6.8) yields $F(2) = \pm 2$, while we know that $T_d(2) = T_d(1+1^{-1}) = 1^d + 1^d = 2$. Hence $F(w) = \pm T_d(w)$, which completes the proof of Lemma 6.10. □

**Lemma 6.11.** *Let $A(w)$ be a polynomial of degree $r \ge 1$ and suppose that $F(w)$ is a polynomial of degree $d \ge 2$ satisfying*

$$A(w)F'(w)^r = d^r A\bigl(F(w)\bigr). \tag{6.10}$$

*Suppose further that $G(w)$ is a polynomial of degree $e \ge 0$ that commutes with $F$, i.e., $F\bigl(G(w)\bigr) = G\bigl(F(w)\bigr)$. Then*

$$A(w)G'(w)^r = e^r A\bigl(G(w)\bigr).$$

*Proof.* Consider the polynomial

$$B(w) = A(w)G'(w)^r - e^r A\bigl(G(w)\bigr).$$

We assume that $B(w) \ne 0$ and derive a contradiction, which will prove the desired result. First we observe that the leading coefficients of $A(w)G'(w)^r$ and $e^r A\bigl(G(w)\bigr)$ cancel, so

$$\deg B < re \qquad \text{(strict inequality).}$$

Next we use the various definitions and given relations to compute

$$\begin{aligned}
&d^r B\bigl(F(w)\bigr)\\
&\quad = d^r A\bigl(F(w)\bigr)G'\bigl(F(w)\bigr)^r - d^r e^r A\bigl(G(F(w))\bigr) && \text{definition of } B,\\
&\quad = d^r A\bigl(F(w)\bigr)G'\bigl(F(w)\bigr)^r - d^r e^r A\bigl(F(G(w))\bigr) && \text{using } F\circ G = G\circ F,\\
&\quad = A(w)F'(w)^r G'\bigl(F(w)\bigr)^r - e^r A\bigl(G(w)\bigr)F'\bigl(G(w)\bigr)^r && \text{using (6.10) twice,}\\
&\quad = A(w)(G\circ F)'(w)^r - e^r A\bigl(G(w)\bigr)F'\bigl(G(w)\bigr)^r && \text{chain rule,}\\
&\quad = A(w)(F\circ G)'(w)^r - e^r A\bigl(G(w)\bigr)F'\bigl(G(w)\bigr)^r && \text{using } F\circ G = G\circ F,\\
&\quad = A(w)F'\bigl(G(w)\bigr)^r G'(w)^r - e^r A\bigl(G(w)\bigr)F'\bigl(G(w)\bigr)^r && \text{chain rule,}\\
&\quad = F'\bigl(G(w)\bigr)^r \bigl[A(w)G'(w)^r - e^r A\bigl(G(w)\bigr)\bigr]\\
&\quad = F'\bigl(G(w)\bigr)^r B(w) && \text{definition of } B.
\end{aligned}$$

Taking degrees of both sides gives

$$(\deg B)(\deg F) = r(\deg F - 1)(\deg G) + (\deg B),$$

and the assumption that $\deg F \ge 2$ means that we can solve for

$$\deg B = r(\deg G) = re.$$

This contradicts the earlier strict inequality $\deg B < re$. Hence $B$ must be the zero polynomial, which completes the proof of Lemma 6.11. □

We now resume the proof of Theorem 6.9(b). Let $\phi(w)$ be as in the statement (c) with $\phi(w)$ a polynomial and let $e = \deg(\phi)$. Lemma 6.10 tells us that

$$(4 - w^2)T_d'(w)^2 = d^2\bigl(4 - T_d(w)^2\bigr).$$

Hence we can apply Lemma 6.11 with $A(w) = 4 - w^2$ and the commuting polynomials $\phi$ and $T_d$ to deduce that

$$(4 - w^2)\phi(w)^2 = e^2\bigl(4 - \phi(w)^2\bigr).$$

Then another application of Lemma 6.10 implies that $\phi(w) = \pm T_e(w)$. □

Using Theorem 6.9, it is easy to describe all of the twists of the Chebyshev polynomials.

**Corollary 6.12.** *Continuing with the notation and assumptions from Theorem* 6.9, *if $d$ is even, then $T_d$ has no nontrivial $\bar{K}/K$-twists, and if $d$ is odd, then each $a \in K^*$ yields a twist*

$$T_{d,a}(w) = \frac{1}{\sqrt{a}} T_d(\sqrt{a}\, w).$$

*Two such twists $T_{d,a}$ and $T_{d,b}$ are $K$-conjugate if and only if $a/b$ is a square in $K^*$.*

*Proof.* We use the description of $\operatorname{Aut}(\phi)$ from (a). If $d$ is even, then the automorphism group $\operatorname{Aut}(\phi)$ is trivial, so Proposition 4.73 says that $\phi$ has no nontrivial twists. For odd $d$ we have $\operatorname{Aut}(\phi) = \{\pm z\}$, so the desired result follows from Example 4.81 (see also Example 4.75). □

*Remark* 6.13. Over $\mathbb{C}$, there is a short proof that $\operatorname{Aut}(T_d) \subset \boldsymbol{\mu}_2$ using the fact (Exercise 1.31) that the Julia set of $T_d$ is $\mathcal{J}(T_d) = [-2, 2]$. Then the assumption that $T_d^f = T_d$ implies that $f$ maps the interval $[-2, 2]$ to itself. Since $f$ is bijective on $\mathbb{P}^1(\mathbb{C})$, it follows in particular that $f$ permutes the endpoints of the interval $[-2, 2]$. Hence $f(2) = \pm 2$ and $f(-2) = \mp 2$. Writing $f$ as $f(w) = aw + b$, this gives two equations to solve for $a$ and $b$, yielding $b = 0$ and $a = \pm 1$. Note that this proof does not carry over to characteristic $p$, since, for example, working over $\mathbb{F}_p$ we have $\operatorname{Aut}(T_p) = \operatorname{PGL}_2(\mathbb{F}_p)$; see Exercise 6.10.

## 6.3 A Primer on Elliptic Curves

The remainder of this chapter is devoted to rational maps associated to elliptic curves. In this section we give some basic definitions and review, without proof, some of the properties of elliptic curves that will be needed later. The reader should also review the summary of elliptic curves over $\mathbb{C}$ given in Section 1.6.3. For further reading on elliptic curves and for the proofs omitted in this section, see for example [96, 198, 248, 250, 254, 257, 410, 412, 420].

### 6.3.1 Elliptic Curves and Weierstrass Equations

**Definition.** An elliptic curve $E$ over a field $K$ (of characteristic different from 2 and 3) is described by a *Weierstrass equation*, which is an equation of the form

$$E : y^2 = x^3 + ax + b \tag{6.11}$$

with $a, b \in K$ and $4a^3 + 27b^2 \neq 0$. Of course, we really mean that $E$ is the projective curve obtained by homogenizing equation (6.11), so $E$ has one extra point "at infinity," which we denote by $\mathcal{O}$. If $K$ has characteristic 2 or 3, then equations of the form (6.11) are insufficient, and indeed they are always singular in characteristic 2, so one uses the generalized Weierstrass equation

$$E : y^2 + a_1xy + a_3y = x^3 + a_2x^2 + a_4x + a_6. \tag{6.12}$$

*Remark* 6.14. Let $E/K$ be an elliptic curve defined over a field $K$. When we write $E$, we mean the geometric points of $E$, i.e., the points in $E(\bar{K})$ for some chosen algebraic closure of $E$. If we want to refer to points defined over $K$, we always explicitly write $E(K)$, and similarly we write $E(L)$ for the points defined over some extension field $L$ of $K$.

More intrinsically, an *elliptic curve* is a pair $(E, \mathcal{O})$ consisting of a smooth algebraic curve $E$ of genus 1 and a point $\mathcal{O} \in E$. For convenience we often call $E$ an elliptic curve, with the understanding that there is a specified point $\mathcal{O}$. We say that $E$ is *defined over a field* $K$ if the curve $E$ is given by equations with $K$-coefficients and the point $\mathcal{O}$ is in $E(K)$.

Using the Riemann–Roch theorem, one can prove that every elliptic curve $E/K$ can be embedded in $\mathbb{P}^2$ by a cubic equation of the form (6.12) with $\mathcal{O}$ mapping to the point at infinity. (See [410, III §3].) Then, if the characteristic of $K$ is neither 2 nor 3, we can complete the square on the left and the cube on the right to obtain the simpler Weierstrass equation (6.11). In order to simplify our discussion, we will generally make this assumption.

The *discriminant* $\Delta(E)$ and *j-invariant* $j(E)$ of the elliptic curve $E$ given by (6.11) are defined by the formulas

$$\Delta(E) = -16(4a^3 + 27b^2), \qquad j(E) = 1728\frac{4a^3}{4a^3 + 27b^2}.$$

**Proposition 6.15.** (a) *Let $a, b \in K$ and let $E$ be the curve given by the Weierstrass equation* (6.11). *Then $E$ is nonsingular, and thus is an elliptic curve, if and only if $\Delta(E) \neq 0$.*

(b) *Two elliptic curves $E$ and $E'$ are isomorphic over $\bar{K}$ if and only if $j(E) = j(E')$. More precisely, $E$ and $E'$ are isomorphic if and only if there is a $u \in \bar{K}^*$ such that $a' = u^4 a$ and $b' = u^6 b$*

*Proof.* See [410, III §1]. □

### 6.3.2 Geometry and the Group Law

There is a natural group structure on the points of $E$ that may be described as follows. Let $L$ be any line in $\mathbb{P}^2$. Then counted with appropriate multiplicities, the cubic curve $E$ and the line $L$ intersect at three points, say

$$E \cap L = \{P, Q, R\},$$

where $P$, $Q$, and $R$ need not be distinct. The group law on $E$ is determined by the requirement that the sum of the points $P, Q, R$ be equal to $\mathcal{O}$,

$$P + Q + R = \mathcal{O}.$$

The point $\mathcal{O}$ serves as the identity element of the group. The inverse of a point $P$, which we denote by $-P$, is the third point on the intersection of $E$ with the line through $P$ and $\mathcal{O}$.

**Theorem 6.16.** *Let $E/K$ be an elliptic curve defined over a field $K$.*

(a) *The addition law described above gives $E = E(\bar{K})$ the structure of an abelian group.*

(b) *The group law is algebraic, in the sense that the addition and inversion maps,*

$$E \times E \xrightarrow{(P,Q)\mapsto P+Q} E, \qquad E \xrightarrow{P\mapsto -P} E,$$

*are morphisms, i.e., are given by everywhere defined rational functions.*

(c) *The subset $E(K)$ consisting of points of $E$ that are defined over $K$ is a subgroup of $E(\bar{K})$.*

*Proof.* See [410, III §§2,3]. □

It is not hard to give explicit formulas for the group law on an elliptic curve, as in the following algorithm.

**Proposition 6.17.** (Elliptic Curve Group Law Algorithm) *Let $E$ be an elliptic curve given by a Weierstrass equation*

$$E : y^2 = x^3 + ax + b,$$

*and let $P_1 = (x_1, y_1)$ and $P_2 = (x_2, y_2)$ be points on $E$.*

*If $x_1 = x_2$ and $y_1 = -y_2$, then $P_1 + P_2 = \mathcal{O}$. Otherwise, define quantities*

$$\lambda = \frac{y_2 - y_1}{x_2 - x_1}, \qquad \nu = \frac{y_1x_2 - y_2x_1}{x_2 - x_1}, \qquad \text{if } x_1 \neq x_2,$$

$$\lambda = \frac{3x_1^2 + a}{2y_1}, \qquad \nu = \frac{-x_1^3 + ax_1 + 2b}{2y_1}, \qquad \text{if } x_1 = x_2.$$

*Then $y = \lambda x + \nu$ is the line through $P_1$ and $P_2$, or tangent to $E$ if $P_1 = P_2$, and the sum of $P_1$ and $P_2$ is given by*

$$P_1 + P_2 = (\lambda^2 - x_1 - x_2, -\lambda^3 + \lambda x_1 + \lambda x_2 - \nu).$$

*As a special case, the* duplication formula *for $P = (x, y)$ is*

$$x([2]P) = \frac{x^4 - 2ax^2 - 8bx + a^2}{4x^3 + 4ax + 4b}.$$

*Proof.* See [410, III.2.3] □

### 6.3.3 Divisors and Divisor Classes

**Definition.** A *divisor* on $E$ is a formal sum of points

$$D = \sum_{P\in E} n_P(P),$$

with $n_P \in \mathbb{Z}$ and all but finitely many $n_P = 0$. The set of divisors under addition forms the *divisor group* $\mathrm{Div}(E)$. The *degree* of a divisor $d$ is

$$\deg(D) = \sum_{P \in E} n_P.$$

The degree map $\deg : \mathrm{Div}(E) \to \mathbb{Z}$ is a group homomorphism.

There is a natural summation map from $\mathrm{Div}(E)$ to $E$ defined by

$$\mathrm{sum} : \mathrm{Div}(E) \longrightarrow E, \qquad \sum_{P \in E} n_P(P) \longmapsto \sum_{P \in E} [n_P](P).$$

(N.B. The two summation signs mean very different things. The first is a formal sum of points in $\mathrm{Div}(E)$. The second is a sum using the complicated addition law on $E$.)

The zeros and poles of a rational function $f$ on $E$ define a divisor

$$\mathrm{div}(f) = \sum_{P \in E} \mathrm{ord}_P(f)(P),$$

where $\mathrm{ord}_P(f)$ is the order of zero of $f$ at $P$ if $f(P) = 0$, and $\mathrm{ord}_P(f)$ is negative the order of the pole of $f$ at $P$ if $f(P) = \infty$. A divisor of the form $\mathrm{div}(f)$ is called a *principal divisor*. The principal divisors form a subgroup of $\mathrm{Div}(E)$, and the quotient group is the *Picard group* $\mathrm{Pic}(E)$. Within $\mathrm{Pic}(E)$ is the important subgroup $\mathrm{Pic}^0(E)$ generated by divisors of degree 0.

The next proposition describes the basic properties of divisors on $E$.

**Proposition 6.18.** *Let E be an elliptic curve.*

(a) *Every principal divisor on E has degree* 0.

(b) *A divisor $D \in \mathrm{Div}(E)$ is principal if and only if both* $\deg(D) = 0$ *and* $\mathrm{sum}(D) = \mathcal{O}$.

(c) *The summation map induces a group isomorphism*

$$\mathrm{sum} : \mathrm{Pic}^0(E) \longrightarrow E.$$

*Proof.* See [410, III.3.4 and III.3.5]. □

### 6.3.4 Isogenies, Endomorphisms, and Automorphisms

**Definition.** An *isogeny* between two elliptic curves $E_1$ and $E_2$ is a surjective morphism $\psi : E_1 \to E_2$ satisfying $\psi(\mathcal{O}) = \mathcal{O}$. (Note that any nonconstant morphism $E_1 \to E_2$ is automatically finite and surjective.) The curves $E_1$ and $E_2$ are said to be *isogenous* if there is an isogeny between them.

*Remark* 6.19. Every nonconstant morphism $\psi : E_1 \to E_2$ is the composition of an isogeny and a translation (cf. [410, III.4.7]). To see this, let $\phi(P) = \psi(P) - \psi(\mathcal{O})$. Then the map $\phi : E \to E$ is a morphism, and $\phi(\mathcal{O}) = \mathcal{O}$, so $\phi$ is an isogeny. Hence

$$\psi(P) = \phi(P) + \psi(\mathcal{O})$$

is the composition of an isogeny and a translation.

*Remark* 6.20. We observe that an isogeny is unramified at all points. This follows from the general Riemann–Hurwitz formula (Theorem 1.5) applied to the map $\psi : E_1 \to E_2$,

$$2g(E_1) - 2 = (\deg \psi)(2g(E_2) - 2) + \sum_{P \in E_1} (e_P(\psi) - 1).$$

The elliptic curves $E_1$ and $E_2$ both have genus 1, and the ramification indices satisfy $e_P(\psi) \geq 1$, so it follows that every $e_P(\psi)$ is equal to 1, so $\psi$ is unramified.

**Theorem 6.21.** *An isogeny $\psi : E_1 \to E_2$ is a homomorphism of groups, i.e.,*

$$\psi(P + Q) = \psi(P) + \psi(Q) \quad \text{for all } P, Q \in E_1(\bar{K}).$$

*Proof.* See [410, III.4.8] □

The *degree* of an isogeny $\psi : E_1 \to E_2$ is the number of points in the inverse image $\psi^{-1}(Q)$ for any point $Q \in E_2$. This number is independent of the point $Q$, since, as noted earlier, $\psi$ is an unramified map. It is clear that if $\deg(\psi) > 1$, then $\psi$ is not invertible, since it is not one-to-one. However, there does exist a dual isogeny that provides a kind of "inverse" for $\psi$.

**Theorem 6.22.** *Let $\psi : E_1 \to E_2$ be an isogeny of degree d. Then there is a unique isogeny $\hat{\psi} : E_2 \to E_1$, called the* dual isogeny of $\psi$, *with the property that*

$$\hat{\psi}(\psi(P)) = [d]P \quad \text{and} \quad \psi(\hat{\psi}(Q)) = [d]Q \quad \text{for all } P \in E_1 \text{ and } Q \in E_2.$$

*Proof.* See [410, III §6]. □

**Definition.** Let $E$ be an elliptic curve. The *endomorphism ring of* $E$, which is denoted by $\operatorname{End}(E)$, is the set of isogenies from $E$ to itself with addition and multiplication given by the rules

$$(\psi_1 + \psi_2)(P) = \psi_1(P) + \psi_2(P), \qquad (\psi_1\psi_2)(P) = \psi_1(\psi_2(P)).$$

(In order to make $\operatorname{End}(E)$ into a ring, we also include the constant map that sends every point to $\mathcal{O}$.) The *automorphism group of* $E$, denoted by $\operatorname{Aut}(E)$, is the set of endomorphisms that have inverses. Equivalently, $\operatorname{Aut}(E) = \operatorname{End}(E)^*$ is the group of units in the ring $\operatorname{End}(E)$.

Every integer $m$ gives a *multiplication-by-$m$* morphism in $\operatorname{End}(E)$. For $m > 0$ this is defined in the natural way as

$$[m] : E \longrightarrow E, \qquad [m](P) = \overbrace{P + P + \cdots + P}^{m \text{ terms}}.$$

For $m < 0$ we set $[m](P) = -[-m](P)$, and of course $[0](P) = \mathcal{O}$. This gives an embedding of $\mathbb{Z}$ into $\operatorname{End}(E)$, and for most elliptic curves (in characteristic 0), there are no other endomorphisms.

**Definition.** An elliptic curve $E$ is said to have *complex multiplication* if $\mathrm{End}(E)$ is strictly larger than $\mathbb{Z}$. The phrase "complex multiplication" is often abbreviated by CM.

*Example* 6.23. The elliptic curve $E : y^2 = x^3 + x$ has CM, since the endomorphism

$$\psi : E \longrightarrow E, \qquad \psi(x, y) = (-x, iy),$$

is not in $\mathbb{Z}$. An easy way to verify this assertion is to note that

$$\psi^2(x, y) = (x, -y) = -(x, y),$$

so $\psi^2 = [-1]$. This gives an embedding of the Gaussian integers $\mathbb{Z}[i]$ into $\mathrm{End}(E)$ via the association $m + ni \mapsto [m] + [n] \circ \psi$, and in fact it is not hard to show that $\mathrm{End}(E)$ is isomorphic to $\mathbb{Z}[i]$.

*Example* 6.24. More generally, there are two special families of elliptic curves that have CM, namely those with $a = 0$ and those with $b = 0$. These are the curves

$$\begin{array}{llll} E'_a : y^2 = x^3 + ax, & j(E'_a) = 1728, & \mathrm{End}(E'_a) = \mathbb{Z}[i], & \mathrm{Aut}(E'_a) = \boldsymbol{\mu}_4, \\ E''_b : y^2 = x^3 + b, & j(E''_b) = 0, & \mathrm{End}(E''_b) = \mathbb{Z}[\rho], & \mathrm{Aut}(E''_b) = \boldsymbol{\mu}_6. \end{array}$$

Here $\rho = (-1+\sqrt{-3})/2$ denotes a cube root of unity and $\boldsymbol{\mu}_n$ is the group of $n^{\text{th}}$ roots of unity.

Of course, all of the $E'_a$ are isomorphic over an algebraically closed field, since they have the same $j$-invariant, and similarly for all of the $E''_b$. However, the curves in each family may not be isomorphic over a field $K$ that is not algebraically closed. This is an example of the phenomenon of twisting as described in Section 4.8 (see also [410, X §5]).

**Proposition 6.25.** *Let $E/K$ be an elliptic curve. Then the endomorphism ring of $E$ is one of the following three kinds of rings:*

(a) $\mathrm{End}(E) = \mathbb{Z}$.

(b) $\mathrm{End}(E)$ *is an order in a quadratic imaginary field $F$. This means that* $\mathrm{End}(E)$ *is a subring of $F$ and satisfies* $\mathrm{End}(E) \otimes \mathbb{Q} = F$. *In particular,* $\mathrm{End}(E)$ *is a subring of finite index in the ring of integers of $F$.*

(c) $\mathrm{End}(E)$ *is a maximal order in a quaternion algebra. (This case can occur only if $E$ is defined over a finite field.)*

*Proof.* See [410, III §9]. □

The automorphisms of an elliptic curve are very easy to describe.

**Proposition 6.26.** *Let $K$ be a field whose characteristic is not equal to 2 or 3 and let $E/K$ be an elliptic curve. Then*

$$\mathrm{Aut}(E) = \begin{cases} \boldsymbol{\mu}_2 & \text{if } j(E) \neq 0 \text{ and } j(E) \neq 1728, \\ \boldsymbol{\mu}_4 & \text{if } j(E) = 1728, \\ \boldsymbol{\mu}_6 & \text{if } j(E) = 0. \end{cases}$$

*Proof.* See [410, III §10]. □

*Remark* 6.27. It is easy to make the description of $\mathrm{Aut}(E)$ in Proposition 6.26 completely explicit. Assuming that $E$ is given by a Weierstrass equation (6.11) as usual, for an appropriate choice of $n$ there is an isomorphism

$$[\cdot] : \boldsymbol{\mu}_n \longrightarrow \mathrm{Aut}(E), \qquad [\xi](x,y) = (\xi^4 x, \xi^3 y). \tag{6.13}$$

Here we take $n = 4$ if $j(E) = 1728$, we take $n = 6$ if $j(E) = 0$, and we take $n = 2$ otherwise. Of course, for $n = 2$ and $n = 4$ the formula simplifies somewhat to $(x, \xi y)$ and $(x, \xi^{-1} y)$, respectively.

### 6.3.5 Minimal Equations and Reduction Modulo $p$

Let $K$ be a local field with ring of integers $R$, maximal ideal $\mathfrak{p}$, and residue field $k = R/\mathfrak{p}$. As in Section 2.3, we write $\tilde{x}$ for the reduction of $x$ modulo $\mathfrak{p}$.

**Definition.** Let $E/K$ be an elliptic curve defined over a local field $K$. A *minimal Weierstrass equation for* $E$ is a Weierstrass equation whose discriminant $\Delta(E)$ has minimal valuation subject to the condition that the coefficients of the Weierstrass equation are all in $R$.

*Example* 6.28. If $k$ does not have characteristic 2 or 3, then a Weierstrass equation

$$E : y^2 = x^3 + ax + b \tag{6.14}$$

for $E$ is minimal if and only if

$$a, b \in R \qquad \text{and} \qquad \min\{3\,\mathrm{ord}_{\mathfrak{p}}(a), 2\,\mathrm{ord}_{\mathfrak{p}}(b)\} < 12.$$

In general, if the residue field $k$ does not have characteristic 2 or 3, then any Weierstrass equation (6.14) can be transformed into a minimal equation by a substitution of the form $(x,y) \mapsto (u^2 x, u^3 y)$ for an appropriate $u \in K^*$.

If $k$ has characteristic 2 or 3, then a minimal Weierstrass equation may require the general form (6.12). There is an algorithm of Tate [412, IV §9] that transforms a given Weierstrass equation into a minimal one.

**Definition.** Fix a minimal Weierstrass equation for $E/K$. Then we can reduce the coefficients of $E$ to obtain a (possibly singular) curve $\tilde{E}/k$. We say that $E$ has *good reduction* if $\tilde{E}$ is nonsingular, which is equivalent to the condition that $\Delta(E) \in R^*$. In any case, we obtain a *reduction modulo* $\mathfrak{p}$ *map* on points,

$$E(K) \longrightarrow \tilde{E}(k), \qquad P \longmapsto \tilde{P}.$$

**Proposition 6.29.** *If $E$ has good reduction, then the reduction modulo $\mathfrak{p}$ map $E(K) \to \tilde{E}(k)$ is a homomorphism.*

*Proof.* See [410, VII.2.1]. □

*Remark* 6.30. For elliptic curves defined over a number field $K$, we say that $E$ has good reduction at a prime $\mathfrak{p}$ of $K$ if it has a Weierstrass equation whose coefficients are $\mathfrak{p}$-adic integers and whose discriminant is a $\mathfrak{p}$-adic unit. Note that one is allowed to use different Weierstrass equations for different primes. If there is a single Weierstrass equation that is simultaneously minimal for all primes, then we say that $E/K$ has a *global minimal Weierstrass equation.* Global minimal equations exist for elliptic curves over $\mathbb{Q}$, and more generally for elliptic curves over any number field of class number 1, but in general the existence of global minimal equations is somewhat subtle; see [410, VIII §8] and [48]. We discussed a related notion of global minimal models of rational maps in Section 4.11.

### 6.3.6 Torsion Points and Reduction Modulo $p$

The kernels of endomorphisms help to determine the arithmetic properties of elliptic curves.

**Definition.** Let $E$ be an elliptic curve. For any endomorphism $\psi \in \operatorname{End}(E)$ we write

$$E[\psi] = \operatorname{Ker}(\psi) = \{P \in E : \psi(P) = \mathcal{O}\}.$$

Of particular importance is the kernel of the multiplication-by-$m$ map,

$$E[m] = \{P \in E : [m]P = \mathcal{O}\}.$$

The group $E[m]$ is called the *$m$-torsion subgroup of $E$*. The union of all $E[m]$ is the *torsion subgroup of $E$*,

$$E_{\text{tors}} = \bigcup_{m \ge 1} E[m].$$

**Theorem 6.31.** *Let $E/K$ be an elliptic curve and assume that either $K$ has characteristic $0$ or else that $K$ has characteristic $p > 0$ and $p \nmid m$. Then as an abstract group,*

$$E[m] = \mathbb{Z}/m\mathbb{Z} \times \mathbb{Z}/m\mathbb{Z}.$$

*In other words, $E[m]$ is the product of two cyclic groups of order $m$.*

*Proof.* See [410, III.6.4]. □

The next result gives conditions that ensure that the reduction modulo $\mathfrak{p}$ map respects the $m$-torsion points. It may be compared with Theorem 2.21, which tells us what reduction modulo $\mathfrak{p}$ does to periodic points of a good-reduction rational map.

**Theorem 6.32.** *Let $K$ be a local field whose residue field has characteristic $p$, let $E/K$ be an elliptic curve with good reduction, and let $m \ge 1$ be an integer with $p \nmid m$. Let $E(K)[m]$ denote the subgroup of $E[m]$ consisting of points defined over $K$, i.e., $E(K)[m] = E[m] \cap E(K)$. Then the reduction map*

$$E(K)[m] \longrightarrow \tilde{E}(k)$$

*is injective. In other words, distinct $m$-torsion points have distinct reductions modulo $\mathfrak{p}$.*

*Proof.* See [410, VII.3.1]. □

Let $E/K$ be an elliptic curve defined over the field $K$. Then the points in $E[m]$ are algebraic over $K$, so their coordinates generate algebraic extensions of $K$. An immediate corollary of the preceding theorem limits the possible ramification of these extensions.

**Corollary 6.33.** *Let $K$ be a local field whose residue field has characteristic $p$, let $E/K$ be an elliptic curve with good reduction, and let $m \ge 1$ be an integer with $p \nmid m$. Then the field $K(E[m])$ obtained by adjoining to $K$ the coordinates of the $m$-torsion points of $E$ is unramified over $K$.*

*Proof Sketch.* Let $K' = K(E[m])$, let $\mathfrak{p}'$ be the maximal ideal of the ring of integers of $K'$, and let $k'$ be the residue field. Suppose that $\sigma \in \operatorname{Gal}(K'/K)$ is in the inertia group. Then $\sigma$ fixes everything modulo $\mathfrak{p}'$, so in particular,

$$\sigma(P) \equiv P \pmod{\mathfrak{p}'} \qquad \text{for all } P \in E[m]. \tag{6.15}$$

But from Theorem 6.32, the reduction map $E[m] \to \tilde{E}(k')$ is injective, so (6.15) implies that $\sigma(P) = P$ for all $P \in E[m]$. The points in $E[m]$ generate $K'/K$, so $\sigma$ fixes $K'$. Hence $\operatorname{Gal}(K'/K)$ has trivial inertia group, so $K'/K$ is unramified. (For further details, see [410, VII.4.1].) □

*Remark* 6.34. The coordinates of the points in $E[m]$ are algebraic over $K$, so the absolute Galois group $G_K = \operatorname{Gal}(\bar{K}/K)$ acts on $E[m]$ compatibly with the group structure. In this way we obtain a representation

$$\rho : G_K \longrightarrow \operatorname{Aut}(E[m]) \cong \operatorname{GL}_2(\mathbb{Z}/m\mathbb{Z}).$$

In order to create a characteristic-0 representation, we fix a prime $\ell$ and combine all of the $\ell$-power torsion to form the *Tate module*

$$T_\ell(E) = \varprojlim E[\ell^n] \cong \mathbb{Z}_\ell \times \mathbb{Z}_\ell.$$

(Here $\mathbb{Z}_\ell$ denotes the ring of $\ell$-adic integers.) Then the *$\ell$-adic representation of $E$* is the homomorphism

$$\rho_{E,\ell} : G_K \longrightarrow \operatorname{Aut}(T_\ell(E)) \cong \operatorname{GL}_2(\mathbb{Z}_\ell).$$

These representations are of fundamental importance in the study of the arithmetic properties of elliptic curves.

### 6.3.7 The Invariant Differential

**Definition.** Let $E : y^2 = x^3 + ax + b$ be an elliptic curve given by a Weierstrass equation. The *invariant differential on $E$* (associated to the given Weierstrass equation) is the differential 1-form

$$\omega_E = \frac{dx}{2y} = \frac{dy}{3x^2 + a}.$$

The next result explains why the invariant differential is so named and shows that it linearizes the group law in a useful way.

**Theorem 6.35.** *Let $E$ be an elliptic curve given by a Weierstrass equation and let $\omega_E$ be the associated invariant differential on $E$.*

(a) *For any given point $Q \in E$, let $\tau_Q : E \to E$ be the translation-by-$Q$ map defined by $\tau_Q(P) = P + Q$. The differential form $\omega_E$ is translation-invariant in the sense that*
$$\tau_Q^*(\omega_E) = \omega_E \qquad \text{for every } Q \in E.$$
(b) *The differential form $\omega_E$ is holomorphic at every point of $E$.*
(c) *Up to multiplication by a nonzero constant, $\omega_E$ is the only holomorphic translation-invariant 1-form on $E$.*
(d) *For every $m \in \mathbb{Z}$ the differential form $\omega_E$ satisfies*
$$[m]^*\omega_E = m\omega_E.$$

*Proof.* See [410, III §5]. □

The invariant differential can also be used to fix an embedding of the endomorphism ring of $E$ into $\mathbb{C}$. Of course, this is of interest only when $E$ has CM, since there is only one way to embed $\mathbb{Z}$ into $\mathbb{C}$.

**Proposition 6.36.** *Let $E/\mathbb{C}$ be an elliptic curve, write $E(\mathbb{C}) \cong \mathbb{C}/L$ for some lattice $L$ as described in Section* 1.6.3, *and let*
$$R = \{\alpha \in \mathbb{C} : \alpha L \subseteq L\}.$$
*Then for each $\alpha \in R$ there is a unique endomorphism $[\alpha] \in \operatorname{End}(E)$ satisfying*
$$[\alpha]^*(\omega) = \alpha\omega, \tag{6.16}$$
*and this association defines a unique ring isomorphism*
$$[\cdot] : R \xrightarrow{\ \sim\ } \operatorname{End}(E).$$
*(Without the normalization* (6.16), *the isomorphism $R \cong \operatorname{End}(E)$ is unique only up to complex conjugation of $R$.)*

*Proof.* We fix the isomorphism $E(\mathbb{C}) \cong \mathbb{C}/L$ so that the invariant differential $\omega$ on $E$ corresponds to $dz$ on $\mathbb{C}/L$. More formally, we use the Weierstrass $\wp$ function to define an isomorphism
$$F : \mathbb{C}/L \longrightarrow E(\mathbb{C}), \qquad F(z) = \left(\wp(z), \frac{1}{2}\wp'(z)\right);$$
see Section 1.6.3 and [410, VI §3]. Then
$$F^*(\omega) = F^*\left(\frac{dx}{2y}\right) = \frac{d\wp(z)}{\wp'(z)} = dz.$$

Let $\psi \in \operatorname{End}(E)$, so using the identification $E(\mathbb{C}) \cong \mathbb{C}/L$, the endomorphism $\psi$ defines a holomorphic map $\psi : \mathbb{C}/L \to \mathbb{C}/L$ satisfying $\psi(0) = 0$. We claim that the analyticity implies that $\psi$ lifts to a map $\bar{\psi} : \mathbb{C} \to \mathbb{C}$ of the form $\psi(z) = \alpha z$ for a unique $\alpha \in \mathbb{C}$, and hence in particular that $\psi$ is a homomorphism. (Compare with the algebraic statement of this fact given in Theorem 6.21.) To prove this, we first observe that the covering map $\mathbb{C} \to \mathbb{C}/L$ is the universal cover of $\mathbb{C}/L$, so we can lift $\psi$ to some holomorphic map $\bar{\psi} : \mathbb{C} \to \mathbb{C}$. Further, the fact that $\bar{\psi}$ lifts $\psi$ means that $\bar{\psi}$ satisfies

$$\bar{\psi}(z+\omega) - \bar{\psi}(z) \in L \qquad \text{for all } z \in \mathbb{C} \text{ and all } \omega \in L.$$

Fixing $\omega \in L$, we find that the map $z \mapsto \bar{\psi}(z+\omega) - \bar{\psi}(z)$ is a holomorphic map from $\mathbb{C}$ to the discrete set $L$, so it must be constant. Thus for each $\omega \in L$ there is a number $c(\omega) \in \mathbb{C}$ such that

$$\bar{\psi}(z+\omega) = \bar{\psi}(z) + c(\omega) \qquad \text{for all } z \in \mathbb{C}. \tag{6.17}$$

Writing $\bar{\psi}(z) = \sum a_i z^i$ as a convergent power series, one easily checks that the relation (6.17) forces $\bar{\psi}$ to be linear, say $\bar{\psi}(z) = \alpha z + \beta$. Then the assumption $\psi(0) = 0$ tells us that $\beta \in L$, so $\alpha z$ and $\alpha z + \beta$ descend to the same map on $\mathbb{C}/L$. Hence $\psi$ lifts to a map of the form $\bar{\psi}(z) = \alpha z$. Further, $\alpha$ is unique, since if $\psi(z)$ also lifts to $\alpha' z$, then the map $z \mapsto (\alpha - \alpha')z$ sends $\mathbb{C}$ to $L$, hence must be constant, so $\alpha = \alpha'$. Finally, we observe that in order for $\alpha z$ to descend to $\mathbb{C}/L$, the complex number $\alpha$ must satisfy $\alpha L \subseteq L$. Thus the association $\psi \mapsto \alpha$ gives a map $\operatorname{End}(E) \to R$, and it is clear that the map $\psi(z) = \alpha z$ satisfies $\psi^*(dz) = \alpha dz$, so with our identifications, we have $\psi^*(\omega) = \alpha\omega$.

Next we check that the resulting map $\operatorname{End}(E) \to R$ is a ring homomorphism. Let $\psi_1, \psi_2 \in \operatorname{End}(E)$. Then on $\mathbb{C}/L$ we have $\psi_1(z) = \alpha_1 z$ and $\psi_2(z) = \alpha_2 z$, so

$$(\psi_1 + \psi_2)(z) = (\alpha_1 + \alpha_2)z \qquad \text{and} \quad (\psi_1 \circ \psi_2)(z) = \alpha_1\alpha_2 z.$$

It remains to check that every $\alpha \in R$ comes from some $\psi \in \operatorname{End}(E)$. By definition any $\alpha \in R$ induces a map $z \mapsto \alpha z$ on $\mathbb{C}$ that descends to a holomorphic homomorphism $\mathbb{C}/L \to \mathbb{C}/L$. Using the theory of elliptic functions (see, e.g., [410, Theorem VI.4.1]), one can show that every such holomorphic map $E(\mathbb{C}) \to E(\mathbb{C})$ is given by rational functions, which shows that $\operatorname{End}(R) \to R$ is surjective. □

### 6.3.8 Maps from $E$ to $\mathbb{P}^1$

The quotient of $E$ by a finite group of automorphisms gives a map from $E$ to $\mathbb{P}^1$. These quotient maps play an important role in dynamics.

**Proposition 6.37.** *Let $\Gamma$ be a nontrivial subgroup of* $\operatorname{Aut}(E)$. *Then the quotient curve $E/\Gamma$ is isomorphic to $\mathbb{P}^1$ and the projection map $\pi : E \to E/\Gamma \cong \mathbb{P}^1$ is given explicitly by*

$$\pi(x,y) = \begin{cases} x & \textit{if } \Gamma = \boldsymbol{\mu}_2 \quad (j(E) \textit{ arbitrary}),\\ x^2 & \textit{if } \Gamma = \boldsymbol{\mu}_4 \quad (j(E) = 1728 \textit{ only}),\\ y & \textit{if } \Gamma = \boldsymbol{\mu}_3 \quad (j(E) = 0 \textit{ only}),\\ x^3 & \textit{if } \Gamma = \boldsymbol{\mu}_6 \quad (j(E) = 0 \textit{ only}). \end{cases}$$

*Proof.* By definition, the quotient curve $E/\Gamma$ is the curve whose function field is the subfield of $K(E) = K(x,y)$ fixed by $\Gamma$. Using the explicit description (6.13) of the action of $\operatorname{Aut}(E)$ on the coordinates of $E$, it is easy to find this subfield. For example, if $\Gamma = \boldsymbol{\mu}_2$, it consists of those functions that are invariant under $(x,y) \mapsto (x,-y)$, so the fixed field $K(E)^\Gamma$ is $K(x,y^2) = K(x)$. As a second example, if $\Gamma = \boldsymbol{\mu}_6$, then we need functions invariant under $(x,y) \mapsto (\rho x, -y)$, where $\rho$ is a primitive cube root of 1. This fixed field is $K(E)^\Gamma = K(x^3,y^2) = K(x^3)$, since in this case the elliptic curve is given by an equation of the form $y^2 = x^3 + b$. The other cases are similar. □

For later use, we prove that the isomorphism class of an elliptic curve $E$ is determined by the critical values of any double cover $E \to \mathbb{P}^1$.

**Lemma 6.38.** *Let $E$ be an elliptic curve defined over a field of characteristic not equal to 2 and let $\pi : E \to \mathbb{P}^1$ be a rational map of degree 2. Then $\pi$ has exactly four critical values and they determine the isomorphism class of $E$.*

*Proof.* The Riemann–Hurwitz formula (Theorem 1.5) for the map $\pi : E \to \mathbb{P}^1$ says that

$$2g(E) - 2 = (\deg \pi)(2g(\mathbb{P}^1) - 2) + \sum_{P \in E} \bigl(e_P(\phi) - 1\bigr).$$

The map $\pi$ has degree 2, the elliptic curve $E$ has genus 1, and $\mathbb{P}^1$ has genus 0, so we find that

$$\sum_{P \in E} \bigl(e_P(\phi) - 1\bigr) = 4.$$

The ramification indices satisfy $1 \le e_P(\phi) \le \deg \pi = 2$, so we conclude that there are exactly four critical points, i.e., four points with $e_P(\phi) = 2$ and all other points satisfy $e_P(\phi) = 1$. Further, these four critical points must have distinct images in $\mathbb{P}^1$, since for any point $t \in \mathbb{P}^1$ we have

$$\sum_{P \in \pi^{-1}(t)} e_P(\phi) = \deg(\pi).$$

Let $t_1, t_2, t_3, t_4 \in \mathbb{P}^1$ be the four critical values of $\pi$, i.e., the images of the critical points, and let $f \in \mathrm{PGL}_2$ be the unique linear fractional transformation satisfying

$$f(t_1) = 0, \qquad f(t_2) = 1, \qquad f(t_3) = \infty.$$

Explicitly,

$$f(t) = \frac{(t_2 - t_3)(t - t_1)}{(t - t_3)(t_2 - t_1)}.$$

(If any of $t_1, t_2, t_3$ equals $\infty$, take the appropriate limit.) The quantity

$$\kappa = f(t_4) = \frac{(t_2 - t_3)(t_4 - t_1)}{(t_4 - t_3)(t_2 - t_1)}$$

is called the *cross-ratio* of $t_1, t_2, t_3, t_4$; cf. Section 2.7, page 71.

We let $x = f \circ \pi$, so $x$ is a rational function of degree 2 on $E$ with critical values 0, 1, $\infty$, and $\kappa$. To ease notation, we let

$$T_0 = \pi^{-1}(0), \quad T_1 = \pi^{-1}(1), \quad T_\kappa = \pi^{-1}(\kappa), \quad \mathcal{O} = \pi^{-1}(\infty).$$

Taking $\mathcal{O}$ to be the identity element for the group law on $E$, we see that

$$\operatorname{div}(x) = 2(T_0) - 2(\mathcal{O}), \quad \operatorname{div}(x-1) = 2(T_1) - 2(\mathcal{O}), \quad \operatorname{div}(x-\kappa) = 2(T_\kappa) - 2(\mathcal{O}),$$

so $T_0, T_1, T_\kappa$ are in $E[2]$, i.e., they are points of order 2. The sum of the three nontrivial 2-torsion points on any elliptic curve is equal to $\mathcal{O}$, so Proposition 6.18 tells us that there is a rational function $y$ on $E$ with divisor

$$\operatorname{div}(y) = (T_0) + (T_1) + (T_\kappa) - 3(\mathcal{O}).$$

After multiplying $y$ by an appropriate constant, it follows that $x$ and $y$ are Weierstrass coordinates for $E$ (cf. [198, IV.4.6] or [410, III.3.1]). More precisely, the rational functions $x$ and $y$ map $E$ isomorphically to the curve with Weierstrass equation

$$E : y^2 = x(x-1)(x-\kappa).$$

It is easy to compute the $j$-invariant of $E$ in terms of $\kappa$ (cf. [410, III.1.7(b)]). We find that

$$j(E) = 2^8 \frac{(\kappa^2 - \kappa + 1)^3}{\kappa^2(\kappa - 1)^2},$$

so in particular $j(E)$ is uniquely determined by $t_1, t_2, t_3, t_4$. Then we apply Proposition 6.15 (or [410, III.1.4(b)]), which says that the isomorphism class of $E$ is determined by its $j$-invariant. This completes the proof of Lemma 6.38. □

### 6.3.9 Complex Multiplication

Let $E/K$ be an elliptic curve with complex multiplication defined over a number field. As described in Proposition 6.25, the endomorphism ring of $E$ is isomorphic to a subring of the ring of integers of a quadratic imaginary field. We briefly recall an analytic proof of this important fact and then discuss the relationship between complex multiplication and the ideal class group of the associated quadratic imaginary field. This material is used only in Section 6.6, so may be omitted at first reading.

**Proposition 6.39.** *Let $E/\mathbb{C}$ be an elliptic curve with complex multiplication, i.e., the endomorphism ring* $\operatorname{End}(E)$ *is strictly larger than $\mathbb{Z}$. Choose a lattice $L \subset \mathbb{C}$ such that $E(\mathbb{C}) \cong \mathbb{C}/L$, let*

$$R = \{\alpha \in \mathbb{C} : \alpha L \subseteq L\},$$

*and let*

$$[\cdot] : R \xrightarrow{\sim} \operatorname{End}(E)$$

*be the isomorphism described in Proposition 6.36. Then $R$ is a subring of the ring of integers of a quadratic imaginary field $F$, and for any $\alpha \in R$, the degree of $[\alpha]$ is given by*

$$\deg[\alpha] = \alpha\bar{\alpha} = \mathrm{N}_{F/\mathbb{Q}}(\alpha),$$

*where $\bar{\alpha}$ denotes the complex conjugate of $\alpha$.*

*Proof.* To describe the ring $R$, we choose a basis for $L$, say $L = \mathbb{Z}\omega_1 + \mathbb{Z}\omega_2$. We have $R \neq \mathbb{Z}$ by assumption, so there exists an $\alpha \in R$ with $\alpha \notin \mathbb{Z}$. Write

$$\alpha\omega_1 = a\omega_1 + b\omega_2 \quad \text{and} \quad \alpha\omega_2 = c\omega_1 + d\omega_2 \quad \text{with} \quad a, b, c, d \in \mathbb{Z}. \tag{6.18}$$

The numbers $\omega_1$ and $\omega_2$ are $\mathbb{R}$-linearly independent, so the relation

$$\begin{pmatrix} \alpha - a & -b \\ -c & \alpha - d \end{pmatrix} \begin{pmatrix} \omega_1 \\ \omega_2 \end{pmatrix} = \begin{pmatrix} 0 \\ 0 \end{pmatrix}$$

implies that the matrix has determinant 0,

$$\alpha^2 - (a + d)\alpha + (ad - bc) = 0. \tag{6.19}$$

Hence $\alpha$ is an algebraic integer in a quadratic field. Further, we must have $\alpha \notin \mathbb{R}$, since if $\alpha$ were real, then the relation $(\alpha - a)\omega_1 = b\omega_2$ (and $\alpha \notin \mathbb{Z}$) would contradict the $\mathbb{R}$-linear independence of $\omega_1$ and $\omega_2$. This proves that every element of $R$ is an algebraic integer in a quadratic imaginary field, and hence $R$ is a subring of the ring of integers of such a field.

Finally, in order to compute the degree of the endomorphism $[\alpha] : E \to E$ corresponding to $\alpha \in R$, we observe that

$$\deg(\alpha) = \#\operatorname{Ker}\left(\mathbb{C}/L \xrightarrow{z \to \alpha z} \mathbb{C}/L\right) = (L : \alpha L).$$

If $\alpha \in \mathbb{Z}$, then it is clear that $(L : \alpha L) = \alpha^2$, since $L = \mathbb{Z}^2$ as an abstract group. Suppose now that $\alpha \notin \mathbb{Z}$. Then continuing with the earlier notation, the transformation formulas (6.18) imply that the index of $\alpha L$ in $L$ is $(L : \alpha L) = ad - bc$. On the other hand, the product $\alpha\bar{\alpha}$ is the constant term in the minimal equation (6.19) for $\alpha$ over $\mathbb{Q}$, hence also equal to $ad - bc$. □

The theory of complex multiplication uses elliptic curves to describe the abelian extensions of a quadratic imaginary field $F$ in a manner analogous to the description of abelian extensions of $\mathbb{Q}$ using torsion points in $\mathbb{G}_m$, i.e., using roots of unity. For a complete introduction to the theory of complex multiplication, see, for example, [257, Part II], [399, Chapter 5], or [412, Chapter II]. We now describe the tiny piece of the theory that will be needed in Section 6.6 in order to prove Theorem 6.62.

Let $F$ be a quadratic imaginary field, let $R_F$ be the ring of integers of $F$, and let $\mathcal{I}_F$ be the group of fractional ideals of $F$. If we fix an embedding $F \subset \mathbb{C}$, then each fractional ideal $\mathfrak{a} \in \mathcal{I}_F$ is a lattice $\mathfrak{a} \subset \mathbb{C}$; hence it determines an elliptic curve $E_\mathfrak{a}$ whose complex points are

$$E_\mathfrak{a}(\mathbb{C}) \cong \mathbb{C}/\mathfrak{a}. \tag{6.20}$$

We observe that $E_\mathfrak{a}$ has complex multiplication by $R_F$, since any $\alpha \in R_F$ has the property that $\alpha\mathfrak{a} \subset \mathfrak{a}$, hence it induces a holomorphic map

$$[\alpha] : \mathbb{C}/\mathfrak{a} \longmapsto \mathbb{C}/\mathfrak{a}, \qquad z \longmapsto \alpha z,$$

which in turn yields an isogeny $[\alpha] : E_\mathfrak{a} \to E_\mathfrak{a}$. In fact, since $R_F$ is the maximal order in $F$, we have $\operatorname{End}(E_\mathfrak{a}) = R_F$. We denote by $\mathcal{E}\ell\ell(R_F)$ the set

$$\mathcal{E}\ell\ell(R_F) = \{\text{isomorphism classes of elliptic curves } E \text{ with } \operatorname{End}(E) \cong R_F\}.$$

We thus have a natural map

$$\mathcal{I}_F \longrightarrow \mathcal{E}\ell\ell(R_F), \qquad \mathfrak{a} \longmapsto (\text{isomorphism class of } E_\mathfrak{a}). \tag{6.21}$$

We also observe that if we multiply $\mathfrak{a}$ by a principal ideal, then the isomorphism class of $E_\mathfrak{a}$ does not change, since for any $c \in K^*$ there is an obvious isomorphism

$$\mathbb{C}/\mathfrak{a} \xrightarrow{\sim} \mathbb{C}/c\mathfrak{a}, \qquad z \longmapsto cz.$$

Hence the map (6.21) induces a natural map from the ideal class group $\mathcal{C}_F = \mathcal{I}_F/F^*$ to elliptic curves with complex multiplication by $R_F$,

$$\mathcal{C}_F \longrightarrow \mathcal{E}\ell\ell(R_F).$$

**Proposition 6.40.** *Let $F$ be a quadratic imaginary field with ring of integers $R_F$ and ideal class group $\mathcal{C}_F$, and let $h_F = \#\mathcal{C}_F$ be the class number of $F$. Then with notation as above, the natural map*

$$\mathcal{C}_F \longrightarrow \mathcal{E}\ell\ell(R_F)$$

*is a bijection. In particular, there are exactly $h_F$ isomorphism classes of elliptic curves whose endomorphism ring is $R_F$.*

*Proof.* See [412, II.1.2]. □

## 6.4 General Properties of Lattès Maps

Chebyshev polynomials arise by restricting the power map $z^n$ to the quotient of $\mathbb{P}^1$ by the finite group of automorphisms $\{z, z^{-1}\}$. As already briefly described in Section 1.6.3, quotients of elliptic curves lead similarly to rational maps called Lattès maps. In this section we define and discuss general properties of these Lattès maps. For an excellent introduction to Lattès maps over $\mathbb{C}$, including historical remarks and proofs of their basic geometric and analytic properties, see [300].

**Definition.** A rational map $\phi : \mathbb{P}^1 \to \mathbb{P}^1$ of degree $d \ge 2$ is called a *Lattès map* if there are an elliptic curve $E$, a morphism $\psi : E \to E$, and a finite separable[1] covering $\pi : E \to \mathbb{P}^1$ such that the following diagram is commutative:

$$\begin{array}{ccc} E & \xrightarrow{\psi} & E \\ \downarrow{\scriptstyle\pi} & & \downarrow{\scriptstyle\pi} \\ \mathbb{P}^1 & \xrightarrow{\phi} & \mathbb{P}^1. \end{array} \tag{6.22}$$

*Example* 6.41. Let $E : y^2 = x^3 + ax + b$ be an elliptic curve. Then the classical formula for $x(2P)$ (Proposition 6.17) and the isomorphism $x : E/\{\pm 1\} \to \mathbb{P}^1$ yield the Lattès map

$$\phi(x) = x(2P) = \frac{x^4 - 2ax^2 - 8bx + a^2}{4x^3 + 4ax + 4b}.$$

Here $\psi$ is the duplication map $\psi(P) = [2]P$, and the projection $\pi$ is given by $\pi(P) = \pi(x, y) = x$.

*Example* 6.42. Let $E$ be the elliptic curve $E : y^2 = x^3 + ax$ with $j(E) = 1728$ and again let $\psi(P) = [2]P$ be the doubling map. If we take $\pi(x, y) = x$, then we are in the $b = 0$ case of Example 6.41, and we obtain the Lattès map

$$\phi(x) = x(2P) = \frac{(x^2 - a)^2}{4x(x^2 + a)}.$$

However, for this curve we may instead take $\pi(x, y) = x^2$. This gives a new Lattès map $\phi_1$. We find a formula for $\phi_1$ using the relation

$$\phi_1(x) = \phi(\sqrt{x})^2 = \left(\frac{(x - a)^2}{4\sqrt{x}(x + a)}\right)^2 = \frac{(x - a)^4}{16x(x + a)^2}.$$

Note that the map $\pi(x, y) = x^2$ corresponds to taking the quotient of $E$ by its automorphism group $\operatorname{Aut}(E) \cong \boldsymbol{\mu}_4$ via the association described in Remark 6.27.

*Example* 6.43. In a similar manner, the doubling map on the elliptic curve

$$E : y^2 = x^3 + 1$$

with $j(E) = 0$ and $\operatorname{Aut}(E) = \boldsymbol{\mu}_6$ gives various Lattès maps corresponding to taking the quotient of $E$ by the different subgroups of $\boldsymbol{\mu}_6$. Explicitly, the Lattès maps corresponding, respectively, to $\boldsymbol{\mu}_2$, $\boldsymbol{\mu}_3$, and $\boldsymbol{\mu}_6$ are

$$\phi_1(z) = \frac{z(z^3 - 8b)}{4(z^3 + b)}, \qquad \phi_2(z) = \frac{z^4 + 18bz^2 - 27b^2}{8z^3}, \qquad \phi_3(z) = \frac{z(z - 8b)^3}{64(z + b)^3}.$$

We leave the verification of these formulas to the reader; see Exercise 6.12.

[1] The assumption that $\pi$ is separable is relevant only when one is working over a field of characteristic $p$, in which case it is equivalent to the assumption that $\pi$ does not factor through the $p$-power Frobenius map.

We begin with an elementary, but useful, characterization of the preperiodic points of a Lattès map (cf. Proposition 1.42).

**Proposition 6.44.** *Let $\phi$ be a Lattès map associated to an elliptic curve $E$. Then*

$$\operatorname{PrePer}(\phi) = \pi(E_{\text{tors}}).$$

*Proof.* Let $\zeta \in \mathbb{P}^1$ and let $P \in E$ be any point satisfying $\pi(P) = \zeta$. We consider the orbits of $\zeta$ and $P$. Thus

$$\begin{aligned}\pi\bigl(\mathcal{O}_\psi(P)\bigr) &= \pi\bigl(\{\psi^n(P) : n \ge 0\}\bigr) = \{\pi\psi^n(P) : n \ge 0\} \\ &= \{\phi^n\pi(P) : n \ge 0\} = \{\phi^n(\zeta) : n \ge 0\} = \mathcal{O}_\phi(\zeta).\end{aligned}$$

The map $\pi$ is finite, so this shows that $\mathcal{O}_\psi(P)$ is finite if and only if $\mathcal{O}_\phi(\zeta)$ is finite. Hence

$$\operatorname{PrePer}(\phi) = \pi(\operatorname{PrePer}(\psi)),$$

and it is left to prove that $\operatorname{PrePer}(\psi) = E_{\text{tors}}$.

We observe that the map $\psi : E \to E$ has the form $\psi(P) = \psi_0(P) + T$ for some $\psi_0 \in \operatorname{End}(E)$ and some point $T \in E$. (See [410, III.4.7].) We are going to prove Proposition 6.44 in the case that $\psi = \psi_0 \in \operatorname{End}(E)$, i.e., assuming that $T = \mathcal{O}$. For the general case, which requires knowing that the point $T$ is a point of finite order, see Exercise 6.14.

Suppose first that $P \in E_{\text{tors}}$, say $[n]P = \mathcal{O}$ for some $n \ge 1$. Consider the images of the iterates $\psi, \psi^2, \psi^3, \dots$ in the quotient ring $\operatorname{End}(E)/n\operatorname{End}(E)$. It follows from the description of $\operatorname{End}(E)$ in Proposition 6.25 that this quotient ring is finite, so we can find iterates $i > j \ge 1$ such that

$$\psi^i \equiv \psi^j \pmod{n\operatorname{End}(E)}.$$

In other words, there is an endomorphism $\beta \in \operatorname{End}(E)$ such that

$$\psi^i = \psi^j + \beta n.$$

Evaluating both sides at $P$ and using the fact that $[n]P = 0$ allows us to conclude that $\psi^i(P) = \psi^j(P)$. Hence $P \in \operatorname{PrePer}(\psi)$, which proves that $E_{\text{tors}} \subset \operatorname{PrePer}(\psi)$.

Next suppose that $P \in \operatorname{PrePer}(\psi)$, say $\psi^i(P) = \psi^j(P)$ for some $i > j$. We rewrite this as $(\psi^i - \psi^j)(P) = \mathcal{O}$ and apply the dual isogeny $\widehat{\psi^i - \psi^j}$ described in Theorem 6.22 to obtain

$$[\deg(\psi^i - \psi^j)](P) = \mathcal{O}.$$

We know that $\psi^i \ne \psi^j$, since $\deg(\psi) = \deg(\phi) \ge 2$ and $i > j$, so $\psi^i - \psi^j$ has positive degree. This proves that $P \in E_{\text{tors}}$, which gives the other inclusion $\operatorname{PrePer}(\psi) \subset E_{\text{tors}}$. □

Many dynamical properties of a rational map can be analyzed by studying the behavior of the critical points under iteration of the map. This is certainly true for Lattès maps, whose postcritical orbits have a simple characterization, which we give after setting some notation.

**Definition.** Let $\phi : C_1 \to C_2$ be a nonconstant rational map between smooth projective curves. The *set of critical points* (also called *ramification points*) of $\phi$ is denoted by

$$\mathsf{CritPt}_\phi = \{P \in C_1 : \phi \text{ is ramified at } P\} = \{P \in C_1 : e_P(\phi) \ge 2\}.$$

The *set of critical values* of $\phi$ is the image of the set of critical points and is denoted by

$$\mathsf{CritVal}_\phi = \phi(\mathsf{CritPt}_\phi).$$

If $\phi : C \to C$ is a map from a curve to itself, the *postcritical set* is the full forward orbit of the critical values and is denoted by

$$\mathsf{PostCrit}_\phi = \bigcup_{n=0}^{\infty} \phi^n(\mathsf{CritVal}_\phi) = \bigcup_{n=1}^{\infty} \mathsf{CritVal}_{\phi^n}.$$

(See Exercise 6.15.)

**Proposition 6.45.** *Let $\phi : \mathbb{P}^1 \to \mathbb{P}^1$ be a Lattès map that fits into a commutative diagram* (6.22). *Then*

$$\mathsf{CritVal}_\pi = \mathsf{PostCrit}_\phi.$$

*In particular, a Lattès map is* postcritically finite.

*Proof.* The key to the proof of this proposition is the fact that the map $\psi : E \to E$ is unramified, i.e., it has no critical points, see Remark 6.20. (In the language of modern algebraic geometry, the map $\psi$ is étale.) More precisely, the map $\psi$ is the composition of an endomorphism of $E$ and a translation (Remark 6.19), both of which are unramified.

For any $n \ge 1$ we compute

$$\begin{aligned}
\mathsf{CritVal}_\pi &= \mathsf{CritVal}_{\pi\psi^n} && \text{because } \psi \text{ is unramified,}\\
&= \mathsf{CritVal}_{\phi^n\pi} && \text{from the commutativity of (6.22),}\\
&= \mathsf{CritVal}_{\phi^n} \cup \phi^n(\mathsf{CritVal}_\pi) && \text{from the definition of critical value,}\\
&\supseteq \mathsf{CritVal}_{\phi^n}.
\end{aligned}$$

This holds for all $n \ge 1$, which gives the inclusion

$$\mathsf{CritVal}_\pi \supseteq \bigcup_{n=0}^{\infty} \phi^n(\mathsf{CritVal}_\phi) = \mathsf{PostCrit}_\phi.$$

In order to prove the opposite inclusion, suppose that there exists a point $P_0 \in E$ satisfying

$$P_0 \in \mathsf{CritPt}_\pi \quad \text{and} \quad \pi(P_0) \notin \mathsf{PostCrit}_\phi. \tag{6.23}$$

Consider any point $Q \in \psi^{-1}(P_0)$. Then $Q$ is a critical point of $\pi\psi$, since $\psi$ is unramified and $\pi$ is ramified at $\psi(Q)$ by assumption. But $\pi\psi = \phi\pi$, so we see that $Q$ is a critical point for $\phi\pi$.

On the other hand,

$$\phi(\pi(Q)) = \pi(P_0) \notin \mathsf{CritVal}_\phi = \phi(\mathsf{CritPt}_\phi),$$

so $\pi(Q)$ is not a critical point for $\phi$. It follows that $Q$ is a critical point of $\pi$. Further, we claim that no iterate of $\phi$ is ramified at $\pi(Q)$. To see this, we use the given fact that $\pi(P_0)$ is not in the postcritical set of $\phi$ to compute

$$\begin{aligned} \pi(P_0) \notin \mathsf{PostCrit}_\phi &\Longrightarrow \pi(P_0) \notin \phi^n(\mathsf{CritPt}_\phi) && \text{for all } n \ge 1, \\ &\Longrightarrow \pi(\psi(Q)) \notin \phi^n(\mathsf{CritPt}_\phi) && \text{for all } n \ge 1, \\ &\Longrightarrow \phi(\pi(Q)) \notin \phi^n(\mathsf{CritPt}_\phi) && \text{for all } n \ge 1, \\ &\Longrightarrow \pi(Q) \notin \phi^n(\mathsf{CritPt}_\phi) && \text{for all } n \ge 0, \\ &\Longrightarrow \pi(Q) \notin \mathsf{PostCrit}_\phi . \end{aligned}$$

To recapitulate, we have now proven that every $Q \in \psi^{-1}(P_0)$ satisfies

$$Q \in \mathsf{CritPt}_\pi \qquad \text{and} \qquad \pi(Q) \notin \mathsf{PostCrit}_\psi .$$

In other words, every point $Q \in \psi^{-1}(P_0)$ satisfies the same two conditions (6.23) that are satisfied by $P_0$. Hence by induction we find that if there is any point $P_0$ satisfying (6.23), then the full backward orbit of $\psi$ is contained in the set of critical points of $\pi$, i.e.,

$$\mathsf{CritPt}_\pi \supset \bigcup_{n=1}^{\infty} \psi^{-n}(P_0).$$

But $\psi$ is unramified and has degree at least 2 (note that $\deg \psi = \deg \phi$), so

$$\#\mathsf{CritPt}_\pi \ge \#\bigl(\psi^{-n}(P_0)\bigr) = (\deg \psi)^n \xrightarrow[n\to\infty]{} \infty.$$

This is a contradiction, since $\pi$ has only finitely many critical points, so we conclude that there are no points $P_0$ satisfying (6.23). Hence

$$P_0 \in \mathsf{CritPt}_\pi \Longrightarrow \pi(P_0) \in \mathsf{PostCrit}_\phi,$$

which gives the other inclusion $\mathsf{CritVal}_\pi \subseteq \mathsf{PostCrit}_\phi$. □

As an application of Proposition 6.45, we show that Lattès maps associated to distinct elliptic curves are not conjugate to one another.

**Theorem 6.46.** *Let $K$ be an algebraically closed field of characteristic not equal to 2 and let $\phi$ and $\phi'$ be Lattès maps defined over $K$ that are associated, respectively, to elliptic curves $E$ and $E'$. Assume further that the projection maps $\pi$ and $\pi'$ associated to $\phi$ and $\phi'$ both have degree 2. If $\phi$ and $\phi'$ are $\mathrm{PGL}_2(K)$-conjugate to one another, then $E$ and $E'$ are isomorphic.*

*Proof.* Let $f \in \mathrm{PGL}_2(K)$ be a linear fractional transformation conjugating $\phi'$ to $\phi$. Then we have a commutative diagram

$$\begin{array}{ccccccc} E & \xrightarrow{\pi} & \mathbb{P}^1 & \xleftarrow{f} & \mathbb{P}^1 & \xleftarrow{\pi'} & E' \\ \downarrow{\scriptstyle\psi} & & \downarrow{\scriptstyle\phi} & & \downarrow{\scriptstyle\phi'} & & \downarrow{\scriptstyle\psi'} \\ E & \xrightarrow{\pi} & \mathbb{P}^1 & \xleftarrow{f} & \mathbb{P}^1 & \xleftarrow{\pi'} & E' \end{array}$$

We let $\pi'' = f \circ \pi'$. Note that since $f$ is an isomorphism, the map $\pi''$ still has degree 2. This yields the simplified commutative diagram

$$\begin{array}{ccccc} E & \xrightarrow{\pi} & \mathbb{P}^1 & \xleftarrow{\pi''} & E' \\ \downarrow{\scriptstyle\psi} & & \downarrow{\scriptstyle\phi} & & \downarrow{\scriptstyle\psi'} \\ E & \xrightarrow{\pi} & \mathbb{P}^1 & \xleftarrow{\pi''} & E' \end{array}$$

showing that $\phi$ is a Lattès map associated to both elliptic curves $E$ and $E'$.

Applying Proposition 6.45, first to $E$ and then to $E'$, we find that

$$\mathsf{CritVal}_\pi = \mathsf{PostCrit}_\phi = \mathsf{CritVal}_{\pi''}. \tag{6.24}$$

In other words, the degree-2 maps $\pi : E \to \mathbb{P}^1$ and $\pi'' : E' \to \mathbb{P}^1$ have the exact same set of critical values. Then Lemma 6.38 tells us that $E$ and $E'$ are isomorphic. □

## 6.5 Flexible Lattès Maps

A Lattès map is a rational map that is obtained by projecting an elliptic curve endomorphism down to $\mathbb{P}^1$. For any integer $m \geq 2$, every elliptic curve has a multiplication-by-$m$ map and a projection $E \to E/\{\pm 1\} \cong \mathbb{P}^1$, so every elliptic curve has a corresponding Lattès map. As $E$ varies, this collection of Lattès maps varies continuously, which prompts the following definition.

**Definition.** A *flexible Lattès map* is a Lattès map $\phi : \mathbb{P}^1 \to \mathbb{P}^1$ that fits into a Lattès commutative diagram (6.22) in which the map $\psi : E \to E$ has the form

$$\psi(P) = [m](P) + T \qquad \text{for some } m \in \mathbb{Z} \text{ and some } T \in E$$

and such that the projection map $\pi : E \to \mathbb{P}^1$ satisfies

$$\deg(\pi) = 2 \qquad \text{and} \qquad \pi(P) = \pi(-P) \text{ for all } P \in E.$$

*Remark* 6.47. The condition that $\pi$ be even, i.e., that it satisfy $\pi(-P) = \pi(P)$, is included for convenience. In general, if $\pi : E \to \mathbb{P}^1$ is any map of degree 2, then there exists a point $P_0 \in E$ such that $\pi(-(P + P_0)) = \pi(P + P_0)$ for all $P \in E$. Thus $\pi$ becomes an even function if we use $P_0$ as the identity element for the group law on $E$. See Exercise 6.16.

*Remark* 6.48. We show in this section that the Lattès maps of a given degree have identical multiplier spectra. This is one reason that these Lattès maps are called "flexible," since they vary in continuous families whose periodic points have identical sets of multipliers. We saw in Section 4.5 that symmetric polynomials in the multipliers give rational functions on the moduli space $M_d$ of rational maps modulo $\mathrm{PGL}_2$-conjugation. Flexible families of rational maps thus cannot be distinguished from one another in $M_d$ solely through the values of their multipliers.

*Example* 6.49. We saw in Example 6.41 that the Lattès function associated to the duplication map $\psi(P) = [2](P)$ on the elliptic curve $E : y^2 = x^3 + ax + b$ is given by the formula

$$\phi_{a,b}(x) = x(2P) = \frac{x^4 - 2ax^2 - 8bx + a^2}{4x^3 + 4ax + 4b}.$$

It is clear that if $a$ and $b$ vary continuously, subject to $4a^3 + 27b^2 \neq 0$, then the Lattès maps $\phi_{a,b}$ vary continuously in the space of rational maps of degree 4.

More precisely, the set of maps $\phi_{a,b}$ is a two-dimensional algebraic family of points in the space $\mathrm{Rat}_4$, given explicitly by

$$\mathbb{A}^2 \longrightarrow \mathrm{Rat}_4 \subset \mathbb{P}^9, \qquad (a,b) \longmapsto [1, 0, -2a, -8b, a^2, 0, 4, 0, 4a, 4b].$$

If we conjugate by $f_u(x) = ux$, the Lattès map $\phi_{a,b}$ transforms into

$$\phi_{a,b}^{f_u}(x) = u^{-1}\phi_{a,b}(ux) = \phi_{u^{-2}a,u^{-3}b}(x).$$

Thus assuming (say) that $ab \neq 0$, we can take $u = b/a$ to transform $\phi_{a,b}$ into

$$\phi_{a,b}^{f_{b/a}} = \phi_{c,c} \quad \text{with } c = a^3/b^2.$$

In other words, the two-dimensional family of Lattès maps $\{\phi_{a,b}\}$ in $\mathrm{Rat}_4$ becomes the one-dimensional family of dynamical systems

$$\mathbb{A}^1 \longrightarrow \mathcal{M}_4, \qquad c \longmapsto \langle \phi_{c,c} \rangle. \tag{6.25}$$

Of course, it is not clear a priori that the map (6.25) is nonconstant. But if the Lattès maps $\phi_{c,c}$ and $\phi_{c',c'}$ are $\mathrm{PGL}_2$-conjugate, then Theorem 6.46 tells us that their associated elliptic curves $E$ and $E'$ are isomorphic. The $j$-invariant of the elliptic curve $E_c : y^2 = x^3 + cx + c$ is

$$j(E_c) = 2^8 \cdot 3^3 \frac{c}{4c + 27},$$

so we see that $j(E_c) = j(E_{c'})$ if and only if $c = c'$. This proves that the map (6.25) is injective, so these flexible Lattès maps do indeed form a one-parameter family of nonconjugate rational maps with identical multiplier spectra, i.e., they are a nontrivial isospectral family.

*Example* 6.50. The elliptic curve $E : y^2 = x^3 + ax^2 + bx$ has the 2-torsion point $T = (0,0)$. To compute the Lattès function $\phi : \mathbb{P}^1 \to \mathbb{P}^1$ associated to the translated duplication map $\psi(P) = [2](P) + T$, we first use the classical duplication formula to compute

$$2P = \left( \frac{x^4 - 2bx^2 + b^2}{4y^2}, \frac{x^6 + 2ax^5 + 5bx^4 - 5b^2x^2 - 2ab^2x - b^3}{8y^3} \right).$$

Then the addition formula and some algebra yield

$$\phi(x) = x(2P + T) = \frac{4b(x^3 + ax^2 + bx)}{x^4 - 2bx^2 + b^2}.$$

As in the previous example, these Lattès maps form a one-dimensional family in $\mathcal{M}_4$.

We begin with a few elementary, but useful, properties of flexible Lattès maps.

**Proposition 6.51.** *Let $\phi : \mathbb{P}^1 \to \mathbb{P}^1$ be a flexible Lattès map whose associated map $\psi : E \to E$ has the form $\psi(P) = [m]P + T$.*

(a) *The map $\phi$ has degree $m^2$.*

(b) *The point $T$ satisfies $[2]T = \mathcal{O}$.*

(c) *Fix a Weierstrass equation* (6.11) *for $E$. Then there is a linear fractional transformation $f \in \mathrm{PGL}_2$ such that $\pi = f \circ x$. Hence $\phi^f$ fits into a commutative diagram*

$$\begin{array}{ccc} E & \xrightarrow{\ \psi\ } & E \\ \big\downarrow{\scriptstyle x} & & \big\downarrow{\scriptstyle x} \\ \mathbb{P}^1 & \xrightarrow{\ \phi^f\ } & \mathbb{P}^1. \end{array} \tag{6.26}$$

*Proof.* (a) The commutativity of the diagram (6.22) tells us that

$$\deg(\phi)\deg(\pi) = \deg(\pi)\deg(\psi).$$

The map $\psi$ has degree $m^2$, since multiplication-by-$m$ has degree $m^2$ and translation-by-$T$ has degree 1. Therefore $\deg(\phi) = m^2$.

(b) We are given that the map $\pi : E \to \mathbb{P}^1$ has degree 2 and satisfies $\pi(P) = \pi(-P)$. It follows that $\pi(P) = \pi(Q)$ if and only if $P = Q$ or $P = -Q$. We use the commutativity of (6.22) to compute

$$\begin{aligned} \pi(-[m]P - T) &= \pi([m]P + T) = \pi(\psi(P)) \\ &= \phi(\pi(P)) = \phi(\pi(-P)) = \pi(\psi(-P)) = \pi(-[m]P + T). \end{aligned}$$

Hence for every $P \in E$ we have either

$$-[m]P - T = -[m]P + T \quad \text{or} \quad -[m]P - T = -(-[m]P + T) = [m]P - T.$$

Simplifying these expressions, we find that every point $P \in E$ satisfies either

$$[2]T = \mathcal{O} \qquad \text{or} \qquad [2m]P = \mathcal{O}.$$

But there are only finitely many points $P \in E$ satisfying $[2m]P = \mathcal{O}$; hence we must have $[2]T = \mathcal{O}$.
(c) The map $\pi$ is a rational function on $E$ satisfying $\pi(-P) = \pi(P)$. It follows that $\pi$ is in the subfield $K(x)$ of the function field $K(E)$; see [410, III.2.3.1]. In other words, there is a rational function $f(z) \in K(z)$ such that $\pi = f(x)$. Equivalently, the map $\pi : E \to \mathbb{P}^1$ factors as

$$\pi : E \xrightarrow{\;x\;} \mathbb{P}^1 \xrightarrow{\;f\;} \mathbb{P}^1.$$

In particular,

$$2 = \deg(\pi) = \deg(f \circ x) = \deg(f)\deg(x) = 2\deg(f),$$

so we see that $\deg(f) = 1$. Hence $f$ is a linear fractional transformation, which proves the first part of (c). Finally, we compute

$$\phi^f \circ x = f^{-1} \circ \phi \circ f \circ x = f^{-1} \circ \phi \circ \pi = f^{-1} \circ \pi \circ \psi = x \circ \psi,$$

which proves the commutativity of (6.26). □

Our next task is to compute the periodic points and multipliers of flexible Lattès maps. For ease of exposition, we do the pure multiplication case, i.e., for maps of the form $\psi(P) = [m](P)$, and leave the general case for the reader.

**Proposition 6.52.** *Let $\phi : \mathbb{P}^1 \to \mathbb{P}^1$ be a flexible Lattès map and assume that $T = \mathcal{O}$, so $\psi(P) = [m](P)$. (See Exercise 6.18 for the case $T \neq \mathcal{O}$.)*
(a) *The set of $n$-periodic points of $\phi$ is*

$$\operatorname{Per}_n(\phi) = \pi\big(E[m^n - 1]\big) \;\cup\; \pi\big(E[m^n + 1]\big).$$

(b) *Let $\zeta$ be a periodic point of $\phi$ of exact period $n$. Then*

$$\lambda_\phi(\zeta) = \begin{cases} m^n & \text{if } \zeta \in \pi\big(E[m^n - 1]\big) \text{ and } \zeta \notin \pi\big(E[2]\big), \\ -m^n & \text{if } \zeta \in \pi\big(E[m^n + 1]\big) \text{ and } \zeta \notin \pi\big(E[2]\big), \\ m^{2n} & \text{if } \zeta \in \pi\big(E[m^n + 1]\big) \cap \pi\big(E[2]\big). \end{cases}$$

*(Notice that $\pi\big(E[2]\big)$ is the set of critical values of $\pi$.)*

*Proof.* (a) Let $\zeta \in \mathbb{P}^1$ be a fixed point of $\phi$ and choose a point $P \in E$ with $\pi(P) = \zeta$. Note that there are generally two choices for $P$, so we simply choose either one of them. Then

$$\pi(P) = \zeta = \phi(\zeta) = \phi(\pi(P)) = \pi(\psi(P)) = \pi([m]P).$$

As noted during the proof of Proposition 6.51(b), we have $\pi(P) = \pi(Q)$ if and only if $P = \pm Q$, so we conclude that either

$$[m]P = P \qquad \text{or} \qquad [m]P = -P.$$

Conversely, if $[m]P = \pm P$, then

$$\phi(\pi(P)) = \pi(\psi(P)) = \pi([m]P) = \pi(\pm P) = \pi(P),$$

so $\pi(P)$ is fixed by $\phi$. This proves that $\zeta = \pi(P) \in \mathrm{Fix}(\phi)$ if and only if

$$[m-1]P = \mathcal{O} \qquad \text{or} \qquad [m+1]P = \mathcal{O},$$

and hence

$$\mathrm{Fix}(\phi) = \pi\bigl(E[m-1]\bigr) \cup \pi\bigl(E[m+1]\bigr). \tag{6.27}$$

In order to find the points of period $n$, we observe that

$$\phi^n(\pi(P)) = \pi(\psi^n(P)) = \pi\bigl([m^n](P)\bigr),$$

so $\phi^n$ is also a flexible Lattès map. It is associated to the map $[m^n] : E \to E$. (For a generalization of this observation, see Exercise 6.17.) Applying (6.27) to the Lattès map $\phi^n$ yields the desired result,

$$\mathrm{Per}_n(\phi) = \mathrm{Fix}(\phi^n) = \pi\bigl(E[m^n-1]\bigr) \cup \pi\bigl(E[m^n+1]\bigr).$$

(b) The multipliers of $\phi$ are invariant under $\mathrm{PGL}_2$-conjugation, so we can use Proposition 6.51(c) to replace $\phi$ by a conjugate satisfying

$$\phi \circ x = x \circ \psi,$$

where $x$ is the $x$-coordinate on a Weierstrass equation

$$E : y^2 = x^3 + ax + b.$$

In order to compute the multipliers of $\phi$, we are going to use the translation-invariant differential form

$$\omega = \frac{dx}{2y} = \frac{dy}{3x^2+a} \tag{6.28}$$

on $E$ described in Theorem 6.35. The invariant differential satisfies the formula

$$\psi^*(\omega) = [m]^*(\omega) = m\omega. \tag{6.29}$$

Substituting $\omega = dx/2y$ and doing some algebra yields

$$\frac{\psi^*(dx)}{dx} = m\frac{y \circ \psi}{y}. \tag{6.30}$$

Let $\phi$ be the Lattès map associated to $\psi$, let $\zeta \in \mathrm{Fix}(\phi)$ with $\zeta \ne \infty$, and let $t$ be a coordinate function on $\mathbb{P}^1$. Then the multiplier $\lambda_\phi(\zeta)$ of $\phi$ at $\zeta$ can be computed using the differential form $dt$ via the equation

$$\left.\frac{\phi^*(dt)}{dt}\right|_{t=\zeta} = \left.\frac{d\phi(t)}{dt}\right|_{t=\zeta} = \phi'(\zeta) = \lambda_\phi(\zeta). \tag{6.31}$$

Using the relation $\phi \circ x = x \circ \psi$ and formula (6.30), we compute

$$\begin{aligned} x^*\left(\frac{\phi^*(dt)}{dt}\right) &= \frac{x^*\phi^*(dt)}{x^*(dt)} = \frac{(\phi \circ x)^*(dt)}{x^*(dt)} \\ &= \frac{(x \circ \psi)^*(dt)}{x^*(dt)} = \frac{\psi^* x^*(dt)}{x^*(dt)} = \frac{\psi^*(dx)}{dx} = m\frac{y \circ \psi}{y}. \end{aligned} \tag{6.32}$$

We lift $\zeta$ to a point $P \in E$ satisfying $x(P) = \zeta$ and evaluate both sides of (6.32) at $P$ to obtain

$$\left.\frac{\phi^*(dt)}{dt}\right|_{t=\zeta} = m\left(\frac{y \circ \psi}{y}\right)(P). \tag{6.33}$$

Equating (6.31) and (6.33) gives the useful formula

$$\lambda_\phi(\zeta) = m\left(\frac{y \circ \psi}{y}\right)(P) \qquad \text{for } \zeta = x(P) \in \operatorname{Fix}(\phi) \text{ with } \zeta \neq \infty. \tag{6.34}$$

Assume first that $[2]P \neq \mathcal{O}$, which ensures that $y(P) \neq 0$ and $y(P) \neq \infty$. Then we can directly evaluate the fraction in (6.34) and conclude that

$$\lambda_\phi(\zeta) = m\frac{(y \circ \psi)(P)}{y(P)}.$$

We are assuming that $\zeta = x(P)$ is a fixed point of $\phi$, so

$$x(P) = \phi(x(P)) = x(\psi(P)).$$

Thus $\psi(P) = \pm P$, and hence $y(\psi(P)) = \pm y(P)$, which proves that $\lambda_\phi(\zeta) = \pm m$. More precisely, $\lambda_\phi(\zeta) = m$ if $mP = P$ and $\lambda_\phi(\zeta) = -m$ if $mP = -P$. To summarize, we have proven that if $\zeta \notin x(E[2])$, then

$$\zeta \in \operatorname{Fix}(\phi) \quad \Longrightarrow \quad \lambda_\phi(\zeta) = \begin{cases} m & \text{if } \zeta \in x(E[m-1]), \\ -m & \text{if } \zeta \in x(E[m+1]). \end{cases} \tag{6.35}$$

Next suppose that $y(P) = 0$, so $[2]P = \mathcal{O}$, but $P \neq \mathcal{O}$. We also have $[m]P = \pm P$ from the assumption that $\zeta = x(P) \in \operatorname{Fix}(\phi)$, so $m$ is odd and $\psi$ fixes $P$. The functions $y$ and $y \circ \psi$ both vanish at $P$, so we can use l'Hôpital's rule to compute

$$\left(\frac{y \circ \psi}{y}\right)(P) = \left(\frac{d(y \circ \psi)}{dy}\right)(P),$$

assuming that the righthand side has a finite value. (Note that this formula is valid algebraically, since what we are really doing is looking at the linear terms in the local expansions of $y$ and $y \circ \psi$ at $P$.) We now use the chain rule to compute

$$\begin{aligned}\frac{d(\psi^*y)}{dy} &= \frac{d(\psi^*y)}{\psi^*(3x^2+a)} \cdot \frac{\psi^*(3x^2+a)}{3x^2+a} \cdot \frac{3x^2+a}{dy} \\ &= \psi^*\left(\frac{dy}{3x^2+a}\right) \cdot \frac{\psi^*(3x^2+a)}{3x^2+a} \cdot \frac{3x^2+a}{dy} \\ &= \frac{\psi^*(\omega)}{\omega} \cdot \frac{\psi^*(3x^2+a)}{3x^2+a} && \text{from (6.28),} \\ &= m\frac{\psi^*(3x^2+a)}{3x^2+a} && \text{from (6.29),} \\ &= m\frac{3(x\circ\psi)^2+a}{3x^2+a}.\end{aligned}$$

We evaluate both sides at $P = (\zeta, 0)$. Note that the quantity $3\zeta^2 + a$ is nonzero, since otherwise $P$ would be a singular point of $E$. Also, $x(\psi(P)) = x(P) = \zeta$. Hence

$$\left(\frac{y\circ\psi}{y}\right)(P) = \frac{d(\psi^*y)}{dy}(P) = m\frac{3(x\circ\psi(P))^2+a}{3x(P)^2+a} = m\frac{3\zeta^2+a}{3\zeta^2+a} = m.$$

Substituting this value into (6.34) yields the desired result $\lambda_\phi(\zeta) = m^2$.

It remains to deal with the case $P = \mathcal{O}$. There are several ways to do this case. First, we could perform an explicit calculation using local coordinates around $\mathcal{O}$. Second, since we are missing only one multiplier, we could use Theorem 1.14, although this would require knowing a priori that $\lambda \neq 1$. Third, at least for odd $m$, we could observe that $\psi(P) = [m]P$ looks the same locally around each of the points in $E[2]$, and we already computed $\lambda = m^2$ for the nonzero points in $E[2]$. We leave as an exercise for the reader (Exercise 6.20) to complete the proof using whichever argument he or she prefers.

Finally, to compute the multiplier of a periodic point $\zeta \in \mathrm{Per}_n(\phi)$, we apply the results that we have just derived to the fixed points of the Lattès map $\phi^n$ satisfying $\phi^n \circ x = x \circ [m^n]$. □

*Remark* 6.53. Let $\phi : \mathbb{P}^1 \to \mathbb{P}^1$ be a flexible Lattès map associated to $\psi(P) = [m]P$ as in Proposition 6.52. If we work over $\mathbb{C}$, then the multiplier of every periodic point $\zeta \in \mathrm{Per}(\phi)$ satisfies

$$\left|\lambda_\phi(\zeta)\right| = m^e > 1 \qquad \text{for some } e = e(\zeta) \geq 1.$$

Hence $\zeta$ is repelling, so $\mathrm{Per}(\phi)$ is contained in the Julia set (cf. Exercise 1.27). Choosing a lattice $L$ and a complex uniformization $\mathbb{C}/L \to E(\mathbb{C})$ as described in Section 1.6.3, it is clear that $E(\mathbb{C})_{\text{tors}}$ is dense in $E(\mathbb{C})$. Therefore $\mathrm{Per}(\phi) = x\bigl(E(\mathbb{C})_{\text{tors}}\bigr)$ is dense in $\mathbb{P}^1(\mathbb{C})$ and is contained in $\mathcal{J}(\phi)$. Further, the Julia set is closed. This proves that $\mathcal{J}(\phi) = \mathbb{P}^1(\mathbb{C})$ and $\mathcal{F}(\phi) = \emptyset$, which is Lattès's Theorem 1.43 discussed in Section 1.6.3.

On the other hand, if we work over a $p$-adic field such as $\mathbb{Q}_p$ or $\mathbb{C}_p$, then every periodic point $\zeta \in \mathrm{Per}(\phi)$ satisfies

$$\left|\lambda_\phi(\zeta)\right| = |m|^e \leq 1 \qquad \text{for some } e = e(\zeta) \geq 1.$$

Thus every periodic point is nonrepelling, so $\operatorname{Per}(\phi) \subset \mathcal{F}(\phi)$ from Proposition 5.20. Further, if $p|m$, then $|\lambda_\phi(\zeta)| = |m|^e < 1$, so in this case every point in $\operatorname{Per}(\phi)$ is attracting.

*Remark* 6.54. Recall that the *multiplier spectrum* of a rational map $\phi : \mathbb{P}^1 \to \mathbb{P}^1$ of degree $d$ is the map that associates to each integer $n \geq 1$ the set

$$\Lambda_n(\phi) = \{\lambda_\phi(\zeta) : \zeta \in \operatorname{Per}_n(\phi)\},$$

where we treat $\operatorname{Per}_n(\phi)$ as the set of $d^n + 1$ (not necessarily distinct) fixed points of $\phi^n$. Two maps with the same multiplier spectrum are called *isospectral*. (See Section 4.5, page 187.) Proposition 6.52(b) shows that flexible Lattès maps of degree $m^2$ are isospectral, since their multiplier spectrum depends only on $m$. A deep theorem of McMullen [294, §2] (Theorem 4.53) says that these are the only isospectral rational maps that vary in a continuous family.

Not surprisingly, good reduction of Lattès maps is closely related to good reduction of the associated elliptic curve.

**Proposition 6.55.** *Let $K$ be a local field, let $R$ be the ring of integers of $K$, and let $\phi : \mathbb{P}^1_K \to \mathbb{P}^1_K$ be a flexible Lattès map of degree $m^2$ associated to an elliptic curve $E/K$. Suppose that $E$ has good reduction and that $m \in R^*$. Then there exists an $f \in \operatorname{PGL}_2(K)$ such that $\phi^f$ has good reduction.*

*Proof.* Since $E$ has good reduction, we can find a Weierstrass equation for $E$ with coefficients in $R$ and discriminant in $R^*$. We then use Proposition 6.51 to replace $\phi$ by $\phi^f$ so that it fits into a Lattès diagram (6.26). In other words, the Lattès projection map $\pi$ is the $x$-coordinate function on a minimal Weierstrass equation for $E$.

For any $n \geq 1$ we can find polynomials $F_n(X), G_n(X) \in R[X]$ such that

$$x([n]P) = \frac{F_n(x(P))}{G_n(x(P))}.$$

This is easily proven by writing out the first few polynomials explicitly and then computing the subsequent ones by a recurrence formula. The recurrence also shows that the leading terms of $F_n$ and $G_n$ are

$$F_n(X) = X^{n^2} + \cdots \qquad \text{and} \qquad G_n(X) = n^2 X^{n^2-1} + \cdots.$$

(See [96, page 133], [410, Exercise III.3.7], or Exercise 6.23.)

We note that the roots of $G_n(X)$ are the $x$-coordinates of the $n$-torsion points, and one can check that $G_n(X)$ factors as

$$G_n(X) = n^2 \prod_{P \in E[n],\, P \neq \mathcal{O}} (X - x(P)).$$

We are assuming that $\psi(P) = [m](P) + T$ for some fixed integer $m$ and some $T \in E[2]$. For simplicity we prove here the case $T = \mathcal{O}$ and leave the general case as an exercise. Then

$$\phi(x(P)) = x(\psi(P)) = x([m](P)), \qquad \text{so} \quad \phi(X) = \frac{F_m(X)}{G_m(X)}.$$

We fix an auxiliary prime $\ell$ satisfying $\ell \nmid m$ and $\ell \in R^*$ and consider the polynomial

$$H(X) = \prod_{Q \in E[\ell],\, Q \neq O} \bigl(F_m(X) - x(Q)G_m(X)\bigr).$$

Notice that $H(X)$ is a monic polynomial, since $F_m(X)$ is monic and $\deg(F_m) > \deg(G_m)$. We also note that all of the $x(Q)$ are integral over $R$, since they are roots of $G_\ell(X)$ and $\ell \in R^*$. It follows that $H(X) \in R[X]$. Further, Proposition 2.13(b) (see also Exercise 2.6) tells us that the resultant of $H(X)$ and $G_m(X)$ is

$$\begin{aligned} \operatorname{Res}\bigl(H(X), G_m(X)\bigr) &= \pm m^{2\deg H} \prod_{G_m(\zeta)=0} H(\zeta) \\ &= \pm m^{2\deg H} \prod_{G_m(\zeta)=0} F_m(\zeta)^{\ell^2-1} \\ &= \operatorname{Res}\bigl(F_m(X), G_m(X)\bigr)^{\ell^2-1}. \end{aligned}$$

Hence in order to show that $\phi$ has good reduction, it suffices to prove that

$$\operatorname{Res}\bigl(H(X), G_m(X)\bigr) \in R^*.$$

Let $K' = K\bigl(E[m\ell]\bigr)$ be the field extension obtained by adjoining the coordinates of the points of order $m\ell$ to $K$, let $R'$ be the ring of integers of $K'$, let $\mathfrak{p}'$ be the maximal ideal in $R'$, and let $k' = R'/\mathfrak{p}'$ be the residue field of $R'$. The extension $K'/K$ is unramified, because we have assumed that $E$ has good reduction and $m\ell$ is a unit in $R$; see Corollary 6.33.

Note that $H(X)$ and $G_m(X)$ factor completely in $K'$, and in fact their roots are in $R'$. This is clear for $G_m(X)$, since its roots are the $x$-coordinates of the points in $E[m]$ and its leading coefficient is $m^2$, which is a unit in $R$. We now analyze $H(X)$ more closely.

**Claim 6.56.** *The roots of $H(X)$ are given by*

$$\{\textit{roots of } H(X)\} = x\bigl(E[m\ell] \smallsetminus E[m]\bigr) \subset R'.$$

*Proof of Claim.* The roots of $H(X)$ are the solutions to

$$\frac{F_m(X)}{G_m(X)} = x(Q) \qquad \text{for some } Q \in E[\ell].$$

Writing a root of $H(X)$ as $x(P)$ for some $P \in E$, this means that

$$x([m]P) = \frac{F_m\bigl(x(P)\bigr)}{G_m\bigl(x(P)\bigr)} = x(Q),$$

and hence $[m]P = \pm Q$. But $Q \in E[\ell]$, so $P \in E[m\ell]$. This shows that the roots of $H(X)$ are contained in $x\bigl(E[m\ell]\bigr)$. Further, if $P \in E[m]$, then $G_m\bigl(x(P)\bigr) = 0$ and $F_m\bigl(x(P)\bigr) \neq 0$, so

$$H\bigl(x(P)\bigr) = F_m\bigl(x(P)\bigr)^{\ell^2-1} \neq 0.$$

This gives the inclusion

$$\{\text{roots of } H(X)\} \subset x\bigl(E[m\ell] \smallsetminus E[m]\bigr).$$

The other inclusion is clear from the definition of $H(X)$, since

$$P \in E[m\ell] \smallsetminus E[m] \quad \Longrightarrow \quad [m]P \in E[\ell] \smallsetminus \{\mathcal{O}\}.$$

Thus $x(P)$ is a root of $F_m(X) - x\bigl([m]P\bigr)G_m(X)$, which is one of the factors in the product defining $H(X)$.

Finally, we note that $K'$ contains the $x$-coordinates of the points in $E[m\ell]$ by construction. Further, these $x$-coordinates are the roots of the polynomial $F_{m\ell}(X) \in R[X]$ whose leading coefficient is $m^2\ell^2 \in R^*$, so the roots are integral over $R$, hence are in $R'$. □

We now resume the proof of Proposition 6.55. We assume that the resultant $\operatorname{Res}\bigl(H(X), G_m(X)\bigr)$ is not a unit in $R$ and derive a contradiction. This assumption means that $H(X)$ and $G_m(X)$ have a common root modulo $\mathfrak{p}'$, so we can find $x_1, x_2 \in R'$ such that

$$H(x_1) = 0, \qquad G_m(x_2) = 0, \qquad \text{and} \qquad x_1 \equiv x_2 \pmod{\mathfrak{p}'}.$$

From our description of the roots of $H(X)$ and $G_m(X)$, this means that we can find points

$$P_1 \in E[m\ell] \smallsetminus E[m] \quad \text{and} \quad P_2 \in E[m] \smallsetminus \{\mathcal{O}\} \quad \text{satisfying} \quad P_1 \equiv P_2 \pmod{\mathfrak{p}'}.$$

(In principle, we might get $P_1 \equiv -P_2$, but if that happens, then just replace $P_2$ by $-P_2$.) Since clearly $P_1 \neq P_2$, this proves that the reduction modulo $\mathfrak{p}'$ map

$$E(K') \longrightarrow \tilde{E}(k')$$

is not injective on $E[m\ell]$. This is a contradiction, since Theorem 6.32 tells us that the prime-to-$p$ torsion injects on elliptic curves having good reduction. □

## 6.6 Rigid Lattès Maps

In general, a Lattès map $\phi : \mathbb{P}^1 \to \mathbb{P}^1$ is defined via the commutativity of a diagram

$$\begin{array}{ccc} E & \xrightarrow{\ \psi\ } & E \\ \downarrow{\scriptstyle \pi} & & \downarrow{\scriptstyle \pi} \\ \mathbb{P}^1 & \xrightarrow{\ \phi\ } & \mathbb{P}^1 \end{array} \tag{6.36}$$

where $\psi$ is a morphism of degree $d \geq 2$ and $\pi$ is a finite separable map. Every morphism of an elliptic curve to itself is the composition of an endomorphism and a translation (Remark 6.19), so $\psi$ has the form $\psi(P) = \alpha(P) + T$ for some $\alpha \in \operatorname{End}(E)$ and some $T \in E$. However, it turns out that the commutativity of (6.36) puts additional constraints on $\phi$, $\psi$, and $\pi$. More precisely, it forces the existence of a similar diagram in which $\pi$ has a special form. We state this important result and refer the reader to [300] for the analytic proof.

**Theorem 6.57.** *Let $K$ be a field of characteristic* 0 *and let $\phi$ be a Lattès map defined over $K$. Then there exists a commutative diagram of the form* (6.36) *such that the map $\pi$ has the form*

$$\pi : E \longrightarrow E/\Gamma \xrightarrow{\sim} \mathbb{P}^1$$

*for some nontrivial finite subgroup $\Gamma \subset \operatorname{Aut}(E)$.*

*Proof.* For a proof over $\mathbb{C}$, see [300, Theorem 3.1]. The general case for characteristic-0 fields follows by the Lefschetz principle, cf. [410, VI §6]. □

**Definition.** Let $\phi$ be a Lattès map. A *reduced Lattès diagram for $\phi$* is a commutative diagram of the form

$$\begin{array}{ccc} E & \xrightarrow{\psi} & E \\ \downarrow{\scriptstyle\pi} & & \downarrow{\scriptstyle\pi} \\ E/\Gamma \cong \mathbb{P}^1 & \xrightarrow{\phi} & \mathbb{P}^1 \cong E/\Gamma \end{array} \qquad (6.37)$$

Theorem 6.57 says that every Lattès map fits into a reduced Lattès diagram.

**Corollary 6.58.** *Let $\phi$ be a Lattès map given by a reduced diagram* (6.37). *Then the point $\psi(\mathcal{O})$ is fixed by every element of $\Gamma$, so in particular, $\psi(\mathcal{O}) \in E_{\text{tors}}$.*

*If further $j(E) \neq 0$ and $j(E) \neq 1728$, then*

$$\Gamma = \boldsymbol{\mu}_2, \quad \deg \pi = 2, \quad \textit{and} \quad \psi(\mathcal{O}) \in E[2].$$

*Proof.* We defer the proof that $\psi(\mathcal{O})$ is fixed by every $\xi \in \Gamma$ until Proposition 6.77(b), where we prove it in a much more general setting. (Cf. the proof for flexible Lattès maps in Proposition 6.51(b).) To see that $\psi(\mathcal{O})$ is a torsion point, let $\xi \in \Gamma$ be a nontrivial element of $\Gamma$. Then $\xi\big(\psi(\mathcal{O})\big) = \psi(\mathcal{O})$, so applying Theorem 6.22 to the isogeny $\xi - 1$, we find that

$$\big[\deg(\xi - 1)\big]\big(\psi(\mathcal{O})\big) = (\widehat{\xi - 1}) \circ (\xi - 1)\big(\psi(\mathcal{O})\big) = \mathcal{O}.$$

For the final statement of the corollary, we note that if $j(E)$ is not equal to 0 or 1728, then Proposition 6.26 tells us that $\operatorname{Aut}(E) = \boldsymbol{\mu}_2$. Hence $\Gamma = \boldsymbol{\mu}_2$ and $\deg \pi = 2$. Further, since $\psi(\mathcal{O})$ is fixed by every element of $\Gamma$, we have $[-1]\psi(\mathcal{O}) = \psi(\mathcal{O})$, so $[2]\psi(\mathcal{O}) = \mathcal{O}$. □

*Remark* 6.59. The proof of Theorem 6.57 in [300] actually shows something a bit stronger. Suppose that $\phi$ is a Lattès map fitting into the commutative diagram (6.36). It need not be true that the map $\pi : E \to \mathbb{P}^1$ is of the form $E \to E/\Gamma$, i.e., the given diagram need not be reduced, and indeed the map $\pi$ may have arbitrarily large degree. However, what is true is that there are an elliptic curve $E'$, an isogeny $E \to E'$, and a finite subgroup $\Gamma' \subset \operatorname{Aut}(E')$ such that $\pi$ factors as

$$E \longrightarrow E' \longrightarrow E'/\Gamma \cong \mathbb{P}^1.$$

Further, this factorization is essentially unique. See [300, Remark 3.3].

*Remark* 6.60. The proof of Theorem 6.57 is analytic and does not readily generalize to characteristic $p$. A full description of Lattès maps in characteristic $p$ is still lacking.

Aside from the curves having $j$-invariant 0 or 1728, every Lattès map has $\Gamma = \boldsymbol{\mu}_2 = \operatorname{Aut}(E)$ and $\deg \pi = 2$, so after a change of coordinates, the projection $\pi : E \to \mathbb{P}^1$ is $\pi(x, y) = x$. For simplicity, we will concentrate on this situation, although we note that the two special cases with $\operatorname{Aut}(E) = \boldsymbol{\mu}_4$ and $\operatorname{Aut}(E) = \boldsymbol{\mu}_6$ have attracted much attention over the years for their interesting geometric, dynamical, and arithmetic properties.

Our next task is to describe the periodic points of (rigid) Lattès maps and to compute their multipliers.

**Proposition 6.61.** *Let $\phi : \mathbb{P}^1 \to \mathbb{P}^1$ be a Lattès map and fix a reduced Lattès diagram* (6.37) *for $\phi$. We assume that $j(E) \neq 0$ and $j(E) \neq 1728$. We further assume that $\psi$ is an isogeny, i.e., with our usual notation $\psi(P) = [\alpha](P) + T$, we are assuming that $T = \mathcal{O}$. (See Exercise* 6.24 *for the other cases.)*

(a) *The set of fixed points of $\phi$ is given by*

$$\operatorname{Fix}(\phi) = \pi\big(E[\alpha + 1] \cup E[\alpha - 1]\big). \tag{6.38}$$

(b) *The intersection satisfies*

$$E[\alpha + 1] \cap E[\alpha - 1] \subset E[2].$$

*If* $\deg(\alpha - 1)$ *is odd, then the intersection is* 0.

(c) *Let $\pi(P) \in \operatorname{Fix}(\phi)$. The multiplier of $\phi$ at $\pi(P)$ is*

$$\lambda_{\pi(P)}(\phi) = \begin{cases} \alpha & \textit{if } P \in E[\alpha - 1] \textit{ and } P \notin E[\alpha + 1], \\ -\alpha & \textit{if } P \in E[\alpha + 1] \textit{ and } P \notin E[\alpha - 1], \\ \alpha^2 & \textit{if } P \in E[\alpha + 1] \cap E[\alpha - 1]. \end{cases} \tag{6.39}$$

*Proof.* (a) We have $\pi(P) \in \operatorname{Fix}(\phi)$ if and only if

$$\pi(P) = \phi(\pi(P)) = \pi(\psi(P)).$$

Our assumption on $j(E)$ means that $\Gamma = \operatorname{Aut}(E) = \boldsymbol{\mu}_2$, so $\pi(P)$ is fixed by $\phi$ if and only if $\psi(P) = \pm P$. Since we are also assuming that $\psi(P) = [\alpha](P)$, this is the desired result.

(b) Let $P \in E[\alpha-1] \cap E[\alpha+1]$. Adding $[\alpha-1](P) = \mathcal{O}$ to $[\alpha+1](P) = \mathcal{O}$ yields $[2]P = \mathcal{O}$, so $P \in E[2]$.

To ease notation, let $m = \deg(\alpha - 1)$. Then using Theorem 6.22, we find that

$$[m](P) = \widehat{[\alpha-1]} \circ [\alpha-1](P) = \widehat{[\alpha-1]}(\mathcal{O}) = \mathcal{O},$$

so $P \in E[m]$. Hence $P \in E[2] \cap E[m]$, so if $m$ is odd, then $P = \mathcal{O}$.
(c) The proof is identical to the proof of Proposition 6.52. The only difference is that $\psi = [\alpha]$ may no longer be multiplication by an integer, but we still have the key formula

$$\psi^*(\omega) = [\alpha]^*(\omega) = \alpha\omega$$

giving the effect of $\psi$ on the invariant differential $\omega$ of $E$. Using this relation in place of formula (6.29) used in proving Proposition 6.52 and tracing through the argument yields the desired result. □

We recall from Section 4.5 that $\sigma_i^{(n)}(\phi)$ denotes the $i^{\text{th}}$ symmetric polynomial of the multipliers of the points in $\operatorname{Per}_n(\phi)$, taken with appropriate multiplicities. For $d \ge 2$ and each $N \ge 1$, we write

$$\boldsymbol{\sigma}_{d,N} : \mathcal{M}_d \longrightarrow \mathbb{A}^k \tag{6.40}$$

for the map defined using all of the functions $\sigma_i^{(n)}$ with $1 \le n \le N$. McMullen's Theorem 4.53 says that for sufficiently large $N$, the map $\boldsymbol{\sigma}_{d,N}$ is finite-to-one away from the locus of the flexible Lattès maps. As noted by McMullen in his paper and stated in Theorem 4.54, rigid Lattès maps can be used to prove that $\boldsymbol{\sigma}_{d,N}$ has large degree. For the convenience of the reader, we restate the theorem before giving the proof.

**Theorem 6.62.** *Define the degree of $\boldsymbol{\sigma}_{d,N}$ to be the number of points in $\boldsymbol{\sigma}_{d,N}^{-1}(P)$ for a generic point $P$ in the image $\boldsymbol{\sigma}_{d,N}(\mathcal{M}_d)$. One can show that the degree of $\boldsymbol{\sigma}_{d,N}$ stabilizes as $N \to \infty$. We write $\deg(\boldsymbol{\sigma}_d)$ for this value. Then for every $\epsilon > 0$ there is a constant $C_\epsilon$ such that*

$$\deg(\boldsymbol{\sigma}_d) \ge C_\epsilon d^{\frac{1}{2}-\epsilon} \qquad \textit{for all } d.$$

*In particular, the multiplier spectrum of a rational function $\phi \in \operatorname{Rat}_d$ determines the conjugacy class of $\phi$ only up to $O_\epsilon(d^{\frac{1}{2}-\epsilon})$ possibilities.*

*Proof.* We prove the theorem in the case that $d$ is squarefree and leave the general case for the reader.

Let $F = \mathbb{Q}(\sqrt{-d})$, let $R_F$ be the ring of integers of $F$, and let $\mathfrak{a}_1, \ldots, \mathfrak{a}_h$ be fractional ideals of $F$ representing the distinct ideal classes of $R_F$. Consider the elliptic curves $E_1, \ldots, E_h$ whose complex points are given by

$$E_i(\mathbb{C}) = \mathbb{C}/\mathfrak{a}_i, \qquad 1 \le i \le h.$$

Each $E_i$ has $\operatorname{End}(R_i) \cong R_F$ (Proposition 6.40), and we normalize an isomorphism

$$[\cdot]_i : R_F \xrightarrow{\sim} \operatorname{End}(E_i)$$

as described in Proposition 6.36. We fix a Weierstrass equation for each $E_i$ and we define a Lattès map $\phi_i$ by

$$\phi_i \circ x = x \circ \left[\sqrt{-d}\right].$$

Then $\deg(\phi_i) = d$ from Proposition 6.39, and Proposition 6.61 tells us that the multipliers of $\phi_i$ are given by (6.39). In particular, they are the same for every $\phi_i$, i.e., the set of maps $\{\phi_1, \dots, \phi_h\}$ is isospectral, so we see that

$$\sigma_{d,N}(\phi_1) = \sigma_{d,N}(\phi_2) = \cdots = \sigma_{d,N}(\phi_h).$$

Next we observe that $\phi_1, \dots, \phi_h$ give distinct points in $\mathcal{M}_d$, because Proposition 6.40 says that $E_1, \dots, E_h$ are pairwise nonisomorphic, and then Theorem 6.46 tells us that $\phi_1, \dots, \phi_h$ are pairwise nonconjugate. This proves that $\sigma_{d,N}$ is generically at least $h$-to-1, where $h$ is the class number of the ring of integers of $\mathbb{Q}(\sqrt{-d})$. (Note that the $\phi_i$ are not flexible Lattès maps, and McMullen's Theorem 4.53 tells us that $\sigma_{d,N}$ is finite-to-one away from the flexible Lattès locus.)

To complete the proof we need an estimate for this class number. Such an estimate is given by the Brauer–Siegel theorem [258, Chapter XVI], which for quadratic imaginary fields says that

$$\lim_{\substack{d\to\infty \\ d \text{ squarefree}}} \frac{\log\left(\text{class number of } \mathbb{Q}(\sqrt{-d})\right)}{\log d} = \frac{1}{2}.$$

(Note that this is where we use the assumption that $d$ is squarefree, since it implies that the discriminant of $\mathbb{Q}(\sqrt{-d})$ is equal to either $d$ or $4d$.) In particular, the class number is larger than $d^{1/2-\epsilon}$ for all sufficiently large squarefree $d$, which completes the proof of Theorem 6.62 for squarefree $d$.

In the general case, there are two ways to proceed. The first, which is sketched in Exercise 6.25, is to find a quadratic imaginary field $F$ whose discriminant is $O(d^{1-\epsilon})$ and whose ring of integers contains an element of norm $d$. The second is to write $d = ab^2$ with $a$ squarefree and use elliptic curves whose endomorphism rings are isomorphic to the order $R_b = \mathbb{Z} + bR_F$ in the field $F = \mathbb{Q}(\sqrt{-a})$. The class number of $R_b$ is equal to $h_F b$ times a small correction factor; see [399, Exercise 4.12]. □

## 6.7 Uniform Bounds for Lattès Maps

A fundamental conjecture in arithmetic dynamics asserts that there is a constant $C = C(d, D)$ such that for all number fields $K/\mathbb{Q}$ of degree $D$ and all rational maps $\phi(z) \in K(z)$ of degree $d \geq 2$, the number of $K$-rational preperiodic points of $\phi$ satisfies

$$\#\operatorname{PrePer}(\phi, \mathbb{P}^1(K)) \leq C(d, D).$$

(See Conjecture 3.15 on page 96.) Aside from monomials and Chebyshev polynomials, the only nontrivial family of rational maps for which Conjecture 3.15 is known is the collection of Lattès maps. The proof uses the following deep theorem, whose demonstration is unfortunately far beyond the scope of this book.

**Theorem 6.63.** (Mazur–Kamienny–Merel) *For all integers $D \geq 1$ there is a constant $B(D)$ such that for all number fields $K/\mathbb{Q}$ of degree at most $D$ and all elliptic curves $E/K$ we have*

$$\#E(K)_{\text{tors}} \leq B(D).$$

*Discussion.* This deep result was first proven by Mazur [292] for $K = \mathbb{Q}$, then by Kamienny [225] for $[K : \mathbb{Q}] = 2$, and then was extended to various specific larger degrees before the proof was completed for all degrees by Merel [297]. The proof uses the theory of modular curves and Jacobians, which do have counterparts in arithmetic dynamics (cf. Sections 4.2–4.6). However, the proof also relies in a fundamental way on the fact that $E$ is a group, and hence that there exist a large number of commuting maps $E \to E$. This is in marked contrast to the situation for a general rational map $\phi : \mathbb{P}^1 \to \mathbb{P}^1$, for which only the iterates of $\phi$ commute with $\phi$. The inclusion $\mathbb{Z} \subset \operatorname{End}(E)$ leads to the existence of Hecke correspondences on elliptic modular curves, and these correspondences provide an essential tool in the proof of Theorem 6.63. Unfortunately, there does not appear to be an analogous theory of correspondences for the dynamical modular curves and varieties attached to non-Lattès maps on $\mathbb{P}^1$. □

**Corollary 6.64.** *For all integers $n \geq 1$, all number fields $K/\mathbb{Q}$, and all elliptic curves $E/K$ we have*

$$\#\Biggl(\bigcup_{[L:K]\leq n} E(L)_{\text{tors}}\Biggr) \leq B\bigl(n[K:\mathbb{Q}]\bigr)^3, \tag{6.41}$$

*where $B(D)$ is the constant appearing in Theorem* 6.63.

*Proof.* To ease notation, we let $D = [K : \mathbb{Q}]$. Every field $L$ appearing in the union in (6.41) satisfies

$$[L : \mathbb{Q}] = [L : K][K : \mathbb{Q}] \leq nD,$$

so Theorem 6.63 tells us that $\#E(L)_{\text{tors}} \leq B(nD)$. In particular, $E(L)$ contains no points of order strictly larger than $B(nD)$. This is true for every such $L$, so we conclude that

$$\bigcup_{[L:K]\leq n} E(L)_{\text{tors}} \subset \bigcup_{1\leq b\leq B(nD)} E[b].$$

Then using $\#E[b] = b^2$ yields

$$\#\Biggl(\bigcup_{[L:K]\leq n} E(L)_{\text{tors}}\Biggr) \leq \sum_{1\leq b\leq B(nD)} \#E[b] = \sum_{1\leq b\leq B(nD)} b^2 \leq B(nD)^3. \qquad \square$$

We now use Theorem 6.63 to prove uniform boundedness of preperiodic points for Lattès maps. This bound is in fact independent of the degree of the Lattès map $\phi$, which may be surprising at first glance. However, it is easily explained by the fact that Lattès maps associated to the same elliptic curve all commute with one another, so they have identical sets of preperiodic points.

**Theorem 6.65.** *Let $D \geq 1$ be an integer. There is a constant $C(D)$ such that for all number fields $K/\mathbb{Q}$ of degree $D$ and all Lattès maps $\phi : \mathbb{P}^1 \to \mathbb{P}^1$ defined over $K$ we have*

$$\#\operatorname{PrePer}(\phi, \mathbb{P}^1(K)) \leq C(D).$$

*Proof.* Without loss of generality we fix a reduced Lattès diagram (6.37) for $\phi$. Then Proposition 6.26 says that the projection map $\pi : E \to \mathbb{P}^1$ has degree at most 6, and indeed if $j(E) \neq 0$ and $j(E) \neq 1728$, then $\deg(\pi) = 2$.

Proposition 6.44 tells us that

$$\operatorname{PrePer}(\phi, \mathbb{P}^1) = \pi(E_{\text{tors}}),$$

so the fact that $\deg(\pi) \leq 6$ yields

$$\operatorname{PrePer}\bigl(\phi, \mathbb{P}^1(K)\bigr) \subset \bigcup_{[L:K]\leq 6} \pi\bigl(E(L)_{\text{tors}}\bigr). \tag{6.42}$$

Corollary 6.64 says that the set on the righthand side of (6.42) has size bounded solely in terms of $D$, hence the same is true of $\#\operatorname{PrePer}(\phi, \mathbb{P}^1(K))$. □

*Example* 6.66. The rational map

$$\phi_{a,b}(x) = \frac{x^4 - 2ax^2 - 8bx + a^2}{4x^3 + 4ax + 4b}$$

is the Lattès map associated to multiplication-by-2 on the elliptic curve

$$E_{a,b} : y^2 = x^3 + ax + b.$$

The $j$-invariant and discriminant of $E_{a,b}$ are given by the usual formulas

$$j(E_a) = 1728\frac{4a^3}{4a^3 + 27b^2} \qquad \text{and} \qquad \Delta(E_a) = -16(4a^2 + 27b^2).$$

Theorem 6.65 tells us that $\#\operatorname{PrePer}\bigl(\phi_{a,b}, \mathbb{P}^1(K)\bigr)$ is bounded solely in terms of the degree $d = [K : \mathbb{Q}]$. In general, the best known bounds are exponential in $d$, but if $j(E_{a,b})$ is an algebraic integer, then much stronger bounds can be proven as in the following result.

**Theorem 6.67.** *Let $K$ be a number field of degree $D \geq 2$, let $E/K$ be an elliptic curve whose $j$-invariant is an algebraic integer, and let $\phi$ be a Lattès map associated to $E$. Then there is an absolute constant $c$ such that*

$$\#\operatorname{PrePer}\bigl(\phi, \mathbb{P}^1(K)\bigr) \leq c(D \log D)^3.$$

*Proof.* The assumption that the elliptic curve $E$ has integral $j$-invariant means that it has everywhere potential good reduction. Replacing $K$ by an extension of bounded degree, we may assume that $E$ has everywhere good reduction. (In fact, it suffices to go to the field $K\bigl(E[3]\bigr)$, a field of degree at most 48 over $K$.) Then a result of Hindry–Silverman [204] implies a bound slightly stronger than

$$\#E(K)_{\text{tors}} \le 2^{21} D \log D.$$

Finally, we note that as in the proof of Corollary 6.64, a bound for $E(K)_{\text{tors}}$ of the form $\#E(K)_{\text{tors}} \le B\bigl([K:\mathbb{Q}]\bigr)$ for all number fields $K$ implies a bound of the form

$$\#\Biggl(\bigcup_{[L:K]\le n} E(L)_{\text{tors}}\Biggr) \le B\bigl(n[K:\mathbb{Q}]\bigr)^3.$$

Hence as in the proof of Theorem 6.65 we have

$$\operatorname{PrePer}(\phi,\mathbb{P}^1(K)) \subset \bigcup_{[L:K]\le 6} \pi\bigl(E(L)_{\text{tors}}\bigr) \le c(D\log D)^3$$

for an absolute constant $c$. □

Theorem 6.65 proves uniformity for rational preperiodic points of Lattès maps. In the other direction, recall that we proved (Theorem 3.43) that the orbits of rational wandering points contain only finitely many integers except in a few precisely specified situations. In particular, Lattès orbits contain only finitely many integer points, since Lattès maps are not polynomial maps. Using deep results from the theory of elliptic curves, it is possible to obtain strong uniformity estimates for the number of integer points lying in Lattès orbits.

For simplicity we state results over $\mathbb{Q}$ and $\mathbb{Z}$, but we note that an appropriately formulated version is true for rings of $S$-integers in number fields.

**Definition.** Let $E/\mathbb{Q}$ be an elliptic curve. Recall that a global minimal Weierstrass equation for $E$ is a Weierstrass equation that is simultaneously minimal at all primes (Remark 6.30). We define a *quasiminimal Weierstrass equation* for $E/\mathbb{Q}$ to be an equation of the form

$$E : y^2 = x^3 + ax + b, \qquad a,b\in\mathbb{Z}, \tag{6.43}$$

such that $|4a^3+27b^2|$ is as small as possible. Equivalently, the equation (6.43) is quasiminimal if there are no primes $p$ such that $p^4|a$ and $p^6|b$.

*Remark* 6.68. Given an arbitrary Weierstrass equation

$$E : y^2 = x^3 + ax + b, \qquad a,b\in\mathbb{Z},$$

it is easy to create a quasiminimal equation. Simply let $u$ be the largest integer such that $u^{12}$ divides $\gcd(a^3,b^2)$, and then

$$E : y^2 = x^3 + au^{-4}x + bu^{-6}$$

is a quasiminimal equation for $E/\mathbb{Q}$. Elliptic curves over $\mathbb{Q}$ have global minimal Weierstrass equations, see Remark 6.30, and it is not hard to show that a quasiminimal equation is minimal at every prime $p \geq 5$, and that it is almost minimal at 2 and 3; see Exercise 6.26.

The following theorem is a conditional resolution of a conjecture of Lang [254, page 140].

**Theorem 6.69.** (Hindry–Silverman) *Let $E/\mathbb{Q}$ be an elliptic curve given by a quasiminimal Weierstrass equation and let $E(\mathbb{Z})$ be the set of points in $E(\mathbb{Q})$ having integer coordinates. Also let $\nu(E)$ be the number of primes dividing the denominator of the $j$-invariant of $E$.*

(a) *There is an absolute constant $C$ such that for any subgroup $\Gamma \subset E(\mathbb{Q})$,*

$$\#\big(\Gamma \cap E(\mathbb{Z})\big) \leq C^{\nu(E)+\operatorname{rank}\Gamma}.$$

(b) *If the "ABC conjecture" is true,[2] then there is an absolute constant $C$ such that for any subgroup $\Gamma \subset E(\mathbb{Q})$,*

$$\#\big(\Gamma \cap E(\mathbb{Z})\big) \leq C^{\operatorname{rank}\Gamma}.$$

*Proof.* The proof of (a) is given in [407] and the proof of (b) is in [202]. □

We can use Theorem 6.69 to prove a uniform bound for integer points in orbits of flexible Lattès maps (cf. Conjecture 3.47).

**Theorem 6.70.** *Let $E/\mathbb{Q}$ be an elliptic curve given by a Weierstrass equation with integer coefficients, let $m \geq 2$ be an integer, and let $\phi(z) \in \mathbb{Q}(z)$ be the Lattès map satisfying*

$$\phi\big(x(P)\big) = x\big([m]P\big) \qquad \text{for all } P \in E.$$

*Assume further that $\phi$ is* affine minimal *in the sense that*

$$\operatorname{Res}(\phi) = \min_{\substack{f \in \mathrm{PGL}_2(\mathbb{Q}) \\ f(z)=az+b}} \operatorname{Res}(\phi^f). \tag{6.44}$$

*(See page 112 for the definition of the resultant $\operatorname{Res}(\phi)$ of a rational map.) Thus the assumption (6.44) says that we cannot reduce the resultant of $\phi$ by conjugation by an affine linear transformation $f(z) = az + b$. Let $\zeta \in \mathbb{Q}$ and consider the orbit $\mathcal{O}_\phi(\zeta)$ of $\zeta$ by $\phi$.*

(a) *There is an absolute constant $C$ such that*

$$\#\big(\mathcal{O}_\phi(\zeta) \cap \mathbb{Z}\big) \leq C^{\nu(E)},$$

*where $\nu(E)$ is the number of primes dividing the denominator of the $j$-invariant of $E$.*

[2] The *ABC conjecture* of Masser and Oesterlé says that if $A, B, C > 0$ are pairwise relatively prime integers satisfying $A + B = C$, then $C \ll_\epsilon \prod_{p \mid ABC} p^{1+\epsilon}$.

(b) *If the ABC conjecture is true, then the number of integer points in* $\mathcal{O}_\phi(\zeta)$ *is bounded by an absolute constant independent of* $E$ *and* $\zeta$.

*Proof.* Write the given Weierstrass equation for $E$ as

$$E : y^2 = x^3 + ax + b \qquad \text{with } a, b \in \mathbb{Z}.$$

We begin by showing that the minimality assumption (6.44) implies that there are no primes $p$ with $p^2|a$ and $p^3|b$. The rational function $\phi(x) = F(x)/G(x)$ is associated to the multiplication-by-$m$ map, so it is given by polynomials

$$F(a, b; x), G(a, b; x) \in \mathbb{Z}[a, b, x]$$

that are weighted homogeneous in the sense that

$$F(t^2a, t^3b; tx) = t^{m^2} F(a, b; x) \quad \text{and} \quad G(t^2a, t^3b; tx) = t^{m^2-1} G(a, b; x).$$

For example, if $m = 2$, then

$$\phi(x) = \frac{x^4 - 2ax^2 - 8bx + a^2}{4x^3 + 4ax + 4b}.$$

Hence if $p^2|a$ and $p^3|b$, then conjugating $\phi(x)$ by $f(x) = px$ yields

$$\phi^f(x) = p^{-1}\frac{F(a, b; px)}{G(a, b; px)} = \frac{F(p^{-2}a, p^{-3}b; x)}{G(p^{-2}a, p^{-3}b; x)}.$$

The assumption that $p^2|a$ and $p^3|b$ implies that these polynomials have integer coefficients, and then homogeneity yields

$$\begin{aligned}
\operatorname{Res}(\phi^f) &= \operatorname{Res}\bigl(F(p^{-2}a, p^{-3}b; x), G(p^{-2}a, p^{-3}b; x)\bigr) \\
&= p^{-m^2(m^2-1)} \operatorname{Res}\bigl(F(a, b; x), G(a, b; x)\bigr) \\
&= p^{-m^2(m^2-1)} \operatorname{Res}(\phi).
\end{aligned}$$

This contradicts (6.44), so we have proven that there are no primes $p$ satisfying $p^2|a$ and $p^3|b$.

We would like to apply Theorem 6.69 to the rank-1 subgroup generated by a point $P = (\zeta, \eta)$ of $E$ lying above $\zeta$. Unfortunately, although $\zeta \in \mathbb{Q}$, there is no reason that $\eta$ need be rational. So it is necessary to move to a twist of $E$.

We are given a point $\zeta \in \mathbb{Q}$ and we choose a point $P = (\zeta, \eta) \in E$ lying above $\zeta$. We do not assume that $\eta = \sqrt{\zeta^3 + a\zeta + b}$ is rational. We write

$$P = P_1 = (x_1, y_1),$$

then we factor

$$x_1^3 + ax_1 + b = u_1 v_1^2 \qquad \text{with } u_1 \text{ squarefree,}$$

and consider the elliptic curve

$$E' : y^2 = x^3 + au_1^2 x + bu_1^3. \tag{6.45}$$

(In the terminology of Section 4.7, $E'$ is a twist of $E$; cf. Example 4.71.) Notice that the point

$$P_1' = (u_1 x_1, u_1^2 v_1) \in E'(\mathbb{Q})$$

is a point of $E'$ having integer coordinates.

We claim that the Weierstrass equation (6.45) for $E'$ is quasiminimal. To prove this claim, let $p$ be any prime. We showed earlier that either

$$\operatorname{ord}_p(a) < 2 \qquad \text{or} \qquad \operatorname{ord}_p(b) < 3.$$

Since $u_1$ is squarefree by construction, it follows that either

$$\operatorname{ord}_p(au_1^2) < 4 \qquad \text{or} \qquad \operatorname{ord}_p(bu_1^3) < 6,$$

which shows that the Weierstrass equation (6.45) is quasiminimal.

This means that we can apply Theorem 6.69 to $E'$ and the rank-1 subgroup generated by $P'$ to conclude that

$$\#\{n \ge 1 : [n]P' \in E'(\mathbb{Z})\} \le C^{\nu(E')+1}. \tag{6.46}$$

Further, if the ABC conjecture is true, then the upper bound may be replaced by $C$.

The two elliptic curves $E$ and $E'$ are isomorphic, although the isomorphism is defined only over $\mathbb{Q}(\sqrt{u_1})$. This isomorphism, which we denote by $F$, is given explicitly by

$$\begin{array}{ccc} E : y^2 = x^3 + ax + b & \xrightarrow{F} & E' : y'^2 = x'^3 + au_1^2 x' + bu_1^3 \\ (x, y) & \longmapsto & (u_1 x, u_1^{3/2} y). \end{array}$$

In particular, $j(E) = j(E')$, so $\nu(E) = \nu(E')$.

In order to relate integers in $\mathcal{O}_\phi(\zeta)$ to integer points in $E'(\mathbb{Q})$, we write

$$P_n = [n]P_1 = (x_n, y_n) \qquad \text{and} \qquad P_n' = [n]P_1' = (x_n', y_n').$$

Since the isomorphism $F$ respects multiplication by $n$, we have

$$F(P_n) = F([n]P_1) = [n](F(P_1)) = [n]P_1' = P_n'.$$

In particular, since $P_1' \in E'(\mathbb{Q})$, it follows that $[n]P_1' = P_n' \in E'(\mathbb{Q})$, so $F$ maps the multiples of $P_1$, which, note, are *not* in $E(\mathbb{Q})$, to points in $E'(\mathbb{Q})$. Further, if $x_n \in \mathbb{Z}$, then it is clear from the definition of $F$ that $x_n' = u_1 x_n \in \mathbb{Z}$, and hence $y_n'$ is also in $\mathbb{Z}$, since we just showed that $y_n'$ is in $\mathbb{Q}$ and the equation of $E'$ shows that $y_n'$ is the square root of an integer.

To summarize, we have proven that

$$x_n \in \mathbb{Z} \implies P_n' = F(P_n) \in E'(\mathbb{Z}). \tag{6.47}$$

By construction we have $\phi^k(\zeta) = x\bigl([m^k]P\bigr) = x_{m^k}$, and hence

$$\begin{aligned}
\#\bigl(\mathcal{O}_\phi(P)\cap\mathbb{Z}\bigr) &= \#\bigl\{k\ge 0 : x_{m^k}\in\mathbb{Z}\bigr\} \\
&\le \#\bigl\{k\ge 0 : P'_{m^k}\in E'(\mathbb{Z})\bigr\} && \text{from (6.47),} \\
&\le \#\bigl\{n\ge 0 : P'_n\in E'(\mathbb{Z})\bigr\} \\
&\le C^{\nu(E')+1} && \text{from (6.46),} \\
&= C^{\nu(E)+1} && \text{since } E \text{ and } E' \text{ are isomorphic.}
\end{aligned}$$

Further, if the ABC conjecture is true, then we may replace the upper bound by $C$. □

*Remark* 6.71. We note that something like the affine minimality of $\phi$ is necessary in the statement of Theorem 6.70. Indeed, without some kind of minimality condition, we saw in Proposition 3.46 that we can make $\#\bigl(\mathcal{O}_\phi(\zeta)\cap\mathbb{Z}\bigr)$ arbitrarily large by replacing $\phi(z)$ with $B\phi(B^{-1}z)$. This conjugation has the effect of multiplying every point in the orbit by $B$, hence allows us to clear an arbitrary number of denominators.

*Remark* 6.72. Continuing with notation from the statement of Theorem 6.70, we note that there is a cutoff value $k_0$ such that

$$\phi^k(\zeta)\in\mathbb{Z}\quad\text{for}\quad 0\le k\le k_0\qquad\text{and}\qquad \phi^k(\zeta)\notin\mathbb{Z}\quad\text{for}\quad k>k_0.$$

This reflects the more general fact that if $x\bigl([n]P\bigr)\in\mathbb{Z}$ and if $r|n$, then $x\bigl([r]P\bigr)\in\mathbb{Z}$ ([410, Exercise 9.12]). Note that no such cutoff statement holds for general rational maps that are not non-Lattès maps.

## 6.8 Affine Morphisms, Algebraic Groups, and Commuting Families of Rational Maps

Power maps and Chebyshev maps are attached to endomorphisms of the multiplicative group $\mathbb{G}_m$ and its quotient $\mathbb{G}_m/\{z = z^{-1}\}$, and similarly Lattès maps are attached to maps of quotients of elliptic curves. In this section we put these constructions into a general context and state a classical theorem on commutativity of one-variable rational maps.

**Definition.** Let $G$ be a commutative algebraic group. An *affine morphism* of $G$ is the composition of a finite endomorphism of degree at least 2 and a translation.

*Remark* 6.73. The reason for this terminology is as follows. Let $G/\mathbb{C}$ be a connected commutative algebraic group of dimension $g$. Then its universal cover is $\mathbb{C}^g$ and every affine morphism $\psi : G\to G$ lifts to an affine map $\mathbb{C}^g\to\mathbb{C}^g$, i.e., there are a matrix $A$ and vector $\mathbf{a}$ such that the following diagram commutes:

$$\begin{array}{ccc}
\mathbb{C}^g & \xrightarrow{\ \mathbf{z}\mapsto A\mathbf{z}+\mathbf{a}\ } & \mathbb{C}^g \\
\downarrow & & \downarrow \\
G & \xrightarrow{\ \psi\ } & G
\end{array}$$

*Example* 6.74. Every affine morphism of the multiplicative group $\mathbb{G}_m$ has the form $\psi(z) = az^d$ for some nonzero $a$ and some $d \in \mathbb{Z}$. More generally, for any commutative group $G$, any $a \in G$, and any $d \in \mathbb{Z}$ there is an affine morphism $\psi(z) = az^d$. Notice that it is easy to compute the iterates of this map,

$$\psi^n(z) = a^{1+d+\cdots+d^{n-1}} z^{d^n}.$$

**Proposition 6.75.** *Let $\psi : G \to G$ be an affine morphism of an algebraic group $G$, so $\psi$ has the form $\psi(z) = a \cdot \alpha(z)$ for some $\alpha \in \mathrm{End}(G)$ and some $a \in G$.*

(a) *The endomorphism $\alpha$ and translation $a$ are uniquely determined by $\psi$.*

(b) *Let $a$ and $\alpha$ be as in* (a)*. Then the iterates of $\psi$ have the form*

$$\psi^n(z) = a \cdot \alpha(a) \cdot \alpha^2(a) \cdots \alpha^{n-1}(a) \cdot \alpha^n(z).$$

*Proof.* The definition of affine morphism tells us that there are an element $a \in G$ and an endomorphism $\alpha$ of $G$ such that the map $\psi$ has the form $\psi(z) = a\alpha(z)$. Evaluating at the identity element $e \in G$ yields $\psi(e) = a\alpha(e) = a$, so $a$ is uniquely determined by $\psi$. Then $\alpha(z) = a^{-1}\psi(z)$ is also uniquely determined by $\psi$. This proves (a). The proof of (b) is an easy induction, using the commutativity of $G$ and the fact that $\alpha$ is a homomorphism. □

**Definition.** A self-morphism of an algebraic variety $\phi : V \to V$ is *dynamically affine* if it is a finite quotient of an affine morphism. What we mean by this is that there are a connected commutative algebraic group $G$, an affine morphism $\psi : G \to G$, a finite subgroup $\Gamma \subset \mathrm{Aut}(G)$, and a morphism $G/\Gamma \to V$ that identifies $G/\Gamma$ with a Zariski dense open subset of $V$ (possibly all of $V$) such that the following diagram is commutative:

$$\begin{array}{ccc} G & \xrightarrow{\ \psi\ } & G \\ \downarrow & & \downarrow \\ G/\Gamma & \longrightarrow & G/\Gamma \\ \downarrow \wr & & \downarrow \wr \\ V & \xrightarrow{\ \phi\ } & V \end{array} \qquad (6.48)$$

*Example* 6.76. Examples of dynamically affine rational maps $\phi : \mathbb{P}^1 \to \mathbb{P}^1$ include the power maps $\phi(z) = z^n$ with $G = \mathbb{G}_m$ and $\Gamma = \{1\}$, the Chebyshev polynomials $T_n(z)$ with $G = \mathbb{G}_m$ and $\Gamma = \{z, z^{-1}\}$, and Lattès maps with $G$ an elliptic curve $E$ and $\Gamma$ a nontrivial subgroup of $\mathrm{Aut}(E)$.

**Proposition 6.77.** *Let $\phi : V \to V$ be a dynamically affine map and let $\psi : G \to G$ and $\Gamma \subset \mathrm{Aut}(G)$ be the associated quantities fitting into the commutative diagram* (6.48).

(a) *For every $\xi \in \Gamma$ there exists a unique $\xi' \in \Gamma$ with the property that $\psi \circ \xi = \xi' \circ \psi$.*

(b) *Write $\psi(z) = a \cdot \alpha(z)$ with $a \in G$ and $\alpha \in \mathrm{End}(G)$ as in* Proposition 6.75. *Then $\xi(a) = a$ for every $\xi \in \Gamma$.*

(c) *Assume that* $\#\Gamma \ge 2$ *and that* $G$ *is simple. (An algebraic group is* simple *if its only connected algebraic subgroups are* $\{1\}$ *and* $G$.) *Then* $a \in G_{\text{tors}}$, *i.e., the translation used to define* $\psi$ *is translation by a point of finite order.*

*Proof.* (a) The uniqueness is clear, since if $\psi \circ \xi = \xi_1 \circ \psi = \xi_2 \circ \psi$, then $\xi_1 = \xi_2$ because the finite map $\psi : G \to G$ is surjective.

We now prove the existence. The commutativity of (6.48) tells us that for all $z \in G$ and all $\xi \in \Gamma$,

$$(\pi \circ \psi \circ \xi)(z) = (\phi \circ \pi \circ \xi)(z) = (\phi \circ \pi)(z) = (\pi \circ \psi)(z).$$

Thus $(\psi \circ \xi)(z)$ and $\psi(z)$ have the same image for the projection map $\pi : G \to G/\Gamma$, so there is an automorphism $\xi' \in \Gamma$ satisfying

$$\psi\bigl(\xi(z)\bigr) = \xi'\bigl(\psi(z)\bigr).$$

We claim that the automorphism $\xi'$, which a priori might depend on both $\xi$ and $z$, is in fact independent of $z$. To see this we fix $\xi$ and write $\psi\bigl(\xi(z)\bigr) = \xi'_z\bigl(\psi(z)\bigr)$ to indicate the possible dependence of $\xi'$ on $z$. In this way we obtain a map (of sets)

$$G \longrightarrow \Gamma, \qquad z \longmapsto \xi'_z.$$

Since $\Gamma$ is finite, there exists some $\xi'' \in \Gamma$ such that $\xi'_z = \xi''$ for a Zariski dense subset of $z \in G$. (Note that a variety cannot be a finite union of Zariski closed proper subsets.) It follows that $\psi \circ \xi$ is equal to $\xi'' \circ \psi$ on a Zariski dense subset of $G$, and hence they are equal on all of $G$.

(b) From (a) we see that there is a permutation of $\Gamma$ defined by the rule

$$\tau : \Gamma \longrightarrow \Gamma, \qquad \psi \circ \xi = \tau(\xi) \circ \psi.$$

Evaluating both sides of $\psi \circ \xi = \tau(\xi) \circ \psi$ at the identity element $1 \in G$ and using the fact that $\xi(1) = 1$ and $\psi(1) = a \cdot \alpha(1) = a$, we find that

$$a = \psi\bigl(\xi(1)\bigr) = \tau(\xi)\bigl(\psi(1)\bigr) = \tau(\xi)(a).$$

But $\tau$ is a permutation of $\Gamma$, so as $\xi$ runs over $\Gamma$, so does $\tau(\xi)$. Hence $a$ is fixed by every element of $\Gamma$.

(c) From (b) and the assumption that $\#\Gamma \ge 2$, there exists a nontrivial $\xi \in \Gamma$ with $\xi(a) = a$. It follows that $a$ is in the kernel of the endomorphism

$$G \longrightarrow G, \qquad z \longmapsto \xi(z) \cdot z^{-1}.$$

The kernel is not all of $G$, since $\xi$ is not the identity map, so the simplicity of $G$ tells us that the kernel is a finite subgroup of $G$. Hence $a$ has finite order. □

*Remark* 6.78. In this book we are primarily interested in dynamically affine maps of $\mathbb{P}^1$, but higher-dimensional analogues, especially of Lattès maps, have also been studied. See for example [68, 134, 145, 439].

The commutativity of (6.48) implies that $\deg(\phi) = \deg(\psi)$. It follows that all dynamically affine maps for the additive group $\mathbb{G}_a$ have degree 1, since every affine morphism of $\mathbb{G}_a$ has the form $\psi(z) = az + b$. Hence nonlinear dynamically affine maps on $\mathbb{P}^1$ are attached to either the multiplicative group $\mathbb{G}_m$ or to an elliptic curve, since these are the only other algebraic groups of dimension 1.

We note that over a field of characteristic 0, the endomorphism ring $\operatorname{End}(G)$ of a one-dimensional algebraic group $G$ is commutative.[3] More precisely, the multiplicative group has endomorphism ring $\operatorname{End}(\mathbb{G}_m) = \mathbb{Z}$, and the endomorphism ring $\operatorname{End}(E)$ of an elliptic curve $E$ is either $\mathbb{Z}$ or an order in a quadratic imaginary field. The commutativity of $\operatorname{End}(G)$ means that dynamically affine maps commute with many other maps. An appropriately formulated converse of this statement is a classical theorem of Ritt.

**Theorem 6.79.** (Ritt and Erëmenko) *Let $\phi, \psi \in \mathbb{C}(z)$ be rational maps of degree at least 2 with the property that $\phi \circ \psi = \psi \circ \phi$. Then one of the following two conditions is true:*

(a) *There are integers $m, n \geq 1$ such that $\phi^n = \psi^m$.*

(b) *Both $\phi$ and $\psi$ are dynamically affine maps, hence they are either power maps, Chebyshev polynomials, or Lattès maps.*

*In all cases, the commuting maps $\phi$ and $\psi$ satisfy*

$$\mathcal{F}(\phi) = \mathcal{F}(\psi), \qquad \mathcal{J}(\phi) = \mathcal{J}(\psi), \qquad \textit{and} \qquad \operatorname{PrePer}(\phi) = \operatorname{PrePer}(\psi).$$

*Proof.* The first part of the theorem, in somewhat different language, is due to Ritt [371]. See Erëmenko's paper [152] for a proof of both parts of the theorem and some additional geometric dynamical properties shared by commuting $\phi$ and $\psi$. A higher-dimensional analogue is discussed in [135]. We remark that the equality $\operatorname{PrePer}(\phi) = \operatorname{PrePer}(\psi)$ is a formal consequence of the commutativity of $\phi$ and $\psi$ and the fact that the preperiodic points of a nonlinear rational map are isolated; see Exercise 1.15. □

Although we do not give a proof of Ritt's theorem, we conclude this section by proving the easier statement that only polynomial maps can commute with polynomial maps. This result was used in our description of the rational maps commuting with the Chebyshev polynomials (Theorem 6.9).

**Theorem 6.80.** *Let $K$ be a field, let $\phi(z) \in K[z]$ be a polynomial of degree $d \geq 2$, and let $\psi(z) \in K(z)$ be a nonconstant rational map. We assume that both $\phi$ and $\psi$ are separable, i.e., neither of the derivatives $\phi'(z)$ and $\psi'(z)$ is identically 0. Suppose further that $\phi$ and $\psi$ commute under composition, $\phi \circ \psi = \psi \circ \phi$. Then one of the following is true:*

(a) *$\psi(z) \in K[z]$, i.e., $\psi$ is also a polynomial.*

[3] Even in characteristic $p$, most elliptic curves have commutative endomorphism ring. However, there are a finite number of elliptic curves whose endomorphism ring is a maximal order in a quaternion algebra. These *supersingular* curves are all defined over $\mathbb{F}_{p^2}$. See [410, V §3].

(b) *After simultaneous conjugation by an affine map* $f(z) = z + \beta$, *the polynomial* $\phi(z)$ *has the form* $\phi(z) = az^d$ *and the rational map* $\psi(z)$ *has the form* $\psi(z) = bz^r$ *for some* $r < 0$.

*Proof.* The proof is an application of ramification theory and the Riemann–Hurwitz formula (Theorem 1.1). By assumption, the map $\phi$ is a polynomial, so $\infty$ is a totally ramified fixed point of $\phi$. Suppose that $\psi(z)$ is not a polynomial. This means that we can find a point $\alpha \in \psi^{-1}(\infty)$ with $\alpha \neq \infty$. We use the commutativity of $\phi$ and $\psi$ to compute

$$\begin{array}{l} e_\alpha(\phi^n \circ \psi) = e_\alpha(\psi) \displaystyle\prod_{i=0}^{n-1} e_{\phi^i\psi(\alpha)}(\phi) = e_\alpha(\psi) e_\infty(\phi)^n = e_\alpha(\psi) d^n \\ \quad \| \\ e_\alpha(\psi \circ \phi^n) = \left( \displaystyle\prod_{i=0}^{n-1} e_{\phi^i(\alpha)}(\phi) \right) e_{\phi^n(\alpha)}(\psi). \end{array} \tag{6.49}$$

Hence

$$\prod_{i=0}^{n-1} \frac{e_{\phi^i(\alpha)}(\phi)}{d} = \frac{e_\alpha(\psi)}{e_{\phi^n(\alpha)}(\psi)} \geq \frac{1}{\deg \psi} \qquad \text{for all } n \geq 1. \tag{6.50}$$

Every ramification index $e_{\phi^i(\alpha)}(\phi)$ is an integer between 1 and $d$, so letting $n \to \infty$, we see that $e_{\phi^i(\alpha)}(\phi) = d$ for all sufficiently large $i$. On the other hand, $\phi$ is a polynomial and $\alpha \neq \infty$, so $\phi^i(\alpha) \neq \infty$ for all $i$. Hence there is at least one point $\beta \neq \infty$ with $e_\beta(\phi) = d$. The Riemann–Hurwitz formula then implies that $\beta$ and $\infty$ are the only two points at which $\phi$ is ramified. It follows that $\phi^i(\alpha) = \beta$ for all sufficiently large $i$, which implies that $\phi(\beta) = \beta$. In other words, $\beta$ and $\infty$ are both totally ramified fixed points of $\phi$, i.e.,

$$\phi^{-1}(\beta) = \{\beta\} \qquad \text{and} \qquad \phi^{-1}(\infty) = \{\infty\}, \tag{6.51}$$

and $\phi$ has no other ramification points. In particular, since by construction we have $\phi^i(\alpha) = \beta$ for some $i$, it follows that $\alpha = \beta$. But $\alpha \neq \infty$ was an arbitrary point in $\psi^{-1}(\infty)$, so we have also proven that $\psi^{-1}(\infty) \subset \{\beta, \infty\}$.

Next let $\gamma \in \psi^{-1}(\beta)$. We use the fact that $\phi^i\psi(\gamma) = \beta$ and $e_\beta(\phi) = d$ to repeat the calculation (6.49) with $\alpha$ replaced by $\gamma$. This again leads to the inequality (6.50), but with $\gamma$ in place of $\alpha$, and hence to the conclusion that $\phi$ is totally ramified at some iterate $\phi^i(\gamma)$. It follows that $\phi^i(\gamma) \in \{\beta, \infty\}$ for some $i$, and hence from (6.51) that $\gamma \in \{\beta, \infty\}$.

We have now proven that

$$\psi^{-1}(\{\beta, \infty\}) \subset \{\beta, \infty\}.$$

Thus $\psi$ is totally ramified at $\beta$ and $\infty$, and since $\psi(\beta) \neq \infty$ by assumption, the map $\psi$ must switch $\beta$ and $\infty$. Since we also know that $\beta$ and $\infty$ are totally ramified fixed points of $\phi$, it follows that $\phi$ and $\psi$ have the form

$$\phi(z) = \beta + a(z - \beta)^d \qquad \text{and} \qquad \psi(z) = \beta + b(z - \beta)^r \quad \text{for some } r < 0.$$

Then conjugation by $f(z) = z + \beta$ puts them into the desired form. □

# Exercises

## Section 6.1. Power Maps and the Multiplicative Group

**6.1.** Let $K$ be a field of positive characteristic $p$.
(a) Let $M_p(z) = z^p$. Prove that the automorphism group of $M_p$ over $K$ equals $\mathrm{PGL}_2(\mathbb{F}_p)$.
(b) More generally, if $q$ is a power of $p$, prove that $\mathrm{Aut}(M_q) = \mathrm{PGL}_2(\mathbb{F}_q)$.
(c) Again let $q$ be a power of $p$, and let $d$ be an integer with $p \nmid d$. Describe $\mathrm{Aut}(M_{qd})$.

**6.2.** Let $K$ be an algebraically closed field, let $d \in \mathbb{Z}$, and let $a \in K^*$. Further, if $K$ has positive characteristic $p$, assume that $p \nmid d$. Describe all rational functions $\phi(z) \in K(z)$ that commute with $az^d$ under composition.

**6.3.** ** Let $M_d(z) = z^d$ be a power map for some $|d| \geq 2$, and if $K$ has positive characteristic $p$, assume that $p \nmid d$. Example 6.5 describes two types of twists of $M_d(z)$. The first type has the form $\phi_a(z) = az^d$ and the second type $\psi_b(z)$ is given by the complicated formula (6.1). Does $M_d(z)$ have any other twists? If so, describe all of the twists of $M_d(z)$.

## Section 6.2. Chebyshev Polynomials

**6.4.** Complete the proof of Proposition 6.6(e) by verifying the identity

$$\frac{1}{d+1-k}\left[(d+1)\binom{d+1-k}{k} + d\binom{d+1-k}{k-1}\right] = \frac{d+2}{d+2-k}\binom{d+2-k}{k}.$$

**6.5.** Let $T_d(w)$ be the $d^{\text{th}}$ Chebyshev polynomial for some $d \geq 2$.
(a) Prove that the fixed points of $T_d(w)$ are as described in Proposition 6.8(a).
(b) Prove that the multipliers of $T_d(w)$ at its fixed points are as described in Proposition 6.8(b).
(*Hint*. For (a) use the trigonometric identity

$$\cos(A) - \cos(B) = \sin\left(\frac{B+A}{2}\right)\sin\left(\frac{B-A}{2}\right),$$

and for (b) differentiate $T_d(z + z^{-1}) = z^d + z^{-d}$ to obtain the identity

$$T_d'(z + z^{-1}) = d\frac{z^d - z^{-d}}{z - z^{-1}}.)$$

**6.6.** Proposition 6.8 describes the multipliers of the fixed points of the Chebyshev polynomial $T_d(w)$. Prove directly that the multipliers satisfy the summation formula described in Theorem 1.14,

$$\sum_{\zeta \in \mathrm{Fix}(T_d)} \frac{1}{1 - \lambda_{T_d}(\zeta)} = 1.$$

**6.7.** Let $K$ be a field of characteristic $p \geq 3$, let $n \geq 1$ be an integer with $p \nmid n$, and let $\boldsymbol{\mu}_n \subset \bar{K}^*$ be the $n^{\text{th}}$ roots of unity. There is no "cosine function" for the field $K$, but we can define a set of cosine values by

$$\mathbf{Cos}_n = \left\{\frac{\alpha + \alpha^{-1}}{2} : \alpha \in \boldsymbol{\mu}_n\right\}.$$

We also let $2\,\mathbf{Cos}_n = \{2\zeta : \zeta \in \mathbf{Cos}_n\}$.

Let $d \geq 2$ be an integer with $p \nmid d(d^2 - 1)$ and let $T_d(w)$ be the $d^{\text{th}}$ Chebyshev polynomial.

(a) Prove that

$$\operatorname{Fix}(T_d) = 2\,\mathbf{Cos}_{d+1} \cup 2\,\mathbf{Cos}_{d-1}.$$

Also compute the intersection $2\,\mathbf{Cos}_{d+1} \cap 2\,\mathbf{Cos}_{d-1}$.

(b) Prove that the multipliers of $T_d$ at its fixed points are given by

$$\lambda_{T_d}(\zeta) = \begin{cases} -d & \text{if } \zeta \in 2\,\mathbf{Cos}_{d+1} \text{ and } \zeta \ne \pm 2, \\ d & \text{if } \zeta \in 2\,\mathbf{Cos}_{d-1} \text{ and } \zeta \ne \pm 2, \\ d^2 & \text{if } \zeta = \pm 2. \end{cases}$$

(c) Give a similar description of the fixed points and their multipliers in the case that $d \equiv \pm 1 \pmod{p}$.

**6.8.** We stated during the proof of Theorem 6.9 that for $d \ge 2$, the Chebyshev polynomial $T_d(w)$ is not equivalent to a monomial, i.e., no conjugate $(f^{-1} \circ T_d \circ f)(w)$ has the form $cw^d$. Prove this assertion.

**6.9.** Prove that the (formal) derivatives of the Chebyshev polynomials satisfy the following identities:

(a) $(4 - w^2)T_d'(w) + dwT_d(w) = 2dT_{d-1}(w)$.

(b) $(4 - w^2)T_d''(w) - wT_d'(w) + d^2T_d(w) = 0$.

**6.10.** Let $K$ be a field of positive characteristic $p$.

(a) Prove that the $p^{\text{th}}$ Chebyshev polynomial $T_p(w)$ is equal to $w^p$ in $K[w]$.

(b) In general, if $q$ is a power of $p$, prove that $T_{qd}(w) = T_d(w)^q = T_d(w^q)$ for all $d \ge 1$.

(c) Again letting $q$ be a power of $p$, deduce that $\operatorname{Aut}(T_q) = \operatorname{PGL}_2(\mathbb{F}_q)$. (Cf. Exercise 6.1.)

**6.11.** Let $K$ be a field of characteristic 2 and let $d \ge 1$ be an odd integer. Prove that $wT_d'(w) = T_d(w)$. What is the derivative $T_d'(w)$ if $d$ is an even integer?

## Section 6.4. Lattès Maps — General Properties

**6.12.** Let $E$ be the elliptic curve $E : y^2 = x^3 + 1$ with $j(E) = 0$, so $\operatorname{Aut}(E) = \boldsymbol{\mu}_6$ is cyclic of order 6. Let $\psi(P) = [2]P$ be the doubling map.

(a) Let $\pi : E \to E/\boldsymbol{\mu}_2 \cong \mathbb{P}^1$. Prove that we can take $\pi(x, y) = x$ and that the Lattès map corresponding to $\psi$ is

$$\phi_1(z) = \frac{z(z^3 - 8b)}{4(z^3 + b)}.$$

(b) Let $\pi : E \to E/\boldsymbol{\mu}_3 \cong \mathbb{P}^1$. Prove that we can take $\pi(x, y) = y$ and that the Lattès map corresponding to $\psi$ is

$$\phi_2(z) = \frac{z^4 + 18bz^2 - 27b^2}{8z^3}.$$

(c) Let $\pi : E \to E/\boldsymbol{\mu}_6 \cong \mathbb{P}^1$. Prove that we can take $\pi(x, y) = x^3$ and that the Lattès map corresponding to $\psi$ is

$$\phi_3(z) = \frac{z(z - 8b)^3}{64(z + b)^3}.$$

Compute the conjugate $\phi_3(z - b) + b$ of $\phi_3(z)$, compare it to $\phi_2(z)$, and explain.

**6.13.** Let $\phi$ be a Lattès map. Prove that there does not exist a linear fractional transformation $f \in \operatorname{PGL}_2$ such that the conjugate $\phi^f$ is a polynomial. (Cf. Exercise 6.8.)

**6.14.** Complete the proof of Proposition 6.44 in the general case that $\psi(P) = \alpha(P) + T$ with $\alpha \in \operatorname{End}(E)$ and $T \in E$ not necessarily equal to $\mathcal{O}$. However, you may assume that $T \in E_{\text{tors}}$, i.e., $T$ is a point of finite order.

**6.15.** Let $\phi : C \to C$ be a nonconstant rational map from a smooth curve to itself. Recall that $\mathsf{CritVal}_\phi$ denotes the set of critical values of $\phi$. Prove that

$$\bigcup_{n=0}^{\infty} \phi^n(\mathsf{CritVal}_\phi) = \bigcup_{n=1}^{\infty} \mathsf{CritVal}_{\phi^n}.$$

## Section 6.5. Elliptic Curves and Flexible Lattès Maps

**6.16.** Let $E$ be an elliptic curve and let $\pi : E \to \mathbb{P}^1$ be a map of degree 2.

(a) Let $R$ be any point on $E$. Show that we can define a new group law (call it $\star$) on $E$ by the rule

$$P \star Q = P + Q - R.$$

Show that $R$ is the identity element for the group $(E, \star)$.

(b) Prove that there exists a point $P_0 \in E$ such that $\pi\bigl(-(P + P_0)\bigr) = \pi(P + P_0)$ for all $P \in E$.

(c) Conclude that after choosing a new identity element for $E$, the map $\pi$ is even, i.e., satisfies $\pi(P) = \pi(-P)$ for all $P \in E$.

**6.17.** Fix an elliptic curve $E$ and and a degree-2 map $\pi : E \to \mathbb{P}^1$ satisfying $\pi(P) = \pi(-P)$. For any integer $m$ and any point $T \in E[2]$, let $\phi_{m,T} : \mathbb{P}^1 \to \mathbb{P}^1$ be the flexible Lattès map associated to the map $\psi(P) = [m](P) + T$ as in the commutative diagram (6.22).

(a) Prove that $\phi_{m,T} \circ \phi_{m',T'} = \phi_{mm',mT'+T}$. In particular, the maps $\phi_{m,\mathcal{O}}$ commute under composition.

(b) Prove that $\phi_{m,T}^n$ is either $\phi_{m^n,T}$ or $\phi_{m^n,\mathcal{O}}$. More precisely, if $m$ is odd and $n$ is even, prove that $\phi_{m,T}^n = \phi_{m^n,\mathcal{O}}$, and prove in all other cases that $\phi_{m,T}^n = \phi_{m^n,T}$. (Of course, if $T = \mathcal{O}$, the cases are all the same.)

(c) It follows from (a) that the collection of maps $\bigl\{\phi_{m,T} : m \ge 1,\ T \in E[2]\bigr\}$ is closed under composition. Prove that $\phi_{1,\mathcal{O}}$ is the identity element and that the associative law holds. Thus this set of flexible Lattès maps for $E$ is a noncommutative monoid.

**6.18.** Let $\phi : \mathbb{P}^1 \to \mathbb{P}^1$ be a flexible Lattès map associated to $\psi(P) = [m]P + T$, where the point $T \in E[2]$ is not necessarily equal to $\mathcal{O}$.

(a) Prove that the set of fixed points of $\phi$ is given by

$$\operatorname{Fix}(\phi) = x\bigl([m-1]^{-1}(T)\bigr) \cup x\bigl([m+1]^{-1}(T)\bigr).$$

(b) Compute the multiplier of $\phi$ at each point in $\operatorname{Fix}(\phi)$.

(c) Use the results from (a) and (b) and the formula for the composition of Lattès maps in Exercise 6.17 to describe the periodic points of $\phi$ and to compute their multipliers.

(*Hint.* Mimic the proof of Proposition 6.52, which dealt with the case $T = \mathcal{O}$.)

**6.19.** Proposition 6.52 describes the multipliers of a flexible Lattès map. Using these values, verify directly that the formula

$$\sum_{\zeta \in \mathrm{Per}_n(\phi)} \frac{1}{1 - \lambda_\phi(\zeta)} = 1$$

from Theorem 1.14 is true for flexible Lattès maps.

**6.20.** Complete the proof of Proposition 6.52(b) by computing the multiplier $\lambda_\phi(\infty)$ at the fixed point $\infty = x(\mathcal{O})$. (*Hint.* Move $\mathcal{O}$ to $(0,0)$ using the change of variables $z = x/y$ and $w = 1/y$. Then write the invariant differential in terms of $z$ and $w$ and mimic the proof in the text.)

**6.21.** Let $K$ be an algebraically closed field and let $\phi$ and $\phi'$ be flexible Lattès maps defined over $K$ that are associated, respectively, to elliptic curves $E$ and $E'$. Suppose that $\phi$ and $\phi'$ are $\mathrm{PGL}_2(K)$-conjugate to one another. We proved (Theorem 6.46) that if the characteristic of $K$ is not equal to 2, then $E$ and $E'$ are isomorphic. What can be said in the case that $K$ has characteristic 2? (Note that in characteristic 2 it is necessary to use a generalized Weierstrass equation (6.12) to define $E$.)

**6.22.** We proved Proposition 6.55 in the case that $\psi(P) = [m](P)$.

(a) Prove Proposition 6.55 for general flexible Lattès maps, i.e., Lattès maps associated to maps of the form $\psi(P) = [m](P) + T$ with $T \in E[2]$.

(b) Formulate and prove a version of Proposition 6.55 for rigid Lattès maps.

(c) ** To what extent is the converse of Proposition 6.55 true? More precisely, if $\phi$ is a Lattès map fitting into a reduced Lattès diagram (6.37) and if $\phi^f$ has bad reduction for every $f \in \mathrm{PGL}_2(K)$, does the elliptic curve $E$ necessarily also have bad reduction?

**6.23.** Let $E$ be an elliptic curve given by a Weierstrass equation

$$E : y^2 = x^3 + ax + b.$$

Let $m \geq 1$ be an integer and write $x\big([m]P\big)$ as a quotient of polynomials

$$x\big([m]P\big) = \frac{F_m\big(x(P)\big)}{G_m\big(x(P)\big)}. \tag{6.52}$$

(a) Prove that $F_m$ and $G_m$ can be taken to be polynomials in $x$, $a$, and $b$. More precisely, prove that there are polynomials $F_m, G_m \in \mathbb{Z}[a,b,x]$ satisfying (6.52) and that they are uniquely determined by the requirement that $F_m$ be monic in the variable $x$.

(b) Prove that $\deg(F_m) = m^2$ and $\deg(G_m) = m^2 - 1$ and that their leading terms are $F_m(x) = x^{m^2} + \cdots$ and $G_m(x) = m^2 x^{m^2-1} + \cdots$.

(c) If $m$ is odd, prove that there is a polynomial $\psi_m(x) \in \mathbb{Z}[a,b,x]$ such that $G_m(x) = \psi_m(x)^2$. Similarly, if $m$ is even, prove that there is a polynomial $\psi_m(x,y) \in \mathbb{Z}[a,b,x,y]$ such that $G_m(x) = \psi_m(x,y)^2$, where in the computation we replace $y^2$ by $x^3 + ax + b$. The polynomial $\psi_m$ is called the $m^{th}$ *division polynomial for* $E$, since its roots are the nontrivial points of order $m$.

(d) Prove that $F_m$ and $G_m$ satisfy

$$F_m(t^2 a, t^3 b; tx) = t^{m^2} F_m(a,b;x) \quad \text{and} \quad G_m(t^2 a, t^3 b; tx) = t^{m^2-1} G_m(a,b;x).$$

Thus $F_m$ and $G_m$ are homogeneous if $x$, $a$, and $b$ are respectively assigned weights 2, 4, and 6.

(e) Let $\Delta(E) = -16(4a^3 + 27b^2)$. Prove that the resultant of $F_m$ and $G_m$ with respect to the variable $x$ is given by

$$\mathrm{Res}(F_m, G_m) = \pm\Delta(E)^{m^2(m^2-1)/6}.$$

## Section 6.6. Elliptic Curves and Rigid Lattès Maps

**6.24.** This exercise extends Proposition 6.61. Let $\phi : \mathbb{P}^1 \to \mathbb{P}^1$ be a Lattès map and fix a reduced Lattès diagram (6.37) for $\phi$. Write $\psi(P) = [\alpha](P) + T$ as usual, where we use the standard normalization described in Proposition 6.36 to identify $\operatorname{End}(E)$ with a subring of $\mathbb{C}$.

(a) Prove that the fixed points of $\phi$ are given by

$$\operatorname{Fix}(\phi) = \bigcup_{\xi \in \Gamma} \{\pi(P) : [\alpha - \xi](P) = -T\}. \tag{6.53}$$

(b) Let $\pi(P) \in \operatorname{Fix}(\phi)$. Prove that $P$ is a critical point for $\pi$ if and only if $P$ is fixed by a nontrivial element of $\xi$. More generally, prove that the ramification index is given by

$$e_P(\pi) = \{\xi \in \Gamma : [\xi]P = P\}.$$

In particular, if $P$ is not a critical point, then there is a unique $\xi \in \Gamma$ that fixes $P$.

(c) Assume that $T = 0$. Let $\pi(P) \in \operatorname{Fix}(\phi)$ and choose some automorphism $\xi \in \Gamma$ such that $\psi(P) = [\xi](P)$. Compute the multiplier of $\phi$ at $\pi(P)$ as in the following table (we have given you the first four values):

$$\lambda_{\pi(P)}(\phi) = \begin{cases} \xi^{-1}\alpha & \text{if } e_P(\pi) = 1, \\ \alpha^2 & \text{if } \Gamma = \boldsymbol{\mu}_2 \text{ and } e_P(\pi) = 2, \\ \alpha^3 & \text{if } \Gamma = \boldsymbol{\mu}_3 \text{ and } e_P(\pi) = 3, \\ \xi^2\alpha^2 & \text{if } \Gamma = \boldsymbol{\mu}_4 \text{ and } e_P(\pi) = 2, \\ \underline{\qquad} & \text{if } \Gamma = \boldsymbol{\mu}_4 \text{ and } e_P(\pi) = 4, \\ \underline{\qquad} & \text{if } \Gamma = \boldsymbol{\mu}_6 \text{ and } e_P(\pi) = 2, \\ \underline{\qquad} & \text{if } \Gamma = \boldsymbol{\mu}_6 \text{ and } e_P(\pi) = 3. \end{cases}$$

**6.25.** In the text we proved Theorem 6.62 under the assumption that $d$ is squarefree. This exercise sketches an argument to eliminate the squarefree hypothesis. We set the notation $\mathsf{S}(b)$ for the squarefree part of the integer $b \geq 1$.

(a) For each integer $d \geq 2$, let $D_d \geq 1$ be an integer with the property that $d$ is a norm from the ring $\mathbb{Z}\big[\sqrt{-D_d}\,\big]$ down to $\mathbb{Z}$. In other words, there are integers $u$ and $v$ such that

$$u^2 + D_d v^2 = d.$$

Then with notation as in the statement of Theorem 6.62, prove that for every $\epsilon > 0$ there is a constant $C_\epsilon$ such that

$$\deg(\boldsymbol{\sigma}_d) \geq C_\epsilon\, \mathsf{S}(D_d)^{\frac{1}{2}-\epsilon} \qquad \text{for all } d.$$

(*Hint.* Use elliptic curves with CM by the ring $\mathbb{Z}\big[\sqrt{-D_d}\,\big]$ and Lattès maps associated to the endomorphism $\big[u + v\sqrt{-D_d}\,\big]$ and follow the proof of Theorem 6.62.)

(b) Prove that for every $\epsilon > 0$ there is a constant $C'_\epsilon > 0$ such that

$$\max_{0 \leq u < \sqrt{d}} \mathsf{S}(d - u^2) \geq C'_\epsilon d^{1-\epsilon} \qquad \text{for all } d.$$

(*Hint.* It suffices to prove that for sufficiently large $d$ the average satisfies

$$\frac{1}{\sqrt{d}} \sum_{0 \le u < \sqrt{d}} \log \mathsf{S}(d-u^2) \ge (1-\epsilon)\log d.$$

Write this as two sums using $\operatorname{ord}_p\bigl(\mathsf{S}(b)\bigr) = \operatorname{ord}_p(b) - 2\left\lfloor \frac{1}{2}\operatorname{ord}_p(b)\right\rfloor$ and show that the first sum is asymptotic to $\log(d)$ and the second is bounded as $d \to \infty$.)

(c) Combine (a) and (b) to complete the proof of Theorem 6.62.

## Section 6.7. Uniform Boundedness for Lattès Maps

**6.26.** Let $E/\mathbb{Q}$ be given by a quasiminimal Weierstrass equation

$$E : y^2 = x^3 + ax + b,$$

i.e., the discriminant $\left|16(4a^3 + 27b^2)\right|$ is minimized subject to the condition that $a$ and $b$ are integers.

(a) Show that the equation for $E$ is minimal at every prime $p \ge 5$.

(b) Let $\Delta_3(E)$ be the discriminant of a general Weierstrass equation (6.12) for $E$ that is minimal at 3. Prove that

$$0 \le \operatorname{ord}_3(4a^3 + 27b^2) - \operatorname{ord}_3\bigl(\Delta_3(E)\bigr) < 6.$$

(c) Let $\Delta_2(E)$ be the discriminant of a general Weierstrass equation (6.12) for $E$ that is minimal at 2. Prove that

$$0 \le \operatorname{ord}_2\bigl(16(4a^3 + 27b^2)\bigr) - \operatorname{ord}_2\bigl(\Delta_2(E)\bigr) < 12.$$

**6.27.** Theorem 6.70 suggests that there should be an absolute upper bound for the number of integer points in the orbits of affine minimal Lattès maps defined over $\mathbb{Q}$.

(a) Let $E$ be the elliptic curve

$$E : y^2 = x^3 - 48907 + 8481094$$

and let $\phi(x)$ be the Lattès map associated to multiplication-by-2, i.e., $f \circ x = x \circ [2]$. Verify that the orbit $\mathcal{O}_\phi(2363)$ contains five integer points, but that $\phi^5(2363) \notin \mathbb{Z}$. Also verify that there is a point in $E(\mathbb{Q})$ with $x$-coordinate 2363.

(b) Let $E$ be the elliptic curve

$$E : y^2 = x^3 - 40467 + 4120274$$

and again let $\phi(x)$ be the Lattès map associated to multiplication-by-2. Verify that the orbit $\mathcal{O}_\phi(193)$ contains five integer points, but that $\phi^5(193) \notin \mathbb{Z}$. In this case $E(\mathbb{Q})$ does not contain a point with $x$-coordinate equal to 193.

(c) ** Find an affine minimal Lattès map $\phi$ and an initial point $\zeta \in \mathbb{Z}$ such that $\phi^5(\zeta) \in \mathbb{Z}$, or prove that none exist. (This will force $\phi^k(\zeta) \in \mathbb{Z}$ for all $0 \le k \le 5$; see Remark 6.72.)

**6.28.** Prove a version of Theorem 6.70 for a Lattès map satisfying

$$\phi\bigl(x(P)\bigr) = x\bigl([m]P + T\bigr) \qquad \text{for all } P \in E,$$

where $T$ is a fixed 2-torsion point of $E$. (You may need a more general version of Theorem 6.69; see [202, 407].)

# Chapter 7

# Dynamics in Dimension Greater Than One

Up to this point our primary focus has been on arithmetic dynamics of rational maps on $\mathbb{P}^1$. In this chapter we take a look at dynamics in higher dimensions. Even over $\mathbb{C}$, although there is now a significant body of knowledge, it seems fair to say that complex dynamics on $\mathbb{P}^N$ is still in its infancy. And arithmetic dynamics in higher dimensions is at present a patchwork of results from which a general theory is yet to emerge. Our goal in this chapter is to provide a glimpse into two aspects of this developing theory by highlighting two ways in which higher-dimensional dynamics differs significantly from the one-dimensional case.

The first difference arises from the fact that rational maps $\mathbb{P}^N \to \mathbb{P}^N$ for $N \geq 2$ need not be everywhere defined, i.e., they need not be morphisms. In Section 7.1 we study the dynamics of rational maps $\phi : \mathbb{P}^N \to \mathbb{P}^N$ having the property that they restrict to automorphisms on $\mathbb{A}^N$. The geometry and arithmetic of such maps can be quite complicated, despite the fact that they are bijective on $\mathbb{A}^N$.

The second difference arises due to the far greater variety of varieties in higher dimensions. Thus in dimension 1, only self-maps of $\mathbb{P}^1$ and of elliptic curves are dynamically interesting, since the only self-maps of a curve of genus greater than 1 are automorphisms of finite order. But even in dimension 2 there is an abundance of varieties that admit self-maps of infinite order, and the dynamical properties of these maps are extremely interesting, although as yet imperfectly understood. In Section 7.4 we study the arithmetic dynamics of certain surfaces that admit a pair of noncommuting involutions $\iota_1$ and $\iota_2$. The composition $\phi = \iota_1 \circ \iota_2$ is an automorphism of infinite order.

The theory of height functions that we developed in Sections 3.1–3.5 provides a powerful tool for studying arithmetic dynamics on $\mathbb{P}^1$. One of the recurring themes of this chapter is the use of height functions in a higher-dimensional setting and on varieties other than projective spaces.

## 7.1 Dynamics of Rational Maps on Projective Space

Recall that a *rational map* $\phi : \mathbb{P}^N \to \mathbb{P}^N$ is described by homogeneous polynomials with no common factor, and that $\phi$ is a *morphism* if the polynomials have no common root in $\mathbb{P}^N(\bar{K})$. (See page 89 in Chapter 3 for the precise definition.) As noted in the introduction to this chapter, height functions are a powerful tool for studying the arithmetic of morphisms $\phi : \mathbb{P}^N \to \mathbb{P}^N$. The situation is considerably more complicated if the map $\phi : \mathbb{P}^N \to \mathbb{P}^N$ is required to be only a rational map. Notice that we did not run into this situation when studying rational functions $\phi(z) \in K(z)$ of one variable, since every rational map $\phi : \mathbb{P}^1 \to \mathbb{P}^1$ is automatically a morphism. But in dimensions 2 and higher, there are many rational maps that are not morphisms.

*Example* 7.1. The rational map

$$\phi : \mathbb{P}^2 \longrightarrow \mathbb{P}^2, \qquad \phi([X_0, X_1, X_2]) = [X_0^2, X_0X_1, X_2^2], \tag{7.1}$$

is not a morphism, since it is not defined at the point $[0,1,0]$. Notice that if we discard $[0,1,0]$, then $\phi$ fixes every point on the line $X_0 = X_2$, and $\phi$ sends every point on the line $X_0 = 0$ to the single point $[0,0,1]$. This kind of behavior is not possible for morphisms $\mathbb{P}^2 \to \mathbb{P}^2$.

Continuing with this example, recall that if $\phi$ were a morphism, then Theorem 3.11 would tell us that $h(\phi(P)) = 2h(P) + O(1)$ for all $P \in \mathbb{P}^2(\bar{\mathbb{Q}})$. But this is clearly false for the map (7.1), since for all $a, b \in \bar{\mathbb{Q}}^*$ we have

$$\phi([a,b,a]) = [a^2, ab, a^2] = [a,b,a].$$

Thus

$$h(\phi([a,b,a])) = h([a,b,a]),$$

so we cannot use Theorem 3.7 to conclude that $\phi$ has only finitely many $\mathbb{Q}$-rational periodic points. Of course, that's good, since in fact $\phi$ has infinitely many $\mathbb{Q}$-rational fixed points!

An initial difficulty in studying the dynamics of a rational map $\phi : \mathbb{P}^N \to \mathbb{P}^N$ arises from the fact that the orbit $\mathcal{O}_\phi(P)$ of a point may "terminate" if some iterate $\phi^n(P)$ arrives at a point where $\phi$ is not defined. This suggests looking first at maps $\phi$ for which there is a large uncomplicated (e.g., quasiprojective, or even affine) subset $U \subset \mathbb{P}^N$ with the property that $\phi(U) \subset U$ and studying the dynamics of $\phi$ on $U$. As a further simplification, we might require that $\phi$ be an automorphism of $U$, since quasiprojective varieties often allow interesting automorphisms.

### 7.1.1 Affine Morphisms and the Locus of Indeterminacy

In this section we study rational maps $\mathbb{P}^N \to \mathbb{P}^N$ with the property that they induce morphisms of affine space $\mathbb{A}^N \to \mathbb{A}^N$. Concretely, an *affine morphism*

$$\phi : \mathbb{A}^N \to \mathbb{A}^N$$

is a map of the form

$$\phi = (F_1, \ldots, F_N) \qquad \text{with} \quad F_1, \ldots, F_N \in K[z_1, \ldots, z_N].$$

To avoid trivial cases, we generally assume that at least one of the $F_i$ is not the zero polynomial.

**Definition.** The degree of a polynomial

$$F(z_1, \ldots, z_N) = \sum_{i_1, \ldots, i_N} a_{i_1 \ldots i_N} z_1^{i_1} \cdots z_N^{i_N} \in K[z_1, \ldots, z_N]$$

is defined to be

$$\deg F = \max\{i_1 + \cdots + i_N : a_{i_1 \ldots i_N} \neq 0\}.$$

In other words, the degree of $F$ is the largest total degree of the monomials that appear in $F$. (By convention the zero polynomial is assigned degree $-\infty$.) The *degree of a morphism* $\phi = (F_1, \ldots, F_N) : \mathbb{A}^N \to \mathbb{A}^N$ is defined to be

$$\deg \phi = \max\{\deg F_1, \ldots, \deg F_N\}.$$

Homogenization of the coordinates of an affine morphism $\phi : \mathbb{A}^N \to \mathbb{A}^N$ of degree $d$ yields a rational map $\bar{\phi} : \mathbb{P}^N \to \mathbb{P}^N$ of degree $d$. For each coordinate function $F_i$ of $\phi$, we let

$$\bar{F}_i(X_0, X_1, \ldots, X_N) = X_0^d F_i\left(\frac{X_1}{X_0}, \frac{X_2}{X_0}, \ldots, \frac{X_N}{X_0}\right).$$

Notice that each $\bar{F}_i$ is a homogeneous polynomial of degree $d$ (or the zero polynomial), so the map

$$\bar{\phi} = [X_0^d, \bar{F}_1, \bar{F}_2, \ldots, \bar{F}_N] : \mathbb{P}^N \to \mathbb{P}^N$$

is a rational map of degree $d$. We call $\bar{\phi}$ the *rational map induced by* $\phi$. A rational map need not be everywhere defined.

**Definition.** Let $\phi : \mathbb{A}^N \to \mathbb{A}^N$ be an affine morphism of degree $d$ and let

$$\bar{\phi} = [X_0^d, \bar{F}_1, \ldots, \bar{F}_N] : \mathbb{P}^N \to \mathbb{P}^N$$

be the rational map that it induces. The *locus of indeterminacy of* $\phi$ is the set

$$Z(\phi) = \{P = [0, x_1, \ldots, x_N] \in \mathbb{P}^N : \bar{F}_1(P) = \cdots = \bar{F}_N(P) = 0\}.$$

(To ease notation, we write $\phi$ and $Z(\phi)$ instead of $\bar{\phi}$ and $Z(\bar{\phi})$.) This is the set of points at which $\bar{\phi}$ is not defined. Notice that $Z(\phi)$ lies in the hyperplane $H_0 = \{X_0 = 0\}$ at infinity, since $\phi$ is well-defined on $\mathbb{A}^N$.

The polynomials $\bar{F}_1, \ldots, \bar{F}_N$ can be used to define a morphism

$$\Phi : \mathbb{A}^{N+1} \longrightarrow \mathbb{A}^{N+1}, \qquad \Phi = (X_0^d, \bar{F}_1, \ldots, \bar{F}_N).$$

The map $\Phi$ is called a *lift of* $\bar{\phi}$. If we let $\pi$ be the natural projection map,

$$\pi : \mathbb{A}^{N+1} \smallsetminus \{0\} \longrightarrow \mathbb{P}^N, \qquad (x_0, \ldots, x_N) \longmapsto [x_0, \ldots, x_N],$$

then $\pi$, $\Phi$, and $\bar{\phi}$ fit together into the commutative diagram

$$\begin{array}{ccc} \mathbb{A}^{N+1} \smallsetminus \{0\} & \xrightarrow{\Phi} & \mathbb{A}^{N+1} \smallsetminus \{0\} \\ \big\downarrow{\scriptstyle \pi} & & \big\downarrow{\scriptstyle \pi} \\ \mathbb{P}^N & \xrightarrow{\bar{\phi}} & \mathbb{P}^N \end{array}$$

*Example* 7.2. The map

$$\phi : \mathbb{A}^2 \longrightarrow \mathbb{A}^2, \qquad \phi(z_1, z_2) = (z_1 z_2, z_1^2),$$

induces the rational map

$$\phi : \mathbb{P}^2 \longrightarrow \mathbb{P}^2, \qquad \bar{\phi}([X_0, X_1, X_2]) = [X_0^2, X_1 X_2, X_1^2],$$

and has indeterminacy locus $Z(\phi) = \{[0,0,1]\}$ consisting of a single point.

### 7.1.2 Affine Automorphisms

Of particular interest are affine morphisms that admit an inverse.

**Definition.** An affine morphism $\phi : \mathbb{A}^N \to \mathbb{A}^N$ is an *automorphism* if it has an inverse morphism. In other words, $\phi$ is an affine automorphism if there is an affine morphism $\phi^{-1} : \mathbb{A}^N \to \mathbb{A}^N$ such that

$$\phi\big(\phi^{-1}(z_1, \ldots, z_N)\big) = (z_1, \ldots, z_N) \quad \text{and} \quad \phi^{-1}\big(\phi(z_1, \ldots, z_N)\big) = (z_1, \ldots, z_N).$$

Somewhat surprisingly, $\phi$ and $\phi^{-1}$ need not have the same degree, nor does $\deg(\phi^n)$ have to equal $(\deg \phi)^n$.

*Example* 7.3. Consider the map $\phi(x, y) = (x, y + x^2)$. It has degree 2 and is an automorphism, since it has the inverse $\phi^{-1}(x, y) = (x, y - x^2)$. The composition $\phi^2$ is

$$\phi^2(x, y) = \phi(x, y + x^2) = (x, y + 2x^2),$$

so $\deg(\phi^2) = 2 = \deg(\phi)$. More generally, $\phi^n(x, y) = (x, y + nx^2)$ has degree 2, so the degree of $\phi^n$ does not grow. This contrasts sharply with what happens for morphisms of $\mathbb{P}^N$.

*Example* 7.4. Let $a \in K^*$ and let $f(y) \in K[y]$ be a polynomial of degree $d \geq 2$. The map

$$\phi : \mathbb{A}^2 \longrightarrow \mathbb{A}^2, \qquad \phi(x, y) = \big(y, ax + f(y)\big),$$

is called a *Hénon map*. It is an automorphism of $\mathbb{A}^2$, since one easily checks that it has an inverse $\phi^{-1}$ given by

$$\phi^{-1} : \mathbb{A}^2 \longrightarrow \mathbb{A}^2, \qquad \phi^{-1}(x, y) = \big(a^{-1}y - a^{-1}f(x), x\big).$$

Hénon maps, especially those with $\deg(f) = 2$, have been extensively studied since Hénon [200] introduced them as examples of maps $\mathbb{R}^2 \to \mathbb{R}^2$ having strange attractors. There are many open questions regarding the real and complex dynamics of Hénon maps; see, for example, [132, §2.9] or [211], as well as [212, 413] for a compactification of the Hénon map.

The rational maps $\mathbb{P}^2 \to \mathbb{P}^2$ induced by $\phi$ and $\phi^{-1}$ are

$$\bar{\phi}([X_0, X_1, X_2]) = [X_0^d,\, X_0^{d-1}X_2,\, aX_0^{d-1}X_1 + \bar{f}(X_0, X_2)],$$
$$\bar{\phi}^{-1}([X_0, X_1, X_2]) = [X_0^d,\, a^{-1}X_0^{d-1}X_2 - a^{-1}\bar{f}(X_0, X_1),\, X_0^{d-1}X_1],$$

where we write $\bar{f}(u, v) = u^d f(v/u)$ for the homogenization of $f$. It is easy to see that the loci of indeterminacy of $\phi$ and $\phi^{-1}$ are

$$Z(\phi) = \{[0,1,0]\} \qquad \text{and} \qquad Z(\phi^{-1}) = \{[0,0,1]\}.$$

In particular, the locus of indeterminacy of $\phi$ is disjoint from the locus of indeterminacy of $\phi^{-1}$. Maps with this property are called *regular*; see Section 7.1.3.

*Example* 7.5. Consider the very simple Hénon map

$$\phi(x, y) = (y, -x + y^2).$$

The extension $\bar{\phi} = [X_0^2, X_0X_2, -X_0X_1 + X_2^2]$ of $\phi$ to $\mathbb{P}^2$ has degree 2, but it is not a morphism, since it is not defined at the point $[0,1,0]$. And just as in Example 7.2, there is no height estimate of the form $h(\phi(P)) = 2h(P) + O(1)$ for $\bar{\phi}$. We can see this by noting that

$$\bar{\phi}([b, a, b]) = [b^2, b^2, -ab + b^2] = [b, b, -a + b],$$

so if $a, b, \in \mathbb{Z}$ with $\gcd(a, b) = 1$ and $b > a > 0$, then $[b, a, b]$ and $\bar{\phi}([b, a, b])$ have the same height. Hence for every $\epsilon > 0$ even the weaker statement

$$h(\bar{\phi}(P)) \geq (1 + \epsilon)h(P) + O(1) \qquad \text{for all } P = (x, y) \in \mathbb{A}^2(\mathbb{Q})$$

is false. It turns out that $\phi$ has only finitely many $\mathbb{Q}$-rational periodic points (Theorem 7.19), but the proof does not follow directly from a simple height argument.

*Example* 7.6. More generally, if $\phi : \mathbb{A}^N \to \mathbb{A}^N$ is an affine automorphism, then it is not possible to have simultaneous estimates of the form

$$\begin{aligned} h(\phi(P)) &\geq (1 + \epsilon)h(P) + O(1), \\ h(\phi^{-1}(P)) &\geq (1 + \epsilon)h(P) + O(1), \end{aligned} \tag{7.2}$$

for some $\epsilon > 0$ and all $P \in \mathbb{A}^N(K)$. To see this, suppose that (7.2) were true. Then we would have for all $P \in \mathbb{A}^N(K)$,

$$h(P) = h(\phi(\phi^{-1}(P))) \geq (1 + \epsilon)h(\phi^{-1}(P)) + O(1) \geq (1 + \epsilon)^2 h(P) + O(1).$$

Thus $h(P)$ would be bounded, leading to the untenable conclusion that $\mathbb{A}^N(K)$ is finite. So it is too much to require that both $\phi(P)$ and $\phi^{-1}(P)$ have heights larger than the height of $P$. However, as we shall see, it is often possible to show that some combination of $h(\phi(P))$ and $h(\phi^{-1}(P))$ is large, which is then sufficient to prove that $\operatorname{Per}(\phi)$ is a set of bounded height.

We conclude this section with two useful geometric lemmas. The first relates the locus of indeterminacy of an affine automorphism and its inverse, and the second characterizes when the degree of a composition is smaller than the product of the degrees.

**Lemma 7.7.** *Let $\phi : \mathbb{A}^N \to \mathbb{A}^N$ be an affine automorphism of degree at least 2 and denote the hyperplane at infinity by $H_0 = \{X_0 = 0\} = \mathbb{P}^N \smallsetminus \mathbb{A}^N$. Then*

$$\bar{\phi}(H_0 \smallsetminus Z(\phi)) \subset Z(\phi^{-1}).$$

*Proof.* Let

$$\Phi = (X_0^d, \bar{F}_1, \bar{F}_2, \dots, \bar{F}_N) \quad \text{and} \quad \Phi^{-1} = (X_0^e, \bar{G}_1, \bar{G}_2, \dots, \bar{G}_N)$$

be the lifts of $\bar{\phi}$ and $\bar{\phi}^{-1}$, respectively. The fact that $\phi$ and $\phi^{-1}$ are inverses of one another implies that there is a homogeneous polynomial $f$ of degree $de - 1$ with the property that

$$(\Phi^{-1} \circ \Phi)(X_0, \dots, X_N) = (f \cdot X_0, f \cdot X_1, \dots, f \cdot X_N).$$

But the first coordinate of the composition is $X_0^{de}$, so we see that $f = X_0^{de-1}$. Thus

$$(\Phi^{-1} \circ \Phi)(X_0, \dots, X_N) = (X_0^{de}, X_0^{de-1}X_1, X_0^{de-1}X_1, \dots, X_0^{de-1}X_N),$$

or equivalently,

$$\bar{G}_j(X_0^d, \bar{F}_1, \dots, \bar{F}_N) = X_0^{de-1}X_j \quad \text{for all } 1 \le j \le N. \tag{7.3}$$

Now let $P = [0, x_1, \dots, x_N] \in H_0 \smallsetminus Z(\phi)$, so $\bar{\phi}(P) = [0, \bar{F}_1(P), \dots, \bar{F}_N(P)]$ with at least one $\bar{F}_i(P) \ne 0$. From (7.3) we see that

$$\bar{G}_j(\Phi(P)) = \bar{G}_j(0, \bar{F}_1(P), \dots, \bar{F}_N(P)) = 0^{de-1}x_j = 0 \quad \text{for all } 1 \le j \le N.$$

Hence

$$\Phi^{-1}(\Phi(P)) = (0, \bar{G}_1(\Phi(P)), \bar{G}_2(\Phi(P)), \dots, \bar{G}_N(\Phi(P)) = (0, 0, 0, \dots, 0),$$

so $\bar{\phi}^{-1}$ is not defined at $\bar{\phi}(P)$. Therefore $\bar{\phi}(P) \in Z(\phi^{-1})$. □

**Lemma 7.8.** *Let $\phi : \mathbb{A}^N \to \mathbb{A}^N$ and $\psi : \mathbb{A}^N \to \mathbb{A}^N$ be affine morphisms, and let $H_0 = \{X_0 = 0\} = \mathbb{P}^N \smallsetminus \mathbb{A}^N$ be the usual hyperplane at infinity. Then*

$$\deg(\psi \circ \phi) < \deg(\psi)\deg(\phi) \quad \textit{if and only if} \quad \bar{\phi}(H_0 \smallsetminus Z(\phi)) \subset Z(\psi).$$

*Proof.* Let $d = \deg(\phi)$, let $e = \deg(\psi)$, and let $\Phi$ and $\Psi$ be lifts of $\bar{\phi}$ and $\bar{\psi}$, respectively. We write $\Phi$ explicitly as

$$\Phi = (X_0^d, \bar{F}_1, \bar{F}_2, \ldots, \bar{F}_N).$$

The composition $\Psi \circ \Phi$ has the form

$$\Psi \circ \Phi = (X_0^{de}, \bar{E}_1, \bar{E}_2, \ldots, \bar{E}_N),$$

where $\bar{E}_1, \ldots, \bar{E}_N$ are homogeneous polynomials of degree $de$. The degree of $\psi \circ \phi$ will be strictly less than $de$ if and only if there is some cancellation in the coordinate polynomials of $\Psi \circ \Phi$. Since the first coordinate is $X_0^{de}$, this shows that

$$\deg(\psi \circ \phi) < \deg(\psi)\deg(\phi) \iff X_0 \text{ divides } \bar{E}_j \text{ for every } 1 \le j \le N.$$

Suppose now that $X_0 | \bar{E}_j$ for every $j$ and let $P = [0, x_1, \ldots, x_N] \in H_0 \smallsetminus Z(\phi)$. Since $\bar{\phi}$ is defined at $P$, some coordinate of

$$\Phi(P) = \bigl(0, \bar{F}_1(P), \ldots, \bar{F}_N(P)\bigr)$$

is nonzero. On the other hand, the assumption that $X_0 | \bar{E}_j$ implies that

$$(\Psi \circ \Phi)(P) = \bigl(0, \bar{E}_1(\Phi(P)), \bar{E}_2(\Phi(P)), \ldots, \bar{E}_N(\Phi(P))\bigr) = (0, 0, 0, \ldots, 0).$$

Hence $\bar{\psi}$ is not defined at $\bar{\phi}(P)$, so $\bar{\phi}(P) \in Z(\psi)$. This completes the proof that if $\deg(\psi \circ \phi) < de$, then $\bar{\phi}(H_0 \smallsetminus Z(\phi)) \subset Z(\psi)$.

For the other direction, suppose that $\bar{\phi}(H_0 \smallsetminus Z(\phi)) \subset Z(\psi)$. This implies that for (almost all) points of the form $(0, x_1, \ldots, x_N)$, the map $\bar{\psi}$ is not defined at the point $\bar{\phi}\bigl([0, x_1, \ldots, x_N]\bigr)$. Hence

$$\Psi\bigl(\Phi(0, X_1, X_2, \ldots, X_N)\bigr) = (0, 0, 0, \ldots, 0),$$

so $\bar{E}_j(0, X_1, X_2, \ldots, X_N) = 0$ for all $j$. Therefore $X_0 | \bar{E}_j$ for all $j$. □

*Example* 7.9. Let $\phi$ be the map $\phi(x, y) = (x, y + x^2)$ that we studied in Example 7.3. Dehomogenizing $\phi$ yields

$$\bar{\phi}([X_0, X_1, X_2]) = [X_0^2,\ X_0X_1,\ X_0X_2 + X_1^2],$$

so the locus of indeterminacy for $\phi$ is $Z(\phi) = \bigl\{[0, 0, 1]\bigr\}$. Notice that

$$\bar{\phi}([0, X_1, X_2]) = [0, 0, X_1^2] = [0, 0, 1] \in Z(\phi).$$

Hence $\bar{\phi}\bigl(H_0 \smallsetminus Z(\phi)\bigr) = Z(\phi)$, so Lemma 7.8 tells us that $\deg(\phi^2) < \deg(\phi)^2$. This is in agreement with Example 7.3, where we computed that $\deg(\phi^2) = 2$.

### 7.1.3 The Geometry of Regular Automorphisms of $\mathbb{A}^N$

In this section we briefly discuss the geometric properties of an important class of affine automorphisms.

**Definition.** An affine automorphism $\phi : \mathbb{A}^N \to \mathbb{A}^N$ is said to be *regular* if the indeterminacy loci of $\phi$ and $\phi^{-1}$ have no points in common,

$$Z(\phi) \cap Z(\phi^{-1}) = \emptyset.$$

The following theorem summarizes some of the geometric properties enjoyed by regular automorphisms of $\mathbb{A}^N$. We sketch the proof of (a) and refer the reader to [401] for (b) and (c).

**Theorem 7.10.** *Let $\phi : \mathbb{A}^N \to \mathbb{A}^N$ be a regular affine automorphism.*

(a) *For all $n \ge 1$,*

$$\phi^n \text{ is regular,} \qquad Z(\phi^n) = Z(\phi), \qquad \text{and} \qquad \deg(\phi^n) = \deg(\phi)^n.$$

(b) *Let*

$$d_1 = \deg\phi, \quad d_2 = \deg\phi^{-1}, \quad \ell_1 = \dim Z(\phi)+1, \quad \ell_2 = \dim Z(\phi^{-1})+1.$$

*Then*

$$\ell_1 + \ell_2 = N \qquad \text{and} \qquad d_2^{\ell_1} = d_1^{\ell_2}.$$

(c) *For all $n \ge 1$ the set of $n$-periodic points $\operatorname{Per}_n(\phi)$ is a discrete subset of $\mathbb{A}^N(\mathbb{C})$. Counted with appropriate multiplicities,*

$$\#\operatorname{Per}_{Nn}(\phi) = d_2^{\ell_1 Nn} = d_1^{\ell_2 Nn}.$$

*Proof.* (a) We first prove by induction on $n$ that

$$Z(\phi^n) \subset Z(\phi) \quad \text{and} \quad Z(\phi^{-n}) \subset Z(\phi^{-1}) \quad \text{for all } n \ge 1.$$

This is trivially true for $n = 1$, so assume now that it is true for $n-1$. Let $P \in Z(\phi^n)$, so in particular $P \in H_0$. Suppose that $P \notin Z(\phi)$. The induction hypothesis tells us that $P \notin Z(\phi^{n-1})$, so applying Lemma 7.7 to the map $\phi^{n-1}$, we deduce that

$$\phi^{n-1}(P) \in \phi^{n-1}(H_0 \setminus Z(\phi^{n-1})) \subset Z(\phi^{-(n-1)}) \subset Z(\phi^{-1}).$$

(For the last equality we have again used the induction hypothesis.) On the other hand, we have that $\phi^{n-1}$ is defined at $P$ and $\phi^n$ is not defined at $P$, which implies that $\phi^{n-1}(P) \in Z(\phi)$. This proves that $\phi^{n-1}(P)$ is in both $Z(\phi^{-1})$ and $Z(\phi)$, contradicting the assumption that $\phi$ is regular. Hence $P \in Z(\phi)$, which completes the proof that $Z(\phi^n) \subset Z(\phi)$. Similarly, we find that $Z(\phi^{-n}) \subset Z(\phi^{-1})$.

Having shown that $Z(\phi^n) \subset Z(\phi)$ and $Z(\phi^{-n}) \subset Z(\phi^{-1})$, we see that the regularity of $\phi$ implies that

$$Z(\phi^n) \cap Z(\phi^{-n}) \subset Z(\phi) \cap Z(\phi^{-1}) = \emptyset,$$

so $\phi^n$ is also regular.

Next suppose that $\deg(\phi^n) < \deg(\phi)^n$ for some $n \geq 2$. We take $n$ to be the smallest value for which this is true, so in particular $\deg(\phi^{n-1}) = \deg(\phi)^{n-1}$, and hence

$$\deg(\phi^n) < \deg(\phi^{n-1})\deg(\phi).$$

We apply Lemma 7.8 with $\psi = \phi^{n-1}$ to conclude that

$$\phi\bigl(H_0 \smallsetminus Z(\phi)\bigr) \subset Z(\phi^{n-1}) \subset Z(\phi),$$

where the last inclusion was proven earlier. On the other hand, Lemma 7.7 says that $\phi\bigl(H_0 \smallsetminus Z(\phi)\bigr) \subset Z(\phi^{-1})$. Hence

$$\phi\bigl(H_0 \smallsetminus Z(\phi)\bigr) \subset Z(\phi) \cap Z(\phi^{-1}) = \emptyset.$$

This is a contradiction, which completes the proof that $\deg(\phi^n) = \deg(\phi)^n$.

It remains to show that $Z(\phi) \subset Z(\phi^n)$. Let

$$\Phi : \mathbb{A}^{N+1} \longrightarrow \mathbb{A}^{N+1}, \qquad \Phi = (X_0^d, F_1, F_2, \ldots, F_N),$$

be a lift of $\phi$, so

$$Z(\phi) = \bigl\{P \in H_0 : F_1(P) = \cdots = F_N(P) = 0\bigr\}.$$

By a slight abuse of notation, we say that $P \in Z(\phi)$ if and only if $\Phi(P) = 0$. (To be precise, we should lift $P$ to $\mathbb{A}^{N+1}$.)

We proved that $\deg(\phi^n) = \deg(\phi)$, which implies that the coordinate functions of $\Phi^n$ have no common factor. Thus $\phi^n$ can be computed by evaluating $\Phi^n$ and mapping down to $\mathbb{P}^N$. Hence just as above we have $P \in Z(\phi^n)$ if and only if $\Phi^n(P) = 0$. Therefore

$$P \in Z(\phi) \quad\Longrightarrow\quad \Phi(P) = 0 \quad\Longrightarrow\quad \Phi^n(P) = 0 \quad\Longrightarrow\quad P \in Z(\phi^n).$$

This proves that $Z(\phi) \subset Z(\phi^n)$ and completes the proof of (a).

(b) See [401, Proposition 2.3.2].

(c) See [401, Theorem 2.3.4]. □

*Remark* 7.11. If $\phi : \mathbb{A}^2 \to \mathbb{A}^2$ is a regular automorphism of the affine plane, then Theorem 7.10(b) tells us that $\ell_1 = \ell_2 = 1$ (which is clear anyway since the indeterminacy locus of a rational map has codimension at least 2) and that $d_1 = d_2$. Thus planar regular automorphisms satisfy $\deg(\phi) = \deg(\phi^{-1})$. In the opposite direction, if $d_1 = d_2$, then Theorem 7.10(b) says that $\ell_1 = \ell_2$, and hence that $N = \ell_1 + \ell_2$ is even. In other words, a regular automorphism $\phi : \mathbb{A}^N \to \mathbb{A}^N$ with $N$ odd always satisfies $\deg(\phi) \neq \deg(\phi^{-1})$.

*Example* 7.12. Let $\phi : \mathbb{A}^3 \to \mathbb{A}^3$ be given by

$$\phi(x, y, z) = (y, z + y^2, x + z^2).$$

One can check that the inverse of $\phi$ is

$$\phi^{-1}(x, y, z) = \bigl(z - (y - x^2)^2, x, y - x^2\bigr).$$

Homogenizing $x = X_1/X_0$, $y = X_2/X_0$, $z = X_3/X_0$, we have the formulas

$$\bar{\phi} = [X_0^2,\ X_0X_2,\ X_0X_3 + X_2^2,\ X_0X_1 + X_3^2],$$
$$\bar{\phi}^{-1} = [X_0^4,\ X_0^3X_3 - (X_0X_2 - X_1^2)^2,\ X_0^3X_1,\ X_0^3X_2 - X_0^2X_1^2],$$

from which it is easy to check that

$$Z(\phi) = \{X_0 = X_2 = X_3 = 0\} = \{[0, 1, 0, 0]\},$$
$$Z(\phi^{-1}) = \{X_0 = X_1 = 0\} = \{[0, 0, u, v]\}.$$

Thus $Z(\phi)$ consists of a single point, while $Z(\phi^{-1})$ is a line. In the notation of Theorem 7.10, we have $N = 3$ and

$$d_1 = \deg\phi = 2, \qquad d_2 = \deg\phi^{-1} = 4,$$
$$\ell_1 = \dim Z(\phi) + 1 = 1, \qquad \ell_2 = \dim Z(\phi^{-1}) = 1 = 2.$$

The map $\phi$ is regular, since $Z(\phi) \cap Z(\phi^{-1}) = \emptyset$.

*Remark* 7.13. Let $\phi : \mathbb{A}^N \to \mathbb{A}^N$ be an affine morphism and let $\Phi : \mathbb{A}^{N+1} \to \mathbb{A}^{N+1}$ be a lift of $\phi$. The map $\phi$ is called *algebraically stable* if

$$\Phi^n\bigl(\{X_0 = 0\}\bigr) \ne \{0\} \quad \text{for all } n \ge 1.$$

In other words, $\phi$ is algebraically stable if for every $n \ge 1$, some coordinate of $\Phi^n(X_0, \dots, X_N)$ is not divisible by $X_0$. Since the first coordinate of $\Phi^n$ is a power of $X_0$, this implies that there can be no cancellation among the coordinates, so an algebraically stable map $\phi$ satisfies

$$\deg(\phi^n) = (\deg\phi)^n.$$

Further, an adaptation of the proof of Theorem 7.10(a) shows that

$$Z(\phi^n) \subset Z(\phi^m) \quad \text{for all } n < m.$$

Regular automorphisms are algebraically stable, but there are algebraically stable automorphisms that are not regular. For a discussion of the complex dynamics of algebraically stable maps, see [174, 187, 401].

*Remark* 7.14. For arbitrary rational maps $\phi : \mathbb{P}^N \to \mathbb{P}^N$, the *dynamical degree of* $\phi$ is defined to be the quantity

$$\operatorname{dyndeg}(\phi) = \lim_{n\to\infty} \deg(\phi^n)^{1/n},$$

and its logarithm $\log \operatorname{dyndeg}(\phi)$ is called the *algebraic entropy of* $\phi$. (One can show that the dynamical degree is in fact the infimum of $\deg(\phi^n)^{1/n}$.) The dynamical degree provides a coarse measure of the stable complexity of the map $\phi$, and presumably it has a major impact on the arithmetic properties of $\phi$. See [10, 199, 290] for an indication of this effect in certain cases.

The dynamical degree need not be an integer, or even a rational number; see Exercise 7.4 for an example. However, Bellon and Viallet [49] have conjectured that it is always an algebraic integer.

The dynamical degree, and more generally the sequence of integers

$$d_n = \deg(\phi^n), \quad n = 0, 1, 2, \ldots,$$

can be quite difficult to describe. See [10, 45, 46, 49, 77, 133, 199] for work on this problem. In many cases the sequence $(d_n)_{n\ge 0}$ satisfies a linear recurrence with rational coefficients, or equivalently, the generating function $\sum_{n\ge 0} d_n T^n$ is in $\mathbb{Q}(T)$. However, see [46] for an example of a birational map $\phi : \mathbb{P}^N \to \mathbb{P}^N$ whose degree generating function is not in $\mathbb{Q}(T)$.

### 7.1.4 A Height Bound for Jointly Regular Affine Morphisms

In this section we prove a nontrivial lower bound for the height of points under regular affine automorphisms. The theorem is an amalgamation of results due to Denis [131], Kawaguchi [230, 231], Marcello [287, 288, 289, 290], and Silverman [413, 418]. Before stating the theorem, we need to define what is meant by the height of a point in affine space.

**Definition.** The height $h(P)$ of a point $P = (x_1, \ldots, x_N) \in \mathbb{A}^N(\bar{\mathbb{Q}})$ in affine space is defined to be the height of the associated point in projective space using the natural embedding $\mathbb{A}^N \to \mathbb{P}^N$,

$$h(P) = h\bigl([1, x_1, \ldots, x_N]\bigr).$$

Eventually we will apply the following height estimate to a regular affine automorphism $\phi$ and its inverse $\phi^{-1}$, but it is no harder to prove the result for any pair of jointly regular maps, and working in a general setting helps clarify the underlying structure of the proof.

**Theorem 7.15.** *Let* $\phi_1 : \mathbb{A}^N \to \mathbb{A}^N$ *and* $\phi_2 : \mathbb{A}^N \to \mathbb{A}^N$ *be affine morphisms with the property that*

$$Z(\phi_1) \cap Z(\phi_2) = \emptyset.$$

(*We say that* $\phi_1$ *and* $\phi_2$ *are* jointly regular.) *Let*

$$d_1 = \deg \phi_1 \qquad \textit{and} \qquad d_2 = \deg \phi_2.$$

*There is a constant* $C = C(\phi_1, \phi_2)$ *such that for all* $P \in \mathbb{A}^N(\bar{\mathbb{Q}})$,

$$\frac{1}{d_1} h(\phi_1(P)) + \frac{1}{d_2} h(\phi_2(P)) \geq h(P) - C. \tag{7.4}$$

*Remark* 7.16. We recall that the upper bound

$$h(\psi(P)) \leq (\deg \psi) h(P) + O(1) \tag{7.5}$$

is valid even for rational maps $\psi : \mathbb{P}^N \to \mathbb{P}^N$ (see Theorem 3.11), since the proof of (7.5) uses only the triangle inequality. Thus Theorem 7.15 may be viewed as providing a nontrivial lower bound complementary to the elementary upper bound

$$\frac{1}{d_1} h(\phi_1(P)) + \frac{1}{d_2} h(\phi_2(P)) \leq 2h(P) + O(1).$$

*Proof of Theorem* 7.15. Write the rational functions $\mathbb{P}^N \to \mathbb{P}^N$ induced by $\phi_1$ and $\phi_2$ as

$$\bar{\phi}_1 = [X_0^{d_1}, \bar{F}_1, \bar{F}_2, \ldots, \bar{F}_N] \qquad \text{and} \qquad \bar{\phi}_2 = [X_0^{d_2}, \bar{G}_1, \bar{G}_2, \ldots, \bar{G}_N],$$

where the $\bar{F}_i$ are homogeneous polynomials of degree $d_1$ and the $\bar{G}_i$ are homogeneous polynomials of degree $d_2$. The loci of indeterminacy of $\phi_1$ and $\phi_2$ are given by

$$Z(\phi_1) = \{X_0 = \bar{F}_1 = \cdots = \bar{F}_N = 0\},$$
$$Z(\phi_2) = \{X_0 = \bar{G}_1 = \cdots = \bar{G}_N = 0\}.$$

We define a rational map $\psi : \mathbb{P}^{2N} \to \mathbb{P}^{2N}$ of degree $d_1 d_2$ by

$$\psi = \left[X_0^{d_1 d_2}, \bar{F}_1^{d_2}, \ldots, \bar{F}_N^{d_2}, \bar{G}_1^{d_1}, \ldots, \bar{G}_N^{d_1}\right].$$

The locus of indeterminacy of $\psi$ is the set

$$Z(\psi) = \{X_0 = \bar{F}_1 = \cdots = \bar{F}_N = \bar{G}_1 = \cdots = \bar{G}_N = 0\} = Z(\phi_1) \cap Z(\phi_2) = \emptyset,$$

since by assumption $Z(\phi_1)$ and $Z(\phi_2)$ are disjoint. Hence $\psi$ is a morphism, so we can apply the fundamental height estimate for morphisms (Theorem 3.11) to deduce that

$$h(\psi(P)) = d_1 d_2 h(P) + O(1) \qquad \text{for all } P \in \mathbb{P}^{2N}(\bar{\mathbb{Q}}). \tag{7.6}$$

The following lemma will give us an upper bound for the height of $\psi(P)$.

**Lemma 7.17.** *Let* $u, a_1, \ldots, a_N, b_1, \ldots, b_N \in \bar{\mathbb{Q}}$ *with* $u \neq 0$. *Then*

$$h([u, a_1, \ldots, a_N, b_1, \ldots, b_N]) \leq h([u, a_1, \ldots, a_N]) + h([u, b_1, \ldots, b_N]).$$

*Proof.* Let $\alpha_i = a_i/u$ and $\beta_i = b_i/u$ for $1 \le i \le N$. Then for any absolute value $v$ we have the trivial estimate

$$\max\{1, |\alpha_1|_v, \dots, |\alpha_N|_v, |\beta_1|_v, \dots, |\beta_N|_v\} \\ \le \max\{1, |\alpha_1|_v, \dots, |\alpha_N|_v\} \cdot \max\{1, |\beta_1|_v, \dots, |\beta_N|_v\}.$$

Raising to an appropriate power, multiplying over all absolute values, and taking logarithms yields

$$h([1, \alpha_1, \dots, \alpha_N, \beta_1, \dots, \beta_N]) \le h([1, \alpha_1, \dots, \alpha_N]) + h([1, \beta_1, \dots, \beta_N]).$$

This is the desired result, since the height does not depend on the choice of homogeneous coordinates of a point. □

We apply Lemma 7.17 to the point

$$\psi(P) = \left[X_0(P)^{d_1 d_2}, \bar{F}_1(P)^{d_2}, \dots, \bar{F}_N(P)^{d_2}, \bar{G}_1(P)^{d_1}, \dots, \bar{G}_N(P)^{d_1}\right]$$

with $P \in \mathbb{A}^N(\bar{\mathbb{Q}})$, which ensures that $X_0(P) \ne 0$. The lemma tells us that

$$\begin{aligned} h(\psi(P)) &\le h\left(\left[X_0(P)^{d_1 d_2}, \bar{F}_1(P)^{d_2}, \dots, \bar{F}_N(P)^{d_2}\right]\right) \\ &\quad + h\left(\left[X_0(P)^{d_1 d_2}, \bar{G}_1(P)^{d_1}, \dots, \bar{G}_N(P)^{d_1}\right]\right) \\ &= d_2 h\left(\left[X_0(P)^{d_1}, \bar{F}_1(P), \dots, \bar{F}_N(P)\right]\right) \\ &\quad + d_1 h\left(\left[X_0(P)^{d_2}, \bar{G}_1(P), \dots, \bar{G}_N(P)\right]\right) \\ &= d_2 h(\phi_1(P)) + d_1 h(\phi_2(P)). \end{aligned}$$

We combine this with (7.6) to obtain

$$d_1 d_2 h(P) + O(1) = h(\psi(P)) \le d_2 h(\phi_1(P)) + d_1 h(\phi_2(P)).$$

Dividing both sides by $d_1 d_2$ completes the proof of Theorem 7.15. □

For regular affine automorphisms, it is conjectured that the height inequality (7.4) in Theorem 7.15 may be replaced by a stronger estimate.

**Conjecture 7.18.** *Let $\phi : \mathbb{A}^N \to \mathbb{A}^N$ be a regular affine automorphism. Then there is a constant $C = C(\phi)$ such that for all $P \in \mathbb{A}^N(\bar{\mathbb{Q}})$,*

$$\frac{1}{d_1} h(\phi(P)) + \frac{1}{d_2} h(\phi^{-1}(P)) \ge \left(1 + \frac{1}{d_1 d_2}\right) h(P) - C. \tag{7.7}$$

Kawaguchi [230] proves Conjecture 7.18 in dimension 2, i.e., for regular affine automorphisms $\phi : \mathbb{A}^2 \to \mathbb{A}^2$; see also [413]. However, for general jointly regular affine morphisms, it is easy to see that (7.4) cannot be improved; see Exercise 7.8. Kawaguchi also constructs canonical heights for maps that satisfy (7.7); see [230] and Exercises 7.17–7.22.

### 7.1.5 Boundedness of Periodic Points for Regular Automorphisms of $\mathbb{A}^N$

Theorem 7.15 applied to a regular affine automorphism $\phi$ and its inverse implies that at least one of $\phi(P)$ and $\phi^{-1}(P)$ has reasonably large height. This suffices to prove that the periodic points of $\phi$ form a set of bounded height, a result first demonstrated by Marcello [287, 288] (see also [131, 418]) using a height bound slightly weaker than the one in Theorem 7.15.

**Theorem 7.19.** (Marcello) *Let $\phi : \mathbb{A}^N \to \mathbb{A}^N$ be a regular affine automorphism of degree at least 2 defined over $\bar{\mathbb{Q}}$. Then* $\operatorname{Per}(\phi)$ *is a set of bounded height in $\mathbb{A}^N(\bar{\mathbb{Q}})$. In particular,*

$$\operatorname{Per}(\phi) \cap \mathbb{A}^N(K) \text{ is finite for all number fields } K.$$

*Proof.* Let

$$d_1 = \deg \phi \qquad \text{and} \qquad d_2 = \deg \phi^{-1}.$$

Applying Theorem 7.15 with $\phi_1 = \phi$ and $\phi_2 = \phi^{-1}$ yields the basic inequality

$$\frac{1}{d_1} h(\phi(P)) + \frac{1}{d_2} h(\phi^{-1}(P)) \geq h(P) - C, \tag{7.8}$$

where $C$ is a constant depending on $\phi$, but not on $P \in \mathbb{A}^N(\bar{\mathbb{Q}})$.

We prove the theorem initially under the assumption that $d_1 d_2 > 4$. Define a function

$$f(P) = \frac{1}{d_1} h(P) - \frac{1}{\alpha d_2} h(\phi^{-1}(P)) - \frac{C}{\alpha - 1}, \tag{7.9}$$

where the real number $\alpha > 1$ will be specified later. Then $f$ satisfies

$$\begin{aligned} f(\phi(P)) - \alpha f(P) &= \left( \frac{1}{d_1} h(\phi(P)) - \frac{1}{\alpha d_2} h(P) - \frac{C}{\alpha - 1} \right) \\ &\quad - \alpha \left( \frac{1}{d_1} h(P) - \frac{1}{\alpha d_2} h(\phi^{-1}(P)) - \frac{C}{\alpha - 1} \right) \\ &= \left( \frac{1}{d_1} h(\phi(P)) + \frac{1}{d_2} h(\phi^{-1}(P)) \right) - \left( \frac{\alpha}{d_1} + \frac{1}{\alpha d_2} \right) h(P) + C \\ &\geq \left( 1 - \frac{\alpha}{d_1} - \frac{1}{\alpha d_2} \right) h(P) \qquad \text{from (7.8).} \end{aligned}$$

Hence if we take

$$\alpha = \frac{d_1 d_2 + \sqrt{(d_1 d_2)^2 - 4 d_1 d_2}}{2 d_2},$$

then

$$1 - \frac{\alpha}{d_1} - \frac{1}{\alpha d_2} = 0,$$

and our assumption that $d_1 d_2 > 4$ ensures that $\alpha > 1$, so for this choice of $\alpha$ we conclude that

$$f(\phi(P)) \geq \alpha f(P) \qquad \text{for all } P \in \mathbb{A}^N(\bar{\mathbb{Q}}).$$

Applying this estimate to the points $P, \phi(P), \phi^2(P), \ldots, \phi^{n-1}(P)$, we obtain the fundamental inequality

$$f(\phi^n(P)) \geq \alpha^n f(P) \qquad \text{for all } P \in \mathbb{A}^N(\bar{\mathbb{Q}}) \text{ and all } n \geq 0. \tag{7.10}$$

Similarly, we define

$$g(P) = \frac{1}{d_2} h(P) - \frac{1}{\beta d_1} h(\phi(P)) - \frac{C}{\beta - 1} \tag{7.11}$$

and take

$$\beta = \frac{d_1 d_2 + \sqrt{(d_1 d_2)^2 - 4 d_1 d_2}}{2 d_1}.$$

Then an analogous calculation, which we leave to the reader, shows that $g$ satisfies

$$g(\phi^{-1}(P)) \geq \beta g(P) \qquad \text{for all } P \in \mathbb{A}^N(\bar{\mathbb{Q}}),$$

from which we deduce that

$$g(\phi^{-n}(P)) \geq \beta^n g(P) \qquad \text{for all } P \in \mathbb{A}^N(\bar{\mathbb{Q}}) \text{ and all } n \geq 0. \tag{7.12}$$

We compute

$$\begin{aligned}
&\alpha^{-n} f(\phi^{n+1}(P)) + \beta^{-n} g(\phi^{-n-1}(P)) \\
&\quad \geq f(\phi(P)) + g(\phi^{-1}(P)) \qquad \text{from (7.10) and (7.12),} \\
&\quad = \left(\frac{1}{d_1} h(\phi(P)) - \frac{1}{\alpha d_2} h(P) - \frac{C}{\alpha - 1}\right) \\
&\quad\quad + \left(\frac{1}{d_2} h(\phi^{-1}(P)) - \frac{1}{\beta d_1} h(P) - \frac{C}{\beta - 1}\right) \\
&\quad\quad\quad \text{from the definition (7.9) and (7.11) of } f \text{ and } g, \\
&\quad \geq \left(1 - \frac{1}{\alpha d_2} - \frac{1}{\beta d_1}\right) h(P) - \left(1 + \frac{1}{\alpha - 1} + \frac{1}{\beta - 1}\right) C \qquad \text{from (7.8).}
\end{aligned}$$

Using the definition of $f$ and $g$ and rearranging the terms, we have proven the inequality

$$\begin{aligned}
&\frac{h(\phi^{n+1}(P))}{\alpha^n d_1} + \frac{h(\phi^{-n-1}(P))}{\beta^n d_2} + \frac{(\alpha\beta - 1)C}{(\alpha - 1)(\beta - 1)} \\
&\quad \geq \left(1 - \frac{1}{\alpha d_2} - \frac{1}{\beta d_1}\right) h(P) + \frac{h(\phi^n(P))}{\alpha^{n+1} d_2} + \frac{h(\phi^{-n}(P))}{\beta^{n+1} d_1}.
\end{aligned} \tag{7.13}$$

Now suppose that $P \in \mathbb{A}^N(\bar{\mathbb{Q}})$ is a periodic point for $\phi$. Then $h(\phi^k(P))$ is bounded independently of $k$, so letting $n \to \infty$ in (7.13) yields

$$\frac{(\alpha\beta - 1)C}{(\alpha-1)(\beta-1)} \geq \left(1 - \frac{1}{\alpha d_2} - \frac{1}{\beta d_1}\right) h(P),$$

where we are using the fact that $\alpha > 1$ and $\beta > 1$. Our assumption that $d_1 d_2 > 4$ also ensures that

$$1 - \frac{1}{\alpha d_2} - \frac{1}{\beta d_1} = \sqrt{1 - \frac{4}{d_1 d_2}} > 0,$$

so the height of $P$ is bounded by a constant depending only on $\phi$. This completes the proof of the first assertion of Theorem 7.19 under the assumption that $d_1 d_2 > 4$, and the second is immediate from Theorem 7.29(f), which says that for any given number field, $\mathbb{P}^N(K)$ contains only finitely many points of bounded height.

In order to deal with the case $d_1 d_2 \leq 4$, i.e., $d_1 = d_2 = 2$, we use Theorem 7.10, which tells us that $\phi^2$ is regular and has degree $d_1^2$. Similarly, $\deg(\phi^{-2}) = d_2^2$. Hence from what we have already proven, the periodic points of $\phi^2$ form a set of bounded height, and since it is easy to see that $\operatorname{Per}(\phi) = \operatorname{Per}(\phi^2)$, this completes the proof in all cases. □

*Remark* 7.20. We observe that Theorem 7.19 applies only to regular maps. It cannot be true for all affine automorphisms, since there are affine automorphisms whose fixed (or periodic) points include components of positive dimension. For example, the affine automorphism $\phi(x, y) = (x, y + f(x))$ fixes all points of the form $(a, b)$ satisfying $f(a) = 0$. Of course, this map $\phi$ is not regular, since one easily checks that

$$Z(\phi) = Z(\phi^{-1}) = \{[0, 1, 0]\}.$$

**Definition.** Let $\phi : V \to V$ be a morphism of a (not necessarily projective) variety $V$. A point $P \in \operatorname{Per}(\phi)$ is *isolated* if $P$ is not in the closure of $\operatorname{Per}_n(\phi) \smallsetminus \{P\}$ for all $n \geq 0$. In particular, if $\operatorname{Per}_n(\phi)$ is finite for all $n$, then every periodic point is isolated.

**Conjecture 7.21.** *Let $\phi : \mathbb{A}^N \to \mathbb{A}^N$ be an affine automorphism of degree at least* 2 *defined over $\bar{\mathbb{Q}}$. Then the set of isolated periodic points of $\phi$ is a set of bounded height in $\mathbb{A}^N(\bar{\mathbb{Q}})$.*

A classification theorem of Friedland and Milnor [176] says that every automorphism $\phi : \mathbb{A}^2 \to \mathbb{A}^2$ of the affine plane is conjugate to a composition of elementary maps and Hénon maps. Using this classification, Denis [131] proved Conjecture 7.21 in dimension 2. (See also [287, 288].)

## 7.2 Primer on Algebraic Geometry

In this section we summarize basic material from algebraic geometry, primarily having to do with the theory of divisors, linear equivalence, and the divisor class group (Picard group). This theory is used to describe the geometry of algebraic varieties and the geometry of the maps between them. We assume that the reader is familiar with basic material on algebraic varieties as may be found in any standard textbook, such as [186, 197, 198, 205].

This section deals with geometry, so we work over an algebraically closed field. Let

$$\begin{aligned} K &= \text{an algebraically closed field,} \\ V &= \text{a nonsingular irreducible projective variety defined over } K, \\ K(V) &= \text{the field of rational functions on } V. \end{aligned}$$

### 7.2.1 Divisors, Linear Equivalence, and the Picard Group

In this section we recall the theory of divisors, linear equivalence, and the divisor class group (Picard group).

**Definition.** A *prime divisor* on $V$ is an irreducible subvariety $W \subset V$ of codimension 1. The *divisor group of* $V$, denoted by $\operatorname{Div}(V)$, is the free abelian group generated by the prime divisors on $V$. Thus $\operatorname{Div}(V)$ consists of all formal sums

$$\sum_W n_W W,$$

where the sum is over prime divisors $W \subset V$, the coefficients $n_W$ are integers, and only finitely many $n_W$ are nonzero. The *support* of a divisor $D = \sum n_W W$ is

$$|D| = \bigcup_{\substack{W \text{ with} \\ n_W \neq 0}} W.$$

If $W$ is a prime divisor of $V$, then the *local ring at* $W$ is the ring

$$\mathcal{O}_{V,W} = \{f \in K(V) : f \text{ is defined at some point of } W\}.$$

It is a discrete valuation ring whose fraction field is $K(V)$. Normalizing the valuation so that $\operatorname{ord}_W(K(V)^*) = \mathbb{Z}$, we say that

$$\operatorname{ord}_W(f) = \text{order of vanishing of } f \text{ along } W.$$

Then $f$ *vanishes on* $W$ if $\operatorname{ord}_W(f) \geq 1$, and $f$ *has a pole on* $W$ if $\operatorname{ord}_W(f) \leq -1$.

**Definition.** Let $f \in K(V)^*$ be a nonzero rational function on $V$. The *divisor of* $f$ is the divisor

$$(f) = \sum_W \operatorname{ord}_W(f) W \in \operatorname{Div}(V).$$

A *principal divisor* is a divisor of the form $(f)$ for some $f \in K(V)$. The principal divisors form a subgroup of $\operatorname{Div}(V)$. The *divisor class group* (or *Picard group*) *of* $V$ is the quotient group

$$\operatorname{Pic}(V) = \frac{\operatorname{Div}(V)}{(\text{principal divisors})}.$$

Two divisors $D_1, D_2 \in \operatorname{Div}(V)$ are *linearly equivalent* if they differ by a principal divisor, $D_1 = D_2 + (f)$, i.e., if their difference is in the kernel of the natural map

$$\operatorname{Div}(V) \longrightarrow \operatorname{Pic}(V).$$

We write $D_1 \sim D_2$ to denote linear equivalence.

The next proposition follows directly from the definitions and the fact that every nonconstant function on a projective variety $V$ has nontrivial zeros and poles.

**Proposition 7.22.** *There is an exact sequence*

$$1 \longrightarrow K^* \longrightarrow K(V)^* \xrightarrow{f \mapsto (f)} \operatorname{Div}(V) \longrightarrow \operatorname{Pic}(V) \longrightarrow 0.$$

*Remark* 7.23. The exact sequence in Proposition 7.22 is analogous to the fundamental exact sequence in algebraic number theory,

$$1 \longrightarrow (\text{units}) \longrightarrow \begin{pmatrix}\text{multiplicative}\\ \text{group}\end{pmatrix} \longrightarrow \begin{pmatrix}\text{fractional}\\ \text{ideals}\end{pmatrix} \longrightarrow \begin{pmatrix}\text{ideal class}\\ \text{group}\end{pmatrix} \longrightarrow 1.$$

**Definition.** Let $\phi : V \to V'$ be a morphism of nonsingular projective varieties and let $W' \subset V'$ be a prime divisor such that $\phi(V)$ is not contained in $W'$. Then $\phi^{-1}(W')$ breaks up into a disjoint union of prime divisors, say

$$\phi^{-1}(W') = W_1 \cup \cdots \cup W_r.$$

Let $f \in K(V')$ be a uniformizer at $W'$, i.e., $\operatorname{ord}_{W'}(f) = 1$. Then the *pullback of* $W'$ *by* $\phi$ is defined to be the divisor

$$\phi^* W' = \sum_{i=1}^{r} \operatorname{ord}_{W_i}(f \circ \phi) W_i \in \operatorname{Div}(V).$$

More generally, if $D' = \sum n_{W'} W' \in \operatorname{Div}(V')$, the pullback of $D'$ is the divisor

$$\phi^* D' = \sum_{W'} n_{W'} \phi^*(W'),$$

provided that all of the terms with $n_{W'} \neq 0$ are well-defined. Thus $\phi^* D'$ is defined if and only if $\phi(V) \not\subset |D'|$.

There is also a way to push divisors forward.

**Definition.** Let $\phi : V \to V'$ be a morphism of nonsingular projective varieties, let $W \subset V$ be a prime divisor, and let $W' = \phi(W)$. If $\dim W' = \dim W$, then the function field $K(W)$ is a finite extension of the function field $K(W')$ via the inclusion

$$\phi^* : K(W') \hookrightarrow K(W), \qquad \phi^*(f) = f \circ \phi,$$

and we define the *pushforward of* $W$ *by* $\phi$ to be the divisor

$$\phi_* W = \bigl[K(W) : K(W')\bigr] W' \in \operatorname{Div}(V').$$

If $\dim W' < \dim W$, we define $\phi_* W = 0$. And in general, for an arbitrary divisor $D = \sum n_W W \in \operatorname{Div}(V)$, the pushforward of $D$ is

$$\phi_* D = \sum_W n_W \phi_*(W).$$

*Example* 7.24. If $\phi : V \to V'$ is a finite map, then

$$\phi_* \phi^* D' = \deg(\phi) D' \quad \text{for all } D' \in \operatorname{Div}(V').$$

**Proposition 7.25.** *Let $\phi : V \to V'$ be a morphism of nonsingular projective varieties.*

(a) *Every $D' \in \operatorname{Div}(V')$ is linearly equivalent to a divisor $D'' \in \operatorname{Div}(V')$ satisfying $\phi(V) \not\subset |D''|$.*

(b) *If $D'$ and $D''$ are linearly equivalent divisors on $V'$ such that $\phi^* D'$ and $\phi^* D''$ are both defined, then $\phi^* D'$ and $\phi^* D''$ are linearly equivalent.*

(c) *Using* (a) *and* (b), *the map*

$$\phi^* : \bigl\{ D' \in \operatorname{Div}(V') : \phi(V) \not\subset |D'| \bigr\} \longrightarrow \operatorname{Div}(V)$$

*extends uniquely to a homomorphism*

$$\phi^* : \operatorname{Pic}(V') \longrightarrow \operatorname{Pic}(V).$$

*Example* 7.26. A prime divisor $W$ of $\mathbb{P}^N$ is the zero set of an irreducible homogeneous polynomial $F \in K[X_0, \ldots, X_N]$. We define the *degree of* $W$ to be the degree of the polynomial $F$ and extend this to obtain a homomorphism

$$\deg : \operatorname{Div}(\mathbb{P}^N) \longrightarrow \mathbb{Z}, \qquad \deg\Bigl(\sum_W n_W W\Bigr) = \sum_W n_W \deg(W).$$

It is not hard to see that a divisor on $\mathbb{P}^N$ is principal if and only if it has degree 0, so the degree map gives an isomorphism

$$\deg : \operatorname{Pic}(\mathbb{P}^N) \xrightarrow{\ \sim\ } \mathbb{Z}.$$

Any hyperplane $H \subset \mathbb{P}^N$ is a generator of $\operatorname{Pic}(\mathbb{P}^N)$.

*Example* 7.27. A prime divisor of $\mathbb{P}^N \times \mathbb{P}^M$ is the zero set of an irreducible bihomogeneous polynomial $F \in K[X_0, \ldots, X_N, Y_0, \ldots, Y_M]$. We say that $F$ and $W$ have *bidegree* $(d,e)$ if $F$ satisfies

$$F(\alpha X_0, \ldots, \alpha X_N, \beta Y_0, \ldots, \beta Y_M) = \alpha^d \beta^e F(X_0, \ldots, X_N, Y_0, \ldots, Y_M).$$

The bidegree map can be extended linearly to give an isomorphism

$$\operatorname{bideg} : \operatorname{Pic}(\mathbb{P}^N \times \mathbb{P}^M) \xrightarrow{\ \sim\ } \mathbb{Z} \times \mathbb{Z}.$$

Let $p_1 : \mathbb{P}^N \times \mathbb{P}^M \to \mathbb{P}^N$ and $p_2 : \mathbb{P}^N \times \mathbb{P}^M \to \mathbb{P}^M$ be the two projections and let $H_1$ be a hyperplane in $\mathbb{P}^N$ and $H_2$ a hyperplane in $\mathbb{P}^M$. Then $\operatorname{Pic}(\mathbb{P}^N \times \mathbb{P}^M)$ is generated by the divisors

$$p_1^* H_1 = H_1 \times \mathbb{P}^M \qquad \text{and} \qquad p_2^* H_2 = \mathbb{P}^N \times H_2.$$

### 7.2.2 Ample Divisors and Effective Divisors

**Definition.** A divisor $D = \sum n_W W$ is said to be *effective* (or *positive*) if $n_W \geq 0$ for all $W$. We write $D \geq 0$ to indicate that $D$ is effective.

The *base locus* of a divisor $D$, denoted by $\operatorname{Base}(D)$, is the intersection of the support of all of the effective divisors in the divisor class of $D$,

$$\operatorname{Base}(D) = \bigcap_{\substack{E \sim D \\ E \geq 0}} |E|.$$

Notice that any divisor is a difference of effective divisors,

$$D = \sum_{\substack{W \text{ with} \\ n_W > 0}} n_W W - \sum_{\substack{W \text{ with} \\ n_W < 0}} (-n_W) W.$$

**Definition.** Let $D \in \operatorname{Div}(V)$. Associated to $D$ is the finite-dimensional $K$-vector space

$$L(D) = \{f \in K(V) : (f) + D \geq 0\} \cup \{0\}.$$

We write $\ell(D) = \dim L(D)$ for the dimension of $L(D)$.

Let $D \in \operatorname{Div}(V)$ be a divisor with $\ell(D) \geq 1$. We choose a basis $f_1, \ldots, f_{\ell(D)}$ for $L(D)$ and use it to define a rational map

$$\phi_D = [f_1, \ldots, f_{\ell(D)}] : V \longrightarrow \mathbb{P}^{\ell(D)-1}.$$

The map $\phi_D$ is well-defined up to a linear change of coordinates on $\mathbb{P}^{\ell(D)-1}$, i.e., up to composition by an element of $\operatorname{PGL}_{\ell(D)}(K)$. Further, if $D$ and $D'$ are linearly equivalent, then $\phi_D$ and $\phi_{D'}$ differ by a change of coordinates.

Conversely, let $i : V \hookrightarrow \mathbb{P}^N$ be a morphism (or even a rational map) and let $H \subset \mathbb{P}^N$ be a hyperplane with $i(V) \not\subset H$. Then $i$ is equal to the composition of $\phi_{i^*H}$ with a change of coordinates and a projection.

**Definition.** A divisor $D \in \operatorname{Div}(V)$ is *very ample* if the map $\phi_D : V \to \mathbb{P}^{\ell(D)-1}$ is an embedding, i.e., an isomorphism onto its image. A divisor $D$ is *ample* if some multiple $nD$ with $n \geq 1$ is very ample. Ampleness and very ampleness are properties of the divisor class of $D$. Notice that if $i : V \hookrightarrow \mathbb{P}^N$ is an embedding and $H \in \operatorname{Div}(\mathbb{P}^N)$ is a hyperplane, then $\phi^* H$ is a very ample divisor on $V$.

*Example* 7.28. Let $p_1^* H_1$ and $p_2^* H_2$ be the generators of $\operatorname{Pic}(\mathbb{P}^N \times \mathbb{P}^M)$ described in Example 7.27. Then $p_1^* H_1 + p_2^* H_2$ is a very ample divisor on $\mathbb{P}^N \times \mathbb{P}^M$. The associated embedding is called the *Segre embedding*. It is given explicitly by the formula

$$\begin{array}{ccc} \mathbb{P}^N \times \mathbb{P}^M & \longrightarrow & \mathbb{P}^{NM+N+M} \\ ([X_0, \ldots, X_N], [Y_0, \ldots, Y_M]) & \longmapsto & [X_0 Y_0, X_0 Y_1, \ldots, X_i Y_j, \ldots, X_N Y_M]. \end{array}$$

Now let $V$ be a subvariety of $\mathbb{P}^N \times \mathbb{P}^M$, say $\phi : V \hookrightarrow \mathbb{P}^N \times \mathbb{P}^M$. Then

$$\phi^*(p_1^* H_1 + p_2^* H_2) = (p_1 \circ \phi)^* H_1 + (p_2 \circ \phi)^* H_2$$

is a very ample divisor on $V$.

## 7.3 The Weil Height Machine

The theory of height functions that we developed in Sections 3.1–3.5 provides a powerful tool for studying the arithmetic of morphisms $\phi : \mathbb{P}^N \to \mathbb{P}^N$ of projective space. For example, if $\phi$ has degree $d \geq 2$ and is defined over a number field $K$, then the fundamental estimate (Theorem 3.11)

$$h(\phi(P)) = d \cdot h(P) + O(1) \qquad \text{for all } P \in \mathbb{P}^N(\bar{\mathbb{Q}}) \tag{7.14}$$

and the fact (Theorem 3.7) that there are only finitely many points in $\mathbb{P}^N(K)$ of bounded height lead immediately to a proof of Northcott's theorem (Theorem 3.12) stating that $\phi$ has only finitely many $K$-rational preperiodic points.

Recall that the height $h(P)$ of a point $P \in \mathbb{P}^N(\bar{\mathbb{Q}})$ is a measure of the arithmetic complexity of $P$. Similarly, the degree of a finite morphism $\phi$ measures the geometric complexity of $\phi$. Thus an enlightening interpretation of (7.14) is that it translates the geometric statement "$\phi$ has degree $d$" into the arithmetic statement "$h(\phi(P))$ is approximately equal to $dh(P)$."

A natural way to define a height function on an arbitrary projective variety is to fix an embedding $\phi : V \hookrightarrow \mathbb{P}^N$ and define $h_V(P)$ to equal $h(\phi(P))$. Unfortunately, different projective embeddings yield different height functions. But letting $H$ denote a hyperplane in $\mathbb{P}^N$, one can show that if the divisors $\phi^*H$ and $\psi^*H$ are linearly equivalent, then the height functions attached to $\phi : V \hookrightarrow \mathbb{P}^N$ and $\psi : V \hookrightarrow \mathbb{P}^M$ differ by a bounded amount.

More intrinsically, the projective embedding $\phi$ determines the divisor class of the very ample divisor $\phi^*H$. This suggests assigning a height function to every divisor on $V$. The Weil height machine provides such a construction. It is a powerful tool that translates geometric facts described by divisor class relations into arithmetic facts described by height relations. As such, the Weil height machine is of fundamental importance in the study of arithmetic geometry and arithmetic dynamics on algebraic varieties of dimension greater than 1.

**Theorem 7.29.** (Weil Height Machine) *For every nonsingular variety $V/\bar{\mathbb{Q}}$ there exists a map*

$$h_V : \operatorname{Div}(V) \longrightarrow \{\text{functions } V(\bar{\mathbb{Q}}) \to \mathbb{R}\}, \qquad D \longmapsto h_{V,D},$$

*with the following properties*:

(a) (Normalization) *Let $H \subset \mathbb{P}^N$ be a hyperplane and let $h : \mathbb{P}^N(\bar{\mathbb{Q}}) \to \mathbb{R}$ be the absolute logarithmic height function on projective space defined in Section* 3.1. *Then*

$$h_{\mathbb{P}^N,H}(P) = h(P) + O(1) \qquad \textit{for all } P \in \mathbb{P}^N(\bar{\mathbb{Q}}).$$

(b) (Functoriality) *Let $\phi : V \to V'$ be a morphism of nonsingular varieties defined over $\bar{\mathbb{Q}}$ and let $D \in \operatorname{Div}(V')$. Then*

$$h_{V,\phi^*D}(P) = h_{V',D}(\phi(P)) + O(1) \qquad \textit{for all } P \in V(\bar{\mathbb{Q}}).$$

(c) (Additivity) *Let $D, E \in \mathrm{Div}(V)$. Then*

$$h_{V,D+E}(P) = h_{V,D}(P) + h_{V,E}(P) + O(1) \qquad \textit{for all } P \in V(\bar{\mathbb{Q}}).$$

(d) (Linear Equivalence) *Let $D, E \in \mathrm{Div}(V)$ with $D$ linearly equivalent to $E$. Then*

$$h_{V,D}(P) = h_{V,E}(P) + O(1) \qquad \textit{for all } P \in V(\bar{\mathbb{Q}}).$$

(e) (Positivity) *Let $D \in \mathrm{Div}(V)$ be an effective divisor. Then*

$$h_{V,D}(P) \ge O(1) \qquad \textit{for all } P \in V(\bar{\mathbb{Q}}) \smallsetminus \mathrm{Base}(D).$$

*That is, $h_{V,D}$ is bounded below for all points not in the base locus of $D$.*

(f) (Finiteness) *Let $D \in \mathrm{Div}(V)$ be ample. Then for all constants $A$ and $B$, the set*

$$\{P \in V(\bar{\mathbb{Q}}) : [\mathbb{Q}(P) : \mathbb{Q}] \le A \text{ and } h_{V,D}(P) \le B\}$$

*is finite. In particular, if $V$ is defined over a number field $K$ and if $L/K$ is a finite extension, then*

$$\{P \in V(L) : h_{V,D}(P) \le B\}$$

*is a finite set.*

(g) (Uniqueness) *The height functions $h_{V,D}$ are determined, up to $O(1)$, by the properties of* (a) *normalization,* (b) *functoriality, and* (c) *additivity. (It suffices to assume functoriality for projective embeddings $V \hookrightarrow \mathbb{P}^N$.)*

*Proof.* See [76, Chapter 2], [205, Theorem B.3.2], or [256, Chapter 4]. □

*Remark* 7.30. All of the $O(1)$ constants appearing in the Weil height machine (Theorem 7.29) depend on the various varieties, divisors, and morphisms. The key fact is that the $O(1)$ constants are independent of the points on the varieties. More precisely, Theorem 7.29 says that it is possible to choose functions $h_{V,D}$, one for each smooth projective variety $V$ and each divisor $D \in \mathrm{Div}(V)$, such that certain properties hold, where those properties involve constants that depend on the particular choice of functions $h_{V,D}$. In principle, one can write down particular functions $h_{V,D}$ and determine specific values for the associated $O(1)$ constants, so the Weil height machine is effective. In practice, the constants often depend on making the Nullstellensatz effective, so they tend to be rather large.

*Remark* 7.31. Many of the properties of the Weil height machine may be succinctly summarized by the statement that there is a unique homomorphism

$$h_V : \mathrm{Pic}(V) \longrightarrow \frac{\{\text{functions } V(\bar{K}) \to \mathbb{R}\}}{\{\text{bounded functions } V(\bar{K}) \to \mathbb{R}\}}$$

such that if $\phi : V \hookrightarrow \mathbb{P}^N$ is a projective embedding, then $h_{V,\phi^*H} = h + O(1)$.

*Example* 7.32. Let $\phi : \mathbb{P}^N \to \mathbb{P}^N$ be a morphism of degree $d$ and let $H \in \operatorname{Div}(\mathbb{P}^N)$ be a hyperplane. Then $\phi^* H \sim dH$, so Theorem 7.29 allows us to compute

$$h_{\mathbb{P}^N,H}\big(\phi(P)\big) = h_{\mathbb{P}^N,\phi^*H}(P)+O(1) = h_{\mathbb{P}^N,dH}(P)+O(1) = dh_{\mathbb{P}^N,H}(P)+O(1).$$

This formula is Theorem 3.11.

*Example* 7.33. Let $V$ be a subvariety of $\mathbb{P}^N \times \mathbb{P}^M$, say $\phi : V \hookrightarrow \mathbb{P}^N \times \mathbb{P}^M$. Continuing with the notation from Examples 7.27 and 7.28, the height of a point $P = [\mathbf{x}, \mathbf{y}] \in V$ with respect to the divisors $\phi^* p_1^* H_1$ and $\phi^* p_2^* H_2$ is given by

$$h_{V,\phi^*p_1^*H_1}(P) = h_{\mathbb{P}^N,H_1}\big(p_1\phi(P)\big) = h(\mathbf{x}),$$
$$h_{V,\phi^*p_2^*H_2}(P) = h_{\mathbb{P}^M,H_2}\big(p_2\phi(P)\big) = h(\mathbf{y}).$$

*Example* 7.34. This example uses properties of elliptic curves; see Sections 1.6.3 and 6.3. Let $E$ be an elliptic curve given by a Weierstrass equation. Then the $x$-coordinate on $E$, considered as a map $x : E \to \mathbb{P}^1$, satisfies

$$x^*(\infty) = 2(\mathcal{O}),$$

so we have

$$h_{E,(\mathcal{O})}(P) = \frac{1}{2} h_{\mathbb{P}^1,(\infty)}\big(x(P)\big) + O(1).$$

Note that the height $h_{\mathbb{P}^1,(\infty)}$ is just the usual height on $\mathbb{P}^1$ from Theorem 7.29(a).

Now let $d \ge 2$, let $[d] : E \to E$ denote the multiplication-by-$d$ map, and let

$$E[d] = \{P \in E : [d]P = \mathcal{O}\}.$$

The map $[d]$ is unramified and

$$\phi^*\big((\mathcal{O})\big) = \sum_{T \in E[d]} (T) \in \operatorname{Div}(E).$$

The group $E[d]$ is isomorphic as an abstract group to $\mathbb{Z}/d\mathbb{Z}\times\mathbb{Z}/d\mathbb{Z}$, so the sum of the points in $E[d]$ is $\mathcal{O}$. It follows from Proposition 6.18 that there is a linear equivalence of divisors

$$\phi^*\big((\mathcal{O})\big) \sim d^2(\mathcal{O}).$$

Hence we can apply Theorem 7.29 to compute

$$\begin{aligned} h_{E,(\mathcal{O})}\big([d]P\big) &= h_{E,[d]^*(\mathcal{O})}(P) + O(1) && \text{from functoriality (b),} \\ &= h_{E,d^2(\mathcal{O})}(P) + O(1) && \text{linear equivalence property (d),} \\ &= d^2 h_{E,(\mathcal{O})}(P) + O(1) && \text{from additivity (c).} \end{aligned}$$

Theorem 3.20 then tells us that there exists a function $\hat{h}_{E,(\mathcal{O})}$ on $E$ satisfying

$$\hat{h}_{E,(\mathcal{O})}\big([d]P\big) = d^2 \hat{h}_{E,(\mathcal{O})}(P) \qquad \text{and} \qquad \hat{h}_{E,(\mathcal{O})}(P) = h_{E,(\mathcal{O})}(P) + O(1).$$

The function $\hat{h}_{E,(\mathcal{O})}$ is called the *canonical height on* $E$. It has many applications, ranging from counting rational points to evaluating $L$-series. For further information about canonical heights on elliptic curves and abelian varieties, see, for example [205, B.5], [256, Chapter 5], or [410, VIII §9].

*Remark* 7.35. The reader should be aware that the theory of heights is often rephrased in the language of metrized line bundles, which offers greater flexibility, albeit at the cost of additional work to set up the general theory.

## 7.4 Dynamics on Surfaces with Noncommuting Involutions

An involution $\iota$ of a variety $V$ is a rational map $\iota : V \to V$ with the property that $\iota^2$ is the identity map on $V$. If we look at the quotient variety

$$W = V/\{\iota(P) = P\},$$

then the natural projection $p : V \to W$ is a double cover , and the effect of $\iota$ on $V$ is to switch the two sheets of the cover. Conversely, any double cover $p : V \to W$ induces an involution $\iota : V \to V$.

The dynamics of a single involution is not very exciting, but some varieties have two (or more) noncommuting involutions $\iota_1$ and $\iota_2$ whose composition $\phi = \iota_1 \circ \iota_2$ is an automorphism of $V$ of infinite order. The dynamics of such maps $\phi$ can be quite interesting. In this section we study in detail an example of this type. The material in this section is taken from [409].

### 7.4.1 K3 Surfaces in $\mathbb{P}^2 \times \mathbb{P}^2$

We consider a surface $S$ contained in $\mathbb{P}^2 \times \mathbb{P}^2$ defined by two bihomogeneous equations, one of bidegree $(1,1)$ and the other of bidegree $(2,2)$. Thus

$$S = \{(\mathbf{x},\mathbf{y}) \in \mathbb{P}^2 \times \mathbb{P}^2 : L(\mathbf{x},\mathbf{y}) = Q(\mathbf{x},\mathbf{y}) = 0\}$$

for bihomogeneous polynomials

$$\begin{aligned} L(\mathbf{x},\mathbf{y}) &= \sum_{i=0}^{2}\sum_{j=0}^{2} A_{ij}x_iy_j, \\ Q(\mathbf{x},\mathbf{y}) &= \sum_{0\le i\le j\le 2}\ \sum_{0\le k\le \ell\le 2} B_{ijk\ell}x_ix_jy_ky_\ell. \end{aligned} \tag{7.15}$$

The surface $S$ is determined by the coefficients

$$\mathbf{A} = [A_{00}, A_{01}, \dots, A_{22}] \in \mathbb{P}^8 \quad \text{and} \quad \mathbf{B} = [B_{0000}, B_{0001}, \dots, B_{2222}] \in \mathbb{P}^{35}$$

of the polynomials $L$ and $Q$. To indicate this dependence, we write $S_{\mathbf{A},\mathbf{B}}$.

There are two natural projections from $S$ to $\mathbb{P}^2$, which we denote by

$$p_1, p_2 : S \longrightarrow \mathbb{P}^2, \qquad p_1(\mathbf{x},\mathbf{y}) = \mathbf{x}, \qquad p_2(\mathbf{x},\mathbf{y}) = \mathbf{y}.$$

These projections are maps of degree 2. To see this, choose a generic point $\mathbf{a} \in \mathbb{P}^2$. Then

$$p_1^{-1}(\mathbf{a}) = \{(\mathbf{a},\mathbf{y}) \in \mathbb{P}^2 \times \mathbb{P}^2 : L(\mathbf{a},\mathbf{y}) = Q(\mathbf{a},\mathbf{y}) = 0\}$$

consists of two points (counted with multiplicity), since it is the intersection of the line $L(\mathbf{a},\mathbf{y}) = 0$ and the conic $Q(\mathbf{a},\mathbf{y}) = 0$ in $\mathbb{P}^2$. And similarly, $p_2$ is a map of degree 2.

In general, a degree-2 map between varieties induces an involution on the domain given by switching the two sheets of the cover. In our situation the maps $p_1$ and $p_2$ induce involutions $\iota_1$ and $\iota_2$ on $S_{\mathbf{A},\mathbf{B}}$. Explicitly, if $P = [\mathbf{a},\mathbf{b}] \in S_{\mathbf{A},\mathbf{B}}$, then $\iota_1(P) = [\mathbf{a},\mathbf{b}']$ is the point satisfying

$$p_1^{-1}(p_1(P)) = \{P, \iota_1(P)\},$$

and similarly, $\iota_2(P) = [\mathbf{a}',\mathbf{b}]$ is the point satisfying

$$p_2^{-1}(p_2(P)) = \{P, \iota_2(P)\}.$$

These involutions are uniquely determined as nonidentity maps $S_{\mathbf{A},\mathbf{B}} \to S_{\mathbf{A},\mathbf{B}}$ satisfying

$$p_1 \circ \iota_1 = p_1 \qquad \text{and} \qquad p_2 \circ \iota_2 = p_2.$$

We note that $\iota_1$ and $\iota_2$ are rational maps on $S$, i.e., they are given by rational functions. To see why this is true, observe that $\mathbf{b}$ and $\mathbf{b}'$ are the intersection points in $\mathbb{P}^2$ of the line and the conic

$$L(\mathbf{a},\mathbf{y}) = 0 \qquad \text{and} \qquad Q(\mathbf{a},\mathbf{y}) = 0.$$

Thus each of $\mathbf{b}$ and $\mathbf{b}'$ can be expressed as a rational function in the coordinates of the other. The following example will help make this clear, or see Exercise 7.25 for explicit formulas to compute $\iota_1$ and $\iota_2$.

*Example* 7.36. We illustrate the involutions on $S_{\mathbf{A},\mathbf{B}}$ using the example

$$\begin{aligned}
L(\mathbf{x},\mathbf{y}) &= x_0y_0 + x_1y_1 + x_2y_2,\\
Q(\mathbf{x},\mathbf{y}) &= x_0^2y_0^2 + 4x_0^2y_0y_1 - x_0^2y_1^2 + 7x_0^2y_1y_2 + 3x_0x_1y_0^2 + 3x_0x_1y_0y_1\\
&\quad + x_0x_1y_2^2 + x_1^2y_0^2 + 2x_1^2y_1^2 + 4x_1^2y_1y_2 - x_0x_2y_1^2\\
&\quad + 5x_0x_2y_0y_2 - 4x_1x_2y_1^2 - 4x_1x_2y_0y_2 - 2x_2^2y_0y_1 + 3x_2^2y_2^2.
\end{aligned}$$

The point $P = ([1,0,0],[0,7,1])$ is in $S(\mathbb{Q})$. In order to compute $\iota_1(P)$, we substitute the value $\mathbf{x} = [1,0,0]$ into $L$ and $Q$ and solve for $\mathbf{y}$. Thus

$$L([1,0,0],\mathbf{y}) = y_0 = 0 \quad \text{and} \quad Q([1,0,0],\mathbf{y}) = y_0^2 + 4y_0y_1 - y_1^2 + 7y_1y_2 = 0,$$

so the solutions are $\mathbf{y} = [0,y_1,y_2]$, where $y_1$ and $y_2$ are the roots of the polynomial $-y_1^2 + 7y_1y_2 = 0$. One solution is $y_1 = 7$, which gives the original point $P$, and the other solution is $y_1 = 0$, which gives $\iota_1(P) = ([1,0,0],[0,0,1])$.

Next we compute $\iota_2(P)$. To do this, we substitute $\mathbf{y} = [0,7,1]$ into $L$ and $Q$ to obtain

$$L(\mathbf{x},[0,7,1]) = 7x_1 + x_2 = 0,$$
$$Q(\mathbf{x},[0,7,1]) = x_0x_1 - 49x_0x_2 + 126x_1^2 - 196x_1x_2 + 3x_2^2 = 0.$$

Substituting $x_2 = -7x_1$ into the second equation gives

$$Q([x_0, x_1, -7x_1], [0,7,1]) = 344x_0x_1 + 1645x_1^2 = 0.$$

The solution $x_1 = 0$ gives back the original point $P$. The other solution is $[x_0, x_1] = [1645, -344]$, and then setting $x_2 = -7x_1 = 2408$ gives

$$\iota_2(P) = \iota_2([1,0,0],[0,7,1]) = ([1645, -344, 2408], [0,7,1]).$$

We could continue this process, but the size of the coordinates grows very rapidly. Indeed, the $\mathbf{y}$-coordinates of $\iota_1(\iota_2(P))$ are already integers with 12 to 13 digits.

*Remark* 7.37. The surface $S$ described by (7.15) is an example of a *K3 surface*. Formally, a K3 surface is a surface $S$ of Kodaira dimension 0 with the property that $H^1(S, \mathcal{O}_S) = 0$. However, all of the information that we will need is contained in the explicit equations (7.15) defining $S$. The reader desiring more information about the geometric properties of K3 surfaces might consult [40, 44, 178, 298]. The dynamics of K3 surfaces with nontrivial automorphisms are studied by Cantat [93] and McMullen [296].

*Remark* 7.38. The collection of K3 surfaces $S_{\mathbf{A},\mathbf{B}}$ is a 43-parameter family, since the coefficients $(\mathbf{A},\mathbf{B})$ vary over $\mathbb{P}^8 \times \mathbb{P}^{35}$. However, many of the surfaces are isomorphic. For example, we can use elements of $\mathrm{PGL}_3$ to change variables in each of two factors of $\mathbb{P}^2$. This reduces the dimension of the parameter space by 16, since $\mathrm{PGL}_3$ has dimension 8. Further, the surface $S_{\mathbf{A},\mathbf{B}}$ really depends only on the ideal generated by the bilinear form $L(\mathbf{x},\mathbf{y})$ and the biquadratic form $Q(\mathbf{x},\mathbf{y})$, so the surface does not change if we replace $Q(\mathbf{x},\mathbf{y})$ by

$$Q(\mathbf{x},\mathbf{y}) + L(\mathbf{x},\mathbf{y}) \cdot M(\mathbf{x},\mathbf{y})$$

for an arbitrary bilinear form $M(\mathbf{x},\mathbf{y})$. The space of such $M$ is 9-dimensional, so we see that the isomorphism classes of K3 surfaces $S_{\mathbf{A},\mathbf{B}}$ constitute a family of dimension at most

$$\underbrace{8}_{\mathbf{A}\in\mathbb{P}^8} + \underbrace{35}_{\mathbf{B}\in\mathbb{P}^{35}} - \underbrace{8}_{\mathrm{PGL}_3} - \underbrace{8}_{\mathrm{PGL}_3} - \underbrace{9}_{M} = 18.$$

One can prove that these are the only isomorphisms between the various $S_{\mathbf{A},\mathbf{B}}$, so there is an 18-parameter family of isomorphism classes of nonsingular surfaces $S_{\mathbf{A},\mathbf{B}}$.

**Definition.** There are several linear, quadratic, and quartic forms that come up naturally when one is working with the surface $S_{\mathbf{A},\mathbf{B}}$. We define linear and quadratic forms by setting

$$\begin{aligned} L_j^x(\mathbf{x}) &= \text{the coefficient of } y_j \text{ in } L(\mathbf{x},\mathbf{y}),\\ L_i^y(\mathbf{y}) &= \text{the coefficient of } x_i \text{ in } L(\mathbf{x},\mathbf{y}),\\ Q_{k\ell}^x(\mathbf{x}) &= \text{the coefficient of } y_k y_\ell \text{ in } Q(\mathbf{x},\mathbf{y}),\\ Q_{ij}^y(\mathbf{y}) &= \text{the coefficient of } x_i x_j \text{ in } Q(\mathbf{x},\mathbf{y}). \end{aligned} \tag{7.16}$$

This notation allows us to write the bilinear form $L$ and the biquadratic form $Q$ as

$$\begin{aligned} L(\mathbf{x},\mathbf{y}) &= \sum_{j=0}^{2} L_j^x(\mathbf{x}) y_j = \sum_{i=0}^{2} L_i^y(\mathbf{y}) x_i,\\ Q(\mathbf{x},\mathbf{y}) &= \sum_{0\le k\le \ell\le 2} Q_{k\ell}^x(\mathbf{x}) y_k y_\ell = \sum_{0\le i\le j\le 2} Q_{ij}^y(\mathbf{y}) x_i x_j. \end{aligned}$$

Then for each triple of distinct indices $i, j, k \in \{0, 1, 2\}$ we define quartic forms

$$\begin{aligned} G_k^x &= (L_j^x)^2 Q_{ii}^x - L_i^x L_j^x Q_{ij}^x + (L_i^x)^2 Q_{jj}^x,\\ G_k^y &= (L_j^y)^2 Q_{ii}^y - L_i^y L_j^y Q_{ij}^y + (L_i^y)^2 Q_{jj}^y,\\ H_{ij}^x &= 2L_i^x L_j^x Q_{kk}^x - L_i^x L_k^x Q_{jk}^x - L_j^x L_k^x Q_{ik}^x + (L_k^x)^2 Q_{ij}^x,\\ H_{ij}^y &= 2L_i^y L_j^y Q_{kk}^y - L_i^y L_k^y Q_{jk}^y - L_j^y L_k^y Q_{ik}^y + (L_k^y)^2 Q_{ij}^y. \end{aligned} \tag{7.17}$$

For some choices of $\mathbf{A}$ and $\mathbf{B}$, there may be points on the surface $S_{\mathbf{A},\mathbf{B}}$ at which $\iota_1$ or $\iota_2$ is not well-defined. The next proposition, which provides a criterion for checking whether $\iota_1$ and $\iota_2$ are defined at a point, shows how the quartic forms (7.17) naturally appear.

**Proposition 7.39.** *Let* $P = [\mathbf{a}, \mathbf{b}] \in S_{\mathbf{A},\mathbf{B}}$.
(a) *The involution* $\iota_1$ *is defined at* $P$ *unless*

$$G_0^x(\mathbf{a}) = G_1^x(\mathbf{a}) = G_2^x(\mathbf{a}) = H_{01}^x(\mathbf{a}) = H_{02}^x(\mathbf{a}) = H_{12}^x(\mathbf{a}) = 0.$$

(b) *The involution* $\iota_2$ *is defined at* $P$ *unless*

$$G_0^y(\mathbf{b}) = G_1^y(\mathbf{b}) = G_2^y(\mathbf{b}) = H_{01}^y(\mathbf{b}) = H_{02}^y(\mathbf{b}) = H_{12}^y(\mathbf{b}) = 0.$$

*Proof.* By symmetry, it is enough to prove (a). The map $\iota_1$ is defined at $P = [\mathbf{a}, \mathbf{b}]$ if and only if the fiber $p_1^{-1}(\mathbf{a})$ consists of exactly two points. That fiber is the set of points $[\mathbf{a}, \mathbf{y}]$ satisfying

$$L(\mathbf{a},\mathbf{y}) = Q(\mathbf{a},\mathbf{y}) = 0,$$

so as long as these two polynomials are not zero, the $\mathbf{y}$ values are given by the intersection of a line and a conic in $\mathbb{P}^2$. If a line and a conic intersect properly, then they intersect in exactly two points, counted with multiplicity. Further, given one solution $\mathbf{y} = \mathbf{b}$, the coordinates of the second solution $\mathbf{b}'$ are rational functions of $\mathbf{b}$ and the coefficients of $L(\mathbf{a}, \mathbf{y})$ and $Q(\mathbf{a}, \mathbf{y})$. Hence $\iota_1$ is a morphism[1] except in the following two situations:

[1]We leave for the reader to check that everything works in a neighborhood of points where the line $L(\mathbf{a}, \mathbf{y}) = 0$ is tangent to the conic $Q(\mathbf{a}, \mathbf{y}) = 0$.

- $L(\mathbf{a},\mathbf{y})$ is identically 0.
- $L(\mathbf{a},\mathbf{y}) = 0$ is a line that is contained in the set where $Q(\mathbf{a},\mathbf{y}) = 0$.

With the notation defined by (7.16) and (7.17), we use the bihomogeneity of $Q$ to write

$$(L_0^x)^2 Q(x_0,x_1,x_2,y_0,y_1,y_2) = Q(x_0,x_1,x_2,L_0^x y_0, L_0^x y_1, L_0^x y_2),$$

and then we eliminate the variable $y_0$ by substituting $L_0^x y_0 = L - L_1^x y_1 - L_2^x y_2$. After some algebra, we obtain an identity of the form

$$\begin{aligned}(L_0^x)^2 Q = G_2^x y_1^2 + H_{12}^x y_1 y_2 + G_1^x y_2^2 \\ + L\{Q_{00}^x L + (L_0^x Q_{01}^x - 2L_1^x Q_{00}^x) y_1 + (L_0^x Q_{02}^x - 2L_2^x Q_{00}^x) y_2\}.\end{aligned}$$

Since we will be interested in studying points $[\mathbf{x},\mathbf{y}]$ satisfying $L(\mathbf{x},\mathbf{y}) = 0$, we write this identity, and the analogous ones obtained by eliminating $y_1$ or $y_2$, as congruences in the polynomial ring $\mathbb{Z}[A_{ij}, B_{ijk\ell}, x_i, y_j]$. Thus

$$L_0^x(\mathbf{x})^2 Q(\mathbf{x},\mathbf{y}) \equiv G_2^x(\mathbf{x}) y_1^2 + H_{12}^x(\mathbf{x}) y_1 y_2 + G_1^x(\mathbf{x}) y_2^2 \pmod{L(\mathbf{x},\mathbf{y})}, \quad (7.18)$$
$$L_1^x(\mathbf{x})^2 Q(\mathbf{x},\mathbf{y}) \equiv G_2^x(\mathbf{x}) y_0^2 + H_{02}^x(\mathbf{x}) y_0 y_2 + G_0^x(\mathbf{x}) y_2^2 \pmod{L(\mathbf{x},\mathbf{y})}, \quad (7.19)$$
$$L_2^x(\mathbf{x})^2 Q(\mathbf{x},\mathbf{y}) \equiv G_1^x(\mathbf{x}) y_0^2 + H_{01}^x(\mathbf{x}) y_0 y_1 + G_0^x(\mathbf{x}) y_1^2 \pmod{L(\mathbf{x},\mathbf{y})}. \quad (7.20)$$

Suppose first that $L(\mathbf{a},\mathbf{y})$ is identically 0. Substituting $\mathbf{x} = \mathbf{a}$ into (7.18), we find that the quadratic form

$$G_2^x(\mathbf{a}) y_1^2 + H_{12}^x(\mathbf{a}) y_1 y_2 + G_1^x(\mathbf{a}) y_2^2$$

is identically 0. Hence $G_2^x(\mathbf{a}) = H_{12}^x(\mathbf{a}) = G_1^x(\mathbf{a}) = 0$. Similarly, substituting $\mathbf{x} = \mathbf{a}$ into (7.19) and (7.20) shows that all of the other values $G_k^x(\mathbf{a})$ and $H_{ij}^x(\mathbf{a})$ are equal to 0, which completes the proof in this case.

We may now suppose that $L(\mathbf{a},\mathbf{y})$ is not identically 0. Then the assumption that $\iota_1$ is not defined at $[\mathbf{a},\mathbf{b}]$ implies that the line $L(\mathbf{a},\mathbf{y}) = 0$ is contained in the zero set of $Q(\mathbf{a},\mathbf{y})$.

If $L_1^x(\mathbf{a}) = L_2^x(\mathbf{a}) = 0$, then the definition (7.17) of $G_0^x$ shows that $G_0^x(\mathbf{a}) = 0$. And if $L_1^x(\mathbf{a})$ and $L_2^x(\mathbf{a})$ are both nonzero, then we let

$$\mathbf{b}' = [0, L_2^x(\mathbf{a}), -L_1^x(\mathbf{a})]$$

and note that $[\mathbf{a},\mathbf{b}'] \in S_{\mathbf{a},\mathbf{b}}$. Hence

$$G_0^x(\mathbf{a}) = L_j^x(\mathbf{a})^2 Q_{ii}^x(\mathbf{a}) - L_i^x(\mathbf{a}) L_j^x(\mathbf{a}) Q_{ij}^x(\mathbf{a}) + L_i^x(\mathbf{a})^2 Q_{jj}^x(\mathbf{a}) = Q(\mathbf{a},\mathbf{b}') = 0.$$

A similar argument shows that also $G_1^x(\mathbf{a}) = G_2^x(\mathbf{a}) = 0$.

Next we evaluate (7.18), (7.19), and (7.20) at $\mathbf{x} = \mathbf{a}$ and use the fact that we now know that $G_0^x(\mathbf{a}) = G_1^x(\mathbf{a}) = G_2^x(\mathbf{a}) = 0$. This yields

$$H_{12}^x(\mathbf{a})y_1y_2 = H_{02}^x(\mathbf{a})y_0y_2 = H_{01}^x(\mathbf{a})y_0y_1 = 0$$
$$\text{for all } \mathbf{y} = [y_0, y_1, y_2] \text{ satisfying } L(\mathbf{a}, \mathbf{y}) = 0. \qquad (7.21)$$

We will prove that $H_{12}^x(\mathbf{a}) = 0$; the others are done similarly. If there is a point on the line $L(\mathbf{a}, \mathbf{y}) = 0$ with $y_1y_2 \neq 0$, then (7.21) immediately implies that $H_{12}^x(\mathbf{a}) = 0$. So we are reduced to the cases in which the line $L(\mathbf{a}, \mathbf{y}) = 0$ is either $y_1 = 0$ or $y_2 = 0$. If it is the line $y_1 = 0$, then $L(\mathbf{a}, \mathbf{y}) = cy_1$ for some constant $c \neq 0$, so $L_0^x(\mathbf{a}) = L_2^x(\mathbf{a}) = 0$, and similarly if it is the line $y_2 = 0$, then $L_0^x(\mathbf{a}) = L_1^x(\mathbf{a}) = 0$. In either case, the definition (7.17) of $H_{12}^x$ yields

$$\begin{aligned} H_{12}^x(\mathbf{a}) &= 2L_1^x(\mathbf{a})L_2^x(\mathbf{a})Q_{00}^x(\mathbf{a}) \\ &\quad - L_1^x(\mathbf{a})L_0^x(\mathbf{a})Q_{20}^x(\mathbf{a}) - L_2^x(\mathbf{a})L_0^x(\mathbf{a})Q_{10}^x(\mathbf{a}) + L_0^x(\mathbf{a})^2Q_{12}^x(\mathbf{a}) = 0. \;\square \end{aligned}$$

*Example* 7.40. We illustrate Proposition 7.39 using the surface described in Example 7.36. The polynomials $G_k^*$ and $H_{ij}^*$ for this example are given in Table 7.1. Proposition 7.39 says that $\iota_1$ is defined at $P = [\mathbf{a}, \mathbf{b}]$ provided that at least one of the six polynomials $G_0^x, G_1^x, G_2^x, H_{01}^x, H_{02}^x, H_{12}^x$ does not vanish at $\mathbf{a}$. For convenience we say that a point $\mathbf{a} \in \mathbb{P}^2$ is *degenerate* if

$$G_0^x(\mathbf{a}) = G_1^x(\mathbf{a}) = G_2^x(\mathbf{a}) = H_{01}^x(\mathbf{a}) = H_{02}^x(\mathbf{a}) = H_{12}^x(\mathbf{a}) = 0.$$

Our first observation is that

$$G_2^x(0, x_1, x_2) = x_1^4 \qquad \text{and} \qquad H_{01}^x(0, x_1, x_2) = 4x_1^2x_2^2 - 2x_2^4.$$

Hence there are no degenerate points with $a_0 = 0$. (We assume that $K$ does not have characteristic 2.) Thus if there exists a degenerate point $\mathbf{a}$, we can dehomogenize $\mathbf{a}$ and write it as $\mathbf{a} = [1, a_1, a_2]$. We use a tilde to indicate the dehomogenization $x_0 = 1$ of the $G_k^*$ and $H_{ij}^*$ polynomials. So for example,

$$\tilde{G}_0^x(x_1, x_2) = G_0^x(1, x_1, x_2) = x_1^3 - 7x_1x_2 - 4x_1^3x_2 - x_2^2 + 5x_1^2x_2^2 - x_2^3 - 4x_1x_2^3.$$

Now suppose that $\tilde{\mathbf{a}} = (a_1, a_2)$ is a degenerate point. Then $x_1 = a_1$ is a common root of the polynomials

$$\tilde{G}_1^x(x_1, a_2), \; \tilde{G}_2^x(x_1, a_2), \; \text{and } \tilde{H}_{01}^x(x_1, a_2).$$

Hence if we take resultants with respect to the $x_2$ variable, then $x_1 = a_1$ is a root of both of the polynomials

$$\begin{aligned} R_1(x_2) &= \operatorname{Res}_{x_1}\big(\tilde{G}_1^x(x_1, x_2), \tilde{H}_{01}^x(x_1, x_2)\big), \\ R_2(x_2) &= \operatorname{Res}_{x_1}\big(\tilde{G}_2^x(x_1, x_2), \tilde{H}_{01}^x(x_1, x_2)\big). \end{aligned}$$

Explicitly, these polynomials are

$$G_0^x = x_0x_1^3 - 7x_0^2x_1x_2 - 4x_1^3x_2 - x_0^2x_2^2 + 5x_1^2x_2^2 - x_0x_2^3 - 4x_1x_2^3$$
$$G_1^x = x_0^3x_1 - x_0^2x_2^2 + 7x_0x_1x_2^2 + x_1^2x_2^2$$
$$G_2^x = -x_0^4 - 4x_0^3x_1 + 3x_0x_1^3 + x_1^4 - x_0^3x_2 - 4x_0^2x_1x_2 + 2x_0x_1x_2^2$$
$$H_{01}^x = 2x_0^2x_1^2 - 7x_0^3x_2 - 4x_0x_1^2x_2 + 4x_0^2x_2^2 + 4x_0x_1x_2^2 + 4x_1^2x_2^2 - 2x_2^4$$
$$\begin{aligned} H_{02}^x = &-7x_0^3x_1 - 4x_0x_1^3 - 2x_0^3x_2 - 4x_0^2x_1x_2 + 6x_0x_1^2x_2 - 4x_1^3x_2 \\ &- 2x_0^2x_2^2 - 8x_0x_1x_2^2 + 2x_1x_2^3 \end{aligned}$$
$$H_{12}^x = 7x_0^4 + 4x_0^2x_1^2 - 4x_0^3x_2 - 6x_0^2x_1x_2 + 10x_0x_1^2x_2 + 2x_1^3x_2 + 2x_0x_2^3$$

$$G_0^y = -2y_0y_1^3 + 4y_1^3y_2 + y_0^2y_2^2 + 4y_0y_1y_2^2 + 5y_1^2y_2^2 + 4y_1y_2^3$$
$$G_1^y = -2y_0^3y_1 + y_0y_1^2y_2 - y_0^2y_2^2 + 4y_0y_1y_2^2 - y_1^2y_2^2 + 7y_1y_2^3$$
$$G_2^y = y_0^4 - 3y_0^3y_1 + 4y_0y_1^3 - y_1^4 + 4y_0^2y_1y_2 + 7y_1^3y_2 - y_0y_1y_2^2$$
$$H_{01}^y = -4y_0^2y_1^2 + 4y_0y_1^2y_2 + y_1^3y_2 + 7y_0^2y_2^2 + 4y_0y_1y_2^2 + y_2^4$$
$$H_{02}^y = 4y_0y_1^3 - y_1^4 + 2y_0^3y_2 + y_0^2y_1y_2 + 6y_0y_1^2y_2 + 8y_0y_1y_2^2 - y_1y_2^3$$
$$H_{12}^y = -4y_0^2y_1^2 + y_0y_1^3 - 7y_0^3y_2 - 6y_0^2y_1y_2 + 8y_0y_1^2y_2 - 2y_1^3y_2 + 14y_1^2y_2^2 - y_0y_2^3$$

Table 7.1: The polynomials $G_k^*$ and $H_{ij}^*$ for the surface in Example 7.36.

$$\begin{aligned} R_1(x_1) = &\ 4x_1^{12} + 112x_1^{11} + 1160x_1^{10} + 5112x_1^9 + 7052x_1^8 - 2271x_1^7 + 18573x_1^6 \\ &+ 2160x_1^5 + 16053x_1^4 - 7304x_1^3 + 1045x_1^2 - 49x_1, \\ R_2(x_1) = &\ 4x_1^{16} + 80x_1^{15} + 600x_1^{14} + 2064x_1^{13} + 2548x_1^{12} - 3616x_1^{11} - 14216x_1^{10} \\ &- 10892x_1^9 + 9856x_1^8 + 21708x_1^7 + 15648x_1^6 + 1000x_1^5 - 13986x_1^4 \\ &- 10462x_1^3 - 3124x_1^2 - 412x_1 - 18, \end{aligned}$$

and the assumption that $R_1$ and $R_2$ have a common root implies that their resultant must vanish. However, when we compute it, we find that

$$\operatorname{Res}(R_1, R_2) = 198929\ldots3830147072 \approx 1.99 \cdot 10^{87}.$$

Hence $\iota_1$ is defined at every point of $S_{\mathbf{A},\mathbf{B}}$ unless the characteristic of $K$ divides this (large) nonzero integer $\operatorname{Res}(R_1, R_2)$.

We can use other resultants to reduce the list of possible bad characteristics. For example, let $R_0(x_2) = \operatorname{Res}_{x_1}(\tilde{G}_0^x, \tilde{H}_{01}^x)$. Then $\iota_1$ is everywhere defined unless both $\operatorname{Res}(R_0, R_2)$ and $\operatorname{Res}(R_1, R_2)$ vanish. We compute

$$\begin{aligned} \gcd\bigl(\operatorname{Res}(R_0, R_2), \operatorname{Res}(R_1, R_2)\bigr) &= 439853213743020234882809856 \\ &= 2^{17} \cdot 3^6 \cdot 317 \cdot 14521485737273461, \quad (7.22) \end{aligned}$$

which proves that $\iota_1$ is everywhere defined unless $p$ is one of the four primes appearing in (7.22).

We sketch a similar calculation for $\iota_2$. Let

$$\begin{aligned} T_0(y_2) &= \operatorname{Res}_{y_1}\bigl(\tilde{G}_0^y(y_1,y_2), \tilde{H}_{01}^y(y_1,y_2)\bigr),\\ T_1(y_2) &= \operatorname{Res}_{y_1}\bigl(\tilde{G}_1^y(y_1,y_2), \tilde{H}_{01}^y(y_1,y_2)\bigr),\\ T_2(y_2) &= \operatorname{Res}_{y_1}\bigl(\tilde{G}_2^y(y_1,y_2), \tilde{H}_{01}^y(y_1,y_2)\bigr). \end{aligned}$$

Then $\operatorname{Res}(T_0,T_2) \approx 2.57 \cdot 10^{97}$ and $\operatorname{Res}(T_1,T_2) \approx 2.75 \cdot 10^{114}$, and

$$\gcd\bigl(\operatorname{Res}(T_0,T_2), \operatorname{Res}(T_1,T_2)\bigr) = 2^{16} \cdot 507593 \cdot 2895545793631.$$

Hence $\iota_2$ is everywhere defined unless the characteristic $p$ of $K$ is one of the three primes appearing in this factorization.

A map $p : V \to W$ of degree 2 between varieties always induces an involution $\iota : V \to V$, but in general $\iota$ is only a rational map, it need not be a morphism. This distinction is quite important. For example, height functions transform well for morphisms, but not for rational maps. We now show that for most choices of $(\mathbf{A},\mathbf{B})$, the involutions on $S_{\mathbf{A},\mathbf{B}}$ are morphisms.

**Proposition 7.41.** *There is a proper Zariski closed set $Z \subset \mathbb{P}^8 \times \mathbb{P}^{35}$ such that if $(\mathbf{A},\mathbf{B}) \notin Z$, then the involutions*

$$\iota_1 : S_{\mathbf{A},\mathbf{B}} \longrightarrow S_{\mathbf{A},\mathbf{B}} \quad \textit{and} \quad \iota_2 : S_{\mathbf{A},\mathbf{B}} \longrightarrow S_{\mathbf{A},\mathbf{B}}$$

*are morphisms.*

*Proof.* According to Proposition 7.39, the involution $\iota_1$ is defined on all of $S_{\mathbf{A},\mathbf{B}}$ provided that the system of equations

$$G_0^x(\mathbf{x}) = G_1^x(\mathbf{x}) = G_2^x(\mathbf{x}) = H_{01}^x(\mathbf{x}) = H_{02}^x(\mathbf{x}) = H_{12}^x(\mathbf{x}) = 0 \tag{7.23}$$

has no solutions in $\mathbb{P}^2$. A general result from elimination theory (see [198, I.5.7A]) says that there are polynomials $f_1,\dots,f_r$ in the coefficients of

$$G_0^x,\ G_1^x,\ G_2^x,\ H_{01}^x,\ H_{02}^x,\ H_{12}^x$$

such that the equations (7.23) have a solution if and only if $f_1 = \cdots = f_r = 0$. The coefficients of $G_0^x,\dots,H_{12}^x$ are themselves polynomials in the coefficients of $L(\mathbf{x},\mathbf{y})$ and $Q(\mathbf{x},\mathbf{y})$, so we can write each $f_i$ as a polynomial in the variables $\mathbf{A}$ and $\mathbf{B}$. There are then two possibilities:

- The set of $f_i$ consists only of the zero polynomial, and hence every surface $S_{\mathbf{A},\mathbf{B}}$ has some point at which $\iota_1$ is not defined.
- The set of $f_i$ contains at least one nonzero polynomial $f_1$, and then $\iota_1$ is well-defined on $S_{\mathbf{A},\mathbf{B}}$ provided that $f_1(\mathbf{A},\mathbf{B}) \neq 0$.

In order to eliminate the first case, it suffices to write down a single surface $S_{\mathbf{A},\mathbf{B}}$ for which the system of equations has no solutions in $\mathbb{P}^2$. We gave such a surface in Example 7.40, at least provided that the characteristic of $K$ is not equal to 2, 3, 317, or 14521485737273461. Similarly, the above argument and Example 7.40 show that $\iota_2$ is defined on a Zariski open set of $(\mathbf{A},\mathbf{B})$ as long as the characteristic of $K$ is not equal to 2, 507593, or 2895545793631. This completes the proof of Proposition 7.41 except for fields having one of these six characteristics. We leave as an exercise for the reader to find other examples to cover the remaining cases. □

*Remark* 7.42. See Exercise 7.28 for a less computational proof of Proposition 7.41.

### 7.4.2 Divisors and Involutions on $S_{\mathbf{A},\mathbf{B}}$

In this section we study how the involutions $\iota_1$ and $\iota_2$ act on divisors on $S_{\mathbf{A},\mathbf{B}}$. Later we use this information to study how iterates of the involutions act on points. This prompts the following definitions.

**Definition.** Let $\iota_1$ and $\iota_2$ be the involutions of the surface $S_{\mathbf{A},\mathbf{B}}$ defined by (7.15). These involutions generate a subgroup (possibly all) of $\operatorname{Aut}(S_{\mathbf{A},\mathbf{B}})$. We denote this subgroup by $\mathcal{A}$. Then for any point $P \in S_{\mathbf{A},\mathbf{B}}$, the *$\mathcal{A}$-orbit of $P$* is the set

$$\mathcal{A}(P) = \{\psi(P) : \psi \in \mathcal{A}\}.$$

Let $H \in \operatorname{Div}(\mathbb{P}^2)$ be a line. As described in Example 7.28, pulling back using the two projections gives divisors on $\mathbb{P}^2 \times \mathbb{P}^2$,

$$H_1 = H \times \mathbb{P}^2 \qquad \text{and} \qquad H_2 = \mathbb{P}^2 \times H,$$

and the Picard group of $\mathbb{P}^2 \times \mathbb{P}^2$ is isomorphic to $\mathbb{Z}^2$ via

$$\mathbb{Z}^2 \xrightarrow{\sim} \operatorname{Pic}(\mathbb{P}^2 \times \mathbb{P}^2), \qquad (n_1,n_2) \longmapsto n_1H_1 + n_2H_2.$$

(By abuse of notation, we write $n_1H_1 + n_2H_2$ for its divisor class.) We note that $n_1H_1 + n_2H_2$ is a very ample divisor on $\mathbb{P}^2 \times \mathbb{P}^2$ if and only if both $n_1$ and $n_2$ are positive.

Next we define two divisors $D_1, D_2 \in \operatorname{Div}(S_{\mathbf{A},\mathbf{B}})$ using the two projections of $S_{\mathbf{A},\mathbf{B}}$ to $\mathbb{P}^2$,

$$D_1 = p_1^*H = S_{\mathbf{A},\mathbf{B}} \cap H_1 \qquad \text{and} \qquad D_2 = p_2^*H = S_{\mathbf{A},\mathbf{B}} \cap H_2.$$

Wehler [448] has proven that the Picard group of a general surface $S_{\mathbf{A},\mathbf{B}}$ satisfies $\operatorname{Pic}(S_{\mathbf{A},\mathbf{B}}) \cong \mathbb{Z}^2$ and that $D_1$ and $D_2$ are generators,[2] but for our purposes it will not matter if $\operatorname{Pic}(S_{\mathbf{A},\mathbf{B}})$ is larger than $\mathbb{Z}^2$, we will simply use the part of $\operatorname{Pic}(S_{\mathbf{A},\mathbf{B}})$ generated by $D_1$ and $D_2$.

We now compute the action of $\iota_1$ and $\iota_2$ on $D_1$ and $D_2$.

[2] What this means is that the set of coefficients $(\mathbf{A},\mathbf{B}) \in \mathbb{P}^8 \times \mathbb{P}^{35}$ for which $\operatorname{Pic}(S_{\mathbf{A},\mathbf{B}})$ is strictly larger than $\mathbb{Z}^2$ forms a countable union of proper Zariski closed subsets of $\mathbb{P}^8 \times \mathbb{P}^{35}$. This implies that most $(\mathbf{A},\mathbf{B})$ in $\mathbb{P}^8(\mathbb{C}) \times \mathbb{P}^{35}(\mathbb{C})$ have Picard group $\mathbb{Z}^2$, but it does not directly imply that there are any such values in $\mathbb{P}^8(\mathbb{Q}) \times \mathbb{P}^{35}(\mathbb{Q})$, since $\mathbb{Q}$ is countable.

**Proposition 7.43.** *Let $D_1 = p_1^* H$ and $D_2 = p_2^* H$. The involutions $\iota_1$ and $\iota_2$ act on the subspace of* $\operatorname{Pic}(S_{\mathbf{A},\mathbf{B}})$ *generated by $D_1$ and $D_2$ according to the following rules:*

$$\iota_1^* D_1 = D_1, \qquad \iota_2^* D_1 = -D_1 + 4D_2, \tag{7.24}$$
$$\iota_1^* D_2 = 4D_1 - D_2, \qquad \iota_2^* D_2 = D_2. \tag{7.25}$$

*Proof.* The involution $\iota_1$ switches the sheets of the projection $p_1$, so it is clear that $p_1 \circ \iota_1 = p_1$. This allows us to compute

$$\iota_1^* D_1 = \iota_1^* p_1^* H = (p_1 \circ \iota_1)^* H = p_1^* H = D_1.$$

This proves the first formula in (7.24).

Next we observe that for any $P \in S_{\mathbf{A},\mathbf{B}}$, the two points in the set $p_1^{-1}(p_1(P))$ are $P$ and $\iota_1(P)$. Thus if we start with a divisor on $S_{\mathbf{A},\mathbf{B}}$, use $p_1$ to push it down to $\mathbb{P}^2$, and then use $p_1$ to pull it back to $S_{\mathbf{A},\mathbf{B}}$, we get back the original divisor plus its translation by $\iota_1$. In other words,

$$p_1^* p_{1*} D = D + \iota_1^* D \quad \text{for all } D \in \operatorname{Pic}(S_{\mathbf{A},\mathbf{B}}).$$

Using this formula with $D = D_2$ allows us to compute

$$\iota_1^* D_2 = p_1^* p_{1*} D_2 - D_2 = p_1^* p_{1*} p_2^* H - D_2. \tag{7.26}$$

The divisor $p_{1*} p_2^* H$ on $\mathbb{P}^2$ is linearly equivalent to some multiple of $H$. For simplicity, let $H$ be the line $y_2 = 0$. Then $p_2^* H$ is the curve in $\mathbb{P}^2 \times \mathbb{P}^2$ (lying on $S_{\mathbf{A},\mathbf{B}}$) given by the equations

$$L(x_0, x_1, x_2, y_0, y_1, 0) = 0, \qquad Q(x_0, x_1, x_2, y_0, y_1, 0) = 0, \qquad y_2 = 0.$$

We solve the linear equation $L = 0$ to express $y_0$ and $y_1$ as linear functions of $[x_0, x_1, x_2]$, and then substituting into the quadratic equation $Q = 0$ yields a homogeneous equation of degree 4 in $x_0, x_1, x_2$. So when we use $p_{1*}$ to push $p_2^* H$ down to $\mathbb{P}^2$, we get a curve of degree 4 in $\mathbb{P}^2$. Hence

$$p_{1*} p_2^* H = 4H,$$

where this is an equality in $\operatorname{Pic}(\mathbb{P}^2)$. Substituting into (7.26) yields the second formula in (7.24).

By symmetry, or by repeating the above argument, the two formulas in (7.25) are also true. □

*Remark* 7.44. It is not hard to prove that the only relations satisfied by compositions of $\iota_1$ and $\iota_2$ are $\iota_1^2 = 1$ and $\iota_2^2 = 1$. In other words, $\mathcal{A}$ is isomorphic to the free product of the groups of order 2 generated by $\iota_1$ and $\iota_2$. An alternative description is that $\mathcal{A}$ is isomorphic to an infinite dihedral group. See Exercise 7.31.

### 7.4.3 Height Functions on $S_{\mathbf{A},\mathbf{B}}$

In this section we use the Weil height machine to translate the divisor relations and transformation formulas from Proposition 7.43 into relations among height functions. We recall from Example 7.33 that the height functions associated to the divisors $D_1$ and $D_2$ are given at a point $P = [\mathbf{x}, \mathbf{y}] \in S_{\mathbf{A},\mathbf{B}}$ by

$$\begin{aligned} h_{D_1}(P) &= h_{p_1^*H}(P) = h_H(p_1P) = h(\mathbf{x}), \\ h_{D_2}(P) &= h_{p_2^*H}(P) = h_H(p_2P) = h(\mathbf{y}). \end{aligned} \tag{7.27}$$

**Proposition 7.45.** *Assume that $S_{\mathbf{A},\mathbf{B}}$ is defined over a number field $K$. Let*

$$\alpha = 2 + \sqrt{3}$$

*and define functions $h^+, h^- : S_{\mathbf{A},\mathbf{B}}(\bar{K}) \to \mathbb{R}$ by the formulas*

$$h^+([\mathbf{x}, \mathbf{y}]) = -h(\mathbf{x}) + \alpha h(\mathbf{y}) \qquad \textit{and} \qquad h^-([\mathbf{x}, \mathbf{y}]) = \alpha h(\mathbf{x}) - h(\mathbf{y}).$$

*Then $h^+$ and $h^-$ transform according to the following rules:*

$$\begin{aligned} h^+ \circ \iota_1 &= \alpha h^- + O(1), & h^- \circ \iota_1 &= \alpha^{-1} h^+ + O(1), \\ h^+ \circ \iota_2 &= \alpha^{-1} h^- + O(1), & h^- \circ \iota_2 &= \alpha h^+ + O(1). \end{aligned}$$

*Remark* 7.46. Before starting the proof of Proposition 7.45, we pause to explain why the number $\alpha$ and the functions $h^+$ and $h^-$ arise naturally. Consider a two-dimensional real vector space $V$ with basis elements $D_1$ and $D_2$, where we view $V$ as a subspace of $\operatorname{Pic}(S_{\mathbf{A},\mathbf{B}}) \otimes \mathbb{R}$. Then the formulas (7.24) and (7.25) in Proposition 7.43 tell us how $\iota_1^*$ and $\iota_2^*$ act on $V$. In terms of the given basis, they are linear transformations that act via the matrices

$$\iota_1^* = \begin{pmatrix} 1 & 4 \\ 0 & -1 \end{pmatrix} \qquad \text{and} \qquad \iota_2^* = \begin{pmatrix} -1 & 0 \\ 4 & 1 \end{pmatrix}.$$

We now look for a new basis $\{E_1, E_2\}$ for $V$ with the property that $\iota_1^*$ and $\iota_2^*$ interchange the basis. More precisely, we ask that

$$\iota_1^* E_1 = aE_2, \qquad \iota_1^* E_2 = bE_1, \qquad \iota_2^* E_1 = cE_2, \qquad \iota_2^* E_2 = dE_1,$$

for some constants $a, b, c, d$. This problem can be solved directly, but it is easier to observe that $\iota_2^*\iota_1^* E_1 = adE_1$ and $\iota_2^*\iota_1^* E_2 = bcE_2$. Thus $E_1$ and $E_2$ must be eigenvectors for the linear transformation $\iota_2^*\iota_1^*$, whose matrix is

$$\iota_2^*\iota_1^* = \begin{pmatrix} 1 & 4 \\ 0 & -1 \end{pmatrix} \begin{pmatrix} -1 & 0 \\ 4 & 1 \end{pmatrix} = \begin{pmatrix} 15 & 4 \\ -4 & -1 \end{pmatrix}.$$

It is easy to check that

$$E_1 = -D_1 + \alpha D_2 \qquad \text{and} \qquad E_2 = \alpha D_1 - D_2$$

are a pair of independent eigenvectors with eigenvalues $\alpha^2$ and $\alpha^{-2}$ respectively. This explains the appearance of $\alpha$, and then one checks that these eigenvectors satisfy

$$\iota_1^* E_1 = \alpha E_2, \quad \iota_1^* E_2 = \alpha^{-1} E_1, \quad \iota_2^* E_1 = \alpha^{-1} E_2, \quad \iota_2^* E_2 = \alpha E_1. \tag{7.28}$$

It is then natural to define height functions $h^+$ and $h^-$ corresponding to the divisors $E_1 = -D_1 + \alpha D_2$ and $E_2 = \alpha D_1 - D_2$, since the divisor relations (7.28) and the Weil height machine should then yield corresponding relations for the height functions.

*Proof of Proposition* 7.45. Having given the motivation, we commence the proof of Proposition 7.45, which is a formal calculation using the additivity and functoriality of height functions (Theorem 7.29(b,c)) and the transformation formulas (7.24) and (7.25) in Proposition 7.43. Note that

$$h(\mathbf{x}) = h_{D_1}\big([\mathbf{x},\mathbf{y}]\big) \quad \text{and} \quad h(\mathbf{y}) = h_{D_2}\big([\mathbf{x},\mathbf{y}]\big)$$

from (7.27). We compute

$$\begin{aligned} h^+ \circ \iota_1 &= -h_{D_1} \circ \iota_1 + \alpha h_{D_2} \circ \iota_1 && \text{by definition of } h^+, \\ &= -h_{\iota_1^* D_1} + \alpha h_{\iota_1^* D_2} + O(1) && \text{from Theorem 7.29(b)}, \\ &= -h_{D_1} + \alpha h_{4D_1 - D_2} + O(1) && \text{from Proposition 7.43}, \\ &= (-1 + 4\alpha) h_{D_1} - \alpha h_{D_2} + O(1) && \text{from Theorem 7.29(c)}, \\ &= \alpha^2 h_{D_1} - \alpha h_{D_2} + O(1) && \text{since } \alpha^2 = 4\alpha - 1, \\ &= \alpha h^- + O(1). \end{aligned}$$

Similarly

$$\begin{aligned} h^+ \circ \iota_2 &= -h_{D_1} \circ \iota_2 + \alpha h_{D_2} \circ \iota_2 && \text{by definition of } h^-, \\ &= -h_{\iota_2^* D_1} + \alpha h_{\iota_2^* D_2} && \text{from Theorem 7.29(b)}, \\ &= -h_{-D_1 + 4D_2} + \alpha h_{D_2} && \text{from Proposition 7.43}, \\ &= h_{D_1} + (-4 + \alpha) h_{D_2} && \text{from Theorem 7.29(c)}, \\ &= h_{D_1} - \alpha^{-1} h_{D_2} + O(1) && \text{since } \alpha^2 = 4\alpha - 1, \\ &= \alpha^{-1} h^- + O(1). \end{aligned}$$

This proves the transformation formulas for $h^+$. The proof for $h^-$ is similar and is left for the reader. □

We can use Proposition 7.45 and the general theory of canonical heights (Theorem 3.20) to construct two heights on $S_{\mathbf{A},\mathbf{B}}$ that are canonical with respect to both $\iota_1$ and $\iota_2$.

**Proposition 7.47.** *Let $S_{\mathbf{A},\mathbf{B}}$ be defined over a number field $K$. There exist unique functions*

$$\hat{h}^+ : S_{\mathbf{A},\mathbf{B}}(\bar{K}) \longrightarrow \mathbb{R} \qquad \textit{and} \qquad \hat{h}^- : S_{\mathbf{A},\mathbf{B}}(\bar{K}) \longrightarrow \mathbb{R}$$

*satisfying both the normalization conditions*

$$\hat{h}^+ = -h_{D_1} + \alpha h_{D_2} + O(1) \qquad \textit{and} \qquad \hat{h}^- = \alpha h_{D_1} - h_{D_2} + O(1) \tag{7.29}$$

*and the canonical transformation formulas*

$$\begin{aligned} \hat{h}^+ \circ \iota_1 &= \alpha \hat{h}^-, & \hat{h}^- \circ \iota_1 &= \alpha^{-1}\hat{h}^+, \\ \hat{h}^+ \circ \iota_2 &= \alpha^{-1}\hat{h}^-, & \hat{h}^- \circ \iota_2 &= \alpha\hat{h}^+. \end{aligned} \tag{7.30}$$

*Proof.* Let $\phi = \iota_1 \circ \iota_2$ be the composition of the two involutions on $S_{\mathbf{A},\mathbf{B}}$ and let

$$h^+ = -h_{D_1} + \alpha h_{D_2} \qquad \text{and} \qquad h^- = \alpha h_{D_1} - h_{D_2}$$

be the functions defined in Proposition 7.45. Then the transformation formulas in Proposition 7.45 allow us to compute

$$h^+ \circ \phi = h^+ \circ \iota_1 \circ \iota_2 + O(1) = \alpha h^- \circ \iota_2 + O(1) = \alpha^2 h^+ + O(1).$$

The constant $\alpha^2$ satisfies $\alpha^2 \approx 13.93 > 1$, so we may apply Theorem 3.20 to the functions $\phi$ and $h^+$ to deduce the existence of a unique function $\hat{h}^+$ satisfying

$$\hat{h}^+ \circ \phi = \alpha^2 \hat{h}^+ \qquad \text{and} \qquad \hat{h}^+ = h^+ + O(1).$$

Repeating this construction with $\phi^{-1} = \iota_2 \circ \iota_1$, we find that

$$h^- \circ \phi^{-1} = h^- \circ \iota_2 \circ \iota_1 + O(1) = \alpha h^+ \circ \iota_1 + O(1) = \alpha^2 h^- + O(1).$$

Applying Theorem 3.20 to the functions $\phi^{-1}$ and $h^-$, we find that there is a function $\hat{h}^-$ satisfying

$$\hat{h}^- \circ \phi^{-1} = \alpha^2 \hat{h}^- \qquad \text{and} \qquad \hat{h}^- = h^- + O(1).$$

The functions $\hat{h}^+$ and $\hat{h}^-$ that we have just constructed satisfy (7.29). In order to check the transformation formulas (7.30), we first note that

$$\hat{h}^+ \circ \iota_1 = h^+ \circ \iota_1 + O(1) = \alpha h^- + O(1) = \alpha \hat{h}^- + O(1).$$

In order to get rid of the $O(1)$, we compose both sides with $\phi^{-n}$ and use the formula

$$\iota_1 \circ \phi^{-n} = \iota_1 \circ (\iota_2 \circ \iota_1)^n = \phi^n \iota_1$$

to compute

$$\alpha^n \hat{h}^+ \circ \iota_1 = \hat{h}^+ \circ \phi^n \circ \iota_1 = \hat{h}^+ \circ \iota_1 \circ \phi^{-n} = \alpha \hat{h}^- \circ \phi^{-n} + O(1) = \alpha^{n+1} \hat{h}^- + O(1).$$

Divide both sides by $\alpha^n$ and let $n \to \infty$ to obtain the desired result $\hat{h}^+ \circ \iota_1 = \alpha \hat{h}^-$. This proves the first of the transformation formulas (7.30). The others are proven similarly.

Finally, in order to prove uniqueness, suppose that $\hat{g}^+$ and $\hat{g}^-$ are functions satisfying (7.29) and (7.30). Then

$$\hat{g}^+ \circ \phi = \hat{g}^+ \circ \iota_1 \circ \iota_2 = \alpha \hat{g}^- \circ \iota_2 = \alpha^2 \hat{g}^+,$$

and similarly $\hat{g}^- \circ \phi^{-1} = \alpha^2 \hat{g}^-$. Hence $\hat{g}^+$ and $\hat{g}^-$ have the same canonical properties as $\hat{h}^+$ and $\hat{h}^-$, so the uniqueness assertion in Theorem 3.20 tells us that $\hat{g}^+ = \hat{h}^+$ and $\hat{g}^- = \hat{h}^-$. □

*Remark* 7.48. In practice, it is infeasible to compute the canonical heights $\hat{h}^+$ and $\hat{h}^-$ to more than a few decimal places using their definition as a limit. As with the other canonical heights studied in Sections 3.4, 3.5, and 5.9, it is possible to decompose $\hat{h}^+$ and $\hat{h}^-$ as sums of local heights that may then be computed using rapidly convergent series (cf. Exercise 5.29). See [89] for details.

### 7.4.4 Properties and Applications of Canonical Heights

The next proposition describes various useful properties of the canonical height functions $\hat{h}^+$ and $\hat{h}^-$ and their sum. As an application, we prove that there are only finitely many $K$-rational points with finite $\mathcal{A}$-orbit. This is the analogue for the K3 surfaces $S_{\mathbf{A},\mathbf{B}}$ of Northcott's Theorem 3.12 on preperiodic points of morphisms on $\mathbb{P}^N$ and of Theorem 7.19 on periodic points of regular affine automorphisms.

**Proposition 7.49.** *Let $S_{\mathbf{A},\mathbf{B}}$ be defined over a number field $K$, let $\hat{h}^+$ and $\hat{h}^-$ be the canonical height functions constructed in Proposition 7.47, and let*

$$\hat{h} = \hat{h}^+ + \hat{h}^-.$$

(a) *The set*

$$\{P \in S_{\mathbf{A},\mathbf{B}}(K) : \hat{h}(P) \le C\}$$

*is finite. (N.B. This is not true if we replace $\hat{h}$ by either of the heights $\hat{h}^+$ and $\hat{h}^-$, see Exercise 7.35.)*

(b) *Let $P \in S_{\mathbf{A},\mathbf{B}}(\bar{K})$. Then*

$$\hat{h}^+(P) = 0 \iff \hat{h}^-(P) = 0 \iff \hat{h}(P) = 0 \iff P \text{ has finite } \mathcal{A}\text{-orbit.}$$

(c) *There are only finitely many points $P \in S_{\mathbf{A},\mathbf{B}}(K)$ with finite $\mathcal{A}$-orbit.*

*Proof.* (a) Using the properties of $\hat{h}^+$ and $\hat{h}^-$, we find that

$$\begin{aligned} \hat{h} &= \hat{h}^+ + \hat{h}^- && \text{by definition of } \hat{h}, \\ &= \bigl(-h_{D_1} + \alpha h_{D_2}\bigr) + \bigl(\alpha h_{D_1} - h_{D_2}\bigr) + O(1) && \text{from Proposition 7.47,} \\ &= (\alpha - 1)\bigl(h_{D_1} + h_{D_2}\bigr) + O(1). \end{aligned}$$

As noted earlier, the heights $h_{D_1}$ and $h_{D_2}$ are given by

$$h_{D_1}([\mathbf{x},\mathbf{y}]) = h(\mathbf{x}) \qquad \text{and} \qquad h_{D_2}([\mathbf{x},\mathbf{y}]) = h(\mathbf{y}),$$

where $h(\mathbf{x})$ and $h(\mathbf{y})$ are the standard heights of $\mathbf{x}$ and $\mathbf{y}$ in $\mathbb{P}^2$. Hence

$$\hat{h}([\mathbf{x},\mathbf{y}]) = (\alpha - 1)\big(h(\mathbf{x}) + h(\mathbf{y})\big) + O(1),$$

so if $\hat{h}([\mathbf{x},\mathbf{y}])$ is bounded, then both $h(\mathbf{x})$ and $h(\mathbf{y})$ are bounded. (Note that $\alpha$ satisfies $\alpha > 1$, which is crucial for the argument to work.) This completes the proof of (a), since Theorem 3.7 tells us that $\mathbb{P}^2(K)$ contains only finitely many points of bounded height.

(b) Since $\hat{h} = \hat{h}^+ + \hat{h}^-$ and both $\hat{h}^+$ and $\hat{h}^-$ are nonnegative, it is clear that

$$\hat{h}(P) = 0 \quad \Longrightarrow \quad \hat{h}^+(P) = \hat{h}^-(P) = 0.$$

Suppose next that $\hat{h}^+(P) = 0$. Let $\phi = \iota_1 \circ \iota_2$ as usual. Then

$$\begin{aligned}\hat{h}(\phi^n(P)) &= \hat{h}^+(\phi^n(P)) + \hat{h}^-(\phi^n(P)) \\ &= \alpha^{2n}\hat{h}^+(P) + \alpha^{-2n}\hat{h}^-(P) \\ &= \alpha^{-2n}\hat{h}^-(P).\end{aligned}$$

The righthand side is bounded (indeed, it goes to 0) as $n \to \infty$, so we see that $\{\phi^n(P) : n \geq 0\}$ is a set of bounded $\hat{h}$-height. It follows from (a) that it is a finite set. Since $\phi$ is an automorphism, we deduce that $P$ is periodic for $\phi$. We now perform a similar calculation using $\phi^{-n}$,

$$\begin{aligned}\hat{h}(\phi^{-n}(P)) &= \hat{h}^+(\phi^{-n}(P)) + \hat{h}^-(\phi^{-n}(P)) \\ &= \alpha^{-2n}\hat{h}^+(P) + \alpha^{2n}\hat{h}^-(P) \\ &= \alpha^{2n}\hat{h}^-(P).\end{aligned}$$

The lefthand side is bounded, since $P$ is periodic for $\phi$, so letting $n \to \infty$ implies that $\hat{h}^-(P) = 0$.

This proves that $\hat{h}^+(P) = 0$ implies $\hat{h}^-(P) = 0$, and a similar argument gives the reverse implication, which completes the proof that

$$\hat{h}^+(P) = 0 \iff \hat{h}^-(P) = 0 \iff \hat{h}(P) = 0.$$

In order to study $\mathcal{A}$-orbits of points, we make further use of the formula

$$\hat{h}(\phi^n(P)) = \hat{h}^+(\phi^n(P)) + \hat{h}^-(\phi^n(P)) = \alpha^{2n}\hat{h}^+(P) + \alpha^{-2n}\hat{h}^-(P). \tag{7.31}$$

Suppose first that $P$ has finite $\mathcal{A}$-orbit. Then $\hat{h}(\phi^n(P))$ is bounded, since it takes on only finitely many values. Letting $n \to \infty$ in (7.31) and using the fact that $\alpha > 1$, we deduce that $\hat{h}^+(P) = 0$.

Finally, suppose that $\hat{h}(P) = 0$. Then $\hat{h}^+(P) = \hat{h}^-(P) = 0$, so (7.31) tells us that $\hat{h}(\phi^n(P)) = 0$ for all $n \in \mathbb{Z}$. In particular, $\{\phi^n(P) : n \in \mathbb{Z}\}$ is a set of bounded $\hat{h}$-height, so (a) tells us that it is a finite set. But the $\mathcal{A}$-orbit of $P$ is equal to

$$\mathcal{A}(P) = \{\phi^n(P) : n \in \mathbb{Z}\} \cup \{(\iota_1 \circ \phi^n)(P) : n \in \mathbb{Z}\},$$

so $\mathcal{A}(P)$ is also finite. □

The canonical height functions on $S_{\mathbf{A},\mathbf{B}}$ can also be used to count the number of points of bounded height in an $\mathcal{A}$-orbit, as in our next result. See Exercise 7.21 for an analogous (conditional) estimate for regular affine automorphisms of $\mathbb{P}^N$.

**Proposition 7.50.** *Let $S_{\mathbf{A},\mathbf{B}}$ be defined over a number field $K$, and for any point $P = [\mathbf{x}, \mathbf{y}] \in S_{\mathbf{A},\mathbf{B}}(K)$, let $h(P)$ be the height function*

$$h(P) = h_{D_1+D_2}(P) = h(\mathbf{x}) + h(\mathbf{y}).$$

*Also let $\alpha = 2 + \sqrt{3}$ as usual. Fix a point $Q \in S_{\mathbf{A},\mathbf{B}}(K)$ with infinite $\mathcal{A}$-orbit and let*

$$\mu(Q) = \#\{\psi \in \mathcal{A} : \psi(Q) = Q\}$$

*be the order of the stabilizer of $Q$. Then*

$$\#\{P \in \mathcal{A}(Q) : h(P) \le B\} = \frac{1}{\mu(Q)} \log_\alpha\left(\frac{B^2}{\hat{h}^+(Q)\hat{h}^-(Q)}\right) + O(1) \quad \text{as } B \to \infty,$$

*where the $O(1)$ constant is independent of both $B$ and $Q$.*

The key to proving Proposition 7.50 is the following elementary counting lemma.

**Lemma 7.51.** *Let $a, b > 0$ and $u > 1$ be real numbers. Then*

$$\#\{n \in \mathbb{Z} : au^n + bu^{-n} \le t\} = \log_u\left(\frac{t^2}{ab}\right) + O(1) \qquad \text{as } t \to \infty,$$

*where the $O(1)$ constant depends only on $u$.*

*Proof.* We start by writing the real number $\log_u(\sqrt{b/a})$ as the sum of an integer and a fractional part,

$$\log_u \sqrt{\frac{b}{a}} = m + r \quad \text{with } m \in \mathbb{Z} \text{ and } |r| \le \frac{1}{2}.$$

(The reason that we do this is because the function $au^x + bu^{-x}$ has a minimum at $x = \log_u(\sqrt{b/a})$.) Then replacing $n$ by $n + m$ in the expression $au^n + bu^{-n}$ yields

$$au^{n+m} + bu^{-n-m} = \sqrt{ab}\,(u^r \cdot u^n + u^{-r} \cdot u^{-n}).$$

Hence

$$\{n \in \mathbb{Z} : au^n + bu^{-n} \le t\} = \left\{n \in \mathbb{Z} : u^r \cdot u^n + u^{-r} \cdot u^{-n} \le \frac{t}{\sqrt{ab}}\right\}.$$

It thus suffices to prove that if $c, d \in \mathbb{R}$ are both between $u^{-1/2}$ and $u^{1/2}$, then

$$\#\{n \in \mathbb{Z} : cu^n + du^{-n} \le t\} = \log(t^2) + O(1) \qquad \text{as } t \to \infty. \tag{7.32}$$

We note that if $n \ge 0$, then

$$\log_u(cu^n + du^{-n}) = n + \log_u(c + du^{-2n}) = n + O(1),$$

and similarly if $n \le 0$, then

$$\log_u(cu^n + du^{-n}) = -n + \log_u(cu^{2n} + d) = -n + O(1).$$

Here the $O(1)$ bounds depend only on $u$, since by assumption $c$ and $d$ are bounded in terms of $u$. Therefore

$$\begin{aligned}\#\{n \in \mathbb{Z} : cu^n + du^{-n} \le t\} &= \#\{n \in \mathbb{Z} : \log_u(cu^n + du^{-n}) \le \log_u(t)\} \\ &= \#\{n \in \mathbb{Z} : |n| + O(1) \le \log_u(t)\} \\ &= 2\log_u(t) + O(1).\end{aligned}$$

This is the desired inequality (7.32), which completes the proof of Lemma 7.51. □

*Proof of Proposition* 7.50. We do the case that $\mu(Q) = 1$ and leave the similar case $\mu(Q) = 2$ to the reader. (It is easy to check that $\mu(Q) \le 2$; see Exercise 7.29.)

Let $\phi = \iota_1 \circ \iota_2$. Every element of $\mathcal{A}$ is given uniquely as an alternating composition of $\iota_1$'s and $\iota_2$'s, so $\mathcal{A}$ splits up as a disjoint union

$$\mathcal{A} = \{\phi^n : n \in \mathbb{Z}\} \cup \{\phi^n \circ \iota_1 : n \in \mathbb{Z}\}.$$

Our assumption that $\mu(Q) = 1$ then implies that the $\mathcal{A}$-orbit of $Q$ is a disjoint union

$$\mathcal{A}(Q) = \mathcal{O}_\phi(Q) \cup \mathcal{O}_\phi(\iota_1 Q). \tag{7.33}$$

Let $\hat{h}^+$ and $\hat{h}^-$ be the canonical height functions constructed in Proposition 7.47 and let $\hat{h} = \hat{h}^+ + \hat{h}^-$. We note that Proposition 7.47 tells us that $\hat{h}^+ \circ \phi = \alpha \hat{h}^+$ and $\hat{h}^- \circ \phi = \alpha^{-1}\hat{h}^-$. This allows us to compute

$$\begin{aligned}&\#\{P \in \mathcal{O}_\phi(Q) : \hat{h}(P) \le B\} \\ &\quad = \#\{n \in \mathbb{Z} : \hat{h}(\phi^n Q) \le B\} && \text{since } \mu(Q) = 1, \\ &\quad = \#\{n \in \mathbb{Z} : \hat{h}^+(\phi^n Q) + \hat{h}^-(\phi^n Q) \le B\} && \text{definition of } \hat{h}, \\ &\quad = \#\{n \in \mathbb{Z} : \alpha^{2n}\hat{h}^+(Q) + \alpha^{-2n}\hat{h}^-(Q) \le B\} && \text{from Proposition 7.47,} \\ &\quad = \frac{1}{2}\log_\alpha\left(\frac{B^2}{\hat{h}^+(Q)\hat{h}^-(Q)}\right) + O(1) && \text{from Lemma 7.51.}\end{aligned}$$

Further, if we replace $Q$ with $\iota_1(Q)$, then we get exactly the same estimate, since

$$\hat{h}^+(\iota_1 Q)\hat{h}^-(\iota_1 Q) = \alpha\hat{h}^-(Q) \cdot \alpha^{-1}\hat{h}^+(Q) = \hat{h}^+(Q)\hat{h}^-(Q).$$

Hence using the decomposition (7.33), we find that

$$\begin{aligned}\#\{P \in \mathcal{A}(Q) : \hat{h}(P) \le B\}&\\ &= \#\{P \in \mathcal{O}_\phi(Q) : \hat{h}(P) \le B\} + \#\{P \in \mathcal{O}_\phi(\iota_1 Q) : \hat{h}(P) \le B\}\\ &= \log_\alpha\left(\frac{B^2}{\hat{h}^+(Q)\hat{h}^-(Q)}\right) + O(1).\end{aligned}$$

Finally, in order to replace the canonical height $\hat{h}$ with the naive height $h$, we note that

$$\begin{aligned}\hat{h} = \hat{h}^+ + \hat{h}^- &= (\alpha h_{D_1} - h_{D_2}) + (-h_{D_1} + \alpha h_{D_2}) + O(1)\\ &= (\alpha - 1)(h_{D_1} + h_{D_2}) + O(1) = (\alpha - 1)h + O(1).\end{aligned}$$

Thus

$$\#\{P \in \mathcal{A}(Q) : h(P) \le B\} = \#\{P \in \mathcal{A}(Q) : \hat{h}(P) \le (\alpha - 1)B + O(1)\},$$

and replacing $B$ with $(\alpha - 1)B$ affects only the $O(1)$, since $\alpha > 1$. □

# Exercises

## Section 7.1. Dynamics of Rational Maps on Projective Space

**7.1.** Let $a, b, c, d, e \in \mathbb{C}$ and let $\phi : \mathbb{A}^3 \to \mathbb{A}^3$ be given by

$$\phi(x, y, z) = \bigl(ax + by^2 + (cx^2 + dz)^2,\ ey + (ax + by^2)^2,\ dz + cx^2\bigr). \tag{7.34}$$

(a) Prove that $\phi$ is invertible if and only if $ade \ne 0$.

(b) Prove that $\phi$ is a regular automorphism if and only if $abcde \ne 0$.

(c) Clearly $(0, 0, 0)$ is a fixed point of $\phi$. Let $b = -1$, $d = 1$, and $e = 1 - t^3$. Prove that $(0, t, t)$ is also a fixed point of $\phi$. Hence there are infinitely many maps $\phi \in \mathbb{Q}[x, y, z]$ of the form (7.34) such that $\operatorname{Fix}(\phi) \cap \mathbb{A}^3(\mathbb{Q})$ contains at least two points.

(d) Let $b = -1$ and $d = 1$. Find all of the (complex) fixed points of $\phi$. If $a, b, c, d, e \in K$, describe the field $K(\operatorname{Fix}(\phi))$. What are its possible Galois groups over $K$? (*Hint*. It is easier to do the computations if you set $e = 1 - t^3$.)

(e) Suppose that $\phi \in \mathbb{R}[x, y, z]$ and that $d = 1$ and $b > 0$. Prove that $\phi$ has only one real fixed point, i.e., show that $\operatorname{Fix}(\phi) \cap \mathbb{A}^3(\mathbb{R}) = \{(0, 0, 0)\}$. In particular, $\phi$ has only one rational fixed point.

**7.2.** Let $\phi : \mathbb{A}^3 \to \mathbb{A}^3$ be the map

$$\phi(x, y, z) = (x^2 z, xy, yz).$$

(a) Calculate the indeterminacy locus of $\phi$.

(b) What are the values of

$$\liminf_{\substack{(x,y,z)\in\mathbb{Z}^3\\ h(x,y,z)\to\infty}} \frac{h\bigl(\phi(x, y, z)\bigr)}{h(x, y, z)} \qquad \text{and} \qquad \limsup_{\substack{(x,y,z)\in\mathbb{Z}^3\\ h(x,y,z)\to\infty}} \frac{h\bigl(\phi(x, y, z)\bigr)}{h(x, y, z)}?$$

(c) Same question as (b), but with the points $(x, y, z) \in \mathbb{Z}^3$ restricted to satisfy $xyz \neq 0$.

**7.3.** Let $\phi : \mathbb{A}^N \to \mathbb{A}^N$ be a regular affine automorphism and let $n \geq 1$. Prove that $\operatorname{Per}_n(\phi)$ is a discrete subset of $\mathbb{A}^N(\mathbb{C})$, and that counted with appropriate multiplicities,

$$\#\operatorname{Per}_{Nn}(\phi) = d_2^{\ell_1 Nn} = d_1^{\ell_2 Nn}.$$

(This is Theorem 7.10(c). *Hint.* Rewrite $\phi^{Nn}(P) = P$ as $\phi^{\ell_2 n}(P) = \phi^{-(N-\ell_2)n}(P)$, show that the homogenizations of $\phi^{\ell_2 n}$ and $\phi^{(N-\ell_2)n}$ have the same degree, and use Bézout's theorem to count the number of solutions.)

**7.4.** Let $\phi : \mathbb{A}^3 \to \mathbb{A}^3$ be the map $\phi(x, y, z) = (y, z, x^2)$.

(a) Find an explicit expression for $\phi^n(x, y, z)$. (There may be more than one case.)

(b) Calculate the dynamical degree of $\phi$,

$$\operatorname{dyndeg}(\phi) = \lim_{n\to\infty} \deg(\phi^n)^{1/n}.$$

(See Remark 7.14 for a discussion of the dynamical degree.)

(c) Let $d_n = \deg(\phi^n)$. Compute the generating function $\sum_{n\geq 0} d_n T^n$ and prove that it is in $\mathbb{Q}(T)$.

(d) Prove that $\operatorname{PrePer}(\phi) \subset \{P \in \mathbb{A}^3(\bar{\mathbb{Q}}) : h(P) = 0\}$.

(e) Let $\delta = \operatorname{dyndeg}(\phi)$ and $P \in \mathbb{A}^3(\bar{\mathbb{Q}})$. Find real numbers $b > a > 0$ such that

$$a\delta^n h(P) \leq h\bigl(\phi^n(P)\bigr) \leq b\delta^n h(P)$$

for all (sufficiently large) integers $n$.

(f) With notation as in (d), if $P \notin \operatorname{PrePer}(\phi)$, prove that

$$\lim_{T\to\infty} \frac{\#\bigl\{n \geq 0 : h\bigl(\phi^n(P)\bigr) \leq T\bigr\}}{\log T} = \frac{1}{\log \delta}.$$

**7.5.** Let $\phi : \mathbb{P}^3 \to \mathbb{P}^3$ be the rational map

$$\phi = [X_1X_3, X_2X_3, X_0^2, X_0X_3].$$

(a) Prove that $\phi$ is a birational map, i.e., find a rational map $\psi$ so that $\phi \circ \psi$ and $\psi \circ \phi$ are the identity map at all points where they are defined.

(b) Compute $Z(\phi)$ and $Z(\phi^{-1})$. Where do they intersect?

(c) * Let $d_n = \deg(\phi^n)$. Prove that the generating function $\sum_{n\geq 0} d_n T^n$ is <u>not</u> in $\mathbb{Q}(T)$.

The map in this exercise and the map in the previous exercise are examples of monomial maps, see [199].

**7.6.** Let $u, a_1, \ldots, a_N, b_1, \ldots, b_N \in \bar{\mathbb{Q}}$ with $u \neq 0$. We proved in Lemma 7.17 that

$$h\bigl([u, a_1, \ldots, a_N, b_1, \ldots, b_N]\bigr) \leq h\bigl([u, a_1, \ldots, a_N]\bigr) + h\bigl([u, b_1, \ldots, b_N]\bigr).$$

Prove that this inequality need not be true if $u = 0$.

**7.7.** This exercise generalizes Theorem 7.15. Let $\phi_1, \ldots, \phi_t : \mathbb{A}^N \to \mathbb{A}^N$ be affine automorphisms with the property that

$$Z(\phi_1) \cap Z(\phi_2) \cap \cdots \cap Z(\phi_t) = \emptyset.$$

Let $d_i = \deg(\phi_i)$ for $1 \le i \le t$. Prove that there is a constant $C = C(\phi_1, \ldots, \phi_t)$ so that for all $P \in \mathbb{A}^N(\bar{\mathbb{Q}})$,

$$\frac{1}{d_1} h\bigl(\phi_1(P)\bigr) + \frac{1}{d_2} h\bigl(\phi_2(P)\bigr) + \cdots + \frac{1}{d_t} h\bigl(\phi_t(P)\bigr) \ge h(P) - C.$$

**7.8.** Let $\phi_1, \phi_2 : \mathbb{A}^2 \to \mathbb{A}^2$ be the maps

$$\phi_1(x,y) = (x^2, xy) \qquad \text{and} \qquad \phi_2(x,y) = (xy, y^2).$$

(a) Prove that $\phi_1$ and $\phi_2$ are jointly regular.
(b) Let $a$ a positive integer and $P = (0,a) \in \mathbb{A}^2(\mathbb{Q})$. Prove that

$$\frac{1}{2} h\bigl(\phi_1(P)\bigr) + \frac{1}{2} h\bigl(\phi_2(P)\bigr) = h(P).$$

This proves that the lower bound in Theorem 7.15 cannot be improved in general for jointly regular affine morphisms.
(c) ** Can the lower bound in Theorem 7.15 be improved for jointly regular affine automorphisms, i.e., if we add the requirement that $\phi_1$ and $\phi_2$ be invertible, although not necessarily inverses of one another?

**7.9.** Let $\phi : \mathbb{A}^N \to \mathbb{A}^N$ be an affine automorphism (not necessarily regular) and let $d_1 = \deg \phi$ and $d_2 = \deg \phi^{-1}$. Prove that

$$\min\left\{ \frac{h\bigl(\phi(P)\bigr)}{d_1}, \frac{h\bigl(\phi^{-1}(P)\bigr)}{d_2} \right\} \ge \frac{h(P)}{d_1 d_2} + O(1) \quad \text{for all } P \in \mathbb{A}^N(\bar{\mathbb{Q}}).$$

**Exercises on Integrability and Reversibility**

The notions of integrability and reversibility play an important role in classical real and complex dynamics. Their algebraic analogues lead to dynamical systems with interesting arithmetic properties, which we explore in Exercises 7.10–7.14.

**Definition.** An affine automorphism $\phi : \mathbb{A}^N \to \mathbb{A}^N$ is said to be *algebraically reversible* if there is linear transformation $g \in \mathrm{GL}_N$ satisfying

$$g^2 = 1, \qquad \det(g) = -1, \qquad \text{and} \qquad g \circ \phi \circ g = \phi^g = \phi^{-1}.$$

The terminology is meant to reflect the idea that conjugation by the involution $g$ has the effect of reversing the flow of the map $\phi$.

**7.10.** Assume that $\phi$ is reversible, say $\phi^g = \phi^{-1}$. Let $\gamma = \phi \circ g$. Prove that $\gamma^2$ is the identity map. Thus $g$ and $\gamma$ are both involutions, so a reversible map can always be written as a composition $\phi = \gamma \circ g$ of two, generally noncommuting, involutions.

**7.11.** Let $a \neq 0$, let $f(y)$ be a polynomial of degree $d \geq 2$, and let

$$\phi : \mathbb{A}^2 \longrightarrow \mathbb{A}^2, \qquad \phi(x,y) = \big(y, ax + f(y)\big)$$

be the associated Hénon map. Suppose that $\phi$ is reversible. Prove that $\phi$ and its reversing involution $g \in \mathrm{GL}_2$ have one of the following forms:

(a) $a = 1$, $\quad g(x,y) = (y,x)$.
(b) $a = 1$, $\quad g(x,y) = (-y,-x)$, $\quad f$ satisfies $f(-y) = f(y)$.
(c) $a = -1$, $\quad g(x,y) = (-y,-x)$, $\quad f$ satisfies $f(-y) = -f(y)$.

**7.12.** The real and complex dynamics of reversible maps are in some ways less chaotic than nonreversible maps. Similarly, reversibility (and integrability) appear to have a significant effect on arithmetic dynamics. For an affine automorphism $\phi : \mathbb{A}^N \to \mathbb{A}^N$, we let

$$C_p(\phi) = \text{number of distinct orbits of } \phi \text{ in } \mathbb{A}^2(\mathbb{F}_p).$$

For each of the following Hénon maps, compute $C_p(\phi)$ for all primes $2 < p < 100$ (or further) and make a graph of $p$ versus $C_p(\phi)$:

(a) $\phi(x,y) = (y, x + y^2)$.
(b) $\phi(x,y) = (y, 2x + y^2)$.
(c) $\phi(x,y) = (y, -x + y^3)$.

Do you see a difference in behavior? Try plotting the ratio $C_p(\phi)/p$. (Notice that Exercise 7.11 says that the maps in (a) and (c) are reversible, while the map in (b) is not reversible.)

**Definition.** Let $\phi : \mathbb{A}^N \to \mathbb{A}^N$ be a *rational automorphism*, by which we mean that $\phi$ is a rational map (but not necessarily a morphism) and that there is an inverse rational map $\phi^{-1} : \mathbb{A}^N \to \mathbb{A}^N$ such that $\phi \circ \phi^{-1}$ is the identity map wherever it is defined. The map $\phi$ is said to be *algebraically integrable* if there is a nonconstant rational function $I : \mathbb{A}^N \to \mathbb{A}^1$ satisfying $I \circ \phi = I$.

**7.13.** Let $\phi : \mathbb{A}^2 \to \mathbb{A}^2$ be the rational map

$$\phi(x,y) = \left(y, -x - \frac{y^2+1}{y+1}\right).$$

(a) Prove that $\phi$ is a rational automorphism.
(b) Let $\bar{\phi}$ and $\bar{\phi}^{-1}$ be the extensions of $\phi$ and $\phi^{-1}$ to maps $\mathbb{P}^2 \to \mathbb{P}^2$. Compute $Z(\phi)$ and $Z(\phi^{-1})$, the sets of point(s) where $\bar{\phi}$ and $\bar{\phi}^{-1}$ are not defined, and verify that $Z(\phi) \cap Z(\phi^{-1}) = \emptyset$.
(c) Prove that $\phi$ is integrable by the function

$$I(x,y) = x^2y + xy^2 + x^2 + y^2 + x + y.$$

In other words, verify that $I \circ \phi(x,y) = I(x,y)$.
(d) Prove that for all but finitely many values of $c \in \mathbb{C}$, the level curve $I(x,y) = c$ is an elliptic curve. Find the exceptional values of $c$ for which the level curve is singular.

**7.14.** This exercise generalizes Exercise 7.13. Let $a, b, c, d, e \in K$ and define a rational map $\phi : \mathbb{P}^2 \to \mathbb{P}^2$ (using dehomogenized coordinates on $\mathbb{A}^2$) by

$$\phi(x,y) = \left(y, -x - \frac{by^2 + dy + e}{ay^2 + by + c}\right).$$

Prove that $\phi$ is integrable by the function

$$I(x,y) = ax^2y^2 + b(x^2y + xy^2) + c(x^2 + y^2) + dxy + e(x + y).$$

**7.15.** Assume that $\phi$ is integrable by the function $I$. For each $c \in K$, the set $I(\mathbf{x}) = c$ is called a level set of $\phi$.

(a) Prove that $\phi$ maps each level set to itself. Thus the dynamics of $\phi$ may be studied by investigating the behavior of the iterates of $\phi$ on the lower-dimensional invariant level sets that give a foliation of $\mathbb{P}^N$.

(b) * Let $N = 2$ and assume that $\phi$ has infinite order, i.e., no iterate of $\phi$ is the identity map. Prove that the level sets of $\phi$ are curves of genus 0 or 1.

The following result will be helpful in doing Exercise 7.16.

**Theorem 7.52.** ([172, Proposition 4.2]) *Let $\phi : \mathbb{P}^N \to \mathbb{P}^N$ be a morphism of degree $d \geq 2$ and let $V \subset \mathbb{P}^N$ be a completely invariant hypersurface, i.e., $\phi^{-1}(V) = V = \phi(V)$. Then $V$ has at most $N+1$ irreducible components.*

**7.16.** Let $\phi = [\phi_0, \ldots, \phi_N]$ be a morphism $\phi : \mathbb{P}^N \to \mathbb{P}^N$ of degree $d$ given by homogeneous polynomials $\phi_i \in \mathbb{C}[X_0, \ldots, X_N]$. We say that such a map is a *polynomial map* if its last coordinate function is equal to $X_N^d$. Equivalently, $\phi$ is a polynomial map if the inverse image of the hyperplane $H = \{X_N = 0\}$ is simply the hyperplane $H$ with multiplicity $d$.

(a) Assume that $\phi$ is a morphism and suppose that there is an $n \geq 1$ such that the iterate $\phi^n$ is a polynomial map. Prove that $\phi^n$ is already a polynomial map for some $n \leq N+1$. This generalizes Theorem 1.7. (*Hint.* Use Theorem 7.52.)

(b) Show that (a) need not be true if we assume only that the map $\phi : \mathbb{P}^N \to \mathbb{P}^N$ is a rational map of degree $d$. More precisely, prove that for all $d \geq 2$ and all $n \geq 2$, there exists a finite rational map $\phi : \mathbb{P}^N \to \mathbb{P}^N$ of degree $d$ such that $\phi^n$ is a polynomial map, but $\phi^i$ is not a polynomial map for all $1 \leq i < n$.

### Exercises on Canonical Heights for Regular Affine Automorphisms

Exercises 7.17–7.22 describe Kawaguchi's construction [230] of canonical heights for regular affine automorphisms assuming the validity of Conjecture 7.18, which is presently known only in dimension 2 [230, 413]. Let $\phi : \mathbb{A}^N \to \mathbb{A}^N$ be a regular affine automorphism of degree at least 2 defined over $\bar{\mathbb{Q}}$ and let

$$d_1 = \deg(\phi) \qquad \text{and} \qquad d_2 = \deg(\phi^{-1}).$$

We assume that Conjecture 7.18 is true, i.e., we assume that there is a constant $C = C(\phi) \geq 0$ such that for all $P \in \mathbb{A}^N(\bar{\mathbb{Q}})$,

$$\textbf{Assumption:} \quad \frac{1}{d_1} h(\phi(P)) + \frac{1}{d_2} h(\phi^{-1}(P)) \geq \left(1 + \frac{1}{d_1 d_2}\right) h(P) - C. \tag{7.35}$$

For any point $P \in \mathbb{A}^N(\bar{\mathbb{Q}})$, Kawaguchi defines *canonical height functions* by the formulas

$$\hat{h}^+(P) = \limsup_{n\to\infty} \frac{1}{d_1^n} h(\phi^n(P)), \qquad \hat{h}^-(P) = \limsup_{n\to\infty} \frac{1}{d_2^n} h(\phi^{-n}(P)), \tag{7.36}$$

$$\hat{h}(P) = \hat{h}^+(P) + \hat{h}^-(P). \tag{7.37}$$

**7.17.** Assuming that (7.35) is true, prove that the canonical height functions $\hat{h}^+$, $\hat{h}^-$, and $\hat{h}$ defined by (7.36) and (7.37) have the following properties:

(a) $\hat{h}^+(P) \leq h(P) + O(1)$ and $\hat{h}^-(P) \leq h(P) + O(1)$.

(b) $h(P) + O(1) \leq \hat{h}(P) \leq 2h(P) + O(1)$.

(c) $\hat{h}^+(P) \geq 0$ and $\hat{h}^-(P) \geq 0$ and $\hat{h}(P) \geq 0$.

(d) $\hat{h}^+(P) = 0 \iff \hat{h}^-(P) = 0 \iff \hat{h}(P) = 0 \iff P \in \mathrm{Per}(\phi)$.

(*Hint*. Before proving (d), you may find it advantageous to do the next exercise.)

**7.18.** Assuming that (7.35) is true, prove that the canonical height functions satisfy the following transformation formulas:

$$\hat{h}^+\big(\phi(P)\big) = d_1\hat{h}^+(P), \qquad \hat{h}^-\big(\phi^{-1}(P)\big) = d_2\hat{h}^-(P), \tag{7.38}$$

$$\frac{1}{d_1}\hat{h}\big(\phi(P)\big) + \frac{1}{d_2}\hat{h}\big(\phi^{-1}(P)\big) = \left(1 + \frac{1}{d_1 d_2}\right)\hat{h}(P). \tag{7.39}$$

**7.19.** Suppose that

$$\hat{h}' : \mathbb{A}^N(\bar{\mathbb{Q}}) \to \mathbb{R} \qquad \text{and} \qquad \hat{h}'' : \mathbb{A}^N(\bar{\mathbb{Q}}) \to \mathbb{R}$$

are two functions satisfying (7.39), and suppose further that

$$\hat{h}' = \hat{h}'' + O(1).$$

Prove that $\hat{h}' = \hat{h}''$.

**7.20.** Let $\phi : \mathbb{A}^N \to \mathbb{A}^N$ be a regular affine automorphism satisfying (7.35) and let $P \in \mathbb{A}^N(\bar{\mathbb{Q}})$ be a wandering point for $\phi$, i.e., $P$ is not a periodic point. Prove that

$$\lim_{n\to\infty} \frac{\frac{1}{d_1}h\big(\phi^{n+1}(P)\big) + \frac{1}{d_2}h\big(\phi^{n-1}(P)\big)}{h\big(\phi^n(P)\big)} = 1 + \frac{1}{d_1 d_2}. \tag{7.40}$$

Hence the constant $1 + \frac{1}{d_1 d_2}$ appearing in the inequality (7.35) cannot be replaced by any larger constant.

**7.21.** Let $\phi : \mathbb{A}^N \to \mathbb{A}^N$ be a regular affine automorphism satisfying (7.35) and let $P \in \mathbb{A}^N(\bar{\mathbb{Q}})$ be a wandering point for $\phi$. We define the (*two-sided*) *orbit-counting function* of $P$ to be

$$N_{\phi,P}(T) = \#\big\{\phi^n(P) : n \in \mathbb{Z} \text{ and } h\big(\phi^n(P)\big) \leq T\big\}.$$

Prove that

$$N_{\phi,P}(T) = \left(\frac{1}{\log d_1} + \frac{1}{\log d_2}\right)\log T - \left(\frac{\log \hat{h}^+(P)}{\log d_1} + \frac{\log \hat{h}^-(P)}{\log d_2}\right) + O(1),$$

where the $O(1)$ constant depends only on the map $\phi$ and is independent of both the point $P$ and the number $T$.

**7.22.** Let $\phi : \mathbb{A}^N \to \mathbb{A}^N$ be a regular affine automorphism satisfying (7.35). Define sequences $(A_n)$ and $(B_n)$ by the formulas

$$A_n = \frac{d_1^n - d_2^{-n}}{d_1 - d_2^{-1}} \qquad \text{and} \qquad B_n = \frac{d_2^n - d_1^{-n}}{d_2 - d_1^{-1}}.$$

Prove that

$$\frac{\hat{h}\big(\phi^n(P)\big)}{d_1} + \frac{\hat{h}\big(\phi^{-n}(P)\big)}{d_2} = \left(\frac{A_{n+1}}{d_1} + \frac{B_{n+1}}{d_2}\right)\hat{h}(P) - B_n\frac{\hat{h}\big(\phi(P)\big)}{d_1} - A_n\frac{\hat{h}\big(\phi^{-1}(P)\big)}{d_2}.$$

(*Hint*. Verify that $A_n$ and $B_n$ satisfy the linear recurrences

$$\begin{aligned} &A_0 = 0, && A_1 = 1, && d_1A_i - (1+d_1d_2)A_{i-1} + d_2A_{i-2} = 0,\\ &B_0 = 0, && B_1 = 1, && d_2B_i - (1+d_1d_2)B_{i-1} + d_1B_{i-2} = 0, \end{aligned}$$

and use a telescoping sum argument.)

**7.23.** ** Let $\phi : \mathbb{A}^N \to \mathbb{A}^N$ be an automorphism defined over $\bar{\mathbb{Q}}$ and denote the dynamical degree of $\phi$ by

$$\delta(\phi) = \lim_{n\to\infty} \deg(\phi^n)^{1/n}.$$

We associate to $\phi$ the number

$$S(\phi) =: \liminf_{\substack{P\in\mathbb{A}^N(\bar{\mathbb{Q}})\\ h(P)\to\infty}} \frac{1}{h(P)}\left(\frac{h(\phi(P))}{\delta(\phi)} + \frac{h(\phi^{-1}(P))}{\delta(\phi^{-1})}\right).$$

Remark 7.16 tells us that $S(\phi)$ satisfies

$$S(\phi) \le 2.$$

If $\phi$ is regular, then $\delta(\phi) = \deg(\phi)$ and $\delta(\phi^{-1}) = \deg(\phi^{-1})$. If in addition $\phi$ satisfies assumption (7.35), then Exercises 7.18 and 7.20 imply that

$$S(\phi) = 1 + \frac{1}{\delta(\phi)\delta(\phi^{-1})}.$$

(a) Do there exist automorphisms $\phi : \mathbb{A}^N \to \mathbb{A}^N$ of degree at least 2 satisfying $S(\phi) = 1$? What if we require that $\phi$ be algebraically stable? (See Remark 7.13.)
(b) Do there exist automorphisms $\phi : \mathbb{A}^N \to \mathbb{A}^N$ of degree at least 2 satisfying $S(\phi) = 2$?
(c) What are the possible values of $S(\phi)$ for automorphisms of $\mathbb{A}^N$?
(d) What are the possible values of $S(\phi)$ for algebraically stable automorphisms of $\mathbb{A}^N$?

**7.24.** Let $K$ be a field that is complete with respect to a nonarchimedean absolute value and let $\phi : \mathbb{P}^N(K) \to \mathbb{P}^N(K)$ be a morphism. Prove that $\phi$ is an open map, i.e., the image of an open set is an open set.

## Section 7.4. Dynamics on Surfaces with Involutions

**7.25.** Let $G_k^*$ and $H_{ij}^*$ be the quartic forms defined by (7.17). Prove that the following algorithm computes $\iota_1$ and $\iota_2$.
(a) Let $[\mathbf{x},\mathbf{y}] \in S_{\mathbf{A},\mathbf{B}}$ and write $\iota_1([\mathbf{x},\mathbf{y}]) = [\mathbf{x},\mathbf{y}']$. Then

$$\mathbf{y}' = \begin{cases} \bigl[y_0G_0^x(\mathbf{x}), -y_0H_{01}^x(\mathbf{x}) - y_1G_0^x(\mathbf{x}), -y_0H_{02}(\mathbf{x}) - y_2G_0^x(\mathbf{x})\bigr] & \text{if } y_0 \ne 0,\\ \bigl[-y_1H_{01}^x(\mathbf{x}) - y_0G_1^x(\mathbf{x}), y_1G_1^x(\mathbf{x}), -y_1H_{12}(\mathbf{x}) - y_2G_1^x(\mathbf{x})\bigr] & \text{if } y_1 \ne 0,\\ \bigl[-y_2H_{02}^x(\mathbf{x}) - y_0G_2^x(\mathbf{x}), -y_2H_{12}^x(\mathbf{x}) - y_1G_2^x(\mathbf{x}), y_2G_2^x(\mathbf{x})\bigr] & \text{if } y_2 \ne 0. \end{cases}$$

(b) Let $[\mathbf{x},\mathbf{y}] \in S_{\mathbf{A},\mathbf{B}}$ and write $\iota_1([\mathbf{x},\mathbf{y}]) = [\mathbf{x}',\mathbf{y}]$. Then

$$\mathbf{x}' = \begin{cases} \bigl[x_0G_0^y(\mathbf{y}), -x_0H_{01}^y(\mathbf{y}) - x_1G_0^y(\mathbf{y}), -x_0H_{02}(\mathbf{y}) - x_2G_0^y(\mathbf{y})\bigr] & \text{if } x_0 \ne 0,\\ \bigl[-x_1H_{01}^y(\mathbf{y}) - x_0G_1^y(\mathbf{y}), x_1G_1^y(\mathbf{y}), -x_1H_{12}(\mathbf{y}) - x_2G_1^y(\mathbf{y})\bigr] & \text{if } x_1 \ne 0,\\ \bigl[-x_2H_{02}^y(\mathbf{y}) - x_0G_2^y(\mathbf{y}), -x_2H_{12}^y(\mathbf{y}) - x_1G_2^y(\mathbf{y}), x_2G_2^y(\mathbf{y})\bigr] & \text{if } x_2 \ne 0. \end{cases}$$

**7.26.** The K3 surface given in Example 7.36 contains the following 12 points of small height:

$$\begin{aligned} P_1 &= \big([0,1,1],[1,1,-1]\big), & P_5 &= \big([0,0,1],[0,1,0]\big), & P_9 &= \big([8,6,9],[-6,5,2]\big),\\ P_2 &= \big([1,0,0],[0,0,1]\big), & P_6 &= \big([0,0,1],[1,0,0]\big), & P_{10} &= \big([1,0,-1],[9,1,9]\big),\\ P_3 &= \big([0,1,0],[0,0,1]\big), & P_7 &= \big([3,1,3],[-3,3,2]\big), & P_{11} &= \big([3,8,11],[1,1,-1]\big),\\ P_4 &= \big([1,0,-1],[0,1,0]\big), & P_8 &= \big([1,0,0],[0,7,1]\big), & P_{12} &= \big([12,1,-20],[2,-4,1]\big). \end{aligned}$$

(a) Which of these 12 points lie in the same $\mathcal{A}$ orbit? How many distinct $\mathcal{A}$ orbits do they generate?
(b) Which of the points in the list are fixed by a nontrivial element of $\mathcal{A}$?
(c) The list includes all points in $S(\mathbb{Q})$ having integer coordinates at most 40. Extend the computation to find all points in $S(\mathbb{Q})$ having integer coordinates at most 100. (*Hint.* Loop over $\mathbf{x}$ with $|x_i| \le 100$, substitute into $L$ and $Q$, eliminate a variable, and check whether the resulting quadratic equation has a rational solution.)

**7.27.** For each of the primes in the set

$$\{2, 3, 317, 507593, 2895545793631, 14521485737273461\}$$

find an example of a surface $S_{\mathbf{A},\mathbf{B}}$ defined over $\mathbb{F}_p$ such that $\iota_1$ and $\iota_2$ are defined at every point of $S_{\mathbf{A},\mathbf{B}}(\bar{\mathbb{F}}_p)$. (These examples can be used to complete the proof of Proposition 7.41.)

**7.28.** This exercise sketches a noncomputational proof of Proposition 7.41 using more advanced methods from algebraic geometry.
(a) Let $S$ and $S'$ be nonsingular projective K3 surfaces and let $\phi : S \to S'$ be a birational map, i.e., a rational map with a rational inverse. Prove that $\phi$ is a morphism. (*Hint.* Find a surface $T$ and birational morphisms $\psi : T \to S$ and $\psi' : T \to S'$ so that $\phi \circ \psi = \psi'$ [198, V.5.5]. Do this so that $\psi$ is a minimal number of blowups and let $E$ be an exceptional curve of the last blowup. Deduce that $q(E)$ is a curve $C$ on $S'$. Then show that the intersection of $C$ with the canonical divisor on $S'$ satisfies $C \cdot K_{S'} \le E \cdot K_S = -1$, which contradicts the fact that $K_{S'} = 0$, since $S'$ is a K3 surface.)
(b) Prove that there is a proper Zariski closed set $Z \subset \mathbb{P}^8 \times \mathbb{P}^{35}$ such that for all $(\mathbf{A}, bfB) \notin Z$, the surface $S_{\mathbf{A},\mathbf{B}}$ is nonsingular. (*Hint.* Elimination theory says that the set of $(\mathbf{A}, bfB) \in \mathbb{P}^8 \times \mathbb{P}^{35}$ such that $S_{\mathbf{A},\mathbf{B}}$ is singular is a Zariski closed set. Thus it suffices to find a single $(\mathbf{A}, bfB)$ for which $S_{\mathbf{A},bfB}$ is nonsingular.)
(c) Combine (a) and (b) to prove Proposition 7.41.

**7.29.** Let $P \in S_{\mathbf{A},\mathbf{B}}$ with infinite $\mathcal{A}$-orbit. Prove that the $\mathcal{A}$-stabilizer of $P$,

$$\{\psi \in \mathcal{A} : \psi(P) = P\},$$

has order either 1 or 2.

**7.30.** This exercise describes intersections on the surface $S_{\mathbf{A},\mathbf{B}}$. For the basics of intersection theory on surfaces, see, for example, [198, V §1].
(a) Let $D_1 = p_1^* H$ and $D_2 = p_2^* H$ be the usual divisors in $\operatorname{Pic}(S_{\mathbf{A},\mathbf{B}})$. Prove that

$$D_1 \cdot D_1 = D_2 \cdot D_2 = 2 \qquad \text{and} \qquad D_1 \cdot D_2 = 4.$$

(b) Let $\alpha = 2 + \sqrt{3}$ and define divisors $E^+$ and $E^-$ in $\operatorname{Pic}(S_{\mathbf{A},\mathbf{B}}) \otimes \mathbb{R}$ by the formulas

$$E^+ = -D_1 + \alpha D_2 \quad \text{and} \quad E^- = \alpha D_1 - D_2.$$

Prove that $\iota_1^* E^{\pm} = \alpha^{\pm 1} E^{\mp}$ and $\iota_2^* E^{\pm} = \alpha^{\mp 1} E^{\mp}$.

(c) Prove that $E^+ \cdot E^+ = E^- \cdot E^- = 0$ and $E^+ \cdot E^- = 12\alpha$.

**7.31.** (a) Prove that under composition, the involutions $\iota_1, \iota_2 \in \operatorname{Aut}(S_{\mathbf{A},\mathbf{B}})$ satisfy no relations other than $\iota_1^2 = \iota_2^2 = 1$. Thus $\mathcal{A}$ is the free product of the subgroups generated by $\iota_1$ and $\iota_2$. (*Hint.* Use Exercise 7.30. Apply a composition of $\iota_1$'s and $\iota_2$'s to $E^+ + E^-$ and intersect with $E^+$.)

(b) Show that $\mathcal{A}$ is isomorphic to the infinite (discrete) dihedral group

$$\mathcal{D}_\infty = \frac{\{s^i t^j : i, j \in \mathbb{Z}\}}{\{t^2 = 1 \text{ and } ts = s^{-1}t\}}$$

via the map

$$\mathcal{D}_\infty \longrightarrow \mathcal{A}, \qquad \begin{pmatrix} s \longmapsto \iota_1\iota_2 \\ t \longmapsto \iota_1 \end{pmatrix}.$$

**7.32.** Let $P \in S_{\mathbf{A},\mathbf{B}}$ be a point whose $\mathcal{A}$-orbit $\mathcal{A}(P)$ is an infinite set. Prove that $\mathcal{A}(P)$ is Zariski dense in $S_{\mathbf{A},\mathbf{B}}$. (*Hint.* If $\mathcal{A}(P)$ is not dense, find a curve $C \subset S_{\mathbf{A},\mathbf{B}}$ fixed by some nontrivial element $\psi \in \mathcal{A}$ and consider the intersection of $C$ with the divisors $E^+$ and $E^-$ defined in Exercise 7.30.)

**7.33.** Let $\phi = \iota_1 \circ \iota_2$ and fix a nonzero integer $n$. Prove that the set

$$\operatorname{Per}_n(\phi) = \bigl\{P \in S_{\mathbf{A},\mathbf{B}} : \phi^n(P) = P\bigr\}$$

is a finite set. (*Hint.* If the set is infinite, find a curve $C \subset S_{\mathbf{A},\mathbf{B}}$ fixed by $\phi^n$ and consider the intersection of $C$ with the divisors $E^+$ and $E^-$ defined in Exercise 7.30.)

**7.34.** Let $D \in \operatorname{Pic}(S_{\mathbf{A},\mathbf{B}})$ be the divisor $D = D_1 + D_2 = p_1^* H + p_2^* H$. Let $\phi = \iota_1 \circ \iota_2$ and $\alpha = 2 + \sqrt{3}$ as usual. Prove that

$$\frac{h_D(\phi^n P) + h_D(\phi^{-n} P)}{\alpha^{2n} + \alpha^{-2n}} = h_D(P) + O(1) \qquad \text{for all } P \in S_{\mathbf{A},\mathbf{B}}(\bar{K}) \text{ and all } n \ge 0.$$

(The $O(1)$ constant depends on the surface $S_{\mathbf{A},\mathbf{B}}$, but is independent of both $P$ and $n$.)

**7.35.** Let $S_{\mathbf{A},\mathbf{B}}$ be defined over a number field $K$ and let $\hat{h}^+$ and $\hat{h}^-$ be the canonical height functions constructed in Proposition 7.47. Assuming that $S_{\mathbf{A},\mathbf{B}}(K)$ is an infinite set, prove that there is a constant $C$ such that both of the sets

$$\bigl\{P \in S_{\mathbf{A},\mathbf{B}}(K) : \hat{h}^+(P) \le C\bigr\} \qquad \text{and} \qquad \bigl\{P \in S_{\mathbf{A},\mathbf{B}}(K) : \hat{h}^-(P) \le C\bigr\}$$

are infinite. This shows that Proposition 7.49 is not true if $\hat{h}$ is replaced by either $\hat{h}^+$ or $\hat{h}^-$.

**7.36.** Let $S_{\mathbf{A},\mathbf{B}}$ be defined over a number field $K$, let $\hat{h}^+$ and $\hat{h}^-$ be the canonical height functions constructed in Proposition 7.47, and let $\hat{h} = \hat{h}^+ + \hat{h}^-$. Fix a point $Q \in S_{\mathbf{A},\mathbf{B}}(\bar{K})$.

(a) Prove that the product $\hat{h}^+(P)\hat{h}^-(P)$ is the same for every point $P \in \mathcal{A}(Q)$. This product measures, in a certain sense, the arithmetic complexity of the $\mathcal{A}$-orbit of $Q$. Notice how $\hat{h}^+(Q)\hat{h}^-(Q)$ naturally appears in Proposition 7.50 counting points of bounded height in the $\mathcal{A}$-orbit of $Q$.

(b) Prove that

$$2\sqrt{\hat{h}^+(Q)\hat{h}^-(Q)} \le \min_{P\in\mathcal{A}(Q)} \hat{h}(P) \le 2\alpha\sqrt{\hat{h}^+(Q)\hat{h}^-(Q)}.$$

(Here $\alpha = 2+\sqrt{3}$ as usual.)

**7.37.** Let $S_{\mathbf{A},\mathbf{B}}$ be defined over a number field $K$ and let $Q \in S_{\mathbf{A},\mathbf{B}}(\bar{K})$ be a point whose $\mathcal{A}$-orbit $\mathcal{A}(Q)$ is infinite. Further, let $\hat{h} = \hat{h}^+ + \hat{h}^-$, and define a *height zeta function* for the $\mathcal{A}$-orbit of $Q$ by the series

$$Z(\mathcal{A}(Q), s) = \sum_{P\in\mathcal{A}(Q)} \frac{1}{\hat{h}(Q)^s}.$$

(a) Prove that the series defining $Z(\mathcal{A}(Q), s)$ converges on the half-plane $\mathrm{Real}(s) > 0$.
(b) Prove that $Z(\mathcal{A}(Q), s)$ has a meromorphic continuation to the entire complex plane.
(c) Find the poles of $Z(\mathcal{A}(Q), s)$.
(d) Find the residues of $Z(\mathcal{A}(Q), s)$ at its poles.

**7.38.** Let $S_{\mathbf{A},\mathbf{B}}$ be defined over a number field $K$ and let $P \in S_{\mathbf{A},\mathbf{B}}(\bar{K})$ be a point whose $\mathcal{A}$-orbit $\mathcal{A}(P)$ is Galois-invariant, i.e., if $Q \in \mathcal{A}(P)$ and $\sigma \in \mathrm{Gal}(\bar{K}/K)$, then $\sigma(Q) \in \mathcal{A}(P)$. Prove that $P$ satisfies one of the following conditions:

(a) $P \in S_{\mathbf{A},\mathbf{B}}(K)$.
(b) $\mathcal{A}(P)$ is finite.
(c) $\big[K(P) : K\big] = 2$.

If $P$ satisfies condition (c), prove that there exist a $\psi \in \mathcal{A}$ and an index $j \in \{1, 2\}$ such that $p_j\big(\psi(P)\big) \in \mathbb{P}^2(K)$.

**7.39.** Let $V \subset \mathbb{P}^N \times \mathbb{P}^N$ be a variety given by the vanishing of $N-1$ bilinear forms and one biquadratic form,

$$L_1(\mathbf{x},\mathbf{y}) = \cdots = L_{N-1}(\mathbf{x},\mathbf{y}) = Q(\mathbf{x},\mathbf{y}) = 0,$$

and let $p_1 : V \to \mathbb{P}^N$ and $p_2 : V \to \mathbb{P}^N$ be the usual projection maps $p_1(\mathbf{x},\mathbf{y}) = \mathbf{x}$ and $p_2(\mathbf{x},\mathbf{y}) = \mathbf{y}$.

(a) Prove that $p_1$ and $p_2$ are generically 2-to-1, so they induce involutions $\iota_1 : V \to V$ and $\iota_2 : V \to V$. In other words, there are rational maps $\iota_1$ and $\iota_2$ such that $\iota_1^2$ and $\iota_2^2$ are the identity map wherever they are defined.
(b) If $N \ge 3$, prove that $\iota_1$ and $\iota_2$ are *not* morphisms.

**7.40.** Let $a \in K^*$. The *Markoff equation*

$$M_a : x^2 + y^2 + z^2 = axyz$$

defines an affine surface in $\mathbb{A}^3$.

(a) Prove that there are involutions $\iota_{12}$, $\iota_{13}$, and $\iota_{23}$ of $M_a$ defined by the formulas

$$\begin{aligned}\iota_{12}(x,y,z) &= (x, y, axy - z),\\ \iota_{13}(x,y,z) &= (x, axz - y, z),\\ \iota_{23}(x,y,z) &= (ayz - x, y, z).\end{aligned}$$

Explain how these involutions correspond to natural double covers $M_a \to \mathbb{A}^2$ induced by projection maps $\mathbb{A}^3 \to \mathbb{A}^2$.

(b) Prove that the involutions $\iota_{ij}$ do *not* extend to morphisms on the projective variety

$$\bar{M}_a = \{x^2 w + y^2 w + z^2 w = axyz\} \subset \mathbb{P}^3,$$

and determine the points at which they fail to be defined.

(c) Find a birational map $\mathbb{P}^2 \to \bar{M}_a$ defined over $\mathbb{Q}$. (A birational map between projective varieties $V$ and $W$ is a rational map from $V$ to $W$ that is an isomorphism from a Zariski open subset of $V$ to a Zariski open subset of $W$.) In particular, this implies that $M_a(\mathbb{Q})$ contains many points.

(d) Prove that every point in $M_3(\mathbb{Z})$ with positive coordinates can be obtained by starting with the point $(1, 1, 1)$ and applying the involutions $\iota_{ij}$. (*Hint.* Define the size of a positive integral point $P = (x, y, z)$ to be the largest of its coordinates and prove that if $P \neq (1, 1, 1)$, then at least one of $\iota_{ij}(P)$ has size strictly smaller than the size of $P$.)

(e) Let $a$ be a positive integer. Prove that if $a \neq 1$ and $a \neq 3$, then $M_a(\mathbb{Z}) = \emptyset$. (*Hint.* Use the same type of descent argument as suggested in (d).)

(f) A *normalized Markoff triple* is a point $(x, y, z) \in M_3(\mathbb{Z})$ with $x \le y \le z$. Let

$$N(T) = \#\{\text{normalized Markoff triples } (x, y, z) \text{ with } z \le T\}.$$

Prove that there are positive constants $c_1$ and $c_2$ such that

$$c_1(\log T)^2 \le N(T) \le c_2(\log T)^2 \quad \text{as } T \to \infty. \tag{7.41}$$

More precisely, prove that there is a constant $c$ such that

$$N(T) = c(\log T)^2 + O\big((\log T)(\log\log T)^2\big). \tag{7.42}$$

(g) ** Let $(x_1, y_1, z_1)$ and $(x_2, y_2, z_2)$ be normalized Markoff triples. Prove that if $z_1 = z_2$, then also $x_1 = x_2$ and $y_1 = y_2$. (This is known as the *unicity conjecture* for Markoff numbers.)

**Exercises on K3 Surfaces with Three Involutions**

Exercises 7.41–7.44 ask you to explore a family of K3 surfaces that admit three noncommuting involutions. These hypersurfaces $S_C \subset \mathbb{P}^1 \times \mathbb{P}^1 \times \mathbb{P}^1$ are described by the vanishing of a trihomogeneous polynomial of degree 2,

$$Q(\mathbf{x}, \mathbf{y}, \mathbf{z}) = \sum_{\substack{0 \le i \le j \le 1 \\ 0 \le k \le \ell \le 1 \\ 0 \le m \le n \le 1}} C_{ijk\ell mn} x_i x_j y_k y_\ell z_m z_n = 0.$$

The surface $S_C$ admits three maps of degree 2 to $\mathbb{P}^1 \times \mathbb{P}^1$,

$$p_{12}(\mathbf{x}, \mathbf{y}, \mathbf{z}) = (\mathbf{x}, \mathbf{y}), \qquad p_{13}(\mathbf{x}, \mathbf{y}, \mathbf{z}) = (\mathbf{x}, \mathbf{z}), \qquad p_{23}(\mathbf{x}, \mathbf{y}, \mathbf{z}) = (\mathbf{y}, \mathbf{z}),$$

and these maps induce corresponding involutions

$$\iota_{12} : S_C \longrightarrow S_C, \qquad \iota_{13} : S_C \longrightarrow S_C, \qquad \iota_{23} : S_C \longrightarrow S_C.$$

We also fix a point $t_0 \in \mathbb{P}^1$, let $\pi_1, \pi_2, \pi_3 : S_C \to \mathbb{P}^1$ be the maps induced by the three projections from $\mathbb{P}^1 \times \mathbb{P}^1 \times \mathbb{P}^1$ to $\mathbb{P}^1$, and define divisors on $S_C$ by setting

$$D_1 = \pi_1^*(t_0), \qquad D_2 = \pi_2^*(t_0), \qquad D_3 = \pi_3^*(t_0).$$

**7.41.** The surface $S_{\mathbf{C}}$ is specified by the 27-tuple

$$\mathbf{C} = [c_{0000}, c_{0001}, \ldots, c_{1111}] \in \mathbb{P}^{26}$$

of coefficients of the trihomogeneous polynomial defining $S_{\mathbf{C}}$.

(a) What is the dimension of the family of surfaces $S_{\mathbf{C}}$ after we identify surfaces that are isomorphic via the action of $\mathrm{PGL}_2$ on each of the three copies of $\mathbb{P}^1$ in $\mathbb{P}^1 \times \mathbb{P}^1 \times \mathbb{P}^1$. (See Remark 7.38 for a similar calculation for the family $S_{\mathbf{A},\mathbf{B}}$.)

(b) Prove that there is a Zariski closed subset $Z \subset \mathbb{P}^{26}$ such that if $\mathbf{C} \notin Z$, then the involutions $\iota_{12}$, $\iota_{13}$, and $\iota_{23}$ are defined at every point of $S_{\mathbf{C}}$.

**7.42.** Let $\mathcal{D}$ be the subspace of $\mathrm{Pic}(S_{\mathbf{C}})$ generated by $D_1$, $D_2$, and $D_3$.

(a) Prove that the action of $\iota_{12}^*$ on $\mathcal{D}$ is given by

$$\iota_{12}^* D_1 = D_1, \qquad \iota_{12}^* D_2 = D_2, \qquad \iota_{12}^* D_3 = 2D_1 + 2D_2 - D_3.$$

Devise analogous formulas for the action of $\iota_{13}^*$ and $\iota_{23}^*$ on $\mathcal{D}$.

(b) Let $M_{ij}$ be the $3 \times 3$ matrix of $\iota_{ij}^*$ acting on $\mathcal{D}$ and let $J$ be the matrix $\left(\begin{smallmatrix} 0 & 2 & 2 \\ 2 & 0 & 2 \\ 2 & 2 & 0 \end{smallmatrix}\right)$. Prove that $M_{ij}^2 = 1$ and that $M_{ij}^t J^{-1} M_{ij} = J^{-1}$.

(c) Prove that double products such at $M_{12}M_{13}$ have all of their eigenvalues equal to 1.

(d) Let $\phi$ be the map $\phi = \iota_{12} \circ \iota_{13} \circ \iota_{23}$ and let $\beta = \frac{1}{2}(3+\sqrt{5})$. Prove that $\beta^3$ is an eigenvalue of $\phi^*$ acting on $\mathcal{D}$ and that a corresponding eigenvector is $\beta^2 D_1 + \beta D_2 + D_3$.

**7.43.** Assume that $S_{\mathbf{C}}$ is defined over a number field $K$. Let $\phi = \iota_{12} \circ \iota_{13} \circ \iota_{23} : S_{\mathbf{C}} \to S_{\mathbf{C}}$ and let $\beta = \frac{1}{2}(3 + \sqrt{5})$.

(a) Define a real-valued function $f : S_{\mathbf{C}}(\bar{K}) \to \mathbb{R}$ by

$$f(P) = \beta^2 h_{D_1}(P) + \beta h_{D_2}(P) + h_{D_3}(P).$$

Prove that

$$f\big(\phi(P)\big) = \beta^3 f(P) + O(1) \qquad \text{for all } P \in S_{\mathbf{C}}(\bar{K}).$$

(b) Prove that there exists a unique function $\hat{f} : S_{\mathbf{C}}(\bar{K}) \to \mathbb{R}$ satisfying

$$\hat{f}(P) = f(P) + O(1) \quad \text{and} \quad \hat{f}\big(\phi(P)\big) = \beta^3 \hat{f}(P) \quad \text{for all } P \in S_{\mathbf{C}}(\bar{K}).$$

**7.44.** Let $S_{\mathbf{C}}$ be the surface given by the equation

$$x_0^2 y_0^2 z_0^2 + x_0^2 y_0 y_1 z_1^2 + 4x_0^2 y_1^2 z_1^2 + x_0 x_1 y_0^2 z_1^2 + x_0 x_1 y_1^2 z_0^2 + x_1^2 y_0 y_1 z_0^2 + x_1^2 y_1^2 z_0 z_1 + 2x_1^2 y_0^2 z_1^2 = 0.$$

(a) Prove that the surface $S_{\mathbf{C}}$ is nonsingular. (*Hint.* Check that it is nonsingular over $\mathbb{F}_2$.)

(b) Prove that the involutions $\iota_{12}$, $\iota_{13}$, and $\iota_{23}$ are defined at every point of $S_{\mathbf{C}}(\mathbb{C})$.

(c) Verify that the point $P_0 = \big([0,1],[-1,1],[-1,1]\big)$ is in $S_{\mathbf{C}}(\mathbb{Q})$. Then compute the "tree" of points starting from $P_0$ and generated by applying the involutions in various orders:

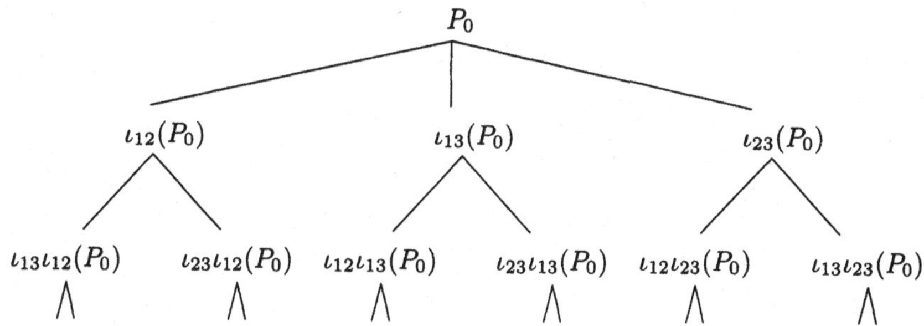


In particular, find two branches of the tree that loop around and reconnect with the top.

# Notes on Exercises

Many of the exercises in this book are standard, or in some cases not so standard, results. These notes thus have a dual purpose: to give credit where due, and to point the reader toward the relevant literature. However, since any attempt to assign credit is bound to be incomplete, the author tenders his apologies to anyone who feels that he or she has been slighted.

## Chapter 1. An Introduction to Classical Dynamics

**1.11.** See [415].
**1.12.** See [415].
**1.18.** (c) is proven in Corollary 4.7.
(d) This is due to I.N. Baker [19], or see [43, §6.8].
**1.24.** See [43, Theorem 3.2.5].
**1.30.** Most of this exercise is proven in Proposition 6.6.
**1.31.** See [43, Section 1.4].

## Chapter 2. Dynamics over Local Fields: Good Reduction

**2.6.** See [436, Section 5.9].
**2.17.** This special case of a theorem of Rivera-Letelier was suggested to the author by Rafe Jones.
**2.19.** See [312, Proposition 3.1].
**2.22.** The first example of this phenomenon is due to Poonen (unpublished). It appears in Zieve's thesis [454, Lemma 6].
**2.24.** This is in Jones's thesis [220].

## Chapter 3. Dynamics over Global Fields

**3.2.** Schanual [391] (or see [256, Theorem 5.3]) proves a general formula for a number field $K$:

$$\lim_{B\to\infty} \frac{\#\{P \in \mathbb{P}^N(K) : H_K(P) \le B\}}{B^{N+1}} = \frac{h_K R_K / w_K}{\zeta_K(N+1)} \left(\frac{2^{r_1}(2\pi)^{r_2}}{D_K^{1/2}}\right)^{N+1} (N+1)^{r_1+r_2-1},$$

where $h_K$, $R_K$, $w_K$, $\zeta_K$,$D_K$, $r_1$, and $r_2$ are, respectively, the class number, regulator, number of roots of unity, zeta function, absolute discriminant, number of real embeddings, and number of complex embeddings of $K$.

**3.4.** See [410, Theorem III.5.9] or [256, Lemma 2.2 in §3.2]. For better estimates, see [256, §3.2] and the references cited there.
**3.9.** See [309].
**3.14.** This exercise was suggested to the author by Rob Benedetto.
**3.38.** See [411].
**3.40.** (b) See [411, Proposition 1.2].
(c) See [411, Theorem B].
**3.46.** See [411].
**3.21.** Sylvester's original article is [428]. See [2, 215] for additional material on Sylvester's and other related sequences.
**3.22.** This exercise was inspired by [2] and [184, Exercise 4.37].
**3.49.** (a) See [306].
(b) $g_e(z) = z^3 - (e-1)z^2/2 - (e^2+2e+9)z/4 + (e^3+e^2+7e-1)/8$ and $\operatorname{Disc}(g_e) = (e^2+e+7)^2$.

## Chapter 4. Families of Dynamical Systems

**4.4.** See [313, Proposition 3.2 and Lemma 3.4].
**4.6.** This exercise was suggested to the author by Michelle Manes.
**4.7.** (b,c) These formulas are due to Morton and Vivaldi [314].
**4.8.** See [313, Theorem 2.1]. For a generalization to morphisms of higher-dimensional varieties, see [214].
**4.12.** (e) See [314].
**4.13.** See [307].
**4.15.** See [132, Proposition 8.6].
**4.20.** (a,b) See [305, 309].
(c,d) See [171].
**4.30.** This exercise was inspired by Milnor's paper [303], which studies the geometry and topology of the spaces that we have denoted by $\operatorname{BiCrit}_d$ and $\mathcal{M}_d^{\mathrm{BiCrit}}$
**4.36.** (a) See [414, example in Section 7].
(b) $\big(\sigma_1(\phi), \sigma_2(\phi)\big) = (-6, 12)$.
**4.45.** See [414, Section 6].
**4.49.** This result is due to Szpiro and Tucker [431]. It is a dynamical version of Faltings' theorem (Shafarevich conjecture) [165, 164] that there are only finitely many principally polarized abelian varieties of given dimension with good reduction outside of a finite set of primes.

## Chapter 5. Dynamics over Local Fields: Bad Reduction

**5.1.** See, for example, [324, §10.1] or [78].
**5.6.** (a) This is wellknown. See [233] for a more general result.
**5.10.** See [436, §19.9] for a proof of the residue theorem over algebraically closed base fields due to Roquette.
**5.11.** (b) Rivera-Letelier shows in his thesis [372] that every indifferent periodic point is contained in a "domain of quasiperiodicity" that contains infinitely many (indifferent) periodic points.

**5.21.** Hsia [208] uses a version of this result in his proof of Montel's theorem with moving targets.

**5.25.** See [56].

**5.26.** This exercise was suggested to the author by Rob Benedetto.

**5.27.** See [234] for the analogous result on $\mathbb{P}^N$.

**5.28.** See [233] for the analogous result on $\mathbb{P}^N$, including explicit values for the Hölder constants.

**5.29.** For elliptic curves, this is due to Tate (unpublished letter to Serre) if $K_v$ is not algebraically closed and to the author [408] for arbitrary $K_v$. See [88, Section 5] for the general dynamical case.

**5.32.** See [88] for a general construction of local canonical heights associated to dynamical systems with eigendivisor classes.

**5.45.** This exercise, which appears in a paper of Rivera-Letelier, was suggested to the author by Rob Benedetto.

## Chapter 6. Dynamics Associated to Algebraic Groups

**6.27.** These examples are due to Noam Elkies [150].

## Chapter 7. Dynamics in Dimension Greater Than One

**7.4.** $\deg \phi^n = 2^{\lfloor (n+2)/3 \rfloor}$ and $\operatorname{dyndeg}(\phi) = \sqrt[3]{2}$.

**7.5.** (c) This example is due to Hasselblatt and Propp [199].

**7.12.** This exercise was inspired by the work of Jogia, Roberts, and Vivaldi [219, 218, 383, 384], who prove results and state conjectures on how reversibility and integrability affect the growth of $C_p(\phi)$ and related quantities.

**7.14.** This family of integrable maps was discovered by McMillan [293]. For an even larger family of integrable maps called the QRT family, see [364, 365].

**7.15.** This result is due to Veselov [439].

**7.16.** This author thanks Shu Kawaguchi for providing a solution to this exercise (private communication).

**7.17–7.21.** These exercises are due to Kawaguchi [230].

**7.20.** See [230, Proposition 4.2] and [413, remark following Theorem 3.1]).

**7.24.** See [233].

**7.26.** See [409, §5].

**7.28.** This exercise was suggested to the author by Shu Kawaguchi.

**7.32.** See [409, Corollary 2.3].

**7.33.** See [409, Corollary 2.4(b)].

**7.36.** See [409].

**7.38.** See [409].

**7.40.** (d) See [210, §11.8].

(f) The estimates (7.41) and (7.42) are due, respectively, to Cohn [108] and Zagier [451]. See also Baragar's articles [32, 33] for a higher-dimensional analogue in which the counting function $N(T)$ grows like $(\log T)^\epsilon$ for an irrational exponent $\epsilon$.

**7.41–7.44.** These exercises were inspired by the work of Baragar, Luijk, and Wang [34, 35, 36, 37, 38, 39, 446], who study the arithmetic and dynamical properties of these triple-involution K3-surfaces.

**7.44.** This example is due to Baragar [38, §4], who notes that $\iota_{13}\iota_{12}\iota_{13}(P_0) = P_0$ and $\iota_{12}\iota_{13}\iota_{12}(P_0) = P_0$.

# List of Notation

| | |
|---|---|
| $\phi^n$ | $n^{\text{th}}$ iterate of the map $\phi$, 1 |
| $\phi^0$ | the identity map, 1 |
| $\mathcal{O}_\phi(\alpha)$ | orbit of $\alpha$ by the map $\phi$, 1 |
| $\operatorname{Per}(\phi, S)$ | set of periodic points of $\phi$ in $S$, 1 |
| $\operatorname{PrePer}(\phi, S)$ | set of preperiodic points of $\phi$ in $S$, 1 |
| $G_{\text{tors}}$ | torsion subgroup of the abelian group $G$, 2 |
| $\operatorname{Aut}(\mathbb{P}^1)$ | automorphism group of the projective line, 10 |
| $\mathrm{PGL}_2$ | projective linear group, 10 |
| $\mathrm{GL}_2$ | general linear group, 10 |
| $\phi^f$ | linear conjugation of $\phi$ by $f$, 11 |
| $\rho$ | chordal metric on $\mathbb{P}^1(\mathbb{C})$, 11 |
| $e_\alpha(\phi)$ | ramification index of $\phi$ at $\alpha$, 12 |
| $\lambda_\phi(\alpha)$ | multiplier of $\phi$ at periodic point $\alpha$, 18 |
| $\operatorname{Per}_n(\phi)$ | set of points of period $n$, 18 |
| $\operatorname{Per}_n^{**}(\phi)$ | set of points of exact period $n$, 18 |
| $\Omega^1_\alpha$ | space of differential one-forms, 19 |
| $\mathcal{M}$ | the Mandelbrot set, 26 |
| $\mathbb{G}_m$ | multiplicative group, 29 |
| $T_d$ | the $d^{\text{th}}$ Chebyshev polynomial, 29 |
| $\mathbb{G}_a$ | the additive group, 30 |
| $\oplus$ | addition on an elliptic curve, 31 |
| $\phi_{E,d}$ | Lattès map associated to multiplication by $d$, 32 |
| $\phi_{E,u}$ | Lattès map associated to an endomorphism $u$, 32 |
| $\psi_E$ | complex uniformization of an elliptic curve $E$, 33 |
| $\wp$ | Weierstrass $\wp$ function, 34 |
| $\iota(\phi, a)$ | residue fixed-point index of $\phi$ at $\alpha$, 38 |
| $\Phi^*_n(z)$ | the $n^{\text{th}}$ dynatomic polynomial, 39 |
| $\rho_v$ | $v$-adic chordal metric on $\mathbb{P}^1$, 45 |
| $\tilde{P}$ | reduction of the point $P$ modulo a prime, 48 |
| $\tilde{\phi}$ | reduction of a rational map $\phi$ modulo a prime, 52 |
| $\operatorname{Res}(A, B)$ | the resultant of $A$ and $B$, 53 |
| $\operatorname{Res}(\phi)$ | resultant of a rational map, 56 |
| $\kappa(P_1, P_2, P_3, P_4)$ | cross-ratio of $P_1, P_2, P_3, P_4$, 71 |
| $\mathcal{K}(\phi)$ | the filled Julia set of $\phi$, 74 |
| $H(P)$ | (multiplicative) height of a rational point, 82 |
| $M_\mathbb{Q}$ | set of standard absolute values on $\mathbb{Q}$, 82 |
| $\lvert \cdot \rvert_\infty$ | the usual absolute value on $\mathbb{R}$, 82 |

| | |
|---|---|
| $\text{ord}_p(a)$ | the exponent of highest power of $p$ dividing $a$, 82 |
| $\lvert\cdot\rvert_p$ | the $p$-adic absolute value on $\mathbb{Q}$, 82 |
| $M_K$ | the standard set of absolute values on $K$, 83 |
| $M_K^\infty$ | the archimedean absolute values on $K$, 83 |
| $M_K^0$ | the nonarchimedean absolute values on $K$, 83 |
| $R_K$ | the ring of integers of $K$, 83 |
| $R_S$ | the ring of $S$-integers of $K$, 83 |
| $n_v$ | local degree at an absolute value $v$, 83 |
| $H_K(P)$ | (multiplicative) height of a $K$-rational point, 84 |
| $H(P)$ | the absolute (multiplicative) height of $P$, 85 |
| $\bar{K}$ | an algebraic closure of the field $K$, 85 |
| $K(P)$ | field of definition of the point $P$, 86 |
| $\sqrt{I}$ | the radical of the ideal $I$, 89 |
| $V(I)$ | the algebraic set attached to the ideal $I$, 90 |
| $I(V)$ | ideal attached to the algebraic set $V$, 90 |
| $\lvert P\rvert_v$ | maximum of absolute values of coordinates of $P$, 90 |
| $\lvert f\rvert_v$ | maximum absolute value of coefficients of a polynomial, 91 |
| $\delta_v(m)$ | equal to $m$ or 1 depending on whether $v$ is archimedean, 91 |
| $h_K$ | logarithmic height, 93 |
| $h$ | absolute logarithmic height, 93 |
| $O(1)$ | a bounded function, 93 |
| $\hat{h}_\phi$ | canonical height associated to morphism $\phi$, 99 |
| $\lambda_v$ | $v$-adic logarithmic distance function, 102 |
| $\hat{\lambda}_{\phi,v}$ | $v$-adic local canonical height, 102 |
| $\text{Res}(\phi)$ | resultant of a rational map, 112 |
| $\text{Per}_n^{**}(\phi)$ | periodic points of $\phi$ of primitive period $n$, 122 |
| $K_{n,\phi}$ | dynatomic field generated by primitive $n$-periodic points, 123 |
| $G_{n,\phi}$ | Galois group of dynatomic field, 123 |
| $\text{Wreath}(H,S)$ | wreath product of $H$ and $S$, 125 |
| $G_{n,\phi}^0$ | subgroup of $G_{n,\phi}$ leaving $\phi$-orbits invariant, 126 |
| $K_{n,\phi}^0$ | fixed field of $G_{n,\phi}^0$, 126 |
| $\mu_\phi$ | the canonical $\phi$-invariant probability measure on $\mathbb{P}^N(\mathbb{C})$, 127 |
| $C(P/K)$ | the set of Galois conjugates of $P$, 128 |
| $\delta_P$ | the Dirac measure supported at $P$, 128 |
| $\mu_P$ | discrete probability measure supported on Galois conjugates of $P$, 128 |
| $S_\phi$ | set of primes of bad reduction for $\phi$, 132 |
| $\varphi$ | Euler's totient function, 137 |
| $\mu$ | the Möbius $\mu$ function, 148 |
| $F_n, G_n$ | coordinate functions of the $n^{\text{th}}$ iterate of $\phi$, 149 |
| $\Phi_{\phi,n}(X,Y)$ | the $n$-period polynomial of $\phi$, 149 |
| $\Phi_{\phi,n}^*(X,Y)$ | that $n^{\text{th}}$ dynatomic polynomial of $\phi$, 149 |
| $\Phi_n(z), \Phi_n^*(z)$ | dehomogenized period and dynatomic polynomials, 149 |
| $\text{Per}_n(\phi)$ | the set of points of period $n$ for $\phi$, 150 |
| $\text{Per}_n^*(\phi)$ | the set of points of formal period $n$ for $\phi$, 150 |
| $\text{Per}_n^{**}(\phi)$ | the set of points of primitive period $n$ for $\phi$, 150 |
| $\nu_d(n)$ | number of points of formal period $n$ for a map of degree $d$, 150 |
| $a_P(n)$ | order of the period polynomial at the point $P$, 151 |
| $a_P^*(n)$ | order of the dynatomic polynomial at the point $P$, 151 |

| | |
|---|---|
| $\phi_c(z)$ | the quadratic polynomial $z^2+c$, 155 |
| $\Phi_n^*(c,z)$ | the $n^{\text{th}}$ dynatomic polynomial for $z^2+c$, 157 |
| $Y_1(n)$ | (affine) dynatomic modular curve for $z^2+c$, 157 |
| $X_1(n)$ | (projective) dynatomic modular curve for $z^2+c$, 157 |
| $\mathrm{Formal}(n)$ | $\mathrm{PGL}_2$-classes of quadratic with point of formal period $n$, 158 |
| $Y_0(n)$ | the quotient of $Y_1(n)$ by $\phi$, 161 |
| $X_0(n)$ | the quotient of $X_1(n)$ by $\phi$, 161 |
| $\theta$ | conformal isomorphism to the complement of Mandelbrot set, 167 |
| $F_{\mathbf{a}}(X,Y)$ | the homogeneous polynomial $a_0X^d+a_1X^{d-1}Y+\cdots+a_dY^d$, 169 |
| $[\mathbf{a},\mathbf{b}]$ | the point $[a_0,\dots,a_d,b_0,\dots,b_d]$ in $\mathbb{P}^{2d+1}$, 169 |
| $\mathrm{Rat}_d$ | the set of rational maps of degree $d$, 169 |
| $\mathbb{P}^1_V$ | the projective line over $V$, 171 |
| $\mathcal{M}_d$ | the moduli space of conjugacy classes of maps of degree $d$, 174 |
| $\langle\cdot\rangle$ | map from $\mathrm{Rat}_d$ to $\mathcal{M}_d$, 174 |
| $\mathbb{Q}[\mathrm{Rat}_d]^{\mathrm{PGL}_2}$ | the ring of $\mathrm{PGL}_2$-invariant functions on $\mathrm{Rat}_d$., 175 |
| $\mathrm{SL}_2$ | the special linear group, 175 |
| $\mathrm{PSL}_2$ | the projective special linear group, 175 |
| $\mathrm{Rat}_d^s$ | set of stable rational maps, 178 |
| $\mathrm{Rat}_d^{ss}$ | set of semistable rational maps, 178 |
| $\mathcal{M}_d^s$ | stable completion of $\mathcal{M}_d$, 178 |
| $\mathcal{M}_d^{ss}$ | semistable completion of $\mathcal{M}_d$, 178 |
| $F_{\phi,n},G_{\phi,n}$ | coordinate functions of the $n^{\text{th}}$ iterate of $\phi$, 181 |
| $\Lambda_n(\phi)$ | the $n$-multiplier spectrum of $\phi$, 182 |
| $\Lambda_n^*(\phi)$ | the formal $n$-multiplier spectrum of $\phi$, 182 |
| $\sigma_i^{(n)}(\phi)$ | symmetric polynomial of $n$-periodic multipliers of $\phi$, 183 |
| $\overset{*}{\sigma}{}_i^{(n)}(\phi)$ | symmetric polynomial of formal $n$-periodic multipliers of $\phi$, 183 |
| $\boldsymbol{\sigma}_{d,N}$ | map of $\mathcal{M}_d$ using $\sigma_i^{(n)}$ with $n\le N$, 187 |
| $\boldsymbol{\sigma}_{d,N}^*$ | map of $\mathcal{M}_d$ using $\overset{*}{\sigma}{}_i^{(n)}$ with $n\le N$, 187 |
| $\Lambda(\phi)$ | the multiplier spectrum of $\phi$, 187 |
| $\boldsymbol{\sigma}$ | map $\mathrm{Rat}_2\to\mathbb{A}^2$ inducing an isomorphism $\mathcal{M}_2\cong\mathbb{A}^2$, 188 |
| $\overline{\mathcal{M}}_2$ | the completion $\mathcal{M}_2^s=\mathcal{M}_2^{ss}$ of $\mathcal{M}_2$, 194 |
| $[\phi]$ | set of rational maps $\bar{K}$-equivalent to $\phi$, 195 |
| $[\phi]_K$ | set of rational maps $K$-equivalent to $\phi$, 195 |
| $\mathrm{Aut}(\phi)$ | the automorphism group of $\phi$, 196 |
| $\mathrm{Twist}(\phi/K)$ | the set of twists of the rational map $\phi$, 197 |
| $\mathrm{Twist}(X/K)$ | the set of twists of the object $X$, 199 |
| $g_\sigma$ | the 1-cocycle associated to a twist, 201 |
| $H^1(\Gamma,A)$ | cohomology set (group), 202 |
| $G_f$ | subgroup of $\mathrm{Gal}(\bar{K}/K)$ associated to $\phi$, 207 |
| $K_f$ | field of moduli of $\phi$, 207 |
| $\mathrm{ord}_{\mathfrak{p}}(P)$ | minimum $\mathrm{ord}_{\mathfrak{p}}$ of the coefficients of the polynomial $P$, 218 |
| $F_A,G_A$ | new coordinate functions for conjugate of $[F,G]$, 218 |
| $\epsilon_{\mathfrak{p}}(\phi)$ | exponent of the minimal resultant of $\phi$, 220 |
| $\mathfrak{R}_\phi$ | the (global) minimal resultant of $\phi$, 220 |
| $\mathfrak{N}_\phi$ | product of the primes of bad reduction for $\phi$, 221 |
| $\mathfrak{a}_{F,G}$ | ideal connecting minimal resultant to resultant of a model, 222 |
| $\bar{\mathfrak{a}}_{\phi/K}$ | the Weierstrass class of $\phi$ over $K$, 223 |
| $\delta_{n,\phi}(x)$ | polynomial whose roots are multipliers of $\phi$, 225 |

$F_{m,n}(X,Y)$ generalized period polynomial, 227
$\Phi^*_{m,n}(X,Y)$ generalized dynatomic polynomial, 227
$(S\phi)(z)$ the Schwarzian derivative of $\phi$, 231
$\mathrm{BiCrit}_d$ the space of bicritical rational maps, 233
$\mathcal{M}_d^{\mathrm{BiCrit}}$ the moduli space of bicritical rational maps, 233
$\mathbb{C}_p$ the completion of the algebraic closure of $\mathbb{Q}_p$, 239
$R = R_K$ the ring of integers of a local field $K$, 242
$R = R_K^*$ the unit group of the ring of integers of a local field $K$, 242
$\mathfrak{M}$ the maximal ideal of the ring of integers of a local field $K$, 242
$D(a,r)$ open disk of radius $r$ at $a$, 242
$\bar{D}(a,r)$ closed disk of radius $r$ at $a$, 242
$S(a,r)$ circle of radius $r$ at $a$, 242
$D(Q,r)$ open disk of radius $r$ in $\mathbb{P}^1(K)$, 243
$\bar{D}(Q,r)$ closed disk of radius $r$ in $\mathbb{P}^1(K)$, 243
$\mathrm{ord}_a(\phi)$ order of vanishing of a holomorphic function, 244
$\|\phi\|_{r,a}$ norm of $\phi$ relative to the disk $\bar{D}(a,r)$, 247
$v(z)$ valuation $-\log_p|z|$, 249
$\mathcal{F}(\phi)$ the Fatou set of $\phi$, 255
$\mathcal{J}(\phi)$ the Julia set of $\phi$, 255
$\mathcal{F}(\phi,K)$ the $K$-valued points of the Fatou set of $\phi$, 255
$\mathcal{J}(\phi,K)$ the $K$-valued points of the Julia set of $\phi$, 255
$\Lambda = \Lambda(\phi)$ set of points with bounded $\left|\phi^n(z)\right|$, 258
$I_0, I_1$ disks $\bar{D}(0,1/p)$ and $\bar{D}(1,1/p)$, 258
$\mathbb{N}$ the set of natural numbers, 258
$S^{\mathbb{N}}$ set of sequences chosen from the set $S$, 258
$\mathcal{D}$ set of "disks" in a topological space, 277
$\mathcal{D}_{\text{closed}}$ standard collection of closed disks in $\mathbb{P}^1(\mathbb{C}_p)$, 277
$\mathcal{D}_{\text{open}}$ standard collection of open disks in $\mathbb{P}^1(\mathbb{C}_p)$, 278
$\mathcal{DC}(\phi)$ set of disk components of the Fatou set, 283
$\pi$ projection map $\mathbb{A}^2 \to \mathbb{P}^1$, 287
$\mathbb{A}^2_*$ the affine plane with $(0,0)$ removed, 287
$\left\|(x,y)\right\|_v$ sup norm of $x$ and $y$, 288
$\mathcal{G}_\Phi(x,y)$ Green function attached to $\Phi : \mathbb{A}^2 \to \mathbb{A}^2$, 288
$\hat{\lambda}_{\phi,E}\bigl([x,y]\bigr)$ local canonical height associated to $\phi$ and $E$, 291
$\bar{D}^{\mathcal{B}}$ the unit Berkovich disk, 295
$\xi_{a,0}$ Type-I point in Berkovich disk $\bar{D}^{\mathcal{B}}$, 295
$\xi_{a,r}$ Type-II or III point in Berkovich disk $\bar{D}^{\mathcal{B}}$, 295
$\xi_{\mathbf{a},\mathbf{r}}$ Type-IV point in Berkovich disk $\bar{D}^{\mathcal{B}}$, 295
$\|f\|_R$ Gauss norm of $f \in \mathbb{C}_p[z]$, 296
$\bar{D}^{\mathcal{B}}$ the closed unit Berkovich disk, 297
$\bar{D}^{\mathcal{B}}_R$ the closed Berkovich disk of radius $R$, 297
$\mathbb{T}_R$ the Tate algebra, 297
$L_{a,r}$ line segment in Berkovich disk, 298
$L_{\mathbf{a},\mathbf{r}}$ line segment in Berkovich disk for Type-IV point, 298
$U(f,B)$ open set in Gel'fond topology on Berkovich disk, 300
$V(f,B)$ open set in Gel'fond topology on Berkovich disk, 300
$\bigwedge_{a,r}$ closed branch of Berkovich disk rooted at $(a,r)$, 300
$\mathring{\bigwedge}_{a,r}$ open branch of Berkovich disk rooted at $(a,r)$, 300

| | |
|---|---|
| $\xi_{0,R}$ | Gauss point of Berkovich disk of radius $R$, 301 |
| $\mathbb{A}^{\mathcal{B}}$ | the Berkovich affine line, 301 |
| $\mathbb{P}^{\mathcal{B}}$ | the Berkovich projective line, 302 |
| $\text{Ann}^{\mathcal{B}}$ | annulus in Berkovich space, 302 |
| $\mathbb{A}^{\circ}_{\infty}$ | branch at infinity of Berkovich projective line, 302 |
| $R(\phi, a, r)$ | radius of the image disk $\phi(\bar{D}(a,r))$, 305 |
| $\mathcal{J}^{\mathcal{B}}(\phi)$ | the Berkovich Julia set of $\phi$, 306 |
| $\mathcal{F}^{\mathcal{B}}(\phi)$ | the Berkovich Fatou set of $\phi$, 306 |
| $\mu_{\phi}$ | canonical measure on Berkovich space associated to $\phi$, 306 |
| $\text{Res}_{z=\alpha}\big(\phi(z)\,dz\big)$ | residue of $\phi(z)$ at $z = a$, 313 |
| $\hat{\mathcal{G}}_{\Phi}$ | modified Green function, 318 |
| $\hat{\lambda}_{\phi,D}$ | local canonical height associated to a divisor, 320 |
| $D_g$ | divisor of the rational function $g$, 320 |
| $\kappa(\xi_{a,r}, \xi_{b,s})$ | path metric on the Berkovich disk, 323 |
| $\delta(\xi_{a,r}, \xi_{b,s})$ | the Hsia kernel on the Berkovich disk, 323 |
| $M_d$ | the power map $z^d$, 325 |
| $\mathbb{G}_m$ | multiplicative group, 325 |
| $\boldsymbol{\mu}_n$ | the group of $n^{\text{th}}$ roots of unity, 326 |
| $T_d$ | the $d^{\text{th}}$ Chebyshev polynomial, 329 |
| $\mathcal{O}$ | point at infinity on an elliptic curve, 336 |
| $E(K)$ | the set of points of $E$ defined over the field $K$, 337 |
| $\text{Div}(E)$ | the group of divisors on $E$, 339 |
| $\deg(D)$ | the degree of the divisor $D$, 339 |
| sum | the summation map on the divisor group of $E$, 339 |
| $\text{div}(f)$ | the divisor associated to the rational function $f$, 339 |
| $\text{ord}_P(f)$ | the order of the zero (pole) of $f$ at $P$, 339 |
| $\text{Pic}(E)$ | the Picard group (group of principal divisors on $E$), 339 |
| $\hat{\psi}$ | the dual isogeny of $\psi$, 340 |
| $\text{End}(E)$ | the endomorphism ring of $E$, 340 |
| $\text{Aut}(E)$ | the automorphism group of $E$, 340 |
| $\tilde{E}$ | the reduction of $E$ modulo $\mathfrak{p}$, 342 |
| $E[\psi]$ | kernel of the endomorphism $\psi$, 343 |
| $E[m]$ | kernel of multiplication-by-$m$ map, 343 |
| $E_{\text{tors}}$ | torsion subgroup of the elliptic curve $E$, 343 |
| $T_{\ell}(E)$ | the Tate module of $E$, 344 |
| $\rho_{E,\ell}$ | the $\ell$-adic representation on an elliptic curve, 344 |
| $\omega_E$ | the invariant differential on $E$, 344 |
| $\mathcal{I}_F$ | group of fractional ideals of $F$, 350 |
| $\mathcal{E}\ell\ell(R_F)$ | elliptic curves with CM by $R_F$, 350 |
| $\mathcal{C}_F$ | ideal class group of $F$, 350 |
| $h_F$ | class number of $F$, 350 |
| $\mathsf{CritPt}_{\phi}$ | set of critical points of $\phi$, 353 |
| $\mathsf{CritVal}_{\phi}$ | set of critical values of $\phi$, 353 |
| $\mathsf{PostCrit}_{\phi}$ | forward orbit of critical values of $\phi$, 353 |
| $E(\mathbb{Z})$ | the set of points in $E(\mathbb{Q})$ with integer coordinates, 372 |
| $\nu(E)$ | number of primes dividing the denominator of $j(E)$, 372 |
| $\bar{F}$ | homogenization of the polynomial $F$, 389 |
| $Z(\phi)$ | locus of indeterminacy of the rational map $\phi$, 389 |

| | |
|---|---|
| $\Phi$ | lift of the rational map $\bar{\phi}$ to $\mathbb{A}^{N+1}$, 389 |
| $\operatorname{Div}(V)$ | the divisor group of $V$, 403 |
| $\mathcal{O}_{V,W}$ | local ring of $V$ at $W$, 403 |
| $\operatorname{ord}_W(f)$ | order of vanishing of $f$ along $W$, 403 |
| $\operatorname{Pic}(V)$ | the Picard group of $V$, 403 |
| $D_1 \sim D_2$ | linear equivalence of divisors, 404 |
| $\phi^* D$ | pullback of the divisor $D$, 404 |
| $f_* D$ | pushforward of the divisor $D$, 404 |
| $\phi^*$ | pullback homomorphism on Picard groups, 405 |
| deg | degree map on $\operatorname{Div}(\mathbb{P}^N)$ and $\operatorname{Pic}(\mathbb{P}^N)$, 405 |
| $\operatorname{Base}(D)$ | base locus of the divisor $D$, 406 |
| $L(D)$ | vector space associated to the divisor $D$, 406 |
| $\ell(D)$ | dimension of the vector space $L(D)$, 406 |
| $\phi_D$ | rational map associated to the divisor $D$, 406 |
| $h_{V,D}$ | height on the variety $V$ associated to the divisor $D$, 407 |
| $S_{\mathbf{A},\mathbf{B}}$ | the K3 surface in $\mathbb{P}^2 \times \mathbb{P}^2$ determined by $\mathbf{A}$ and $\mathbf{B}$, 410 |
| $p_1, p_2$ | projections of a K3 surface to $\mathbb{P}^2$, 410 |
| $\iota_1, \iota_2$ | involutions on a K3 surface $S_{\mathbf{A},\mathbf{B}}$, 411 |
| $L_j^x, L_i^y$ | linear forms associated to a K3 surface, 412 |
| $Q_{k\ell}^x, Q_{ij}^y$ | quadratic forms associated to a K3 surface, 412 |
| $G_k^x, G_k^y, H_{ij}^x, H_{ij}^y$ | quartic forms associated to a K3 surface, 413 |
| $\mathcal{A}$ | subgroup of $\operatorname{Aut}(S_{\mathbf{A},\mathbf{B}})$ generated by $\iota_1$ and $\iota_2$, 418 |
| $\mathcal{A}(P)$ | the $\mathcal{A}$-orbit of $P$ on the K3 surface $S_{\mathbf{A},\mathbf{B}}$, 418 |
| $D_1, D_2$ | divisors on the K3 surface $S_{\mathbf{A},\mathbf{B}}$, 418 |
| $h^+, h^-$ | height functions on $S_{\mathbf{A},\mathbf{B}}$, 420 |
| $\hat{h}^+, \hat{h}^-$ | canonical heights on the K3 surface $S_{\mathbf{A},\mathbf{B}}$, 421 |
| $\hat{h}$ | the sum $\hat{h}^+ + \hat{h}^-$ on the K3 surface $S_{\mathbf{A},\mathbf{B}}$, 423 |
| $\mu(Q)$ | the order of the stabilizer of a point on a K3 surface, 425 |
| $C_p(\phi)$ | number of distinct $\mathbb{F}_p$-orbits of $\phi$, 430 |
| $\hat{h}^+, \hat{h}^-, \hat{h}$ | canonical heights on $\mathbb{A}^N$ for a regular affine automorphism, 431 |
| $E^+, E^-$ | eigendivisors in $\operatorname{Pic}(S_{\mathbf{A},\mathbf{B}}) \otimes \mathbb{R}$, 435 |
| $\mathcal{D}$ | subspace of $\operatorname{Pic}(S_{\mathbf{C}})$ generated by $D_1$, $D_2$, and $D_3$, 438 |

# References


[1] L. V. Ahlfors. *Complex Analysis*. McGraw-Hill Book Co., New York, 1978.

[2] A. V. Aho and N. J. A. Sloane. Some doubly exponential sequences. *Fibonacci Quart.*, 11(4):429–437, 1973.

[3] W. Aitken, F. Hajir, and C. Maire. Finitely ramified iterated extensions. *IMRN*, 14:855–880, 2005.

[4] S. Albeverio, M. Gundlach, A. Khrennikov, and K.-O. Lindahl. On the Markovian behavior of $p$-adic random dynamical systems. *Russ. J. Math. Phys.*, 8(2):135–152, 2001.

[5] S. Albeverio, B. Tirotstsi, A. Y. Khrennikov, and S. de Shmedt. $p$-adic dynamical systems. *Teoret. Mat. Fiz.*, 114(3):349–365, 1998.

[6] N. Ali. Stabilité des polynômes. *Acta Arith.*, 119(1):53–63, 2005.

[7] F. Amoroso and R. Dvornicich. A lower bound for the height in abelian extensions. *J. Number Theory*, 80(2):260–272, 2000.

[8] F. Amoroso and U. Zannier. A relative Dobrowolski lower bound over abelian extensions. *Ann. Scuola Norm. Sup. Pisa Cl. Sci. (4)*, 29(3):711–727, 2000.

[9] V. Anashin. Ergodic transformations in the space of $p$-adic integers. In *$p$-adic mathematical physics*, volume 826 of *AIP Conf. Proc.*, pages 3–24. Amer. Inst. Phys., Melville, NY, 2006.

[10] J.-C. Anglès d'Auriac, J.-M. Maillard, and C. M. Viallet. On the complexity of some birational transformations. *J. Phys. A*, 39(14):3641–3654, 2006.

[11] T. M. Apostol. *Introduction to Analytic Number Theory*. Springer-Verlag, New York, 1976. Undergraduate Texts in Mathematics.

[12] J. Arias de Reyna. Dynamical zeta functions and Kummer congruences. *Acta Arith.*, 119(1):39–52, 2005.

[13] D. K. Arrowsmith and F. Vivaldi. Some $p$-adic representations of the Smale horseshoe. *Phys. Lett. A*, 176(5):292–294, 1993.

[14] D. K. Arrowsmith and F. Vivaldi. Geometry of $p$-adic Siegel discs. *Phys. D*, 71(1-2):222–236, 1994.

[15] P. Autissier. Hauteur des correspondances de Hecke. *Bull. Soc. Math. France*, 131(3):421–433, 2003.

[16] P. Autissier. Dynamique des correspondances algébriques et hauteurs. *Int. Math. Res. Not.*, (69):3723–3739, 2004.

[17] M. Ayad and D. L. McQuillan. Irreducibility of the iterates of a quadratic polynomial over a field. *Acta Arith.*, 93(1):87–97, 2000.

[18] M. Ayad and D. L. McQuillan. Corrections to: "Irreducibility of the iterates of a quadratic polynomial over a field" [Acta Arith. **93** (2000), no. 1, 87–97]. *Acta Arith.*, 99(1):97, 2001.

[19] I. N. Baker. Fixpoints of polynomials and rational functions. *J. London Math. Soc.*, 39:615–622, 1964.

[20] M. Baker. A finiteness theorem for canonical heights attached to rational maps over function fields, 2005. `ArXiv:math.NT/0601046`.

[21] M. Baker. A lower bound for average values of dynamical Green's functions. *Math. Res. Lett.*, 13(2-3):245–257, 2006.

[22] M. Baker. Uniform structures and Berkovich spaces, 2006. `ArXiv:math.NT/0606252`.

[23] M. Baker and L.-C. Hsia. Canonical heights, transfinite diameters, and polynomial dynamics, 2005.

[24] M. Baker and S.-i. Ih. Equidistribution of small subvarieties of an abelian variety. *New York J. Math.*, 10:279–285 (electronic), 2004.

[25] M. Baker, S.-I. Ih, and R. Rumely. A finiteness property of torsion points, 2005. `ArXiv:math.NT/0509485`.

[26] M. Baker and R. Rumely. Analysis and dynamics on the Berkovich projective line, 2004. `ArXiv:math.NT/0407433`.

[27] M. Baker and R. Rumely. Montel's theorem for the Berkovich projective line and the Berkovich Julia set of a rational map, 2005. Preprint.

[28] M. Baker and R. Rumely. Equidistribution of small points, rational dynamics, and potential theory. *Ann. Inst. Fourier (Grenoble)*, 56(3):625–688, 2006.

[29] M. Baker and R. Rumely. Potential theory on the Berkovich projective line, 2006. `http://www.math.gatech.edu/~mbaker/pdf/BerkBook.pdf`, in preparation.

[30] M. Baker and R. Rumely. Harmonic analysis on metrized graphs. *Canadian J. Math.*, 2007. To appear.

[31] T. F. Banchoff and M. I. Rosen. Periodic points of Anosov diffeomorphisms. In *Global Analysis (Proc. Sympos. Pure Math., Vol. XIV, Berkeley, Calif., 1968)*, pages 17–21. Amer. Math. Soc., Providence, R.I., 1970.

[32] A. Baragar. Asymptotic growth of Markoff-Hurwitz numbers. *Compositio Math.*, 94(1):1–18, 1994.

[33] A. Baragar. Integral solutions of Markoff-Hurwitz equations. *J. Number Theory*, 49(1):27–44, 1994.

[34] A. Baragar. Rational points on $K3$ surfaces in $\mathbb{P}^1 \times \mathbb{P}^1 \times \mathbb{P}^1$. *Math. Ann.*, 305(3):541–558, 1996.

[35] A. Baragar. Rational curves on $K3$ surfaces in $\mathbb{P}^1 \times \mathbb{P}^1 \times \mathbb{P}^1$. *Proc. Amer. Math. Soc.*, 126(3):637–644, 1998.

[36] A. Baragar. Canonical vector heights on algebraic $K3$ surfaces with Picard number two. *Canad. Math. Bull.*, 46(4):495–508, 2003.

[37] A. Baragar. Orbits of curves on certain $K3$ surfaces. *Compositio Math.*, 137(2):115–134, 2003.

[38] A. Baragar. Canonical vector heights on $K3$ surfaces with Picard number three—an argument for nonexistence. *Math. Comp.*, 73(248):2019–2025 (electronic), 2004.

[39] A. Baragar and R. van Luijk. $K3$ surfaces with picard number three and canonical heights. *Math. Comp.* To appear.

[40] W. P. Barth, K. Hulek, C. A. M. Peters, and A. Van de Ven. *Compact complex surfaces*, volume 4 of *Ergebnisse der Mathematik und ihrer Grenzgebiete. 3. Folge. A Series of Modern Surveys in Mathematics*. Springer-Verlag, Berlin, second edition, 2004.

[41] A. Batra and P. Morton. Algebraic dynamics of polynomial maps on the algebraic closure of a finite field. I. *Rocky Mountain J. Math.*, 24(2):453–481, 1994.

[42] A. Batra and P. Morton. Algebraic dynamics of polynomial maps on the algebraic closure of a finite field. II. *Rocky Mountain J. Math.*, 24(3):905–932, 1994.

[43] A. F. Beardon. *Iteration of Rational Functions*, volume 132 of *Graduate Texts in Mathematics*. Springer-Verlag, New York, 1991. Complex analytic dynamical systems.

[44] A. Beauville. *Complex Algebraic Surfaces*, volume 34 of *London Mathematical Society Student Texts*. Cambridge University Press, Cambridge, second edition, 1996.

[45] E. Bedford and K. Kim. On the degree growth of birational mappings in higher dimension. *J. Geom. Anal.*, 14(4):567–596, 2004.

[46] E. Bedford and K. Kim. Degree growth of matrix inversion: birational maps of symmetric, cyclic matrices, 2005. `ArXiv:math.DS/0512507`.

[47] M. B. Bekka and M. Mayer. *Ergodic Theory and Topological Dynamics of Group Actions on Homogeneous Spaces*, volume 269 of *London Mathematical Society Lecture Note Series*. Cambridge University Press, Cambridge, 2000.

[48] E. Bekyel. The density of elliptic curves having a global minimal Weierstrass equation. *J. Number Theory*, 109(1):41–58, 2004.

[49] M. P. Bellon and C.-M. Viallet. Algebraic entropy. *Comm. Math. Phys.*, 204(2):425–437, 1999.

[50] S. Ben-Menahem. $p$-adic iterations. Preprint, TAUP 1627–88, Tel-Aviv University, 1988.

[51] R. L. Benedetto. *Dynamics in one p-adic variable*. In preparation, 2007.

[52] R. L. Benedetto. Preperiodic points of polynomials over global fields. *J. Reine Angew. Math.* To appear.

[53] R. L. Benedetto. *Fatou components in p-adic dynamics*. PhD thesis, Brown University, 1998.

[54] R. L. Benedetto. $p$-adic dynamics and Sullivan's no wandering domains theorem. *Compositio Math.*, 122(3):281–298, 2000.

[55] R. L. Benedetto. An elementary product identity in polynomial dynamics. *Amer. Math. Monthly*, 108(9):860–864, 2001.

[56] R. L. Benedetto. Hyperbolic maps in $p$-adic dynamics. *Ergodic Theory Dynam. Systems*, 21(1):1–11, 2001.

[57] R. L. Benedetto. Reduction, dynamics, and Julia sets of rational functions. *J. Number Theory*, 86(2):175–195, 2001.

[58] R. L. Benedetto. Components and periodic points in non-Archimedean dynamics. *Proc. London Math. Soc. (3)*, 84(1):231–256, 2002.

[59] R. L. Benedetto. Examples of wandering domains in $p$-adic polynomial dynamics. *C. R. Math. Acad. Sci. Paris*, 335(7):615–620, 2002.

[60] R. L. Benedetto. Non-Archimedean holomorphic maps and the Ahlfors Islands theorem. *Amer. J. Math.*, 125(3):581–622, 2003.

[61] R. L. Benedetto. Heights and preperiodic points of polynomials over function fields. *Int. Math. Res. Not.*, (62):3855–3866, 2005.

[62] R. L. Benedetto. Wandering domains and nontrivial reduction in non-Archimedean dynamics. *Illinois J. Math.*, 49(1):167–193 (electronic), 2005.

[63] R. L. Benedetto. Wandering domains in non-archimedean polynomial dynamics. *Bull. London Math. Soc.*, 38(6):937–950, 2006.

[64] V. G. Berkovich. *Spectral Theory and Analytic Geometry over Non-Archimedean Fields*, volume 33 of *Mathematical Surveys and Monographs*. American Mathematical Society, Providence, RI, 1990.

[65] V. G. Berkovich. Étale cohomology for non-Archimedean analytic spaces. *Inst. Hautes Études Sci. Publ. Math.*, (78):5–161 (1994), 1993.

[66] V. G. Berkovich. The automorphism group of the Drinfel'd half-plane. *C. R. Acad. Sci. Paris Sér. I Math.*, 321(9):1127–1132, 1995.

[67] V. G. Berkovich. $p$-adic analytic spaces. In *Proceedings of the International Congress of Mathematicians, Vol. II (Berlin, 1998)*, number Extra Vol. II, pages 141–151 (electronic), 1998.

[68] F. Berteloot and J.-J. Loeb. Une caractérisation géométrique des exemples de Lattès de $\mathbb{P}^k$. *Bull. Soc. Math. France*, 129(2):175–188, 2001.

[69] E. A. Bertram. Polynomials which commute with a Tchebycheff polynomial. *Amer. Math. Monthly*, 78:650–653, 1971.

[70] J.-P. Bézivin. Sur les ensembles de Julia et Fatou des fonctions entières ultramétriques. *Ann. Inst. Fourier (Grenoble)*, 51(6):1635–1661, 2001.

[71] J.-P. Bézivin. Sur les points périodiques des applications rationnelles en dynamique ultramétrique. *Acta Arith.*, 100(1):63–74, 2001.

[72] J.-P. Bézivin. Fractions rationnelles hyperboliques $p$-adiques. *Acta Arith.*, 112(2):151–175, 2004.

[73] J.-P. Bézivin. Sur la compacité des ensembles de Julia des polynômes $p$-adiques. *Math. Z.*, 246(1-2):273–289, 2004.

[74] P. Blanchard. Complex analytic dynamics on the Riemann sphere. *Bull. Amer. Math. Soc. (N.S.)*, 11(1):85–141, 1984.

[75] P. E. Blanksby and H. L. Montgomery. Algebraic integers near the unit circle. *Acta Arith.*, 18:355–369, 1971.

[76] E. Bombieri and W. Gubler. *Heights in Diophantine Geometry*. Number 4 in New Mathematical Monographs. Cambridge University Press, Cambridge, 2006.

[77] A. M. Bonifant and J. E. Fornæss. Growth of degree for iterates of rational maps in several variables. *Indiana Univ. Math. J.*, 49(2):751–778, 2000.

[78] A. I. Borevich and I. R. Shafarevich. *Number Theory*. Translated from the Russian by Newcomb Greenleaf. Pure and Applied Mathematics, Vol. 20. Academic Press, New York, 1966.

[79] G. Boros, M. Joyce, and V. Moll. A transformation of rational functions. *Elem. Math.*, 58(2):73–83, 2003.

[80] G. Boros, J. Little, V. Moll, E. Mosteig, and R. Stanley. A map on the space of rational functions. *Rocky Mountain J. Math.*, 35(6):1861–1880, 2005.

[81] S. Bosch, U. Güntzer, and R. Remmert. *Non-Archimedean analysis*, volume 261 of *Grundlehren der Mathematischen Wissenschaften*. Springer-Verlag, Berlin, 1984. A systematic approach to rigid analytic geometry.

[82] D. Bosio and F. Vivaldi. Round-off errors and $p$-adic numbers. *Nonlinearity*, 13(1):309–322, 2000.

[83] T. Bousch. *Sur quelques problèmes de dynamique holomorphe*. PhD thesis, Université de Paris-Sud, Centre d'Orsay, 1992.

[84] J. Bryk and C. E. Silva. Measurable dynamics of simple $p$-adic polynomials. *Amer. Math. Monthly*, 112(3):212–232, 2005.

[85] A. Buium. Complex dynamics and invariant forms mod $p$. *Int. Math. Res. Not.*, (31):1889–1899, 2005.

[86] A. Buium and K. Zimmerman. Differential orbit spaces of discrete dynamical systems. *J. Reine Angew. Math.*, 580:201–230, 2005.

[87] G. S. Call and S. W. Goldstine. Canonical heights on projective space. *J. Number Theory*, 63(2):211–243, 1997.

[88] G. S. Call and J. H. Silverman. Canonical heights on varieties with morphisms. *Compositio Math.*, 89(2):163–205, 1993.

[89] G. S. Call and J. H. Silverman. Computing the canonical height on $K3$ surfaces. *Math. Comp.*, 65(213):259–290, 1996.

[90] J. K. Canci. Cycles for rational maps with good reduction outside a prescribed set. *Monatsh. Math.*, 149(4):265–287, 2007.

[91] J. K. Canci. Finite rational orbits for rational functions. *Indag. Math. (N.S.)*, 2007. To appear.

[92] J. K. Canci. Rational periodic points for quadratic maps, 2007. Preprint, 2007.

[93] S. Cantat. Dynamique des automorphismes des surfaces $K3$. *Acta Math.*, 187(1):1–57, 2001.

[94] D. C. Cantor and E. G. Straus. On a conjecture of D. H. Lehmer. *Acta Arith.*, 42(1):97–100, 1982/83.

[95] L. Carleson and T. W. Gamelin. *Complex Dynamics*. Universitext: Tracts in Mathematics. Springer-Verlag, New York, 1993.

[96] J. W. S. Cassels. *Lectures on Elliptic Curves*, volume 24 of *London Mathematical Society Student Texts*. Cambridge University Press, Cambridge, 1991.

[97] J. W. S. Cassels and A. Fröhlich, editors. *Algebraic Number Theory*, London, 1986. Academic Press Inc. [Harcourt Brace Jovanovich Publishers]. Reprint of the 1967 original.

[98] A. Chambert-Loir. Mesures et équidistribution sur les espaces de Berkovich. *J. Reine Angew. Math.*, 595:215–235, 2006.

[99] A. Chambert-Loir and A. Thuillier. Formule de Mahler et équidistribution logarithmique, 2006. `ArXiv:math.NT/0612556`.

[100] G. Chassé. Combinatorial cycles of a polynomial map over a commutative field. *Discrete Math.*, 61(1):21–26, 1986.

[101] V. Chothi, G. Everest, and T. Ward. $S$-integer dynamical systems: periodic points. *J. Reine Angew. Math.*, 489:99–132, 1997.

[102] W.-S. Chou and I. E. Shparlinski. On the cycle structure of repeated exponentiation modulo a prime. *J. Number Theory*, 107(2):345–356, 2004.

[103] A. Chowla. Contributions to the analytic theory of numbers (II). *J. Indian Math. Soc.*, 20:120–128, 1933.

[104] Z. Coelho and W. Parry. Ergodicity of $p$-adic multiplications and the distribution of Fibonacci numbers. In *Topology, Ergodic Theory, Real Algebraic Geometry*, volume 202 of *Amer. Math. Soc. Transl. Ser. 2*, pages 51–70. Amer. Math. Soc., Providence, RI, 2001.

[105] H. Cohen. *A Course in Computational Algebraic Number Theory*, volume 138 of *Graduate Texts in Mathematics*. Springer-Verlag, Berlin, 1993.

[106] S. D. Cohen and D. Hachenberger. Actions of linearized polynomials on the algebraic closure of a finite field. In *Finite Fields: Theory, Applications, and Algorithms (Waterloo, ON, 1997)*, volume 225 of *Contemp. Math.*, pages 17–32. Amer. Math. Soc., Providence, RI, 1999.

[107] S. D. Cohen and D. Hachenberger. The dynamics of linearized polynomials. *Proc. Edinburgh Math. Soc. (2)*, 43(1):113–128, 2000.

[108] H. Cohn. Minimal geodesics on Fricke's torus-covering. In *Riemann surfaces and related topics: Proceedings of the 1978 Stony Brook Conference (State Univ. New York, Stony Brook, N.Y., 1978)*, volume 97 of *Ann. of Math. Stud.*, pages 73–85, Princeton, N.J., 1981. Princeton Univ. Press.

[109] O. Colón-Reyes, A. Jarrah, R. Laubenbacher, and B. Sturmfels. Monomial dynamical systems over finite fields. *Complex Systems*, 16(4):333–342, 2006.

[110] C. Consani and M. Marcolli. Noncommutative geometry, dynamics, and $\infty$-adic Arakelov geometry. *Selecta Math. (N.S.)*, 10(2):167–251, 2004.

[111] J.-M. Couveignes. Calcul et rationalité de fonctions de Belyĭ en genre 0. *Ann. Inst. Fourier (Grenoble)*, 44(1):1–38, 1994.

[112] D. Cox, J. Little, and D. O'Shea. *Ideals, Varieties, and Algorithms*. Undergraduate Texts in Mathematics. Springer-Verlag, New York, 1997.

[113] J. E. Cremona. On the Galois groups of the iterates of $x^2+1$. *Mathematika*, 36(2):259–261 (1990), 1989.

[114] P. D'Ambros, G. Everest, R. Miles, and T. Ward. Dynamical systems arising from elliptic curves. *Colloq. Math.*, 84/85(, part 1):95–107, 2000. Dedicated to the memory of Anzelm Iwanik.

[115] L. Danielson and B. Fein. On the irreducibility of the iterates of $x^n - b$. *Proc. Amer. Math. Soc.*, 130(6):1589–1596 (electronic), 2002.

[116] S. De Smedt and A. Khrennikov. A $p$-adic behaviour of dynamical systems. *Rev. Mat. Complut.*, 12(2):301–323, 1999.

[117] P. Dèbes and J.-C. Douai. Algebraic covers: field of moduli versus field of definition. *Ann. Sci. École Norm. Sup. (4)*, 30(3):303–338, 1997.

[118] P. Dèbes and J.-C. Douai. Local-global principles for algebraic covers. *Israel J. Math.*, 103:237–257, 1998.

[119] P. Dèbes and J.-C. Douai. Gerbes and covers. *Comm. Algebra*, 27(2):577–594, 1999.

[120] P. Dèbes and D. Harbater. Fields of definition of $p$-adic covers. *J. Reine Angew. Math.*, 498:223–236, 1998.

[121] L. DeMarco. Iteration at the boundary of the space of rational maps, 2005.

[122] L. DeMarco. The moduli space of quadratic rational maps, 2007.

[123] L. DeMarco and R. Rumely. Transfinite diameter and the resultant, 2007. To appear.

[124] C. Deninger. Some analogies between number theory and dynamical systems on foliated spaces. In *Proceedings of the International Congress of Mathematicians, Vol. I (Berlin, 1998)*, number Extra Vol. I, pages 163–186 (electronic), 1998.

[125] C. Deninger. On dynamical systems and their possible significance for arithmetic geometry. In *Regulators in Analysis, Geometry and Number Theory*, volume 171 of *Progr. Math.*, pages 29–87. Birkhäuser Boston, Boston, MA, 2000.

[126] C. Deninger. Number theory and dynamical systems on foliated spaces. *Jahresber. Deutsch. Math.-Verein.*, 103(3):79–100, 2001.

[127] C. Deninger. A note on arithmetic topology and dynamical systems. In *Algebraic Number Theory and Algebraic Geometry*, volume 300 of *Contemp. Math.*, pages 99–114. Amer. Math. Soc., Providence, RI, 2002.

[128] C. Deninger. Arithmetic geometry and analysis on foliated spaces, 2005. unpublished, `ArXiv:math.NT/0505354`.

[129] C. Deninger. A dynamical systems analogue of Lichtenbaum's conjectures on special values of Hasse-Weil zeta functions, 2006. `ArXiv:math.NT/0605724`.

[130] C. Deninger. $p$-adic entropy and a $p$-adic Fuglede-Kadison determinant, 2006. `ArXiv:math.DS/0608539`.

[131] L. Denis. Points périodiques des automorphismes affines. *J. Reine Angew. Math.*, 467:157–167, 1995.

[132] R. Devaney. *An Introduction to Chaotic Dynamical Systems*. Addison-Wesley, Redwood City, CA, 2nd edition, 1989.

[133] J. Diller and C. Favre. Dynamics of bimeromorphic maps of surfaces. *Amer. J. Math.*, 123(6):1135–1169, 2001.

[134] T.-C. Dinh. Sur les applications de Lattès de $\mathbb{P}^k$. *J. Math. Pures Appl. (9)*, 80(6):577–592, 2001.

[135] T.-C. Dinh and N. Sibony. Sur les endomorphismes holomorphes permutables de $\mathbb{P}^k$. *Math. Ann.*, 324(1):33–70, 2002.

[136] T.-C. Dinh and N. Sibony. Dynamique des applications polynomiales semi-régulières. *Ark. Mat.*, 42(1):61–85, 2004.

[137] Z. Divišová. On cycles of polynomials with integral rational coefficients. *Math. Slovaca*, 52(5):537–540, 2002.

[138] E. Dobrowolski. On a question of Lehmer and the number of irreducible factors of a polynomial. *Acta Arith.*, 34:391–401, 1979.

[139] M. M. Dodson and J. A. G. Vickers, editors. *Number Theory and Dynamical Systems*, Cambridge, 1989. Cambridge University Press. Papers from the meeting held at the University of York, York, March 30–April 15, 1987.

[140] V. Dolotin and A. Morozov. Algebraic geometry of discrete dynamics. The case of one variable. ITEP-TH-02/05.

[141] A. Douady and J. H. Hubbard. Itération des polynômes quadratiques complexes. *C. R. Acad. Sci. Paris Sér. I Math.*, 294(3):123–126, 1982.

[142] A. Douady and J. H. Hubbard. *Étude dynamique des polynômes complexes. Partie I*, volume 84 of *Publications Mathématiques d'Orsay [Mathematical Publications of Orsay]*. Université de Paris-Sud, Département de Mathématiques, Orsay, 1984.

[143] A. Douady and J. H. Hubbard. *Étude dynamique des polynômes complexes. Partie II*, volume 85 of *Publications Mathématiques d'Orsay [Mathematical Publications of Orsay]*. Université de Paris-Sud, Département de Mathématiques, Orsay, 1985. With the collaboration of P. Lavaurs, Tan Lei and P. Sentenac.

[144] B.-S. Du, S.-S. Huang, and M.-C. Li. Newton, Fermat, and exactly realizable sequences. *J. Integer Seq.*, 8(1):Article 05.1.2, 8 pp. (electronic), 2005.

[145] C. Dupont. Exemples de Lattès et domaines faiblement sphériques de $\mathbb{C}^n$. *Manuscripta Math.*, 111(3):357–378, 2003.

[146] R. Dvornicich and U. Zannier. Cyclotomic Diophantine problems (Hilbert irreducibility and invariant sets for polynomial maps), 2006. Preprint, June 2006.

[147] M. Einsiedler, G. Everest, and T. Ward. Entropy and the canonical height. *J. Number Theory*, 91(2):256–273, 2001.

[148] M. Einsiedler, G. Everest, and T. Ward. Morphic heights and periodic points. In *Number Theory (New York, 2003)*, pages 167–177. Springer, New York, 2004.

[149] M. Einsiedler and T. Ward. Fitting ideals for finitely presented algebraic dynamical systems. *Aequationes Math.*, 60(1-2):57–71, 2000.

[150] N. Elkies. Nontorsion points of low height on elliptic curves over $\mathbb{Q}$, 2002. www.math.harvard.edu/~elkies/low_height.html.

[151] P. Erdős, A. Granville, C. Pomerance, and C. Spiro. On the normal behavior of the iterates of some arithmetic functions. In *Analytic number theory (Allerton Park, IL, 1989)*, volume 85 of *Progr. Math.*, pages 165–204. Birkhäuser Boston, Boston, MA, 1990.

[152] A. È. Erëmenko. Some functional equations connected with the iteration of rational functions. *Algebra i Analiz*, 1(4):102–116, 1989.

[153] G. Everest. On the elliptic analogue of Jensen's formula. *J. London Math. Soc. (2)*, 59(1):21–36, 1999.

[154] G. Everest and B. N. Fhlathúin. The elliptic Mahler measure. *Math. Proc. Cambridge Philos. Soc.*, 120(1):13–25, 1996.

[155] G. Everest and C. Pinner. Bounding the elliptic Mahler measure. II. *J. London Math. Soc. (2)*, 58(1):1–8, 1998.

[156] G. Everest and C. Pinner. Corrigendum: "Bounding the elliptic Mahler measure. II" [J. London Math. Soc. (2) **58** (1998), no. 1, 1–8.]. *J. London Math. Soc. (2)*, 62(2):640, 2000.

[157] G. Everest, A. van der Poorten, Y. Puri, and T. Ward. Integer sequences and periodic points. *J. Integer Seq.*, 5(2):Article 02.2.3, 10 pp. (electronic), 2002.

[158] G. Everest, A. van der Poorten, I. Shparlinski, and T. Ward. *Recurrence Sequences*, volume 104 of *Mathematical Surveys and Monographs*. American Mathematical Society, Providence, RI, 2003.

[159] G. Everest and T. Ward. A dynamical interpretation of the global canonical height on an elliptic curve. *Experiment. Math.*, 7(4):305–316, 1998.

[160] G. Everest and T. Ward. *Heights of polynomials and entropy in algebraic dynamics*. Springer-Verlag London Ltd., London, 1999.

[161] N. Fagella and J. Llibre. Periodic points of holomorphic maps via Lefschetz numbers. *Trans. Amer. Math. Soc.*, 352(10):4711–4730, 2000.

[162] N. Fakhruddin. Boundedness results for periodic points on algebraic varieties. *Proc. Indian Acad. Sci. Math. Sci.*, 111(2):173–178, 2001.

[163] N. Fakhruddin. Questions on self maps of algebraic varieties. *J. Ramanujan Math. Soc.*, 18(2):109–122, 2003.

[164] G. Faltings. Endlichkeitssätze für abelsche Varietäten über Zahlkörpern. *Invent. Math.*, 73(3):349–366, 1983.

[165] G. Faltings. Finiteness theorems for abelian varieties over number fields. In *Arithmetic geometry (Storrs, Conn., 1984)*, pages 9–27. Springer, New York, 1986. Translated from the German original [Invent. Math. **73** (1983), no. 3, 349–366; ibid. **75** (1984), no. 2, 381] by Edward Shipz.

[166] P. Fatou. Sur les équations fonctionnelles. *Bull. Soc. Math. France*, 47:161–271, 1919.

[167] P. Fatou. Sur les équations fonctionnelles. *Bull. Soc. Math. France*, 48:33–94 and 208–314, 1920.

[168] C. Favre and J. Rivera-Letelier. Théorème d'équidistribution de Brolin en dynamique $p$-adique. *C. R. Math. Acad. Sci. Paris*, 339(4):271–276, 2004.

[169] C. Favre and J. Rivera-Letelier. Équidistribution quantitative des points de petite hauteur sur la droite projective. *Math. Ann.*, 335(2):311–361, 2006.

[170] G. Fernandez. Wandering Fatou components on $p$-adic polynomial dynamics, 2004. `ArXiv:math.DS/0503720`.

[171] E. V. Flynn, B. Poonen, and E. F. Schaefer. Cycles of quadratic polynomials and rational points on a genus-2 curve. *Duke Math. J.*, 90(3):435–463, 1997.

[172] J. E. Fornæss and N. Sibony. Complex dynamics in higher dimension. I. *Astérisque*, (222):5, 201–231, 1994.

[173] J. E. Fornæss and N. Sibony. Complex dynamics in higher dimensions. In *Complex Potential Theory (Montreal, PQ, 1993)*, volume 439 of *NATO Adv. Sci. Inst. Ser. C Math. Phys. Sci.*, pages 131–186. Kluwer Acad. Publ., Dordrecht, 1994.

[174] J. E. Fornaess and N. Sibony. Complex dynamics in higher dimension. II. In *Modern Methods in Complex Analysis (Princeton, NJ, 1992)*, volume 137 of *Ann. of Math. Stud.*, pages 135–182. Princeton Univ. Press, Princeton, NJ, 1995.

[175] J. Fresnel and M. van der Put. *Rigid analytic geometry and its applications*, volume 218 of *Progress in Mathematics*. Birkhäuser Boston Inc., Boston, MA, 2004.

[176] S. Friedland and J. Milnor. Dynamical properties of plane polynomial automorphisms. *Ergodic Theory Dynam. Systems*, 9(1):67–99, 1989.

[177] J. A. C. Gallas. Units: remarkable points in dynamical systems. *Phys. A*, 222(1-4):125–151, 1995.

[178] *Géométrie des surfaces K3: modules et périodes*. Société Mathématique de France, Paris, 1985. Papers from the seminar held in Palaiseau, October 1981–January 1982, Astérisque No. 126 (1985).

[179] S. Getachew. *Galois Theory of Polynomial Iterates*. PhD thesis, Brown University, 2000.

[180] D. Ghioca. Equidistribution for torsion points of a Drinfeld module. *Math. Ann.*, 336(4):841–865, 2006.

[181] D. Ghioca and T. Tucker. A dynamical version of the Mordell-Lang conjecture for the additive group, 2006. `arXiv:0704.1333`.

[182] H. Glockner. Equidistribution and integral points for Drinfeld modules, 2006. `ArXiv: math.NT.0609120`.

[183] F. Q. Gouvêa. *p-adic numbers*. Universitext. Springer-Verlag, Berlin, second edition, 1997. An introduction.

[184] R. L. Graham, D. E. Knuth, and O. Patashnik. *Concrete Mathematics*. Addison-Wesley Publishing Company, Reading, MA, second edition, 1994. A foundation for computer science.

[185] B. Green and M. Matignon. Order $p$ automorphisms of the open disc of a $p$-adic field. *J. Amer. Math. Soc.*, 12(1):269–303, 1999.

[186] P. Griffiths and J. Harris. *Principles of Algebraic Geometry*. Wiley Classics Library. John Wiley & Sons Inc., New York, 1994. Reprint of the 1978 original.

[187] V. Guedj and N. Sibony. Dynamics of polynomial automorphisms of $\mathbb{C}^k$. *Ark. Mat.*, 40(2):207–243, 2002.

[188] M. Gundlach, A. Khrennikov, and K.-O. Lindahl. On ergodic behavior of $p$-adic dynamical systems. *Infin. Dimens. Anal. Quantum Probab. Relat. Top.*, 4(4):569–577, 2001.

[189] M. Gundlach, A. Khrennikov, and K.-O. Lindahl. Topological transitivity for $p$-adic dynamical systems. In *p-adic functional analysis (Ioannina, 2000)*, volume 222 of *Lecture Notes in Pure and Appl. Math.*, pages 127–132. Dekker, New York, 2001.

[190] N. B. Haaser and J. A. Sullivan. *Real Analysis*. Dover Publications Inc., New York, 1991. Revised reprint of the 1971 original.

[191] F. Halter-Koch and P. Konečná. Polynomial cycles in finite extension fields. *Math. Slovaca*, 52(5):531–535, 2002.

[192] F. Halter-Koch and W. Narkiewicz. Finiteness properties of polynomial mappings. *Math. Nachr.*, 159:7–18, 1992.

[193] F. Halter-Koch and W. Narkiewicz. Polynomial cycles in finitely generated domains. *Monatsh. Math.*, 119(4):275–279, 1995.

[194] F. Halter-Koch and W. Narkiewicz. Polynomial cycles and dynamical units. In *Proceedings of a Conference on Analytic and Elementary Number Theory (Wien 1996)*, pages 70–80. 1997. `www.boku.ac.at/math/proc.html`.

[195] F. Halter-Koch and W. Narkiewicz. Scarcity of finite polynomial orbits. *Publ. Math. Debrecen*, 56(3-4):405–414, 2000. Dedicated to Professor Kálmán Győry on the occasion of his 60th birthday.

[196] B. Harris. Probability distributions related to random mappings. *Ann. Math. Statist.*, 31:1045–1062, 1960.

[197] J. Harris. *Algebraic Geometry*, volume 133 of *Graduate Texts in Mathematics*. Springer-Verlag, New York, 1995. A first course, Corrected reprint of the 1992 original.

[198] R. Hartshorne. *Algebraic Geometry*. Springer-Verlag, New York, 1977. Graduate Texts in Mathematics, No. 52.

[199] B. Hasselblatt and J. Propp. Monomial maps and algebraic entropy, 2006. `ArXiv: math.DS/0604521`.

[200] M. Hénon. A two-dimensional mapping with a strange attractor. *Comm. Math. Phys.*, 50(1):69–77, 1976.

[201] M. Herman and J.-C. Yoccoz. Generalizations of some theorems of small divisors to non-archimedean fields. In *Geometric Dynamics*, volume 1007 of *Lecture Notes in Mathematics*, pages 408–447. Springer-Verlag, 1983. Rio de Janairo (1981).

[202] M. Hindry and J. H. Silverman. The canonical height and integral points on elliptic curves. *Invent. Math.*, 93(2):419–450, 1988.

[203] M. Hindry and J. H. Silverman. On Lehmer's conjecture for elliptic curves. In *Séminaire de Théorie des Nombres, Paris 1988–1989*, volume 91 of *Progr. Math.*, pages 103–116. Birkhäuser Boston, Boston, MA, 1990.

[204] M. Hindry and J. H. Silverman. Sur le nombre de points de torsion rationnels sur une courbe elliptique. *C. R. Acad. Sci. Paris Sér. I Math.*, 329(2):97–100, 1999.

[205] M. Hindry and J. H. Silverman. *Diophantine Geometry*, volume 201 of *Graduate Texts in Mathematics*. Springer-Verlag, New York, 2000. An introduction.

[206] L.-C. Hsia. A weak Néron model with applications to $p$-adic dynamical systems. *Compositio Math.*, 100(3):277–304, 1996.

[207] L.-C. Hsia. On the dynamical height zeta functions. *J. Number Theory*, 63(1):146–169, 1997.

[208] L.-C. Hsia. Closure of periodic points over a non-Archimedean field. *J. London Math. Soc. (2)*, 62(3):685–700, 2000.

[209] L.-C. Hsia. $p$-adic equidistribution theorems. manuscript, 2003.

[210] L. K. Hua. *Introduction to Number Theory*. Springer-Verlag, Berlin, 1982. Translated from the Chinese by Peter Shiu.

[211] J. Hubbard. The Hénon mapping in the complex domain. In *Chaotic Dynamics and Fractals (Atlanta, Ga., 1985)*, volume 2 of *Notes Rep. Math. Sci. Engrg.*, pages 101–111. Academic Press, Orlando, FL, 1986.

[212] J. Hubbard, P. Papadopol, and V. Veselov. A compactification of Hénon mappings in $\mathbb{C}^2$ as dynamical systems. *Acta Math.*, 184(2):203–270, 2000.

[213] J. Hubbard and D. Schleicher. The spider algorithm. In *Complex Dynamical Systems (Cincinnati, OH, 1994)*, volume 49 of *Proc. Sympos. Appl. Math.*, pages 155–180. Amer. Math. Soc., Providence, RI, 1994.

[214] B. Hutz. *Arithmetic Dynamics on Varieties in Dimension Greater Than One*. PhD thesis, Brown University, 2007.

[215] E. Ionascu and P. Stanica. Effective asymptotics for some nonlinear recurrences and almost doubly-exponential sequences. *Acta Math. Univ. Comenian. (N.S.)*, 73(1):75–87, 2004.

[216] K. Ireland and M. Rosen. *A Classical Introduction to Modern Number Theory*, volume 84 of *Graduate Texts in Mathematics*. Springer-Verlag, New York, second edition, 1990.

[217] K. Jänich. *Topology*. Undergraduate Texts in Mathematics. Springer-Verlag, New York, 1984. With a chapter by Theodor Bröcker, Translated from the German by Silvio Levy.

[218] D. Jogia, J. A. G. Roberts, and F. Vivaldi. The Hasse-Weil bound and integrability detection in rational maps. *J. Nonlinear Math. Phys.*, 10(suppl. 2):166–180, 2003.

[219] D. Jogia, J. A. G. Roberts, and F. Vivaldi. An algebraic geometric approach to integrable maps of the plane. *J. Phys. A*, 39(5):1133–1149, 2006.

[220] R. Jones. *Galois Martingales and the p-adic Hyperbolic Mandelbrot Set*. PhD thesis, Brown University, 2005.

[221] R. Jones. The density of prime divisors in the arithmetic dynamics of quadratic polynomials, 2006. `ArXiv:math.NT/0612415`.

[222] R. Jones. Iterated Galois towers, their associated martingales, and the $p$-adic Mandelbrot set. *Compositio Math.*, 2007. To appear.

[223] G. Julia. Mémoire sur l'itération des fonctions rationelles. *Journal de Math. Pures et Appl.*, 8:47–245, 1918.

[224] G. Julia. Mémoire sur la permutabilité des fractions rationnelles. *Ann. Sci. École Norm. Sup. (3)*, 39:131–215, 1922.

[225] S. Kamienny. Torsion points on elliptic curves and $q$-coefficients of modular forms. *Invent. Math.*, 109(2):221–229, 1992.

[226] A. Katok and B. Hasselblatt. *Introduction to the Modern Theory of Dynamical Systems*, volume 54 of *Encyclopedia of Mathematics and its Applications*. Cambridge University Press, Cambridge, 1995. With a supplementary chapter by Katok and Leonardo Mendoza.

[227] S. Kawaguchi. Some remarks on rational periodic points. *Math. Res. Lett.*, 6(5-6):495–509, 1999.

[228] S. Kawaguchi. Canonical heights for random iterations in certain varieties, 2005. Preprint.

[229] S. Kawaguchi. Projective surface automorphisms of positive topological entropy from an arithmetic viewpoint, 2005. Preprint.

[230] S. Kawaguchi. Canonical height functions for affine plane automorphisms. *Math. Ann.*, 335(2):285–310, 2006.

[231] S. Kawaguchi. Canonical heights, invariant currents, and dynamical eigensystems of morphisms for line bundles. *J. Reine Angew. Math.*, 597:135–173, 2006.

[232] S. Kawaguchi and J. H. Silverman. Arithmetic complexity of morphisms, 2006. Preprint.

[233] S. Kawaguchi and J. H. Silverman. Nonarchimedean Green functions and dynamics on projective space, 2006. Preprint.

[234] S. Kawaguchi and J. H. Silverman. Dynamics of projective morphisms having identical canonical heights. *Proc. Lond. Math. Soc., II. Ser.*, 2007. To appear.

[235] L. Keen. Julia sets of rational maps. In *Complex Dynamical Systems*, volume 49 of *Proceedings of Symposia in Applied Mathematics*, pages 71–90. American Mathematical Society, 1994. Cincinnati (1994).

[236] M. Khamraev and F. Mukhamedov. On a class of rational $p$-adic dynamical systems. *J. Math. Anal. Appl.*, 315(1):76–89, 2006.

[237] A. Khrennikov. $p$-adic dynamical systems: description of concurrent struggle in a biological population with limited growth. *Dokl. Akad. Nauk*, 361(6):752–754, 1998.

[238] A. Khrennikov. $p$-adic discrete dynamical systems and collective behaviour of information states in cognitive models. *Discrete Dynamics in Nature and Society*, 5(1):59–69, 2000.

[239] A. Khrennikov. Ergodic and non-ergodic behaviour for dynamical systems in rings of $p$-adic integers. In *Fourth Italian-Latin American Conference on Applied and Industrial Mathematics (Havana, 2001)*, pages 404–409. Inst. Cybern. Math. Phys., Havana, 2001.

[240] A. Khrennikov. Small denominators in complex $p$-adic dynamics. *Indag. Math. (N.S.)*, 12(2):177–189, 2001.

[241] A. Khrennikov, K.-O. Lindahl, and M. Gundlach. Ergodicity in the $p$-adic framework. In *Operator Methods in Ordinary and Partial Differential Equations (Stockholm, 2000)*, volume 132 of *Oper. Theory Adv. Appl.*, pages 245–251. Birkhäuser, Basel, 2002.

[242] A. Khrennikov and M. Nilson. *p-adic Deterministic and Random Dynamics*, volume 574 of *Mathematics and Its Applications*. Kluwer Academic Publishers, Dordrecht, 2004.

[243] A. Khrennikov and M. Nilsson. On the number of cycles of $p$-adic dynamical systems. *J. Number Theory*, 90(2):255–264, 2001.

[244] A. Khrennikov and M. Nilsson. Behaviour of Hensel perturbations of $p$-adic monomial dynamical systems. *Anal. Math.*, 29(2):107–133, 2003.

[245] A. Khrennikov, M. Nilsson, and R. Nyqvist. The asymptotic number of periodic points of discrete polynomial $p$-adic dynamical systems. In *Ultrametric functional analysis (Nijmegen, 2002)*, volume 319 of *Contemp. Math.*, pages 159–166. Amer. Math. Soc., Providence, RI, 2003.

[246] J. Kiwi. Puiseux series polynomial dynamics and iteration of complex cubic polynomials. *Ann. Inst. Fourier (Grenoble)*, 56(5):1337–1404, 2006.

[247] D. Kleinbock, N. Shah, and A. Starkov. Dynamics of subgroup actions on homogeneous spaces of Lie groups and applications to number theory. In *Handbook of Dynamical Systems, Vol. 1A*, pages 813–930. North-Holland, Amsterdam, 2002.

[248] A. W. Knapp. *Elliptic Curves*, volume 40 of *Mathematical Notes*. Princeton University Press, Princeton, NJ, 1992.

[249] N. Koblitz. *p-adic Numbers, p-adic Analysis, and Zeta-Functions*, volume 58 of *Graduate Texts in Mathematics*. Springer-Verlag, New York, second edition, 1984.

[250] N. Koblitz. *Introduction to Elliptic Curves and Modular Forms*, volume 97 of *Graduate Texts in Mathematics*. Springer-Verlag, New York, second edition, 1993.

[251] B. Kra. Ergodic methods in additive combinatorics, 2006. `ArXiv:math.DS/0608105`.

[252] J. C. Lagarias. Number theory and dynamical systems. In *The Unreasonable Effectiveness of Number Theory (Orono, ME, 1991)*, volume 46 of *Proc. Sympos. Appl. Math.*, pages 35–72. Amer. Math. Soc., Providence, RI, 1992.

[253] J. C. Lagarias. The $3x+1$ problem: An annotated bibliography, I & II, 2006. `ArXiv:math.NT/0309224`, `ArXiv:math.NT/0608208`.

[254] S. Lang. *Elliptic Curves: Diophantine Analysis*, volume 231 of *Grundlehren der Mathematischen Wissenschaften*. Springer-Verlag, Berlin, 1978.

[255] S. Lang. *Introduction to Algebraic and Abelian Functions*. Springer-Verlag, Berlin, 2 edition, 1982.

[256] S. Lang. *Fundamentals of Diophantine Geometry*. Springer-Verlag, New York, 1983.

[257] S. Lang. *Elliptic Functions*, volume 112 of *Graduate Texts in Mathematics*. Springer-Verlag, New York, second edition, 1987. With an appendix by J. Tate.

[258] S. Lang. *Algebraic Number Theory*, volume 110 of *Graduate Texts in Mathematics*. Springer-Verlag, New York, second edition, 1994.

[259] S. Lang. *Algebra*, volume 211 of *Graduate Texts in Mathematics*. Springer-Verlag, New York, third edition, 2002.

[260] S. Lattès. Sur l'iteration des substitutions rationelles et les fonctions de Poincaré. *Comptes Rendus Acad. Sci. Paris*, 166:26–28, 1918.

[261] E. Lau and D. Schleicher. Internal addresses in the Mandelbrot set and irreducibility of polynomials. Technical Report 1994/19, December 1994.

[262] F. Laubie, A. Movahhedi, and A. Salinier. Systèmes dynamiques non archimédiens et corps des normes. *Compositio Math.*, 132(1):57–98, 2002.

[263] M. Laurent. Minoration de la hauteur de Néron-Tate. In *Séminaire de Théorie des Nombres*, Progress in Mathematics, pages 137–151. Birkhäuser, 1983. Paris 1981–1982.

[264] D. H. Lehmer. Factorization of certain cyclotomic functions. *Ann. of Math. (2)*, 34(3):461–479, 1933.

[265] D. Lewis. Invariant set of morphisms on projective and affine number spaces. *Journal of Algebra*, 20:419–434, 1972.

[266] H.-C. Li. Counting periodic points of $p$-adic power series. *Compositio Math.*, 100(3):351–364, 1996.

[267] H.-C. Li. $p$-adic dynamical systems and formal groups. *Compositio Math.*, 104(1):41–54, 1996.

[268] H.-C. Li. $p$-adic periodic points and Sen's theorem. *J. Number Theory*, 56(2):309–318, 1996.

[269] H.-C. Li. When is a $p$-adic power series an endomorphism of a formal group? *Proc. Amer. Math. Soc.*, 124(8):2325–2329, 1996.

[270] H.-C. Li. Isogenies between dynamics of formal groups. *J. Number Theory*, 62(2):284–297, 1997.

[271] H.-C. Li. $p$-adic power series which commute under composition. *Trans. Amer. Math. Soc.*, 349(4):1437–1446, 1997.

[272] H.-C. Li. On dynamics of power series over unramified extensions of $\mathbb{Q}_p$. *J. Reine Angew. Math.*, 545:183–200, 2002.

[273] H.-C. Li. On heights of $p$-adic dynamical systems. *Proc. Amer. Math. Soc.*, 130(2):379–386 (electronic), 2002.

[274] H.-C. Li. $p$-typical dynamical systems and formal groups. *Compositio Math.*, 130(1):75–88, 2002.

[275] R. Lidl and H. Niederreiter. *Finite Fields*, volume 20 of *Encyclopedia of Mathematics and Its Applications*. Cambridge University Press, Cambridge, second edition, 1997. With a foreword by P. M. Cohn.

[276] D. Lind and K. Schmidt. Symbolic and algebraic dynamical systems. In *Handbook of Dynamical Systems, Vol. 1A*, pages 765–812. North-Holland, Amsterdam, 2002.

[277] D. Lind, K. Schmidt, and T. Ward. Mahler measure and entropy for commuting automorphisms of compact groups. *Invent. Math.*, 101(3):593–629, 1990.

[278] D. Lind and T. Ward. Automorphisms of solenoids and $p$-adic entropy. *Ergodic Theory Dynam. Systems*, 8(3):411–419, 1988.

[279] K.-O. Lindahl. On Siegel's linearization theorem for fields of prime characteristic. *Nonlinearity*, 17(3):745–763, 2004.

[280] J. Lubin. Non-Archimedean dynamical systems. *Compositio Math.*, 94(3):321–346, 1994.

[281] J. Lubin. Sen's theorem on iteration of power series. *Proc. Amer. Math. Soc.*, 123(1):63–66, 1995.

[282] J. Lubin. Formal flows on the non-Archimedean open unit disk. *Compositio Math.*, 124(2):123–136, 2000.

[283] J. Lubin. Seminar on $p$-adic time in nonarchimedean dynamical systems. Seminar at Brown University, prepared April 25, 1996.

[284] J. Lubin and G. Sarkis. Extrinsic properties of automorphism groups of formal groups, 2007. To appear.

[285] K. Mahler. On the lattice points on curves of genus 1. *Proc. Lond. Math. Soc., II. Ser.*, 39:431–466, 1935.

[286] M. Manes. *Arithmetic Dynamics and Moduli Spaces of Rational Maps*. PhD thesis, Brown University, 2007.

[287] S. Marcello. *Sur la dynamique arithmétique des automorphismes affines*. PhD thesis, Université Paris 7, 2000.

[288] S. Marcello. Sur les propriétés arithmétiques des itérés d'automorphismes réguliers. *C. R. Acad. Sci. Paris Sér. I Math.*, 331(1):11–16, 2000.

[289] S. Marcello. Géométrie, points rationnels et itérés des automorphismes de l'espace affine, 2003. ArXiv:math.NT/0310434.

[290] S. Marcello. Sur la dynamique arithmétique des automorphismes de l'espace affine. *Bull. Soc. Math. France*, 131(2):229–257, 2003.

[291] D. W. Masser. Counting points of small height on elliptic curves. *Bull. Soc. Math. France*, 117(2):247–265, 1989.

[292] B. Mazur. Modular curves and the Eisenstein ideal. *Inst. Hautes Études Sci. Publ. Math.*, (47):33–186 (1978), 1977.

[293] E. M. McMillan. A problem in the stability of periodic systems. In *Topics in Modern Physics: A Tribute to E. U. Condon*, pages 219–244. Colorado Assoc. Univ. Press, Boulder, CO, 1971.

[294] C. T. McMullen. Families of rational maps and iterative root-finding algorithms. *Ann. of Math. (2)*, 125(3):467–493, 1987.

[295] C. T. McMullen. From dynamics on surfaces to rational points on curves. *Bull. Amer. Math. Soc. (N.S.)*, 37(2):119–140, 2000.

[296] C. T. McMullen. Dynamics on $K3$ surfaces: Salem numbers and Siegel disks. *J. Reine Angew. Math.*, 545:201–233, 2002.

[297] L. Merel. Bornes pour la torsion des courbes elliptiques sur les corps de nombres. *Invent. Math.*, 124(1-3):437–449, 1996.

[298] J.-Y. Mérindol. Propriétés élémentaires des surfaces $K3$. *Astérisque*, (126):45–57, 1985. Geometry of $K3$ surfaces: moduli and periods (Palaiseau, 1981/1982).

[299] J. S. Milne. *Étale Cohomology*, volume 33 of *Princeton Mathematical Series*. Princeton University Press, Princeton, N.J., 1980.

[300] J. Milnor. On Lattès maps. ArXiv:math.DS/0402147, Stony Brook IMS Preprint #2004/01.

[301] J. Milnor. Geometry and dynamics of quadratic rational maps. *Experiment. Math.*, 2(1):37–83, 1993. With an appendix by the author and Lei Tan.

[302] J. Milnor. *Dynamics in One Complex Variable*. Friedr. Vieweg & Sohn, Braunschweig, 1999. Introductory lectures.

[303] J. Milnor. On rational maps with two critical points. *Experiment. Math.*, 9(4):481–522, 2000.

[304] D. W. Morris. *Ratner's Theorems on Unipotent Flows*. Chicago Lectures in Mathematics. University of Chicago Press, Chicago, IL, 2005.

[305] P. Morton. Arithmetic properties of periodic points of quadratic maps. *Acta Arith.*, 62(4):343–372, 1992.

[306] P. Morton. Characterizing cyclic cubic extensions by automorphism polynomials. *J. Number Theory*, 49(2):183–208, 1994.

[307] P. Morton. On certain algebraic curves related to polynomial maps. *Compositio Math.*, 103(3):319–350, 1996.

[308] P. Morton. Periods of maps on irreducible polynomials over finite fields. *Finite Fields Appl.*, 3(1):11–24, 1997.

[309] P. Morton. Arithmetic properties of periodic points of quadratic maps. II. *Acta Arith.*, 87(2):89–102, 1998.

[310] P. Morton. Galois groups of periodic points. *J. Algebra*, 201(2):401–428, 1998.

[311] P. Morton and P. Patel. The Galois theory of periodic points of polynomial maps. *Proc. London Math. Soc. (3)*, 68(2):225–263, 1994.

[312] P. Morton and J. H. Silverman. Rational periodic points of rational functions. *Internat. Math. Res. Notices*, (2):97–110, 1994.

[313] P. Morton and J. H. Silverman. Periodic points, multiplicities, and dynamical units. *J. Reine Angew. Math.*, 461:81–122, 1995.

[314] P. Morton and F. Vivaldi. Bifurcations and discriminants for polynomial maps. *Nonlinearity*, 8(4):571–584, 1995.

[315] P. Moussa. Ensembles de Julia et propriétés de localisation des entiers algébriques. In *Seminar on Number Theory, 1984–1985 (Talence, 1984/1985)*, pages Exp. No. 21, 10. Univ. Bordeaux I, Talence, 1985.

[316] P. Moussa. Diophantine properties of Julia sets. In *Chaotic Dynamics and Fractals (Atlanta, Ga., 1985)*, volume 2 of *Notes Rep. Math. Sci. Engrg.*, pages 215–227. Academic Press, Orlando, FL, 1986.

[317] P. Moussa, J. S. Geronimo, and D. Bessis. Ensembles de Julia et propriétés de localisation des familles itérées d'entiers algébriques. *C. R. Acad. Sci. Paris Sér. I Math.*, 299(8):281–284, 1984.

[318] F. Mukhamedov and J. F. Mendes. On chaos of a cubic $p$-adic dynamical system, 2006. `ArXiv:math.DS/0608573`.

[319] F. Mukhamedov and J. F. Mendes. On the chaotic behavior of a generalized logistic $p$-adic dynamical system. *J. Differential Equations*, 2007. To appear.

[320] F. Mukhamedov and U. Rozikov. On rational $p$-adic dynamical systems. *Methods Funct. Anal. Topology*, 10(2):21–31, 2004.

[321] D. Mumford. *Abelian Varieties*. Tata Institute of Fundamental Research Studies in Mathematics, No. 5. Published for the Tata Institute of Fundamental Research, Bombay, 1970.

[322] D. Mumford, J. Fogarty, and F. Kirwan. *Geometric Invariant Theory*, volume 34 of *Ergebnisse der Mathematik und ihrer Grenzgebiete (2) [Results in Mathematics and Related Areas (2)]*. Springer-Verlag, Berlin, third edition, 1994.

[323] D. Mumford and K. Suominen. Introduction to the theory of moduli. In *Algebraic Geometry, Oslo 1970 (Proc. Fifth Nordic Summer-School in Math.)*, pages 171–222. Wolters-Noordhoff, Groningen, 1972.

[324] M. R. Murty. *Problems in Analytic Number Theory*, volume 206 of *Graduate Texts in Mathematics*. Springer-Verlag, New York, 2001. Readings in Mathematics.

[325] W. Narkiewicz. Polynomial cycles in algebraic number fields. *Colloq. Math.*, 58(1):151–155, 1989.

[326] W. Narkiewicz. *Polynomial Mappings*, volume 1600 of *Lecture Notes in Mathematics*. Springer-Verlag, Berlin, 1995.

[327] W. Narkiewicz. Arithmetics of dynamical systems: a survey. *Tatra Mt. Math. Publ.*, 11:69–75, 1997. Number theory (Liptovský Ján, 1995).

[328] W. Narkiewicz. Finite polynomial orbits. A survey. In *Algebraic Number Theory and Diophantine Analysis (Graz, 1998)*, pages 331–338. de Gruyter, Berlin, 2000.

[329] W. Narkiewicz. Polynomial cycles in certain rings of rationals. *J. Théor. Nombres Bordeaux*, 14(2):529–552, 2002.

[330] W. Narkiewicz. Polynomial cycles in cubic fields of negative discriminant. *Funct. Approx. Comment. Math.*, 35:261–270, 2006.

[331] W. Narkiewicz and R. Marszalek. Finite polynomial orbits in quadratic rings. *J. Ramanujan Math. Soc.*, 12(1):91–130, 2006.

[332] W. Narkiewicz and T. Pezda. Finite polynomial orbits in finitely generated domains. *Monatsh. Math.*, 124(4):309–316, 1997.

[333] A. Néron. Quasi-fonctions et hauteurs sur les variétés abéliennes. *Ann. of Math. (2)*, 82:249–331, 1965.

[334] M. Nevins and T. Rogers. Quadratic maps as dynamical systems on the $p$-adic numbers. unpublished, `www.maths.ex.ac.uk/~mwatkins/zeta/nevins.pdf`, March 2000.

[335] P. E. Newstead. *Introduction to Moduli Problems and Orbit Spaces*, volume 51 of *Tata Institute of Fundamental Research Lectures on Mathematics and Physics*. Tata Institute of Fundamental Research, Bombay, 1978.

[336] H. Niederreiter and I. E. Shparlinski. Dynamical systems generated by rational functions. In *Applied Algebra, Algebraic Algorithms and Error-Correcting Codes (Toulouse, 2003)*, volume 2643 of *Lecture Notes in Comput. Sci.*, pages 6–17. Springer, Berlin, 2003.

[337] M. Nilsson. Cycles of monomial and perturbated monomial $p$-adic dynamical systems. *Ann. Math. Blaise Pascal*, 7(1):37–63, 2000.

[338] M. Nilsson. Distribution of cycles of monomial $p$-adic dynamical systems. In *$p$-adic Functional Analysis (Ioannina, 2000)*, volume 222 of *Lecture Notes in Pure and Appl. Math.*, pages 233–242. Dekker, New York, 2001.

[339] M. Nilsson. Fuzzy cycles of $p$-adic monomial dynamical systems. *Far East J. Dyn. Syst.*, 5(2):149–173, 2003.

[340] M. Nilsson and R. Nyqvist. The asymptotic number of periodic points of discrete $p$-adic dynamical systems. *Tr. Mat. Inst. Steklova*, 245(Izbr. Vopr. $p$-adich. Mat. Fiz. i Anal.):210–217, 2004.

[341] K. Nishizawa, K. Sekiguchi, and K. Yoshino. Location of algebraic integers and related topics. In *Dynamical Systems and Related Topics (Nagoya, 1990)*, volume 9 of *Adv. Ser. Dyn. Syst.*, pages 422–450. World Sci. Publishing, 1991.

[342] I. Niven. The iteration of certain arithmetic functions. *Canadian J. Math.*, 2:406–408, 1950.

[343] D. G. Northcott. Periodic points on an algebraic variety. *Ann. of Math. (2)*, 51:167–177, 1950.

[344] R. Nyqvist. Some dynamical systems in finite field extensions of the $p$-adic numbers. In *$p$-adic functional analysis (Ioannina, 2000)*, volume 222 of *Lecture Notes in Pure and Appl. Math.*, pages 243–253. Dekker, New York, 2001.

[345] R. W. K. Odoni. The Galois theory of iterates and composites of polynomials. *Proc. London Math. Soc. (3)*, 51(3):385–414, 1985.

[346] R. W. K. Odoni. Realising wreath products of cyclic groups as Galois groups. *Mathematika*, 35(1):101–113, 1988.

[347] R. W. K. Odoni. On the Galois groups of iterated generic additive polynomials. *Math. Proc. Cambridge Philos. Soc.*, 121(1):1–6, 1997.

[348] R. Oselies and H. Zieschang. Ergodische Eigenschaften der Automorphismen $p$-adischer Zahlen. *Arch. Math. (Basel)*, 26:144–153, 1975.

[349] A. Pal. On the torsion of Drinfeld modules of rank two, 2007. Preprint.

[350] A. Peinado, F. Montoya, J. Muñoz, and A. J. Yuste. Maximal periods of $x^2 + c$ in $\mathbb{F}_q$. In *Applied Algebra, Algebraic Algorithms and Error-Correcting Codes (Melbourne, 2001)*, volume 2227 of *Lecture Notes in Comput. Sci.*, pages 219–228. Springer, Berlin, 2001.

[351] I. Percival and F. Vivaldi. Arithmetical properties of strongly chaotic motions. *Phys. D*, 25(1-3):105–130, 1987.

[352] J. Pettigrew, J. A. G. Roberts, and F. Vivaldi. Complexity of regular invertible $p$-adic motions. *Chaos*, 11(4):849–857, 2001.

[353] T. Pezda. Cycles of polynomial mappings in several variables. *Manuscripta Math.*, 83(3-4):279–289, 1994.

[354] T. Pezda. Cycles of polynomials in algebraically closed fields of positive characteristic. *Colloq. Math.*, 67(2):187–195, 1994.

[355] T. Pezda. Polynomial cycles in certain local domains. *Acta Arith.*, 66(1):11–22, 1994.

[356] T. Pezda. Cycles of polynomials in algebraically closed fields of positive chracteristic. II. *Colloq. Math.*, 71(1):23–30, 1996.

[357] T. Pezda. Cycles of rational mappings in algebraically closed fields of positive characteristics. *Ann. Math. Sil.*, 12:15–21, 1998. Number theory (Cieszyn, 1998).

[358] T. Pezda. On cycles and orbits of polynomial mappings $\mathbb{Z}^2 \mapsto \mathbb{Z}^2$. *Acta Math. Inform. Univ. Ostraviensis*, 10(1):95–102, 2002.

[359] T. Pezda. Cycles of polynomial mappings in several variables over rings of integers in finite extensions of the rationals. *Acta Arith.*, 108(2):127–146, 2003.

[360] J. Pineiro, L. Szpiro, and T. J. Tucker. Mahler measure for dynamical systems on $\mathbb{P}^1$ and intersection theory on a singular arithmetic surface. In *Geometric Methods In Algebra and Number Theory*, volume 235 of *Progr. Math.*, pages 219–250. Birkhäuser Boston, Boston, MA, 2005.

[361] B. Poonen. The classification of rational preperiodic points of quadratic polynomials over $\mathbb{Q}$: a refined conjecture. *Math. Z.*, 228(1):11–29, 1998.

[362] Y. Puri and T. Ward. Arithmetic and growth of periodic orbits. *J. Integer Seq.*, 4(2):Article 01.2.1, 18 pp. (electronic), 2001.

[363] Y. Puri and T. Ward. A dynamical property unique to the Lucas sequence. *Fibonacci Quart.*, 39(5):398–402, 2001.

[364] G. R. W. Quispel, J. A. G. Roberts, and C. J. Thompson. Integrable mappings and soliton equations. *Phys. Lett. A*, 126(7):419–421, 1988.

[365] G. R. W. Quispel, J. A. G. Roberts, and C. J. Thompson. Integrable mappings and soliton equations. II. *Phys. D*, 34(1-2):183–192, 1989.

[366] U. Rausch. On a theorem of Dobrowolski about the product of conjugate numbers. *Colloq. Math.*, 50(1):137–142, 1985.

[367] M. Raynaud. Courbes sur une variété abélienne et points de torsion. *Invent. Math.*, 71(1):207–233, 1983.

[368] M. Raynaud. Sous-variétés d'une variété abélienne et points de torsion. In *Arithmetic and Geometry, Vol. I*, volume 35 of *Progr. Math.*, pages 327–352. Birkhäuser Boston, Boston, MA, 1983.

[369] M. Rees. A partial description of the parameter space of rational maps of degree two (1). *Acta Math.*, 168:11–87, 1992.

[370] M. Rees. A partial description of the parameter space of rational maps of degree two (2). *Proc. London Math. Soc.*, 70:644–690, 1995.

[371] J. F. Ritt. Periodic functions with a multiplication theorem. *Trans. Amer. Math. Soc.*, 23(1):16–25, 1922.

[372] J. Rivera-Letelier. *Dynamique des fonctions rationnelles sur des corps locaux*. PhD thesis, Universite de Paris XI, 2000.

[373] J. Rivera-Letelier. Sur la structure des ensembles de Fatou $p$-adiques, 2002. `ArXiv: math.DS/0412180`.

[374] J. Rivera-Letelier. Une caractérisation des fonctions holomorphes injectives en analyse ultramétrique. *C. R. Math. Acad. Sci. Paris*, 335(5):441–446, 2002.

[375] J. Rivera-Letelier. Dynamique des fonctions rationnelles sur des corps locaux. *Astérisque*, (287):xv, 147–230, 2003. Geometric methods in dynamics. II.

[376] J. Rivera-Letelier. Espace hyperbolique $p$-adique et dynamique des fonctions rationnelles. *Compositio Math.*, 138(2):199–231, 2003.

[377] J. Rivera-Letelier. Points périodiques des fonctions rationnelles dans l'espace hyperbolique $p$-adique. *Comment. Math. Helv.*, 80(3):593–629, 2005.

[378] J. Rivera-Letelier. Wild recurrent critical points. *J. London Math. Soc. (2)*, 72(2):305–326, 2005.

[379] J. Rivera-Letelier. Notes sur la droite projective de Berkovich, 2006. `ArXiv:math.MG/0605676`.

[380] J. Rivera-Letelier. Polynomials over $\mathbb{C}_p$ with wandering domains, after R. Benedetto, 2006. Preprint, `www.math.sunysb.edu/~rivera/mypapers/wand.ps`.

[381] J. Rivera-Letelier. Théorie de Fatou et Julia dans la droite projective de Berkovich, 2007. In preparation.

[382] A. Robert. *Elliptic curves*. Springer-Verlag, Berlin, 1973.

[383] J. A. G. Roberts and F. Vivaldi. Arithmetical method to detect integrability in maps. *Phys. Rev. Lett.*, 90(3):034102, 4, 2003.

[384] J. A. G. Roberts and F. Vivaldi. Signature of time-reversal symmetry in polynomial automorphisms over finite fields. *Nonlinearity*, 18(5):2171–2192, 2005.

[385] T. D. Rogers. The graph of the square mapping on the prime fields. *Discrete Math.*, 148(1-3):317–324, 1996.

[386] M. Ru and E. Yi. Nevanlinna theory and iteration of rational maps. *Math. Z.*, 249(1):125–138, 2005.

[387] W. Rudin. *Real and Complex Analysis*. McGraw-Hill Book Co., New York, third edition, 1987.

[388] P. Russo and R. Walde. Rational periodic points of the quadratic function $Q_c(x) = x^2 + c$. *Am. Math. Monthly*, 101:318–331, 1994.

[389] G. Sarkis. *Formal Groups and p-adic Dynamical Systems*. PhD thesis, Brown University, 2001.

[390] G. Sarkis. On lifting commutative dynamical systems. *J. Algebra*, 293(1):130–154, 2005.

[391] S. Schanuel. Heights in number fields. *Bull. Soc. Math. France*, 107:443–449, 1979.

[392] K. Schmidt. *Dynamical Systems of Algebraic Origin*, volume 128 of *Progress in Mathematics*. Birkhäuser Verlag, Basel, 1995.

[393] W. Schmidt. *Diophantine Approximation*, volume 785 of *Lecture Notes in Mathematics*. Springer, Berlin, 1980.

[394] W. Schmidt and N. Steinmetz. The polynomials associated with a Julia set. *Bull. London Math. Soc.*, 27(3):239–241, 1995.

[395] A. Schweizer. On periodic points under the iteration of additive polynomials. *Manuscripta Math.*, 113(1):25–34, 2004.

[396] J.-P. Serre. *Local Fields*, volume 67 of *Graduate Texts in Mathematics*. Springer-Verlag, New York, 1979. Translated from the French by Marvin Jay Greenberg.

[397] J.-P. Serre. *Lectures on the Mordell-Weil Theorem*. Aspects of Mathematics. Friedr. Vieweg & Sohn, Braunschweig, third edition, 1997. Translated from the French and edited by Martin Brown from notes by Michel Waldschmidt, with a foreword by Brown and Serre.

[398] G. Shimura. On the field of definition for a field of automorphic functions. I, II, III. *Ann. of Math. (2)*, 80, 81, 83:160–189, 124–165, 377–385, 1964, 1965, 1966.

[399] G. Shimura. *Introduction to the Arithmetic Theory of Automorphic Functions*, volume 11 of *Publications of the Mathematical Society of Japan*. Princeton University Press, Princeton, NJ, 1994. Reprint of the 1971 original, Kanô Memorial Lectures, 1.

[400] I. Shparlinski. On some dynamical systems in finite fields and residue rings. *Discrete Contin. Dyn. Syst.*, 17:901–917, 2007.

[401] N. Sibony. Dynamique des applications rationnelles de $\mathbb{P}^k$. In *Dynamique et géométrie complexes (Lyon, 1997)*, volume 8 of *Panor. Synthèses*, pages ix–x, xi–xii, 97–185. Soc. Math. France, Paris, 1999.

[402] N. Sidorov. Arithmetic dynamics. In *Topics in Dynamics and Ergodic Theory*, volume 310 of *London Math. Soc. Lecture Note Ser.*, pages 145–189. Cambridge Univ. Press, Cambridge, 2003.

[403] C. Siegel. The integer solutions of the equation $y^2 = ax^n + bx^{n-1} + \cdots + k$. *J. London Math. Soc.*, 1:66–68, 1926.

[404] C. Siegel. Über einige Anwendungen diophantischer Approximationen. In *Collected Works*, pages 209–266. Springer, Berlin, 1966.

[405] J. H. Silverman. Integer points on curves of genus 1. *J. London Math. Soc. (2)*, 28(1):1–7, 1983.

[406] J. H. Silverman. Arithmetic distance functions and height functions in Diophantine geometry. *Math. Ann.*, 279(2):193–216, 1987.

[407] J. H. Silverman. A quantitative version of Siegel's theorem: integral points on elliptic curves and Catalan curves. *J. Reine Angew. Math.*, 378:60–100, 1987.

[408] J. H. Silverman. Computing heights on elliptic curves. *Math. Comp.*, 51(183):339–358, 1988.

[409] J. H. Silverman. Rational points on $K3$ surfaces: a new canonical height. *Invent. Math.*, 105(2):347–373, 1991.

[410] J. H. Silverman. *The Arithmetic of Elliptic Curves*, volume 106 of *Graduate Texts in Mathematics*. Springer-Verlag, New York, 1992. Corrected reprint of the 1986 original.

[411] J. H. Silverman. Integer points, Diophantine approximation, and iteration of rational maps. *Duke Math. J.*, 71(3):793–829, 1993.

[412] J. H. Silverman. *Advanced Topics in the Arithmetic of Elliptic Curves*, volume 151 of *Graduate Texts in Mathematics*. Springer-Verlag, New York, 1994.

[413] J. H. Silverman. Geometric and arithmetic properties of the Hénon map. *Math. Z.*, 215(2):237–250, 1994.

[414] J. H. Silverman. The field of definition for dynamical systems on $\mathbb{P}^1$. *Compositio Math.*, 98(3):269–304, 1995.

[415] J. H. Silverman. Rational functions with a polynomial iterate. *J. Algebra*, 180(1):102–110, 1996.

[416] J. H. Silverman. The space of rational maps on $\mathbb{P}^1$. *Duke Math. J.*, 94(1):41–77, 1998.

[417] J. H. Silverman. A zeta function over a recurrent sequence. *Amer. Math. Monthly*, 106(7):686–688, 1999. Problem 10486 with solutions.

[418] J. H. Silverman. Height bounds and preperiodic points for families of jointly regular affine maps. *Quart. J. Pure Appl. Math.*, 2:135–145, 2006.

[419] J. H. Silverman and N. Stephens. The sign of an elliptic divisibility sequence. *J. Ramanujan Math. Soc.*, 21(1):1–17, 2006.

[420] J. H. Silverman and J. Tate. *Rational Points on Elliptic Curves*. Undergraduate Texts in Mathematics. Springer-Verlag, New York, 1992.

[421] C. J. Smyth. On the product of the conjugates outside the unit circle of an algebraic integer. *Bull. London Math. Soc.*, 3:169–175, 1971.

[422] C. J. Smyth. The Mahler measure of algebraic numbers: A survey. In *Number theory and polynomials (Univ. Bristol April 2006)*, LMS Lecture Notes. London Mathematical Society, 2007. `ArXiv:math.NT/0701397`.

[423] A. N. Starkov. *Dynamical systems on homogeneous spaces*, volume 190 of *Translations of Mathematical Monographs*. American Mathematical Society, Providence, RI, 2000. Translated from the 1999 Russian original by the author.

[424] C. L. Stewart. Algebraic integers whose conjugates lie near the unit circle. *Bull. Soc. Math. France*, 106(2):169–176, 1978.

[425] M. Stoll. Galois groups over $\mathbb{Q}$ of some iterated polynomials. *Arch. Math. (Basel)*, 59(3):239–244, 1992.

[426] D. Sullivan. Quasiconformal homeomorphisms and dynamics. I. Solution of the Fatou-Julia problem on wandering domains. *Ann. of Math. (2)*, 122(3):401–418, 1985.

[427] P.-A. Svensson. Perturbed dynamical systems in $p$-adic fields. *Tr. Mat. Inst. Steklova*, 245(Izbr. Vopr. $p$-adich. Mat. Fiz. i Anal.):264–272, 2004.

[428] J. J. Sylvester. On a point in the theory of vulgar fractions. *Amer. J. Math.*, 3(4):332–335, 1880.

[429] L. Szpiro and T. Tucker. Equidistribution and generalized Mahler measures, 2006. `ArXiv:math.NT/0510404`.

[430] L. Szpiro and T. Tucker. One half log discriminant, 2006. Diophantine Geometry Proceedings, Publications of the Scuola Normale Superiore de Pisa, `ArXiv:math.NT/0510404`.

[431] L. Szpiro and T. Tucker. A Shafarevich-Faltings theorem for rational functions. *Pure Appl. Math. Q.*, 2:37–48, 2006.

[432] L. Szpiro, E. Ullmo, and S. Zhang. Équirépartition des petits points. *Invent. Math.*, 127(2):337–347, 1997.

[433] E. Thiran, D. Verstegen, and J. Weyers. $p$-adic dynamics. *J. Statist. Phys.*, 54(3-4):893–913, 1989.

[434] A. Thuillier. *Théorie du potentiel sur les courbes en géométrie analytique non archimédienne. Applications à la théorie d'Arakelov*. PhD thesis, Université Rennes, 2005.

[435] E. Ullmo. Positivité et discrétion des points algébriques des courbes. *Ann. of Math. (2)*, 147(1):167–179, 1998.

[436] B. van der Waerden. *Algebra*. Frederick Ungar Publ. Co., New York, 7th edition, 1970.

[437] T. Vasiga and J. Shallit. On the iteration of certain quadratic maps over $\mathrm{GF}(p)$. *Discrete Math.*, 277(1-3):219–240, 2004.

[438] D. Verstegen. $p$-adic dynamical systems. In *Number Theory and Physics (Les Houches, 1989)*, volume 47 of *Springer Proc. Phys.*, pages 235–242. Springer, Berlin, 1990.

[439] A. P. Veselov. Integrable mappings. *Uspekhi Mat. Nauk*, 46(5(281)):3–45, 190, 1991.

[440] F. Vivaldi. The arithmetic of chaos. In *Chaos, Noise and Fractals (Como, 1986)*, volume 3 of *Malvern Phys. Ser.*, pages 187–199. Hilger, Bristol, 1987.

[441] F. Vivaldi. Arithmetical theory of Anosov diffeomorphisms. *Proc. Roy. Soc. London Ser. A*, 413(1844):97–107, 1987.

[442] F. Vivaldi. Algebraic number theory and Hamiltonian chaos. In *Number Theory and Physics (Les Houches, 1989)*, volume 47 of *Springer Proc. Phys.*, pages 294–301. Springer, Berlin, 1990.

[443] F. Vivaldi. Dynamics over irreducible polynomials. *Nonlinearity*, 5(4):941–960, 1992.

[444] F. Vivaldi. Geometry of linear maps over finite fields. *Nonlinearity*, 5(1):133–147, 1992.

[445] P. Voutier. An effective lower bound for the height of algebraic numbers. *Acta Arith.*, 74(1):81–95, 1996.

[446] L. Wang. Rational points and canonical heights on $K3$-surfaces in $\mathbb{P}^1 \times \mathbb{P}^1 \times \mathbb{P}^1$. In *Recent Developments in the Inverse Galois Problem (Seattle, WA, 1993)*, volume 186 of *Contemp. Math.*, pages 273–289. Amer. Math. Soc., Providence, RI, 1995.

[447] T. B. Ward. Almost all $S$-integer dynamical systems have many periodic points. *Ergodic Theory Dynam. Systems*, 18(2), 1998.

[448] J. Wehler. $K3$-surfaces with Picard number 2. *Arch. Math. (Basel)*, 50(1):73–82, 1988.

[449] C. F. Woodcock and N. P. Smart. $p$-adic chaos and random number generation. *Experiment. Math.*, 7(4):333–342, 1998.

[450] X. Yuan. Big line bundles over arithmetic varieties, 2006. `ArXiv:math.NT/0612424`.

[451] D. Zagier. On the number of Markoff numbers below a given bound. *Math. Comp.*, 39(160):709–723, 1982.

[452] S.-W. Zhang. Equidistribution of small points on abelian varieties. *Ann. of Math. (2)*, 147(1):159–165, 1998.

[453] S.-W. Zhang. Distributions in algebraic dynamics. In *Differential Geometry: A Tribute to Professor S.-S. Chern, Surv. Differ. Geom., Vol. X*, pages 381–430. Int. Press, Boston, MA, 2006.

[454] M. Zieve. *Cycles of Polynomial Mappings*. PhD thesis, University of California at Berkeley, 1996.

[455] K. Zimmerman. Commuting polynomials and self-similarity. *New York J. Math.*, 13:89–96 (electronic), 2007.

# Index

# Graduate Texts in Mathematics

*(continued from p. ii)*

125 BERENSTEIN/GAY. Complex Variables: An Introduction.
126 BOREL. Linear Algebraic Groups. 2nd ed.
127 MASSEY. A Basic Course in Algebraic Topology.
128 RAUCH. Partial Differential Equations.
129 FULTON/HARRIS. Representation Theory: A First Course. *Readings in Mathematics*
130 DODSON/POSTON. Tensor Geometry.
131 LAM. A First Course in Noncommutative Rings. 2nd ed.
132 BEARDON. Iteration of Rational Functions.
133 HARRIS. Algebraic Geometry: A First Course.
134 ROMAN. Coding and Information Theory.
135 ROMAN. Advanced Linear Algebra. 2nd ed.
136 ADKINS/WEINTRAUB. Algebra: An Approach via Module Theory.
137 AXLER/BOURDON/RAMEY. Harmonic Function Theory. 2nd ed.
138 COHEN. A Course in Computational Algebraic Number Theory.
139 BREDON. Topology and Geometry.
140 AUBIN. Optima and Equilibria. An Introduction to Nonlinear Analysis.
141 BECKER/ WEISPFENNING/KREDEL. Gröbner Bases. A Computational Approach to Commutative Algebra.
142 LANG. Real and Functional Analysis. 3rd ed.
143 DOOB. Measure Theory.
144 DENNIS/FARB. Noncommutative Algebra.
145 VICK. Homology Theory. An Introduction to Algebraic Topology. 2nd ed.
146 BRIDGES. Computability: A Mathematical Sketchbook.
147 ROSENBERG. Algebraic *K*-Theory and Its Applications.
148 ROTMAN. An Introduction to the Theory of Groups. 4th ed.
149 RATCLIFFE. Foundations of Hyperbolic Manifolds.
150 EISENBUD. Commutative Algebra with a View Toward Algebraic Geometry.
151 SILVERMAN. Advanced Topics in the Arithmetic of Elliptic Curves.
152 ZIEGLER. Lectures on Polytopes.
153 FULTON. Algebraic Topology: A First Course.
154 BROWN/PEARCY. An Introduction to Analysis.
155 KASSEL. Quantum Groups.
156 KECHRIS. Classical Descriptive Set Theory.
157 MALLIAVIN. Integration and Probability.
158 ROMAN. Field Theory.
159 CONWAY. Functions of One Complex Variable II.
160 LANG. Differential and Riemannian Manifolds.
161 BORWEIN/ERDÉLYI. Polynomials and Polynomial Inequalities.
162 ALPERIN/BELL. Groups and Representations.
163 DIXON/MORTIMER. Permutation Groups.
164 NATHANSON. Additive Number Theory: The Classical Bases.
165 NATHANSON. Additive Number Theory: Inverse Problems and the Geometry of Sumsets.
166 SHARPE. Differential Geometry: Cartan's Generalization of Klein's Erlangen Program.
167 MORANDI. Field and Galois Theory.
168 EWALD. Combinatorial Convexity and Algebraic Geometry.
169 BHATIA. Matrix Analysis.
170 BREDON. Sheaf Theory. 2nd ed.
171 PETERSEN. Riemannian Geometry. 2nd ed.
172 REMMERT. Classical Topics in Complex Function Theory.
173 DIESTEL. Graph Theory. 2nd ed.
174 BRIDGES. Foundations of Real and Abstract Analysis.
175 LICKORISH. An Introduction to Knot Theory.
176 LEE. Riemannian Manifolds.
177 NEWMAN. Analytic Number Theory.
178 CLARKE/LEDYAEV/STERN/WOLENSKI. Nonsmooth Analysis and Control Theory.
179 DOUGLAS. Banach Algebra Techniques in Operator Theory. 2nd ed.
180 SRIVASTAVA. A Course on Borel Sets.
181 KRESS. Numerical Analysis.
182 WALTER. Ordinary Differential Equations.
183 MEGGINSON. An Introduction to Banach Space Theory.
184 BOLLOBAS. Modern Graph Theory.
185 COX/LITTLE/O'SHEA. Using Algebraic Geometry. 2nd ed.
186 RAMAKRISHNAN/VALENZA. Fourier Analysis on Number Fields.
187 HARRIS/MORRISON. Moduli of Curves.
188 GOLDBLATI. Lectures on the Hyperreals: An Introduction to Nonstandard Analysis.

189 Lam. Lectures on Modules and Rings.
190 Esmonde/Murty. Problems in Algebraic Number Theory. 2nd ed.
191 Lang. Fundamentals of Differential Geometry.
192 Hirsch/Lacombe. Elements of Functional Analysis.
193 Cohen. Advanced Topics in Computational Number Theory.
194 Engel/Nagel. One-Parameter Semigroups for Linear Evolution Equations.
195 Nathanson. Elementary Methods in Number Theory.
196 Osborne. Basic Homological Algebra.
197 Eisenbud/Harris. The Geometry of Schemes.
198 Robert. A Course in $p$-adic Analysis.
199 Hedenmalm/Korenblum/Zhu. Theory of Bergman Spaces.
200 Bao/Chern/Shen. An Introduction to Riemann–Finsler Geometry.
201 Hindry/Silverman. Diophantine Geometry: An Introduction.
202 Lee. Introduction to Topological Manifolds.
203 Sagan. The Symmetric Group: Representations, Combinatorial Algorithms, and Symmetric Functions.
204 Escofier. Galois Theory.
205 Felix/Halperin/Thomas. Rational Homotopy Theory, 2nd ed.
206 Murty. Problems in Analytic Number Theory. *Readings in Mathematics*
207 Godsil/Royle. Algebraic Graph Theory.
208 Cheney. Analysis for Applied Mathematics.
209 Arveson. A Short Course on Spectral Theory.
210 Rosen. Number Theory in Function Fields.
211 Lang. Algebra. Revised 3rd ed.
212 Matoušek. Lectures on Discrete Geometry.
213 Fritzsche/Grauert. From Holomorphic Functions to Complex Manifolds.
214 Jost. Partial Differential Equations.
215 Goldschmidt. Algebraic Functions and Projective Curves.
216 D. Serre. Matrices: Theory and Applications.
217 Marker. Model Theory: An Introduction.
218 Lee. Introduction to Smooth Manifolds.
219 Maclachlan/Reid. The Arithmetic of Hyperbolic 3-Manifolds.
220 Nestruev. Smooth Manifolds and Observables.
221 Grünbaum. Conv0ex Polytopes. 2nd ed.
222 Hall. Lie Groups, Lie Algebras, and Representations: An Elementary Introduction.
223 Vretblad. Fourier Analysis and Its Applications.
224 Walschap. Metric Structures in Differential Geometry.
225 Bump: Lie Groups.
226 Zhu. Spaces of Holomorphic Functions in the Unit Ball.
227 Miller/Sturmfels. Combinatorial Commutative Algebra.
228 Diamond/Shurman. A First Course in Modular Forms.
229 Eisenbud. The Geometry of Syzygies.
230 Stroock. An Introduction to Markov Processes.
231 Björner/Brenti. Combinatorics of Coxeter Groups.
232 Everest/Ward. An Introduction to Number Theory.
233 Albiac/Kalton. Topics in Banach Space Theory.
234 Jorgensen. Analysis and Probability: Wavelets, Signals, Fractals.
235 Sepanski. Compact Lie Groups.
236 Garnett. Bounded Analytic Functions.
237 Martinez-Avendano/Rosenthal. An Introduction to Operators on the Hardy-Hilbert Space.
238 Aigner. Course in Enumeration.
239 Cohen. Number Theory 1.
240 Cohen. Number Theory 2.
241 Silverman. The Arithmetic of Dynamical Systems.
242 Grillet. Abstract Algebra.
243 Givant. Simple Relation Algebras.